Bottom Gravity Currents and Overflows in Deep Channels of the Atlantic Ocean

Eugene G. Morozov • Roman Y. Tarakanov
Dmitry I. Frey

Bottom Gravity Currents and Overflows in Deep Channels of the Atlantic Ocean

Observations, Analysis, and Modeling

With a contribution by Walter Zenk

Springer

Eugene G. Morozov
Physical Department
Shirshov Institute of Oceanology
Moscow, Russia

Roman Y. Tarakanov
Physical Department
Shirshov Institute of Oceanology
Moscow, Russia

Dmitry I. Frey
Physical Department
Shirshov Institute of Oceanology
Moscow, Russia

ISBN 978-3-030-83073-1 ISBN 978-3-030-83074-8 (eBook)
https://doi.org/10.1007/978-3-030-83074-8

Cover illustration: Chart of the Vema Channel region and Santos Plateau. The bathymetry source is the GEBCO 2019 database.

This Springer imprint is published by the registered company Springer Nature Switzerland AG
The registered company address is: Gewerbestrasse 11, 6330 Cham, Switzerland

Foreword

This book is dedicated to the study of structure and transport of deep and bottom waters in underwater channels of the Atlantic Ocean and spreading of these waters in the deep basins of the Atlantic. Thus, we study the bottom circulation in the ocean. Our research is based on recent observations, analysis of historical data, modeling, and literature reviews. This approach allows us to understand the bottom water transport and analyze how deep currents and water mass properties have changed over the last years and decades. Our study is focused on the propagation of bottom waters in the Atlantic Ocean based on new field data at key points in the channels.

At the end of the 1920s, the first integral study of water masses and bottom topography of the Central and South Atlantic was carried out from the German research vessel *Meteor*. This German Atlantic Expedition was one of the first cruises equipped with the newly developed echo sounder (fathometer): an obligatory prerequisite for the investigation of bottom topography in the deep sea. The results of the expedition were published by Wüst, Defant, and colleagues in the multivolume METEOR publication series starting with the cruise report by the ship's commander (Spiess 1928; 1932). Historically, this series of publications, intermittently interrupted by World War II, was the basis for many years of research into the development of modern concepts about Atlantic water masses and their circulation Schemes. A history of this expedition has been recently published by Emery and Zenk (2019).

Since the 1920s, many national and international programs were initiated to study the properties of bottom water in the Atlantic Ocean and their propagation through narrow channels in submarine ridges. This book summarizes the results of many field experiments and describes our contribution to this research. This book is a continuation of our research published in many journals and in the book by E.G. Morozov, A.N. Demidov, R.Yu. Tarakanov, and W. Zenk (2010), *Abyssal*

With a contribution by Walter Zenk, GEOMAR Helmholtz Centre for Ocean Research Kiel, Germany

Channels in the Atlantic Ocean: Water Structure and Flows. Dordrecht: Springer. 266 p. https://doi.org/10.1007/978-90-481-9358-5.

Recently, deep-water fractures, troughs, and channels between large oceanic basins became interesting objects of research. These investigations have allowed us to characterize specific features of water exchange in deep-water parts of the ocean through narrow channels. Such features cannot be studied even using modern global circulation models with a high resolution. The peculiarities of abyssal water exchange through narrow channels are extremely difficult to integrate in numerical global circulation models with a sufficiently high resolution. Long-term observations and profiling measurements from the ships indicate that high velocities exceeding half a meter per second are not unusual at selected choke points in the deepest layers of the ocean.

Intense research into the deep ocean was carried out during the field works that were a part of the World Ocean Circulation Experiment (WOCE) in the 1990s. The deepest fracture zones and channels of different origins in the ocean provide water exchange between deep basins. Since the beginning of the World Ocean Circulation Experiment in the early 1990s, sustained observations of Antarctic Bottom Water characteristics and transport were conducted at selected sites around the Brazil Basin. One of the WOCE core projects, the Deep Basin Experiment, was focused on the key passages that allow an equatorward flow of bottom water from the south and inter-basin exchange. After the WOCE decade, field measurements in the Atlantic abyssal channels continued to broaden our understanding of these phenomena. Since then, we have been continuing our research in the deep channels and deep basins, widening the regions of investigation to the entire ocean basin.

Physical Department
Shirshov Institute of Oceanology
Moscow, Russia

Eugene G. Morozov
Roman Y. Tarakanov
Dmitry I. Frey

References

Emery WJ, Zenk W (2019) Captain F. Spiess and the German Meteor Expedition of 1925–27. Brown Walker Press. Irvine, CA. 275 pp.

Spiess F (1928) The Meteor Expedition. Scientific results of the German Atlantic Expedition, 1925–1927. D. Reimer Verlag, Berlin. Translated by W. J. Emery. Amerind Publishing Co. Pvt. Ltd., New Delhi, 429 pp.

Spiess F (1932) Das Forschungsschiff und seine Reise. Dt. Atl. Exp. "Meteor" 1925–1927, Bd. 1. 442 pp. [in German]

Preface

Our book reviews the studies carried out in different abyssal channels of the Atlantic Ocean, combined with our recent oceanographic research and analysis of observations in the main underwater channels of the Atlantic Ocean. We also analyze properties of water masses involved in the flows through the channels and their variability on the decadal time scale. This book presents a continuation of our research published by E. Morozov, A. Demidov, R. Tarakanov, and W. Zenk, *Abyssal Channels in the Atlantic Ocean: Water Structure and Flows*, Springer, 2010. The authors strongly regret that Walter Zenk could not participate in writing this book, but we are very grateful to him for discussions. Many ideas about abyssal flows and flows in the abyssal channels are reported by W. Zenk in (Zenk 2019). In this book, we left part of the previous text that was written together with him.

The following abyssal channels are the most important in bottom water exchange in the Atlantic Ocean: Falkland Gap, Vema Channel, Hunter Channel, Vema Fracture Zone, Romanche and Chain fracture zones, Equatorial Channel, Kane Gap, Charlie Gibbs Fracture Zone, and a group of fractures in the Southwest Indian Ridge at 25° E. Of course, numerous channels in the Scotia Sea and Drake Passage, as well as passages connecting the Weddell Sea with the Georgia Basin, which provide the initial outflow of Antarctic Bottom Water from the Weddell Sea to the Atlantic abyssal zone, also play a significant role. The fractures in the northern part of the Mid-Atlantic Ridge play an important role in the propagation of bottom waters from the West Atlantic to the East Atlantic. Here, the main role belongs to the Vema Fracture Zone; however, the other channels jointly provide a pathway for almost the same amount of bottom water to the East Atlantic: Strakhov, Bogdanov, Nameless (07°28′ N), Vernadsky, Doldrums, Arkhangelsky, Ten Degree, Marathon, Fifteen Twenty, and Kane fracture zones.

With a contribution by Walter Zenk, GEOMAR Helmholtz Centre for Ocean Research Kiel, Germany

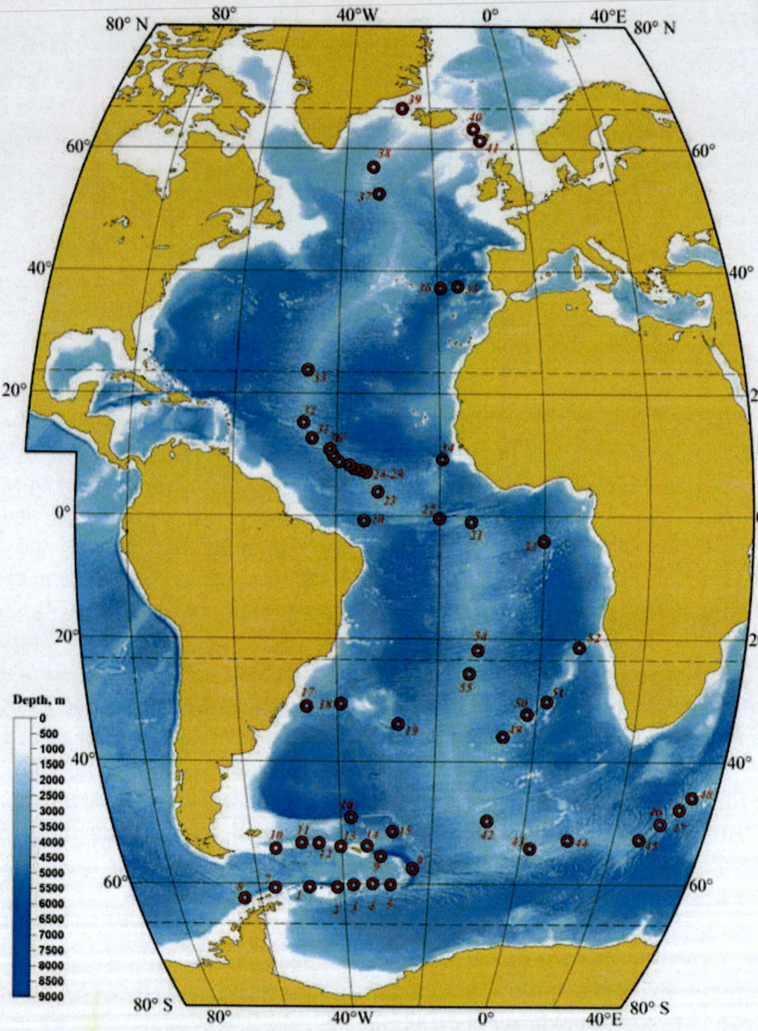

Fig. 1 Chart of the Atlantic Ocean showing the location of the major abyssal channels usually associated with fracture zones (FZ). 1: Philip Passage, 2: Orkney Passage, 3: Bruce Passage, 4: Discovery Passage, 5: Hoyer Passage, 6: South Sandwich Trench, 7: Shackleton Passage, 8: Hero Passage, 9: Georgia Passage, 10: 54–54 Passage, 11: Shag Rocks Passage, 12: Black Rock Passage, 13: Unnamed Passage, 14: Northeastern Georgia Passage, 15: Nameless Passage east of the Islas Orcadas Rise, 16: Falkland Passage, 17: Lower Santos Plateau and Santos Basin), 18: Vema Channel, 19: Hunter Channel, 20: Equatorial Channel, 21: Chain FZ, 22: Romanche FZ, 23: Strakhov FZ, 24: Bogdanov FZ, 25: 7°30′ N FZ, 26: Vernadsky FZ, 27: Doldrums FZ, 28: Arkhangelsky Fracture Zone, 29: Ten Degree FZ, 30: Vema FZ, 31: Marathon FZ, 32: Cabo Verde FZ (Fifteen Twenty FZ), 33: Kane FZ, 34: Kane Gap, 35: Discovery Gap, 36: Western Gap, 37: Charlie Gibbs FZ, 38: Bight FZ, 39: Denmark Strait, 40: Iceland-Faroe Passage, 41: Faroe Bank Channel, 42: 50°S FZ in the South Atlantic Ridge, 43: Conrad-Bouvet FZ, 44: 10° E FZ in the Southwest Indian Ridge, 45: Du Toit FZ, 46: Andrew Bain FZ, 47: Marion FZ, 48: Prince Edward FZ, 49: Walvis Passage, 50: Unnamed Passage at 2° W in the Walvis Ridge, 51: Unnamed Passage at 2° E in the Walvis Ridge, 52: Namib Col, 53: Angola Gap, 54: Rio-de-Janeiro FZ, 55: Rio Grande FZ

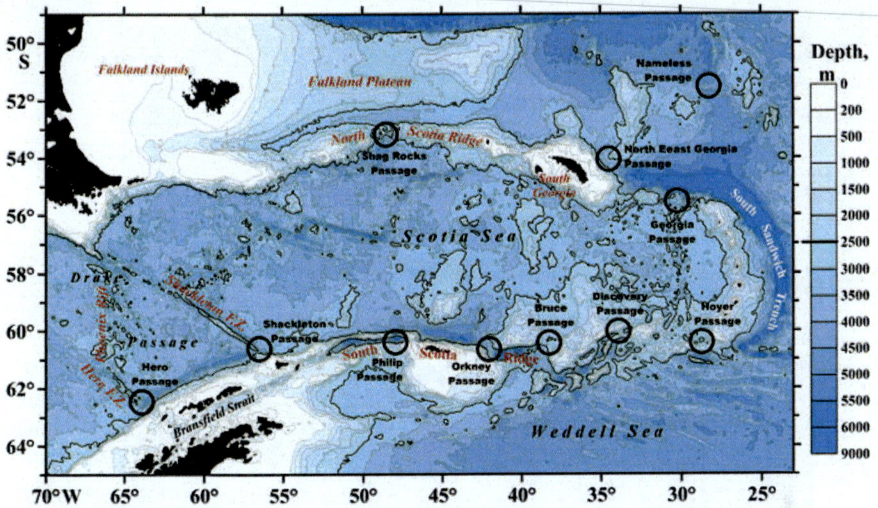

Fig. 2 Chart of the Scotia Sea and Drake Passage showing numerous channels that allow bottom water spreading

The water in the Weddell Sea is the source of the bottom waters in the Atlantic and Southern oceans, while the Scotia Sea is one of the basins responsible for the first appreciable transformation of bottom water. Locations of the main abyssal channels are shown in Fig. 1. Detailed charts of bottom topography in the regions of major channels are shown in individual topographic maps in Figs. 2, 3, 4, 5, 6, 7 and 8.

Most of the Atlantic abyssal channels provide a unidirectional flow of bottom water. Progressive vector diagrams based on moored measurements of currents in the channels give a clear illustration of the prevailing flow directions, current intensity, and steadiness. Figure 9 shows ten progressive vector diagrams in the major channels around the Argentine and Brazil basins of the South Atlantic.

Images on the left side of Fig. 9 are inferred from observations across the Rio Grande Rise. Diagram 1 is related to the Deep Western Boundary Current on the continental rise off the Santos Plateau. The near-bottom flows through the Vema Channel were recorded on the moorings near the west bank of the Vema Sill (Diagrams 2 and 3) and on its east bank (Diagrams 4 and 5). Diagram 6 represents the main vein in the Hunter Channel. Diagram 7 shows examples of inflow for bottom water from the Shag Rocks Passage (Fig. 2) at the northern rim of the Scotia Sea. Diagram 8 shows the outflow of bottom water to the north and the Vema Extension (northern part of the Vema Channel). Diagrams 9 and 10 show outflows across the Mid-Atlantic Ridge in the Romanche and Chain fracture zones. All curves are barely 1.5 years long (527 days) except diagram 7, which is about half this

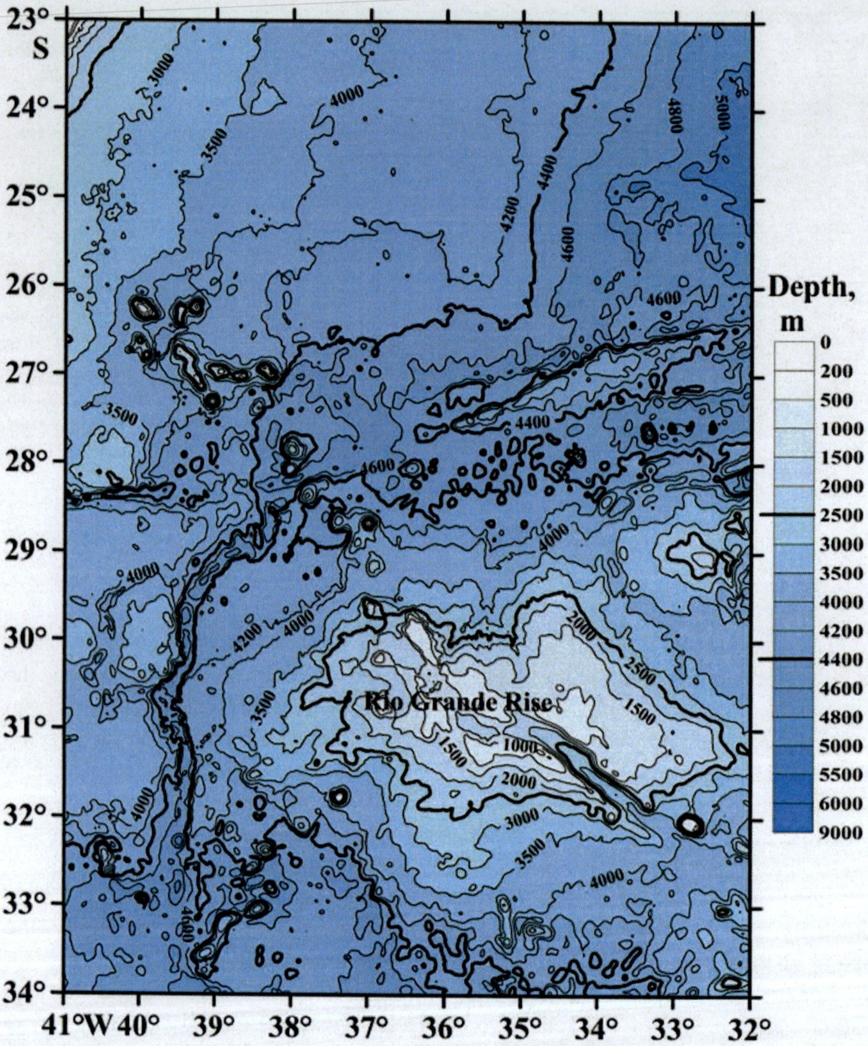

Fig. 3 Bathymetric chart of the Vema Channel

duration time. Dots are 30 days apart. No filters were applied to the data sets. The sampling interval is 2 h. The topographic steering effect is obvious at all locations.

Moreover, specific similarity can be found between flows in different abyssal channels (Zenk et al. 1999). The authors constructed daily vertical profiles based on measurements in several channels. The vertical axis was normalized by the level, where the velocities decreased to 14% of their maximum (unit depth). The zero depth was defined as the level of the deepest current meter (15–50 m above the

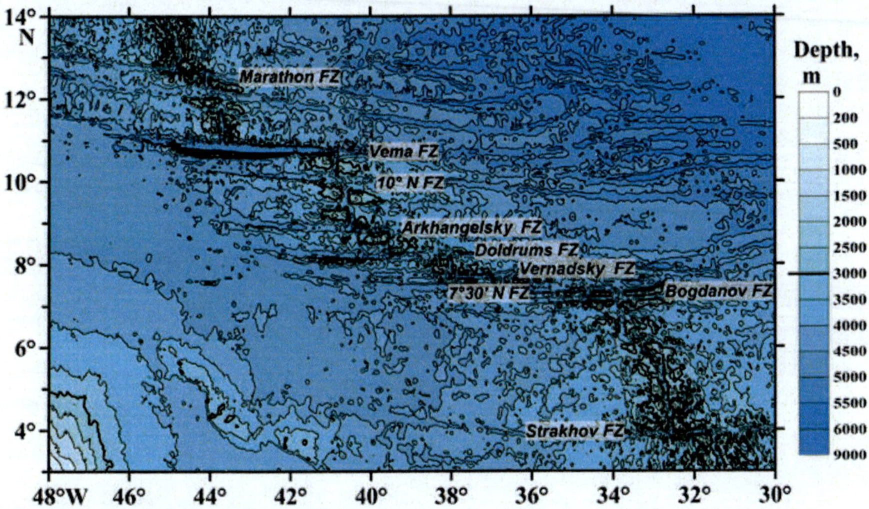

Fig. 4 Bathymetric chart of the Vema Fracture Zone and other fractures of the northern part of the Mid-Atlantic Ridge

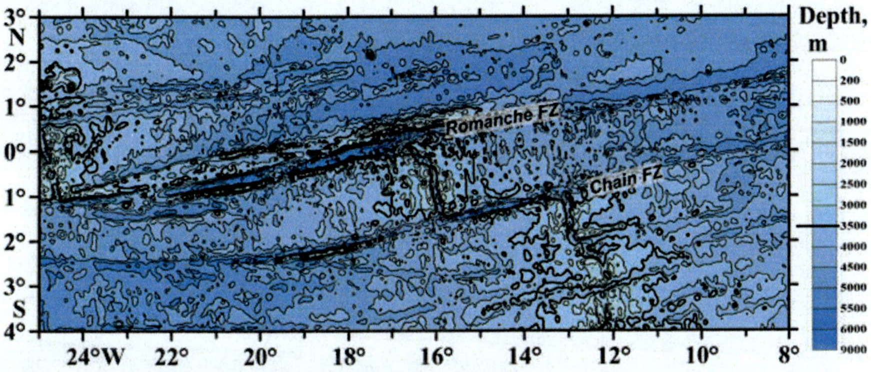

Fig. 5 Bathymetric chart of the Romanche and Chain fracture zones

bottom). Mean velocity maxima served as scaling factors on the abscissa axis. Figure 10 shows time-averaged profiles from different abyssal channels: Vema Channel, Hunter Channel, Romanche Fracture Zone (Mercier and Speer 1998), and Vema Fracture Zone at 11°N (Vangriesheim 1980).

The general common property of abyssal flows in deep channels is location of the maximum velocity at a distance above the bottom. Velocities decrease in both upward and downward directions. This property of the flow was found from the moored observations and later confirmed by measurements with lowered acoustic

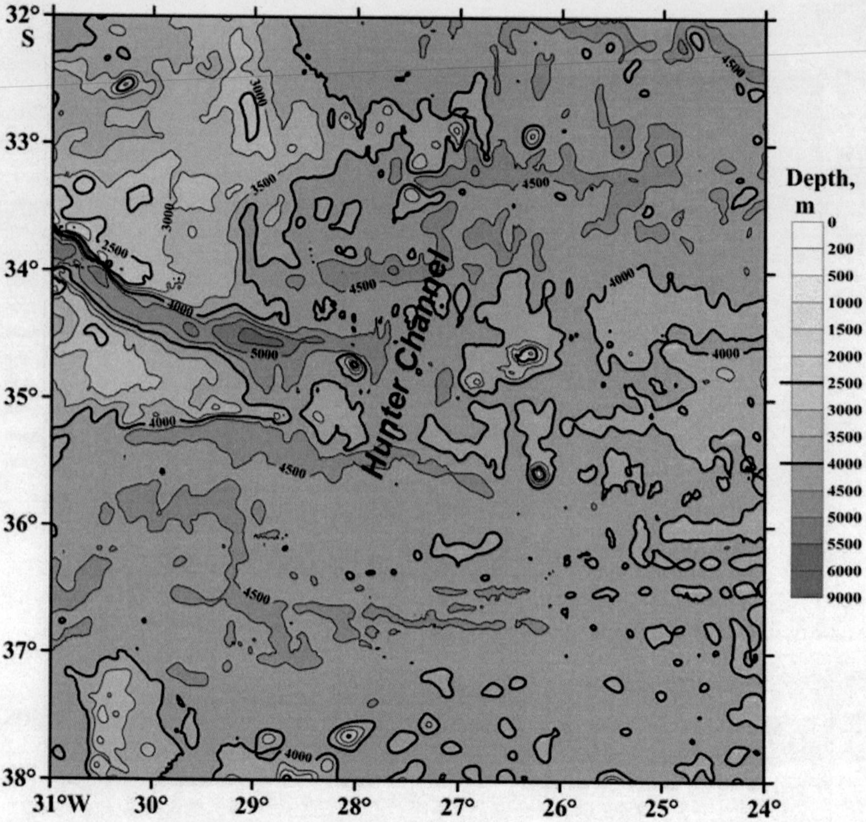

Fig. 6 Bathymetric chart of the Hunter Channel

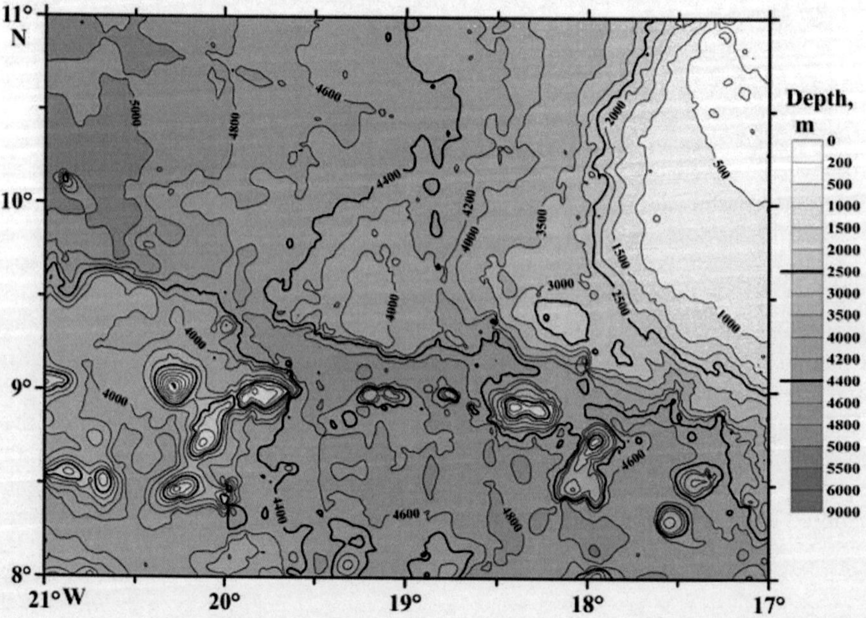

Fig. 7 Bathymetric chart of the Kane Gap

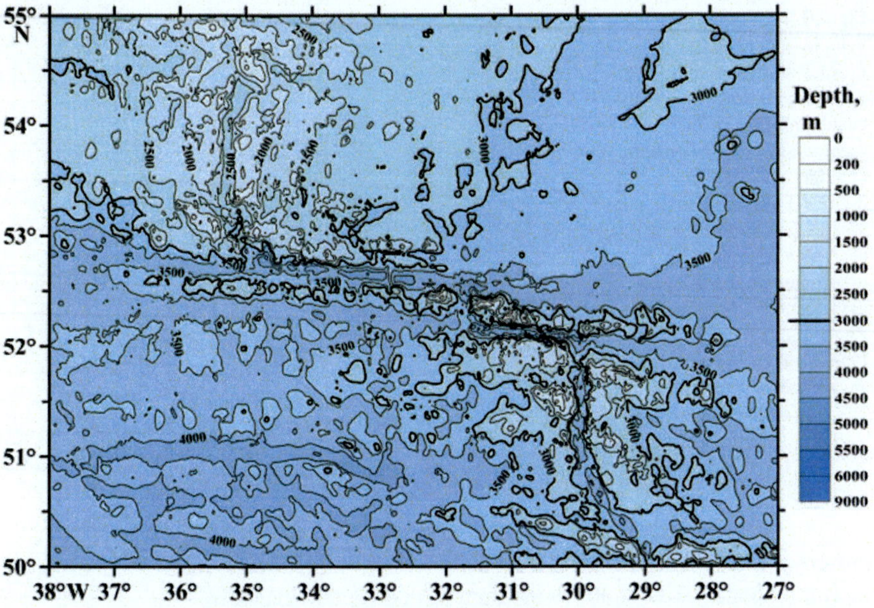

Fig. 8 Bathymetric chart of the Charlie Gibbs Fracture Zone

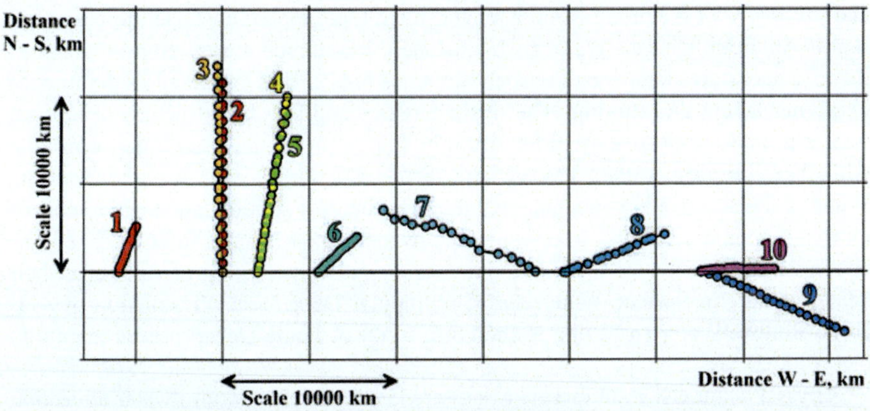

Fig. 9 Progressive diagrams based on 527-days long measurements in different abyssal passages around the Argentine and Brazil basins: (1) Santos Plateau; (2–3) west bank of the Vema Channel; (4–5) east bank of the Vema Channel; (6) Hunter Channel; (7) Shag Rocks Passage (one year only); (8) Vema Extension; (9) Romanche Fracture Zone; (10) Chain Fracture Zone. Tick marks are 30 days apart. See Fig. 1 for the location of passages. The figures are based on archive data of the Shirshov Institute of Oceanology, IFM-GEOMAR, and WOD18

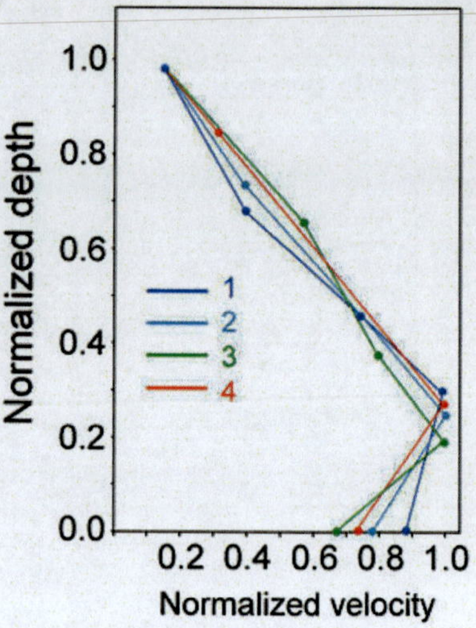

Fig. 10 Similarity of velocity versus depth profiles in different channels. Averaged vertical current profiles of the recorded meridional current components in different regions (plotted based on the idea in (Zenk et al. 1999)): (1) Vema Channel, (2) Hunter Channel, (3) Romanche Fracture Zone, and (4) Vema Fracture Zone. The diagram shows velocities between 2 °C potential temperature isotherm and the bottom

current profilers. They demonstrated that the core of the flows is located at a distance of approximately 100 m above the bottom in different channels.

The major part of the book is dedicated to the propagation of Antarctic Bottom Water (AABW) and different modifications of this water in the Atlantic abyssal channels. Antarctic Bottom Water occupies a bottom position over the major part of the Atlantic Ocean. The bottom topography plays a crucial role in the propagation and properties of Antarctic Bottom Water. Each topographic obstacle limits spreading of Antarctic Bottom Water with extremely low temperatures. Bottom water at these temperatures can overflow such obstacles only if it is entrained in the flow of overlying layers and mixing. The characteristics and specific structure of abyssal waters in these regions may differ strongly due to entrainment and intense mixing with the overlying waters. Everywhere in the Atlantic Ocean, Antarctic Bottom Water is characterized by low potential temperature, low salinity, and high content of nutrients compared to the overlying deep waters. It is formed in several regions above the continental slope of Antarctica as the result of mixing of Antarctic Shelf Water, Antarctic Surface Water, and Circumpolar Deep Water. Formation of Antarctic Bottom Water spreading to the north in the Atlantic Ocean occurs mainly in the Weddell Sea.

Abyssal channels in submarine ridges are key points for bottom water advection in the ocean. Locations of abyssal channels are geographically distant from each other. Therefore, the specific structure of waters is different in these regions. In addition, it is determined by the complex processes of interaction between the waters of the North Atlantic and Antarctic origin. However, abyssal channels play an exclusive role in the formation of deep circulation, because they are the main pathways for Antarctic Bottom Water spreading to the north.

Although we are aware of a geographically unequal coverage of individual channels in the ocean, we believe that the Vema Channel is a test laboratory and the best example for analysis of the flows in underwater channels. This is the only channel that could be investigated thoroughly and deeply because many other abyssal channels are located very far from the leading oceanographic centers of the world. Many of our efforts were concentrated on the measurements in the Vema Channel, which is probably the best example of strong flow in an abyssal channel. Experiments in the Vema Channel constitute the most important part of our research. Our measurements in the Vema Channel provide insights into physical processes that occur in the deep narrow channels with high velocities and intense mass transport. However, we studied the currents in many other abyssal channels, which also play an important role in the bottom water transport.

Physical Department
Shirshov Institute of Oceanology
Moscow, Russia

Eugene G. Morozov
Roman Y. Tarakanov
Dmitry I. Frey

References

Emery WJ, Zenk W (2019) Captain F. Spiess and the German Meteor Expedition of 1925–27. Brown Walker Press. Irvine. CA. 275 pp.

Mercier H, Speer KG (1998) Transport of bottom water in the Romanche Fracture Zone and the Chain Fracture Zone. J Phys Oceanogr 28(5):779–790. https://doi.org/10.1175/1520-0485(1998)028<0779:TOBWIT>2.0.CO;2

Morozov EG, Demidov AN, Tarakanov RYu, Zenk W (2010) Abyssal Channels in the Atlantic Ocean: Water Structure and Flows. Dordrecht: Springer. 266 p. https://doi.org/10.1007/978-90-481-9358-5

Orsi AH, Johnsson GC, Bullister JL (1999) Circulation, mixing, and production of Antarctic Bottom Water. Prog Oceanogr 43:55–109. https://doi.org/10.1016/S0079-6611(99)00004-X

Reid JL, Nowlin WD, Patzert WC (1977) On the characteristics and circulation of the Southwestern Atlantic Ocean. J Phys Oceanogr 7(1):62–91

Siedler G, Church J, Gould J (Eds.) (2001) Ocean Circulation and Climate, Academic Press. San Diego

Sievers HA, Nowlin WD Jr. (1984) The stratification and water masses at Drake Passage. J Geophys Res Oceans 89(C6):10489–10514

Spiess F (1928) The Meteor Expedition. Scientific results of the German Atlantic Expedition, 1925–1927. D. Reimer Verlag, Berlin. Translated by W. J. Emery. Amerind Publishing Co. Pvt. Ltd., New Delhi, 429 pp.

Spiess F (1932) Das Forschungsschiff und seine Reise. Dt. Atl. Exp. "Meteor" 1925–1927, Bd. 1. 442 pp. [in German]

Vangriesheim A. (1980) Antarctic Bottom Water flow through the Vema Fracture Zone. Oceanol Acta 3:199–207

Wüst G (1936) Schichtung und Zirkulation des Atlantischen Ozeans, Das Bodenwasser und die Stratosphäre. In: A. Defant (Ed.), Wissenschaftliche Ergebnisse, Deutsche Atlantische Expedition auf dem Forschungs - und Vermessungsschiff "Meteor" 1925–1927, 6(1), Walter de Gruyter & Co, Berlin, 411 pp.

Zenk W, Siedler G, Lenz B, Hogg NG (1999) Antarctic Bottom Water flow through the Hunter Channel. J Phys Oceanogr 29(11):2785–2801

Zenk W (2019) Abyssal Currents, Encyclopedia of Ocean Sciences. (3rd Edition) Ed by K. Cochran, H. Bokuniewicz, P. Yager, Academic Press, Vol. 3, pp. 189–203. https://doi.org/10.1016/B978-0-12-409,548-9.11343-0

Acknowledgments

The expeditions onboard research vessels *Akademik Ioffe* and *Akademik Sergey Vavilov* were carried out as part of the projects "Southern Ocean" and "Meridian," and other Russian scientific research projects. The study was supported by the Russian Foundation for Basic Research, Departments of Earth Sciences and Engineering Sciences (projects nos. 11-08-00076, 14-08-00107, 17-08-00085, 18-05-00283, 20-08-00246), and NWO-RFBR, project 047.017.2006.003; Federal Purposeful Program of Russia "World Ocean"; Programs 17 and 49 of the Presidium of the Russian Academy of Sciences; and the Russian Science Foundation (projects 16-17-10149 and 21-77-20004 that continue the previous project). Investigations within these projects have been continuing for many years.

We are extremely grateful to W. Zenk from GEOMAR for the fruitful discussions when he was a co-author of our previous book on abyssal currents. We have been consulting with him in our further research after the book was published.

The authors are grateful to numerous colleagues who participated in the field work and helped to perform observations in the ocean: H. van Haren, L. Maas, T.A. Demidova, N.I. Makarenko, S.V. Pisarev, I.S. Podymov, T. Gerkema, A.V. Remeslo, V.P. Shevchenko, S.G. Skolotnev, V.S. Zapotylko, Yu.A. Zyulyaeva, V.S. Ledenev, N.N. Dmitrevsky, and A.N. Demidov. We are also grateful to the captains of research vessels for their contributions to the success of our expeditions: G.A. Poskonny, L.V. Sazonov, V.V. Beluga, and Yu.N. Gorbach. We thank our colleagues who helped in the organization of expeditions: R.I. Nigmatulin, A.V. Sokov, S.S. Lappo, V.B. Lapshin, S.M. Shapovalov, L.I. Tolpygin, T.V. Ershova, V.P. Tereshchenkov. Comments by professors H. van Haren, M.N. Koshlyakov, N.I. Makarenko, and V.Yu. Liapidevsky were extremely helpful. Thanks go to colleagues who participated in data processing, analytical works, and preparation of the manuscript: D.R. Sakya and L.P. Filatova.

We also thank the crews of research vessels *Meteor, Polarstern, Akademik Ioffe, Akademik Sergey Vavilov, and Akademik Mstislav Keldysh* for their contribution to this research.

Abstract

This book is dedicated to the analysis of bottom water flows through underwater channels of the Atlantic Ocean. The study is based on recent observations of the authors, analysis of historical data, numerical modeling, and literature review.

Seawater is cooled at high latitudes, descends to the ocean bottom, and slowly flows to the tropical latitudes and further. This current is slow in the deep basins but intensifies in the abyssal channels connecting the basins. The current overflows submarine topographic structures and sometimes forms deep spillways when water descends over slopes by several hundred meters. The flow of Antarctic Bottom Water (AABW) is studied on the basis of CTD sections combined with LADCP profiling carried out annually at a few locations, and long-term moored measurements. The main channels that provide the flow of AABW are the Vema Channel (30°S), Romanche Fracture Zone (equator), and Vema Fracture Zone (11°N). Many other fractures and deep channels in the Atlantic are also the pathways for AABW. In addition to AABW, cold water is formed in the Arctic Ocean. It overflows the submarine ridge in the Denmark Strait and descends to the depths over the layer of AABW forming an opposing current. The mean speed of the flow in abyssal channels is 10–20 cm s^{-1}, but in some places the velocity is as high as 60 cm s^{-1}. The temperature of bottom water flow increases in the course of its propagation.

Antarctic Bottom Water in the Atlantic is formed in the Weddell Sea. After passing the channels in the South Scotia Ridge and other channels in the Scotia Sea, it circulates in the Argentine Basin and then flows to the Brazil Basin through the Vema Channel. The further flow of bottom water in the Brazil Basin splits in the northern part of the basin. Part of water flows to the East Atlantic basins through the Romanche and Chain fracture zones. The other part is a northwestern flow to the North American Basin. Part of the northwesterly flow propagates into the Northeastern Atlantic basins through the Vema Fracture Zone (11° N) and other fracture zones in the Northern Mid-Atlantic Ridge.

This book is a collection of oceanographic data, interpretation, and analysis, which can be used by field oceanographers, specialists in numerical modeling, and students who specialize in oceanography.

Contents

About the Authors

Eugene G. Morozov graduated in oceanography in 1970 from the Moscow Institute of Physics and Technology, and then received a Ph.D. in 1975 and a doctoral degree in 1990. He is a field oceanographer and a specialist in the observations in the open ocean. His career developed at the Shirshov Institute of Oceanology in Moscow where he heads a laboratory. His research focuses on mesoscale dynamics, internal waves, and circulation. His research is also related to the Arctic and Antarctic oceanography. He participated in more than 50 oceanic expeditions all over the globe and was the leading scientist in many of them. From 2011 to 2015, he was the president of the International Association for the Physical Sciences of the Ocean (IAPSO).

Roman Y. Tarakanov graduated in oceanography in 1996 from the Moscow Institute of Physics and Technology, and then received a Ph.D. in 2006. In 2015, he defended his doctoral dissertation. He is a field oceanographer and his career developed at the Shirshov Institute of Oceanology in Moscow. He participated in more than 20 oceanic expeditions. His research focuses on water masses, ocean fronts, currents, and abyssal ocean circulation. He developed some methods for analyzing satellite altimetry data in relation to the identification of jet currents, closed circulations, and estimates of the barotropic component of transport through transoceanic hydrophysical sections. He is a professor at the Moscow Institute of Physics and Technology, where he lectures on physical oceanography. Since 2015, he is a professor at the Russian Academy of Sciences.

Dmitry I. Frey graduated in applied physics and mathematics in 2014 from the Moscow Institute of Physics and Technology, and then received a Ph.D. in 2019. Since 2015, he has been working at the Laboratory of Hydrological Processes of the Shirshov Institute of Oceanology as a senior scientist. His research focuses on different aspects of ocean circulation including bottom gravity flows, western boundary currents, Arctic and Antarctic oceanography, and equatorial undercurrents. He is a field oceanographer and specialist in numerical modeling.

Abbreviations

AABW	Antarctic Bottom Water. Note: AABW = WSBW + WSDW in classification of (Orsi et al. 1999); AABW = WSDW + LCPW in classification of (Wüst 1936) and (Reid et al. 1977)
AAIW	Antarctic Intermediate Water
AASW	abbreviation used for both Antarctic Surface Water and Antarctic Shelf Water (AASurW and AAShW, respectively, in this book)
ACC	Antarctic Circumpolar Current
ACCbw	ACC Bottom Water
ADCP	Acoustic Doppler Current Profiler
ADT	absolute dynamic topography
BaA	Argentine Basin
BaG	Georgia Basin
BaO	Ona Basin
BaPh	Phoenix Basin
BaPo	Powell Basin
BaPr	Protector Basin
BaY	Yagan Basin
BBr	Bruce Bank
BBu	Burdwood Bank
BF	Brazil Front
BFZ	Bight Fracture Zone
BME	Maurice Ewing Bank
BP	Pirie Bank
BS	Shackleton Bank
BSX	Barents Sea Exit
BSO	Barents Sea Opening
CBW	Circumpolar Bottom Water (term used in this book). Original term introduced in (Orsi et al. 1999) is ACCbw (ACC Bottom Water)
CFC	Chlorofluorocarbon

CGFZ	Charlie Gibbs Fracture Zone
CORE	Coordinated Ocean-ice Reference Experiments (International Program)
CTD-profiler	Conductivity-Temperature-Depth profiler
DF	Deep Front (in the northern part of the Argentine Basin near the Vema Channel)
DSOW	Denmark Strait Overflow Water
DWBC	Deep Western Boundary Current
FSCBW	Faroe Shetland Channel Bottom Water
FZ	Fracture Zone
FZE	Endurance Fracture Zone
FZH	Hero Fracture Zone
FZQ	Quest Fracture Zone
FZS	Shackleton Fracture Zone
FZT	Tehuelche Fracture Zone
GEBCO	General Bathymetric Chart of the Oceans
GO	Ona Gap
GWS	West Scotia Gap
HFU	Heat Flux Unit; 1 HFU = 42 mW/m^2
IDW	Indian Ocean Deep Water
INMOM	Institute of Numerical Mathematics Model
ISOW	Iceland Scotland Overflow Water
LADCP	Lowered Acoustic Doppler Current Profiler
LCDW	Lower Circumpolar Deep Water (in the ACC zone)
LCPW	Lower Circumpolar Water (in the Argentine Basin (Reid et al. 1977)) (Note: LCPW is not the same as LCDW)
lNADW	Lower North Atlantic Deep Water
LOIW	Lower Intermediate Water
LSW	Labrador Sea Water
LWSDW	Lower Weddell Sea Deep Water
MAR	Mid-Atlantic Ridge
MEIW	Modified East Icelandic Water
mNADW	Middle North Atlantic Deep Water
NADW	North Atlantic Deep Water
NAO	North Atlantic Oscillation
NBSO	Northern Barents Sea Opening
NSAIW	Norwegian Sea Arctic Intermediate Water
NSDW	Norwegian Sea Deep Water
NWO	Netherland Research Council (Nederlandse Orgsanisatir voor Wetenschappelijk)
PB	Bruce Passage
PD	Discovery Passage
PDW	Pacific Deep Water (sometimes SPDSW designates South Pacific Deep Slope Water)

PF	Polar Front
PFl	Passage Falkland
PG	Georgia Passage
PH	Hoyer Passage
PHe	Hero Passage
PNEG	North East Georgia Passage
PO	Orkney Passage
PP	Philip Passage
PS	Shackleton Passage
PSR	Shag Rocks Passage
PSU	Practical Salinity Unit
PVD	Progressive Vector Diagrams
RBr	Bruce Ridge
RF	Falkland Ridge
RFBR	Russian Foundation for Basic Research
RNEG	North East Georgia Ridge
RNWG	North West Georgial Ridge
RPh	Phoenix Rift
RSF	Russian Science Foundation
RV	Research Vessel
SAC	Subantacrtic Current
SACCB	South Boundary of ACC
SACCF	South ACC Front
SAF	Subantarctic Front
SAMW	Subantarctic Mode Water
SB	South Boundary of ACC in some figures
SEPDW	South East Pacific Deep Water. Its regional modification CBW in the Drake Passage, introduced by Sievers and Nowlin (1984)
SF	South ACC Front in some figures
SlAC	Slope Antarctic Current
SPC	South Polar Current
SPDW	Southeast Pacific Deep Water
SPF	South Polar Front
STD-profiler	Salinity-Temperature-Depth profiler
STF	Subtropical Front
SthAC	Southern Jet of the Antarctic Circumpolar Current (Southern ACC Front)
Sv	Sverdrup (1 Sv = 10^6 m^3 s^{-1})
TSO	South Orkney Trough
TSSa	South Sandwich Trench
TSSh	South Shetland Trench
UCDW	Upper Circumpolar Deep Water in the Antarctic Circumpolar Current zone
UCPW	Upper Circumpolar Water in the Argentine Basin (Reid et al. 1977)

uNADW	Upper North Atlantic Deep Water
UPIW	Upper Intermediate Water
USlTFC	Upper Slope Tierra del Duego Current
UWSDW	Upper Weddell Sea Deep Water
WDW	Warm Deep Water (in the Weddell Gyre) (Note: WDW = LCDW + CBW if we use the approach in (Orsi et al. 1999))
WOCE	World Ocean Circulation Experiment in the 1990s (Siedler et al. 2001)
WOD	World Ocean Database
WSBW	Weddell Sea Bottom Water
WSDW	Weddell Sea Deep Water
WF	Weddell Front

Chapter 1
Deep Water Masses of the South and North Atlantic

Abstract This chapter summarizes our knowledge about the structure and propagation of waters in the deep and bottom layers of the Atlantic Ocean, description of the properties of the deep, and bottom waters and conditions of their formation. Despite the fact that almost the whole book is related to the spreading of Antarctic Bottom Water, the dynamics of this water is correlated with the other water masses of the Atlantic. We also describe the properties of several water masses in the deep and bottom layers of the Atlantic. We analyze the mechanisms of the formation of these waters and their circulation. We compare different definitions of Antarctic Bottom Water (as well as some other water masses) and criteria to identify their boundaries.

Keywords Deep water masses · Antarctic bottom water · Mechanisms of formation · Circumpolar water · North Atlantic deep water

1.1 General Description

The Southern Ocean and the Antarctic Circumpolar Current isolate the Antarctic continent from the other regions of the Earth. Thus, Antarctic and Sub-Antarctic conditions provide the formation of a special water structure around Antarctica, whereas water structure in the northern regions is determined by the interactions between waters of the North Atlantic and those of the Arctic origin.

Deep and bottom waters are formed as a result of thermohaline interaction in both polar and subpolar regions of the oceans. Deep and bottom waters are formed in several regions around Antarctica in each of the oceans. In the Northern Hemisphere, they are formed only in the Atlantic Ocean. Winter cooling in the North Atlantic basins (North European Basin, Norwegian and Greenland seas, and Labrador Sea) results in the descent of water from the ocean surface to the deep layers. The density of waters of the North Atlantic origin is not high enough to occupy the bottom layer over the major part of the World Ocean (Orsi et al. 1999) because density of waters of the Antarctic origin is greater. Conditions of water formation in the Arctic Ocean are similar to those in the Antarctic region, but the dense waters that fill the bottom

E. G. Morozov et al., *Bottom Gravity Currents and Overflows in Deep Channels of the Atlantic Ocean*, https://doi.org/10.1007/978-3-030-83074-8_1

layer of the Arctic Ocean cannot overflow the two thresholds between Greenland and Iceland and between Iceland and the Faroe-Shetland Islands without decreasing the density due to mixing with the surrounding Atlantic waters because the depths of these sills are shallower than 650 and 850 m, respectively. As a consequence, all waters in the bottom layer of the Atlantic have a component of the Antarctic origin.

1.2 Global Overturning Circulation

The deep and bottom waters of the oceans are formed mainly at high latitudes in the Southern Ocean and North Atlantic. The descent of the deep and bottom waters of the oceans at high latitudes of the Southern Ocean and North Atlantic is only a part of the global overturning circulation in the World Ocean. From the modern point of view, this circulation is a wind-thermohaline one: simultaneous forcing of wind and thermohaline factors. The existing configuration of the continents in the current geological era is fundamentally important for the existence of this circulation. One of the recent schemes of this circulation was published in (Koshlyakov and Tarakanov 2011). A modified scheme of circulation is shown in Fig. 1.1. It is based on the analysis of the authors and a summary of a large number of publications (Schmitz and McCartney 1993; Dickson and Brown 1994; Schmitz 1996b; Neiman et al. 1997; Döös and Coward 1997; Orsi et al. 1999; Stramma and England 1999; Lebedev and Yaremchuk 2000; Stephens and Marshall 2000; Anisimov et al. 2002; Stammer et al. 2002; Koshlyakov and Tarakanov 2003b; Drijfhout et al. 2003; Koshlyakov and Tarakanov 2004; Sen Gupta and England 2004; Koshlyakov and Tarakanov 2005; Davis 2005; Friocourt et al. 2005; Hallberg and Gnanadesikan 2006; Lumpkin and Speer 2007; Sen Gupta and England 2007; Iudicone et al. 2008; Richardson 2008; Morozov et al. 2010; Naveira Garabato et al. 2014; Badin et al. 2013).

Forcing by strong westerly winds on the ocean surface over the Antarctic Circumpolar Current (ACC) free of continental barriers generates a pure drift current in a thin surface layer up to several tens of meters thick, the total flux of which is directed across the ACC over the entire circumpolar ring of currents around Antarctica. The eastward Antarctic Circumpolar Current exists as a result of the adjustment of the density field to the sea surface height variation caused by pure drift currents in the upper layer of the ocean. Then, the ACC penetrates to the deep layers of the ocean. Thus, the ACC prevents the development of meridional currents in the upper 1500–2000 m layer of the ocean, which could compensate for the equatorward flow of the pure drift currents. As a result, such compensation occurs due to the meridional transfrontal eddy transport across the ACC and due to the geostrophic bottom currents below the ACC penetration depth. The reality of the existence of such a balance based on the field data was first shown in (Koshlyakov and Tarakanov 2011) along the line of the Subantarctic Front, which is the northernmost jet of the ACC. In this case, the southward transport compensating for the pure drift currents is an isopycnal ascent of water from the mid-oceanic depths north of the ACC almost to

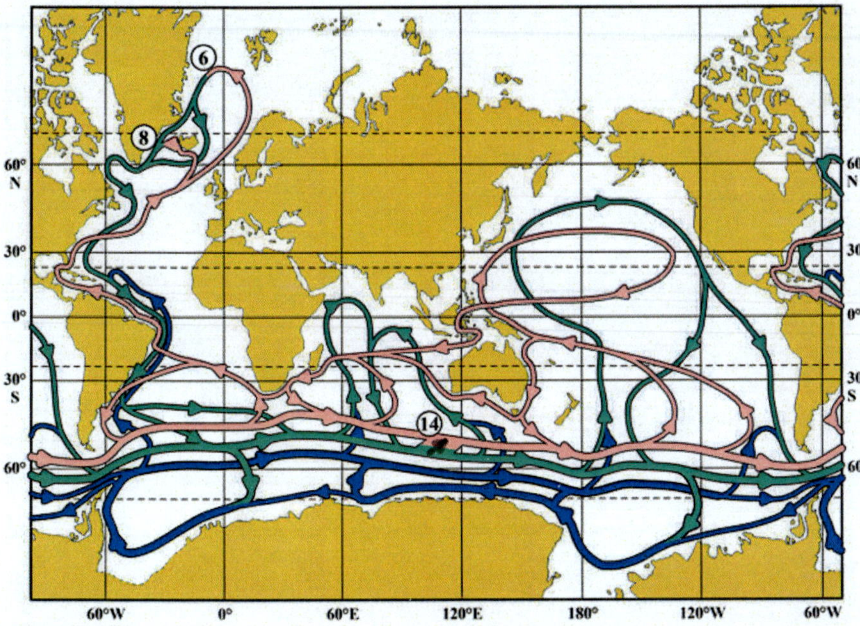

Fig. 1.1 Global ocean conveyor belt that includes water ascent from the mid-depths in the Southern Ocean and descent in the North Atlantic and near the Antarctic slope. Pink arrows show water propagation in the upper layer; green arrows show water propagation in the isopycnal range corresponding to the water descending in the North Atlantic, and blue arrows show propagation of Antarctic waters in the bottom layer of the World Ocean. The water transport from the deep layer to the thermocline, shown by one arrow, is actually distributed along the entire perimeter of the Antarctic ring. The numbers in circles indicate the intensity of water sinking into the depths in the Norwegian and Greenland Seas (6 Sv) and Labrador Sea (8 Sv), and the compensating ascent from the depths into the thermocline layer in the Southern Ocean (14 Sv) based on (Koshlyakov and Tarakanov 2011) and many other publications

the surface on the southern periphery of the current and south of it. This ascent from the mid-depths is compensated by the descent of the North Atlantic waters from the surface to the mid-oceanic depths as well as the descent of additionally cooled and denser waters along the continental slope in several regions around Antarctica. A decrease in the density of water ascending from the depth in the ACC zone occurs in the northward pure drift currents due to heating, prevalence of precipitation over evaporation, and prevalence of ice melting over ice formation. Then, these waters flow to the North Atlantic as part of the circulation in the upper 1500–2000 m of the ocean at low latitudes. The waters that descended in the Antarctic spread to the north at the bottom and gradually mix with the overlying deep waters.

In the Atlantic, this mixing closes the Antarctic cell of the global overturning circulation. In the Pacific and Indian oceans, the bottom waters from the Southern Ocean mix with the local intermediate waters and thereby form mid-deep waters, which actually return to the Southern Ocean where they transfer to the pure drift current to circulate in the upper layer of the low latitude oceans. Note that the

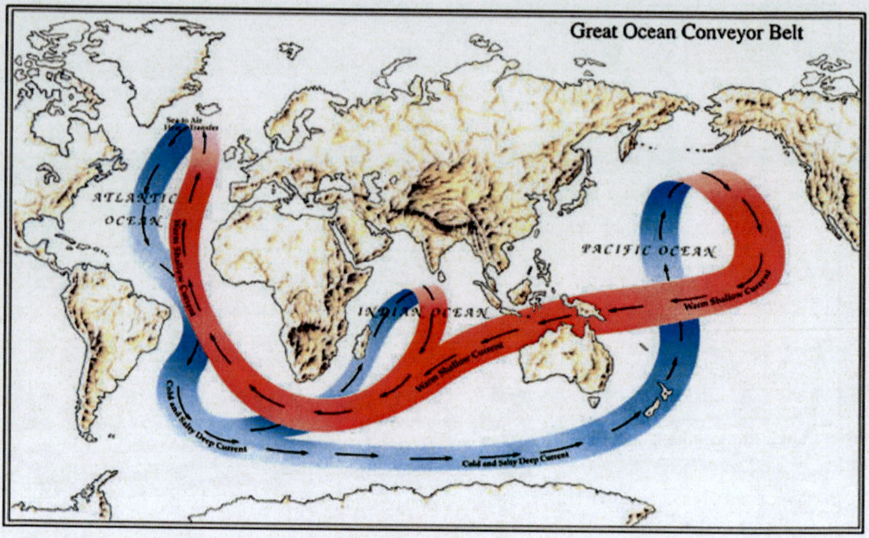

Fig. 1.2 The great ocean conveyor belt logo shown by Broecker (1987) and Richardson (2008) illustrated by J. Le Monnier, Natural History Magazine

described circulation differs significantly from the well-known, but becoming out of date Global Ocean Conveyor Belt scheme (Broecker 1991) (Fig. 1.2), in which the global overturning circulation has a purely thermohaline nature and does not take into account the wind forcing factor. The idea of this scheme of inter-oceanic exchange has been considered in (Broecker and Peng 1982; Lappo 1984; Gordon 1986; Broecker 1987). We emphasize that in the scheme suggested in (Koshlyakov and Tarakanov 2011) compensation for lowering of waters to the depths in the North Atlantic does not occur in the Pacific and Indian oceans as in the scheme suggested by Broecker (1991), but occurs in the Southern Ocean. The idea of such compensation was suggested earlier (Toggweiler and Samuels 1995; Gnanadesikan 1999). Based on this idea, Marshall and Speer (2012) published a more simplified scheme of the Global Overturning Circulation than the one shown in Fig. 1.1. Later, numerous publications on the Global Overturning Circulation appeared (e.g., Wolfe and Cessi 2011; Nikurashin and Vallis 2011, 2012; Shakespeare and Hogg 2012; Talley 2013; Jansen and Nadeau 2016; Thompson et al. 2016; Ferrari et al. 2017). Detailed reviews of these studies can be found in (Vallis 2017; Cessi 2019; Baker et al. 2020). We also note review works on the upper cell of the Global Overturning Circulation in the Atlantic associated with water descent in the Atlantic (Buckley and Marshall 2016; Frajka-Williams et al. 2019; Johnson et al. 2020; Srokosz et al. 2021).

1.3 Mechanisms of the Formation of the Deep and Bottom Waters

Two main mechanisms of the formation of the deep and bottom waters in the ocean exist: deep winter convection of the open ocean and convection on the shelf and continental slope. In the first case, downwelling occurs directly from the ocean surface by means of strong cooling and salinity increase caused by ice formation and associated increased density (Coachman and Aagaard 1974; Schmitz and McCartney 1993; Koltermann et al. 1999; Dobrolyubov and Falina 2002; Dye et al. 2007; Sarafanov et al. 2009). In the second case, the water density on the shelf increases (the water is almost cooled to the freezing point) and water salinity increases. At the same time, in Antarctica, salinization occurs not only at the surface, but also under shelf glaciers (Patterson and Whitworth 1990). These very cold and dense shelf waters then descend along the mainland slope mixing with the less dense surrounding (both deep and surface!) waters. The resulting waters reach the depth of their density in the open ocean, the densest waters reach the ocean bed (Foster and Carmack 1976; Baines and Condie 1998). Note that the process described is not exactly descending of water from the surface into the interior of the ocean. In this case, density increase of deep and surface waters by mixing with the shelf waters is essential (Foster and Carmack 1976; Orsi et al. 2002). Only surface and shelf waters are actually in contact with the atmosphere.

In the North Atlantic and Nordic seas, the lowering of water from the surface layer to the depths of the ocean occurs mainly as a result of deep winter convection in the open ocean. This process is localized in the Greenland Sea and in the Labrador and Irminger seas (Fig. 1.1). Estimates of transport of this downwelling based on numerical modeling of ocean circulation (Marsh et al. 2000; Blanke et al. 2002; Iudicone et al. 2008; Mazloff et al. 2010) using inverse (box) models of the ocean (Ganachaud and Wunsch 2000; Sloyan and Rintoul 2001b; Lumpkin and Speer 2007), reanalyses (Cessi 2019), and mainly on the analysis of observations (Schmitz 1996b; Smethie and Fine 2001; Talley et al. 2003; Dye et al. 2007; Talley 2008; Koshlyakov and Tarakanov 2011; Talley 2013; McCarthy et al. 2015; Meinen et al. 2018) yielded the values of this transport fairly close to each other in the range of 14–18 Sv. In the Arctic, slope convection with the entrainment of extremely cold shelf waters accounts for a comparatively lower value, not exceeding 0.1 Sv (Smethie et al. 1988). The coldest and densest waters formed in the Greenland Sea due to the topographic isolation of the Nordic seas from the Atlantic Ocean reach the ocean bed almost exclusively in the Arctic Ocean. Traces of the flow of these waters south over the thresholds in the Iceland region separating the Arctic Ocean from the Atlantic Ocean, as far as one can judge from the WOD18 database (WOD18 2018; Boyer et al. 2018) of CTD measurements can be traced at the bottom in the Atlantic only in a limited region of the ocean adjacent to the thresholds. In the major part of the ocean, Antarctic Bottom Water occupies the bottom layer.

In the Southern Ocean, downwelling into the depths of the ocean is mainly confined to the Antarctic slope. According to modern estimates, the proportion of shelf Antarctic waters in this downwelling (which leads to the formation of bottom and renewal of deep waters) is not greater than 30%, while the share of surface waters is about 25%, and the share of deep waters is about 45% (Orsi et al. 2002) (similar values were previously estimated in the individual regions of the Antarctic, for example, in the Weddell Sea (Foster and Carmack 1976). This process occurs only in a few regions around the perimeter of Antarctica: in the Weddell Sea (in the western and southern parts of the sea) in the Atlantic, in the western and eastern parts of the Ross Sea in the Pacific Ocean, off the Adélie Land at the boundary between the Indian and Pacific oceans, and in the Commonwealth Sea in the Indian Ocean (Whitworth et al. 1998). It is noteworthy that this process is non-uniform in time due to the seasonal nature of the formation of shelf waters. The occurrence of downwelling (in the form of plumes and individual portions) is sporadic along the Antarctic slope (Baines and Condie 1998).

The transport estimates of downwelling in the Antarctic vary much wider and may differ several times depending on the region. In particular, the transport estimates based on the balance of tracers are 15–20 Sv (CFCs, radiocarbon, quasi-conservative phosphate complex, etc.) in the Antarctic (Broecker et al. 1998; Peacock et al. 2000; Orsi et al. 2002). Recent publications report about the rate of formation of the Antarctic waters in the bottom layer based on the estimates of diapycnal mixing (de Lavergne et al. 2016a, b; Kunze 2017). The maximum transports are estimated at 10–15 Sv (de Lavergne et al. 2016b) and 20 Sv (Kunze 2017). In most studies, the intensity of downwelling is determined, ultimately, by the transport of Antarctic waters in the bottom layer approximately at 30° S. (Schmitz 1996b; Ganachaud and Wunsch 2000; Sloyan and Rintoul 2001b; Talley et al. 2003; Lumpkin and Speer 2007; Iudicone et al. 2008; Mazloff et al. 2010; Katsumata et al. 2013; Talley 2013; Cessi 2019). Among these works, there is a group of publications that report about the transport of Antarctic waters at 8–12 Sv. (Schmitz 1996b; Iudicone et al. 2008; Mazloff et al. 2010; Katsumata et al. 2013); the other authors (Ganachaud and Wunsch 2000; Sloyan and Rintoul 2001a; Talley et al. 2003; Lumpkin and Speer 2007; Talley 2008; Talley 2013; Cessi 2019) estimate the transport at 20–29 Sv.

Note that this method of determining the transport does not take into account the diapycnal mixing of the Antarctic waters with the overlying deep waters south of 30° S, which, apparently, seems to be significant for the Southern Ocean (Badin et al. 2013). In addition to the downwelling of waters at 32° S, Lumpkin and Speer (2007) report about the intensity of downwelling south of 62° S, which is approximately 13 Sv.

In the Antarctic, deep winter convection can also occur sporadically in the open ocean. In the early 1970s, such convection in the polynyas of mesoscale eddies was recorded in the vicinity of the Maud Rise in the region of the Prime meridian (Gordon 1978). Satellite observations of the ice cover show that the necessary conditions for the occurrence of this mechanism exist only during 11% of the time of such observations, which can ensure the formation of about 2.4 Sv of deep waters (Gordon 1978). This estimate is almost one order of magnitude smaller than the estimates of the intensity of formation and renewal of deep waters on the Antarctic slope. However, it was shown in (Whitworth et al. 1998) that this mechanism should be considered as potentially possible over most of the perimeter of the Antarctic slope. Among the possible mechanisms of bottom water formation, one should consider cabbeling (density increase during mixing) (Fofonoff 1956) and internal wave breaking at the shelf edge (Alekseev et al. 1978). These issues are described in more detail in monographs (Sarukhanyan and Smirnov 1986; Morozov et al. 2010).

1.4 Classifications of Deep and Bottom Waters in the Atlantic

A quantitative study of the individual elements of the global overturning circulation requires a detailed and at the same time unified classification of water masses in the World Ocean. Taking into account the specific problems considered here, it is important to classify deep water masses. However, such a single generally accepted classification does not exist. Two different classifications of water masses are known in the Atlantic Ocean. Table 1.1 provides information about these classifications. The sense of the table is to show the application of these classifications on a geographical basis; however, it does not cover all existing publications. Figure 1.3 shows a diagram of the deep-water masses in the Atlantic for convenience of presentation and approximate locations of the extreme values of hydrophysical and hydrochemical properties characterizing these water masses.

In the first of the classifications suggested by Wüst (1936), only two deep water masses were distinguished in the Atlantic: relatively saline North Atlantic Deep Water (NADW) spreads generally in the southern direction. In the underlying bottom layer, colder and less saline Antarctic Bottom Water (AABW) propagates to the north. This classification is very often applied to the greater part of the Atlantic Ocean located north of the Antarctic Circumpolar Current (Table 1.1). In the classical definition (Wüst 1936), North Atlantic Deep Water and Antarctic Bottom Water are delimited by the potential temperature $\theta = 2.0$ °C isotherm. However, in some regional studies, other potential temperature values from 1.7 °C to 2.3 °C are also used as the boundary value, while in general, the selected temperatures increase towards the North Atlantic (Table 1.2). The values of potential density $\sigma_4 = 45.90$, neutral density $\gamma^n = 28.11$, and other criteria are also used (see Table 1.2). Neutral density γ^n is determined from the condition that liquid particles, when moving along its iso-surface, do not work against the buoyancy force (Jacket and McDougall 1997).

The second classification is closely related to the existence of the Antarctic Circumpolar Current (ACC) around the Antarctic continent. Circumpolar Deep Water (CDW) is the main water mass in the ACC (Sverdrup et al. 1942). The entire volume of this water together with Antarctic Surface Water (AASW), Antarctic Intermediate Water (AAIW), and Subantarctic Mode Water (SAMW) located above it forms the entire eastward circumpolar transport around Antarctica. Since the ACC is a geostrophic current, the entire thickness of these waters creates an obstacle to the

Table 1.1 Terminology of water masses used in different publications

	Name of near bottom water mass	Comments
Classification by (Wüst 1936), θ < 2 °C		
Wüst (1936)	AABW	
Sverdrup et al. (1942)		
Hogg et al. (1982)		Vema Channel
Harvey and Arhan (1988)		Hydrography of the Northern MAR in the North Atlantic
Broecker (1991)		Global Ocean Conveyor
Speer and McCartney (1992)		Southwest of the North Atlantic
Friedrichs and Hall (1993)		Section along 11° N in the North Atlantic
McCartney and Curry (1993)		Equatorial Channel
Rhein et al. (1996)		Equatorial Channel
Hall et al. (1997)		Equatorial Channel
Stramma and Rhein (2001)		Equatorial Channel
Schott et al. (2003)		Equatorial Channel
Lavin et al. (2003)		Section along 24.5° N in the North Atlantic
Lherminier et al. (2007)		West European Basin
Hybrid Wüst's Classification (Reid et al. 1977)		
WSDW (< 0.2 °C), LCPW (< 2 °C)		
Reid et al. (1977)	LCPW + WSDW (without AABW term)	Argentine Basin
Mantyla and Reid (1983)	LCPW + WSDW	Abyss of the World Ocean
Reid (1989)	Circumpolar layers, upper layers of Weddell Sea waters	Whole Atlantic
Speer and Zenk (1993)	AABW = LCDW + WSDW	Rio Grande Plateau
Tsuchiya et al. (1994)	LCPW + WSDW (without AABW term)	Quasi-Meridional Section A16
Zenk and Hogg (1996)	AABW = LCDW + WSDW	Vema Channel
Larque et al. (1997)	LCDW + WSDW (without AABW term)	Atlantic
Rhein et al. (1998)	AABW = LCDW + WSDW	Equatorial Channel, Vema Fracture Zone
Arhan et al. (1998)	AABW = LCPW + WSDW	Zonal sections between 4° S and 7° N
Arhan et al. (1999)	SPDW + LCDW + WSDW (without AABW term)	Argentine Basin
Stramma and England (1999)	AABW = LCDW + WSDW	Atlantic
Messias et al. (1999)	AABW = LCDW + WSDW	Romanche and Chain fracture zones
Wienders et al. (2000)	AABW = LCPW + WSDW	Quasi-Meridional Section A17
Memery et al. (2000)	AABW = LCPW + WSDW	Quasi-Meridional Section A17

(continued)

Table 1.1 (continued)

	Name of near bottom water mass	Comments
Sandoval and Weatherly (2001)	AABW = LCPW + WSDW	Brazil Basin
Vanicek and Siedler (2002)	AABW = LCDW + WSDW	South Atlantic
McDonagh et al. (2002)	AABW = LCPW + WSDW	Vema Channel
Zenk and Morozov (2007)	AABW = CDW + WSDW	Vema Channel
Special Opinion		
Fu (1981)	LCPW only	Subtropics of the Atlantic (28° S–28° N), Box-model
Roemmich (1983)	LCPW only	Zonal sections between 24° S и 24° N
Onken (1995)	LCDW consisting from WSDW and LCDW passed through the Drake Passage	Model of the LCDW inflow
Johnson (2008)	WSBW, Bottom Water of Adelie Land, Bottom Water of the Ross Sea, Weddell Sea Shelf Water as modifications of AABW	Atlantic, Optimum Multiparameter Analyses
Southern Ocean Classification:		
AABW ($\gamma^n > 28.27$), CBW ($28.27 > \gamma^n > 28.18$), LCDW ($28.18 > \gamma^n > 28.00$)		
Mantyla and Reid (1983), Patterson and Whitworth (1990)	LCPW(=LCDW+CBW) = NADW, iso- and diapycnal mixed with AABW	World Ocean
Orsi et al. (1999)	LCDW + ACCbw (=CBW) + WSDW	Southern Ocean
	AABW = WSDW + WSBW	
Hoppema et al. (2001)	WDW (=LCDW) + WSDW + WSBW	Weddell Sea
Naveira Garabato et al. (2002b)	AABW = WSDW + WSBW	Scotia Sea
Klatt et al. (2002)	WDW (=LCDW) + WSDW + WSBW	Weddell Sea
Klatt et al. (2005)	WDW (=LCDW) + WSDW + WSBW	Weddell Sea
Barre et al. (2008)	AABW = WSDW + WSBW	Drake Passage
Southern Ocean Classification applicated to the South Atlantic		
Coles et al. (1996)	LCDW (=CBW) + AABW (=WSDW)	Argentine Basin
Smythe-Wright and Boswell (1998)	LCDW (=CBW) + AABW (=WSDW)	Argentine Basin
van Aken (2007)	LCDW (=CBW) + AABW (=WSDW)	World Ocean

meridional water exchange by geostrophic currents between low and high latitudes of the Southern Hemisphere. The possibility of such advective water exchange appears only in the layer of denser waters below Circumpolar Deep Water, which in the framework of this classification is called Antarctic Bottom Water. Given that the main threshold in the series of basins surrounding Antarctica is in the region of the Drake Passage and the Scotia Sea, Antarctic Bottom Water includes all waters that cannot overflow this threshold with the flow from west to east (Orsi et al. 1999). Antarctic Bottom Water and Circumpolar Deep Water are separated by

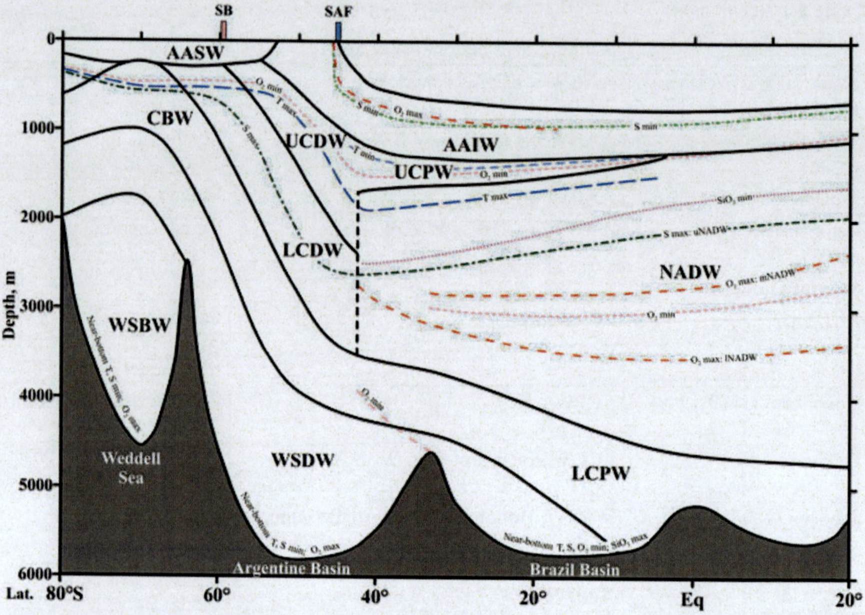

Fig. 1.3 Layout of water masses in the Atlantic based on many publications; positions of the extrema of hydrophysical and hydrochemical properties characterizing individual water masses are shown. AASW: Antarctic Surface Water, AAIW: Antarctic Intermediate Water, UCDW: Upper Circumpolar Deep Water, UCPW: Upper Circumpolar Water, LCDW: Lower Circumpolar Deep Water, LCPW: Lower Circumpolar Water, NADW: North Atlantic Deep Water (uNADW, mNADW, lNADW are Upper, Middle, and Lower NADW, respectively), WSDW: Weddell Sea Deep Water, WSBW: Weddell Sea Bottom Water, CBW: Circumpolar Bottom Water; LCPW +WSDW≈AABW in the Wüst's (1936) classification, WSDW+WSBW ≈AABW in the Southern Ocean classification; SAF: Subantarctic Front, SB: Southern Boundary of the Antarctic Circumpolar Current

approximately the $\theta = 0.2$ °C isotherm (Sievers and Nowlin 1984). In the units of neutral density, the existing estimates for this boundary yield $\gamma^n = 28.26$ (Naveira Garabato et al. 2002b; Koshlyakov and Tarakanov 2003a), which approximately corresponds to the previously mentioned $\theta = 0.2$ °C isotherm in the Drake Passage. However, it is noteworthy that, as the classification for the South Ocean develops, the boundary between Antarctic Bottom Water and Circumpolar Deep Water associates with the other values of temperature and density (see the review in Sect. 3.1 of this book, Weddell Gyre). In particular, isopycnal $\gamma^n = 28.27$ in (Orsi et al. 1999) was associated with this boundary, while isotherm $\theta = 0.0$ °C was used in (Gordon 1971; Gordon 1972). The latter version is most often found in literature sources.

Within the described South Ocean classification of water masses, two layers are distinguished in the Circumpolar Deep Water, which are characterized by extreme values of hydrophysical or hydrochemical parameters (Gordon 1967). The minimum concentration of dissolved oxygen and the maximum concentration of nutrients are the characteristics of the upper layer of Circumpolar Deep Water called Upper

Table 1.2 Variants of interpretation of the upper boundary of AABW in the Wüst's approach

Authors	Upper AABW boundary	Region
Macdonald 1998	$\theta = 1.7$ °C	Brazil Basin (27° S)
Hall et al. 1997 Friedrichs and Hall 1993 Macdonald 1998	$\theta = 1.8$ °C	Equatorial Channel Guinea Basin in the West Atlantic (11°N) Guinea Basin in the West Atlantic (11°N)
Weatherly and Kelley 1982	$\theta = 1.82$ °C	North American Basin (40° N)
Tucholke et al. 1973	$\theta = 1.84$ °C	North American Basin
Mercier and Morin 1997 Mercier and Speer 1998 Whitehead and Worthington 1982 McCartney 1992 McCartney and Curry 1993 Macdonald 1998	$\theta = 1.9$ °C	Romanche Fracture Zone (Equator) Romanche and Chain fracture zones (Equator) Guinea Basin in the West Atlantic Guinea Basin in the West Atlantic Guinea Basin in the West Atlantic (4° N), Equatorial Channel North American Basin (24° N, 36° N)
Wright 1970 McCartney et al. 1991 Speer and Zenk 1993 Fischer et al. 1996 Lenz 1997 Macdonald 1998 Hogg et al. 1999 Holfort and Siedler 2001	$\theta = 2.0$ °C	Guinea Basin in the West Atlantic (8° N) Vema Fracture Zone (11° N) Vema and Hunter channels, lower Santos Plateau Vema Fracture Zone (11° N) Vema Channel North American Basin (48° N) Vema Channel Vema Channel
Saunders 1987	$\theta = 2.05$ °C	East Azores Ridge (37° N)
Schlitzer 1987	$\theta = 2.07$ °C	East Atlantic
Vanicek and Siedler 2002	$\gamma^n = 28.12$ ($\theta \approx 1.2°–1.3$ °C)	Southwest Atlantic, Box-model
Ganachaud 2003	$\gamma^n = 28.11$ (from 1.3 °C in the South Atlantic to 2.3 °C in the North Atlantic)	Global box-model
Saunders and King 1995	$\sigma_4 = 45.95$ ($\theta \approx 0.8°–1.0$ °C)	Argentine Basin (45° S)
Fu 1981	$\sigma_4 = 45.92$ ($\theta \approx 1.5$ °C)	8°–24° S, Box-model
Fischer et al. 1996 Rhein et al. 1998 Schott et al. 2003	$\sigma_4 = 45.9$ ($\theta \approx 1.7°–1.8$ °C)	Vema Fracture Zone (11° N) Equatorial Channel Brazil Basin (5° S)
Lherminier et al. 2007	$\sigma_4 = 45.85$ ($\theta \approx 2.18°–2.3$ °C)	Northeast Atlantic

Circumpolar Deep Water (UCDW). These are the properties of waters of the Indo- and Pacific origin and are associated with the great age of the deep waters of the Indian (Indian Deep Water) and Pacific (Pacific Deep Water) oceans. The inflow of these waters into the ACC zone is related to their predominantly isopycnal mixing with circumpolar waters at the northern periphery of the ACC. South of the Polar Front, the maximum temperature induced by the lower temperatures of the overlying Antarctic Surface Water and the underlying layers of circumpolar waters is also observed in the Upper Circumpolar Deep Water layer (Patterson and Whitworth 1990). Based on this property Kuksa (1978, 1983) identified this layer as Intermediate water of higher temperature. Lower Circumpolar Deep Water (LCDW) underlying Upper Circumpolar Deep Water is characterized by the salinity maximum. This characteristic of Lower Circumpolar Deep Water is associated with the predominantly isopycnal mixing of Circumpolar Deep Water with more saline North Atlantic Deep Water at the northern periphery of the ACC in the Atlantic Ocean (Gordon 1967; Stramma and England 1999). The boundary between Upper Circumpolar Deep Water and Lower Circumpolar Deep Water is associated with neutral density values in the range 27.95–28.04 (Table 1.3). Significant non-isopycnicity of this boundary in the transversal direction to the current streamlines in the ACC is shown in (Koshlyakov and Tarakanov 2003b, 2004). Separation of Circumpolar Deep Water into only two components (Upper Circumpolar Deep Water and Lower Circumpolar Deep Water) has been reported in the publications. With this approach, Lower Circumpolar Deep Water is the product of both iso- and diapycnal mixing of North Atlantic Deep Water and Antarctic Bottom Water.

It was emphasized in (Reid et al. 1977; Mantyla and Reid 1983) that not extremely cold and dense Antarctic waters, which form as a result of slope convection at the Antarctic shelf break, spread from the Southern Ocean to the north. The major part of the bottom current is formed from the lower layers of circumpolar waters, which are a product of diapycnal mixing of Lower Circumpolar Deep Water and the previously mentioned Antarctic waters caused by various mechanisms. To develop this idea in its application to the Atlantic Reid et al. (1977) suggested the terminology based on the fact that North Atlantic Deep Water wedges into the Circumpolar Deep Water layer in the Southwest Atlantic (in the Argentine Basin) and divides it into two parts. These two parts were called Upper Circumpolar Water (UCPW) and Lower Circumpolar Water (LCPW). Another terminology is also used

Table 1.3 Boundary between Lower Circumpolar Deep Water and Upper Circumpolar Deep Water

Authors	Boundary
Arhan et al. 1999	$\sigma_4 = 45.79$ ($\gamma^n \approx 28.02$)
Sloyan and Rintoul 2001a, b	$\gamma^n = 28.00$
Orsi et al. 2002	$\gamma^n = 27.98$
Ganachaud 2003	$\gamma^n = 27.903$
Naveira Garabato et al. 2003	$\gamma^n = 28.00$
Koshlyakov and Tarakanov 2003b, 2004	$\gamma^n = 27.95$–28.04

in other sectors of the Southern Ocean in relation to the lowest layer of circumpolar waters (roughly corresponding to the Lower Circumpolar Water in the Atlantic): Southeast Pacific Deep Water (Sievers and Nowlin 1984), Southwest Pacific Bottom Water (Downes et al. 2012), Lower Deep Water (Arhan et al. 2003), and others. To complete the development of the South Ocean classification of deep waters, Orsi et al. (1999) combined all these regional modifications into a single circumpolar water mass named ACC Bottom Water (ACCbw), which is not renewed by isopycnal mixing with North Atlantic Deep Water. Term Circumpolar Bottom Water for this layer was suggested in the publications by (Koshlyakov and Tarakanov 2003a, b) devoted to the study of the structure of water masses in the Pacific Antarctic. We shall keep to this term within the Southern Ocean classification. The upper boundary of this water (including its regional modifications) in the Southern Ocean is estimated by different authors in a rather wide range of values of neutral density from 28.20 to 28.11 and potential temperature (see Table 1.4). The water layer equivalent in its density range to the Circumpolar Bottom Water layer occupies the largest volume; it is distributed over the largest area among all other water masses of the World Ocean (van Aken 2007).

Circumpolar Bottom Water does not have any extremum of hydrophysical (temperature, salinity, density) or hydrochemical parameters almost in the entire circumpolar circle. The only exception is the Drake Passage, in which the water of this isopycnal range is characterized by a maximum of silicates at a distance from the bottom (Sievers and Nowlin 1984). Initially, it was believed that Circumpolar Bottom Water is formed mainly due to diapycnal mixing of overlying and underlying waters. However, relatively recently it was shown in (Orsi et al. 2002; Orsi 2010) that slope convection on the Antarctic shelf plays a significant role in the formation of Circumpolar Bottom Water. According to (Orsi et al. 2002), the isopycnal range of this water and partially the range of Lower Circumpolar Deep Water account for about half (10 Sv!) of all water that is a product of densification in the Southern Ocean; the other half is Antarctic Bottom Water.

It was indicated in the previously mentioned work (Reid et al. 1977) that the Southern Ocean waters spreading in the bottom layer of the Atlantic to the north are formed mainly in the Weddell Sea, while in the Antarctic regions these waters do not occupy a bottom position. Based on this fact Reid et al. (1977) decided to abandon

Table 1.4 Boundary between Circumpolar Bottom Water and Lower Circumpolar Deep Water

Authors	Waters	Boundary
Sievers and Nowlin 1984	SEPDW/LCDW	$\gamma^n \approx 28.20$
Coles et al. 1996	LCDW(≈CBW)/NADW	$\theta = 1.4$ °C ($\gamma^n \approx 28.11$)
Arhan et al. 1999	SEPDW/LCDW-2	$\sigma_4 = 46.00$ ($\gamma^n \approx 28.22$)
Orsi et al. 1999	ACCbw/LCDW	$\gamma^n = 28.18$
Arhan et al. 2003	LDW/MDW	$\sigma_4 = 45.90$ ($\gamma^n \approx 28.11$)
Koshlyakov and Tarakanov 2003a, b	CBW/LCDW	$\gamma^n = 28.16$

Notes: SEPDW is South East Pacific Deep Water, LDW and MDW are Lower Deep Water and Middle Deep Water in the Southeast Atlantic the term "Antarctic Bottom Water" and proposed a more detailed classification of Antarctic bottom waters in the Atlantic. In their classification, Weddell Sea Deep Water and Weddell Sea Bottom Water were distinguished among the waters from the Weddell Sea in addition to Lower Circumpolar Water. In this concept, Weddell Sea Bottom Water is the water formed as a result of convection on the Antarctic slope in the Weddell Sea, and Weddell Sea Deep Water is basically a result of diapycnal mixing of Weddell Sea Bottom Water and overlying waters of the circumpolar origin entering the Weddell Gyre from the ACC zone through a group of fractures in the Southwest Indian Ridge in the region of 25°–32° E and then spreading westward along with the southern part of this gyre. Weddell Sea Bottom Water unlike the overlying Weddell Sea Deep Water does not spread beyond the Weddell Gyre (Weddell Abyssal Plain) due to the topographic restrictions (Reid et al. 1977; Orsi et al. 1999; Naveira Garabato et al. 2002b). Based on the recent concepts (Orsi et al. 1993; Fahrbach et al. 1995; Weppernig et al. 1996; Whitworth et al. 1998; Orsi et al. 1999) convection on the Antarctic slope in the Weddell Sea is an important factor for the formation of the Weddell Sea Deep Water layer. Part of Weddell Sea Deep Water is formed from Antarctic Bottom Water spreading westward along the Antarctic slope from its Indian Ocean sources (Meredith et al. 2000; Hoppema et al. 2001; Klatt et al. 2002). The boundary between Weddell Sea Deep Water and Weddell Sea Bottom Water is most often associated with isotherms: $\theta = -0.7$ °C (Fahrbach et al. 1995) or $\theta = -0.8$ °C (Foldvik et al. 1985), which approximately corresponds to neutral density $\gamma^n = 28.40$ (Orsi et al. 1999; Naveira Garabato et al. 2002a). A more detailed review of the options for this boundary is given in Subsection 3.1 of this book on Weddell Sea and Weddell Gyre.

We note that rejection of the term "Antarctic Bottom Water" was neither supported in the Wust's classification nor in the Southern Ocean water mass classification. Both Weddell Sea Deep Water and Weddell Sea Bottom Water actually make up Antarctic Bottom Water in the Atlantic in the Southern Ocean classification, while Weddell Sea Deep Water and Lower Circumpolar Water make up Antarctic Bottom Water in the Wust's classification. It should also be emphasized that term Lower Circumpolar Water introduced in (Reid et al. 1977) does not correspond to the term Lower Circumpolar Deep Water in the Southern Ocean classification of water masses.

In the Weddell Gyre, a layer of water of the circumpolar origin is traditionally called Warm Deep Water (Orsi et al. 1993; Naveira Garabato et al. 2002b). As Warm Deep Water spreads along the southern part of the Weddell Gyre its salinity decreases and the water is cooled due to mixing with overlying Antarctic Surface Water and underlying Antarctic Bottom Water and shelf waters of the Weddell Sea while maintaining its distinctive feature: salinity maximum. In the terms of circumpolar classification, Warm Deep Water combines Circumpolar Bottom Water and Lower Circumpolar Deep Water. In the southern part of the Drake Passage, Warm Deep Water together with underlying Weddell Sea Deep Water feeds the westward Antarctic Slope Current. In the southern part of the Scotia Sea south of the ACC zone (in the region east of the Drake Passage) meridional minima of salinity and temperature are observed in the density range of Warm Deep Water, which, according to

(Whitworth et al. 1994), are associated with the eastward transport of Antarctic Shelf Water from the Bransfield Strait around the tip of the Antarctic Peninsula. In the eastward direction, these minima become diffuse and almost disappear in the eastern part of the Scotia Sea (Whitworth et al. 1994).

In some regions of the ocean, even more detailed classifications of deep-water masses are used. For example, in the Scotia Sea region, Weddell Sea Deep Water is divided into Upper Weddell Sea Deep Water and Lower Weddell Sea Deep Water (Arhan et al. 1999; Meredith et al. 2008), which is associated with the topographic restrictions on the distribution of Lower Weddell Sea Deep Water in this region. North Atlantic Deep Water in the South Atlantic is divided into three layers (Fig. 1.3) characterized by the presence of different extreme values of hydrophysical (temperature, salinity, density) and hydrochemical properties.

In this book, we do not analyze in detail the properties of Antarctic Surface Water, Antarctic Intermediate Water, and Subantarctic Mode Water (Fig. 1.3) located above the Circumpolar Deep Water layer. Nevertheless, we note that the relatively fresh and cold Antarctic Surface Water is formed as a result of winter cooling and excess of precipitation and ice melting over evaporation and ice formation. This water occupies the surface layer south of the Polar Front. In summer, a winter minimum of temperature in the Antarctic Surface Water layer (Winter Water) is covered by a heated layer and a seasonal thermocline (Patterson and Whitworth 1990). Antarctic Intermediate Water is characterized by an intermediate salinity minimum extended northward from the Subantarctic Front (Reid 1965; Kuksa 1978; Kuksa 1983; Patterson and Whitworth 1990; Schmid et al. 2000).

This minimum is formed as a result of convergence (in the Subantarctic Front region) of the pure drift current directed northward transporting Antarctic Surface Water from the south (Sloyan and Rintoul 2001a). Downwelling of low salinity water from the upper layer of the ocean in the zone south of the Subantarctic Front to the intermediate layer in the zone north of the Subantarctic Front is caused mainly due to the separation of the cyclonic meanders of the Subantarctic Front and further decay of the formed quasi-geostrophic cyclonic eddies. During this process, the water with relatively low salinity follows the descent of the domes of isopycnal surfaces in the eddies from the upper layer of the ocean to the intermediate layer. At the same time, mesoscale isopycnal mixing of this water with the surrounding ocean water occurs (Golivets and Koshlyakov 2003, 2004). Subantarctic Mode Water located above Antarctic Intermediate Water is formed during deep winter convection on the northern side of the Subantarctic Front (Hanawa and Talley 2001).

There are also alternative hypotheses about the formation of Antarctic Intermediate Water. McCartney (1977) suggested another hypothesis about non-circumpolar formation of Antarctic Intermediate Water. According to this hypothesis, the entire volume of Antarctic Intermediate Water is formed as a result of deep winter convection north of the Subantarctic Front in the southeastern part of the Pacific Ocean. When transported to the Atlantic Ocean through the Drake Passage this water is transformed in the Drake Passage and western part of the Scotia Sea. In this case, Antarctic Intermediate Water consists only of the coldest, freshest, and densest type of Subantarctic Mode Water. Martineau (1953) put forward a hypothesis about the third source of Antarctic Intermediate Water near the South American coast extending from 45° to 25° S. Waters of the Brazil and Falkland currents are mixed up in this region. Waters formed as a result of mixing descend below lighter waters of the Brazil Current and join Antarctic Intermediate Water formed near the South Polar Front.

1.5 Upper Circumpolar Water and Upper Circumpolar Deep Water

It was previously noted that two terms are used in the literature: Upper Circumpolar Water and Upper Circumpolar Deep Water. Upper Circumpolar Deep Water (UCDW) is a product of isopycnal and diapycnal mixing of Indian Ocean Deep Water (IDW) and Pacific Deep Water (PDW), which spread into the Antarctic Circumpolar Current zone from the north, with Antarctic Surface Water, Antarctic Intermediate Water, and Lower Circumpolar Deep Water. A scheme of Upper Circumpolar Deep Water circulation together with its parental Pacific and Indian Deep Waters in the Southern Ocean is shown in Fig. 1.4.

Pacific Deep Water underlying Antarctic Intermediate Water is formed in the North Pacific as a result of mixing of Lower Circumpolar Deep Water and Circumpolar Bottom Water with North Pacific Intermediate Water and Subarctic Surface Water. Then, it propagates to the South Pacific (Tomczak and Godfrey 1994; Koshlyakov and Tarakanov 2004).

Indian Ocean Deep Water, in turn, is related to mixing of Lower Circumpolar Deep Water and Circumpolar Bottom Water with local intermediate waters. Together with Antarctic Intermediate Water and Red Sea Water, the former waters spread in the bottom layer from the south to the Subtropical zone of the Indian Ocean in its western and eastern deep basins (Tomczak and Godfrey 1994; Neiman et al. 1997). Transport of Indian Ocean Deep Water and Pacific Deep Water to the Antarctic Circumpolar Current zone at 1500–3000 m and their transformation to Upper Circumpolar Deep Water occur predominantly at the western and eastern margins of the Indian and Pacific oceans (Tomczak and Godfrey 1994; Orsi et al. 1995; Neiman et al. 1997; Sloyan and Rintoul 2001b). According to (Sloyan and Rintoul 2001b), transport of Upper Circumpolar Deep Water with the Antarctic Circumpolar Current is 60 Sv at the Tasmania meridian and increases to 69 Sv in the Drake Passage, mainly owing to the additional transport of Pacific Deep Water from the north. Lower values of oxygen concentration and higher content of nutrients in Indian Ocean Deep Water and Pacific Deep Water are caused by oxidizing processes combined with the old age of these waters. Like Indian Ocean Deep Water and Pacific Deep Water, Upper Circumpolar Deep Water is characterized by a minimum of oxygen concentration and maxima of phosphates and nitrates. In the zone south of the South Polar Front, which is at the same time the southern boundary of Antarctic Intermediate Water, Upper Circumpolar Deep Water gets into contact with Antarctic

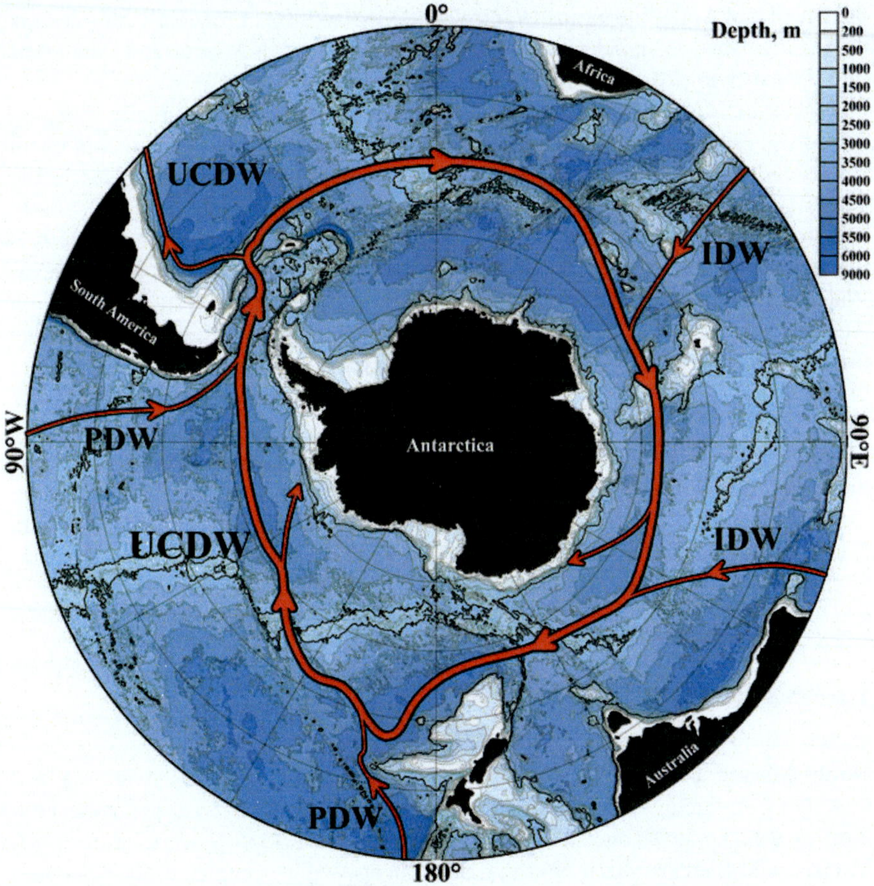

Fig. 1.4 Scheme of Upper Circumpolar Water spreading in the Southern Ocean (thick lines) based on many publications. Thin lines show the Indian Ocean Deep Water (IDW) and Pacific Deep Water (PDW) inflow to the Antarctic Circumpolar Current region

Surface Water. In this region, Upper Circumpolar Deep Water is characterized by a temperature maximum at the intermediate depth, related to the low temperature of the overlying Antarctic Surface Water. The term 'Upper Circumpolar Deep Water' is used by the authors of papers devoted to the Southern Ocean (Patterson and Whitworth 1990; Naveira Garabato et al. 2002a; Well et al. 2003), and the Southern Atlantic (Arhan et al. 1998; Ruth et al. 2000; van Aken 2007; and others).

Oxygen minimum and maximum of nutrients are distinguishing properties of Upper Circumpolar Water (UCPW) in the Atlantic Ocean (Saunders and King 1995; Larque et al. 1997; Vanicek and Siedler 2002). Another distinguishing property of this layer in the Atlantic north of the Antarctic Circumpolar Current is a temperature minimum between two temperature maxima in North Atlantic Deep Water from below and Antarctic Intermediate Water from above. This feature is observed only in the Brazil and Argentine basins (Vanicek and Siedler 2002). Frequently, this layer is not distinguished as an independent water mass in the South Atlantic and is included into the overlying Antarctic Intermediate Water.

In the South Atlantic, this water spreads in the western basin, gradually changing its propagation to the southeastern direction (Larque et al. 1997). These authors do not exclude that a small portion of Upper Circumpolar Water propagates in the system of the Deep Western Boundary Current. According to (Memery et al. 2000), it is not likely that it spreads north of 20° S. Nevertheless, Messias et al. (1999) distinguish Upper Circumpolar Deep Water (in our context: Upper Circumpolar Water) near the equator. The term 'upper branch of Circumpolar Water' in the South Atlantic was used in (Reid et al. 1977) for the upper layer of circumpolar waters dividing North Atlantic Deep Water into the upper and lower parts, i.e., Upper Circumpolar Water (UCPW) and Lower Circumpolar Water (LCPW). This term was later used in (Tsuchiya et al. 1994; Arhan et al. 1998; Outdot et al. 1998; Memery et al. 2000). In recent years, Upper Circumpolar Water has ceased to be used in the literature; following the work by Ganachaud (2003) it is being replaced by Upper Circumpolar Deep Water (e.g., Jullion et al. 2010; Meinen et al. 2012; Naveira Garabato et al. 2014; Evans et al. 2017; Meinen et al. 2017; Valla et al. 2018; Hernández-Guerra et al. 2019; Kersalé et al. 2019; Manta et al. 2021).

1.6 North Atlantic Deep Water

North Atlantic Deep Water (NADW) is distinguished from the overlying and underlying Antarctic waters by higher salinity, temperature, and oxygen concentration, as well by low concentration of nutrients (Wüst 1936; Larque et al. 1997; Vanicek and Siedler 2002). North Atlantic Deep Water occupies a layer between 1200 and 4000 m (Baum 2004). A scheme of North Atlantic Deep Water propagation in the North Atlantic based on compilation of (Schmitz and McCartney 1993; Schott and Brandt 2007; Sutherland and Pickart 2008; Lherminier et al. 2010; Mercier et al. 2015; Daniault et al. 2016; Sarafanov et al. 2018) is shown in Fig. 1.5. North Atlantic Deep Water is formed according to the following scheme (Koltermann et al. 1999). The densest waters of the Northern Hemisphere are located in the bottom layers of the Arctic Ocean, but they cannot propagate to the Atlantic owing to limited depths (650–850 m) of the straits and thresholds separating the Arctic Ocean from the Atlantic Ocean. Waters located above the bottom waters flow to the Atlantic through the Denmark Strait (Denmark Strait Overflow Water, DSOW) (Macrander et al. 2007) and Iceland Scotland thresholds (Iceland-Scotland Overflow Water, ISOW) (Olsen et al. 2008; van Aken and de Boer 1995). Formation of North Atlantic Deep Water is related not only to the inflow of waters from the Arctic Basin, but also to the deep convection in the Labrador Sea and Irminger Sea (Pickart et al. 2003). Among them, Labrador Sea Water (LSW) has a lower density and waters from the Denmark Strait are the densest ones. Iceland-Scotland Overflow Water flows to the Irminger Basin through the Charlie Gibbs Fracture Zone

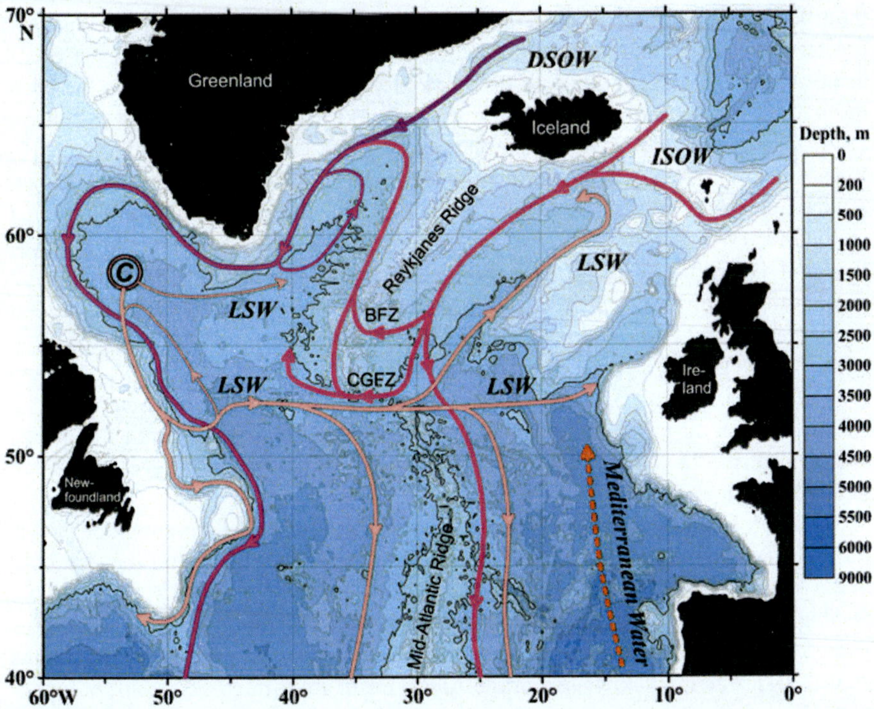

Fig. 1.5 Circulation layers in the North Atlantic based on many publications. Currents in the intermediate layers are shown with white lines. The Mediterranean Water pathway is shown with a dashed line. Bottom flows are shown with black and gray lines. The region of convection is indicated as "C". Other notations: BFZ: Bight Fracture Zone; CGFZ: Charlie Gibbs Fracture Zone; LSW: Labrador Sea Water; DSOW: Denmark Strait Overflow Water; ISOW: Iceland Scotland Overflow Water

(Worthington and Volkmann 1965; Harvey 1980). Beyond the fracture zone, this water is sometimes called Charlie Gibbs Fracture Zone Water (Mamayev 1992; Koltermann et al. 1999). The region near 60° N is the southernmost region of final formation of North Atlantic Deep Water (Mamayev 1992). On the other hand, North Atlantic Deep Water is transformed strongly by the influence of Mediterranean Water far to the south of 60° N (Keeling and Peng 1995).

Labrador Sea Water is the water of lowest density among North Atlantic Deep Water components. It is formed as a result of winter convection in the ice-free central and southern parts of the Labrador Sea and possibly in the Irminger Basin (Pickart et al. 2003; Bacon et al. 2003; Våge et al. 2011; Piron et al. 2017).

Iceland-Scotland Overflow Water is formed when water from the Norwegian Sea overflows the thresholds east of Iceland: Iceland Faroe Ridge and Faroe Shetland gateways (Dickson and Brown 1994; van Aken and de Boer 1995; Fleischmann et al. 2001; Lankhorst and Zenk 2006) with depths of 480 and 850 m, respectively. Since the warmer and more saline intermediate waters are mixed and entrained when they descend to greater depths, salinity of Iceland-Scotland Overflow Water is greater than that in the Northwest Atlantic. Iceland-Scotland Overflow Water propagates along the eastern slope of the Reykjanes Ridge. Then, it flows to the west through the deep Charlie Gibbs Fracture Zone and Bight Fracture Zone (Saunders 1994; Chang et al. 2009; Xu et al. 2010; Bower and Furey 2017; Zou et al. 2017; Racapé et al. 2019) and to the north along the western slope of the Reykjanes Ridge. After reaching the northern boundary of the Labrador Basin, it flows along the coast of America (Worthington 1970). In the western basin of the North Atlantic, Iceland-Scotland Overflow Water is identified by local salinity, silicate maximum, and oxygen minimum (van Aken and de Boer 1995). Recent investigations show that the densest waters in the straits are related to mixing of deep and intermediate waters of the Norwegian Sea, respectively), but the ratio of their contribution is variable (Fogelqvist et al. 2003). The estimates of spreading of Labrador Sea Water in (Koltermann et al. 1999) indicate that it takes less than two years for Labrador Sea Water to arrive to the Newfoundland Basin and 18–20 years to reach the Deep Western Basin interior at 24.5° N.

Denmark Strait Overflow Water is the densest and fresher water component. It is formed during the flow of intermediate waters of the Arctic Basin over the threshold in the Strait of Denmark with a depth of 650 m. Spatial and temporal structure of the Denmark Strait Overflow Water was revealed by acoustic observations (Macrander et al. 2007), field measurements, and models (Käse and Oschlies 2000; Jungclaus et al. 2001). Intermediate waters of the Arctic Basin are formed as a result of winter convection in the Greenland Sea (Swift et al. 1980; McCartney 1992).

The share of the densest waters does not exceed 10% of the total volume of Denmark Strait Overflow Water (Swift et al. 1980). After overflowing the threshold, Denmark Strait Overflow Water propagates along the continental slope of Greenland to the south (Girton et al. 2000; Köhl et al. 2007), then flows along the western boundary of the Labrador Sea, and eventually continues its motion in the general direction to the southwest (Worthington 1970). It can easily be identified in the Irminger Basin and Labrador Sea by the bottom oxygen maximum and minimum of temperature, salinity, and silicates. Denmark Strait Overflow Water propagates at greater depths and cannot spread into the eastern part of the Atlantic even through the Charlie Gibbs Fracture Zone (Mantyla and Reid 1983; Dickson et al. 1990).

A significant contribution of Mediterranean Sea Water to the North Atlantic Deep Water is discussed in many publications (Wüst 1936; Keeling and Peng 1995; Schmitz 1996a; Iorga and Lozier 1999). Thus, we can consider that North Atlantic Deep Water is formed as a result of mixing between Labrador Sea Water, Iceland-Scotland Overflow Water, Denmark Strait Overflow Water, and Mediterranean Sea Water shown in (Schmitz 1995).

According to the summary publication (Schmitz 1996a), quantitative estimates of deep-water transport (density range $\sigma_0 = 27.5$–27.8) are as follows: 3 Sv through the Strait of Denmark, 2 Sv east of Iceland, and 8 Sv is contributed by waters of the intermediate layer. According to (Schmitz 1996a), the amounts of deep waters transported from the Newfoundland Bank to a latitude of 30° S with the Deep Western Boundary Current is 14–18 Sv.

The fastest pathway for North Atlantic Deep Water components to the Southern Hemisphere is the Deep Western Boundary Current in the Atlantic (Rhein et al. 2015), although most of the southward transport occurs through the basin interior pathways (Gary et al. 2012). In the South Atlantic, components of North Atlantic Deep Water are usually not distinguished based on the regions of their formation. Here, North Atlantic Deep Water is divided into three layers: Upper (uNADW), Middle (mNADW), and Lower (lNADW) (Molinari et al. 1990; Tsuchiya et al. 1994). Sometimes, it is divided into two layers: upper and lower (in this case, Middle North Atlantic Deep Water is related to the lower layer) (Fu 1981; Roemmich 1983; Macdonald 1993).

High salinity is a distinguishing indicator of the upper layer of North Atlantic Deep Water. Local maxima of oxygen concentration indicate the middle and lower layers (Tsuchiya et al. 1994). According to (Molinari et al. 1992), the minimum chlorofluorocarbon (CFC) concentration can be used for distinguishing Middle North Atlantic Deep Water from Lower and Upper North Atlantic Deep Waters, which have local CFC maxima.

Vanicek and Siedler (2002) divide North Atlantic Deep Water into four components. They distinguish: Upper NADW, Labrador Sea Water, old Lower NADW, and overflowing Lower NADW and indicate that Upper North Atlantic Deep Water approximately corresponds to shallower Upper NADW according to the classification in (Rhein et al. 1996). The components are also characterized by maxima of salinity, temperature, concentration of CFC, and minimum of nutrients concentration. Labrador Sea Water, the core of which is associated with the oxygen maximum, and old Lower NADW are a part of Middle North Atlantic Deep Water. They are characterized by minimal concentration of oxygen and CFC, in contrast to Lower North Atlantic Deep Water. The North Atlantic Deep Water component can also be distinguished by the minimal concentration of the terrigenous helium (Vanicek and Siedler 2002).

There is no commonly accepted opinion about the origin of North Atlantic Deep Water components. For example, according to (Arhan et al. 1998; Larque et al. 1997), Upper North Atlantic Deep Water originates mainly from Mediterranean Water (Mediterranean Sea Water). According to (Pickart 1992; Andrie et al. 1998), Upper North Atlantic Deep Water originates from waters of the southern Labrador Sea. Distinguishing indicators of this water such as salinity maximum and minimum of silicate concentration are typical for Mediterranean Water, while high oxygen concentration is a characteristic feature of Labrador Sea Water.

Some authors (Tsuchiya 1989; Messias et al. 1999) associate Middle North Atlantic Deep Water with the maximum oxygen concentration, while others (Molinari et al. 1992; Andrie et al. 1998) believe that a minimum of CFC concentration is the best indicator of these waters, which does not coincide with the maximum of oxygen. This fact explains a wide spectrum of opinions about the origin of this water. For example, Arhan et al. (1998) consider that Middle North Atlantic Deep Water is formed in the Labrador Sea. In addition, Andrie et al. (1998) suggest that these waters are mixed with circumpolar waters and Iceland-Scotland Overflow Water.

Lower North Atlantic Deep Water originates from Denmark Strait Overflow Water (Wüst 1936; Tsuchiya et al. 1994; Andrie et al. 1998; Arhan et al. 1998). Speer and McCartney (1992) assume that Lower North Atlantic Deep Water is formed from the mixture of Iceland-Scotland Deep Water and Denmark Strait Overflow Water. A minimum of nutrients is also a characteristic property of Lower North Atlantic Deep Water formed from the waters of the Strait of Denmark transported to the Equatorial Atlantic by the Deep Western Boundary Current (Speer and McCartney 1991; Tsuchiya et al. 1992). In the Southwest Atlantic, salinity maximum south of 25° S is continued from the layer of Lower North Atlantic Deep Water (Tsuchiya et al. 1994).

A scheme of North Atlantic Deep Water spreading in the South Atlantic is shown in Fig. 1.6. This scheme is based on compilation of publications (Friedrichs et al. 1994; Schmitz 1996a; Larque et al. 1997; Müller et al. 1998; Zangenberg and Siedler 1998; Arhan et al. 1999; Arhan et al. 2003; Dengler et al. 2004; van Sebille et al. 2012; Garzoli et al. 2015; Hummels et al. 2015). Distinguishing properties of North Atlantic Deep Water are given in Table 1.5.

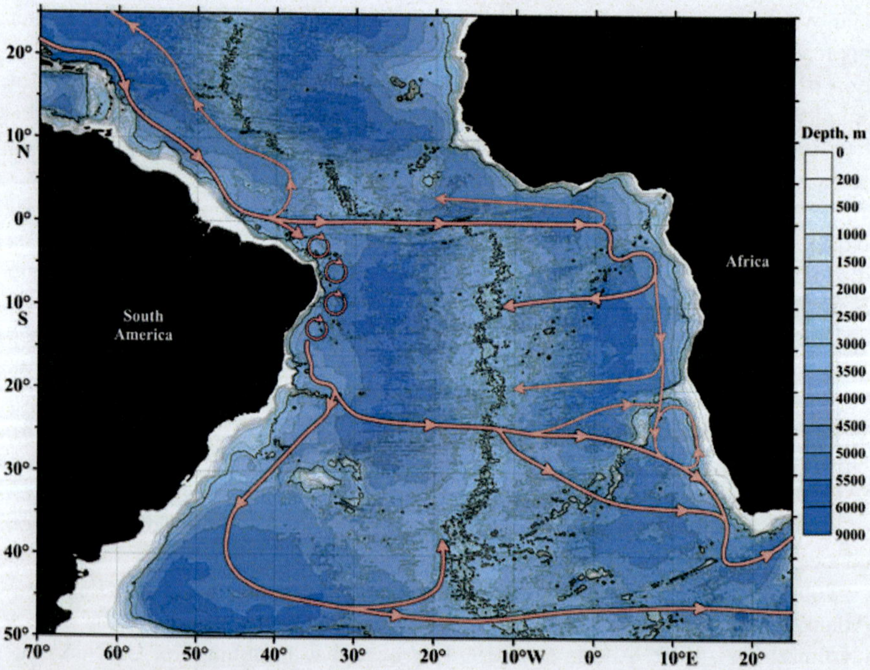

Fig. 1.6 Spreading of North Atlantic Deep Water in the South Atlantic compiled from different publications. Circles designate predominantly eddy transport of NADW along the Deep Western Boundary Current path in the Atlantic. (Dengler et al. 2004)

Table 1.5 Difference between extrema of different NADW components

Wüst 1936; Molinari et al. 1990			Rhein et al. 1996; Vanicek and Siedler 2002		
	max	min		max	min
UNADW	S,CFC		SUNADW	T, S, CFC	Si, P
MNADW	O_2	CFC	LSW	O_2	
			LNADW-old		O_2, CFC
LNADW	O_2, CFC		OLNADW	O_2, CFC	

1.7 Lower Circumpolar Water and Lower Circumpolar Deep Water, Circumpolar Bottom Water, Southeast Pacific Deep Water, and Warm Deep Water

We noted in the beginning of the chapter that Lower Circumpolar Deep Water and Lower Circumpolar Water are different objects in the Atlantic water structure. North Atlantic Deep Water is the main component of Lower Circumpolar Deep Water (Patterson and Whitworth 1990). Intense, mainly isopycnal mixing of North Atlantic Deep Water with deep waters of the Antarctic Circumpolar Current occurs at the northern periphery of the Antarctic Circumpolar Current in the Atlantic sector of the Southern Ocean (Patterson and Whitworth 1990; Stramma and England 1999). The resultant new water mass was called Lower Circumpolar Deep Water (Gordon 1967; Patterson and Whitworth 1990). It is the opinion of Mantyla and Reid (1983) that Lower Circumpolar Deep Water is a mixture of Antarctic Bottom Water and North Atlantic Deep Water. Unlike North Atlantic Deep Water, it is characterized by smoother extrema found in North Atlantic Deep Water (salinity maximum and minima of the concentrations of phosphates and nitrates). During the further easterly transport of Lower Circumpolar Deep Water by the Antarctic Circumpolar Current, these extrema become weaker, but they are still clearly pronounced along the entire pathway of these waters around Antarctica up to the next renewal of Lower Circumpolar Deep Water by North Atlantic Deep Water in the Southwest Atlantic.

The main distinguishing property of Lower Circumpolar Water at the northern periphery of the Antarctic Circumpolar Current in the Atlantic Ocean is an oxygen minimum caused by the existence of higher oxygen concentrations in the overlying and underlying layers in Weddell Sea Deep Water and North Atlantic Deep Water (Reid et al. 1977). This feature allows us to distinguish Lower Circumpolar Water only in the Argentine Basin. Already in the Brazil Basin, characteristics of waters of this density range are intermediate between the bottom water and North Atlantic Deep Water. In recent years, term Lower Circumpolar Water has not been used in the literature and following Ganachaud (2003) it is replaced by Lower Circumpolar Deep Water.

If we use the two-layer classification of circumpolar waters (Gordon 1971, 1972), Lower Circumpolar Deep Water would include Lower Circumpolar Water. However, if we use the three-layer classification with supplements by (Orsi et al. 1999), which is likely more reasonable, density ranges of these waters would only partially overlap. Hereinafter, we shall keep to the three-layer classification of circumpolar waters in the Antarctic Circumpolar Current region. A scheme of Lower Circumpolar Deep Water propagation in the Southern Ocean is shown in Fig. 1.7.

Circumpolar Bottom Water (CBW) (Bottom Water of the Antarctic Circumpolar Current according to (Orsi et al. 1999) classification) occupies a layer between Antarctic Bottom Water and Lower Circumpolar Deep Water in the Southern Ocean. During the motion with the Antarctic Circumpolar Current, Lower Circumpolar Deep Water is partly mixed with the underlying colder and denser Antarctic Bottom Water. Circumpolar Bottom Water is actually formed as a result of this mixing and also due to isopycnal inflow of waters from Antarctica (Orsi et al. 1999; Orsi et al. 2002). This water can overflow the underwater thresholds in the Drake

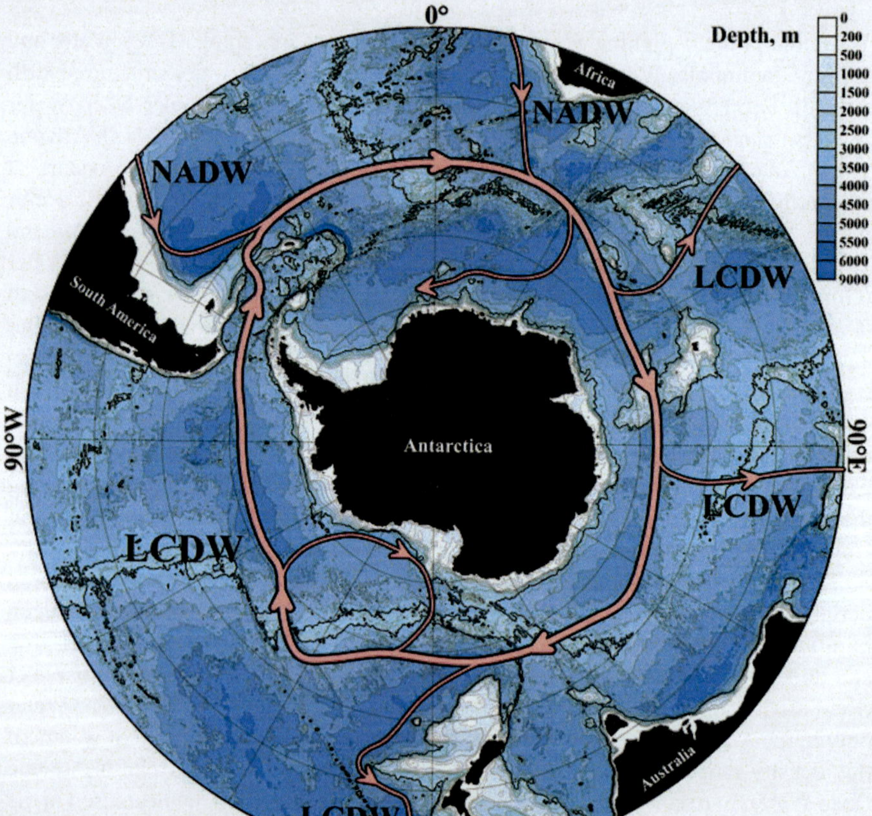

Fig. 1.7 Scheme of Lower Circumpolar Deep Water propagation in the Southern Ocean (thick line) based on many publications. Thin lines show inflow of North Atlantic Deep Water to the Antarctic Circumpolar Current region and outflow of Lower Circumpolar Deep Water to the Indian and Pacific oceans

Passage during the motion from west to east because its density is low. However, its density is high enough to occupy the isopycnal interval of North Atlantic Deep Water renewing Lower Circumpolar Deep Water in the South Atlantic. Because of such specific properties of this water, relative to Antarctic Bottom Water and Lower Circumpolar Deep Water, it should be distinguished as an independent water mass, as was stated for the first time in (Orsi et al. 1999). Scheme of propagation of Circumpolar Bottom Water is shown in Fig. 1.8.

According to (Tarakanov 2012; Koshlyakov et al. 2013), Circumpolar Bottom Water has a near-zero total transport through the Drake Passage. It means that Circumpolar Bottom Water outflowing from the Weddell Sea to the Scotia Sea maintains northward direction in the Southwest Atlantic and is a constituent part of the Deep Western Boundary Current transporting Antarctic waters in this part of the Atlantic. The validity of this conclusion is confirmed by two independent

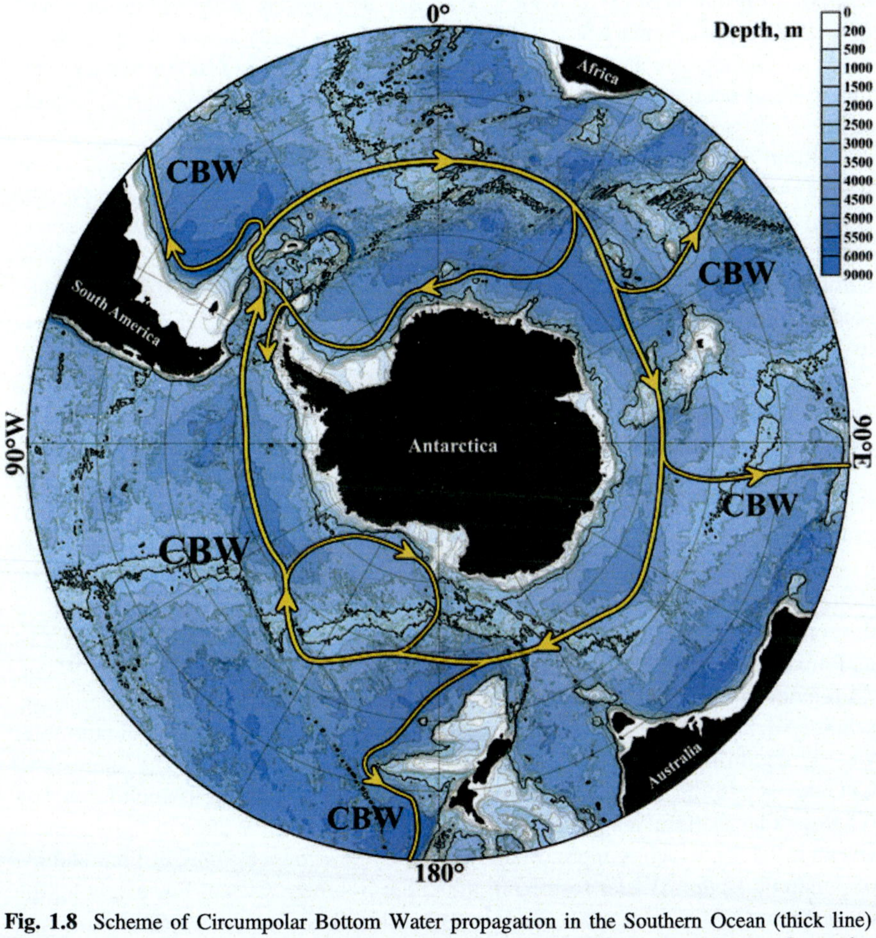

Fig. 1.8 Scheme of Circumpolar Bottom Water propagation in the Southern Ocean (thick line) based on many publications. Thin lines show outflow of Circumpolar Bottom Water to the north

considerations: by changes in θ,S-properties of water flowing from the Southeast Pacific to the Scotia Sea and by joint analyses of altimetry maps and digital maps of bottom topography (Tarakanov 2010, 2012). This issue is discussed in more detail in Sect. 3.3 of this book (Drake Passage). This conclusion does not contradict the circumpolar distribution of Circumpolar Bottom Water, as shown in Fig. 1.8, although it causes doubts related to the statement in (Orsi et al. 1999) about the repeated circumnavigation of this water around Antarctica with the Antarctic Circumpolar Current.

Circumpolar Bottom Water is the most uniform water layer along the pathway of the Antarctic Circumpolar Current. It is subjected to least variability of properties in the course of its propagation in the circumpolar circle compared with the overlying and underlying layers (Orsi et al. 1999). Circumpolar Bottom Water rather than waters of the Weddell and Ross seas propagates to the equator in the Pacific and Indian oceans. According to (Orsi et al. 1999), the neutral density range for Circumpolar Bottom Water is $28.18 < \gamma^n < 28.27$. Later, it was accepted that isopycnal $\gamma^n = 28.26$ is the boundary between Weddell Sea Deep Water and circumpolar water (Naveira Garabato et al. 2002a). Koshlyakov and Tarakanov (2003a) showed that this value of neutral density gives the best indicator for this boundary in the Pacific sector of the Southern Ocean.

Gordon (1966) was the first to demonstrate the existence of the Pacific origin waters in the Atlantic Ocean (in the Argentine Basin). Later, Sievers and Nowlin (1984) distinguished Southeast Pacific Deep Water (formed in the southeastern sector of the Pacific Ocean) in the Drake Passage as a mixture of overlying Lower Circumpolar Deep Water with bottom waters of the Ross Sea. According to (Orsi et al. 1999), Circumpolar Bottom Water in the Drake Passage is actually equivalent to Southeast Pacific Deep Water. A high age of the mixing waters in the Pacific Ocean explains high silicate concentration in Southeast Pacific Deep Water in the Drake Passage compared to a relatively low content in the younger Antarctic Bottom Water of the Weddell Sea.

Koshlyakov and Tarakanov (2003b) analyzed the properties of deep waters in the Pacific Ocean and showed that Circumpolar Bottom Water in the southeastern part of the Pacific Ocean is a product of diapycnal mixing of the overlying Lower Circumpolar Deep Water and the underlying Antarctic Bottom Water. Antarctic Bottom Water in this region is represented by Weddell Sea Deep Water that passed the route with the Antarctic Circumpolar Current from the Weddell Sea to the east and mixed along this pathway with other modifications of Antarctic Bottom Water. As previously mentioned, formation of Antarctic Bottom Water based on the classic scheme of Deacon (1937) is accompanied by isopycnal renewal of circumpolar waters. According to the estimates in (Orsi et al. 2002) based on the analysis of CFC balance in the World Ocean, the total volume of renewed circumpolar waters (Circumpolar Bottom Water and Lower Circumpolar Deep Water) in the Southern Ocean is 9.4 Sv, which is approximately equal to the known estimates of the volume of Antarctic Bottom Water formed in the entire Southern Ocean. For example, the estimate reported in (Orsi et al. 1999) is 10 Sv. It is likely that formation of Antarctic Bottom Water and renewal of Circumpolar Deep Water are confined to specific regions around Antarctica. The northwestern periphery of the Weddell Gyre (60°–35° W) is one of the regions where this process is intense (Whitworth et al. 1994). The other region is George V Coast (120°–160° E) (Foster 1995; Rintoul 1998; Rintoul and Bullister 1999).

Traditionally, the relatively warm and saline layer in the Weddell Gyre, which occupies an intermediate position between Antarctic Bottom Water and Antarctic Surface Water, is called Warm Deep Water (see, for example, (Orsi et al. 1993)). Salinity maximum in this water indicates its circumpolar origin, precisely from Lower Circumpolar Deep Water. According to (Orsi et al. 1993; Gordon 1998), Lower Circumpolar Deep Water propagates to the Weddell Sea approximately at 25° E. It flows along the southern arc and later along the western arc of the Weddell Gyre. Gradual cooling and freshening of its core are observed during this propagation. If we use classification in (Orsi et al. 1999) this water joins Circumpolar Bottom Water and lower layers of Lower Circumpolar Deep Water. Thus, renewal of Circumpolar Deep Water at the northwestern periphery of the Weddell Gyre is related to the inflow of the cooler and fresher Warm Deep Water from the Weddell Sea.

Hence, several water masses formed or renewed as a result of convection in the Weddell Sea and Antarctic Circumpolar Current owing to their interaction with North Atlantic Deep Water are distinguished in the Antarctic bottom zone. We shall consider peculiarities of their northward propagation in the Atlantic Ocean.

1.8 Antarctic Bottom Water

The definition of Antarctic Bottom Water suggested by Wüst (1936) is generally accepted for water mass analysis in the Atlantic Ocean. Wüst defines Antarctic Bottom Water as water with potential temperature below $\theta = 2$ °C. Wüst (1936) was the first to distinguish waters of the Antarctic origin in the bottom layer of the Atlantic Ocean based on the data of the German Atlantic Expedition on the R/V *Meteor*. In this elaborated benchmark book, the bottom water was named Antarctic Bottom Water. It was found that this water of the so-called oceanic stratosphere is formed in the Weddell Sea. Its potential temperature is below $\theta = -0.8$ °C in the Weddell Sea, increases to +0.6 °C at the equator, and reaches +1.8 °C near the Bermuda Islands. The terminal point of Antarctic Bottom Water spreading is the Discovery Gap at 37° N.

Definitions of Antarctic Bottom Water are different for research in the Southern Ocean and Antarctic region. According to the definition of Antarctic Bottom Water given in (Whitworth et al. 1998; Orsi et al. 1999), the limiting density value of neutral density is $\gamma^n = 28.27$. This property does not allow Antarctic Bottom Water to overflow the meridional ridges in the Drake Passage in the eastern direction together with the Antarctic Circumpolar Current (Reid et al. 1977; Orsi et al. 1999). Thus, the term 'Antarctic Bottom Water' covers all bottom waters with density $\gamma^n > 28.27$ (Whitworth et al. 1998; Orsi et al. 1999). This density approximately corresponds to the water with potential temperatures lower than $\theta = 0.2$ °C. An additional criterion was suggested in (Whitworth et al. 1998) to distinguish Antarctic Bottom Water from Antarctic Shelf Water: potential temperature $\theta > -1.7$ °C.

Since a large part of our book is dedicated to the review of literature, we shall keep to both definitions of Antarctic Bottom Water and try to follow the definitions accepted for each individual Atlantic region. Each definition will be specially emphasized if not clear from the text.

According to the modern concepts, using the definition by (Orsi et al. 1999), Antarctic Bottom Water is formed over the Antarctic slope as a result of mixing of the cold and dense Antarctic Shelf Water with the lighter, warmer, and more saline Circumpolar Deep Water. In its turn, Antarctic Shelf Water is formed in the autumn-winter season over the Antarctic shelf due to cooling of the relatively fresh Antarctic Surface Water to nearly freezing point temperature and salinification due to ice formation (Fig. 1.9). Formation of Antarctic Shelf Water, its further downwelling along the Antarctic slope to the ocean abyssal, and formation of Antarctic Bottom Water occur only in the case of a combination of specific conditions: existence of permanent large polynya supported by strong offshore winds, leading to the withdrawal of heat to the atmosphere; sufficient width of the shelf that provides significant time for the interaction with the atmosphere; and existence of crossing canyons that allow outflow of the heaviest forms of Antarctic Shelf Water from the shelf. Cooling and freshening at the bottom of shelf glaciers also take place during the formation (cabbeling – mixing with densification) of Antarctic Shelf Water. This is especially characteristic of the western part of the Ross Sea, where the temperature under the ice column can reach extremely low values $\theta < -2.0$ °C (Patterson and Whitworth 1990). If the salinity of the mixed waters is relatively high and the temperature is low, the resulting water mass reaches the ocean floor, thus forming Antarctic Bottom Water. Because of these limitations, formation of Antarctic Bottom Water is confined to a few regions along the perimeter of the Antarctic slope. At present, one can identify several regions (Baines and Condie 1998; Whitworth et al. 1998; Orsi et al. 1999): southern and western parts of the Weddell Sea; Commonwealth Sea in the Indian Ocean sector of Antarctica; George V Coast (or Adélie Land) at the boundary between the Indian Ocean and Pacific sectors; western and eastern parts of the Ross Sea (Fig. 1.10). The Weddell Sea is a region marked by the formation of 60–65% of Antarctic Bottom Water (Carmack 1977; Rintoul 1998; Orsi et al. 1999). There are a few other regions of lesser importance for Antarctic Bottom Water formation (Whitworth et al. 1998) including the Drake Passage (Meredith et al. 2003). Waters formed in these sources are also called Antarctic Bottom Water, but their properties can strongly differ from those of waters from the Weddell and Ross seas and other sources in the Southern Ocean. Antarctic Bottom Water formation is not uniform in time because of the seasonal character of Antarctic Shelf Water formation and sporadic flow of Antarctic Shelf Water down the Antarctic slope in the form of plumes and individual portions (Baines and Condie 1998). Some inter-seasonal and inter-annual differences in the characteristics (including density) of the waters flowing from the shelf are quite natural.

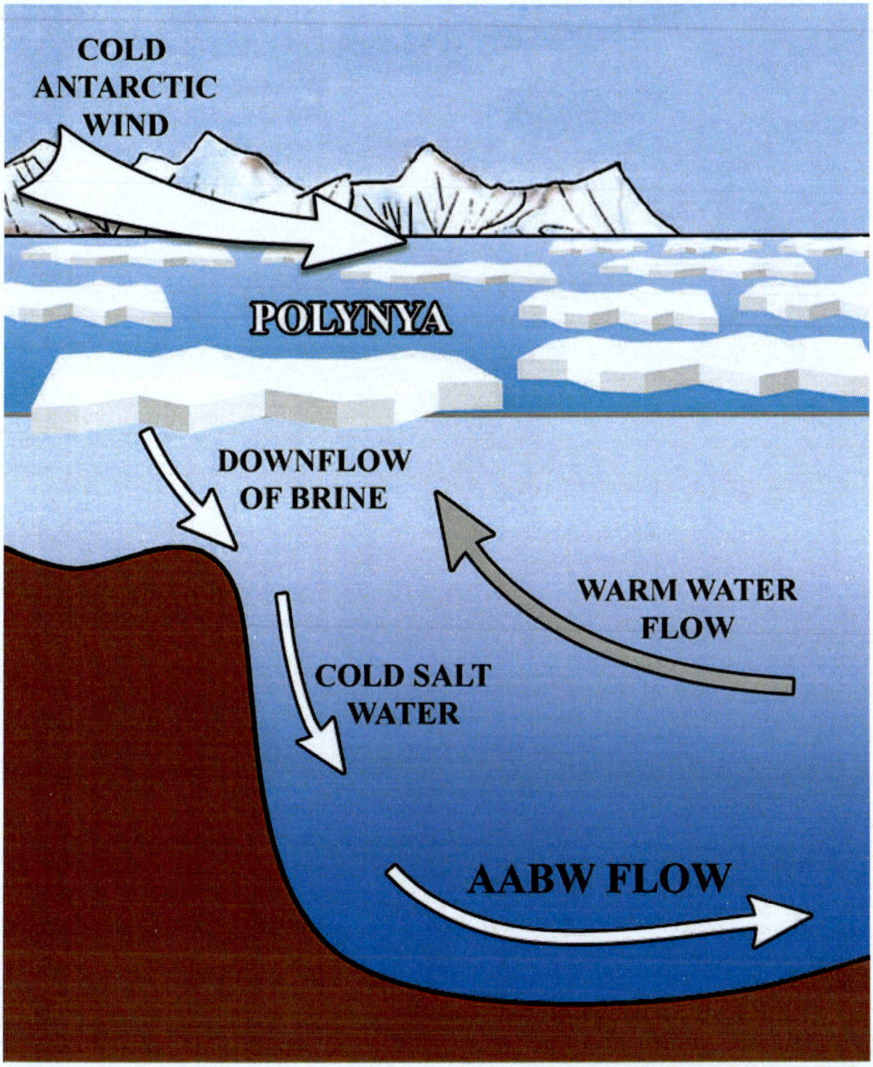

Fig. 1.9 Scheme of Antarctic Bottom Water formation

Hypotheses about the formation of cold bottom waters in the Antarctic region as a result of mixing of cold shelf waters with warmer deep waters were first put forward by Brennecke (1921) and Mosby (1934). Later, Fofonoff (1956) indicated the possibility of cabbeling during mixing of these waters that changes the water density. Gordon (1978) and Whitehead (1989) suggested another mechanism of Antarctic Bottom Water formation: deep convective mixing in polynyas within mesoscale eddies in the open ocean intensified by strong cooling in the surface layer. In the early 1970s, such convection was observed in the Southern Ocean in the Maud Rise

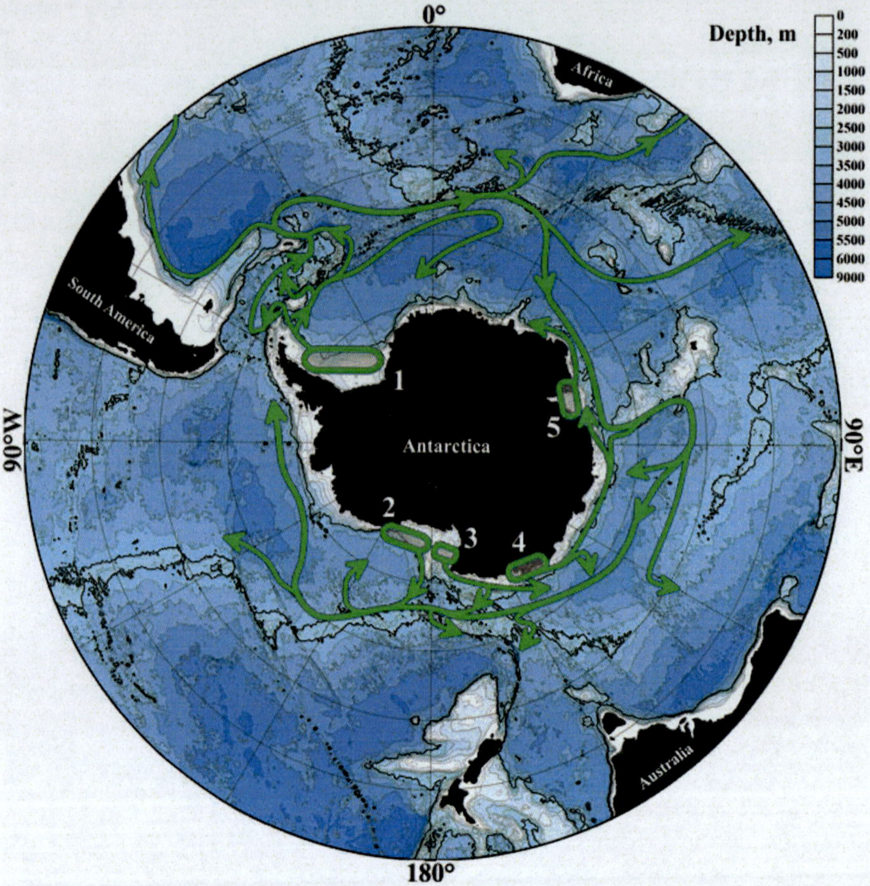

Fig. 1.10 Scheme of Antarctic Bottom Water propagation in the Southern Hemisphere according to (Orsi et al. 1999). Thick lines denote quasi-circumpolar flow of Antarctic Bottom Water. Thin lines correspond to outflow of this water from the sources and pathways for further Antarctic Bottom Water propagation from the Southern Ocean. Numerals indicate main regions of AABW formation. (1) Weddell Sea; (2) Eastern part of the Ross Sea; (3) Western part of the Ross Sea; (4) Adélie Land; (5) Commonwealth Sea

region (Prime or Greenwich meridian). However, satellite observations of ice cover in the ocean demonstrated that necessary conditions for this mechanism existed only during 11% of the time of such observations, which could provide the formation of approximately 2.4 Sv of deep waters (Gordon 1978). The mean estimate of the rate of deep-water formation in the Weddell Sea due to deep convection is more than one order of magnitude smaller than estimates of the rate of Antarctic Bottom Water formation over the Antarctic slope. However, Whitworth et al. (1998) demonstrated that this process should be considered as potentially possible over the major part of the perimeter of the Antarctic slope.

According to the formation mechanism in all regions of the Antarctic Bottom Water spreading, it is characterized by low temperature values. In the major part of regions of Antarctic Bottom Water formation mentioned here, Antarctic Shelf Water is also characterized by lower salinity compared to Circumpolar Deep Water. Therefore, Antarctic Bottom Water corresponding to different types of shelf waters also differs by lower salinity compared to the overlying waters. Antarctic Shelf Water and, correspondingly, Antarctic Bottom Water from the western part of the Ross Sea are exceptions. According to (Patterson and Whitworth 1990), this peculiarity is caused by quasi-isolated location of the cyclonic gyre near the Wilkes Coast in the northwestern part of the Ross Sea shelf, which, in turn, leads to extremely strong salinification of the local Antarctic Shelf Water (salinity reaches 35.0 PSU). Antarctic Bottom Water within the Antarctic Circumpolar Current is distinguished for increased content of oxygen. Concentrations of oxygen and nutrients depend on the age of the bottom waters and trajectory of their flow. In some cases, these features allow us to distinguish Antarctic Bottom Water formed in different Antarctic regions (Patterson and Whitworth 1990; Koshlyakov and Tarakanov 1999; 2003a). Potential temperature of Antarctic Bottom Water in the Weddell Sea can be as low as −1.5 °C but even lower temperatures are recorded; salinity is approximately 34.65 PSU, and silicate concentration is nearly 100 μmol/kg. The type of Antarctic Bottom Water from the Ross Sea has a slightly higher temperature (−0.2 °C) and salinity (34.75 PSU), while salinity is slightly smaller (34.4–34.7 PSU) near the George V Coast (Adélie Land). Mean temperature and salinity values of Antarctic Bottom Water in the region of 60° S are equal to −0.4 °C and 34.66 PSU, respectively (Mamayev 1992), while the silicate concentration is approximately 125 μmol/kg (Ostlund et al. 1987). Propagation of Antarctic Bottom Water around Antarctica is shown in Fig. 1.10.

References

Alekseev GV, Afanasyev BV, Zykov ID, Smirnov NP (1978) Thermal structure of Antarctic waters above the continental slope in the area of Molodezhnaya station. Proc AARI 345:138–141 [in Russian]

Andrie C, Ternon JF, Messias MJ, Memery L, Bourlès B (1998) Chlorofluormethane distribution in the deep equatorial Atlantic during January-March 1993. Deep-Sea Res I 45:903–930

Anisimov MV, Ivanov YA, Subbotina MM (2002) Abyssal part of the Global Ocean Conveyor. Oceanology 42(3):305–316

Arhan M, Mercier H, Bourlès B, Gouriou Y (1998) Hydrographic section across the Atlantic at 7°30′ N and 4°30′ S. Deep-Sea Res I 45:829–872

Arhan M, Heywood KJ, King BA (1999) The deep waters from the Southern Ocean at the entry to the Argentine Basin. Deep-Sea Res II 46:475–499

Arhan M, Mercier H, Park Y-H (2003) On the deep water circulation of the eastern South Atlantic Ocean. Deep-Sea Res I 50:889–916

Bacon S, Gould WJ, Jia Y (2003) Open-ocean convection in the Irminger Sea. Geophys Res Lett 30(5):1246. https://doi.org/10.1029/2002GL016271

Badin G, Williams RG, Jing Z, Wu L (2013) Water mass transformations in the Southern Ocean diagnosed from observations: contrasting effects of air–sea fluxes and diapycnal mixing. J Phys Oceanogr 43(7):1472–1484

Baines PG, Condie S (1998) Observation and modelling of Antarctic downslope flows: a review, ocean, ice, and atmosphere: interactions at the Antarctic continental margin. Antarctic Res Ser 75:29–49. Washington DC: AGU. https://doi.org/10.1029/AR075p0029

Baker JA, Watson JA, Vallis GK (2020) Meridional overturning circulation in a multibasin model. Part I: dependence on Southern Ocean buoyancy forcing. J Phys Oceanogr 50(5):1159–1178

Barre N, Provost C, Sennechael N, Lee JH (2008) Circulation in the Ona Basin, southern Drake Passage. J Geophys Res Oceans 113:C04033. https://doi.org/10.1029/2007JC004549

Baum SK (2004) Glossary of physical oceanography and related disciplines. http://stommel.tamu.edu/~baum/paleo/ocean/ocean.html

Blanke B, Arhan M, Speich S, Pailler K (2002) Diagnosing and picturing the North Atlantic segment of the global Conveyer Belt by means an ocean general circulation model. J Phys Oceanogr 32(5):1430–1451

Bower AS, Furey H (2017) Iceland-Scotland overflow water transport variability through the Charlie-Gibbs fracture zone and the impact of the North Atlantic current. J Geophys Res: Oceans 122(9):6989–7012

Boyer TP, Baranova OK, Coleman C, Garcia HE, Grodsky A, Locarnini RA, Mishonov AV, Paver CR, Reagan JR, Seidov D, Smolyar IV, Weathers K, Zweng MM (2018) World Ocean database 2018. AV Mishonov, technical Ed., NOAA Atlas NESDIS 87. https://www.ncei.noaa.gov/sites/default/files/2020-04/wod_intro_0.pdf

Brennecke WS (1921) Die ozeanographischen Arbeiten der Deutchen Antarktischen Expedition, 1911–1912. Arch Dtsch Seewarte 39(1):1–216 [in German]

Broecker WS (1987) The biggest chill. Nat Hist 96:74–82

Broecker WS (1991) The great ocean conveyor. Oceanography 4(2):79–89

Broecker WS, Peng T-H (1982) Tracers in the sea. Eldigio Press, 690 pp

Broecker WS, Peacock SL, Walker S, Weiss R, Fahrbach E, Schröder M, Mikolajewicz U, Heinze C, Key R, Peng T-H, Rubin S (1998) How much deep water is formed in the Southern Ocean? J Geophys Res Oceans 103(C8):15833–15843

Buckley MW, Marshall J (2016) Observations, inferences, and mechanisms of Atlantic meridional overturning circulation variability: a review. Rev Geophys 54:5–63. https://doi.org/10.1002/2015RG000493

Carmack EC (1977) Water characteristics of the Southern Ocean south of the polar front. In: Deacon G, Angel M (eds) A voyage of discovery, 70th anniversary volume. Pergamon Press, Oxford, pp 15–41

Cessi P (2019) The global overturning circulation. Annu Rev Mar Sci 11:249–270. https://doi.org/10.1146/annurevmarine-010318-095241

Chang YS, Garraffo ZD, Peters H, Özgökmen TM (2009) Pathways of Nordic overflows from climate model scale and eddy resolving simulations. Ocean Model 29:66–84

Coachman KK, Aagaard K (1974) Physical oceanography of Arctic and Subarctic seas. In: Herman Y (ed) Marine geology and oceanography of the Arctic seas. Springer, Berlin, pp 1–74

Coles VJ, McCartney MS, Olson BD, Smethie WJ Jr (1996) Changes in Antarctic bottom water properties in the western South Atlantic in the late 1980s. J Geophys Res Oceans 101(C4): 8957–8970

Daniault N, Mercier H, Lherminier P, Sarafanov A, Falina A, Zunino P et al (2016) The northern North Atlantic Ocean mean circulation in the early 21st century. Progress Oceanogr 146:142–158. https://doi.org/10.1016/j.pocean.2016.06.007

Davis RE (2005) Intermediate-depth circulation of the Indian and South Pacific oceans measured by autonomous floats. J Phys Oceanogr 35(5):683–707

de Lavergne C, Madec G, Sommer JL, Nurser A, Naveira Garabato AC (2016a) The impact of a variable mixing efficiency on the abyssal overturning. J Phys Oceanogr 46:663–681 de Lavergne C, Madec G, Le Sommer J, Nurser AG, Naveira Garabato AC (2016b) On the consumption of Antarctic bottom water in the abyssal ocean. J Phys Oceanogr 46:635–661

Deacon GER (1937) The hydrology of the Southern Ocean. Discovery Reports 15:1–124

Dengler M, Schott FA, Eden C, Brandt P, Fischer J, Zantopp RJ (2004) Break-up of the Atlantic deep Western boundary current into eddies at 8°S. Nature 432(7020):1018–1020. https://doi.org/10.1038/nature03134

Dickson RR, Brown J (1994) The production of North Atlantic deep water: sources, rates, and pathways. J Geophys Res Oceans 99:12,319–12,342. https://doi.org/10.1029/94JC00530

Dickson RR, Gmitrowicz EM, Watson AJ (1990) Deep-water renewal in the northern North Atlantic. Nature 344:848–850

Dobrolyubov SA, Falina AS (2002) Interannual variability of the Intermediate North Atlantic Water circulation from a multiparameter analysis of the water masses. Oceanology 42:619–627

Döös K, Coward A (1997) The Southern Ocean as the major upwelling zone of North Atlantic deep water. International WOCE Newsletter, No. 27, WOCE international project office, Southampton, United Kingdom, 3–4

Downes SM, Key RM, Orsi AH, Speer KG, Swift JH (2012) Tracing Southwest Pacific bottom water using potential vorticity and helium-3. J Phys Oceanogr 42(12):2153–2168

Drijfhout SS, de Vries P, Döös K, Coward AC (2003) Impact of eddy-induced transport of the lagrangian structure of the upper branch of the thermohaline circulation. J Phys Oceanogr 33(10):2141–2155

Dye S, Hansen B, Østerhus S, Quadfasel D, Rudels B (2007) The overflow of dense water across the Greenland-Scotland Ridge. Exchanges № 40, pp 20–22

Evans GR, McDonagh EL, King BA, Bryden HL et al (2017) South Atlantic interbasin exchanges of mass, heat, salt and anthropogenic carbon. Prog Oceanogr 151:62–82. https://doi.org/10.1016/j.pocean.2016.11.005

Fahrbach E, Rohardt G, Scheele N, Schröder M, Strass V, Wisotzki A (1995) Formation and discharge of deep and bottom water in the Northwestern Weddell Sea. J Mar Res 53:515–538

Ferrari F, Nadeau LP, Marshall DP, Allison LC, Johnson HL (2017) A model of the ocean overturning circulation with two closed basins and a reentrant channel. J Phys Oceanogr 47:2887–2906. https://doi.org/10.1175/JPO-D-16-0223.1

Fischer J, Rhein M, Schott F, Stramma L (1996) Deep water masses and transports in the Vema fracture zone. Deep-Sea Res I 43(7):1067–1074

Fleischmann U, Hildebrandt H, Putzka A, Bayer R (2001) Transport of newly ventilated deep water from the Iceland Basin to the Westeuropean Basin. Deep Sea Res I 48:1793–1819

Fofonoff NP (1956) Some properties of sea water influencing the formation of Antarctic bottom water. Deep-Sea Res I 4:32–35

Fogelqvist E, Blindheim J, Tanhua T, Osterhus S, Buch E, Rey F (2003) Greenland-Scotland overflow studied by hydro-chemical multivariate analysis. Deep-Sea Res I 50:73–102

Foldvik A, Gammelsrød T, Tørresen T (1985) Circulation and water masses on the southern Weddell Sea shelf. In: Jacobs SS (ed) Oceanology of the Antarctic continental shelf, Antarctic research series, vol 43. AGU, Washington, DC, pp 5–20

Foster TD (1995) Abyssal water mass formation off the eastern Wilkes Island coast of Antarctica. Deep-Sea Res I 42:501–522

Foster TD, Carmack EC (1976) Frontal zone mixing and Antarctic bottom water formation in the southern Weddell Sea. Deep-Sea Res I 23:301–317. https://doi.org/10.1016/0011-7471(76)90872-X

Frajka-Williams E, Ansorge IG, Baehr J, Bryden HL, Paz Chidichimo M, Cunningham SA et al (2019) Atlantic meridional overturning circulation: observed transport and variability. Front Mar Sci 6:260. https://doi.org/10.3389/fmars.2019.00260

Friedrichs MA, Hall MM (1993) Deep circulation in the tropical North Atlantic. J Mar Res 51(4):697–736

Friedrichs MA, McCartney MS, Hall MM (1994) Hemispheric asymmetry of deep water transport modes in the western Atlantic. J Geophys Res Oceans 99(C12):25165–25179

Friocourt Y, Drijfhourt S, Blanke B, Speich S (2005) Water mass export from Drake Passage to the Atlantic, Indian, and Pacific oceans: a lagrangian model analysis. J Phys Oceanogr 35(7): 1206–1222

Fu LL (1981) The general circulation and meridional heat transport of the subtropical South Atlantic determined by inverse methods. J Phys Oceanogr 11(9):1171–1193

Ganachaud A (2003) Large-scale mass transports, water mass formation, and diffusivities estimated from World Ocean circulation experiment (WOCE) hydrographic data. J Geophys Res Oceans 108(C7):3213. https://doi.org/10.1029/2002JC001565

Ganachaud A, Wunsch C (2000) Improved estimates of global ocean circulation, heat transport and mixing from hydrographic data. Nature 408:453–457. https://doi.org/10.1038/35044048

Gary SF, Lozier MS, Biastoch A, Böning CW (2012) Reconciling tracer and float observations of the export pathways of Labrador Sea water. Geophys Res Lett 39(24):1–5. https://doi.org/10.1029/2012GL053978

Garzoli SL, Dong S, Fine R, Meinen CS, Perez RC, Schmid C et al (2015) The fate of the deep Western boundary current in the South Atlantic. Deep Sea Research I 103:125–136. https://doi.org/10.1016/j.dsr.2015.05.008

Girton JB, Sanford TB, Käse R (2000) Synoptic sections of the Denmark Strait overflow. Geophys Res Lett 28(8):1619–1622

Gnanadesikan A (1999) A simple predictive model for the structure of the oceanic pycnocline. Science 283:2077–2079

Golivets SV, Koshlyakov MN (2003) Cyclonic eddies of the Subantarctic front and formation of the Antarctic intermediate water. Oceanology 43(3):305–317

Golivets SV, Koshlyakov MN (2004) Eddy formation at the Subantarctic front based on the data of satellite observations and formation of Antarctic intermediate water. Oceanology 44(4): 485–494

Gordon AL (1966) Potential temperature, oxygen and circulation of bottom water in the Southern Ocean. Deep-Sea Res I 13:1125–1138

Gordon AL (1967) Structure of Antarctic waters between 20° W and 170° W. In: Bushell VC (ed) Antarctic map folio series (folio 6). American Geographic Society, New York

Gordon AL (1971) Oceanography of Antarctic waters. In: Reid JL (ed) Antarctic oceanography, Antarctic research series, vol 15. American Geophysical Union, Washington, DC, pp 169–203

Gordon AL (1972) Spreading of Antarctic bottom waters II. In: Studies in physical oceanography – a tribute of George Wüst on his 80th birthday, vol 2. Gordon and Breach, New York, pp 1–17

Gordon AL (1978) Deep Antarctic convection west of Maud rise. J Phys Oceanogr 8:600–612

Gordon AL (1986) Interocean exchange of thermocline water. J Geophys Res 91:5037–5046. https://doi.org/10.1029/JC091iC04p05037

Gordon AL (1998) Western Weddell Sea thermohaline stratification. In: Jacobs SS, Weiss RF (eds) Ocean, ice, and atmosphere: interactions at the Antarctic continental margin, Antarctic research series 75, pp 215–240

Hall MM, McCartney MS, Whitehead JA (1997) Antarctic bottom water flux in the equatorial Western Atlantic. J Phys Oceanogr 27(9):1903–1926. https://doi.org/10.1175/1520-0485(1997)027<1903:ABWFIT>2.0.CO;2

Hallberg RW, Gnanadesikan A (2006) The role of eddies in determining the structure and response of the wind driven Southern Hemisphere overturning: results from the modeling eddies in the Southern Ocean project. J Phys Oceanogr 36(12):2232–2252

Hanawa K, Talley LD (2001) Mode waters. In: Siedler G, Church J, Gould J (eds) Ocean circulation and climate. Academic, San Diego, pp 373–386

Harvey JG (1980) Deep and bottom water in the Charlie-Gibbs fracture zone. J Mar Res 38:172–173

Harvey J, Arhan M (1988) The water masses of the Central North Atlantic in 1983–1984. J Phys Oceanogr 18(12):1855–1874

Hernández-Guerra A, Talley LD, Pelegríc JL, Vélez-Belchíd P, Baringer MO, Macdonald AM, McDonagh EL (2019) The upper, deep, abyssal and overturning circulation in the Atlantic Ocean at 30°S in 2003 and 2011. Progr Oceanogr 176:102136

Hogg N, Biscaye P, Gardener W, Schmitz WJ (1982) On the transport and modification of Antarctic bottom water in the Vema Channel. J Mar Res 40(suppl):231–263

Hogg N, Siedler G, Zenk W (1999) Circulation and variability at the southern boundary of the Brazil Basin. J Phys Oceanogr 29:145–157

Holfort J, Siedler G (2001) The meridional oceanic transport of heat and nutrients in the South Atlantic. J Phys Oceanogr 31(1):5–28

Hoppema M, Klatt O, Roether W, Fahrbach E, Bulsiewicz K, Rodehacke C, Rohardt G (2001) Prominent renewal of Weddell Sea deep water from a remote source. J Mar Res 59:257–279

Hummels R, Brandt P, Dengler M, Fischer J, Araujo M, Veleda D, Durgadoo JV (2015) Interannual to decadal changes in the western boundary circulation in the Atlantic at 11°S. Geophys Res Lett 42:7615–7622. https://doi.org/10.1002/2015GL065254

Iorga M, Lozier MS (1999) Signatures of the Mediterranean outflow from a North Atlantic climatology 2. Diagnostic velocity fields. J Geophys Res Oceans 104(26):26011–26029

Iudicone D, Speich S, Madec G, Blanke B (2008) The global Conveyor Belt from a Southern Ocean perspective. J Phys Oceanogr 38(7):1401–1425

Jacket DR, McDougall TJ (1997) A neutral density variable for the World's ocean. J Phys Oceanogr 27(2):237–263

Jansen MF, Nadeau LP (2016) The effect of Southern Ocean surface buoyancy loss on the deep-ocean circulation and stratification. J Phys Oceanogr 46:3455–3470. https://doi.org/10.1175/JPO-D-16-0084.1

Johnson GC (2008) Quantifying Antarctic bottom water and North Atlantic deep water volumes. J Geophys Res Oceans 113:C05027. https://doi.org/10.1029/2007JC004477

Johnson HL, Cessi P, Marshall DP, Schloesser F, Spall MA (2020) Recent contributions of theory to our understanding of the Atlantic meridional overturning circulation. J Geophys Res 124:5376–5399. https://doi.org/10.1029/2019JC015330

Jullion L, Heywood KJ, Naveira Garabato AC, Stevens DP (2010) Circulation and water mass modification in the Brazil-Malvinas confluence. J Phys Oceanogr 40:845–864. https://doi.org/10.1175/2009JPO4174.1

Jungclaus JH, Hauser J, Käse RH (2001) Cyclogenesis in the Denmark Strait overflow plume. J Phys Oceanogr 31(11):3214–3229

Käse RH, Oschlies A (2000) Flow through Denmark Strait. J Geophys Res Oceans 105(C12): 28527–28546. https://doi.org/10.1029/2000JC900111

Katsumata K, Sloyan BM, Masuda S (2013) Diapycnal and isopycnal transports in the Southern Ocean estimated by a box inverse model. J Phys Oceanogr 43(11):2270–2287

Keeling RF, Peng T (1995) Transport of heat, CO2, and O2 by the Atlantic's thermohaline circulation. Philos Trans R Soc London B 348:133–142

Kersalé M, Perez RC, Speich S, Meinen CS, Lamont T, Le Hénaff M et al (2019) Shallow and deep eastern boundary currents in the South Atlantic at 34.5°S: mean structure and variability. J Geophys Res Oceans 124:1634–1659. https://doi.org/10.1029/2018JC014554

Klatt O, Roether W, Hoppema M, Bulsiewicz K, Fleischmann U, Rodehacke C, Fahrbach E, Weiss RF, Bullister JL (2002) Repeated CFC sections at the Greenwich Meridian in the Weddell Sea. J Geophys Res Oceans 107(C4):3030. https://doi.org/10.1029/2000JC000731

Klatt O, Fahrbach E, Hoppema M, Rohardt G (2005) The transport of the Weddell gyre across the prime Meridian. Deep-Sea Res II 52:513–528

Köhl A, Käse RH, Stammer D (2007) Causes of changes in the Denmark Strait overflow. J Phys Oceanogr 37(6):1678–1696

Koltermann KP, Sokov AV, Tereschenkov VP, Dobroliubov SA, Lorbacher K, Sy A (1999) Decadal changes in the thermohaline circulation of the North Atlantic. Deep-Sea Res II 46:109–138

Koshlyakov MN, Tarakanov RY (1999) Water masses of the Pacific Antarctic. Oceanology 39(1):5–15

Koshlyakov MN, Tarakanov RY (2003a) Antarctic bottom water in the Pacific sector of the Southern Ocean. Oceanology 43(1):5–20

Koshlyakov MN, Tarakanov RY (2003b) Antarctic circumpolar water in the southern part of the Pacific Ocean. Oceanology 43(5):607–621

Koshlyakov MN, Tarakanov RY (2004) Pacific deep water in the Southern Ocean. Oceanology 44(3):325–340

Koshlyakov MN, Tarakanov RY (2005) Intermediate water masses in the southern part of the Pacific Ocean. Oceanology 45(4):455–473

Koshlyakov MN, Tarakanov RY (2011) Water transport across the Subantarctic front and the Global Ocean Conveyer Belt. Oceanology 51(5):721–735

Koshlyakov MN, Gladyshev SV, Ryu T, Fedorov DA (2013) Currents in the Drake Passage by observations in October—November of 2011. Oceanology 53(1):1–12

Kuksa VI (1978) Atlas of intermediate and Subsurface intermediate waters of the World Ocean. Moscow. Gidrometeoizdat, 1978. 83 p. [in Russian]

Kuksa VI (1983) Intermediate waters of the World Ocean. Gidrometeoizdat, Leningrad. [in Russian]

Kunze E (2017) The internal-wave-driven meridional overturning circulation. J Phys Oceanogr 47(11):2673–2688

Lankhorst M, Zenk W (2006) Lagrangian observations of the mid-depth and deep velocity fields of the northeastern Atlantic Ocean. J Phys Oceanogr 36:43–63

Lappo SS (1984) On the origin of the northward cross-equatorial heat advection in the Atlantic Ocean. In: Ocean-atmosphere interaction studies, Gidrometeoizdat, Moscow, pp 125–129 [in Russian]; translated in Clivar Exchanges (2006) 11(2):28–10

Larque L, Maamaatuaiahutapu K, Garcon VC (1997) On the intermediate and deep-water flow in the South Atlantic Ocean. J Geophys Res Oceans 102(C6):12425–12440

Lavin AM, Bryden HL, Parrilla G (2003) Mechanisms of heat, freshwater, oxygen and nutrient transports and budgets at 24.5° N in the subtropical North Atlantic. Deep-Sea Res I 50:1099–1128

Lebedev KV, Yaremchuk MI (2000) A diagnostic study of the Indonesian Throughflow. J Geophys Res Oceans 105(5):11243–11258

Lenz B (1997) Bodenwassertransporte über die Rio-Grande-Schwelle. Diplom thesis, Kiel University, 68 pp. [Available from Library, Christian-Albrechts-Universität, D-24098 Kiel, Germany

Lherminier P, Mercier H, Gourcuff C, Alvarez M, Bacon S, Kermabon C (2007) Transports across the 2002 Greenland-Portugal Ovide section and comparison with 1997. J Geophys Res Oceans 112:C07003. https://doi.org/10.1029/2006JC003716

Lherminier P, Mercier H, Huck T, Gourcuff C, Perez FF, Morin P et al (2010) The Atlantic meridional overturning circulation and the subpolar gyre observed at the A25–Ovide section in June 2002 and 2004. Deep Sea Res I 57:1374–1391. https://doi.org/10.1016/j.dsr.2010.07.009

Lumpkin R, Speer K (2007) Global Ocean meridional overturning. J Phys Oceanogr 37:2550–2562. https://doi.org/10.1175/JPO3130.1

Macdonald A (1993) Property fluxes at 30° S and their implications for the Pacific-Indian throughflow and the global heat budget. J Geophys Res Oceans 98(4):6851–6868

Macdonald A (1998) The global ocean circulation a hydrographic estimate and regional analysis. Prog Oceanogr 41:281–382

Macrander A, Käse RH, Send U, Valdimarsson H, Jónsson S (2007) Spatial and temporal structure of the Denmark Strait overflow revealed by acoustic observations. Ocean Dyn 57:75–89. https://doi.org/10.1007/s10236-007-0101-x

Mamayev OI (1992) Abyssal waters of the World Ocean. IRO Publishers, Moscow. [in Russian]

Manta G, Speich S, Karstensen J, Hummels R, Kersalé M, Laxenaire R et al (2021) The South Atlantic meridional overturning circulation and mesoscale eddies in the first GO-SHIP section at 34.5°S. J Geoph Res Oceans 126:e2020JC016962. https://doi.org/10.1029/2020JC016962

Mantyla AW, Reid JL (1983) Abyssal characteristics of the World Ocean waters. Deep-Sea Res I 30(8):805–833. https://doi.org/10.1016/0198-0149(83)90002-X

Marsh R, Nurser AJG, Megann AP, New AL (2000) Water mass transformation in the Southern Ocean of a global isopycnal coordinate GCM. J Phys Oceanogr 30(5):1013–1045

Marshall J, Speer K (2012) Closer of the meridional overturning circulation through Southern Ocean upwelling. Nat Geosci 5:171–180

Martineau DP (1953) The influence of the current systems and lateral mixing upon Antarctic intermediate water in the South Atlantic. Ref. № 53–72. WHOI. pp 12

Mazloff MR, Heimbach P, Wunsch C (2010) An eddy-permitting Southern Ocean state estimate. J Phys Oceanogr 40(5):880–899

McCarthy GD, Smeed DA, Johns WE, Frajka-Williams E, Moat BI et al (2015) Measuring the Atlantic meridional overturning circulation at 26°N. Prog Oceanogr 130:91–111

McCartney MS (1977) Mode water. In: Angel M (ed) A Voyage of Discovery, George Deacon 70th Anniversary Volume. Pergamon Press, Oxford, pp 103–119

McCartney MS (1992) Recirculating components to the deep boundary current of the northern North Atlantic. Prog Oceanogr 29:283–382

McCartney MS, Curry RA (1993) Trans-equatorial flow of Antarctic bottom water in the western Atlantic Ocean: abyssal geostrophy at the equator. J Phys Oceanogr 23:1264–1276

McCartney MS, Bennet SL, Woodgate-Jones ME (1991) Eastward flow through the Mid-Atlantic ridge at 11° N and its influence on the abyss of the Eastern basin. J Phys Oceanogr 21(8): 1089–1121. https://doi.org/10.1175/1520-0485(1991)021<1089:EFTTMA>2.0.CO;2

McDonagh EL, Arhan M, Heywood KJ (2002) On the circulation of bottom water in the region of the Vema Channel. Deep-Sea Res I 49:1119–1139

Meinen CS, Piola AR, Perez RC, Garzoli SL (2012) Deep Western boundary current transport variability in the South Atlantic: preliminary results from a pilot array at 34.5 S. Ocean Sci 8:1041–1054. https://doi.org/10.5194/os-8-1041-2012

Meinen CS, Garzoli SL, Perez RC, Campos E, Piola AR, Chidichimo MP et al (2017) Characteristics and causes of deep Western boundary current transport variability at 34.5°S during 2009–2014. Ocean Sci 13(1):175–194. https://doi.org/10.5194/os-13-175-2017

Meinen CS, Speich S, Piola AR, Ansorge I, Campos E et al (2018) Meridional overturning circulation transport variability at 34.5°S during 2009–2017: baroclinic and barotropic flows and the dueling influence of the boundaries. Geophys Res Lett 45:4810–4888

Memery L, Arhan M, Alvarez-Salgado XA, Messias M-J, Mercier H, Castro CG, Rios AF (2000) The water masses along the western boundary of the south and Equatorial Atlantic. Prog Oceanogr 47:69–98

Mercier H, Morin P (1997) Hydrography of the Romanche and chain fracture zones. J Geophys Res Oceans 102(C5):10,373–10,389. https://doi.org/10.1029/97JC00229

Mercier H, Speer KG (1998) Transport of bottom water in the Romanche fracture zone and the chain fracture zone. J Phys Oceanogr 28(5):779–790. https://doi.org/10.1175/1520-0485(1998)028<0779:TOBWIT>2.0.CO;2

Mercier H, Lherminier P, Sarafanov A, Gaillard F, Daniault N, Desbruyères D, Falina A, Ferron B, Gourcuff C, Huck T, Thierry V (2015) Variability of the meridional overturning circulation at the Greenland–Portugal OVIDE section from 1993 to 2010. Prog Oceanogr 132:250–261

Meredith MP, Locarnini RA, van Scoy KA, Watson AJ, Heywood KJ, King BA (2000) On the sources of Weddell gyre Antarctic bottom water. J Geophys Res Oceans 105(1):1093–1104

Meredith MP, Hughes CW, Foden PR (2003) Downslope convection north of Elephant Island, Antarctica: influence on deep waters and dependence on ENSO. Geophys Res Lett 30(9):1462. https://doi.org/10.1029/2003GL017074

Meredith MP, Naveira Garabato AC, Gordon AL, Johnson GC (2008) Evolution of the deep and bottom waters of the Scotia Sea, Southern Ocean, during 1995–2005. J Clim 21:3327–3343

Messias M-J, Andrie C, Memery L, Mercier H (1999) Tracing the North Atlantic deep water through the Romanche and chain fracture zones with chlorofluoromethanes. Deep-Sea Res I 46:1247–1278. https://doi.org/10.1016/S0967-0637(99)00005-9

Molinari RL, Johns E, Festa JF (1990) The annual cycle of meridional heat flux in the Atlantic Ocean at 26° N. J Phys Oceanogr 20(3):476–482

Molinari RL, Fine RA, Johns E (1992) The deep Western boundary current in the tropical North Atlantic. Deep-Sea Res I 39:1967–1984

Morozov EG, Demidov AN, Tarakanov RYu, Zenk W (2010) Abyssal channels in the Atlantic Ocean: water structure and flows. Dordrecht: Springer. 266 p. https://doi.org/10.1007/978-90-481-9358-5

Mosby H (1934) The waters of the Atlantic Antarctic Ocean. In Scientific Results of the Norwegian Antarctic Expedition 1927–1928, pp 1–131

Müller TJ, Ikeda Y, Zangenberg N, Nonato LV (1998) Direct measurements of western boundary currents off Brazil between 20°S and 28°S. J Geophys Res 103(C3):5429–5437. https://doi.org/10.1029/97JC03529

Naveira Garabato AC, Heywood KJ, Stevens DP (2002a) Modification and pathways of Southern Ocean deep waters in the Scotia Sea. Deep-Sea Res I 49:681–705

Naveira Garabato AC, McDonagh EL, Stevens DP, Heywood KJ, Sanders RJ (2002b) On the export of Antarctic bottom water from the Weddell Sea. Deep-Sea Research II 49:4715–4742

Naveira Garabato AC, Stevens DP, Heywood KJ (2003) Water mass conversion, fluxes, and mixing in the Scotia Sea diagnosed by an inverse model. J Phys Oceanogr 33(12):2565–2587

Naveira Garabato AC, Williams AP, Bacon S (2014) The three-dimensional overturning circulation of the Southern Ocean during the WOCE era. Prog Oceanogr 120:41–78

Neiman VG, Burkov VA, Shcherbinin AD (1997) Dynamics of the Indian Ocean waters. Nauchnyi Mir, Moscow. [in Russian]

Nikurashin M, Vallis G (2011) A theory of deep stratification and overturning circulation in the ocean. J Phys Oceanogr 41:485–502. https://doi.org/10.1175/2010JPO4529.1

Nikurashin M, Vallis G (2012) A theory of the interhemispheric meridional overturning circulation and associated stratification. J Phys Oceanogr 42:1652–1667. https://doi.org/10.1175/JPO-D-11-0189.1

Olsen SM, Hansen B, Quadfasel D, Østerhus S (2008) Observed and modelled stability of overflow across the Greenland–Scotland ridge. Nature 455:519–522. https://doi.org/10.1038/nature07302

Onken R (1995) The spreading of lower circumpolar water in the Atlantic Ocean. J Phys Oceanogr 25(12):3051–3063

Orsi AH (2010) Recycling bottom waters. Nat Geosci 3:307–309

Orsi AH, Nowlin WD Jr, Whitworth T III (1993) On the circulation and stratification of the Weddell gyre. Deep-Sea Res I 40:169–203

Orsi AH, Whitworth T III, Nowlin WD Jr (1995) On the meridional extent and fronts of the Antarctic circumpolar current. Deep-Sea Res I 42(5):641–673

Orsi AH, Johnsson GC, Bullister JL (1999) Circulation, mixing, and production of Antarctic bottom water. Prog Oceanogr 43:55–109. https://doi.org/10.1016/S0079-6611(99)00004-X

Orsi AH, Smethie WM, Bullister JL (2002) On the total input of Antarctic waters to the deep ocean: a preliminary estimate from chlorofluorocarbon measurements. J Geophys Res Oceans 107(C8): 3122. https://doi.org/10.1029/2001JC000976

Ostlund HG, Broecker WS, Spencer D (1987) GEOSECS Atlantic, Pacific and Indian Ocean expeditions. Shorebased data and graphics. U.S. Govt. Printing Office, Washington DC, vol 17

Outdot C, Morin P, Baurand F, Wafar M, Le Corre P (1998) Northern and southern water masses in the equatorial Atlantic: distribution of nutrients on the WOCE A6 and A7 lines. Deep-Sea Res I 45:873–902

Patterson SL, Whitworth T (1990) Physical oceanography. In: Glasby GP (ed) Antarctic Sector of the Pacific, Elsevier oceanography series, vol 51, pp 55–93

Peacock S, Visbeck M, Broecker WS (2000) Deep water formation rates inferred from global tracer distributions: an inverse model. In: Kasibhatla P et al (eds) Inverse methods in global biogeochemical cycles, Geophysical monograph series, vol 114. AGU, Washington, DC, pp 185–195

Pickart RS (1992) Water mass components of the North Atlantic deep western boundary current. Deep-Sea Res I 39:1553–1572

Pickart RS, Straneo F, Moore GWK (2003) Is Labrador Sea water formed in the Irminger basin? Deep-Sea Res I 50:23–52

Piron A, Thierry V, Mercier H, Caniaux G (2017) Gyre-scale deep convection in the subpolar North Atlantic Ocean during winter 2014–2015. Geophys Res Lett 44(3):1439–1447

Racapé V, Thierry V, Mercier H, Cabanes C (2019) ISOW spreading and mixing as revealed by deep-Argo floats launched in the Charlie-Gibbs Fracture Zone. J Geophys Res Oceans 124. https://doi.org/10.1029/2019JC015040

Reid JL (1965) Intermediate waters of the Pacific Ocean. The Johns Hopkins Press, Baltimore

Reid JL (1989) On the total geostrophic circulation of the South Atlantic Ocean: flow pattern, tracers and transports. Prog Oceanogr 23:149–244

Reid JL, Nowlin WD, Patzert WC (1977) On the characteristics and circulation of the Southwestern Atlantic Ocean. J Phys Oceanogr 7(1):62–91

Rhein M, Schott F, Stramma L, Fischer J, Plahn O, Send U (1996) The deep Western boundary current in the tropical Atlantic: deep water distribution and circulation off Brazil. Int WOCE Newslett 23:11–14

Rhein M, Stramma L, Krahmann G (1998) The spreading of Antarctic bottom water in the tropical Atlantic. Deep-Sea Res I 45:507–527

Rhein M, Kieke D, Steinfeldt R (2015) Advection of North Atlantic deep water from the Labrador Sea to the southern hemisphere. J Geophys Res Oceans 120(4):2471–2487. https://doi.org/10.1002/2014jc010605

Richardson PL (2008) On the history of meridional overturing circulation schematic diagrams. Progr Oceanogr 76:466–486

Rintoul SR (1998) On the origin and influence of Adelie Land bottom water, ocean, ice, and atmosphere: interactions at the Antarctic continental margin. Antarctic Res Ser 75:151–171

Rintoul SR, Bullister JL (1999) A late winter hydrographic section from Tasmania to Antarctica. Deep-Sea Res I 46:1417–1454

Roemmich D (1983) The balance of geostrophic and Ekman transports in the tropical Atlantic Ocean. J Phys Oceanogr 13(8):1534–1539

Ruth C, Well R, Roether W (2000) Primordial ^{3}He in South Atlantic deep waters from sources on the mid-Atlantic ridge. Deep-Sea Res I 47:1059–1075

Sandoval FJ, Weatherly GL (2001) Evolution of the deep western boundary current of Antarctic bottom water in the Brazil Basin. J Phys Oceanogr 31(6):1440–1460. https://doi.org/10.1175/1520-0485(2001)031<1440:EOTDWB>2.0.CO;2

Sarafanov AA, Sokov AV, Falina AS (2009) Warming and salinification of Labrador Sea water and deep waters in the subpolar North Atlantic at 60°N in 1997–2006. Oceanology 49(2):193–204

Sarafanov AA, Falina AS, Sokov AV, Zapotylko VS, Gladyshev SV (2018) Ship-based monitoring of the Northern North Atlantic Ocean by the Shirshov Institute of Oceanology. The main results. In: Velarde G, Tarakanov RY, Marchenko AV (eds) The ocean in motion. Springer, Dordrecht, pp 415–427

Sarukhanyan EI, Smirnov NP (1986) Water masses and circulation of the Southern Ocean. Leningrad, Gidrometeoizdat. 288 p. [in Russian]

Saunders PM (1987) Flow through discovery gap. J Phys Oceanogr 17(5):631–643

Saunders PM (1994) The flux of overflow water through the Charlie-Gibbs fracture zone. J Geophys Res Oceans 99(C6):12343–12355

Saunders PM, King B (1995) Oceanic fluxes on the WOCE A11 section. J Phys Oceanogr 25(9):1942–1958

Schlitzer R (1987) Renewal rates of East Atlantic deep water estimated by inversion of 14C data. J Geophys Res Oceans 92:2953–2980

Schmid C, Siedler G, Zenk W (2000) Dynamics of intermediate water circulation in the subtropical South Atlantic. J Phys Oceanogr 30(12):3191–3211
Schmitz WJ (1995) On the interbasin-scale thermohaline circulation. Rev Geophys 33(2):151–173
Schmitz WJ (1996a) On the World Ocean circulation: volume I, some global features. North Atlantic circulation (technical report WHOI–96–03). Woods Hole, Woods Hole Oceanographic Institution, 150 p
Schmitz WJ (1996b) On the world ocean circulation: volume II. The Pacific and Indian oceans, a global update. Technical report WHOI-96-08. Woods Hole Oceanographic Institution, USA
Schmitz WJ Jr, McCartney MS (1993) On the North Atlantic circulation. Rev Geophys 31:29–49
Schott FA, Brandt P (2007) Circulation and deep-water export of the subpolar North Atlantic during the 1990s. In: Schmittner A, Chiang J, Hemmings S (eds) Ocean circulation: mechanisms and impacts, vol 173. Geophysical Monograph Series, pp 91–118. https://doi.org/10.1029/173gm08
Schott FA, Dengler M, Brandt P, Affler K et al (2003) The zonal currents and transports at 35° W in the tropical Atlantic. Geophys Res Lett 30(7):1349. https://doi.org/10.1029/2002GL016849
Sen Gupta A, England M (2004) Evaluation of interior circulation in a high–resolution global ocean model. Part I: deep and bottom waters. J Phys Oceanogr 34(12):2592–2614
Sen Gupta A, England M (2007) Evaluation of interior circulation in a high–resolution global ocean model. Part II: Southern Hemisphere intermediate, mode, and thermocline waters. J Phys Oceanogr 37(1):2612–2636
Shakespeare CJ, Hogg AM (2012) An analytical model of the response of the meridional overturning circulation to changes in wind and buoyancy forcing. J Phys Oceanogr 42:1270–1287. https://doi.org/10.1175/JPO-D-11-0198.1
Sievers HA, Nowlin WD Jr (1984) The stratification and water masses at Drake Passage. J Geophys Res Oceans 89(C6):10489–10514
Sloyan BM, Rintoul SR (2001a) Circulation, renewal, and modification of Antarctic mode and intermediate water. J Phys Oceanogr 31(4):1005–1030
Sloyan BM, Rintoul SR (2001b) The Southern Ocean limb of the global deep overturning circulation. J Phys Oceanogr 31(1):143–173
Smethie WM, Fine R (2001) Rates of North Atlantic deep water formation calculated from chlorofluorocarbon inventories. Deep Sea Res I 48(1):189–215
Smethie WM, Chipman DW, Swift JH, Koltermann KP (1988) Chlorofluoromethanes in the Arctic Mediterranean seas: evidence for formation of bottom water in the Eurasian Basin and deep-water exchange through Fram Strait. Deep-Sea Res I 35:347–369
Smythe-Wright D, Boswell S (1998) Abyssal circulation in the argentine basin. J Geophys Res Oceans 103(C8):15845–15851
Speer KG, McCartney MS (1991) Tracing lower North Atlantic deep water across the equator. J Geophys Res Oceans 96(C11):20443–20448
Speer KG, McCartney MS (1992) Bottom water circulation in the Western North Atlantic. J Phys Oceanogr 22(1):83–92
Speer KG, Zenk W (1993) The flow of Antarctic bottom water into the Brazil Basin. J Phys Oceanogr 23:2667–2682
Srokosz M, Danabasoglu G, Patterson M (2021) Atlantic meridional overturning circulation: reviews of observational and modeling advances – an introduction. J Geophys Res Oceans 126(1):e2020JC016745
Stammer D, Wunsch C, Giering R, Heimbach P, Marotzke J, Adcroft A et al (2002) The global ocean circulation during 1992–1997, estimated from ocean observations and a general circulation model. J Geophys Res Oceans 107(9):3118. https://doi.org/10.1029/2001JC000888
Stephens JC, Marshall DP (2000) Dynamical pathways of Antarctic bottom water in the Atlantic. J Phys Oceanogr 30(3):622–640
Stramma L, England M (1999) On the water mass and mean circulation of the South Atlantic Ocean. J Geophys Res Oceans 104(C9):20863–20883
Stramma L, Rhein M (2001) Variability in the deep Western boundary current in the equatorial Atlantic at 44° W. Geophys Res Lett 28:1623–1626

Sutherland DA, Pickart RS (2008) The East Greenland coastal current: structure, variability, and forcing. Prog Oceanogr 78:58–77. https://doi.org/10.1016/j.pocean.2007.09.006

Sverdrup H, Johnson M, Fleming R (1942) The oceans. Their physics, chemistry and general biology. Prentice Hall Inc, New York

Swift JH, Aagaard K, Malmberg S-A (1980) The contribution of the Denmark Strait overflow to the deep North Atlantic. Deep-Sea Res I 27:29–42

Talley LD (2008) Freshwater transport estimates and the global overturning circulation: shallow, deep and throughflow components. Prog Oceanogr 78:257–303. https://doi.org/10.1016/j.pocean.2008.05.001

Talley LD (2013) Closure of the global overturning circulation through the Indian, Pacific, and southern oceans: schematics and transports. Oceanography 26:80–97. https://doi.org/10.5670/oceanog.2013.07

Talley LD, Reid JL, Robbins PE (2003) Data-based meridional overturning streamfunctions for the Global Ocean. J Clim 16(10):3213–3226

Tarakanov RY (2010) Circumpolar bottom water in the Scotia Sea and Drake Passage. Oceanology 50(1):5–22

Tarakanov RY (2012) The Drake Passage and the Scotia Sea as an orographic barrier for the Antarctic circumpolar current. Oceanology 52:157–170

Thompson AF, Stewart A, Bischoff T (2016) A multibasin residual-mean model for the global overturning circulation. J Phys Oceanogr 46:2583–2604. https://doi.org/10.1175/JPO-D-15-0204.1

Toggweiler JR, Samuels B (1995) Effect of Drake Passage on the global thermohaline circulation. Deep-Sea Res I 42(4):477–500

Tomczak M, Godfrey SR (1994) Regional oceanography: an introduction. Pergamon Press, p 422

Tsuchiya M (1989) Circulation of the Antarctic intermediate water in the North Atlantic Ocean. J Mar Res 47(4):747–755

Tsuchiya V, Talley LD, McCartney MS (1992) An eastern Atlantic section from Iceland southward across the equator. Deep Sea Res I 39:1885–1918

Tsuchiya V, Talley LD, McCartney MS (1994) Water mass distribution in the western South Atlantic: a section from South Georgia Island (54°S) northward across the equator. J Mar Res 52:55–81

Tucholke BE, Wright WR, Hollister CD (1973) Abyssal circulation over the greater Antilles outer ridge. Deep Sea Res I 20:973–995

Våge K, Pickart RS, Sarafanov A, Knutsen O, Mercier H, Lherminier P et al (2011) The Irminger gyre: circulation, convection, and interannual variability. Deep-Sea Res I 58(5):590–614. https://doi.org/10.1016/j.dsr.2011.03.001

Valla D, Piola AR, Meinen CS, Campos E (2018) Strong mixing and recirculation in the North-western argentine basin. J Geophys Res Oceans 123:4624–4648. https://doi.org/10.1029/2018JC013907

Vallis GK (2017) Atmospheric and oceanic fluid dynamics. Cambridge University Press, 946 pp

van Aken HM (2007) The oceanic thermohaline circulation: an introduction. Atmospheric and Oceanographic Sciences Library, vol 39. New York, Springer

van Aken HM, de Boer CJ (1995) On the synoptic hydrography of intermediate and deep-water masses in the Iceland Basin. Deep-Sea Res I 42(2):165–189

van Sebille E, Johns WE, Beal LM (2012) Does the vorticity flux from Agulhas rings control the zonal pathway of NADW across the South Atlantic? J Geophys Res 117:C05037. https://doi.org/10.1029/2011JC007684

Vanicek M, Siedler G (2002) Zonal fluxes in the deep water layers of the western South Atlantic Ocean. J Phys Oceanogr 32(8):2205–2235

Weatherly GL, Kelley EA (1982) 'Too cold' bottom layers at the base of the Scotian rise. J Mar Res 40(4):985–1012

Well R, Roether W, Stevens DP (2003) An additional deep-water mass in Drake Passage as revealed by 3He data. Deep-Sea Res I 50:1079–1098

Weppernig R, Schlosser P, Khatiwala S, Fairbanks RG (1996) Isotope data from Ice Station Weddell: implications for deep water formation in the Weddell Sea. J Geophys Res Oceans 101(C9):25723–25739

Whitehead JA (1989) Surges of Antarctic bottom water into the North Atlantic. J Phys Oceanogr 19(6):853–861

Whitehead JA, Worthington LV (1982) The flux and mixing rates of Antarctic bottom water within the North Atlantic. J Geophys Res Oceans 87(C10):7903–7924. https://doi.org/10.1029/jc087ic10p07903

Whitworth T, Nowlin WD, Orsi AH, Locarnini RA, Smith SG (1994) Weddell Sea shelf water in the Bransfield Strait and Weddell-scotia confluence. Deep-Sea Res I 41:629–641

Whitworth T, Orsi AH, Kim S-J, Nowlin WD, Locarnini RA (1998) Water masses and mixing near the Antarctic slope front. In: Jacobs S, Weiss R (eds) Ocean, ice, and atmosphere: interactions at the Antarctic continental margin, Antarctic Research Series 75. American Geophysical Union, Washington, DC, pp 1–27

Wienders N, Arhan M, Mercier H (2000) Circulation at the western boundary of the south and equatorial Atlantic: exchanges with the ocean interior. J Mar Res 58:1007–1039

WOD18 (2018) World Ocean Database 2018 https://www.nodc.noaa.gov/OC5/WOD/pr_wod.html

Wolfe CL, Cessi P (2011) The adiabatic pole-to-pole overturning circulation. J Phys Oceanogr 41:1795–1810

Worthington LV (1970) The Norwegian Sea as a mediterranean basin. Deep-Sea Res I 17:77–84

Worthington LV, Volkmann GH (1965) The volume transport of the Norwegian Sea overflow water in the North Atlantic. Deep-Sea Res I 12:667–676

Wright WR (1970) Northward transport of Antarctic bottom water in the Western Atlantic Ocean. Deep-Sea Res I 17:367–371

Wüst G (1936) Schichtung und Zirkulation des Atlantischen Ozeans, Das Bodenwasser und die Stratosphäre. In: A. Defant (Ed) Wissenschaftliche Ergebnisse, Deutsche Atlantische Expedition auf dem Forschungs - und Vermessungsschiff “Meteor” 1925–1927, 6(1), Walter de Gruyter & Co, Berlin, 411pg

Xu X, Schmitz WJ, Hurlburt HE, Hogan PJ, Chassignet EP (2010) Transport of Nordic seas overflow water into and within the Irminger Sea: an eddy-resolving simulation and observations. J Geophys Res 115:C12048

Zangenberg N, Siedler G (1998) Path of the North Atlantic deep water in the Brazil Basin. J Geophys Res 103(C3):5419–5428. https://doi.org/10.1029/97JC03287

Zenk W, Hogg NG (1996) Warming trend in Antarctic bottom water flowing into the Brazil Basin. Deep-Sea Res I 43(9):1461–1473

Zenk W, Morozov EG (2007) Decadal warming of the coldest Antarctic bottom water flow through the Vema Channel. Geophys Res Lett 34:L14607. https://doi.org/10.1029/2007GL030340

Zou S, Lozier S, Zenk W, Bower A, Johns W (2017) Observed and modeled pathways of the Iceland Scotland overflow water in the eastern North Atlantic. Prog Oceanogr 159:211–222. https://doi.org/10.1016/j.pocean.2017.10.003

Chapter 2
General Overview of Abyssal Pathways, and Channels (for Waters of the Antarctic Origin)

Abstract This chapter describes a general scheme of the distribution of Antarctic waters in the abyssal of the Atlantic Ocean after propagating from their sources in the Southern Ocean. The chapter also summarizes the quantitative indicators of bottom flows in the abyssal channels. Comparison of the limits of distribution of Antarctic waters is given depending on the specific characteristics used to identify them.

Keywords Abyssal pathways · Deep cataracts · Indicators of bottom flows · Limits of spreading

2.1 Propagation of Deep and Bottom Waters as Series of Deep Cataracts

Cold and dense Antarctic waters descending to the bottom at high latitudes of the Southern Ocean spread to the north flowing from one basin to another. The ridges separating the abyssal basins block spreading of waters with the lowest temperatures. Warmer bottom waters flowing over the crests of the ridges spill into the abyss of the next downstream basin. This worldwide spilling called the giant cataracts is often accompanied by intense turbulent mixing with the surrounding warmer waters that increases the water temperature in the bottom layer (Whitehead 1989). Deep cataracts of the Antarctic waters in the Atlantic Ocean from Antarctica to the North Atlantic are included in the lower limb of the Antarctic cell of the Global Overturning Circulation in this ocean: flows over the South Orkney Ridge into the Scotia Sea, over the Rio Grande Rise, Ceara Rise (4° N), through the equatorial Romanche and Chain fracture zones, through the Vema Fracture Zone (11° N), and the ridge of the East Azores Fracture Zone (Whitehead 1989). The bottom flows over the Falkland Fracture Zone and over the Walvis Ridge are also related to these cataracts (Fig. 2.1).

There are several cataracts at relatively high latitudes of the North Atlantic, which are related only to the lower branch of the North Atlantic cell: the flows over thresholds separating the Nordic Arctic seas from the Atlantic Ocean (the Denmark

E. G. Morozov et al., *Bottom Gravity Currents and Overflows in Deep Channels of the Atlantic Ocean*, https://doi.org/10.1007/978-3-030-83074-8_2

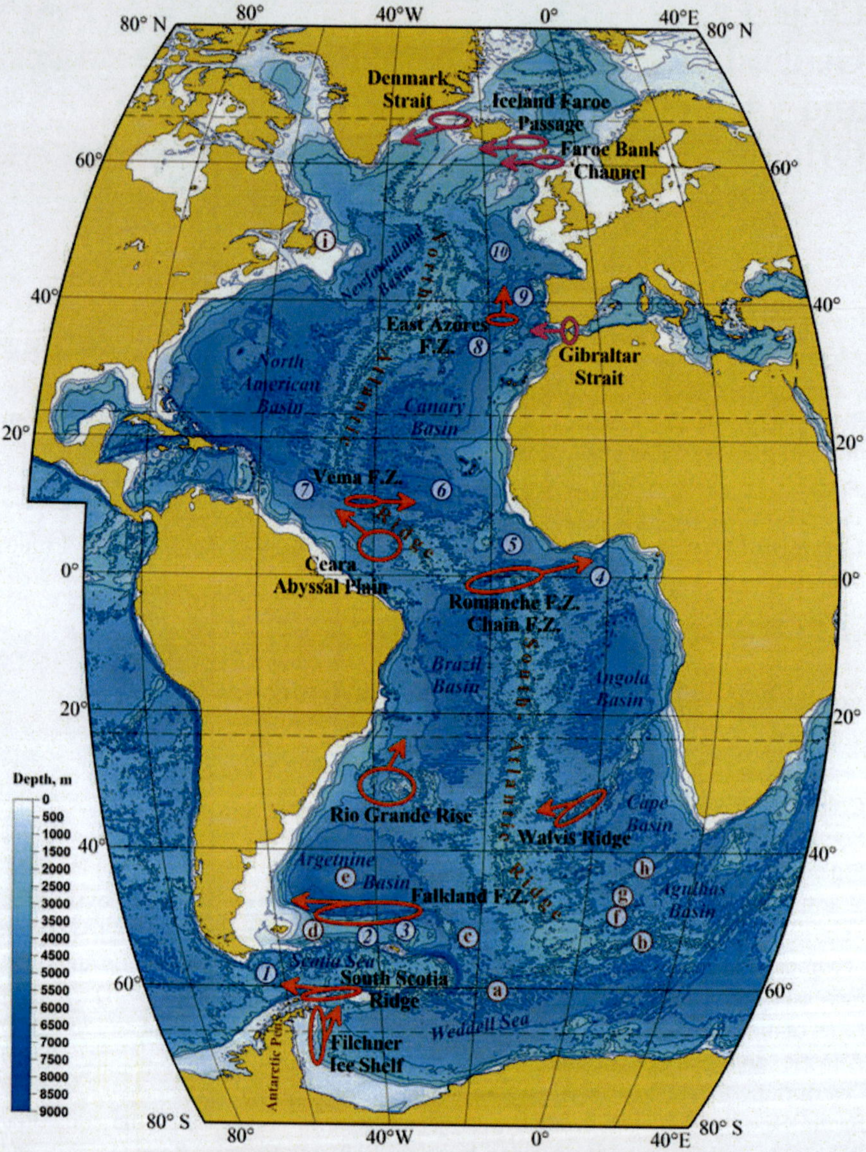

Fig. 2.1 Bottom topography and main deep cataracts of the Atlantic Ocean. Red and magenta ellipses show the cataracts related and not related to the propagation of Antarctic waters in the abyss of the World Ocean. The arrows show the general direction of the flow in the cataracts. Deep blue numerals in circles indicate: (1) Drake Passage, (2) Falkland Trough (3) Georgia Basin, (4) Guinea Basin, (5) Sierra Leone Basin, (6) Cape Verde Basin, (7) Guiana Basin, (8) Madeira Basin, (9) Iberian Basin, (10) West European Basin. Brown letters in circles indicate: (a) North Weddell Ridge, (b) Southwest Indian Ridge, (c) Islas Orcadas Rise, (d) Falkland Plateau, (e) Zapiola Ridge, (f) Shona Ridge, (g) Meteor Rise, (h) Agulhas Ridge (i) Newfoundland Bank

Strait and Faroe-Iceland Channel). The Strait of Gibraltar can also be attributed to the channels of the North Atlantic meridional circulation cell. Each cataract can consist of a series of local spillways associated with critical sections (narrows and/or thresholds) along the pathway of the abyssal flow. The flow through such sections, which are usually located in deep-sea channels, occurs in the so-called hydraulically controlled flow mode. In this mode, the flow is accelerated when passing through the critical section due to the available potential energy of the gravitational field, followed by shock dissipation of the released kinetic energy in the hydraulic jump zone and mixing due to this dissipation including the entrainment of overlying water layers in the bottom flow (Pratt and Whitehead 2007), and also due to the generation of internal waves over the sill (Clément et al. 2017).

Investigations of hydraulically controlled overflow in deep-water channels are generally based on indirect indications following from the theory: channel width in its narrowest parts and sill height, density stratification that restructures when the current overflows the threshold, correspondence of measured velocities and transports in the critical section to the velocities and transports predicted by the theory. Such a theory is well developed for the case of a two-layer inviscid incompressible fluid (Baines 1984; Armi 1986; Lawrence 1993). The so-called 1½-layer model of hydraulically controlled fluid flow between two basins through a channel connecting them with a rectangular cross section is of key importance for the study of flows in real channels in the oceans (Whitehead et al. 1974). Based on this model, Whitehead (1998) compared the transport estimates in a number of ocean channels obtained on the basis of field measurements (direct velocity measurements, geostrophic estimates) and those predicted by theory. Later, the same model was used to study hydraulically controlled flow in a number of studies, in particular, along passages in the Iceland-Shetland Ridge in the North Atlantic (Nikolopoulos et al. 2003; Girton et al. 2006).

According to the well-known concept (Faller 1966; Munk 1966; Munk and Wunsch 1998), the dissipation rate of the available potential energy of the gravitational field on the global scale is negligible compared to the two other energy sources, which establish the deep density stratification in the oceans: barotropic tide and wind energy. Energy dissipation of the barotropic tide in the ocean occurs when tidal currents flow over sloping bottom topography of underwater ridges and internal tidal waves are generated (Morozov 1995). According to laboratory experiments in (Whitehead and Wang 2008), deep ocean circulation can be driven by turbulent mixing. In the case of wind, the transport of wind energy to the abyssal occurs by means of the baroclinic instability of the jet currents of the upper, thermocline layer of the ocean with the formation of meanders and eddies and their further barotropization. This mechanism performs energy transport to the ocean interior. However, in contrast to this previously mentioned concept, Wright (1972) showed that the potential energy released by sinking of cold-water masses is approximately the same magnitude as the wind stress over the North Atlantic. Moreover, there are theoretical publications (Bryden and Nurser 2003) and observations in the ocean (Polzin et al. 1996; Mercier and Morin 1997; Ferron et al. 1998; Tarakanov et al. 2013b; Tarakanov et al. 2018; Tarakanov et al. 2020), which show

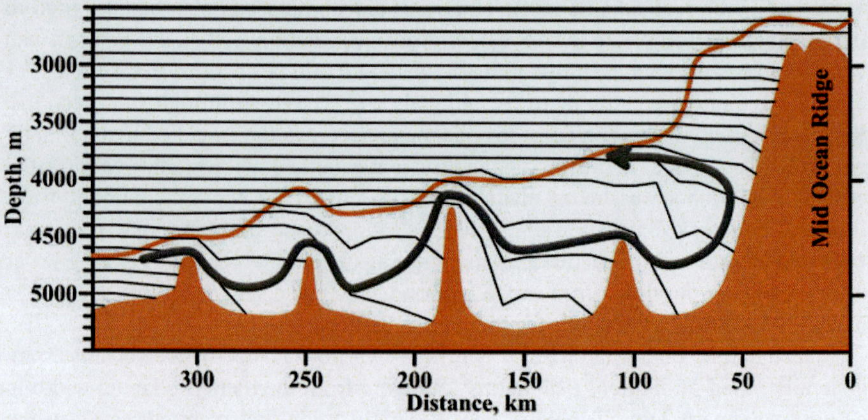

Fig. 2.2 Scheme of density distribution along a dead-end zonal fracture zone in the mid-ocean ridge; west is on the left side. Bottom topography along the thalweg is shown in brown color; crest depth of the shallowest of two walls of the fracture zone is shown with a brown line above the thalweg. Broken black lines denote contour lines of neutral density illustrating virtual data from a hydrographic section. Contour lines are initially spaced 100 m apart in the westernmost profile. Gray arrow shows a scheme of circulation in the fracture zone

that despite the localization of the hydraulically controlled flow process, mixing in the narrows and over the sills can play a significant role in the transformation of the flow along the deep-water channels, as well as in the formation of thermohaline characteristics and establishing the density stratification in the basins downstream.

It was reported in the publications (Thurnherr and Speer 2003; Thurnherr et al. 2005) that hydraulically controlled overflow that takes place on a whole set of narrows and sills of the mid-oceanic transform ridges is responsible for bottom mixing in a vast domain above the ridge slope. In this case, in each channel (fracture zone) with a hydraulically controlled flow through a critical section (narrowness or sill), a local overturning circulation cell is formed, created by the pressure drop through it, with the lower branch directed in the channel toward the ridge crest, upward motion due to mixing with overlying layers and reverse horizontal motion over the slope of the ridge (Thurnherr and Speer 2003; Thurnherr et al. 2020). The presence of a similar critical section (sill) with a hydraulically controlled flow towards the crest of the South Mid-Atlantic Ridge has recently been established in one of the "dead-end" (for the eastward near-bottom flow) transform fractures in the Brazil Basin at 21° S (Clément et al. 2017). A scheme is shown in Fig. 2.2.

Targeted field studies of hydraulically controlled flow through narrows and over sills for the investigation of the structure of bottom currents including its acceleration and revealing the hydraulic jump require specific work, in particular, the specific configuration of the study site of CTD/LADCP stations, application of the so-called "tow-yo" sections when a vessel is moving at a low speed and rises and lowers the profiling device at a distance from the bottom, as well as deploying mooring stations. Measurements using microstructural probes can also be performed to provide an instrumental estimate of the energy dissipation rate. Such works are not numerous today and started not long ago. We cite several publications with such a set of measurements in the rift valleys of the North Mid-Atlantic Ridge (Thurnherr et al. 2002), near the South Mid-Atlantic Ridge (Tippenhauer et al. 2015), in the transform fracture zone in the Brazil Basin at 21° S (Clément et al. 2017).

2.2 Propagation of Antarctic Waters in the Abyss of the Atlantic

Spreading of deep waters from the regions of their formation in Antarctica and the North Atlantic towards the equator due to the western intensification is concentrated in the relatively narrow Deep Western Boundary Currents (DWBC), which are mainly confined to the eastern slopes of the continents (or submarine ridges or lines of islands) and oriented in the meridional direction (Stommel 1958; Stommel and Arons 1960). Since the bottom waters of Antarctic origin are colder compared with those in the North Atlantic, the tongues of lower temperature bottom flows in the Atlantic (Fig. 2.3) show the main pathways of Antarctic waters from the sources of their formation in the Southern Ocean. A qualitative distribution scheme of these waters is shown in Fig. 2.4 taking into account the detailed regional patterns of abyssal circulation given in this book. The scheme in Fig. 2.4 shows the pathways located only at the bottom. Due to mixing isotherms gradually terminate at the bottom, and warmer waters appear at the bottom. The actual pathways of these warmer waters may differ when the flow is not at the bottom. Sometimes, such pathways strongly differ from this scheme. In this section of the book, we restrict ourselves to describing only the general abyssal circulation in the Atlantic and mentioning only key publications.

We emphasize that narrow depressions, fractures, and channels in submarine ridges play the key role in the flow of abyssal waters between individual deep basins. Table 2.1 shows the estimates of the Antarctic Bottom Water transport in the abyssal channels of the Atlantic based on publications. In this table, we selected the estimates of transport based on direct measurements of the current velocity either on moorings or lowered velocity profilers (LADCP). More detailed overview tables of transport estimates in abyssal channels and across individual sections of deep-sea basins of the Atlantic are given in the corresponding sections of this book. Table 2.1 also shows estimates of the minimum bottom potential temperature in these channels based on the World Ocean DataBase 2018 (WOD18 2018). In addition, the data of measurements of the Shirshov Institute of Oceanology of the Russian Academy of Sciences (both temperature and velocity of abyssal currents) were also used.

Cold Antarctic waters from the sources in the Indian Ocean flowing westward along the Antarctic slope in the southern periphery of the Weddell Gyre (Meredith et al. 2000; Hoppema et al. 2001; Klatt et al. 2002) mix with the waters that were formed and renewed on the shelf of the southern and western parts of the Weddell

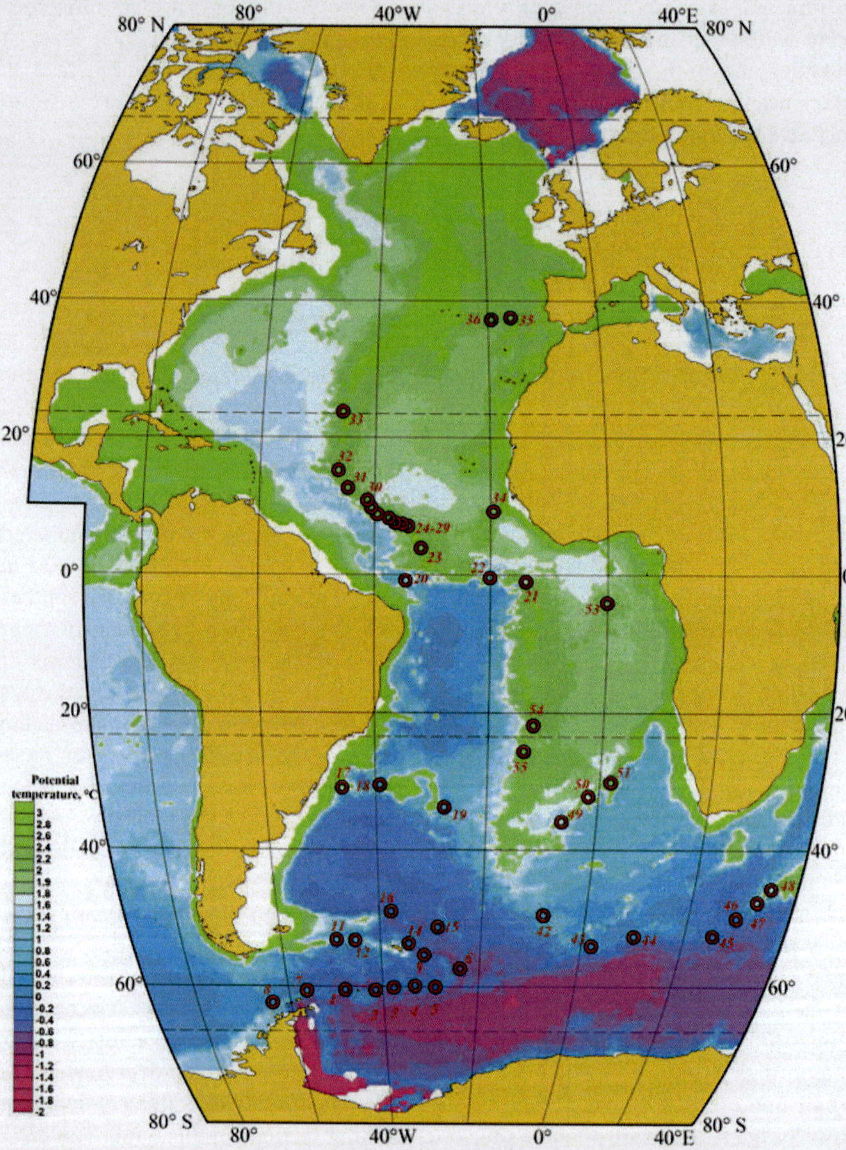

Fig. 2.3 Climatic distribution of the bottom potential temperature (°C) and deep channels (Table 2.1) in the Atlantic Ocean based on the WOCE Climatology (Gouretski and Koltermann 2004). The channels are indicated with circles. The numbers at the circles correspond to the numbers in Table 2.1

Sea and reach the northwestern periphery of the Weddell Gyre (Fig. 2.4). The coldest and densest Weddell Sea Deep Water ($\gamma^n > 28.40$) is blocked in the Weddell Gyre from spreading to the north by the South Scotia Ridge (the southern arc of the South Antilles), the South American Antarctic Ridge, and Southwest Indian Ridge

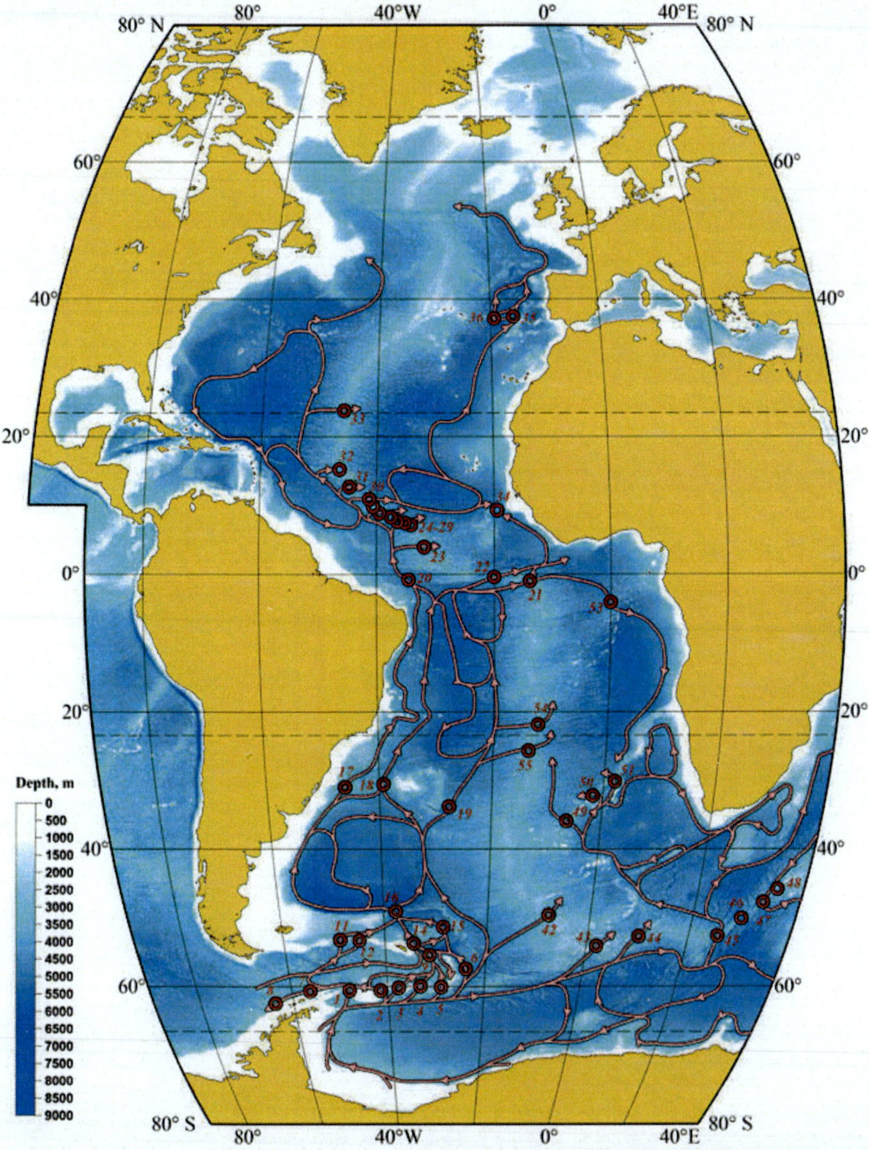

Fig. 2.4 Sketch of propagation of Antarctic waters in the bottom layer and deep channels (Table 2.1) in the Atlantic Ocean. The channels are indicated with circles. The numbers at the circles correspond to the numbers in Table 2.1

(Figs. 2.3 and 2.4) (Orsi et al. 1999; Naveira Garabato et al. 2002b). The densest layers of the overlying Weddell Sea Deep Water moving eastward along these ridges reach a deep depression between the Crozet Islands and the Kerguelen Ridge at 55° E, through which they penetrate into the Crozet Basin and further into the

Table 2.1 Minimal potential temperature near bottom and transport estimates in the channels of the Atlantic Ocean

№	Name of the channel	Minimum potential temperature, °C	Transport, Sv	Data	Comments
1	Philip Passage	<−0.60	5.0–8.4 for θ < 0.2 °C	Naveira Garabato et al. (2003)	Inverse model (box-model) of the Scotia Sea, LADCP-measurements
2	Orkney Passage				
3	Bruce Passage				
4	Discovery Passage				
5	Hoyer Passage				
6	South Sandwich Trench	<−0.70	5.0 for θ < 0.2 °C	Naveira Garabato et al. (2002b)	LADCP-measurements, Inverse model of the Weddell Sea
7	Shackleton Passage	−0.11	1.0 for θ < 0.4 °C	Gladyshev et al. (2008)	LADCP-measurements
8	Hero Passage	0.10			
9	Georgia Passage	<−0.30			
11	Shag Rocks Passage	0.20			
12	Black Rock Passage				
14	Northeastern Georgia Passage	<−0.40			
15	Unnamed Passage east of the Islas Orcadas Rise	<−0.70			
16	Falkland Passage	<−0.40			
17	Lower Santos Plateau (with Santos Basin)	0.12	0.8–0.0 for θ < 2.0 °C	Hogg et al. (1999)	Moorings + geostrophic calculations
			2.0 for θ < 2.0 °C	Speer and Zenk (1993)	Geostrophic calculations with reference level at θ = 2.0 °C
18	Vema Channel (southern and standard sections)	< −0.20	1.3–3.2 for θ < 0.2 °C	Denker (2007), Zenk (2008)	Moorings
			1.5–2.0 for θ < 0.2 °C	Morozov et al. (2010)	LADCP-measurements
			1.5–5.0 for θ < 2.0 °C	Holfort and Siedler (2001), Denker (2007) and Zenk (2008)	Moorings

(continued)

Table 2.1 (continued)

№	Name of the channel	Minimum potential temperature, °C	Transport, Sv	Data	Comments
			1.5–3.2 for θ < 2.0 °C	Morozov et al. (2010)	LADCP-measurements
			4.5 for θ < 2.0 °C	Hogg et al. (1999)	Moorings + geostrophic calculations with reference level at θ = 2.0 °C
			3.9 for θ < 2.0 °C	Speer and Zenk (1993)	Geostrophic calculations with reference level at θ = 2.0 °C
			1.1–3.1 for θ < 2.0 °C	Morozov et al. (2019)	LADCP-measurements
19	Hunter Channel	0.17	1.4–4.1 for θ < 2.0 °C	Zenk et al. (1999)	Moorings
			0.7 for θ < 2.0 °C	Speer and Zenk (1993)	Geostrophic calculations with reference level at θ = 2.0 °C
20	Equatorial Channel	<0.60	2.1 for θ < 1.8 °C	Hall et al. (1997) and Limeburner et al. (2005)	Moorings
21	Chain Fracture Zone	0.40 at the entrance	0.39–0.73 for θ < 1.9 °C	Mercier and Speer (1998)	Moorings
		1.70 at the exit			
22	Romanche Fracture Zone	0.48 at the entrance	0.52–0.80 for θ < 1.9 °C	Mercier and Speer (1998)	Moorings
		1.40 at the exit			
			0.88 for θ < 2.0 °C	Morozov et al. (2010), van Haren et al. (2014) and Tarakanov et al. (2018)	LADCP-measurements
			0.69 for θ < 1.9 °C		
			0.62 for θ < 1.8 °C		
			0.43 for θ < 1.5 °C		
23	Strakhov Fracture Zone	1.74	0.5 for θ < 2.0 °C	Tarakanov et al. (2017), Morozov et al. (2017)	LADCP-measurements
24	Bogdanov Fracture Zone	1.97			
25	7°30'N Fracture Zone	1.20 at the entrance			
		~1.80 at the exit			

(continued)

Table 2.1 (continued)

№	Name of the channel	Minimum potential temperature, °C	Transport, Sv	Data	Comments
26	Vernadsky Fracture Zone	1.87			
27	Doldrums Fracture Zone	1.87			
28	Arkhangelsky Fracture Zone	1.85			
29	10°N Fracture Zone	2.08			
30	Vema Fracture Zone	<1.35	0.05–0.46 for $\theta < 1.5$ °C	Vangriesheim (1980a, b)	Mooring
			1.8–2.0 for $\theta < 2.0$ °C	Fischer et al. (1996)	LADCP-measurements + geostrophic calculations
			2.1–2.4 for $\sigma_4 > 45.9$		
			0.01–0.04 for $\theta < 1.5$ °C	Demidov et al. (2007)	LADCP-measurements
			0.12–0.37 for $\theta < 1.8$ °C		
			0.13–0.51 for $\theta < 1.9$ °C		
			0.15–0.64 for $\theta < 2.0$ °C		
			0.7–1.2 for $\theta < 2.0$ °C	Morozov et al. (2017)	LADCP-measurements
31	Marathon Fracture Zone	1.92		Morozov et al. (2018)	
32	Cape Verde Fracture Zone	2.02			
33	Kane Fracture Zone	1.77			
34	Kane Gap	1.80	0.0 (amplitude 0.2 Sv) for $\theta < 1.9$ °C	Morozov et al. (2013)	Mooring, LADCP-measurements
35	Discovery Gap	2.01	0.2 for $\theta < 2.05$ °C	Saunders (1987)	Moorings

(continued)

Table 2.1 (continued)

№	Name of the channel	Minimum potential temperature, °C	Transport, Sv	Data	Comments
36	Western Discovery Gap	1.99	0.3 for $\theta < 2.05$ °C	Tarakanov et al. (2013a)	LADCP-measurements
42	50°S Fracture Zone in the South Atlantic Ridge	<0.05			
43	Conrad-Bouvet fracture zones	<−0.10			
44	10° E Fracture Zone in the Southwest Indian Ridge	<−0.20			
45	Du Toit Fracture Zone	<−0.50			
46	Andrew Bain Fracture Zone				
47	Marion Fracture Zone				
48	Prince Edward Fracture Zone				
49	Walvis Passage	<1.40		Connary and Ewing (1974)	
50	Unnamed Passage at 2° W in the Walvis Ridge				
51	Unnamed Passage at 2° E in the Walvis Ridge				
53	Angola Gap			Connary and Ewing (1974)	
54	Rio de Janeiro Fracture Zone	<2.1	0.5	Mercier et al. (2000)	Geostrophic calculations with reference level at $\theta = 2.1$ °C
55	Rio Grande Fracture Zone	<2.01			

The numbers correspond to the charts in Figs. 2.1, 2.3 and 2.4

western basins of the Indian Ocean (Haine et al. 1998). In the Atlantic sector of the Southern Ocean, Weddell Sea Deep Water spreads to the north from the Weddell Gyre zone along several pathways, both in the western and eastern parts of the Atlantic. However, the major part of the flow is concentrated in the western part of the Atlantic.

The upper layers of Weddell Sea Deep Water in the West Atlantic penetrate from the Weddell Sea into the Scotia Sea through the passages in the South Scotia Ridge (Philip, Orkney, Bruce, Discovery, and Hoyer) (Seabrooke et al. 1971; Nowlin and Zenk 1988; Locarnini et al. 1993; Orsi et al. 1999; Naveira Garabato et al. 2002a; Schodlock et al. 2002) where they divide into two streams (Fig. 2.4). One stream follows westward in the Deep Western Boundary Current system along the north-western slope of the Antarctic Peninsula (Mann 1977; Treshnikov et al. 1978; Nowlin and Zenk 1988; Naveira Garabato et al. 2002a). Moreover, according to (Orsi et al. 1999), this flow does not reach the Pacific Ocean basins, although, it is reported in (Tarakanov 2009) that tracks of Weddell Sea Deep Water had been found on the Antarctic slope in the Pacific Ocean at 72° W. This flow in the western part of the Scotia Sea separates from the Antarctic slope along the eastern slope of the Shackleton Ridge while crossing the Drake Passage, flows around the Scotia Sea in the west, and then transforms and flows to the eastern part of the Scotia Sea at the depths below the Antarctic Circumpolar Current (Tarakanov 2009). Another stream from the Weddell Sea turns east from the mentioned passages in the South Scotia Ridge, merges with the stream from the western part of the sea, and then passing through the George Passage merges with the Weddell Sea Deep Water stream, which propagates directly from the Weddell Sea through the South Sandwich Trough and transports the denser forms of Weddell Sea Deep Water (Locarnini et al. 1993; Orsi et al. 1999; Naveira Garabato et al. 2002a). Then, this combined stream flows into the Georgia Basin through the Northeastern Georgia Passage and the passage adjacent to the Islas Orcadas Rise from the west. The overlying Circumpolar Bottom Water flows along another pathway from the Scotia Sea through the Shag Rocks and Nameless passages and further through the Falkland Trough joins the previously described flow of Weddell Sea Deep Water flow (Zenk 1981; Wittstock and Zenk 1983; Walkden et al. 2008; Smith et al. 2010). Overflow of Weddell Sea Deep Water through the two latter passages occurs only sporadically (Zenk 1981; Wittstock and Zenk 1983; Walkden et al. 2008).

Antarctic waters (Circumpolar Bottom Water and Weddell Sea Deep Water) enter the Argentine Basin from the Georgia Basin through the depression (hereinafter, the Falkland Passage) in the Falklands Fracture Zone (hereinafter, the Falkland Ridge) (Fig. 2.4) (Whitworth et al. 1991; Arhan et al. 1999). In addition, this flow into the Argentine Basin probably occurs east of the Islas Orcadas Rise (Whitworth et al. 1991, Orsi et al. 1999). In the Argentine Basin, part of the Antarctic water flow that passed through the Falkland Passage joins the Deep Western Boundary Current system, flowing first to the west along the Falklands Escarpment (i.e., the northern edge of the Falkland Plateau), and then to the northeast along the continental slope of South America (Reid et al. 1977; Georgi 1981; Mantyla and Reid 1983; Whitworth et al. 1991; Coles et al. 1996; Memery et al. 2000). The other part entrains the eastward ACC flow and joins thc abyssal anticyclonic circulation around the Zapiola Rise (Whitworth et al. 1991; Coles et al. 1996; Smythe-Wright and Boswell 1998).

Further northward spreading of Antarctic waters into the Brazil Basin occurs along three pathways: through the Hunter and Vema channels and over the lower Santos Plateau (Fig. 2.4) (Speer and Zenk 1993; Hogg et al. 1999; McDonagh et al. 2002; Morozov et al. 2010). Moreover, the coldest part of these waters ($\theta < 0.2$ °C, i.e., Weddell Sea Deep Water) propagates only through the Vema Channel (Reid et al. 1977; Sandoval and Weatherly 2001). In the Brazil Basin, Antarctic waters spread northward mainly in the Deep Western Boundary Current system along the continental slope of South America (Stommel and Arons 1960; Rhein et al. 1995), splitting into two cores, one of which is located at the foot of the slope, and the second is above the slope (Sandoval and Weatherly 2001). A complex system of recirculation of these waters is formed over the abyssal of the Brazil Basin (Fig. 2.4) (De Madron and Weatherly 1994; Sandoval and Weatherly 2001). In the equatorial region, the Antarctic water flow, which consists here only of Lower Circumpolar Deep Water (≈ Circumpolar Bottom Water), is divided into two parts. One part continues in the West Atlantic, where after passing through the relatively narrow quasi-zonal Equatorial Channel, eventually reaches the Newfoundland Bank (Mantyla and Reid 1983; Dobrovolsky and Zalogin 1992; Schmitz 1996). Moreover, in the Guiana and North American basins, recirculation of these waters occurs, the intensity of which significantly exceeds the mean flow (Mantyla and Reid 1983). In the region of 11° N, there is a branch of this stream that crosses the Mid-Atlantic Ridge through the Vema Fracture Zone (Heezen et al. 1964; Vangriesheim 1980a, b; Mantyla and Reid 1983; McCartney et al. 1991). Another part of the Antarctic waters from the Brazil Basin flows into the East Atlantic through the equatorial Romanche and Chain fracture zones (Heezen et al. 1964; Mantyla and Reid 1983; Mercier and Speer 1998; Messias et al. 1999; Tarakanov et al. 2013b; Tarakanov et al. 2018). After the flow crosses the equator in the West Atlantic through the equatorial fractures in the Mid-Atlantic Ridge, a significant increase in the bottom temperature occurs ranging from ~0.4 ° C to more than 1.4 °C (Fig. 2.3). Note that the flow of Antarctic waters through the fractures in the Mid-Atlantic Ridge near the equator and at 8–11° N was suggested by Wüst (1936), although the Chain and Vema fracture zones were not yet discovered at that time.

After Antarctic waters pass through the Romanche and Chain equatorial fracture zones in the East Atlantic (Fig. 2.4) they fill the abyssal of the equatorial basins and partly the Angola Basin, following south along the African continental slope (Mantyla and Reid 1983). In the region of the Kane Gap (9° N) between the Sierra Leone Basin in the south and Cabo Verde Basin in the north, the waters from the equatorial fracture zones merge with the waters that passed through the Vema Fracture Zone (Hobart et al. 1975). Before a series of measurements have been performed by the scientists of the Shirshov Institute of Oceanology in 2009–2012, the direction of the bottom water motion in the Kane Gap was unknown. The only source of information about this at that time was the oriented bottom photographs (Hobart et al. 1975) that showed the northwestern direction of the abyssal current.

Our measurements revealed seasonal variations in the flow direction through the Kane Gap with almost zero annual mean transport (Morozov et al. 2013).

The low-latitude Vema Fracture Zone (Figs. 2.3 and 2.4) is the source of the coldest bottom waters for almost all northern basins of the East Atlantic (Mantyla and Reid 1983; van Aken 2000). In addition, the flow of Antarctic waters into the East Atlantic, but with slightly higher potential temperatures, occurs through a series of fractures in the northern tropical part of the Mid-Atlantic Ridge (Tarakanov et al. 2017, Morozov et al. 2018). These bottom waters spread to the north after they pass the Vema Fracture Zone and other fractures. The main part of the flow is concentrated at the foot of the slope of the African continent (Saunders 1987). Potential temperature of bottom waters of the Antarctic origin north of the quasi-zonal ridge in the East Azores Fracture Zone (hereinafter the East Azores Ridge) is slightly higher than 2.0 °C (Table 2.1, Fig. 2.3). The two flows of the cold Antarctic water through this ridge occur through the deep-water Western Gap and Discovery Gap in the central and eastern parts of the East Azores Ridge (Fig. 2.4) (Saunders 1987; Tarakanov et al. 2013a). In the Northeast Atlantic, Antarctic waters are distinguished in the Iberian (van Aken 2000) and even in the West European basins (Lherminier et al. 2007).

The flow of Antarctic waters (Weddell Sea Deep Water) from the Weddell Gyre in the Southeast Atlantic occurs mainly through several fractures in the Southwest Indian Ridge, in particular, the Conrad-Bouvet Fracture Zone (Fig. 2.4) (Sarukhanyan 1980; Tucholke and Embley 1984; Reid 1989). However, the main location of the propagation of these waters into the southern basins of the East Atlantic is considered to be a group of faults at 25°–31° E (Du Toit, Andrew Bain, Marion, and Prince Edward fracture zones, Fig. 2.4) (Tucholke and Embley 1984; Reid 1989; Orsi et al. 1993). The coldest part of the waters propagating here (colder than $\theta = 0.0$ °C) is localized in the southeastern Agulhas Basin and in the Mozambique Basin in the Southwest Indian Ocean (Fig. 2.3). Warmer waters propagate westward along the Southwest Indian Ridge and reach a quasi-meridional series of elevations southwest of the southern tip of the African continent (Agulhas Ridge - Meteor Rise - Shona Ridge). They cross these elevations but already as warmer waters ($\theta > 0.3$ °C) and spread into the Cape Basin (Reid 1989). From the Cape Basin, the main outflow of bottom waters of Antarctic origin ($\theta > 1.4$ °C) occurs northward into the Angola Basin through the Walvis Passage in the Walvis Ridge (Fig. 2.4) (Connary and Ewing 1974). An additional weak outflow also exists in two narrow passages in the Walvis Ridge at 31–32° S, 2° E (Shannon and van Rijswijck 1969; Shannon and Chapman 1991) and presumably at 32°40′ S, 2° W (Needham et al. 1986; Warren and Speer 1991). In the Angola Basin, traces of these overflows at the bottom quickly disappear in the northward direction (Fig. 2.3). Antarctic waters also flow into the Angola Basin from the west (Mercier et al. 2000) through the Rio de Janeiro and Rio Grande fracture zones (Fig. 2.4, Table 2.1) in the Mid-Atlantic Ridge.

2.3 Comparison of Spreading of Water Masses

Now, we shall describe the structure and circulation of bottom waters of Antarctic origin with quantitative estimates in each of the basins. Figure 2.5 shows contour lines of different characteristics, which are associated in literature with the boundaries of water masses of Antarctic origin in the bottom layer of the Atlantic. The figure is based on the data of quasi-meridional section A17 occupied in 2003 from the R/V *Akademik Sergey Vavilov*. The section crossed the Brazil and Argentine basins from north to south. Figure 2.6 shows the corresponding regions of the spreading of bottom waters. The scheme is based on climatic WOCE dataset. We do not show all possible boundaries used in literature not to overload the figure.

The group of contour lines $\sigma_2 = 37.11$ (brown line in Fig. 2.5), $\theta = 2.0$ °C (red line), $\gamma^n = 28.11$ (orange line), $\sigma_4 = 45.87$ (yellow line) corresponds to the upper boundary of Lower Circumpolar Water (Wüst 1936; Reid et al. 1977; Ganachaud 2003). One can see that contour lines $\gamma^n = 28.11$, $\sigma_4 = 45.87$ are very close to each other (Fig. 2.5) and the spreading regions bounded by these lines practically coincide (Fig. 2.6). However, it is not the same with other characteristics. The $\theta = 2.0$ °C

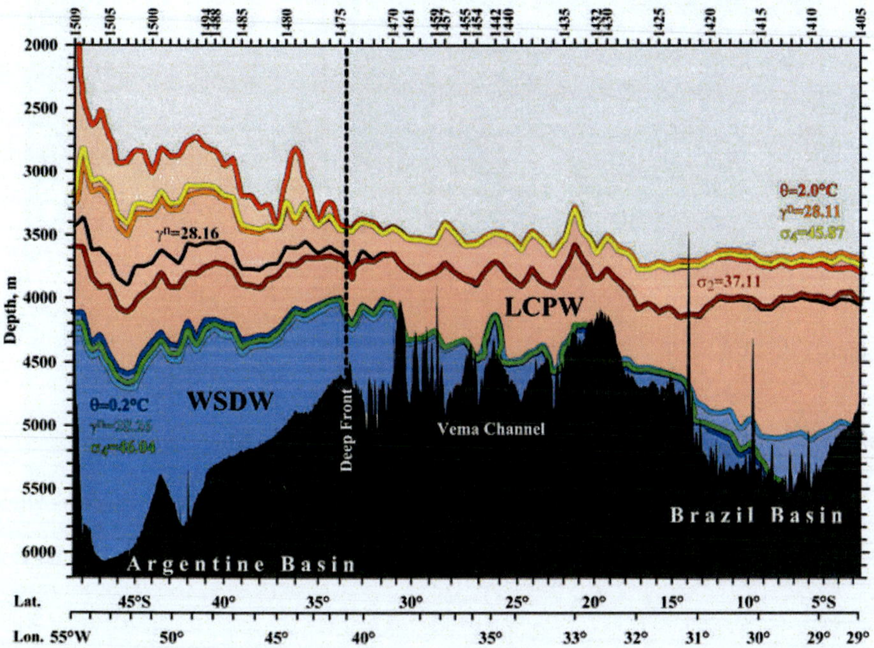

Fig. 2.5 Location of isopycnals related to the boundaries between water masses based on the different characteristics used in various publications. LCPW is Lower Circumpolar Water approximately corresponding by the depth of isopycnals to the lowest layer (Circumpolar Bottom Water) of the Antarctic Circumpolar Current, WSDW is Weddell Sea Deep Water, which is the upper layer of Antarctic Bottom Water in the Antarctic classification of water masses, and γ^n is neutral density (Jacket and McDougall 1997)

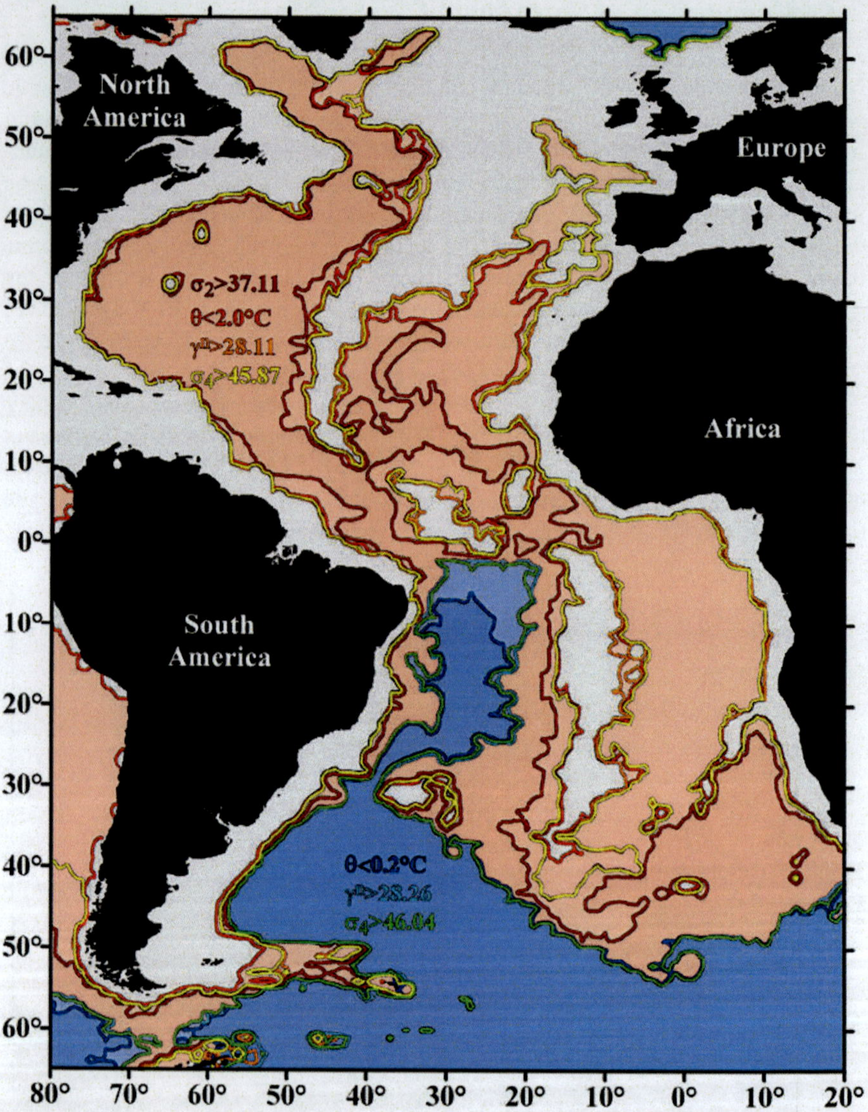

Fig. 2.6 Spreading regions of Antarctic waters in the bottom layer of the Atlantic according to different selection of boundaries. (See Fig. 2.5)

isotherm almost coincides with $\gamma^n = 28.11$ and $\sigma_4 = 45.87$ in the Brazil Basin and northern part of the Argentine Basin (Fig. 2.5). The regions of their spreading are also very close excluding the northern part of the Northeast Atlantic, where the isotherm spreads to the north only up to 37° N (Fig. 2.6). The isotherms and isopycnals diverge strongly in the southern part of the Argentine Basin. This divergence is clearly seen at the Deep Front (34° S, Fig. 2.5), which according to

(McDonagh et al. 2002), is related to the change in the dominating propagation of North Atlantic Deep Water from the meridional to zonal directions. Hogg and Thurnherr (2005) studied this peculiarity of circulation in detail. The depth and spreading region of isopycnal $\sigma_2 = 37.11$ differ strongly from the other three characteristics. In the Brazil Basin, this isopycnal is located deeper than $\theta = 2.0$ °C, $\gamma^n = 28.11$, and $\sigma_4 = 45.87$ approximately by 300–500 m (Fig. 2.5). Unlike the similar selection of Lower Circumpolar Water boundary, its spreading is strictly limited to the region east of the Mid-Atlantic Ridge. It spreads only in the Cape Basin and basins located in the equatorial and tropical parts of the Atlantic Ocean (Fig. 2.6). We emphasize that north of the Deep Front, this isopycnal is close to contour line $\gamma^n = 28.16$ (Fig. 2.5). The latter is assumed the best approximation of the boundary between Lower Circumpolar Deep Water and Circumpolar Bottom Water on the basis of the analysis of temperature and salinity gradients in the southern part of the Pacific Ocean (Koshlyakov and Tarakanov 2003b). Thus, it is our opinion that $\sigma_2 = 37.11$ is the best choice of the boundary that corresponds to the circumpolar water mass classification of the Antarctic waters (Orsi et al. 1999), where the authors identified this boundary as $\gamma^n = 28.18$. In the literature, the upper boundary of the Antarctic waters in the bottom layer of the Atlantic Ocean is also identified as isobaths (3200 and 4000 m) and isotherms $\theta = 2.07$ °C, $\theta = 1.9$ °C, $\theta = 1.8$ °C.

The group of contour lines $\theta = 0.2$ °C (blue line in Fig. 2.6), $\gamma^n = 28.26$ (light blue line), $\sigma_4 = 46.04$ (green line) corresponds to the upper boundary of Weddell Sea Deep Water (Reid et al. 1977; Naveira Garabato et al. 2002b; Koshlyakov and Tarakanov 2003a). The boundaries determined from these contour lines are quite close to each other, although in the Brazil Basin the spreading region of water with $\theta = 0.2$ °C and cooler is smaller than the spreading regions bounded by other characteristics (Fig. 2.6). This boundary is also identified with contour lines $\gamma^n = 28.27$ (Orsi et al. 1999) and $\sigma_4 = 46.06$ (Arhan et al. 1999).

Each of the groups described above determines the upper boundary of Antarctic waters according to two approaches. The first approach (Wüst 1936) (isotherm $\theta = 2.0$ °C) unites all bottom waters in the Atlantic, which have Antarctic rather than North Atlantic origin, under the general term Antarctic Bottom Water. The second approach divides these waters into the waters of circumpolar and Antarctic origin (Reid et al. 1977). In this case, following the approach developed in (Orsi et al. 1999) the term Antarctic Bottom Water unites all Antarctic waters in the Southern Ocean with high density that does not enable these waters to overflow the threshold in the Drake Passage (isotherms close to $\theta = 0.2$ °C). Naturally, the estimates of transport and regions of spreading of these waters are principally different. Even more, differences can be notable even within one approach to the classification of the Antarctic waters. This is especially related to the approach of Wüst (1936) because the authors of different publications applying this approach use a wide range of characteristics and their values to determine the upper boundary of Antarctic Bottom Water. Thus, ambiguity in classification, different (but sometimes the same) terminology, and different boundaries complicate comparison of quantitative characteristics of waters in publications by different authors. We shall make an attempt to solve this problem applying mainly the data of our measurements.

References

Arhan M, Heywood KJ, King BA (1999) The deep waters from the Southern Ocean at the entry to the argentine basin. Deep-Sea Res II 46:475–499

Armi L (1986) The hydraulics of two layers with different densities. J Fluid Mech 163:27–58

Baines PG (1984) A unified description of two-layer flow over topography. J Fluid Mech 146:127–167

Bryden HL, Nurser AJG (2003) Effect of strait mixing on ocean stratification. J Phys Oceanogr 33(8):1870–1872. https://doi.org/10.1175/1520-0485(2003)033<1870:EOSMOO>2.0.CO;2

Clément L, Thurnherr AM, St Laurent LC (2017) Turbulent mixing in a deep fracture zone on the mid-Atlantic ridge. J Phys Oceanogr 47(8):1873–1896. https://doi.org/10.1175/JPO-D-16-0264.1

Coles VJ, McCartney MS, Olson BD, Smethie WJ Jr (1996) Changes in Antarctic bottom water properties in the western South Atlantic in the late 1980s. J Geophys Res Oceans 101(C4): 8957–8970

Connary SD, Ewing M (1974) Penetration of Antarctic bottom water from the Cape Basin into the Angola Basin. J Geophys Res Oceans 79:463–469

De Madron XD, Weatherly G (1994) Circulation, transport and bottom boundary layers of the deep currents in the Brazil Basin. J Mar Res 52:583–638

Demidov AN, Dobrolyubov SA, Morozov EG, Tarakanov RY (2007) Transport of bottom waters through the Vema fracture zone in the mid-Atlantic ridge. Dokl Earth Sci 416:1120–1124

Denker C (2007) Schwankungen von Wassermasseneigenschaften an der Schwelle des Vema-Kanals. Diploma Thesis, Christian-Albrechts-Universitaet, Kiel, 63 pp

Dobrovolsky AD, Zalogin BS (1992) Regional oceanography. MGU Publishers, Moscow. [in Russian]

Faller A (1966) Sources of energy for the ocean circulation and a theory for the mixed layer. Proceedings of Fifth U.S. National Congress of Applied Mechanics. American Society of Mechanical Engineers, Minneapolis, pp 651–672

Ferron B, Mercier H, Speer K, Gargett A, Polzin K (1998) Mixing in the Romanche fracture zone. J Phys Oceanogr 28:1929–1945. https://doi.org/10.1175/1520-0485(1998)028<1929:MITRFZ>2.0.CO;2

Fischer J, Rhein M, Schott F, Stramma L (1996) Deep water masses and transports in the Vema fracture zone. Deep-Sea Res I 43(7):1067–1074

Ganachaud A (2003) Large-scale mass transports, water mass formation, and diffusivities estimated from World Ocean circulation experiment (WOCE) hydrographic data. J Geophys Res Oceans 108(C7):3213. https://doi.org/10.1029/2002JC001565

Georgi DT (1981) Circulation of bottom waters in the southwestern South Atlantic. Deep-Sea Res 28:959–979

Girton JB, Pratt LJ, Sutherland DA Price JF (2006) Is the Faroe Bank Channel overflow hydraulically controlled? J Phys Oceanogr 36(12):2340–2349

Gladyshev SV, Koshlyakov MN, Tarakanov RY (2008) Currents in the Drake Passage based on observations in 2007. Oceanology 48(6):759–770

Gouretski VV, Koltermann KP (2004) WOCE global hydrographic climatology. Ber Bundesamtes Seeschiffahrt Hydrogr 35:1–52

Haine TWN, Watson AJ, Liddicoat MI, Dickson RR (1998) The flow of Antarctic bottom water to the Southwest Indian Ocean estimated using CFCs. J Geophys Res Oceans 103(C12): 27637–27653

Hall MM, McCartney MS, Whitehead JA (1997) Antarctic bottom water flux in the equatorial Western Atlantic. J Phys Oceanogr 27(9):1903–1926. https://doi.org/10.1175/1520-0485(1997)027<1903:ABWFIT>2.0.CO;2

Heezen B, Bunce E, Hersey J, Tharp M (1964) Chain and Romanche fracture zones. Deep Sea Res I 11:30–33. https://doi.org/10.1016/0011-7471(64)91079-4

Hobart MA, Bunce ET, Sclater JG (1975) Bottom water flow through the Kane gap, Sierra Leone rise, Atlantic Ocean. J Geophys Res Oceans 80:5083–5088

Hogg NG, Thurnherr M (2005) A zonal pathway for NADW in the South Atlantic. J Oceanogr 61:493–507

Hogg N, Siedler G, Zenk W (1999) Circulation and variability at the southern boundary of the Brazil Basin. J Phys Oceanogr 29:145–157

Holfort J, Siedler G (2001) The meridional oceanic transport of heat and nutrients in the South Atlantic. J Phys Oceanogr 31(1):5–28

Hoppema M, Klatt O, Roether W, Fahrbach E, Bulsiewicz K, Rodehacke C, Rohardt G (2001) Prominent renewal of Weddell Sea deep water from a remote source. J Mar Res 59:257–279

Jacket DR, McDougall TJ (1997) A neutral density variable for the World's ocean. J Phys Oceanogr 27(2):237–263

Klatt O, Roether W, Hoppema M, Bulsiewicz K, Fleischmann U, Rodehacke C, Fahrbach E, Weiss RF, Bullister JL (2002) Repeated CFC sections at the Greenwich Meridian in the Weddell Sea. J Geophys Res Oceans 107(C4):3030. https://doi.org/10.1029/2000JC000731

Koshlyakov MN, Tarakanov RY (2003a) Antarctic bottom water in the Pacific sector of the Southern Ocean. Oceanology 43(1):5–20

Koshlyakov MN, Tarakanov RY (2003b) Antarctic circumpolar water in the southern part of the Pacific Ocean. Oceanology 43(5):607–621

Lawrence GA (1993) The hydraulics of steady two-layer flow over a fixed obstacle. J Fluid Mech 254(C5):605–633

Lherminier P, Mercier H, Gourcuff C, Alvarez M, Bacon S, Kermabon C (2007) Transports across the 2002 Greenland-Portugal Ovide section and comparison with 1997. J Geophys Res Oceans 112:C07003. https://doi.org/10.1029/2006JC003716

Limeburner R, Whitehead JA, Cenedese C (2005) Variability of Antarctic bottom water flow into the North Atlantic. Deep Sea Res II 52:495–512. https://doi.org/10.1016/j.dsr2.2004.12.012

Locarnini RA, Whitworth T III, Nowlin WD Jr (1993) The importance of the Scotia Sea on the outflow of Weddell see deep water. J Mar Res 51:135–153

Mann CR (1977) Currents and water masses in the vicinity of Drake Passage. Proceedings 1974 Polar Oceans Conference. Ed. Dunbar MJ. AINA. Calgary. Canada. pp 121–128

Mantyla AW, Reid JL (1983) Abyssal characteristics of the World Ocean waters. Deep-Sea Res I 30(8):805–833. https://doi.org/10.1016/0198-0149(83)90002-X

McCartney MS, Bennet SL, Woodgate-Jones ME (1991) Eastward flow through the mid-Atlantic ridge at 11° N and its influence on the abyss of the eastern basin. J Phys Oceanogr 21(8): 1089–1121. https://doi.org/10.1175/1520-0485(1991)021<1089:EFTTMA>2.0.CO;2

McDonagh EL, Arhan M, Heywood KJ (2002) On the circulation of bottom water in the region of the Vema Channel. Deep-Sea Res I 49:1119–1139

Memery L, Arhan M, Alvarez-Salgado XA, Messias M-J, Mercier H, Castro CG, Rios AF (2000) The water masses along the western boundary of the south and equatorial Atlantic. Prog Oceanogr 47:69–98

Mercier H, Morin P (1997) Hydrography of the Romanche and chain fracture zones. J Geophys Res Oceans 102(C5):10,373–10,389. https://doi.org/10.1029/97JC00229

Mercier H, Speer KG (1998) Transport of bottom water in the Romanche fracture zone and the chain fracture zone. J Phys Oceanogr 28(5):779–790. https://doi.org/10.1175/1520-0485(1998)028<0779:TOBWIT>2.0.CO;2

Mercier H, Weatherly G, Arhan M (2000) Bottom water throughflows at the Rio de Janeiro and Rio Grande fracture zones. Geophys Res Lett 27:1503–1506
Meredith MP, Locarnini RA, van Scoy KA, Watson AJ, Heywood KJ, King BA (2000) On the sources of Weddell gyre Antarctic bottom water. J Geophys Res Oceans 105(1):1093–1104
Messias M-J, Andrie C, Memery L, Mercier H (1999) Tracing the North Atlantic deep water through the Romanche and chain fracture zones with chlorofluoromethanes. Deep-Sea Res I 46:1247–1278. https://doi.org/10.1016/S0967-0637(99)00005-9
Morozov EG (1995) Semidiurnal internal wave global field. Deep-Sea Res I 42(1):135–148
Morozov EG, Demidov AN, Tarakanov RY, Zenk W (2010) Abyssal channels in the Atlantic Ocean: water structure and flows. Springer. 266 p, Dordrecht. https://doi.org/10.1007/978-90-481-9358-5
Morozov EG, Tarakanov RY, van Haren H (2013) Transport of AABW through the Kane gap, tropical NE Atlantic Ocean. Ocean Sci 9:825–835. https://doi.org/10.5194/os-9-825-2013
Morozov EG, Tarakanov RY, Demidova TA, Makarenko NI (2017) Measurements of bottom currents in the western part of the Romanche fracture zone on cruises 37 and 38 of the R/V *Akademik Sergey Vavilov*. Oceanology 53(5):751–752. https://doi.org/10.1134/S0001437017050125
Morozov EG, Tarakanov RY, Frey DI, Demidova TA, Makarenko NI (2018) Bottom water flows in the tropical fractures of the northern mid-Atlantic ridge. J Oceanogr 74(2):147–167. https://doi.org/10.1007/s10872-017-0445-x
Morozov EG, Frey DI, Neiman VG, Makarenko NI, Tarakanov RY (2019) Extreme transport velocities of Antarctic bottom water flow through the deep Vema Channel. Dokl Earth Sci 486:659–662
Munk WH (1966) Abyssal recipes. Deep-Sea Res I 13:707–730
Munk WH, Wunsch C (1998) Abyssal recipes II: energetics of tidal and wind mixing. Deep-Sea Res I 45:1977–2010. https://doi.org/10.1016/S0967-0637(98)00070-3
Naveira Garabato AC, Heywood KJ, Stevens DP (2002a) Modification and pathways of Southern Ocean deep waters in the Scotia Sea. Deep-Sea Res I 49:681–705
Naveira Garabato AC, McDonagh EL, Stevens DP, Heywood KJ, Sanders RJ (2002b) On the export of Antarctic bottom water from the Weddell Sea. Deep-Sea Research II 49:4715–4742
Naveira Garabato AC, Stevens DP, Heywood KJ (2003) Water mass conversion, fluxes, and mixing in the Scotia Sea diagnosed by an inverse model. J Phys Oceanogr 33(12):2565–2587
Needham HD, Carré D, Sibuet J-C (1986) Carte Bathymétrique de la Ride de Walvis: Océan Atlantique Sud. Echelle 1/4. 382, 832. IFREMER, Direction de l'Environnement et des Recherches Océaniques, Département Géosciences Marines
Nikolopoulos A, Borena K, Hietala R, Lundberg P (2003) Hydraulic estimates of Denmark Strait overflow. J Geophys Res Oceans 108(C3):3095
Nowlin WD, Zenk W (1988) Westward bottom currents along the margin of the south Shetland island arc. Deep-Sea Res I 35:269–301
Orsi AH, Nowlin WD Jr, Whitworth T III (1993) On the circulation and stratification of the Weddell gyre. Deep-Sea Res I 40:169–203
Orsi AH, Johnsson GC, Bullister JL (1999) Circulation, mixing, and production of Antarctic bottom water. Prog Oceanogr 43:55–109. https://doi.org/10.1016/S0079-6611(99)00004-X
Polzin KL, Speer KG, Toole JM, Schmitt RW (1996) Intense mixing of Antarctic bottom water in the equatorial Atlantic Ocean. Nature 380:54–57. https://doi.org/10.1038/380054a0
Pratt LJ, Whitehead JA (2007) Rotating hydraulics: nonlinear topographic effects in the ocean and atmosphere. Springer. 550 p, New York. https://doi.org/10.1007/978-0-387-49572-9
Reid JL (1989) On the total geostrophic circulation of the South Atlantic Ocean: flow pattern, tracers and transports. Prog Oceanogr 23:149–244
Reid JL, Nowlin WD, Patzert WC (1977) On the characteristics and circulation of the Southwestern Atlantic Ocean. J Phys Oceanogr 7(1):62–91
Rhein M, Stramma L, Send U (1995) The Atlantic deep western boundary current: water masses and transports near the equator. J Geophys Res Oceans 100:2441–2457

Sandoval FJ, Weatherly GL (2001) Evolution of the deep western boundary current of Antarctic bottom water in the Brazil Basin. J Phys Oceanogr 31(6):1440–1460. https://doi.org/10.1175/1520-0485(2001)031<1440:EOTDWB>2.0.CO;2

Sarukhanyan EI (1980) On the spatial structure of the Antarctic circumpolar current between Africa and Antarctica. Doklady AN SSSR 250(3):738–741

Saunders PM (1987) Flow through discovery gap. J Phys Oceanogr 17(5):631–643

Schmitz WJ (1996) On the World Ocean circulation: volume I, some global features. North Atlantic circulation (Technical report WHOI–96–03). Woods Hole Oceanographic Institution, Woods Hole. 142 p

Schodlock MP, Hellmer HH, Beckmann A (2002) On the transport, variability and origin of dense water masses crossing the south scotia ridge. Deep-Sea Res II 49:4807–4825

Seabrooke JD, Hufford GL, Elder RB (1971) Formation of Antarctic bottom water in the Weddell Sea. J Geophys Res Oceans 76:2164–2178

Shannon LV, Chapman P (1991) Evidence of Antarctic bottom water in the Angola Basin at 32°S. Deep-Sea Res I 38(10):1299–1304

Shannon LV, van Rijswijck M (1969) Physical oceanography of the Walvis ridge region. Investigational report no 70, division of sea fisheries of RSA [Available from the library of the University of Cape Town]

Smith IJ, Stevens DP, Heywood KJ, Meredith MP (2010) The flow of the Antarctic circumpolar current over the north scotia ridge. Deep-Sea Res I 57(1):14–28

Smythe-Wright D, Boswell S (1998) Abyssal circulation in the argentine basin. J Geophys Res Oceans 103(C8):15845–15851

Speer KG, Zenk W (1993) The flow of Antarctic bottom water into the Brazil Basin. J Phys Oceanogr 23:2667–2682

Speer KG, Siedler G, Talley L (1995) The Namib col current. Deep-Sea Res I 42:1933–1950

Stommel H (1958) The abyssal circulation. Deep-Sea Res (1953) 5(1):80–82

Stommel H, Arons AB (1960) On the abyssal circulation of the World Ocean. II. An idealized model of the circulation pattern and amplitude in oceanic basins. Deep-Sea Res (1953) 6(3): 217–233

Tarakanov RY (2009) Antarctic bottom water in the Scotia Sea and the Drake Passage. Oceanology 49(5):607–621

Tarakanov RY, Morozov EG, Gritsenko AM, Demidova TA, Makarenko NI (2013a) Transport of Antarctic bottom water through passages in the East Azores ridge (37° N) in the East Atlantic. Oceanology 53(4):432–441

Tarakanov RY, Makarenko NI, Morozov EG (2013b) Antarctic bottom water flow in the western part of the Romanche fracture zone based on the measurements in October of 2011. Oceanology 53(6):655–667. https://doi.org/10.1134/S0001437013050147

Tarakanov RY, Morozov EG, Makarenko NI, Frey DI, Demidova TA (2017) Measurements of bottom currents in fractures of the southern part of the mid-Atlantic ridge on cruises 39, 40, and 41 of the R/V *Akademik Sergey Vavilov*. Oceanology 57(5):753–755

Tarakanov RY, Morozov EG, van Haren H, Makarenko NI, Demidova TA (2018) Structure of the deep spillway in the western part of the Romanche fracture zone. J Geophys Res Oceans 123:8508–8531. https://doi.org/10.1029/2018JC013961

Tarakanov RY, Morozov EG, Frey DI (2020) Hydraulic continuation of the abyssal flow from the Vema Channel in the southwestern part of the Brazil Basin. J Geophys Res Oceans 125: e2020JC016232. https://doi.org/10.1029/2020JC016232

Thurnherr AM, Speer KG (2003) Boundary mixing and topographic blocking on the mid-Atlantic ridge in the South Atlantic. J Phys Oceanogr 33(4):848–862

Thurnherr AM, Richards KJ, German CR, Lane-Serff GF, Speer KG (2002) Flow and mixing in the rift valley of the mid-Atlantic ridge. J Phys Oceanogr 32(6):1763–1778

Thurnherr AM, Laurent LC, Speer KG, Toole JM, Ledwell JR (2005) Mixing associated with sills in a canyon on the Midocean ridge flank. J Phys Oceanogr 35(8):1370–1381

Thurnherr AM, Clément L, St Laurent LC, Ferrari R, Ijichi T (2020) Transformation and upwelling of bottom water in fracture zone valleys. J Phys Oceanogr 50(3):715–726

Tippenhauer S, Dengler M, Fischer T, Kanzow T (2015) Turbulence and finestructure in a deep ocean channel with sill overflow on the mid-Atlantic ridge. Deep Sea Res I 99:10–22

Treshnikov AF, Sarukhanyan EI, Smirnov NP (1978) Researches on the polar experiment program. Gidrometeoizdat, Leningrad. 9p. [in Russian]

Tucholke BE, Embley RW (1984) Cenozoic regional erosion of the abyssal sea floor off South Africa. In: Schlee JS (ed) Interregional unconformities and hydrocarbon accumulation, AAPG memoir 36, Tulsa, pp 145–164

van Aken HM (2000) The hydrography of the mid-latitude Northeast Atlantic Ocean I: the deep-water masses. Deep-Sea Res I 47:757–788

van Haren H, Gostiaux L, Morozov E, Tarakanov R (2014) Extremely long kelvin–Helmholtz billow trains in the Romanche fracture zone. Geophys Res Lett 41(23):8445–8451. https://doi.org/10.1002/2014GL062421

Vangriesheim A (1980a) Antarctic bottom water flow through the Vema fracture zone. Oceanol Acta 3:199–207

Vangriesheim A (1980b) Antarctic bottom water flow through the Vema fracture zone. Oceanol Acta 3:199–207

Walkden GJ, Heywood KJ, Stevens DP (2008) Eddy heat fluxes from direct current measurements of the Antarctic, polar front in shag rocks passage. Geophys Res Lett 35:L06602

Warren BA, Speer KG (1991) Deep circulation in the eastern South Atlantic Ocean. Deep-Sea Res I 38(Suppl. 1):281–322

Whitehead JA (1989) Giant Ocean cataracts. Sci Am 260:50–57. https://doi.org/10.1038/scientificamerican0289-50

Whitehead JA (1998) Topographic control of oceanic flows in deep passages and straits. Rev Geophys 36(3):423–440. https://doi.org/10.1029/98RG01014

Whitehead JA, Wang W (2008) A laboratory model of vertical ocean circulation driven by mixing. J Phys Oceanogr 38(5):1091–1106

Whitehead JA, Leetma A, Knox RA (1974) Rotating hydraulics of strait and sill flows. Geophys Fluid Dyn 6(2):101–125. https://doi.org/10.1080/03091927409365790

Whitworth T, Nowlin WD, Pillsbury RD, Moore MI, Weiss RF (1991) Observations of the Antarctic circumpolar current and deep boundary current in the Southwest Atlantic. J Geophys Res Oceans 96(15):105–118

Wittstock R-R, Zenk W (1983) Some current observations and surface T/S distribution from the Scotia Sea and the Bransfield Strait during early austral summer 1980/81. Meteor Forschung-Ergebnisse, Reihe A/B. No 24, pp 77–86

WOD18 (2018) World Ocean Database 2018. https://www.nodc.noaa.gov/OC5/WOD/pr_wod.html

Wright WR (1972) Northern sources of energy for the deep Atlantic. Deep-Sea Res 19:865–877

Wüst G (1936) Schichtung und Zirkulation des Atlantischen Ozeans, Das Bodenwasser und die Stratosphäre. In: A Defant (Ed) Wissenschaftliche Ergebnisse, Deutsche Atlantische Expedition auf dem Forschungs - und Vermessungsschiff "Meteor" 1925–1927, 6(1), Walter de Gruyter & Co, Berlin, 411pg

Zenk W (1981) Detection of overflow events in the shag rocks passage, scotia ridge. Science 213(4512):1113–1114

Zenk W (2008) Temperature fluctuations and current shear in Antarctic bottom water at the Vema sill. Prog Oceanogr 77:276–284

Zenk W, Siedler G, Lenz B, Hogg NG (1999) Antarctic bottom water flow through the Hunter Channel. J Phys Oceanogr 29(11):2785–2801

Chapter 3
Source Regions

Abstract In this chapter we consider formation of bottom water in the Weddell Sea and further spreading of bottom water to the north and around the Antarctic continent. The distribution paths of Antarctic waters between the basins of the Atlantic are confined to the depressions of the bottom topography. From the Weddell Sea, Antarctic waters in the abyss spread to the north through four passages in the South Scotia Ridge as well as through the South Sandwich Trench. Then, Antarctic waters propagate further to the west in the Drake Passage and to the north into the Argentine Basin through the Falkland Passage in the Falkland Ridge. Further, waters of Antarctic origin are transported from the Argentine Basin to the Brazil Basin along three pathways: through the Vema Chanel, Hunter Channel, and over the Santos Plateau. Another part of this flow is entrained into the Antarctic Circumpolar Current and propagates to the east around Antarctica. A small upper part of Antarctic waters spreads through the Agulhas and Cape basins to the Angola Basin.

Keywords Source regions of Antarctic Bottom Water · Weddell Sea · Vema Channel · Hunter Channel · Santos Plateau

3.1 Weddell Sea and Weddell Gyre

The cyclonic Weddell Gyre spreading to the bottom occupies a vast region from Antarctica in the south to the South Scotia Ridge and the America-Antarctic (North Weddell) and Southwest Indian (African Antarctic) ridges in the north (Fig. 3.1). The Antarctic Peninsula is the western boundary of the gyre. The eastern boundary of the gyre is associated with the propagation of circumpolar waters from the north, which is confined to a wide gap in the Southwest Indian Ridge at 25°–32° E (Orsi et al. 1993; Vernet et al. 2019). The measurements evidence penetration of this gyre further to the east approximately to 53° E (Park et al. 2001). Northern branching of the gyre observed in the South Sandwich Islands region is associated with the corresponding deflection of the Antarctic Circumpolar Current from the zonal direction during the flow of this current around the South Sandwich Arc of islands (Naveira Garabato et al. 2002a). The Weddell Gyre at the surface is a unique

E. G. Morozov et al., *Bottom Gravity Currents and Overflows in Deep Channels of the Atlantic Ocean*, https://doi.org/10.1007/978-3-030-83074-8_3

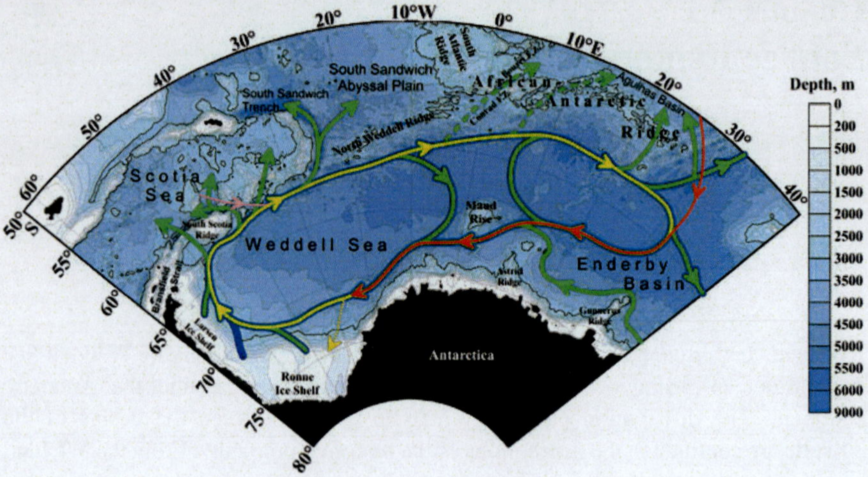

Fig. 3.1 Scheme of cyclonic Weddell Gyre. Deep and bottom water circulation in the Weddell Sea based on (Orsi et al. 1993; Klatt et al. 2005; Ryan et al. 2016; Reeve et al. 2019). The thick blue line shows circulation of Weddell Sea Bottom Water. The green line shows circulation of Weddell Sea Deep Water. The red, pink, and deep yellow lines show the main inflow of Circumpolar Deep Water, water exchange of the Weddell Gyre with the Antarctic Circumpolar Current at its northwestern periphery, and Warm Deep Water circulation, respectively. The deep yellow dotted line shows the inflow of the modified Warm Deep Water into the Antarctic Shelf. The inflow of Antarctic Bottom Water along the Antarctic continental slope from the Indian Ocean is also shown with a green line

cyclonic system elongated from southwest to northeast. In the deep layers, its central part splits into two independent sub-gyres divided at 0–20° W depending on the depth (Orsi et al. 1993).

The thermohaline properties of waters in the Weddell Sea are described in (Fahrbach et al. 1995, 2001, 2004). Cold Winter Water occupies the upper layer up to the depths of 100–300 m. In summer, the upper layer up to 50–100 m is warmed and mixed. A layer of Warm Deep Water is located below this water, which is typical for the major part of the Weddell Sea. This layer of relatively warm (θ = 0.0–0.5 °C) water spreads almost to 1000 m. This water determines the maximum temperature and salinity at intermediate depths of about 600 m. A layer of Antarctic Bottom Water is located deeper. Beneath it, there is a layer of Weddell Sea Bottom Water with a temperature cooler than −0.7 °C. This water does not overflow the sills in the South Scotia Ridge.

Researchers of water masses of the Antarctic region agree (Baines and Condie 1998; Orsi et al. 1999; Whitworth et al. 1998) that 60–65% of AABW is formed in the southern and western parts of the Weddell Sea (Carmack and Foster 1975; Orsi et al. 1999; Rintoul 1998). It is shown in (Carmack and Foster 1975; Naveira Garabato et al. 2002a) that Weddell Sea Deep Water (WSDW) can overflow the South Scotia Ridge, while the denser bottom water, Weddell Sea Bottom Water (WSBW) cannot overflow.

Antarctic Bottom Water in the region described here is divided into Weddell Sea Bottom Water and the overlying Weddell Sea Deep Water. According to early concept (e.g., Reid et al. 1977), Weddell Sea Bottom Water is a layer, in which Antarctic Bottom Water is predominantly renewed as a result of convection over the Antarctic slope. At the same time, Weddell Sea Deep Water is a product of mixing of Weddell Sea Bottom Water with the surrounding waters: Antarctic Shelf Water and overlying subsurface Warm Deep Water. The last name is traditionally used for the Circumpolar Deep Water flowing into the Weddell Gyre over its eastern flank east of 25° E (e.g.: Deacon 1979; Orsi et al. 1993; Gouretski and Danilov 1993; Ryan et al. 2016; Reeve et al. 2019). Additional inflow of Circumpolar Deep Water at the northwestern periphery of the Weddell Gyre is related to the water exchange with Antarctic Circumpolar band by eddies (Klatt et al. 2005). At the southern flank of the Weddell Gyre, Warm Deep Water is modified by upwelling and mixing with the surface layer to compensate the heat loss at the surface. Modified Warm Deep Water flows to the Antarctic Shelf and transforms to the densest Antarctic Shelf Water due to cooling and brine rejection, which then feeds Antarctic Bottom Water. According to the modern concept, the slope convection during which Antarctic Shelf Water mixes with Warm Deep Water and Antarctic Surface Water leads to renovation not only of Weddell Sea Bottom Water but also Weddell Sea Deep Water (Orsi et al. 1993; Fahrbach et al. 1995; Weppernig et al. 1996). Antarctic Bottom Water in the Weddell Sea is formed at the southern (Ronne Ice Shelf) (Nicholls et al. 2009) and western (Larsen Ice Shelf) (Huhn et al. 2008; van Caspel et al. 2015a) peripheries of the Weddell Gyre.

Sources of Antarctic Bottom Water in the Indian Ocean sector of the Southern Ocean form a contribution to Weddell Sea Deep Water (Meredith et al. 2000; Hoppema et al. 2001; Klatt et al. 2002). At the northwestern periphery of the Weddell Gyre, Weddell Sea Deep Water of lowest density penetrates to the Bransfield Strait (Gordon et al. 2000; von Gyldenfeldt et al. 2002) and the Scotia Sea (Whitworth et al. 1994). Heavier forms of Weddell Sea Deep Water outflow from the Weddell Sea through the passages in the South Scotia Ridge (Phillip, Orkney, Bruce, and Discovery passages, and maybe Hoyer Passage) to the Scotia Sea and Georgia Basin through the South Sandwich Trench (Nowlin and Zenk 1988; Locarnini et al. 1993; Orsi et al. 1999; Naveira Garabato et al. 2002a; Schodlok et al. 2002; Franco et al. 2007; Palmer et al. 2012; Jullion et al. 2014; van Caspel et al. 2015b). The densest forms of Weddell Sea Deep Water flow through the latter of these passages, leaving the Weddell Gyre in the Southwest Atlantic.

Bathymetric map of the northern part of the Weddell Sea is shown in Fig. 3.2. The line of ridge crests including islands is shown with a thick line. Profile of depths along the crest of the South Scotia Ridge is shown in Fig. 3.3.

Further spreading of AABW from the Scotia Sea to the north into the Argentine Basin occurs through the Falkland Passage in the Falkland Ridge (Whitworth et al. 1991). Part of this stream follows along the southern and western margins of the Argentine Basin. The other part is entrained by the Subantarctic Front and flows to the east together with the Antarctic Circumpolar Current (Whitworth et al. 1991).

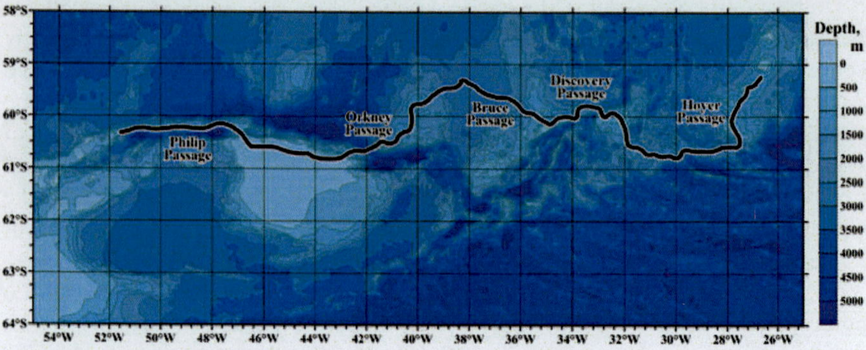

Fig. 3.2 Bathymetric map of the northern part of the Weddell Sea. Thick black line shows the South Scotia Ridge, the profile of which is shown in Fig. 3.3. (Reproduced from Morozov et al. 2020, with permission from Pleiades Publishers)

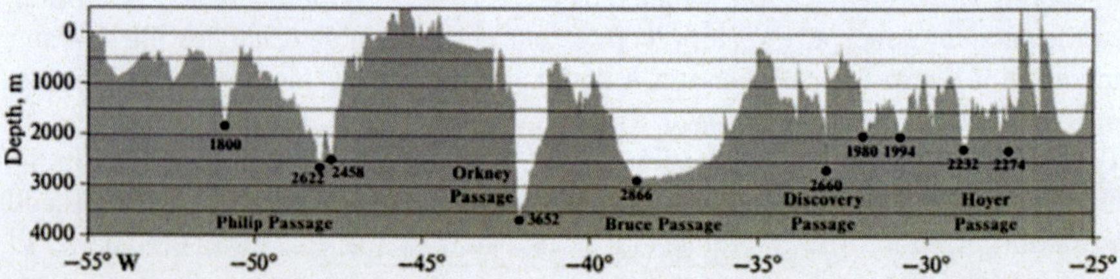

Fig. 3.3 Profile of depths and heights along the crest line of the South Scotia Ridge and islands. Numerals indicate maximum depth of passages in meters. (Reproduced from Morozov et al. 2020, with permission from Pleiades Publishers)

Since all five passages in the South Scotia Ridge are deeper than 1000 m, Weddell Sea Deep Water can flow through the major narrow passages in the ridge. The total deep-water transport of the Weddell Sea through the South Scotia Ridge is estimated at 6.7 ± 1.7 Sv (Schodlok et al. 2002) based on the BRIOS-1 numerical model. The major part of AABW flows northward from the Weddell Sea through the Orkney Passage, which is the deepest passage. The estimate through the Orkney Passage is about 5 Sv (Schodlok et al. 2002).

The major flow of Antarctic Bottom Water to the Agulhas Basin is confined to a topographic depression in the Southwest Indian Ridge approximately at 25–30° E (Reid 1989; Orsi et al. 1993). The minor flow occurs at 20° E and, probably, through the Conrad – Bouvet fracture zones (Tucholke and Embley 1984). The densest forms of Weddell Sea Deep Water reach the gap between the Crozet Islands and Kerguelen Ridge at 55° E, through which Antarctic Bottom Water penetrates to the Crozet Basin in the Indian Ocean (not shown in Fig. 3.1) (Haine et al. 1998). Propagation of Weddell Sea Bottom Water is confined to the Weddell Gyre including the South Sandwich Abyssal Plain (Orsi et al. 1999; Naveira Garabato et al. 2002b). We also note that upper layers of Weddell Sea Deep Water propagate to the east and other sectors of the Southern Ocean, thus forming a quasi-circumpolar bottom circulation, which includes Antarctic Bottom Water from different Antarctic basins, and forms a single water mass (Orsi et al. 1999).

As compared to the Weddell Sea Deep Water, the Younger Weddell Sea Bottom Water is characterized by lower temperature, salinity, and silicate concentration, and by higher oxygen concentration. Generally, temperature and salinity of Weddell Sea Deep Water are intermediate between Weddell Sea Bottom Water and Warm Deep Water (Orsi et al. 1993). At the same time, several maxima and minima in silicate concentration are found in Weddell Sea Deep Water. This peculiarity of Weddell Sea Deep Water is related to its mixing at the northwestern periphery of the Weddell Gyre with circumpolar silicate-rich waters, solutions of sediment-containing silicates, and transport of Antarctic Bottom Water from the Indian Ocean sector of the Southern Ocean, which is characterized by lower concentration of silicates.

The $\theta = -0.7$ °C isotherm is most frequently accepted as the upper boundary of Weddell Sea Bottom Water (Fahrbach et al. 1995; Weppernig et al. 1996; Gordon 1998; Fahrbach et al. 2001; Gordon et al. 2001; Harms et al. 2001; Naveira Garabato et al. 2002b), which approximately corresponds to isopycnal $\gamma^n = 28.40$ or the $\theta = -0.8$ °C isotherm (Foldvik et al. 1985; Yaremchuk et al. 1998; Gordon et al. 1993; Fahrbach et al. 1994; Muench and Gordon 1995; Foldvik et al. 2004; Mensch et al. 1998). Other methods of determination of this boundary are applied rarely (Carmack and Foster 1975; Foster and Carmack 1976; Weiss et al. 1979; Mensch et al. 1996) (Table 3.1). Following (Gordon 1971, 1972), a potential temperature isotherm of 0 °C is most frequently accepted as the upper boundary of Weddell Sea Deep Water and Antarctic Bottom Water, in general (Carmack 1977; Yaremchuk et al. 1998; Fahrbach et al. 1994; Meredith et al. 2001b; Naveira Garabato et al. 2002b). In (Orsi et al. 1999), $\gamma^n = 28.27$ is assumed the upper boundary of Antarctic Bottom Water. Sievers and Nowlin (1984) associated this boundary in the Drake Passage with the maximum of hydrostatic stability (close to $\theta = 0.2$ °C), which approximately corresponds to $\gamma^n = 28.26$. This value of neutral density was assumed in (Naveira Garabato et al. 2002b) as the upper boundary of Antarctic Bottom Water. The same value was also accepted for this boundary in the Pacific sector of the Southern Ocean in (Koshlyakov and Tarakanov 2003a) based on the analysis of hydrophysical and hydrochemical characteristics over the section in the entire Pacific sector of the Southern Ocean.

Large amounts of Antarctic Bottom Water are formed in the Weddell Sea. Then, Weddell Sea Deep Water flows to the north through the passages in the South Scotia Ridge. Cooling of Shelf Water and Antarctic Surface Water generates convective plumes that descend down the continental slope of Antarctica (Naveira Garabato et al. 2002b). The cold water flows along and simultaneously down the continental slope and mixes with Circumpolar Deep Water. The mass of water resulting from mixing of these waters reaches the ocean floor, and thus, forms Antarctic Bottom Water.

The rate of Antarctic Bottom Water formation is determined by researchers on the basis of different methods (Table 3.1). Estimates of the rate of Weddell Sea Bottom Water formation range from 1 to 5 Sv. The rate of Antarctic Bottom Water formation in the Weddell Sea ranges from 3 to 8 Sv. Note that 60–65% of Antarctic Bottom

Table 3.1 Estimates of the formation rates Antarctic Bottom Water and Weddell Sea Bottom Water in the Weddell Sea

Source and method of estimating	Water mass and boundaries	Transport, Sv
Transport of the Deep Western Boundary Current in the Weddell Sea		
Carmack and Foster (1975)	WSBW ($-1.4 < \theta < -1.2$ °C)	2.0–5.0
Foster and Carmack (1976)	WSBW ($\theta = -1.3$ °C)	3.6
Foldvik et al. (1985)	WSBW ($\theta < -0.8$ °C)	2.0–5.0
Gordon et al. (1993)	WSBW ($\theta < -0.8$ °C)	3.0
Fahrbach et al. (1995)	WSBW ($\theta < -0.7$ °C)	1.0–4.0
Muench and Gordon (1995)	WSBW ($\theta < -0.8$ °C)	2.5–3.0
Gordon (1998)	WSBW ($\theta < -0.7$ °C)	4.0–4.8
Fahrbach et al. (2001)	WSBW ($\theta < -0.7$ °C)	1.3 ± 0.4
Gordon et al. (2001)	WSBW ($\theta < -0.7$ °C)	5.0
Foldvik et al. (2004)	WSBW ($\theta < -0.8$ °C)	4.3 ± 1.4
Transport in the Weddell Sea Gyre		
Fahrbach et al. (1994)	AABW ($\theta <0.0$ °C)	3.3–5.6
	WSBW ($\theta < -0.8$ °C)	2.6–2.8
Klatt et al. (2005)	WSBW ($\theta < -0.7$ °C)	3.0
Balance of shelf waters		
Gill (1973)	AABW ($\theta = -0.6$ °C)	6.0–9.0
Carmack (1977)	AABW ($\theta <0.0$ °C)	5.0–10.0*
Application of tracers		
Weiss et al. (1979)	AABW ($\theta <0.4$ °C)	8.0, 4.5
	WSBW ($\theta < -0.9$ °C)	5.0, 3.0
Weppernig et al. (1996)	WSBW ($\theta < -0.7$ °C)	5.0
Mensch et al. (1996)	AABW ($\theta = -0.5$ °C)	11.0
Mensch et al. (1998)	WSBW ($\theta < -1.0$ °C)	3.5
	WSBW ($\theta < -0.8$ °C)	5.0
Broecker et al. (1998)	AABW ($PO_4 = 1.95$ mMol/kg)	15.0
	AABW ($\gamma^n > 28.27$)	
Orsi et al. (1999)	AABW ($\gamma^n > 28.27$)	8.1*
	AABW ($\theta <0.0$ °C)	4.9
Meredith et al. (2001b)		<6.6, 3.7 ± 1.6
Orsi et al. (2002)	AABW ($\gamma^n > 28.27$)	4.9
Mass balance		
Orsi et al. (1999)	AABW ($\gamma^n > 28.27$)	10.0*
Freshwater balance		
Harms et al. (2001)	WSBW ($\theta < -0.7$ °C, $S > 34.64$)	2.6
Numerical models		
Hellmer and Beckmann (2001)	AABW ($\sigma_2 > 37.16$)	11.0
Schodlok et al. (2002)	WSDW ($\theta <0.0$ °C)	6.4 ± 0.6
Wang et al. (2009)	WSBW ($\theta < -0.8$ °C)	2.2
Kerr et al. (2012)	AABW ($\gamma^n > 28.27$)	10.6 ± 3.1

(continued)

Table 3.1 (continued)

Source and method of estimating	Water mass and boundaries	Transport, Sv
Inverse (Box) models		
Yaremchuk et al. (1998)	WSDW ($-0.8 < \theta < 0.0$ °C)	2.6 ± 1.3
	WSBW ($\theta < -0.8$ °C)	2.5 ± 1.9
Sloyan and Rintoul (2001b)	AABW ($\gamma^n > 28.30$)	11 ± 1
Naveira Garabato et al. (2002a)	AABW ($\gamma^n > 28.26$)	9.7 ± 3.7
	($\theta < 0.0$ °C)	10.0 ± 3.7
	($\gamma^n > 28.27$)	9.8 ± 3.7
	($\sigma_2 > 37.16$)	9.8 ± 3.7
	($\gamma^n > 28.30$)	10.2 ± 3.8
	WSBW ($\gamma^n > 28.40$)	3.9 ± 0.8
	($\theta < -0.7$ °C)	4.5 ± 0.9
Jullion et al. (2014)	WSBW ($\gamma^n > 28.40$)	4.0 ± 2.0
	WSDW ($28.40 > \gamma^n > 28.27$)	2.0 ± 2.0
	AABW ($\gamma^n > 28.27$)	6.0 ± 2.0
Optimum Multiparameter Analysis		
Huhn et al. (2008)	WSBW ($\theta < -0.8$ °C)	5.0 ± 1.3

Adapted from Naveira Garabato et al. (2002b) with supplements
Note: The estimates marked with asterisk correspond to the circumpolar rate of AABW formation

Water is formed in the southern and western parts of the Weddell Sea (Carmack and Foster 1975; Orsi et al. 1999; Rintoul 1998). The rate of Antarctic Bottom Water formation in the Southern Ocean, as a whole, ranges from 5 to 10 Sv. The total intensity of water densification including formation of Antarctic Bottom Water and renewal of Warm Deep Water in the Weddell Sea is estimated at 8.4 ± 2.0 Sv (Naveira Garabato et al. 2016). The estimate of (Schodlok et al. 2002) is 6.4 Sv. Model estimates in (Jullion et al. 2014) are even smaller, 2 Sv.

The amount of Antarctic Bottom Water formation was estimated in (Gordon et al. 2001) at 30 Sv; not more than 6.8 Sv of AABW propagates into the Atlantic Ocean north of the Antarctic Circumpolar Current (McCartney and Curry 1993; Coles et al. 1996). It was estimated in (Gordon et al. 2001) that the transports of Weddell Sea Deep and Weddell Sea Bottom Water crossing 44° W are 25 and 5 Sv, respectively; but they do not report that Weddell Sea Bottom Water flows further to the north.

The flow of Antarctic Bottom Water through the passages in the South Scotia Ridge and variations in this flow influence the properties of bottom water in the Scotia Sea, in the Argentine Basin, and further affect the variability of water temperature in the Vema Channel. This influence is explained by the mechanism suggested in (Coles et al. 1996; Meredith et al. 2008). Variations in the cyclonic wind of atmospheric Weddell Gyre displace isopycnal surfaces in the sea. They become either flatter or more inclined depending of the wind stress at the surface caused by strengthening or weakening of the atmospheric gyre (spinup or spindown). Vertical motions of isopycnals near the South Scotia Ridge changes the density range of Weddell Sea Deep Water that can overflow the passages in the

South Scotia Ridge and spread into the Scotia Sea. If the winds are strong, the coldest water remains trapped south of the ridge, and only less dense water overflows. At light or moderate winds, the isopycnal surfaces become flatter and denser water from the deeper layers can flow through the passages in the ridge. Thus, volumes of water of different densities enter the Scotia Sea and flow further into the Argentine Basin modulating the process of further spreading of bottom water to the north.

It was shown in (Abrahamsen et al. 2019; Gordon et al. 2020; Purkey and Johnson 2012, 2013) that recently Antarctic Bottom Water has warmed, freshened and declined in volume. The most pronounced decrease in the volume of Antarctic Bottom Water was observed from 1990 to 2014; then it stabilized. Abrahamsen et al. (2019) show based on observations in the Scotia Sea that this process stopped recently and related these changes to the variations in the densest waters.

It is shown in (Naveira Garabato et al. 2002b) that the northward transport of Weddell Sea Deep Water through the passages in the South Scotia Ridge occurs in the western parts of the passages, while in the eastern parts of the passages, it is transported to the south.

Abrahamsen et al. (2019) define Lower Weddell Sea Deep Water (LWSDW) as water with neutral density (γ^n) between 28.31 and 28.40 kg m^{-3}. In the Orkney Passage, this water occupies a layer below 2200–2300 m. The major northward flow was recorded in the western part of the passage, west of 42° W. The monthly mean transport ranges between 2 and 3 Sv (the mean value is 2.4 Sv). The velocities are of the order of 20 cm s^{-1}. Gordon et al. (2010) revealed the annual cycle in the fluctuations of temperature, salinity, and velocity from the data of two moorings in the Orkney Passage in 1999–2007. The temperatures were highest in October–November and lower by 0.1 °C in April–May. Salinity was changing approximately by 0.01 PSU similar to temperature. Velocities were low (2–4 cm s^{-1}) in December–January and high (12–15 cm s^{-1}) in July–August. Gordon et al. (2010) associate these cycles with the spinup and spindown of the wind stress in the atmospheric Weddell Gyre. It was shown in Abrahamsen et al. (2019) that northward currents in the Orkney Passage from 2011 to 2017 decreased in 2015–2016 to less than 2 Sv compared to ~2.5 Sv in 2013–2014. A decrease in the sea water density at the sill depth was observed induced by the Weddell Gyre spin-up, thus preventing water of higher density to overflow the ridge into the Scotia Sea (Meredith et al. 2008).

The transport of deep water in the Weddell Sea through the Philip Passage based on the data of LADCP profiling data is estimated at 0.7 ± 0.4 Sv (Naveira Garabato et al. 2002b). Another calculation by these authors based on the geostrophic ratio with the reference surface determined from the LADCP data yields in a transport of 0.2 ± 0.4 Sv, which in both calculations is close to zero. The maximum speeds in the Philip Passage are about 20 cm s^{-1}. The transport through the Orkney Passage is maximum (6.0 ± 0.6 Sv) and accounts for almost all transport over the South Scotia Ridge. The velocities there exceed 50 cm s^{-1}. A small amount of water flows through the Bruce and Discovery passages: 1.4 and 0.5 Sv, respectively (Naveira Garabato et al. 2002b).

Similar results are reported in (Franco et al. 2007) based on the Brazilian DOVETAIL 2000 and 2001 field experiments. In the summer of 2000 the transports of Weddell Sea Deep Water through the Philip Passage, Orkney Passage, and southwestern Bruce Passage were 0.12, 3.48, and 1.20 Sv.

Two prominent features of bottom topography are known in this region: the South Scotia Ridge, approximately along 60°00–60°10′ S and the Philip Ridge at 60°30′ S. The deep Hesperides Trench is located between them.

We will consider our measurements of currents in the Philip Passage, which were carried out on cruise 79 of the R/V *Akademik Mstislav Keldysh* in January 2020. In (Naveira Garabato et al. 2002b), the Philip Passage is considered as a wide space between the South Orkney Islands in the east (47° W) and Shishkov (Clarence) Island in the west (55° W).

There are several depressions in this wide interval. We carried our CTD/LADCP measurements only in the eastern narrow part of this interval at 60°08′ S between 48°06′ and 47°42′ W. This passage is 14 km wide at a depth of the 2500 m isobath. In (Palmer et al. 2012), this passage is called the Hesperides Gap.

Field works in the expedition onboard the R/V *Akademik Mstislav Keldysh* in the Philip Passage were carried out on January 28–29, 2020. First, an echo sounder survey was performed. The position of the transverse ridge was found and depths were measured along this ridge. A map of depths, stations, and velocity vectors at the bottom are shown in Fig. 3.4 The instantaneous velocities measured by the profilers were processed to remove tidal currents estimated using the TPXO9 model of Oregon State University (USA) (Egbert and Erofeeva 2002). Three stations across

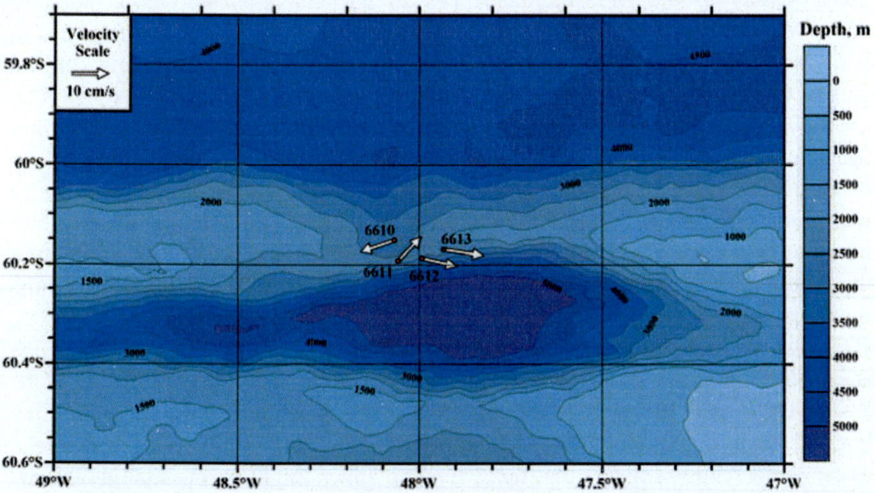

Fig. 3.4 Bottom topography of the eastern part of the Philip Passage. Sketch map of stations and bottom current vectors. (Reproduced from Morozov et al. 2020, with permission from Pleiades Publishers)

the passage and one station on the northern slope, presumably downstream, were occupied.

In 2020 during the cruise of the R/V *Akademik Mstislav Keldysh*, a current of about 8 cm s^{-1} to the northeast was measured at bottom depths at station 6611 in the western part of the Philip Passage at a depth of 2376 m (Fig. 3.4). This station was located exactly on the sill of the transverse ridge across the passage. Eastern-northeastern transport was recorded with a bottom current velocity of about 9 cm s^{-1} at two eastern stations (6612 and 6613). These stations were located on the southern slope of the transverse ridge.

The mean northward velocities of the Weddell Sea Deep Water layer located deeper than 1000 m at stations in the western part of the passage are estimated at 4 cm s^{-1}, which results in the northward transport of deep water within 0.1–0.2 Sv. These estimates generally confirm the previous studies that the flow of Antarctic Bottom Water through the Philip Passage is low.

3.2 Agulhas and Cape Basins

Cyclonic circulation in the bottom layer of these basins was found quite long ago. It existed even in the previous ages, which was revealed on the basis of sedimentary rocks (Tucholke and Embley 1984). The bottom circulation based on the analysis in (Reid 1989) and other publications is shown in Fig. 3.5. Weddell Sea Deep Water flows mainly to the Mozambique Basin in the Indian Ocean propagating to the southeastern part of the Agulhas Basin from the south through the fractures in the Southwest Indian Ridge approximately at 25°–32° E (Reid 1989). Comparing the data of CTD-sections in the Southeast Atlantic with the classification in (Orsi et al. 1999), we can state that there is only a small inflow of Weddell Sea Deep Water into the southwestern part of the Agulhas Basin, while Circumpolar Bottom Water occupies the bottom part of the Cape Basin. Thus, taking into account the existence of cyclonic circulation, we can state that the lowest layer of circumpolar waters flows into the Cape Basin not with the jets of the Antarctic Circumpolar Current crossing the Mid-Atlantic Ridge, but flows from the south through the fractures in the Southwest Indian Ridge. According to WOD18 (World Ocean Data Base 2018), CTD-measurements in the Fracture Zones at 50° S in the South Mid-Atlantic Ridge at 49°45′ S, 8°45′ W (November 1989) and in the Fracture Zone at 10° E in the Southwest Indian Ridge at 51°45′ S, 12°04′ E (December 1976) show that the bottom potential temperatures there are <0.05 °C and < −0.25 °C, respectively. These temperatures suggest that these fracture zones might be possible additional pathways of the inflow of Antarctic Bottom Water into the region north of the Southwest Indian Ridge. Anticyclonic circulation in the northern part of the Cape Basin is shown in Fig. 3.5 according to (Arhan et al. 2003). The authors of this

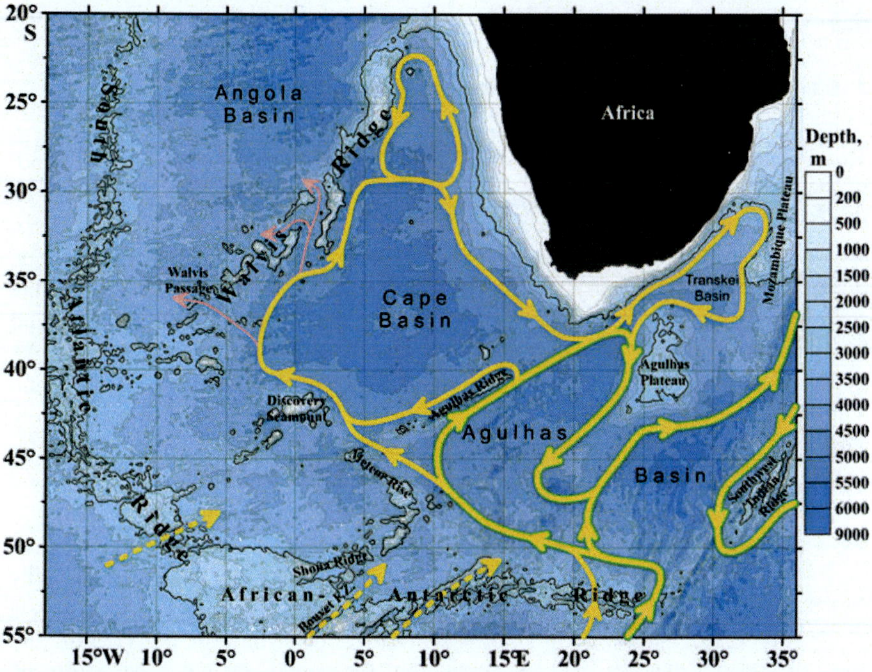

Fig. 3.5 Bottom circulation in the Cape and Agulhas basins based on (Shannon and van Rijswijck 1969; Connary and Ewing 1974; Tucholke and Embley 1984; Needham et al. 1986; Reid 1989; Shannon and Chapman 1991; Warren and Speer 1991; Arhan et al. 2003) and database WOD18 (WOD18 2018) compiled by the authors. The thick green lines show circulation of the densest waters in the bottom layer. The thin deep yellow lines show circulation of waters in the bottom layer of lower density. The dashed deep yellow lines show possible pathways of Antarctic Bottom Water inflow through the Bouvet Fracture Zone from the Weddell Gyre. The outflow of water with a temperature of approximately 1.5 °C through the Walvis Passage and unnamed passages at 2° W and 2° E in the Walvis Ridge is also shown by pink arrows

publication estimate intensity of bottom water circulation in the Cape Basin at 8 ± 3 Sv assuming the upper boundary as $\sigma_4 = 46.0$ ($\theta < 1.5$ °C, $S < 34.80$ PSU). They also estimate weak anticyclonic circulation in the northern part of the basin at 2 Sv. From the Cape Basin, the main outflow of bottom waters of Antarctic origin ($\theta > 1.4$ °C) to the north into the Angola Basin occurs through the Walvis Passage in the (Connary and Ewing 1974). An additional weak outflow also exists in two narrow passages in the Walvis Ridge at 31–32° S, 2° E (Shannon and van Rijswijck 1969; Shannon and Chapman 1991) and presumably at 32°40′ S, 2° W (Needham et al. 1986; Warren and Speer 1991).

3.3 Drake Passage, Scotia Sea, and Georgia Basin

3.3.1 General Description and Bottom Topography

The southwestern part of the Atlantic Ocean, which includes the Scotia Sea, South Sandwich Abyssal Plain, and Georgia Basin, is the first basin in the pathway of the bottom water flow to the North Atlantic. Basins of this region are separated from each other and from the adjacent basins by lines of islands and high submarine ridges: e.g., the ridge in the Shackleton Fracture Zone and its continuation to the continental slope of South America (hereinafter, the Shackleton Ridge), Falkland Ridge and Southwest Indian Ridge, and other smaller ridges, as well as the arc of the Southern Antilles, which includes the North Scotia Ridge, South Georgia, South Sandwich Islands, South Orkney Islands, and South Scotia Ridge. A few relatively narrow passages through these ridges and between islands play a key role in water exchange of the deep and abyssal waters between the South Atlantic basins and surrounding waters of the World Ocean. Moreover, jets of the Antarctic Circumpolar Current are confined to some of these passages.

The Scotia Sea is a basin bounded by a series of high ridges and islands, the western part of which is much deeper than the eastern one. The western part is characterized by depths of 4000 m and deeper, the depths in the eastern part generally do not exceed 3500 m; hereinafter, the depths are given based on the digital database (Smith and Sandwell 1997). The Burdwood Bank, North Scotia Ridge, South Georgia, South Sandwich Islands, South Scotia Ridge, South Shetland Islands, a ridge in the Shackleton Fracture Zone and its extension to the continental slope of South America (hereinafter Shackleton Ridge) form a boundary of the Scotia Sea. In the Drake Passage, the Shackleton Ridge on the one hand, the ridge in the Hero Fracture Zone (hereinafter the Hero Ridge) and the Phoenix Rift on the other, are the borders of the Phoenix Basin from the northeast and from the west, respectively (Fig. 3.6; decoding of all abbreviations in Fig. 3.6 is given in Table 3.2). The greatest depths in the basin between the previously mentioned ridges are found in the South Shetland Trench (deeper than 4500 m), oriented from southwest to northeast and located in the southern part of the basin (Fig. 3.6); local depressions up to 5500 m are located on the southwestern side of the Shackleton Ridge, in its southern part.

The South Scotia Ridge and the South Sandwich Islands arc form a barrier that limits water exchange in the deep layers of the Scotia and Weddell seas. Only a few passages in this ridge are important for water exchange: Philip, Orkney, Bruce, and Discovery, as well as the Georgia Passage between the northern tip of the arc of islands and South Georgia Island (Fig. 3.6). Water exchange through these straits (up to 2500 m deep) between the islands of this arc connecting the Scotia Sea basin with the South Sandwich Trench is possible. The threshold depths of passages in the South Scotia Ridge are about 2000 m in the Phillip Passage, 3650 m in the Orkney Passage, 2850 m in the Bruce Passage, and 2650 m in the Discovery Passage. We also note the depth of another pass in the South Scotia Ridge, the Hoyer Passage

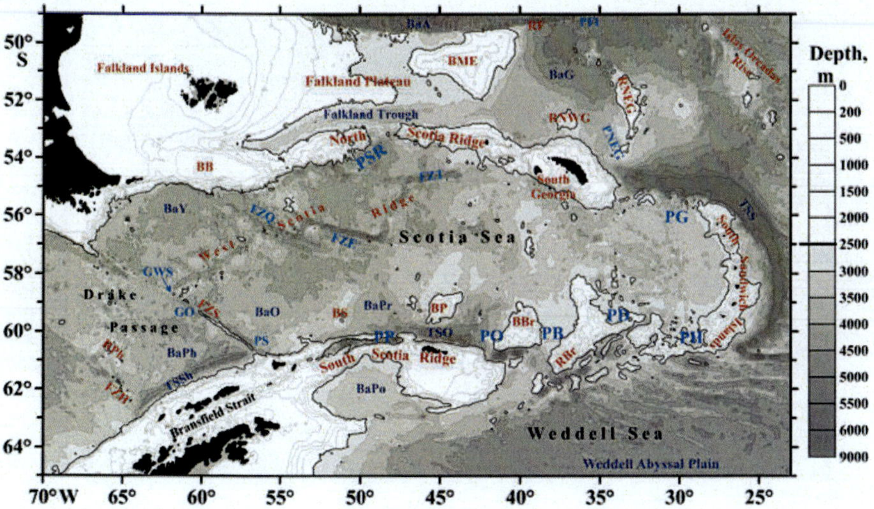

Fig. 3.6 Bottom topography in the Drake Passage and the Scotia Sea. Decoding of the geographical objects is given in Table 3.2. (Reproduced from Tarakanov 2009, with permission from Pleiades Publishers)

Table 3.2 Decoding of geographical abbreviations in Fig. 3.6

		PFl	Falkland Passage
		PB	Bruce Passage
BBu	Burdwood Bank	PD	Discovery Passage
BME	Maurice Ewing Bank	PG	Georgia Passage
BP	Pirie Bank	PH	Hoyer Passage
BS	Shackleton Bank	PNEG	North East Georgia Passage
		PO	Orkney Passage
TSO	South Orkney Trough	PP	Phillip Passage
		PS	Shackleton Passage
TSSa	South Sandwich Trench		
TSSh	South Shetland Trench	FZE	Endurance Fracture
		FZH	Hero Fracture
BaA	Argentine Basin	FZQ	Quest Fracture
BaG	Georgia Basin	FZS	Shackleton Fracture
BaO	Ona Basin	FZT	Tehuelche Fracture
BaPh	Phoenix Basin		
BaPo	Powell Basin		
BaPr	Protector Basin	RBr	Bruce Ridge
BaY	Yaghan Basin	RF	Falkland Ridge
		RNEG	North East Georgia Ridge
GO	Ona Gap	RNWG	North West Georgia Ridge
GWS	West Scotia Gap	RPh	Phoenix Rift

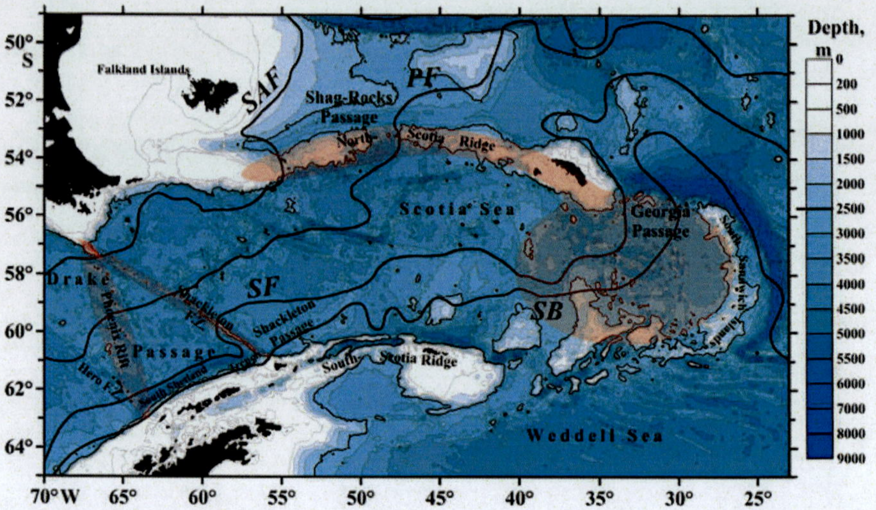

Fig. 3.7 Orographic barrier in the Drake Passage and in the Scotia Sea. Individual thresholds of this barrier are highlighted in red. The fronts of the Antarctic Circumpolar Current are also shown: SAF is the Subantarctic Front, PF is the Polar Front, SF is the Southern Front of the ACC, SB is the Southern Boundary of the ACC. (Reproduced from Tarakanov 2012, with permission from Pleiades Publishers)

(about 2250 m), which is adjacent from the south to the arc of the South Sandwich Islands. We also note that the threshold depth of the South Scotia Ridge is about 3350 m. This threshold point is located in an elongated depression from east to west south of the Bruce Bank (Fig. 3.6).

The ridges in the Drake Passage and in the Scotia Sea form an orographic barrier for the Antarctic Circumpolar Current (ACC). There are at least three thresholds in this region that form this barrier (Fig. 3.7). Figure 3.7 also shows the locations of the Sub-Antarctic, Polar, and Southern fronts of the ACC in the generally accepted classification (Orsi et al. 1995). The first of the barriers along the ACC is the Hero Ridge and the Phoenix Rift (Fig. 3.7). The depths above the Hero Ridge vary from 2000 to 3000 m. The ridge is separated from the Antarctic slope by a depression with depths exceeding 4000 m (hereinafter the Hero Pass). The Phoenix Rift is oriented from north to south and is actually a series of uplands starting from the northern end of the Hero Ridge and deep (up to 4500 m and deeper) depressions (transform faults). The next thresholds are the Shackleton Ridge (Fig. 3.7). In the southern part of the ridge, the depths are shallower than 2500 m with a minimum depth of about 750 m. The depths in the northern part are generally 2500–3200 m, and the greatest depth (~3600 m) was found in the passage at the foot of the continental slope of South America. The Shackleton Ridge is separated from the Antarctic slope by a passage (hereinafter the Shackleton Pass) with a depth of about 3300 m. Finally, the third and shallowest threshold is the North Scotia Ridge together with the shallow eastern part of the Scotia Sea (Fig. 3.7). The greatest depth above the ridge is observed in the Shag Rocks Passage in the central part of the ridge, which is slightly deeper than 3000 m. We also note the depths of three more passages in the North Scotia Ridge, which are important from the point of view of water exchange in the Scotia Sea with the adjacent basins. The depth of Passage 54–54 in the western part of the ridge is about 1800 m, the depth of the Nameless Passage east of the Shag Rocks is about 2000 m, and finally, the Black Rock Passage is 1600 m deep. The threshold depth in the eastern part of the Scotia Sea is about 3200 m. Hereinafter, the listed thresholds would be referred to as the Phoenix Thresholds, Shackleton Thresholds, and North Scotia Thresholds, respectively.

3.3.2 *Deep and Bottom Water Masses and Previous Concepts of Circulation*

Antarctic Bottom Water (in the Southern Ocean classification) occupies the bottom layer in the major part of the Scotia Sea, excluding its northwestern sector (Yagan Basin). Antarctic Bottom Water is represented here by Weddell Sea Deep Water. This water is also found in the Drake Passage in the South Shetland Trench west of the Shackleton Ridge (Nowlin and Zenk 1988). In the other parts of the Scotia Sea and the Drake Passage, the lower layers of Circumpolar Bottom Water are located at the bottom. Figure 3.8 shows a scheme of the distribution of these waters in the

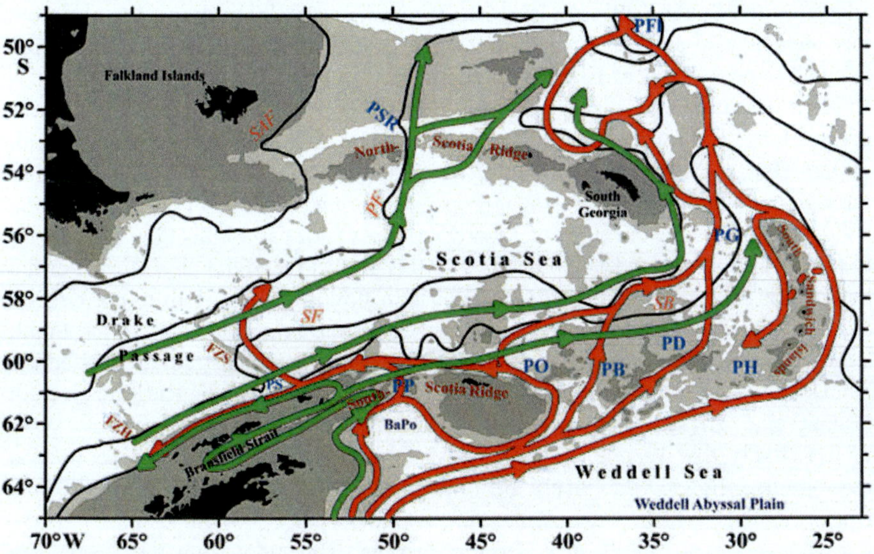

Fig. 3.8 Circulation diagram of Weddell Sea Deep Water (red arrows), Circumpolar Deep Water, and Warm Deep Water (green arrows) based on publications. The regions shallower than 3000 m and 1500 m, respectively, are shaded in dark and light gray tones. The rest is as in Fig. 3.7

Scotia Sea and Drake Passage based on (Naveira Garabato et al. 2002a; Nowlin and Zenk 1988; Locarnini et al. 1993).

For a long time, the South Sandwich Trench was considered the only transit pathway for Weddell Sea Deep Water to the northern regions of the Atlantic Ocean. After passing this trench, the bottom waters propagate to the Georgia Basin and are then transported to the Argentine Basin through the Falkland Passage in the Falkland Ridge (Fig. 3.8). Although penetration of Weddell Sea Deep Water to the Scotia Sea through the passages in the South Scotia Ridge has been known for a long time (Gordon 1966, 1967), the role of the Scotia Sea in the transport of Weddell Sea Deep Water to the north was underestimated. It was considered that the Scotia Sea is shallower than the adjacent basins of the Atlantic (Weddell Sea, South Sandwich Trench, and Georgia Basin) and, therefore, cannot serve as a transit pathway for Weddell Sea Deep Water. The essential role of the Scotia Sea in this process was reported first in (Locarnini et al. 1993). Possibility of the northerly transport of Weddell Sea Deep Water from the Weddell Sea through the South Sandwich Abyssal Plain is shown in (Orsi et al. 1999). The South Sandwich Abyssal Plain is located east of the South Sandwich Islands.

According to present-day concepts, Weddell Sea waters propagate to the Scotia Sea through four passages in the South Scotia Ridge: Phillip, Orkney, Bruce, and Discovery. The Orkney Passage is the deepest one (3500 m) and, therefore, considered the main passage (Naveira Garabato et al. 2002a). Additional inflow of these waters to the eastern part of the sea can also occur through the Georgia Passage between the South Sandwich Islands and South Georgia (Meredith et al. 2008), as well as from the South Sandwich Trench through the passages between individual islands of the South Sandwich Islands (approximately 2300 m deep). We also note that there is also an additional local AABW source (Meredith et al. 2003) in the Scotia Sea near Elephant Island at the tip of the Antarctic Peninsula.

After passing the passages in the South Scotia Ridge, Weddell Sea Deep Water is divided into two flows in the southern part of the Scotia Sea (Naveira Garabato et al. 2002a) (Fig. 3.8). The western flow, which includes the waters from the Phillip Passage and partly Orkney Passage, propagates along the slope of the South Scotia Ridge and then along the slope of Antarctica. The major part of this flow is deflected to the north by the topographic features. It fills almost the entire abyssal part of the Scotia Sea (Locarnini et al. 1993). The eastern flow propagating from the Orkney Passage includes Weddell Sea Deep Water from the other passages in the South Scotia Ridge (Naveira Garabato et al. 2002a). A more detailed structure of Weddell Sea Deep Water overflow through the passages in the South Scotia Ridge is given in (Palmer et al. 2012).

It is likely that Weddell Sea Deep Water is transported from the Scotia Sea through the following previously mentioned passages (Locarnini et al. 1993): Shag Rocks Passage (Wittstock and Zenk 1983; Walkden et al. 2008), Shackleton Passage (Nowlin and Zenk 1988), and Georgia Passage. This flow from the Scotia Sea through the first passage (Shag Rocks Passage) (with the upper boundary of Weddell Sea Deep Water determined as $\theta = 0.2$ °C) was considered insignificant or zero. Lack of Weddell Sea Deep Water in this passage was shown on the section occupied along the crest of the North Scotia Ridge (Smith et al. 2010). The North Scotia Ridge at the northern periphery of the Scotia Sea is only occasionally permeable for the upper layers of Weddell Sea Deep Water. Thus, the northern part of the Scotia Sea is actually a dead end for Weddell Sea Deep Water. Nowlin and Zenk (1988) showed that Weddell Sea Deep Water propagates through the Shackleton Passage (second just mentioned passage) to the South Shetland Trench in the southern part of the Drake Passage with the Antarctic Slope Current, which continues in the Pacific Ocean at least up to 120° W (Hollister and Heezen 1967). As to the third passage (Georgia Passage), the transport of Weddell Sea Deep Water through this passage is currently considered comparable with the transport through the South Sandwich Trench and South Sandwich Abyssal Plain (Naveira Garabato et al. 2002b).

The Weddell Sea Deep Water transport through the passages in the South Scotia Ridge is estimated at 4.7 ± 0.7 Sv (Naveira Garabato et al. 2002b) and 5.9 ± 1.5 Sv (Naveira Garabato et al. 2003). A part of this water (0.4–0.8 Sv) is transported westward to the Drake Passage by the Antarctic Slope Current (Locarnini et al. 1993). The total transport of Antarctic Bottom Water to the north through the South Sandwich Trench and South Sandwich Abyssal Plain is estimated at 5.0 ± 4.3 Sv (Naveira Garabato et al. 2002b), including 1.1 ± 3.5 Sv transported to this region in the density range of Weddell Sea Deep Water and 3.9 ± 0.8 Sv transported initially in the layer of Weddell Sea Bottom Water and then under conditions of intense diapycnal mixing as Weddell Sea Deep Water. Northward transport through the South Sandwich Trench and South Sandwich Abyssal Plain is estimated as part of Antarctic Bottom Water transport from the Indian Ocean sector of the Southern Ocean (Naveira Garabato et al. 2002b). According to (Meredith et al. 2000), this inflow takes place in the density range of Weddell Sea Deep Water equal to 2.7 ± 0.9 Sv (Hoppema et al. 2001), which agrees with the results of calculations (2.5 Sv) reported in (Schodlok et al. 2002). Rintoul (1998) argued that not more than 25% of the circumpolar production of Antarctic Bottom Water occurs in the Indian Ocean, although it is not clear what fraction of this water is exported to the Weddell Gyre.

The arc formed by the South Sandwich Islands and South Georgia (Fig. 3.8) is an orographic barrier for the densest forms of Weddell Sea Deep Water. Therefore, only its upper layers ($\gamma^n < 28.31$) can overflow through the Georgia Passage, although denser layers of Weddell Sea Deep Water exist at both sides of the arc of islands (Arhan et al. 1999). In the basins north of the Scotia Sea, properties of the upper and lower layers of Weddell Sea Deep Water, which propagated along different pathways through the Scotia Sea and South Sandwich Trench, respectively, are notably different. Based on this fact Weddell Sea Deep Water has been recently divided into Lower Weddell Sea Deep Water and Upper Weddell Sea Deep Water) (Arhan et al. 1999; Naveira Garabato et al. 2002b; Tarakanov 2009; Abrahamsen et al. 2019; Meredith et al. 2008). The value of neutral density $\gamma^n = 28.31$ is considered the threshold value (in the climatic sense) for overflowing the threshold in the Georgia Passage (Arhan et al. 1999).

It is shown in (Meredith et al. 2001a, 2008) that traces of more saline and warmer Weddell Sea Deep Water with density $\gamma^n > 28.31$, which penetrates through the

Georgia Passage, were found on the quasi-meridional section in the eastern part of the Scotia Sea. This inflow is irregular. It was detected only over one of the three sections (Meredith et al. 2008). According to (Meredith et al. 2008), such irregularity is related to the variability of the wind stress over the Weddell Gyre. Baroclinic properties of waters in the cyclonic Weddell Gyre increase in the periods of high wind stress. Thus, inclination of isopycnals to the gyre periphery increases, and penetration of the densest layers of Weddell Sea Deep Water through the passages in the South Scotia Ridge is hampered. At the same time, a flow of the denser Weddell Sea Deep Water exists, i.e., the flow of Lower Weddell Sea Deep Water from the South Sandwich Trench through the Georgia Passage to the Scotia Sea. On the contrary, when the wind stress weakens, the basins of the southern part of the Scotia Sea are filled with the densest water, and inflow of Lower Weddell Sea Deep Water from the South Sandwich Trench terminates.

Spreading of the lower layers of Circumpolar Deep Water including Circumpolar Bottom Water in the Scotia Sea and Drake Passage correlates with the eastward ACC branches (Fig. 3.8) (Naveira Garabato et al. 2002a). We note that the upper boundary of Circumpolar Bottom Water is assumed as $\gamma^n = 28.16$ (Koshlyakov and Tarakanov 2003b, 2004), while Circumpolar Bottom Water has the same isopycnal sources as Antarctic Bottom Water. In the northern part of the basin, the transport of Circumpolar Deep Water includes the flow of relatively fresh and cold Pacific Deep Water along the continental slope of South America. This water plays a specific role in smoothing the salinity maximum in the Lower Circumpolar Deep Water layer (Well et al. 2003). Cold and fresh waters from the Weddell Sea flow through the South Scotia Ridge into the southern part of the Scotia Sea and then get mixed with Circumpolar Deep Water. It is clear that strong eddy generation at the fronts of the Antarctic Circumpolar Current plays a significant role in this mixing, but the mechanism is not clear yet. We also note that part of Weddell Sea waters in the Scotia Sea is entrained into the Slope Antarctic Current directed to the west.

The southern part of the Scotia Sea is located south of the southern boundary of the Antarctic Circumpolar Current, thus it is included into the northwestern periphery of the Weddell Gyre (Orsi et al. 1999). It is noteworthy that waters of the circumpolar origin in the Weddell Gyre are called Warm Deep Water. Thus, Warm Deep Water, which includes the density ranges of Circumpolar Bottom Water and Lower Circumpolar Deep Water, spreads in the southern part of the Scotia Sea south of the southern Antarctic Circumpolar Current boundary. Meridional minima of salinity and temperature are found in the Scotia Sea in this density range due to the easterly transport of Antarctic Shelf Water from the Bransfield Strait to the east of the Antarctic Peninsula (Whitworth et al. 1994). These minima become smoother with increasing distance in the eastern direction and almost disappear in the eastern part of the Scotia Sea.

3.3.3 Analysis of Recent Data

In recent years, detailed studies of the internal circulation of bottom layers in the Scotia Sea and in the Drake Passage (Tarakanov 2009, 2010, 2012; Koshlyakov et al. 2010, 2013) were carried out, first of all, on the basis of measurement performed by the Shirshov Institute of Oceanology in 2003–2011, as well as by summarizing the data of hydrographic sections, climatology, bottom topography, and satellite altimetry available in open sources. The results of these studies led to a significant revision of the previous concepts about the abyssal circulation in the Scotia Sea. The main findings can be summarized as follows:

1. The ACC in the Drake Passage does not penetrate deeper than 2700 m in the axial part of the current; approximately along the line of the classical Polar Front. At the same time, south of the Polar Front, the lower boundary of the ACC is located not deeper than the neutral density surface $\gamma^n = 28.16$ (the upper boundary of Circumpolar Bottom Water) ascending to the south to about 1000 m (the upper boundary of Circumpolar Bottom Water), in the northern direction it ascends across the isopycnals approximately to a depth of 1600–1800 m on the northern periphery of the ACC (Tarakanov 2012).
2. At depths greater than the previously mentioned lower boundary of the ACC, a circulation is formed that is fundamentally different from the transport by the ACC jets (Tarakanov 2012; Koshlyakov et al. 2013). In this layer, very intense bottom circulations are formed both stationary and non-stationary, which are steered by the bottom topography. An important feature of this circulation in the Scotia Sea is the existence of northwardly directed streamlines in the Circumpolar Bottom Water layer transverse to the ACC. This transport is a constituent part of the Deep Western Boundary Current of the South Atlantic, which transports deep and bottom waters of the Antarctic origin from the high southern latitudes of the Atlantic to the north. The existence of this transport is in good agreement with the total quasi-zero transport of Circumpolar Bottom Water through the Drake Passage (i.e., in the ACC direction) based on the direct measurements of current velocities. This result is also in good agreement with the Antarctic water propagation in the system of deep cataracts in the Atlantic (Whitehead 1989a, b), in which there is westward retroflection of the Antarctic water flow into the Scotia Sea. This retroflection of the pathways was shown based on indirect observational data.
3. The orographic barrier of the region of the Drake Passage and the Scotia Sea plays an important role in limiting the ACC in depth and in the formation of abyssal circulation in the region. This barrier consists of at least three successive thresholds from west to east: the Hero Ridge together with the Phoenix Rift, the Shackleton Ridge, and the North Scotia Ridge and the relatively shallow eastern part of the Scotia Sea (Tarakanov 2012).
4. The western part of the Scotia Sea is not a dead end for the propagation of Weddell Sea Deep Water, which has passed through the passages in the South Scotia Ridge (Tarakanov 2009). Weddell Sea Deep Water propagates westward

along the continental slope and turns to the north governed by the topographic barriers. It flows at least partly to the eastern part of the Scotia Sea forced by the jets of the Antarctic Circumpolar Current and exits the Scotia Sea through the Georgia Passage. Thus, the western part of the Scotia Sea is another transit route for Weddell Sea Deep Water from the Weddell Sea to the low latitudes of the Atlantic.

The brief summary presented here requires a detailed explanation, which is given in the following sections of this book.

3.3.3.1 Data of Measurements in the Drake Passage and Methods of Processing

Analysis is performed on the basis of a large dataset in the oceanic sector, which is limited by the arc of the South Sandwich Islands and sections A21_99, S04_99, A23_99, Falk_99 occupied in March – April 1999 within the ALBATROSS program (Antarctic Large-Scale Box Analysis and The Role Of the Scotia Sea), (Fig. 3.9 and Table 3.3). Namely:

1. Data from hydrographic sections of the World Ocean Circulation Experiment (WOCE) program, CTD and STD data from the World Ocean Database 2018

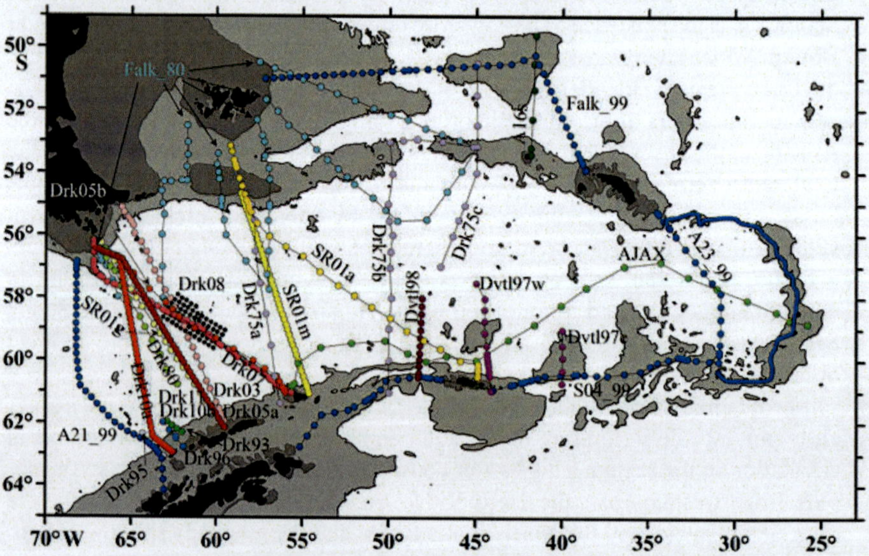

Fig. 3.9 Locations of hydrographic sections in the study region (Table 3.3). Regions shallower than 2500 and 200 m are shown in light gray and dark gray, respectively. The bold blue line is the eastern border of the sector under study. (Reproduced from Tarakanov 2009, with permission from Pleiades Publishers)

Table 3.3 Hydrographic sections in the Drake Passage and Scotia Sea (Fig. 3.9)

Section (Fig. 3.9)	Notation (Fig. 3.9)	Time	Ship	Country
SR01a		September 1992	R/V "PolarStern"	Germany
SR01g		November 1994	R/V "Vidal Gormaz"	Chilie
SR01m		December 1997–January 1998	RRS "James Clark Ross"	USA
A16s		February 1989	R/V "Melville"	USA
A21_99		March–April 1999	RRS "James Clark Ross"	UK
S04_99				
A23_99				
Falk_99				
Drk75a		February 1975	R/V "Atlantis II"	USA
Drk75b				
Drk75c				
Falk_80		August–September 1980	R/V "Atlantis II"	USA
Drk80		October 1980	R/V "Atlantis II"	USA
AJAX		February 1984	R/V "Atlantis II"	USA
Drk93		May 1993	R/V "Nathaniel B. Palmer"	USA
Drk95		December 1995	R/V "Hesperides"	Spain
Drk96		January 1996	R/V "Hesperides"	Spain
Dvtl97e		August 1997	R/V "Nathaniel B. Palmer"	USA
Dvtl97w				
Dvtl98		January 1998	R/V "Hesperides"	Spain
Drk03		December 2003	R/V "Akademik Sergey Vavilov"	Russia
Drk05a		November 2005	R/V "Akademik Ioffe"	Russia
Drk05b		November 2005	R/V "Akademik Ioffe"	Russia
Drk07		November 2007	R/V "Akademik Ioffe"	Russia
Drk08		October–November 2008	R/V "Akademik Sergey Vavilov"	Russia
Drk10a		January 2010	R/V "Akademik Ioffe"	Russia
Drk10b		November 2010	R/V "Akademik Sergey Vavilov"	Russia
Drk11		October–November 2011	R/V "Akademik Ioffe"	Russia

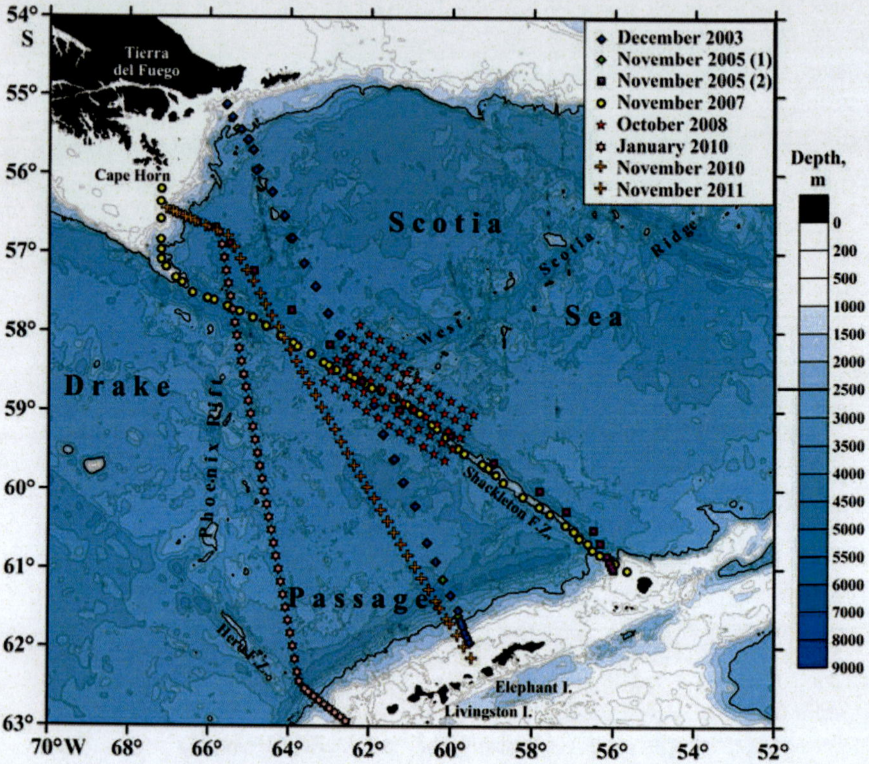

Fig. 3.10 Locations of Russian hydrographic sections in the Drake Passage. (Reproduced from Tarakanov 2012, with permission from Pleiades Publishers)

(WOD18 2018). In this case, only the data of those sections were used, in which the high-resolution CTD or STD casts were performed to the bottom.

2. Data from expeditions onboard the vessels of the Shirshov Institute of Oceanology of the Russian Academy of Sciences *Akademik Sergei Vavilov* and *Akademik Ioffe* in 2003, 2005, and 2007–2011. (Figs. 3.9 and 3.10 and Table 3.3). At the stations of the major part of these sections (except for the November 2010 section) the currents were measured with a lowered Doppler current profiler (LADCP) in addition to the CTD-casts to the bottom. The data were processed using the approach in (Visbeck 2002). The barotropic tidal velocities in the region were calculated from the satellite data using the NASA 0.25 × 0.25 database (TPXO9 model) available at the Oregon State University (Egbert and Erofeeva 2002).
3. Data from the WOCE climatic dataset (Gouretski and Koltermann 2004) containing data on temperature, salinity, and oxygen, as well as on some other hydrochemical characteristics at the nodes of the half-degree grid.

4. Data on the bottom topography on a one-minute grid (Smith and Sandwell 1997). In addition, data were used from echo sounding measurements in the cruises of the Russian ships.
5. Data on absolute dynamic topography (ADT) of the French agency CLS (Collecte Localization Satellites) on the dates of the Russian sections. The ADT version was used, based on the results of simultaneous measurements of up to four altimetry satellites. The weekly ADT data (DT-Global-MADT-Upd product) over $1^{\circ}/3 \times 1^{\circ}/3_{\cos\varphi}$ grid (φ is latitude).
6. In addition to the CTD data in the ALBATROSS sector, the data from two sections of the WOCE program in the Southeast Pacific Ocean were used: S4P (Russia, R/V *Akademik Ioffe*, February – March 1992) and P19 (USA, R/V *Knorr*, January 1993).

3.3.3.2 Depth of the Antarctic Circumpolar Current Penetration in the Drake Passage

3.3.3.2.1 On Quasi-Zero Transport of Abyssal Waters

Figure 3.11 shows cumulative curves of water transport increasing from the bottom calculated from the CTD data and LADCP measurements of current velocities over four sections across the Drake Passage. Three of them (Drk03, Drk10a, Drk11) crossed the Shackleton Ridge, and the line of the fourth (Drk07) was along the crest of this ridge. The summation was carried out over isopycnic layers, but the vertical scale was maintained in meters in order to give an idea of the thickness of these layers. The idea of quasi-isopycnic propagation of water masses is exactly the basis for this method of calculating the cumulative transport. The bounding curves of shaded areas in Fig. 3.11 reflect the length of the corresponding isopycnic layer along the section at each depth. Thus, the square of this region is equivalent to the square of the cross-section. It is important to note that all four series of measurements can be considered uncorrelated with each other because LADCP measures all currents regardless of their physical nature, and, in addition, the sections were made along different trajectories.

It is noteworthy that the total cumulative transport over all four sections increases from the bottom and fluctuates around zero in the Antarctic Bottom Water and Circumpolar Bottom Water layers (within a few Sverdrups); a steady increase in transport is noted only starting from the upper boundary of Circumpolar Bottom Water. In this case, the amplitude of water exchange, i.e., the purely eastern or purely western transport in the Circumpolar Bottom Water + AABW layer over three sections crossing the Shackleton Ridge is about 22–28 Sv, and over one section, which was located almost along its ridge, it is about 8 Sv (Table 3.4). The latter value is most likely overestimated due to the resulting displacement of the Drk07 section line to the southwest of the ridge crest. The indicated values turn out to be more than one order of magnitude larger than the total transport in this layer. The almost threefold difference in the amplitudes of water exchange between the Drk07 section

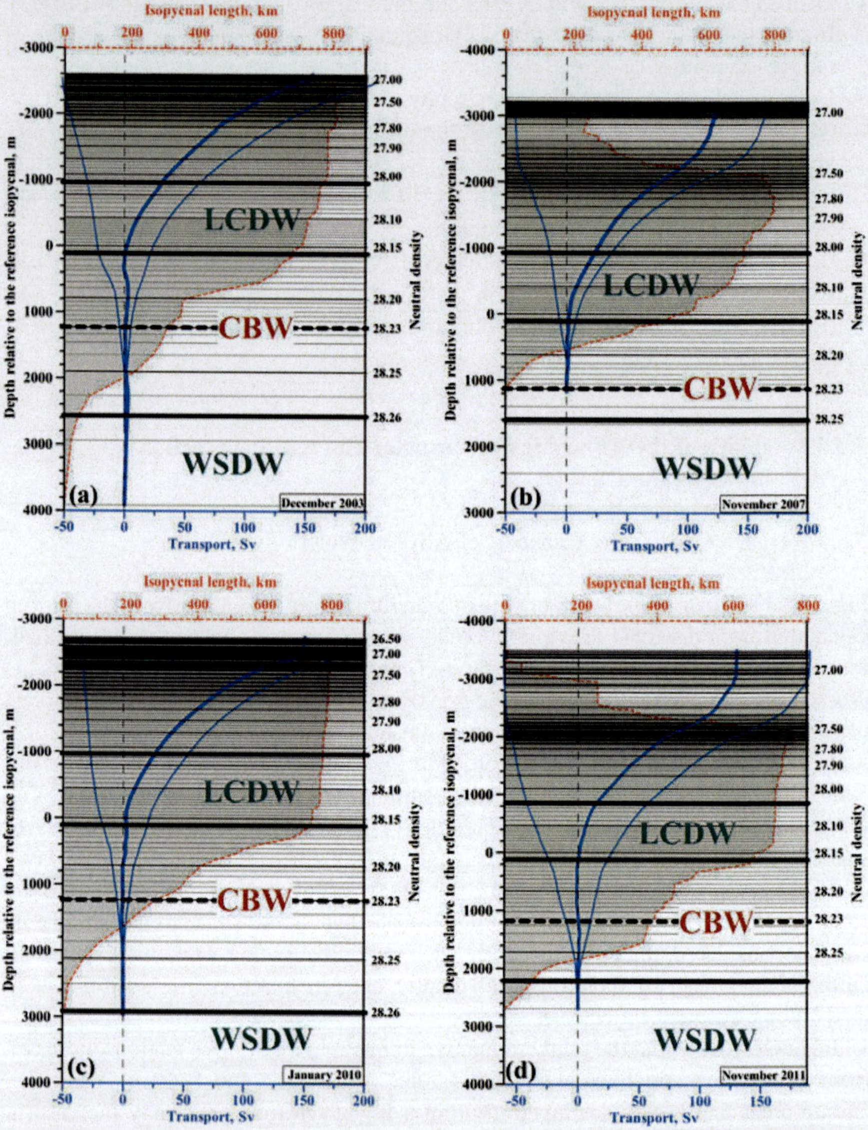

Fig. 3.11 Cumulative curves of water transport in the Drake Passage increasing from the bottom (blue bold curve) calculated from the LADCP data over the sections in December 2003 (**a**), November 2007 (**b**), January 2010 (**c**), and November 2011 (**d**). The thin blue curves show the actual positive and negative cumulative transports. The summation method is described in the text. The shaded area bounded by the red dashed line corresponds to the square of the cross-section. (Figures **a**–**c** are reproduced from Tarakanov 2012, with permission from Pleiades Publishers)

Table 3.4 Transport of Circumpolar Bottom Water (Sv) through the sections near the Shackleton Ridge based on LADCP-measurements

Section	Zone	Transport to the east, Sv	Transport to the west, Sv	Total
Drk07	SlAC	0.0	−0.8	−0.8
	SthAC	2.7	−4.5	−1.8
	SPC	4.7	−2.0	2.7
	SAC	1.1	−0.6	0.5
	Total	8.5	−7.9	**0.6**
Drk03	South of the Shackleton Ridge	18.2	−16.5	1.3
	North of the Shackleton Ridge	8.8	−10.3	−1.5
	Total	27.0	−26.8	**0.2**
Drk10a	South of the Shackleton Ridge	17.9	−15.4	2.5
	North of the Shackleton Ridge	3.1	−2.8	0.3
	Total	21.0	−18.2	**2.8**
Drk11	South of the Shackleton Ridge	25.4	−23.3	−1.3
	North of the Shackleton Ridge	2.4	−2.8	−0.4
	Total	27.8	−26.1	**−1.7**

and the Drk03, Drk10a, and Drk11 sections clearly indicates the formation of abyssal circulations in the basins adjacent to the Shackleton Ridge that are more intense than the water exchange over the ridge.

It is noteworthy that the western currents in the abyssal of the Drake Passage are not concentrated at the continental slope of Antarctica in the Antarctic Slope Current (ASC) system, but are distributed over the entire cross-section of the strait. This feature is seen, for example, in the distribution of the measured normal component of the current velocity on the Drk07 section along the Shackleton Ridge (Fig. 3.12). Moreover, as can be seen from Fig. 3.12, all abyssal currents (both in the western and eastern directions) in the Circumpolar Bottom Water + AABW layer are characterized by bottom intensification. All the previously mentioned features show that a very intense circulation is formed in the abyssal of the Drake Passage, at least in the AABW and Circumpolar Bottom Water layers, which is strongly different from the transport by the ACC jets intensified at the surface. The abyssal circulation in the Scotia Sea and Drake Passage in these layers is considered in more detail in Sects. 3.3.3.3 and 3.3.3.4. Tics on the upper axis in Fig. 3.12 show the cores of the ACC jets in accordance with the classification (Sokolov and Rintoul 2009b): SAC-n, SAC-m, SAC-s are the northern, middle, and southern jets of the Subantarctic Current; SPC-n, SPC-m, SPC-s are the northern, middle, and southern jets of the South Polar Current; SthAC-n and SthAC-s are the northern and southern jets of the

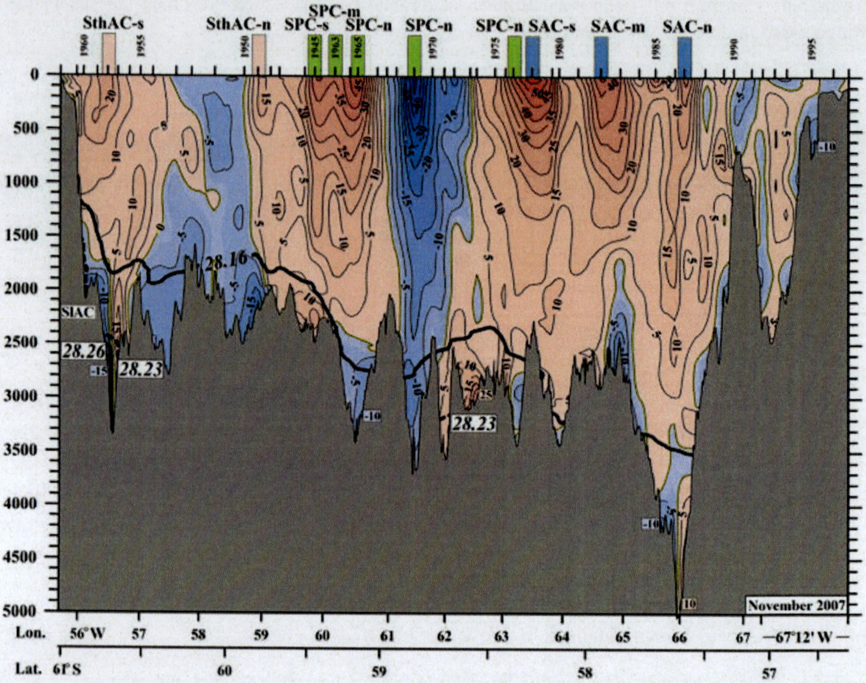

Fig. 3.12 Distribution of the velocity component normal to the section based on the LADCP measurements over the 2007 section (Fig. 3.9 and Table 3.3). Locations of isopycnals $\gamma^n = 28.26$, $\gamma^n = 28.23$, $\gamma^n = 28.16$ are shown. The tics at the top mark the position of stations and cores of the current's jets of the ACC according to the 8-jet classification (Sokolov and Rintoul 2009b): SAC-n, SAC-m, and SAC-s are the northern, middle, and southern jets of the Subantarctic Current; SPC-n, SPC-m, SPC-s are the northern, middle, and southern jets of the South Polar Current; SthAC-n and SthAC-s are the northern and southern jets of the Subantarctic Current, respectively. SAC-m and SPC-n correspond to classical (Orsi et al. 1995) Subantarctic and Polar fronts; SthAC is the classical Southern jet of the ACC. (Reproduced from Tarakanov 2012, with permission from Pleiades Publishers)

South Antarctic Current, respectively. The classical Subantarctic Front and Polar Front correspond to SAC-m and SPC-n; the SthAC jet is equivalent to the Southern ACC Front.

3.3.3.2.2 Transformation of Circumpolar Bottom Water from the Southeast Pacific Ocean to the Western Scotia Sea

Abyssal currents of different directions, which are intensified at the bottom in the Drake Passage, are interpreted in the literature as local disturbances of the geostrophic velocity field (stationary disturbances, internal lee waves, inertial oscillations) when the ACC interacts with the rough bottom topography in the strait (Naveira Garabato et al. 2004; Koshlyakov et al. 2007; Nikurashin and Ferrari

2010a, b; Sheen et al. 2013). In this case, the thermohaline properties of each perturbation (i.e., a pair of oppositely directed flows) should be the same. However, in reality, in the Drake Passage, such currents may differ significantly by their thermohaline properties, and these differences may have the character of isopycnic intrusions, which indicates the transport of waters with different properties in opposite directions, and, consequently, the lack of the previously mentioned explanation for the existence of these abyssal currents. In order to demonstrate this, let us consider in detail the transformation of waters in the bottom layer from the southeastern part of the Pacific Ocean to the western part of the Scotia Sea.

Figure 3.13 shows the θ,*S*-curves in the bottom layer over several sections from the Southeast Pacific Ocean to the Scotia Sea illustrating such a transformation. Measurements were made with different instruments, and periods of research were divided by the time exceeding 15 years. Therefore, we cannot formulate an exact concept about the transformation of waters in the course of their easterly propagation on the basis of these data. Variation in the form (slope) of the θ,*S*-curve during the motion of water with the Antarctic Circumpolar Current is significant in this case. On section P19 (R/V *Knorr*, (US), January and February, 1993) in the southeastern part of the Pacific Ocean in the Subantarctic Front zone and north of this front, the θ,*S*-curves corresponding to the Circumpolar Bottom Water layer are straight lines; they are bent to low salinities south of this front (Fig. 3.13a). Further to the east, the form of θ,*S*-curves in the northern part of section A21_99 does not almost change as compared with section P19, with the only difference that waters with $\theta < 0.55$ °C (Fig. 3.13b) do not appear on the θ,*S*-curves. In the central part of the section, the slope of the curves slightly changes to lower salinities at temperatures $\theta < 0.35$ °C, showing the influence of fresher cold waters, which flow to the west in the Slope Antarctic Current. Finally, in the southern part of section A21_99 located in the depression of the Hero Fracture Zone, we find significant freshening and cooling of waters due to the influence of waters of the Slope Antarctic Current that covers the lower layers of Circumpolar Bottom Water with $\theta < 0.5$ °C. Further to the east, the same form of θ,*S*-curves is conserved in the part of the Drk03 section located south of the Shackleton Ridge (Fig. 3.13c). In the northern part of this section, we find not only strong variation in the slope of θ,*S*-curves, but also lower salinities as compared with the southern part of the section (Fig. 3.13c), which was not found over sections P19 and A21_99 (Fig. 3.13a, b). This can only be explained by the transport of fresher waters from the Weddell Sea across the Antarctic Circumpolar Current.

The shape of the θ,*S*-curves, which is the same as in the northern part of the Drk03 section, is repeated in the northern part of the Drk05b section, located to the east of the Shackleton Ridge (Fig. 3.13d). More saline water (from the region south of the ridge) is found in the Polar Front Zone and at one station in the southern part of the section (Fig. 3.13d). Decreased salinity and temperature are observed in the southern part of the section along with the layering of the *θ,S*-curves (Fig. 3.13d). This fact undoubtedly indicates an intense isopycnic mixing of Circumpolar Bottom Water with Warm Deep Water to the east of the ridge. On the SR01a section located further to the east (Table 3.3), the *θ,S*-curves of the northern and southern parts of the section, with the exception of the intrusions of less saline water in the south, are

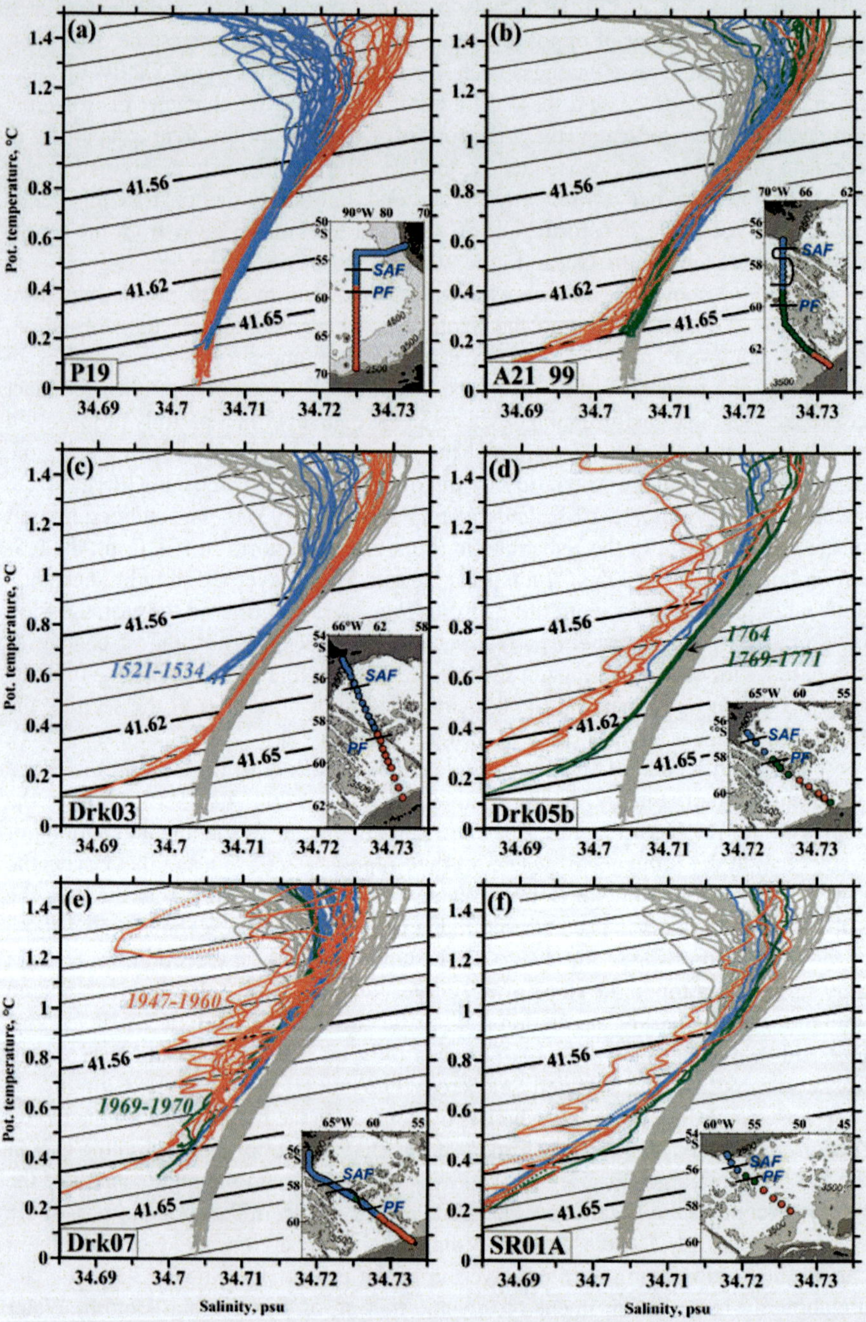

Fig. 3.13 Curves of θ,*S*-distributions for sections P19 (**a**), A21_99 (**b**), Drk03 (**c**), Drk05b (**d**), Drk07 (**e**), SR01a (**f**). Bold lines show isopycnals $\sigma_3 = 41.56$ ($\sim\gamma^n = 28.16$), $\sigma_3 = 41.62$ ($\sim\gamma^n = 28.23$), and $\sigma_3 = 41.65$ ($\sim\gamma^n = 28.26$). Schemes of stations, locations of PF and SAF, and bottom topography are shown in the insets. (Reproduced from Tarakanov 2010, with permission from Pleiades Publishers)

almost the same (Fig. 3.13f), and, in addition, their slope follows the slope of the θ,*S*-curves of the northern sections of the Drk03 and Drk05b sections (Fig. 3.13c, d, f), which indicates the isopycnic nature of the Circumpolar Bottom Water desalination in the northwestern Scotia Sea. These intrusions show continued mixing of Circumpolar Bottom Water with Warm Deep Water at the southern periphery of the ACC. Slightly higher salinity in the Polar Front zone is related to the remains of more saline water flowing along the Polar Front from the west (Fig. 3.13f). We conclude that the Shackleton Ridge creates conditions for the transformation (cooling and desalination) of Circumpolar Bottom Water due to the presence of the Weddell Sea waters, and this transformation reaches the northern edge of the ACC.

We emphasize the importance of the Drk07 section along the crest of the Shackleton Ridge. The southern section of this section was slightly displaced to the west from the crest of the Shackleton Ridge (Gladyshev et al. 2008). Numerous intrusions of less saline waters of the Weddell Sea into Circumpolar Bottom Water and even into Lower Circumpolar Deep Water (Fig. 3.13e) have been observed here. In the central part of the Shackleton Ridge, intrusions of fresher waters into Circumpolar Bottom Water layer were noted at two stations of the section (1969 and 1970) located over the southwestern slope of the ridge. No such intrusions were found in the northern part of the section. However, they were recorded on sections Falk_1980 and SR01g (Table 3.3 and Fig. 3.9) at several stations adjacent to the slope (Fig. 3.14).

Analysis of the *θ*,*S*-curves shows that there is a line in the Drake Passage (along the crest of the Shackleton Ridge) where the opposing abyssal currents in the layer of Circumpolar Bottom Water normal to the ridge do indeed transport water with different thermohaline properties. This transport is observed over the entire width of the strait. Thus, these currents are not just a disturbance of the velocity field during

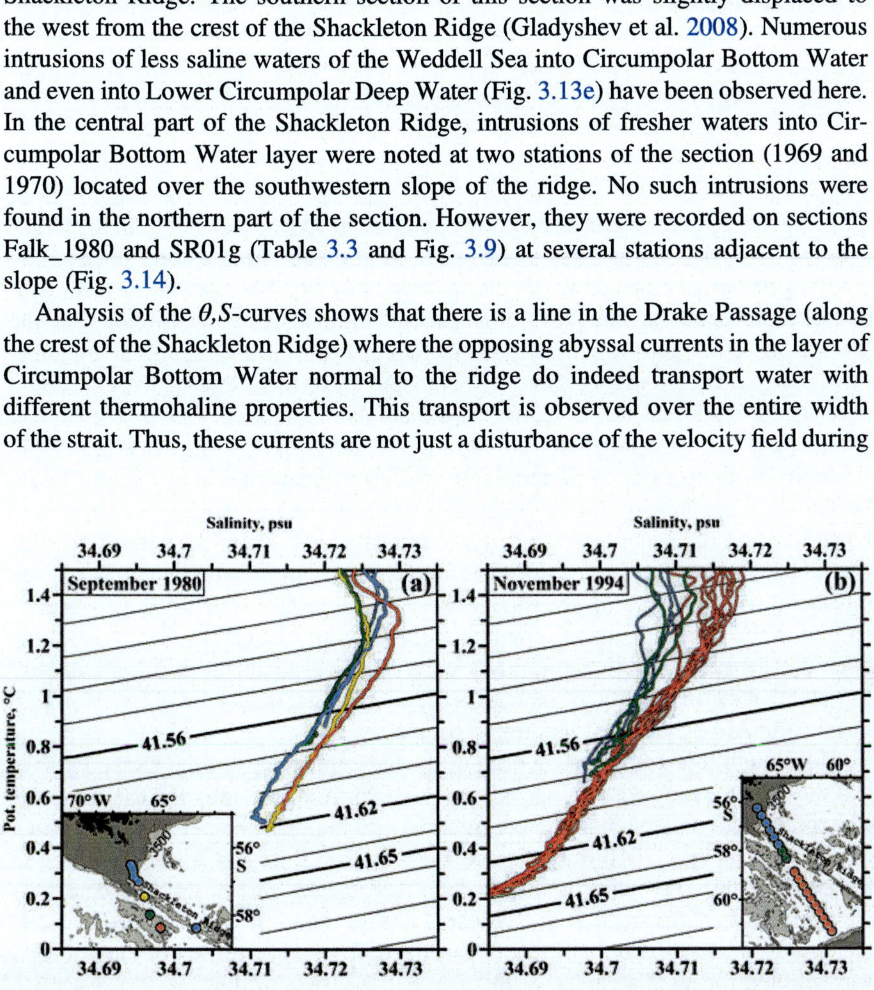

Fig. 3.14 Curves of θ,*S*-distributions in the deep layer over sections Falk_1980 (**a**) and SR01g (**b**). The other notations are the same as in Fig. 3.13

the interaction of the ACC jets with the rough bottom topography of the Drake Passage, but most likely these currents are of gravitational nature. Bottom intensification is in good agreement with the gravitational nature of the currents (Fig. 3.12). Moreover, it seems that these currents are confined to the narrow passages in the Shackleton Ridge. Hence, the balance of forces responsible for these abyssal currents is not limited to the geostrophic balance. The longitudinal pressure gradient along these passages is a significant or even the governing factor. The gravitational nature of these abyssal currents combined with the quasi-zero transport in the Circumpolar Bottom Water layer (Table 3.4 and Fig. 3.11) completely excludes the existence of currents in this layer from the ACC structure as a current forced by wind and thermohaline factors not only over the Shackleton Ridge, but in the entire ocean area adjacent to the ridge.

3.3.3.2.3 Role of the Phoenix, Shackleton, and North Scotia Thresholds in the Formation of the Orographic Barrier for the ACC Waters

Conclusions of Sects. 3.3.3.1 and 3.3.3.2 indicate the location of the upper limit of Circumpolar Bottom Water as the lower limit of the ACC propagation in the Drake Passage. Note that this is an estimate of the ACC depth "from below" obtained by studying the circulation and properties of abyssal waters. The idea of the method to estimate the depth of the ACC propagation "from above" is as follows. If the streamline of the geostrophic current (isohypse of absolute dynamic topography, ADT) on the ocean surface has a circumpolar form, then, due to the equivalent barotropic structure of the ACC, the isopycnals should be horizontal on the vertical plane along this line, reflecting the absence of geostrophic transport across it. In other words, in the equivalent barotropic ACC, the streamlines at any depth should be similar to the streamlines on the ocean surface. Moreover, such a pattern should be conserved not deeper than the depth of the thresholds along the path of this line (Koshlyakov and Tarakanov 2011). Thus, if the bottom topography along the thresholds is shifted along the streamlines of the surface-intensified geostrophic current to the section plane of a hydrographic section with CTD casts to the ocean bottom, then the data from this section can be used to estimate the maximum density of water that has the ability to circulate in the circumpolar manner along each of such lines in the ACC equivalent barotropic structure.

The bottom topography along each of the three thresholds in the Drake Passage and the Scotia Sea (Phoenix, Shackleton, and North Scotia) was shifted along the isohypses (ADT isolines) on the ocean surface to the plane of the Drk10a section, occupied in January 2010 in the Drake Passage. The shift was made according to the data of the ADT map of the French CLS agency for the middle date over the period when the section was occupied (Fig. 3.15). If the isohypse crossed the threshold line several times, the largest depth from the entire set of intersections was selected for it. The choice of this particular section to illustrate the described method was determined by two facts. First, the section line was almost entirely located west of the Shackleton Ridge. A transversal transport to the ACC

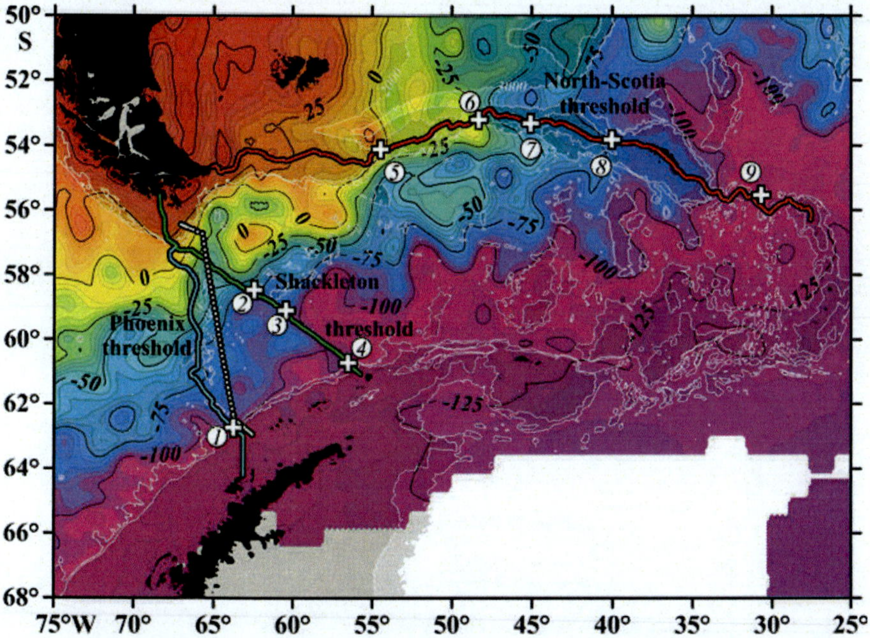

Fig. 3.15 Map of absolute dynamic topography (cm) on January 6, 2010 in the Drake Passage and in the Scotia Sea and the location of the section in 2010. Light and dark gray shaded regions highlighted by white isolines are the regions with depths shallower than 3500 and 2500 m, respectively. The three thresholds are shown with lines: Phoenix, Shackleton, and North Scotia. Crosses and numbers in circles indicate several passages in these thresholds: (*1*) Hero Passage, (*2*) West Scotia Gap, (*3*) Ona Gap, (*4*) Shackleton Passage, (*5*) 54–54 Passage, (*6*) Shag Rocks Passage, (*7*) Nameless passage, (*8*) Black Rock Passage, (*9*) Georgia Passage. (Reproduced from Tarakanov 2012, with permission from Pleiades Publishers)

should be manifested in the layer of Circumpolar Bottom Water and should appear exactly at this location of the section. This transport exists between the Shackleton and North Scotia ridges. Second, the high horizontal resolution of the section by the stations (separated by not more than 10 miles) made it possible to reliably identify all the details of the synoptic pattern of jet streams along the section.

Figure 3.16a shows the distribution of the normal component of velocity over the section in January 2010, measured using LADCP profilers. Figure 3.16b–d show the profiles of bottom topography along three thresholds superimposed onto the section plane. In addition, these figures show the position of the upper and lower boundaries of Circumpolar Bottom Water ($\gamma^n = 28.16$ and $\gamma^n = 28.26$, respectively), as well as the $\gamma^n = 28.23$ isopycnal delimiting the Circumpolar Bottom Water layer into the upper and lower density ranges. The sense of this division is explained below in Sect. 3.3.3.4.

The tics on the top axis of Fig. 3.16 mark the cores of the ACC jets in the multi-jet classification (Sokolov and Rintoul 2009a, b) over the section: three jets of the

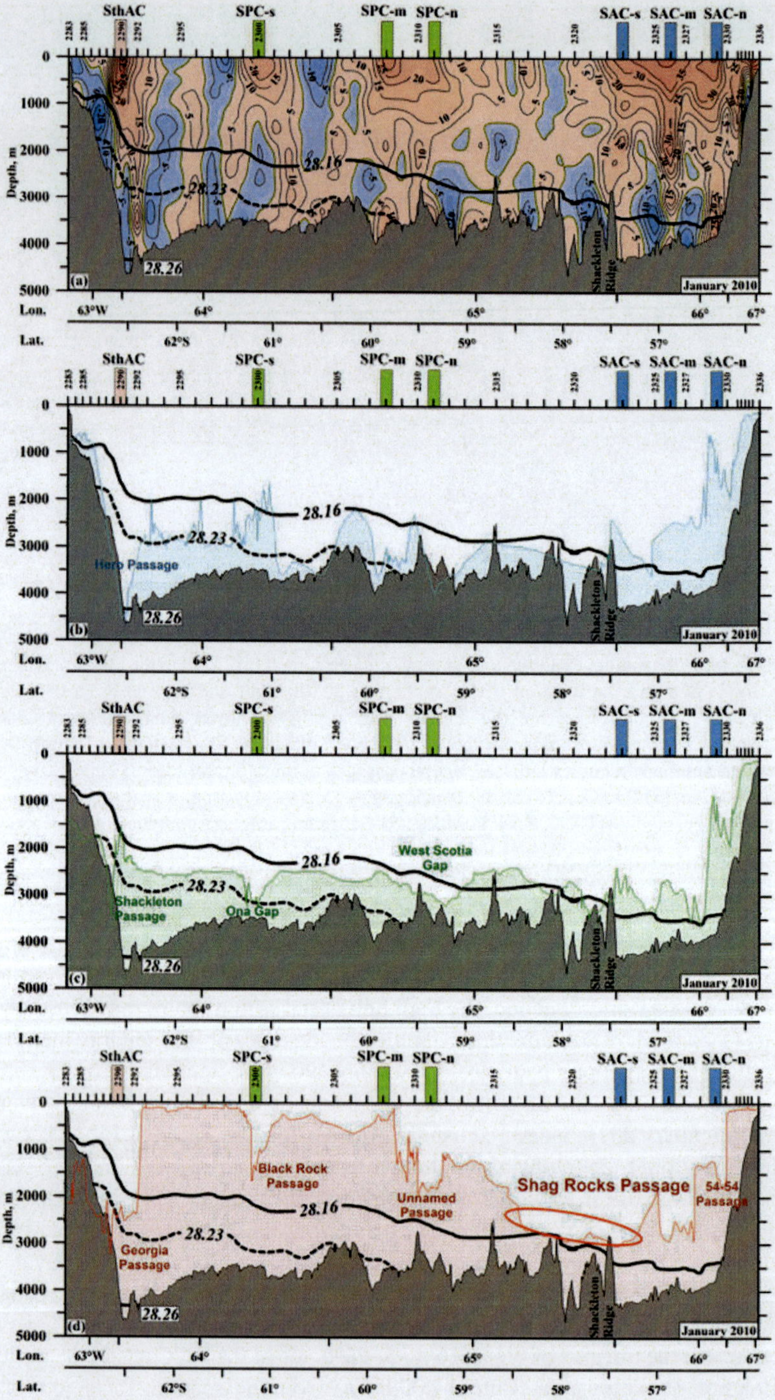

Fig. 3.16 Distribution of the velocity component normal to the section based on the LADCP measurements over the section in 2010 (**a**) and the projection (along the isohypse of absolute

Subantarctic Current (SAC): North (SAC-n), Middle (SAC-m), and South (SAC-s); three jets of the South Polar Current (SPC): North (SPC-n), Middle (SPC-m), and South (SPC-s); South Antarctic Current (SthAC) merged from two streams: North (SthAC-n) and South (SthAC-s). Classical Polar Front (PF), Subantarctic Front (SAF), and Southern Front of the ACC in the classification (Orsi et al. 1995) correspond to SPC-n, SAC-m, and SthAC, respectively.

Figure 3.16 shows that the mapping of the bottom topography of the North Scotia thresholds onto the cross-sectional plane of the 2010 section covers the Circumpolar Bottom Water layer almost completely, with the exception of the narrow South Antarctic Current zone, while the Shackleton and Phoenix thresholds cover this layer only partially. The estimates of threshold mappings on the 2010 section using synoptic charts for any other dates, as well as the application of this method to other sections located in the Drake Passage west of the Shackleton Ridge principally repeats the described pattern. Thus, the waters in the Circumpolar Bottom Water layer could potentially follow the surface streamlines and overflow the Phoenix and Shackleton thresholds, but at the same time they cannot overflow the North Scotia threshold. It is known that Circumpolar Bottom Water is observed in its entire density range over the North Scotia Ridge in the Shag-Rocks Passage (Fig. 3.16d), through which this water flows from the Scotia Sea (Smith et al. 2010). Obviously, this transport can be provided only due to the flow of Circumpolar Bottom Water normal to the ACC jets in the zone between the two thresholds: Shackleton and North Scotia. Note that due to the quasi-zero estimate of the transport of Circumpolar Bottom Water through the Drake Passage this transport should be part of the system of the Deep Western Boundary Current of the South Atlantic Ocean, which is predominantly thermohaline and transports the deep and bottom waters of the Antarctic origin from the high southern latitudes of the Atlantic to the north. In addition, the thermohaline nature of the northward transport of Circumpolar Bottom Water in the Scotia Sea is demonstrated by the measurements on moorings in the Shag Rocks Passage, when the bottom intensification of currents with velocities up to 80 cm s^{-1} has been periodically recorded (Sarukhanyan and Smirnov 1979).

It is seen from Fig. 3.16d that the Circumpolar Bottom Water layer in the northern part of the Drake Passage (northwestern part of the Yagan Basin) deepens below the threshold depth of the Shag Rocks Passage. In this case, the maximum depth of the ACC propagation can be estimated as the maximum depth of the upper boundary of

Fig. 3.16 (continued) dynamic topography, Fig. 3.15) of the bottom topography at the Phoenix (**b**), Shackleton (**c**), and North Scotia (**d**) thresholds onto the plane of this section. The passages in the corresponding thresholds are marked in Fig. 3.16b–d. Positions of isopycnals $\gamma^n = 28.26$, $\gamma^n = 28.23$, $\gamma^n = 28.16$ are shown. The tics at the top mark the position of stations and cores of the current. Explanations of abbreviations are given in the text. The ellipse in Fig. 3.13d marks the region in the Shag Rocks Passage, in which the Circumpolar Bottom Water layer is observed (Smith et al. 2010). The other notations are the same as in Fig. 3.12. (Reproduced from Tarakanov 2012, with permission from Pleiades Publishers)

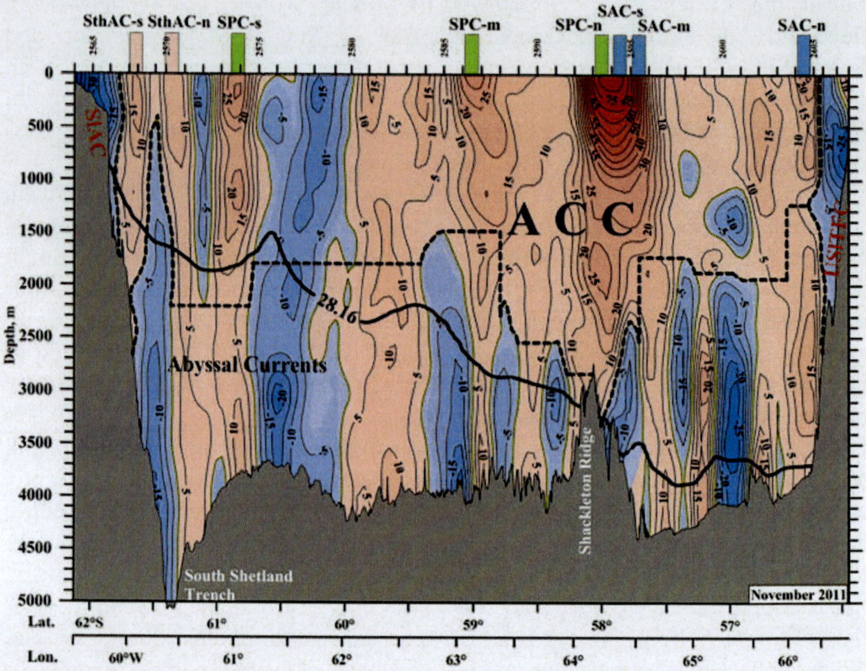

Fig. 3.17 Distribution of the velocity component normal to the section (cm s^{-1}) over the Drk11 section (October 28–November 4, 2011) based on the LADCP data (Koshlyakov et al. 2011). Isopycnal $\gamma^n = 28.16$ is shown. The thick dashed line separates the ACC zone from the zones of abyssal and slope currents. SlAC is Slope Antarctic Current, USlTFC is the Upper Slope Tierra del Fuego Current. The other notations are the same as in Fig. 3.12

Circumpolar Bottom Water in the Shag Rocks Passage. This depth is about 2700 m approximately along the line of the Polar Front (Fig. 3 in Smith et al. 2010). Thus, north of the Polar Front, the ACC jets are limited in depth by a stronger restriction than the upper boundary of Circumpolar Bottom Water. Note that the mapping of a part of the North Scotia Ridge located west of the Shag Rocks Passage on the Drk10a section line in the SAC zone (north of the Polar Front) closes the section sector from the bottom to 2700 m in the SAC-s zone for an equivalent barotropic transport up to 1600 m for the SAC-n jet (Fig. 3.16d). However, it is impossible to judge similarly unambiguously about the degree of limitation on the depth of the SAC jets due to the presence of significant errors in the satellite ADT in the sense that this limitation can be even stronger.

Indeed, it can be seen in Fig. 3.17a that in the SAC zone the ocean layer below ~1500 m to the bottom is occupied by multidirectional currents of intermediate intensification with velocities up to 30 cm s^{-1} and greater. These currents are classified as deep unsteady ocean eddies that do not extend into the main thermocline. For the first time, monitoring of several such eddies of both signs, was carried out by American oceanographers in 2007–2008 in the Yagan Basin (Fig. 3.6) by means of annual measurements of pressure and currents at the network of bottom-moored instruments (Chereskin et al. 2009). In the same work, an explanation of the physical nature of unsteady abyssal eddies was given assuming their formation due to meandering of the ACC jets in the upper layer of the ocean (thermocline) and associated compression and extension of vertical water columns in the lower layer (abyssal), which is similar to the mechanism of formation of atmospheric cyclones and anticyclones in the lower troposphere in the mid-latitude zone of the Earth. In the two-layer ocean model, the linearized equation of conservation of potential vorticity in the abyssal can be written as follows (Gill 1982; Chereskin et al. 2009):

$$\xi_t = \frac{f}{H_0} h_t. \tag{3.1}$$

Here, f is the Coriolis parameter, H_0 is the mean abyssal thickness, h is the perturbation of this thickness, $\xi = (v_x - u_y)$ is relative vorticity; the x-axis is directed to the east, y-axis is directed to the north, u and v are the zonal and meridional velocity components; the lower indices denote partial derivatives with respect to the corresponding variables; subscripts denote partial derivatives with respect to the corresponding variables. Before the thermocline meander passes by a specific point in the ocean, the following equation $h = \xi = 0$ is true at this point. Passing of the frontal part of anticyclonic meander indicates that $h_t < 0$ because in this case the boundary between the thermocline and abyssal descends; hence, $\xi_t > 0$ ($f < 0$ in the Southern Hemisphere). Thus, an anticyclonic eddy is generated in the abyssal. Passing of the frontal part of cyclonic meander indicates that $h_t > 0$, $\xi_t < 0$, and a cyclonic eddy is generated in the abyssal. The transition from a two-layer to continuous density stratification does not change the essence of the process.

The structure of currents over the Drk11 section across the Drake Passage, carried out by the Shirshov Institute of Oceanology in October – November 2011 was analyzed in (Koshlyakov et al. 2013), taking into account the concept of limiting the penetration of the ACC into the ocean depth by the orographic barrier of the Drake Passage and analysis of the results in (Chereskin et al. 2009) about the formation of abyssal eddies in the Yagan Basin. Figure 3.17 shows the distribution of the measured velocity component normal to the section over this section. The delineation of the velocity field in Fig. 3.17 is simplified compared to the interpretation in (Koshlyakov et al. 2013). Here (Fig. 3.17), we distinguished the entire ACC zone, zones of abyssal circulation to the south and north of the Shackleton Ridge, and the zones of slope currents in Antarctica and South America. The abyssal zone north of the Shackleton Ridge covered a layer from approximately 1800 m to the bottom and included abyssal eddies with an estimated transport of 20.4 Sv (22.0 Sv) westward (eastward). In the abyssal zone south of the Shackleton Ridge, the corresponding estimates were 30.1 Sv (29.4 Sv) westward (eastward); the upper boundary of this zone generally turned out to be somewhat higher than the $\gamma^n = 28.16$ isopycnal. Thus, the total transport in both abyssal zones turned out to be almost zero.

Summarizing the previous discussion, we can say that the ACC in the Drake Passage is limited to a depth not exceeding 2700 m. At the same time, approximately to the south of the classic Polar Front, the lower boundary of the ACC does not penetrate deeper than the $\gamma^n = 28.16$ isopycnal. North of the Polar Front, this boundary is sharply non-isopycnic and varies approximately from a depth not exceeding 2700 m at the southern periphery of the SAC zone to 1600–1800 m at the northern one. The orographic barrier of the Scotia Sea and Drake Passage plays an important role in this limitation; it consists of at least three successive thresholds: the Hero Ridge together with the Phoenix Rift, the Shackleton Ridge, and the North Scotia Ridge together with the relatively shallow eastern part of the Scotia Sea (Fig. 3.7).

3.3.3.3 Weddell Sea Deep Water in the Scotia Sea and Drake Passage

This section of the book examines and quantifies the Upper Weddell Sea Deep Water and Lower Weddell Sea Deep Water pathways in the Scotia Sea and in the Drake Passage.

Figures 3.18, 3.19, 3.20 and 3.21 show the distributions of potential temperature, salinity and neutral density γ^n, and the boundaries between water masses over several sections in the studied ocean sector. To determine the boundaries of Lower Weddell Sea Deep Water and Upper Weddell Sea Deep Water, we generally used the isolines of neutral density γ^n: $\gamma^n = 28.31$ for the upper boundary of Lower Weddell Sea Deep Water and $\gamma^n = 28.26$ for the upper boundary of Upper Weddell Sea Deep Water. The first of the indicated γ^n values is the threshold for overflowing the Georgia Passage (Arhan et al. 1999), the second is used in a number of works in the Atlantic as the upper boundary of Weddell Sea Deep Water (Naveira Garabato et al. 2002a, b; Meredith et al. 2008). It is also the best approximation for the upper limit of Antarctic Bottom Water in the Pacific sector of the Southern Ocean (Koshlyakov and Tarakanov 2003a). The boundaries of the overlying water layers were determined using an original method linking the position of the boundary with the maximum vertical gradient of any hydrophysical or hydrochemical characteristic. This approach was previously used to analyze the structure of water masses in the Pacific sector of the Southern Ocean and is described in more detail in (Koshlyakov and Tarakanov 1999, 2003a). The spatial position of the boundaries in the region between the sections was determined using the interpolation algorithm described in (Koshlyakov and Tarakanov 2003a) based on the WOCE climatic array (Gouretski and Koltermann 2004).

Table 3.5 presents the estimates of quantitative characteristics of Upper Weddell Sea Deep Water and Lower Weddell Sea Deep Water in the Drake Passage and in the Scotia Sea; these estimates were obtained using the procedure for determining the spatial position of boundaries between water masses indicated in the introduction to the section. Figure 3.22 shows the boundaries of Upper Weddell Sea Deep Water and Lower Weddell Sea Deep Water as well as the boundaries of the sectors for the calculation in Table 3.5. Figure 3.23 shows the climatic distribution of salinity at the

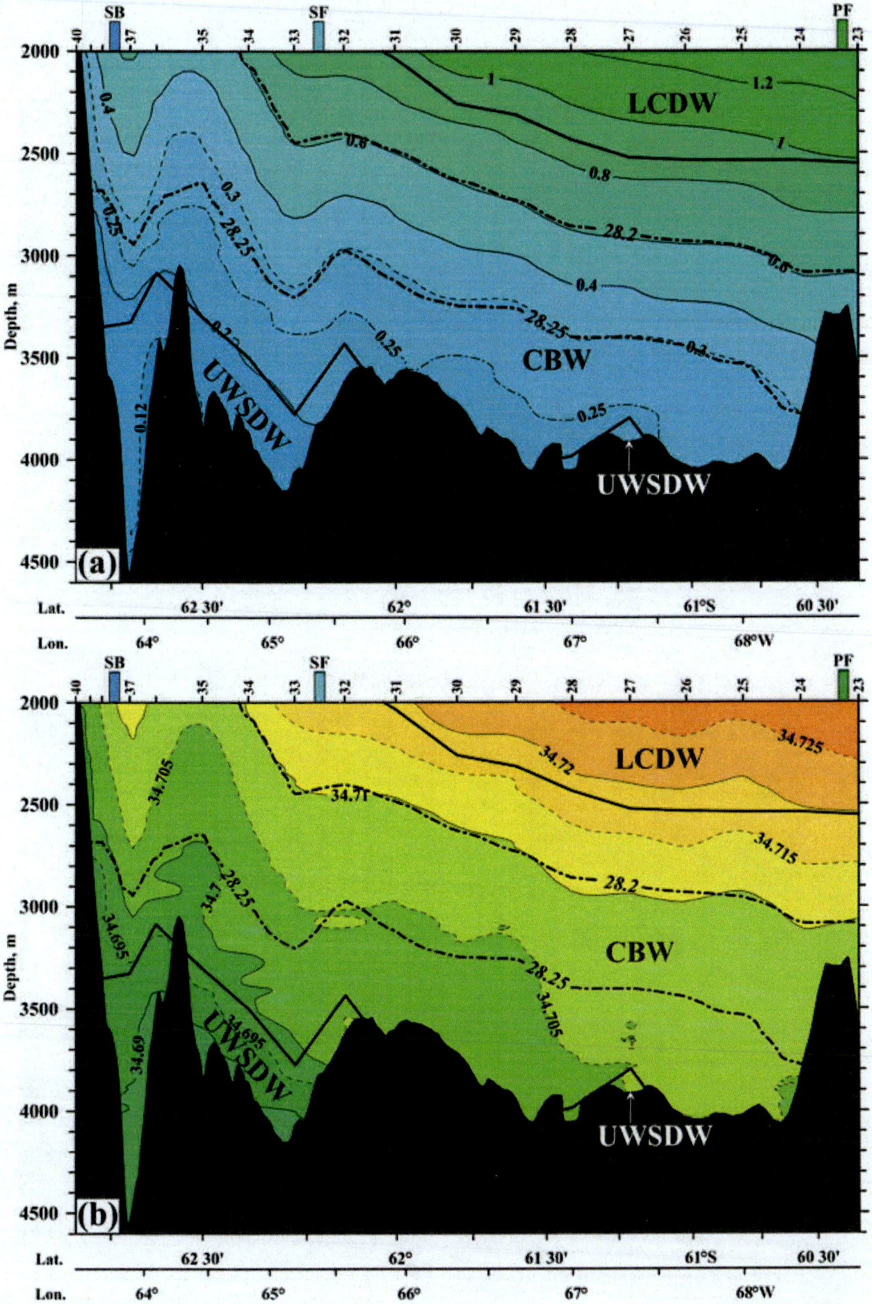

Fig. 3.18 Potential temperature, °C (**a**) and salinity, PSU (**b**) over section A21_1999 (Figs. 3.7 and 3.9). Solid bold lines show the boundaries between water masses. Dash-dotted lines show contours of neutral density γ^n. Numbers of stations are shown along the upper axis. Notations: (SB) Southern boundary of the ACC, (SF) Southern Front of the ACC, (PF) Polar Front, (LCDW) Lower Circumpolar Deep Water, (CBW) Circumpolar Bottom Water, (UWSDW) Upper Weddell Sea Deep Water. (Reproduced from Tarakanov 2009, with permission from Pleiades Publishers)

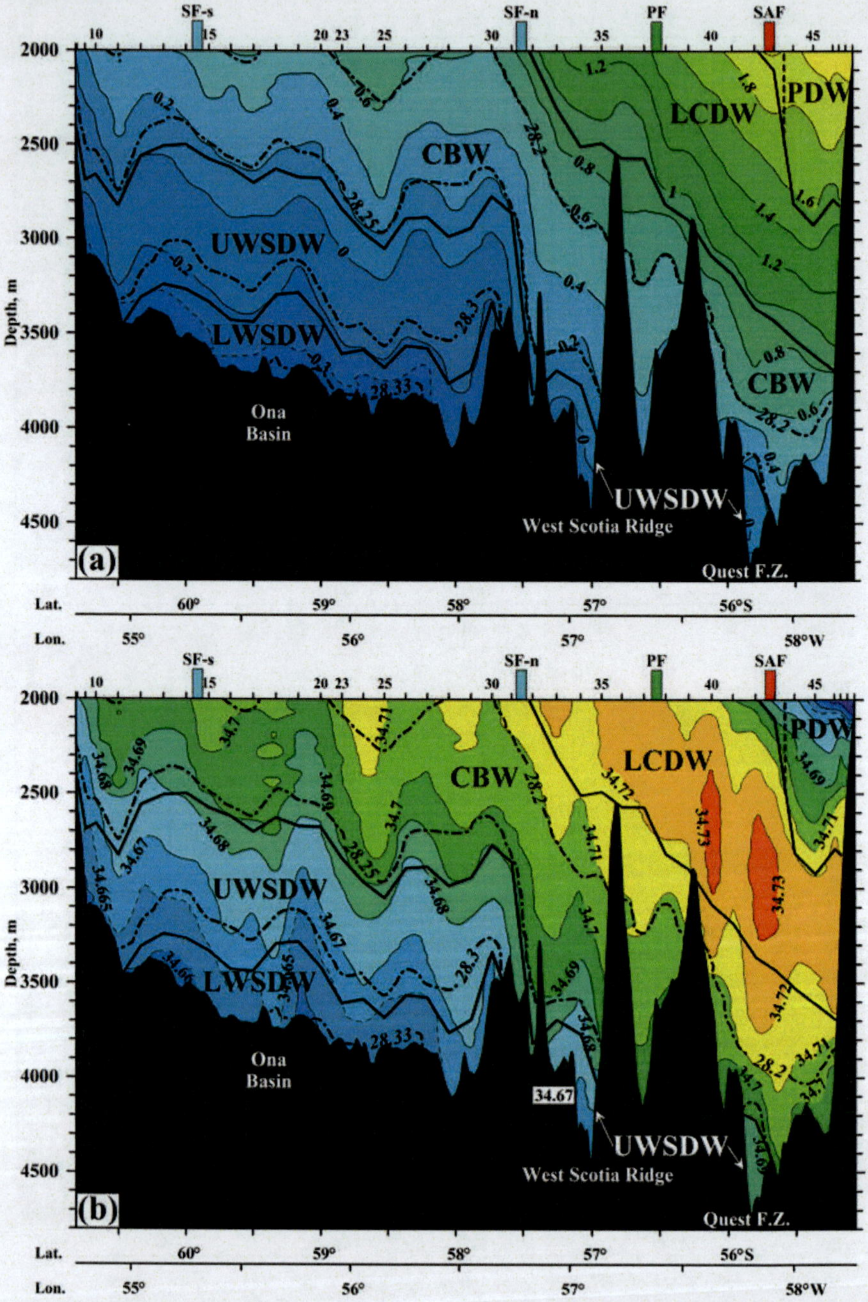

Fig. 3.19 Potential temperature, °C (a) and salinity, PSU (b) over section SR01m (Figs. 3.7 and 3.9). SF-s and SF-n are the southern and northern ACC branches, respectively, in the Southern Front of the ACC, (SAF) Subantarctic Front, (PDW) Pacific Deep Water, (LWSDW) Lower Weddell Sea Deep Water. See Fig. 3.18 for other notations. (Reproduced from Tarakanov 2009, with permission from Pleiades Publishers)

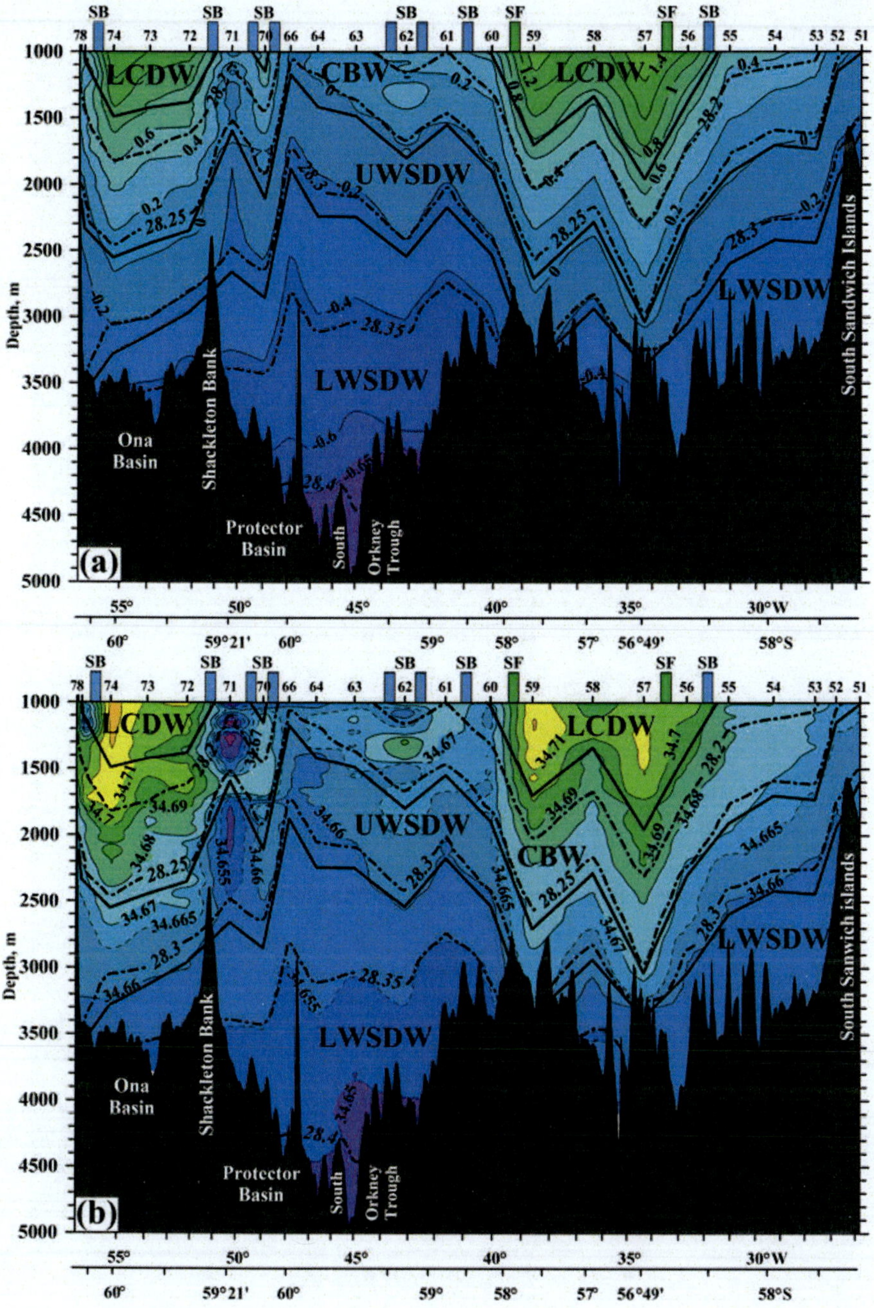

Fig. 3.20 Potential temperature, °C (**a**) and salinity, PSU (**b**) over section AJAX (Figs. 3.7 and 3.9). See Figs. 3.18 and 3.19 for other notations. (Reproduced from Tarakanov 2009, with permission from Pleiades Publishers)

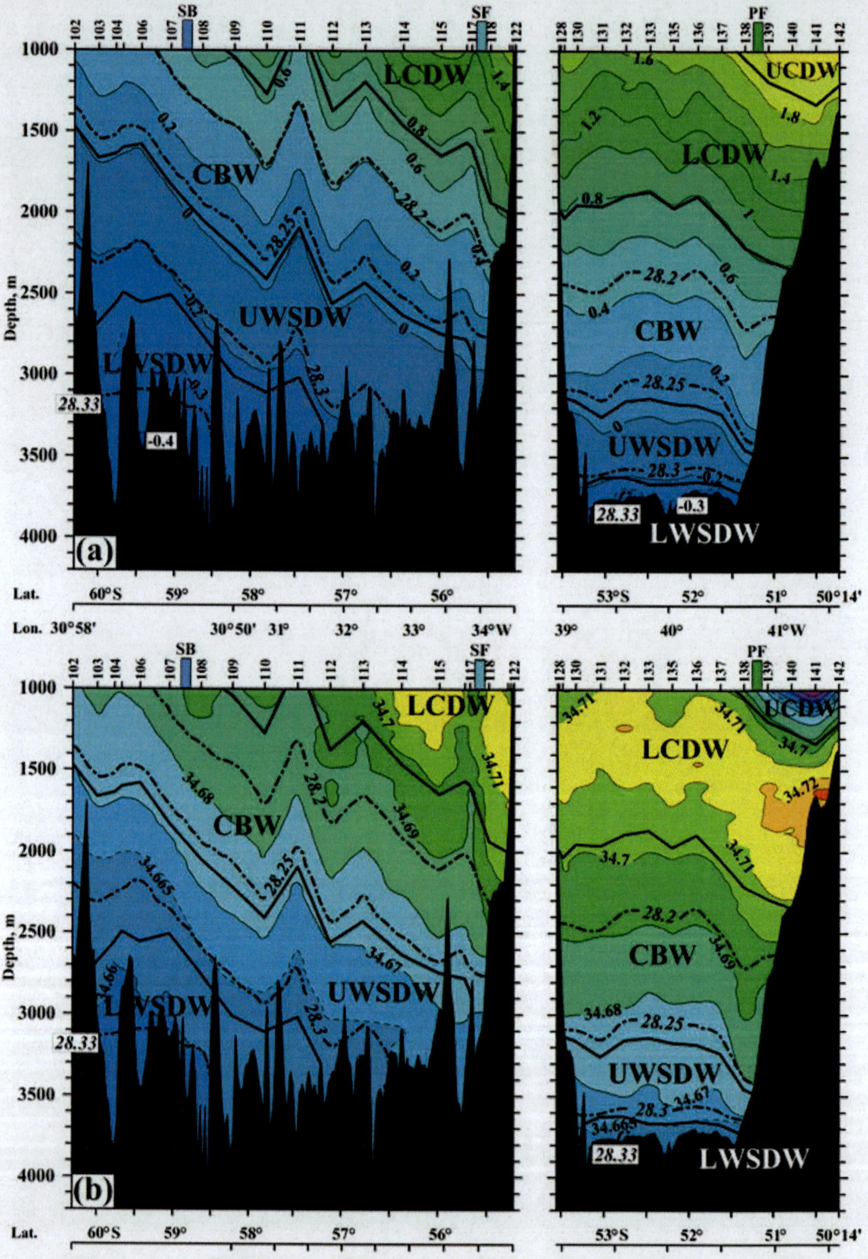

Fig. 3.21 Potential temperature, °C (**a**) and salinity, PSU (**b**) over section A23_1999 (left) and quasi-meridional part of section Falk_99 (right panel) (Figs. 3.7 and 3.9). UCDW is Upper Circumpolar Deep Water, See Fig. 3.18 for other notations. (Reproduced from Tarakanov 2009, with permission from Pleiades Publishers)

Table 3.5 Volume, range of variations, and mean values of Upper Weddell Sea Deep Water and Lower Weddell Sea Deep Water characteristics in the Scotia Sea and Drake Passage

Ocean region	Volume, 10^4 km^3	Square, 10^5 km^2	–	θ, °C		S, PSU		γ^n, kg/m^3		O_2, ml/l	
LWSDW (2)	13.8	2.5	Mean value	−0.30		34.660		28.324		5.32	
			Range of variations	−0.70	0.00	34.650	34.675	28.260	28.410	5.05	5.85
LWSDW (3)	24.8	4.1	Mean value	−0.29		34.659		28.320		5.39	
			Range of variations	−0.75	0.00	34.645	34.675	28.265	28.420	5.15	5.90
LWSDW (4)	0.1	0.1	Mean value	−0.28		34.663		28.319		5.23	
			Range of variations	−0.45	−0.15	34.655	34.670	28.300	28.345	5.20	5.25
LWSDW (total)	38.7	6.7	Mean value	−0.30		34.659		28.322		5.36	
			Range of variations	−0.75	0.00	34.645	34.675	28.260	28.420	5.05	5.90
UWSDW (1)	2.4	0.5	Mean value	0.12		34.686		28.265		4.99	
			Range of variations	−0.15	0.45	34.665	34.705	28.240	28.295	4.80	5.25
UWSDW (2)	31.5	5.7	Mean value	0.01		34.671		28.267		5.11	
			Range of variations	−0.50	0.50	34.650	34.695	28.205	28.330	4.85	5.40
UWSDW (3)	26.3	5.1	Mean value	−0.08		34.666		28.276		5.18	
			Range of variations	−0.35	0.20	34.655	34.685	28.230	28.320	4.95	5.40
UWSDW (4)	1.0	0.3	Mean value	−0.07		34.671		28.281		5.14	
			Range of variations	−0.30	0.15	34.660	34.680	28.245	28.325	5.00	5.25
UWSDW (total)	61.2	11.6	Mean value	−0.02		34.670		28.271		5.14	
			Range of variations	−0.50	0.50	34.650	34.705	28.205	28.330	4.80	5.40

Numbers in the first column indicate regions in Fig. 3.22

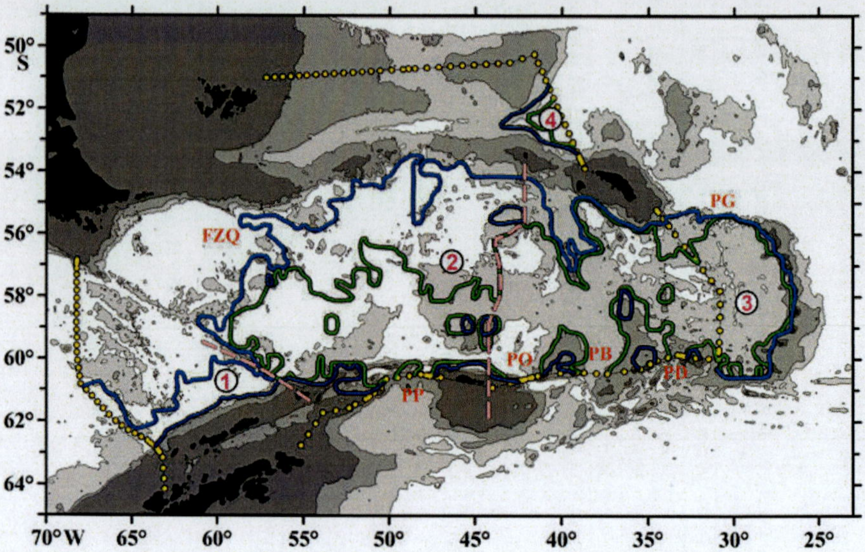

Fig. 3.22 Limits of Upper and Lower Weddell Sea Deep Water spreading in the Drake Passage and Scotia Sea (blue and green solid lines, respectively). Dashed pink lines and numerals in circles denote boundaries and numbers of sectors (Table 3.5). Isobaths 1000, 2500, and 3500 m are shown. Numerals 1, 2, 3 4 denote regions indicated in Table 3.5. FZQ is Quest Fracture Zone. See Fig. 3.6 for other notations. (Reproduced from Tarakanov 2009, with permission from Pleiades Publishers)

ocean floor, and Fig. 3.24 shows the topography of the upper boundary of Upper Weddell Sea Deep Water. Distribution patterns of Weddell Sea Deep Water shown in Figs. 3.23 and 3.24 were constructed taking into account the known literature data of these distributions given in (Nowlin and Zenk 1988; Locarnini et al. 1993; Naveira Garabato et al. 2002a, b) as well as the features of bottom topography and direct measurements of velocity using LADCP. We note that the horizontal scale of the synoptic variability of the ocean is comparable to the horizontal dimensions of individual basins in the study sites. Therefore, schemes shown in Figs. 3.23 and 3.24 should be understood as propagation schemes rather than the mean currents of Weddell Sea Deep Water.

As shown in Fig. 3.22, Lower Weddell Sea Deep Water in the Scotia Sea does not spread west of the Shackleton Ridge (the elevated part of the Shackleton Fracture Zone). In the northern direction, it spreads much farther than the Southern Front of the Antarctic Circumpolar Current. The Lower Weddell Sea Deep Water region is divided by the Pirie Bank (BP in Fig. 3.6) into two virtually isolated sectors, which are connected only by the South Orkney Trough. Weddell Sea Deep Water is found almost in the entire Scotia Sea excluding the Yagan Basin (Figs. 3.6 and 3.22). West of the Shackleton Ridge, Weddell Sea Deep Water is found only in the South Shetland Trench and a deep depression in the Hero Fracture Zone (Figs. 3.6 and 3.22).

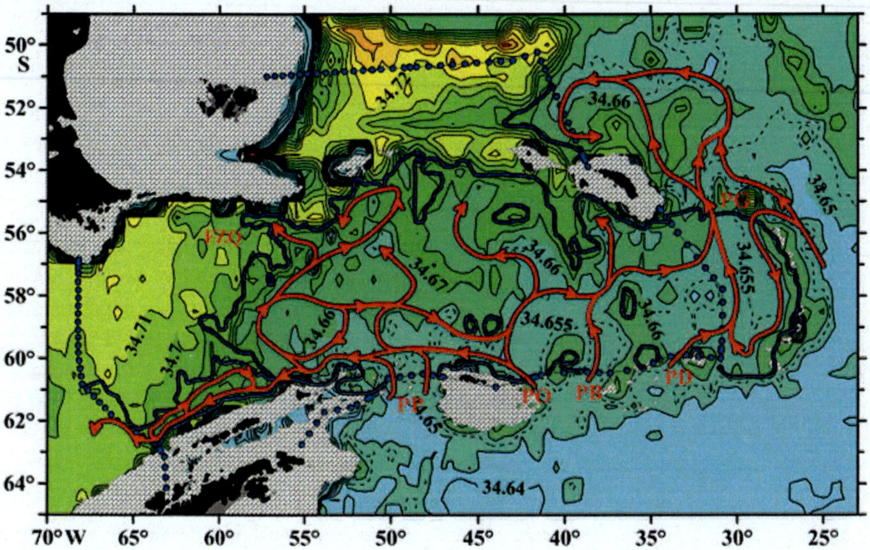

Fig. 3.23 Distribution of salinity (PSU) at the ocean bottom in the Drake Passage and Scotia Sea. Regions with depths shallower than 1000 m are hashed. Arrows show the authors' scheme of Weddell Sea Deep Water spreading in the bottom layer. The blue solid line shows the limits of this water spreading in the Scotia Sea and Drake Passage. FZQ is Quest Fracture Zone. See Fig. 3.6 for other notations. (Reproduced from Tarakanov 2009, with permission from Pleiades Publishers)

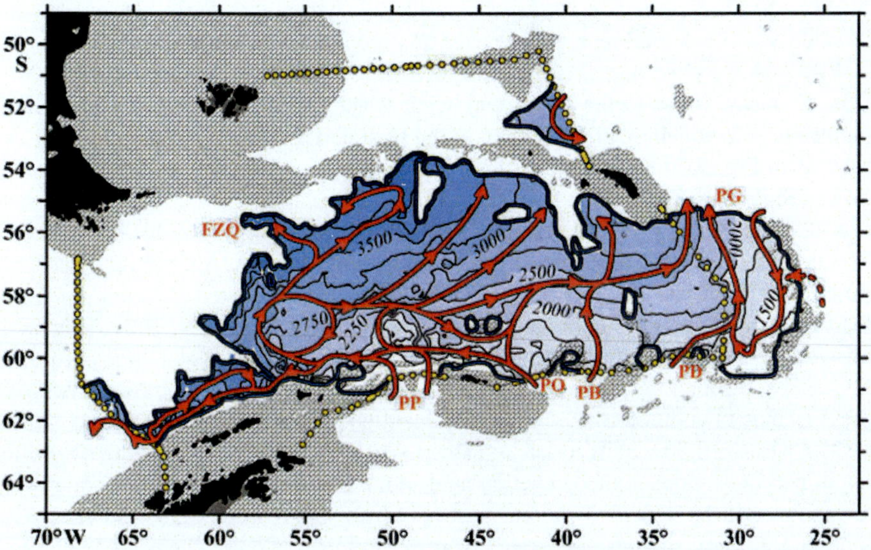

Fig. 3.24 Topography of the upper boundary surface of Upper Weddell Sea Deep Water (m) in the Drake Passage and Scotia Sea. Regions with depths shallower than 2000 m are hashed. Arrows show the authors' scheme of Upper Weddell Sea Deep Water spreading near the upper boundary surface. The blue solid line shows the limits of this water spreading in the Scotia Sea and Drake Passage. FZQ is Quest Fracture Zone. See Figs. 3.6 and 3.22 for other notations. (Reproduced from Tarakanov 2009, with permission from Pleiades Publishers)

3.3.3.3.1 Weddell Sea Deep Water in the Scotia Sea

Since Weddell Sea Deep Water is fresher than the overlying waters, the tongues of low salinity at the bottom should indicate the directions of Weddell Sea Deep Water spreading from the sources of its propagation to the Scotia Sea. Together with the chart of bottom topography (Fig. 3.6), the topography of the upper boundary surface (Fig. 3.24) gives information about the thickness of Weddell Sea Deep Water layer in the Scotia Sea. Minimum depths (1250–1750 m) of the upper boundary surface and bottom salinity (<34.655 PSU) in the southern part of the Scotia Sea (Figs. 3.23 and 3.24) correspond to Weddell Sea Deep Water propagating through the passages in the South Scotia Ridge. Tongues of low salinity at the bottom (<34.66 PSU) extending from this region to the east and west (Fig. 3.23) show the main directions of Weddell Sea Deep Water propagation in the Scotia Sea. The tongues of higher salinity (but less than 34.670 PSU) in the northern part of the sea mark the propagation of Weddell Sea Deep Water to the southeast and north in the zone between the South Polar Front and Southern Front of the Antarctic Circumpolar Current. Deflection to the west from this direction in the northwestern part of the Scotia Sea (Figs. 3.23 and 3.24) fits the tongue of low salinity related to the deep depressions in the Endurance and Quest fracture zones (locations of fractures are shown in Fig. 3.6). Salinity distribution at the bottom in the Georgia Basin is associated with anticyclonic bottom circulation in this region (Fig. 3.23). It is worth noting that this conclusion agrees with the results of LADCP measurements over the quasi-meridional part of section Falk_99 (Arhan et al. 2002). Distributions of salinity and potential temperature over this section are shown in Fig. 3.21.

The AJAX section passes along the entire part of the Scotia Sea from west to east from Elephant Island to the South Sandwich Islands and follows almost along the pathways of Weddell Sea Deep Water to the west and east (Figs. 3.9 and 3.20). It is seen from Fig. 3.20 that thermohaline properties of two Weddell Sea Deep Water flows propagating to the west from the Protector Basin (BaPr in Fig. 3.6) and to the east from the Orkney Passage are almost similar. Upwelling of Weddell Sea Deep Water up to 1000 m is observed in the eastern part of the AJAX section near the South Sandwich Islands (Fig. 3.20) corresponding to an isolated elevation (to less than 1500 m) at the upper boundary of Weddell Sea Deep Water (Fig. 3.24. This fact provides evidence about a possibility of Weddell Sea Deep Water inflow to the Scotia Sea, through the Hoyer Passage with depths exceeding 2000 m southwest of these islands in addition to the previously mentioned pathways (Fig. 3.6).

The westward flow of Weddell Sea Deep Water has a bimodal structure over the major part of sections in the Drake Passage and in the western part of the Scotia Sea (Dvtl97w, SR01a, Drk75b, western part of AJAX, SR01m, Drk05b, and A21_99 sections). Several stations are located near the Antarctic slope, where Weddell Sea Deep Water is characterized by low temperature and salinity compared to the offshore stations (Figs. 3.18, 3.19 and 3.20). According to (Naveira Garabato et al.

2002a), such structure of the western flow is related to the propagation of two types of Weddell Sea Deep Water with different scenarios of formation to the Scotia Sea. The young (cold and fresh) variety of Weddell Sea Deep Water formed in the western part of the Weddell Sea propagates to the Scotia Sea through the Philip Passage and partly through the Orkney Passage. The older (warm and saline) variety of Weddell Sea Deep Water, which propagates closer to the central part of the Weddell Gyre, propagates to the Scotia Sea through the Orkney, Bruce, and Discovery passages, and is later divided into the western and eastern flows (Naveira Garabato et al. 2002a). It was shown in (Meredith et al. 2003) that convection was developing over the Antarctic slope north of Elephant Island (South Shetland Islands) as a result of ice formation during winter in the 1990s, resulting in the formation of water with a temperature of $\theta = -0.9$ °C at a depth of 1040 m in the years marked by the extreme conditions. The density of descending water can reach the density range of Weddell Sea Deep Water. This water flowing to the west as plumes can also form a bimodal structure in the westward Weddell Sea Deep Water flow in the Drake Passage.

As was previously mentioned, the major part of the westward flow of Weddell Sea Deep Water deflects to the north due to the topographic obstacles in the western part of the Scotia Sea. Then the flow turns to the east and in the return current of the ACC fills the major part of the abyssal in the Scotia Sea (Locarnini et al. 1993). Such topographic obstacles for the westward flow of Weddell Sea Deep Water are the Shackleton Bank in the western part of the Protector Basin and the Shackleton Ridge (the depth over the ridge is generally shallower than 2500 m), and only in the narrow Shackleton Passage separating the ridge from the Antarctic slope the sill depth is slightly over 3000 m (Fig. 3.6). Indeed, the northwestern direction of Weddell Sea Deep Water motion along the northeastern slope of the Shackleton Ridge was found in 2005 over section Drk05b (Table 3.3 and Figs. 3.9 and 3.10) based on LADCP measurements, and then confirmed in the 2008 study site in the central part of the Drake Passage on the northeastern side of the Shackleton Ridge (Koshlyakov et al. 2010).

Two main topographic obstacles for the Antarctic Circumpolar Current exist along its pathway through the Scotia Sea in the Southern ACC Front (South Antarctic Current) zone. Their depths only slightly exceed 3000 m. The first is the Shackleton Ridge (Fig. 3.6). Two jets of the South Antarctic Current overflow it. The southern jet passes through the Shackleton Passage (Barre et al. 2008; Zhou et al. 2010). The second (quite large) region of the sea is located between the Bruce Bank and Georgia Passage (Fig. 3.6). Thus, because the upper boundary of Weddell Sea Deep Water in the South Antarctic Current region on the sections in the Scotia Sea is located at a depth of 2500–3000 m (Figs. 3.20 and 3.21), we can suppose that the upper layers of Weddell Sea Deep Water transported by the westerly Antarctic Slope Current are entrained by the southern branch of the Antarctic Circumpolar Current and transported from the western part of the Scotia Sea to its eastern part. Naturally, Weddell Sea Deep Water could mix in the course of its motion with the overlying warm and saline circumpolar water. As a result, this flow of Weddell Sea Deep

Water should be characterized by higher values of temperature and salinity by the time when it merges with Weddell Sea Deep Water flowing from the passages in the South Scotia Ridge directly to the east. Indeed, the warmer and more saline form of Weddell Sea Deep Water is observed in the eastern part of the Scotia Sea on sections AJAX and A23_99 at the northern periphery of the eastern flow of Weddell Sea Deep Water (Figs. 3.20 and 3.21). Based on this fact, the eastern direction of Weddell Sea Deep Water transport in the entire Scotia Sea in the Southern ACC Front zone is shown in Fig. 3.24. This pathway of Weddell Sea Deep Water motion is an additional (previously not described) transit pathway of transport in the Scotia Sea to the northern basins of the Atlantic Ocean.

Propagation of Weddell Sea Deep Water into the northern part of the Scotia Sea is likely governed both by the South Polar Front and bottom topography. Deep depressions of the bottom play important roles in this process: the rift valley of the West Scotia Ridge and Tehuelche, Endurance, and Quest fractures (Fig. 3.6 and Table 3.2). Quasi-meridional section SR01m located in the western part of the Scotia Sea (Fig. 3.9) crosses the northwesterly flow of Weddell Sea Deep Water in the Quest Fracture Zone (Fig. 3.19).

3.3.3.3.2 Weddell Sea Deep Water in the South Shetland Trench

Weddell Sea Deep Water was found only at one station on section Drk07 occupied along the Shackleton Ridge crest (Gladyshev et al. 2008) (Fig. 3.10). The station was located at the southern side of the Shackleton Passage. The upper boundary of Weddell Sea Deep Water with a temperature of 0.00 °C and salinity of 34.666 PSU was found at a depth of 2500 m. The westerly transport of Weddell Sea Deep Water was estimated at 0.5 Sv. The reverse easterly transport in the northern part of the passage occurs in the circumpolar water layer. We also note that on all sections in the South Shetland Trench the depth of the upper boundary of Weddell Sea Deep Water location exceeds the threshold depth of the Shackleton Passage. This result agrees with the modern concepts that Weddell Sea Deep Water is transferred to the density range of circumpolar waters due to diapycnal mixing in the Drake Passage and only then transported to the east by the Antarctic Circumpolar Current (Orsi et al. 1999).

It is likely that Weddell Sea Deep Water circulation in the South Shetland Trench is cyclonic and quasi-isolated. This can be seen from a number of direct and indirect indications:

1. In addition to the westerly flow of Weddell Sea Deep Water, an easterly reverse current of Weddell Sea Deep Water was found on the northern slope of the trench from the LADCP data on sections Drk03, Drk05a, and Drk11 (Fig. 3.10). The currents in the easterly flow reached 10 cm s^{-1}. In January 2010, Weddell Sea Deep Water was found only in the deepest part of the trench on the Drk10a section in the westernmost part of the section. However, even in this case, in the overlying layer, bottom flows of the western and eastern directions were observed

on the southern and northern slopes, respectively. No LADCP data are available from the Drk10b section.

2. If we assume that only the westerly flow of Weddell Sea Deep Water exists in the Southern Shetland Trench and take into account the available estimates of the volume transport (Table 3.5) and also take into account the Weddell Sea Deep Water transport through the Shackleton Passage, the renewal time of this water in the trench would be equal to 1.5 years. The length of the trench is approximately 400 km. Thus, the mean velocity of the westerly flow should be less than 1 cm s^{-1}. At the same time, velocities of the mean current along the Antarctic Slope in the Southern Shetland Trench measured on moorings at depths of 2700 and 3590 m were equal to 10–20 cm s^{-1} (Nowlin and Zenk 1988).
3. The South Shetland Trench with a threshold depth of about 4000 m (Fig. 3.6) is a region of closed geostrophic contours f/H (f is the Coriolis parameter; H is the ocean depth). Under these conditions, any upwelling over the trench would stimulate intense cyclonic circulation in the weakly stratified abyssal waters of the trench along the geostrophic contours (Johnson 1998). Upwelling is compensated by a water flow crossing the geostrophic contours in the bottom Ekman friction layer. We note that Weddell Sea Deep Water in the South Shetland Trench is distinguished for high density homogeneity compared with the Ona Basin. Density homogeneity can be characterized by quasi-conservative parameter fN^2 (N is the Brunt-Väisälä frequency). The value of fN^2 is 1–2·10^{-10} s^{-3} based on the stations in the South Shetland Trench and fN^2 is 3–6·10^{-10} s^{-3} in the same isopycnal interval in the Ona Basin.

3.3.3.3.3 Weddell Sea Deep Water West of the Hero Ridge

The next obstacle on the pathway of the westerly flow of Weddell Sea Deep Water after the Shackleton Ridge is a deep bank with depths of 3000–3500 m located west of the Hero Fracture Zone (Fig. 3.6). Since the depth of the upper boundary of Weddell Sea Deep Water in the South Shetland Trench exceeds 3000 m, the slope flow of Weddell Sea Deep Water should be deflected to the north along the isobaths. This property of the Weddell Sea Deep Water slope current was found on section A21_99 (Figs. 3.9 and 3.18) that was occupied along the trough of the Hero Fracture Zone (Fig. 3.9) dividing this bank and the ridge of the fracture zone, the depths over which are shallower than 3000 m.

As was mentioned above, it is commonly accepted that the westerly flow of Weddell Sea Deep Water does not reach the Pacific Ocean. However, traces of the cooler and fresher Weddell Sea Deep Water, as compared to the warmer and more saline Pacific type of Antarctic Bottom Water, were found, for example, in the eastern part of section S04P (see inset in Fig. 3.25) occupied in February and March of 1992 during the cruise of the R/V *Akademik Ioffe* (Koshlyakov and Sazhina 1995). Water in the layer of $\gamma^n > 28.26$ with salinity lower by 0.001 PSU and cooler by 0.005 °C than the surrounding waters of this density range was found at station 690 close to Adelaide Island at a depth greater than 3000 m (Fig. 3.25). The

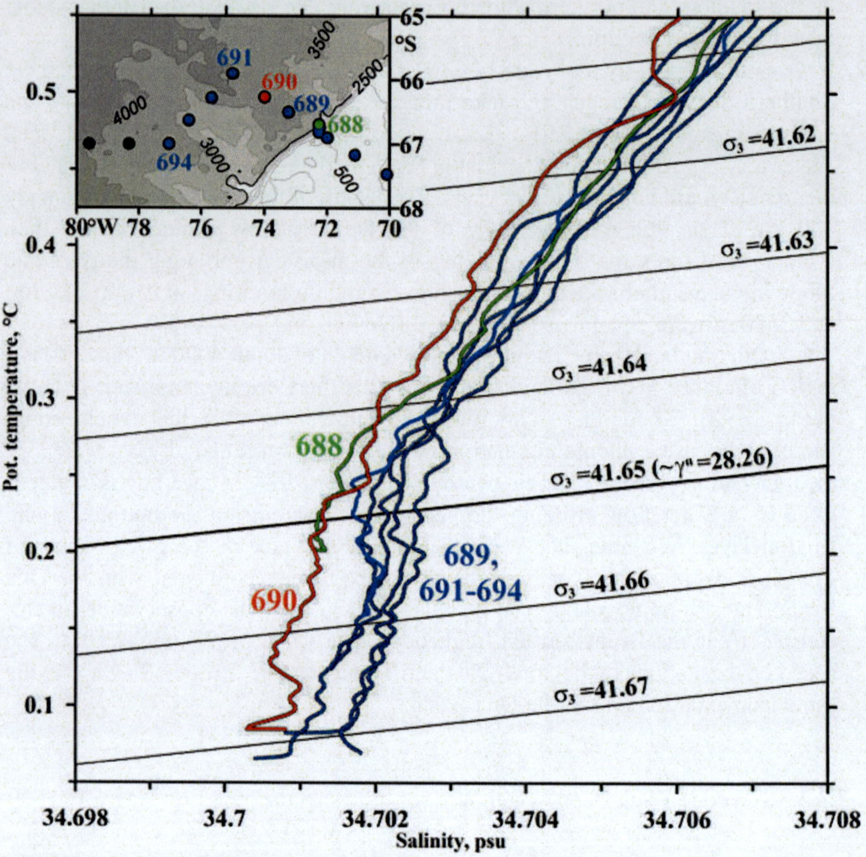

Fig. 3.25 Curves of θ,*S*-distributions for the eastern part of section S04P (stations 682–696) near Adelaide Island in the Pacific Ocean. Bold line emphasizes isopycnal $\sigma_3 = 41.65$ (~$\gamma^n = 28.26$) assumed as the upper boundary of Weddell Sea Deep Water. Scheme of stations along section S04P and bottom topography are shown in the inset. (Reproduced from Tarakanov 2009, with permission from Pleiades Publishers)

westerly direction of the flow at the bottom with a velocity of about 2 cm s^{-1} in this part of section S04P was reported in (Koshlyakov and Sazhina 1995).

3.3.3.3.4 Quantitative Characteristics of Weddell Sea Deep Water in the Scotia Sea and in the Drake Passage

Data in Table 3.5 indicate that characteristics of Lower Weddell Sea Deep Water in the western (sector 2, Fig. 3.22) and eastern (sector 3) parts of the Scotia Sea are very close to each other. The estimate of the total volume of Lower Weddell Sea Deep Water (Table 3.5) is slightly greater than $3.1 \pm 0.8 \cdot 10^4$ km^3 given in (Meredith et al.

2008). Renewal time of Lower Weddell Sea Deep Water in the Scotia Sea at the known transport through the South Scotia Ridge (3.5 ± 1.2 Sv) (Meredith et al. 2008) is approximately 3.5 years, which takes into account the isolation of this water in the Scotia Sea from the Atlantic basins. This is the total time of Lower Weddell Sea Deep Water transition to the density range of Upper Weddell Sea Deep Water. Characteristics of Upper Weddell Sea Deep Water are slightly different in the western and eastern parts of the Scotia Sea. Higher values of salinity and potential temperature in the western part of the sea are likely because the spreading region of Upper Weddell Sea Deep Water in this part of the sea is much greater than the spreading region of Lower Weddell Sea Deep Water. In addition, the region of spreading of the densest layers of Upper Weddell Sea Deep Water is strongly limited. This is because the neutral density is lower in the eastern part of the Scotia Sea. Significantly higher values of potential temperature and salinity in the Southern Shetland Trench (sector 1, in Fig. 3.22) almost at the same values of neutral density as in the western part of the Scotia Sea (sector 2) provide evidence of strong mixing of Upper Weddell Sea Deep Water in the trench with the overlying warmer and more saline circumpolar waters. Quasi-isolated cyclonic circulation in the trench facilitates efficient mixing of these waters. Estimate of the total volume of Weddell Sea Deep Water (UWSDW + LWSDW) in the Scotia Sea and Drake Passage is $9.9 \cdot 10^4$ km^3 is also slightly greater than $8.7 \pm 1.9 \cdot 10^4$ km^3 given in (Meredith et al. 2008). The total renewal time of Weddell Sea Deep Water in the Scotia Sea and Drake Passage is approximately 5 years.

3.3.3.4 Circumpolar Bottom Water in the Scotia Sea and Drake Passage

In this section of the book, we analyze the quantitative characteristics of Circumpolar Bottom Water and Warm Deep Water in the Scotia Sea and Drake Passage: volumes, their mean characteristics, and ranges of variation. We also update the scheme of distribution of Circumpolar Bottom Water and Warm Deep Water in these regions. The main attention is focused on the analysis of data from a series of Russian sections made in the region of the Shackleton Fracture Zone (Figs. 3.9 and 3.10). Figures 3.26, 3.27 and 3.28 show the distributions of potential temperature, salinity, and neutral density γ^n, as well as the boundaries between water masses over the Drk03, Drk05b, and Drk07 sections. The lower and upper boundaries of Circumpolar Bottom Water were determined from the neutral density isolines $\gamma^n = 28.26$ and $\gamma^n = 28.16$, respectively, which is the best approximation of these boundaries in the Pacific sector of the Southern Ocean (Koshlyakov and Tarakanov 2003b).

Figures 3.29a–c show the velocity vectors based on the data of the measurements of currents at the stations of the sections using LADCP profiler, averaged over the Circumpolar Bottom Water layer; these figures also show maps of the absolute dynamic sea level available at http://www.aviso.oceanobs.com (NRT-MADT product).

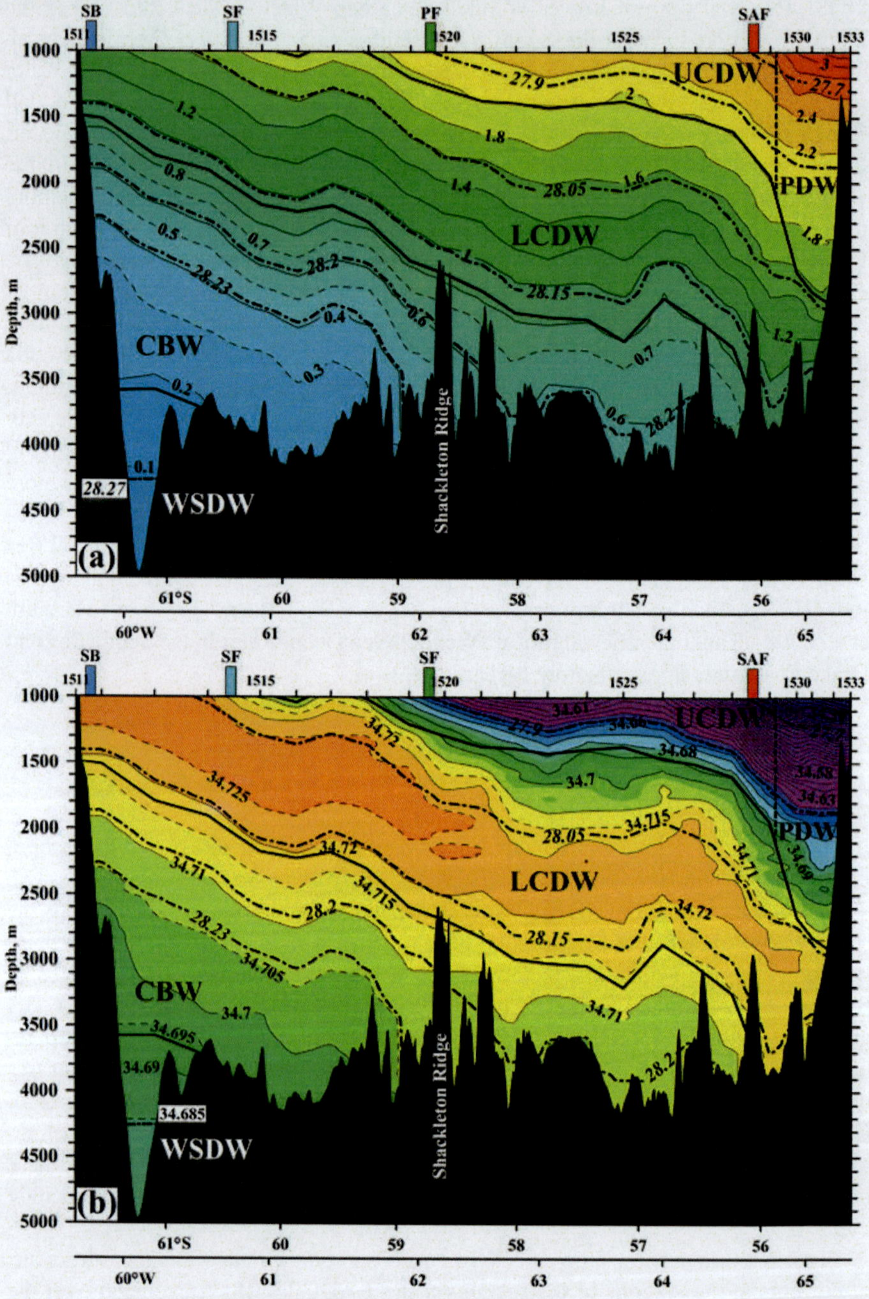

Fig. 3.26 Potential temperature, °C (**a**) and salinity, PSU (**b**) over section Drk03 (See Fig. 3.10). See Figs. 3.18, 3.19, and 3.21 for other notations. (Reproduced from Tarakanov 2010, with permission from Pleiades Publishers)

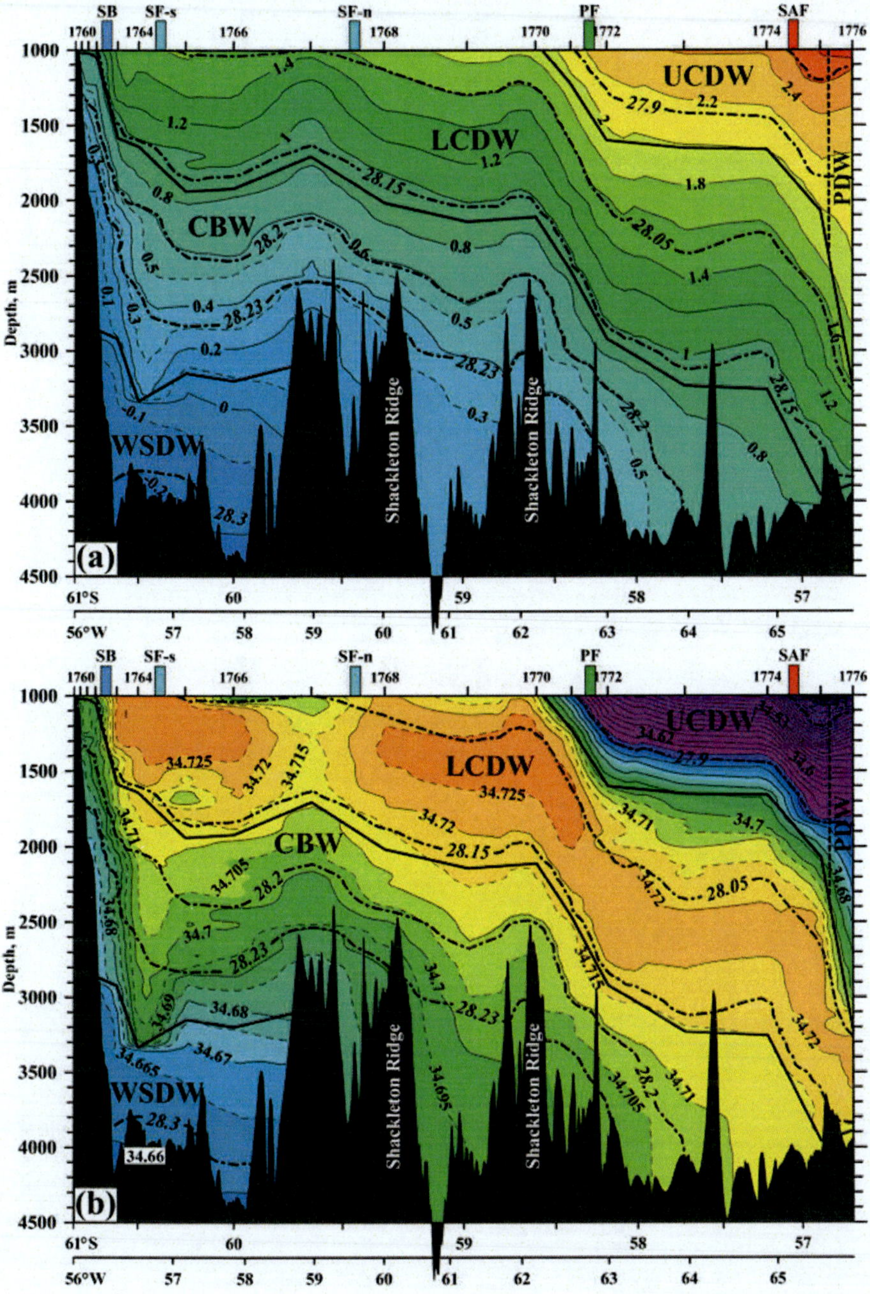

Fig. 3.27 Potential temperature, °C (**a**) and salinity, PSU (**b**) over section Drk05b (See Fig. 3.10). SF-s and SF-n are the southern and northern branches in the Southern Front of the ACC. See Fig. 3.26 for other notations. (Reproduced from Tarakanov 2010, with permission from Pleiades Publishers)

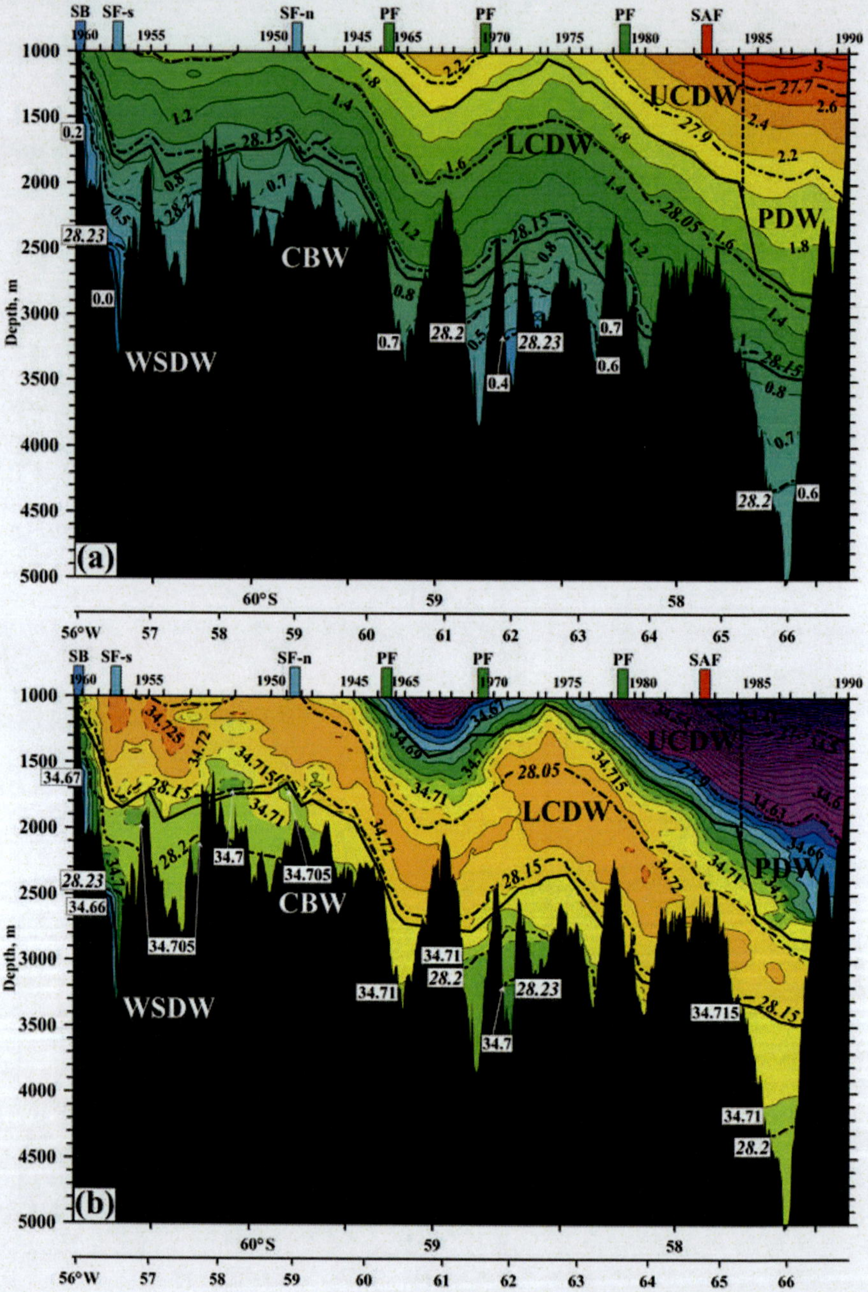

Fig. 3.28 Potential temperature, °C (**a**) and salinity, PSU (**b**) over section Drk07 (Fig. 3.10). See Fig. 3.26 for other notations. (Reproduced from Tarakanov 2010, with permission from Pleiades Publishers)

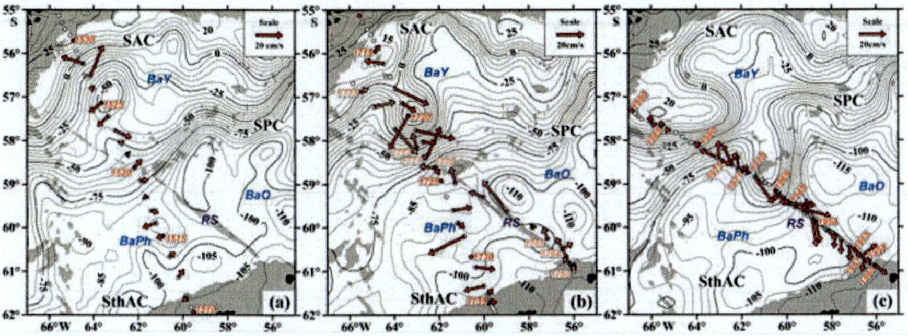

Fig. 3.29 Chart of absolute dynamic topography (cm) on December 13, 2003 (**a**); November 12, 2005 (**b**); and November 15, 2007 (**c**). Arrows show velocity vectors averaged over the layer of Circumpolar Bottom Water based on LADCP data from section Drk03 (**a**), Drk05a and Drk05b (**b**), and Drk07 (**c**). Notations: (BaO) Ona Basin, (BaPh) Phoenix Basin, (BaY) Yagan Basin, (RS) Shackleton Ridge. Abbreviations SthAC, SPC, and SAC denote South Antarctic Current, South Polar Current, and Subantarctic Current, respectively that are related to the corresponding fronts of the Antarctic Circumpolar Current. (Reproduced from Tarakanov 2010, with permission from Pleiades Publishers)

Table 3.6 shows the quantitative estimates of Circumpolar Bottom Water and Warm Deep Water in the Scotia Sea and Drake Passage obtained using the same procedure for determining the spatial position of the boundaries between water masses as we used before for Weddell Sea Deep Water. Figure 3.30 shows the distribution limits of Circumpolar Bottom Water and Warm Deep Water, as well as the boundaries of the sectors for calculating Table 3.6. Figure 3.31 shows a distribution diagram and topography of the upper boundary surface of Circumpolar Bottom Water and Warm Deep Water. The slope of this surface to the north is related to the corresponding slope of the isopycnal surfaces characteristic of the ACC zone. The boundary between Warm Deep Water and Circumpolar Bottom Water is defined by the climatic location of the ACC south boundary (Orsi et al. 1995). The scheme was constructed on the basis of detailed analysis of the data of sections in this region of the ocean, the features of the bottom topography, and literature sources (Nowlin and Zenk 1988; Orsi et al. 1993; Arhan et al. 1999; Orsi et al. 1999; Arhan et al. 2002; Naviera Garabato 2002a; Chereskin et al. 2009; Smith et al. 2010). The circulation patterns of the upper ($28.23 < \gamma^n < 28.16$) and lower ($28.26 < \gamma^n < 28.23$) layers of Circumpolar Bottom Water are basically similar, and the differences between them are associated with different degrees of blocking of water exchange in these layers by the thresholds in the Scotia Sea and Drake Passage. Thinning of the lower layer of Circumpolar Bottom Water to the bottom at the northern periphery of the study site is also seen.

To investigate the properties of Circumpolar Bottom Water in the Shackleton Ridge region, in addition to Fig. 3.13 the θ,*S*-curves are shown in Fig. 3.31 for this layer based on two sections in 2005, one of which crossed the Shackleton Ridge, and on a survey in 2008 in the central part of the ridge (Fig. 3.10) crossing the ridge in its northern part. Similarly to Fig. 3.13, these *θ*,*S*-curves show the diagrams for the P19

Table 3.6 Volume, range of variations, and other characteristics of Circumpolar Bottom Water and Warm Deep Water in the Scotia Sea and Drake Passage

Ocean region	Volume, 10^4 km^3	Square, 10^5 km^2	–	θ, °C		S, psu		γ^n, kg/m^3		O_2, ml/l	
CBW (1)	28.3	2.3	Mean value	0.54		34.706		28.208		4.80	
			Range of variations	0.05	1.25	34.655	34.740	28.060	28.275	4.75	5.10
CBW (2)	75.1	7.4	Mean value	0.57		34.698		28.194		4.84	
			Range of variations	0.00	1.25	34.650	34.725	28.080	28.265	4.45	5.20
CBW (3)	32.5	3.3	Mean value	0.37		34.682		28.205		4.94	
			Range of variations	−0.10	0.95	34.665	34.710	28.125	28.275	4.60	5.20
CBW (4)	9.6	1.7	Mean value	0.55		34.697		28.197		4.81	
			Range of variations	−0.10	1.20	34.670	34.720	28.110	28.265	4.50	5.15
CBW (total)	145.5	14.7	Mean value	0.52		34.696		28.199		4.85	
			Range of variations	−0.10	1.25	34.650	34.740	28.060	28.275	4.45	5.20
WDW (5)	7.4	0.6	Mean value	0.28		34.654		28.167		5.02	
			Range of variations	−0.05	0.75	34.600	34.685	28.060	28.255	4.75	5.30
WDW (6)	20.6	1.3	Mean value	0.29		34.661		28.180		4.95	
			Range of variations	−0.10	0.85	34.585	34.680	28.030	28.270	4.70	5.20
WDW (7)	15.5	1.1	Mean value	0.36		34.667		28.178		4.88	
			Range of variations	−0.05	1.10	34.590	34.690	28.030	28.270	4.65	5.10
WDW (total)	43.5	3.0	Mean value	0.31		34.662		28.177		4.94	
			Range of variations	−0.10	1.10	34.585	34.690	28.030	28.270	4.75	5.30

Numbers in the first column indicate regions in Fig. 3.30.

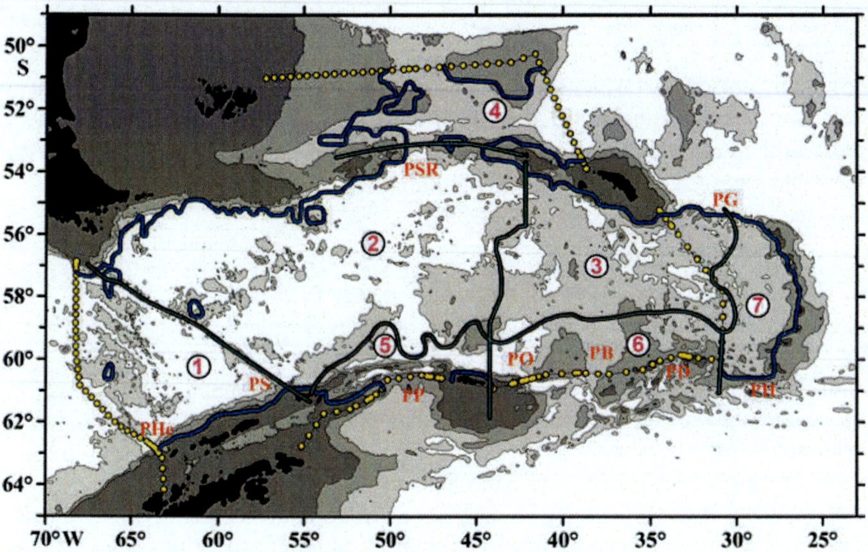

Fig. 3.30 Limits of Circumpolar Bottom Water and Warm Deep Water spreading in the Drake Passage and Scotia Sea are shown with blue solid bold lines. Green lines and numerals in circles denote boundaries and numbers of sectors (Table 3.6). Isobaths 1000, 2500, and 3500 m are shown. PB: Bruce Passage, PD: Discovery Passage, PG: Georgia Passage, PH: Hoyer Passage, PHe: Hero Passage, PO: Orkney Passage, PP: Philip Passage, PS: Shackleton Passage, PSR: Shag Rocks Passage. (Reproduced from Tarakanov 2010, with permission from Pleiades Publishers)

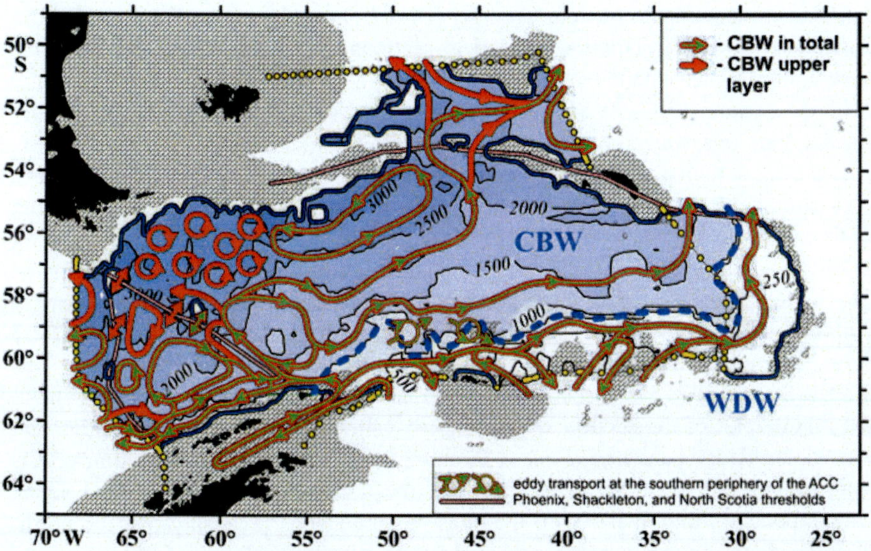

Fig. 3.31 Topography of the upper boundary surface of Circumpolar Bottom Water and Warm Deep Water (m) in the Drake Passage and Scotia Sea. Regions with depths shallower than 2000 m are dashed. Arrows show the authors' scheme of Circumpolar Bottom Water and Warm Deep Water spreading in the region. The blue solid line shows the limits of this water spreading in the Scotia Sea and Drake Passage. Thick blue dashed line divides the spreading regions of Circumpolar Bottom Water and Warm Deep Water. See Fig. 3.30 for other notations. (Reproduced from Tarakanov 2010, with permission from Pleiades Publishers)

section in 1993 from the continental slope of South America to the Antarctic slope to compare the properties of Circumpolar Bottom Water in the Drake Passage with those in the Pacific Ocean. The meridional part of the section passed along 88° W; the zonal part of the section passed approximately along 54° S.

3.3.3.4.1 Shackleton and Hero Ridges as Partially Permeable Barriers for Circumpolar Bottom Water

A thick quasi-homogeneous layer in the lower part of Circumpolar Bottom Water in the southern part of section Drk03 is worth attention (Fig. 3.26). Its upper boundary, determined from the maximum of $|d^2\sigma_4/dz^2|$, approximately corresponds to isopycnal $\gamma^n = 28.23$ ($\theta \approx 0.4$ °C). Such stratification of the lower layer of Circumpolar Bottom Water is almost missing on the sections adjacent to the eastern ridge from the east. In contrast, stratification is observed at all stations west of the ridge. This is illustrated by the distribution of fN^2 averaged over the lower layer of Circumpolar Bottom Water ($28.23 < \gamma^n < 28.26$) (Fig. 3.33) that characterizes homogeneity of the vertical density stratification. This fact allows us to suppose that waters of the lower layer of Circumpolar Bottom Water at both sides of the Shackleton Ridge are strongly isolated by this ridge, in the sense that intensity of circulation at both sides of the ridge strongly exceeds water exchange over the ridge.

Comparison of water transport in this Circumpolar Bottom Water layer based on LADCP measurements on four sections (Drk07 along the crest of the Shackleton Ridge and the Drk03, Drk10a, and Drk11 sections crossing the ridge) confirms this assumption. Water with density $28.23 < \gamma^n < 28.26$ was found on section Drk07 only at three stations 1957, 1958, and 1971 (Fig. 3.28). An estimate of the easterly transport of this water is 0.5 Sv and the westerly transport is less than 0.1 Sv, which is more than one order of magnitude smaller than the corresponding estimates on section Drk03 (8.9 Sv to the east and 9.6 Sv to the west) (Table 3.4). In the region of this section, water with such density was found only south of the ridge (Fig. 3.26).

We note that stations 1957 and 1958 on section Drk07 were located in the Shackleton Passage in the southern part of the section (Fig. 3.28). The easterly current was recorded only in the northern part of the passage at station 1957 (Fig. 3.29c). The transport in the lower layer of Circumpolar Bottom Water was 0.3 Sv. Station 1971 was occupied in the Polar Front (South Polar Current) zone in the middle part of the section, where the rift valley of the West Scotia Ridge and Shackleton Ridge converge (Figs. 3.28 and 3.29c). The westerly water transport is estimated at 0.2 Sv. It is noteworthy that the traces of overflow of more saline Circumpolar Bottom Water with the quasi-homogeneous lower layer to the north-eastern side of the ridge were found in the same places on section Drk05b (stations 1764, 1771, 1772) in 2005 (Figs. 3.27 and 3.13d), indicating stability of these deep currents. It is noteworthy that the current was directed to the northeast at station 1764 displaced to the east from the Shackleton Ridge (Fig. 3.29b) and directed to the northwest approximately along the slope of the ridge at stations 1771 and 1772 in the South Polar Current zone (Fig. 3.29b).

It is likely that the Hero Ridge, like the Shackleton Ridge, significantly blocks the exchange of Circumpolar Bottom Water, in particular, it blocks the exchange between its lower layers. This is evident from the differences in the slopes of θ,S-curves, as well as from the values of potential temperature and salinity corresponding to the lower quasi-homogeneous layer in the southern part of section Drk03 and central part of section A21_99 that are located east and west of the Hero Ridge, respectively (Fig. 3.13b, c). All the previously mentioned facts provide evidence that the Hero and Shackleton ridges facilitate the formation of a few quasi-stable and quasi-isolated circulations of abyssal and deep waters in the adjacent regions.

3.3.3.4.2 Circumpolar Bottom Water Exchange over the Shackleton Ridge

As shown in Sect. 3.3.3.2, a strong transformation (cooling and desalination) of the Circumpolar Bottom Water layer along the entire length of the Shackleton Ridge is related to the transition of the ACC across the ridge. Such a change in the properties of Circumpolar Bottom Water precisely over the Shackleton Ridge is obviously associated with its shape and location in the Drake Passage (Figs. 3.7 and 3.10). The shallower southern part of the ridge creates a partially permeable barrier for both the easterly transport of the relatively saline Circumpolar Bottom Water along with the ACC jets, and for the westerly transport along the Antarctic slope for the fresher and colder waters of the Weddell Sea and Bransfield Strait. The Shackleton Ridge diverts some of the western less saline water flow, which is then transported across the ACC as the Deep Western Boundary Current on the northeastern slope of the ridge. This Deep Western Boundary Current transporting at least Circumpolar Bottom Water and Weddell Sea Deep Water was first instrumentally revealed in the southern part of the Shackleton Ridge over the section in 2005 along the slope of the ridge (Gladyshev et al. 2008). Later, in 2008, the same current was measured during a survey in the central part of the ridge (Koshlyakov et al. 2010). Along with the Deep Western Boundary Current, the freshening of the Circumpolar Bottom Water layer in the Scotia Sea may be related to the eddy transport at the southern periphery of the ACC (Fig. 3.31). The southern component of this transport may be responsible for salinization of the Warm Bottom Water layer at the northwestern periphery of the Weddell Gyre (Naveira Garabato et al. 2003). The partial permeability of the Shackleton Ridge to the waters of the Circumpolar Bottom Water layer, the existence of which was already indicated in Sect. 3.3.3.2, leads to salinization and warming of these waters on the eastern side of the ridge and, conversely, to desalination and cooling on its western side. At the same time, due to the different degree of blocking of water exchange by the ridge, the degree of this transformation is different in the lower and upper layers of Circumpolar Bottom Water.

Lower Layer of Circumpolar Bottom Water (28.26 > γ^n > 28.23) Water exchange over the Shackleton Ridge in the lower part of the Circumpolar Bottom Water layer, is extremely complicated; hence, there are significant differences in the θ,S-characteristics of this layer between the basins separated by the ridge (Fig. 3.32a, b). Such water exchange was recorded only in two deep-water passages: in 2005 and 2007 in the Shackleton Passage (Figs. 3.32a and 3.13d) and in 2005 in the depression in the convergence zone of the Shackleton Ridge with the rift valley of the West Scotia Ridge (hereinafter, the West Scotia Gap, Figs. 3.6 and 3.32a). The easterly transport of the more saline type of Circumpolar Bottom Water within this water exchange manifests itself in the form of intrusions localized in the region of the previously mentioned passages on the eastern side of the ridge (Fig. 3.32a). It is impossible to judge about the degree of influence of these intrusions on the θ,S properties of the lower layer of Circumpolar Bottom Water in the basin east of the ridge based on the available data. The intensity of the westerly transport of the less saline type of Circumpolar Bottom Water is sufficient for a notable decrease in salinity compared to the Pacific Ocean type of water in the lower layer of Circumpolar Bottom Water in the Phoenix Basin located between the Shackleton and Hero ridges. This is clearly seen from the comparison of the forms of the θ, S-diagrams at stations in the basin with the θ,S-curves at stations over the WOCE P19 section (Figs. 3.32a and 3.13d). The clearly pronounced influence of the less saline type of Circumpolar Bottom Water in this basin is due to the fact that water exchange in this layer is hampered by both ridges.

Upper Layer of Circumpolar Bottom Water (28.23 > γ^n > 28.16) Water exchange across the Shackleton Ridge in the upper layer of Circumpolar Bottom Water (28.23 > γ^n > 28.16) occurs along the entire length of the ridge. This is seen in its central and southern parts (zones of the South Polar Current and Southern Antarctic Current, respectively) from the distribution of the normal component of velocity over the 2007 section along the ridge crest (Fig. 3.12). However, in the southern part of the ridge, the differences in θ,S characteristics in this layer between the basins separated by the ridge remain as significant as in the lower layer of Circumpolar Bottom Water (Fig. 3.32a), which implies the difficult nature of heat and salt exchange in the southern part of the ridge in general in the Circumpolar Bottom Water layer. This fact combined with a rather high estimate of the intensity of water exchange (i.e., in this case, westerly and easterly transport) in the Southern Antarctic Current zone in the Drk07 (2–5 Sv) section (Table 3.4) indicates the formation of local circulation around the high southern part of the Shackleton Ridge (Fig. 3.31). This circulation acts as a buffer zone hampering efficient (i.e., exchange of characteristics) water exchange between the basins in the upper layer of Circumpolar Bottom Water. The hypothesis about the existence of anticyclonic circulation in this region was put forward in (Gladyshev et al. 2008).

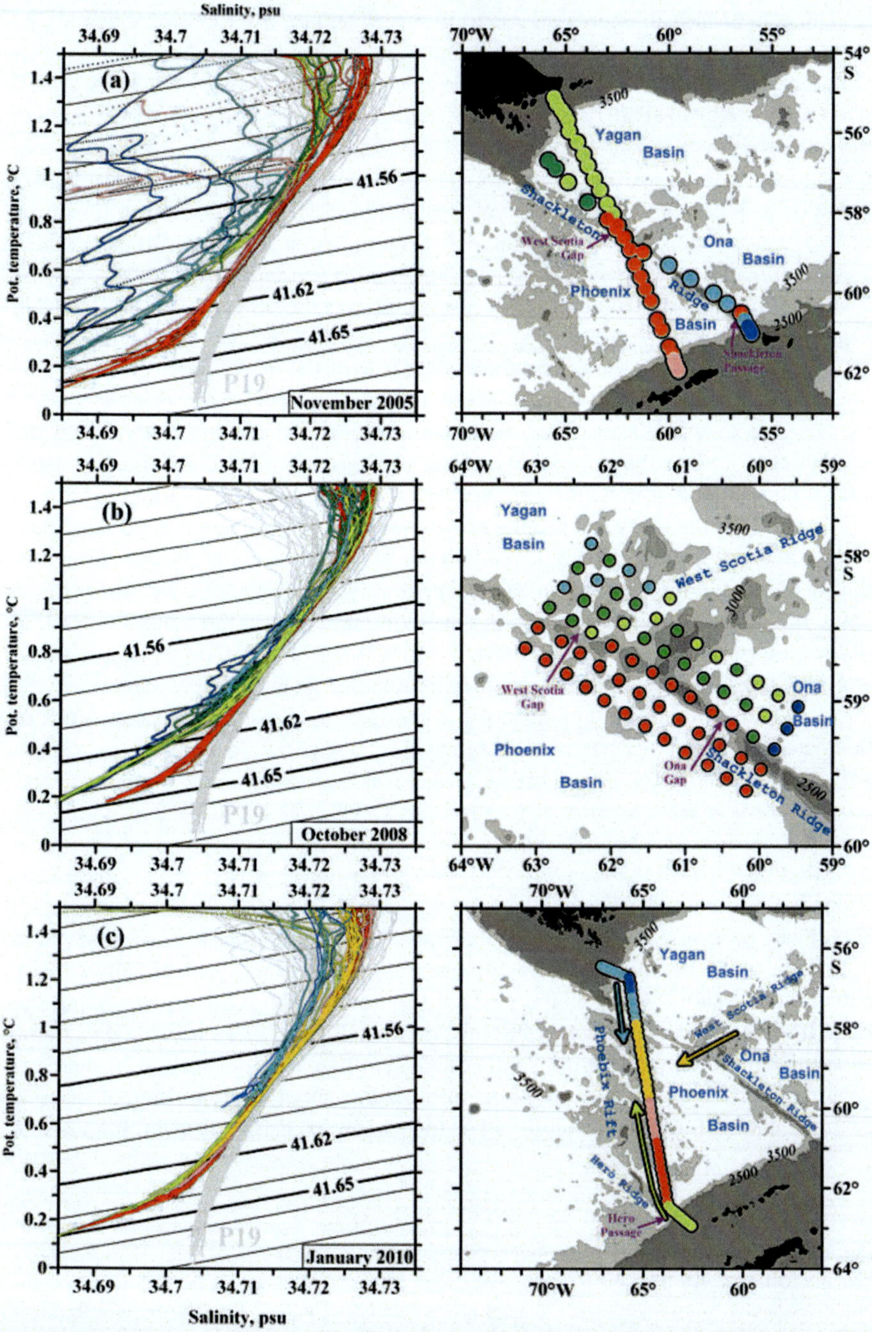

Fig. 3.32 Curves of θ,*S*-distributions for the Circumpolar Bottom Water layer (left panels) and schemes of locations of stations and bottom topography (right panels) along two hydrographic sections in 2005 made in the region of the Shackleton Ridge from the R/V *Akademik Ioffe* (**a**), 2008 regional survey in the central part of the Shackleton Ridge from the R/V *Akademik Sergey Vavilov*

The data of measurements on the 2008 survey and on the 2005 sections provide evidence of strong transport of more saline type of this water in the upper layer of Circumpolar Bottom Water on the eastern side of the Shackleton Ridge in its central part (Koshlyakov et al. 2010), which is manifested in a decrease in the differences in θ,*S* properties between the basins separated by the ridge (Fig. 3.32a, b). In this case, the stations with almost the same *θ,S* characteristics as those west of the ridge are noted (Fig. 3.32a, b). The locations of these stations correspond to the cores of the flow of more saline type of Circumpolar Bottom Water from west to east, which were confined to the West Scotia Gap (in 2005 and 2008) and to the depression in the Shackleton Ridge located at 59°05′ S (measured over the section in 2008, Ona Gap, Figs. 3.6 and 3.32b).

Other stations of the survey on the eastern side of the ridge, either indicating the presence of intrusions of two types of Circumpolar Bottom Water, or a completely fresher type of water were characterized by the lowest salinities observed on the southeastern and northern periphery of the survey (Fig. 3.32b). On the one hand, such a horizontal thermohaline structure of the upper layer of Circumpolar Bottom Water indicates the isopycnal mixing of two water types observed in 2008 east of the ridge. On the other hand, Fig. 3.32b shows the unsteadiness of the abyssal currents in the central part of the Shackleton Ridge, apparently due to synoptic variability in the entire ocean depth, which can manifest itself, in particular, in the periodic cessation of the eastern transport of Circumpolar Bottom Water over the ridge, as a result of which the less saline modification of Circumpolar Bottom Water can reach the northern end of the ridge without notable changes. In the latter case, intrusions of this type of Circumpolar Bottom Water can also be expected to appear in the central part of the Shackleton Ridge on its western side. Indeed, such intrusions were recorded on the 2007 section at two stations (1969 and 1970) located within the survey zone (Figs. 3.12 and 3.13e).

Based on the measurements in 2007 along the northern part of the Shackleton Ridge in the Subantarctic Front (Subantarctic Current) zone, all Circumpolar Bottom Water was concentrated near a deep narrow trough west of the ridge (Figs. 3.6 and 3.28). The upper boundary of Circumpolar Bottom Water was located only slightly above the threshold depth of the passage (approximately 3600 m) separating the ridge from the continental slope of South America (Fig. 3.28). Hence, the data of the Circumpolar Bottom Water transport (Table 3.4) based on the section in the Subantarctic Current zone give us more information about the intensity of abyssal cyclonic circulation in this trough rather than water exchange over the ridge. Note

Fig. 3.32 (continued) (**b**), and the section in 2010 located on the eastern side of the Hero Ridge and Phoenix Rift occupied from the R/V *Akademik Ioffe* (**c**). The θ,*S*-curves for the P19 section in the Southeast Pacific Ocean are shown in light gray (see text). The bold lines show the isopycnals $\sigma_3 = 41.56$ ($\sim\gamma^n = 28.16$), $\sigma_3 = 41.62$ ($\sim\gamma^n = 28.23$), and $\sigma_3 = 41.65$ ($\sim \gamma^n = 28.26$). Arrows in Fig. 3.32c indicate possible inflow paths of the less saline type of Circumpolar Bottom Water into the northern part of the Phoenix Basin. (Reproduced from Tarakanov 2012, with permission from Pleiades Publishers) that, despite the actual absence of Circumpolar Bottom Water flow through the Shackleton Ridge in the South Antarctic Current zone in 2007, the available data do not exclude the possibility of such water exchange, including the westward flow of Circumpolar Bottom Water, which can be realized in the meandering process of the ACC jets. The presence of such overflows was shown in Sect. 3.3.3.2. Based on the available data, it is impossible to establish the presence of overflows of Circumpolar Bottom Water, which could occur in this part of the ridge in the opposite direction.

3.3.3.4.3 Circumpolar Bottom Water in the Phoenix Basin

One can see from the LADCP measurements in 2003 and 2005 (Fig. 3.29a, b) that at least two cyclonic gyres of Circumpolar Bottom Water exist in the southern part of the Drake Passage between the Hero and Shackleton ridges (Fig. 3.31). Circulation in the South Shetland Trough is the upper part of the Upper Weddell Sea Deep Water gyre described in Sect. 3.3.3.3. Its northern arc is related to the southern branch of the Antarctic Circumpolar Current. The southern arc is the Slope Antarctic Current, which is seen in Fig. 3.31 over the shallow depths of the upper boundary surface of Circumpolar Bottom Water and closer location of isobaths. It is likely that the other circulation pattern located in the northern zone between the South Polar Front and Southern Front of the Antarctic Circumpolar Current does not correlate with the surface currents, which are mapped from the absolute dynamic topography of the ocean (Fig. 3.29).

Results of measurement over the January 2010 section showed that less saline Circumpolar Bottom Water intrusions in the northern and central parts of the Shackleton Ridge can leave a pronounced trace in the properties of Circumpolar Bottom Water in the northern periphery of the Phoenix Basin (Fig. 3.32c). In the section from the Antarctic slope to the Shackleton Ridge, θ,S-curves in the Circumpolar Bottom Water layer form four groups (Fig. 3.32c), each of which apparently corresponded to a separate abyssal circulation cell. These cells can be both quasi-stationary elements of the circulation and represent deep eddies (Koshlyakov and Tarakanov 2011). The group of stations with the lowest salinity in the Circumpolar Bottom Water layer in the 2010 section was located above the Antarctic slope, while those with the highest salinity are in the southern part of the basin seaward of the slope. These two groups apparently correspond to the circulations in the South Shetland Trench and in the central part of the Phoenix Basin. Salinity decreased again in the two groups of stations further to the north indicating the influence of less saline waters on the eastern side of the Shackleton Ridge (Fig. 3.32c). Since this salinity decrease was found in the lower and partly upper layers of Circumpolar Bottom Water, the intrusions leading to this decrease could contain water from both density ranges of Circumpolar Bottom Water. It is noteworthy that the described change in the thermohaline properties of the Circumpolar Bottom Water layer along the 2010 section admits another explanation, which is in the existence of the episodic or permanent flow of the Deep Western Boundary Current along the northeastern slope of the Hero Ridge and further along the elevation in the southern part of the

Phoenix Rift and transport of the more saline form of Circumpolar Bottom Water from the Slope Antarctic Current zone (3.29c). Since the Hero Ridge hinders water exchange in the lower layer of Circumpolar Bottom Water, this assumption seems justified. However, to verify it, additional measurements are needed in this region.

The unsteadiness of Circumpolar Bottom Water exchange in the central and northern parts of the Shackleton Ridge, in particular, the possibility of both eastern and western transport across the ridge in the Subantarctic Current and South Polar Current zones, suggests that Circumpolar Bottom Water exchange in this zone through the Phoenix threshold is similar. Partial blocking of the Circumpolar Bottom Water layer by this part of the sill leads to the formation of several quasi-isolated abyssal circulations west of the Phoenix Rift (Fig. 3.31). Their anticyclonic nature is hypothetical. It is also possible that in the northern part of the Phoenix Basin and west of the Phoenix Rift, an eddy field can be observed similar to that in the Yagan Basin in the western part of the Scotia Sea.

3.3.3.4.4 Circumpolar Bottom Water in the Scotia Sea

Circumpolar Bottom Water in the Ona Basin According to the LADCP data in 2005, the northwesterly flow of Circumpolar Bottom Water, freshened due to mixing with Warm Deep Water, was observed along the northeastern slope of the Shackleton Ridge. This flow was confined to the zone between the Shackleton Passage and Polar Front, which is the outflow region of more homogeneous and saline Circumpolar Bottom Water from the basin west of the ridge (Fig. 3.29b). This fact combined with the satellite altimetry data (Fig. 3.29) provides evidence that cyclonic circulation is formed in the western part of the Ona Basin in the water column from the surface to the bottom. This gyre is a meander of the southern branch of the Southern ACC Front (South Antarctic Current), which crosses the ridge near the Shackleton Passage. Satellite altimetry data in the Ona Basin indicate that this meander periodically separates from the main flow and splits off into smaller eddies (Fig. 3.29b). The existence of such an eddy circulation was suggested in (Barre et al. 2008; Zhou et al. 2010).

Circumpolar Bottom Water in the Northern Scotia Sea According to the latest studies (Chereskin et al. 2009), intense abyssal eddies exist in the abyssal in the western part of the Yagan Basin, an oceanic region adjacent to the northern part of the Shackleton Ridge from the east. The eddy field in this region is shown in Fig. 3.31. The upper layer of Circumpolar Bottom Water with a density of $\gamma^n < 28.205$ ($\theta > 0.55$ °C), which occupies the bottom position here, can exhibit, depending on the synoptic situation, the properties close both to more and less saline type of Circumpolar Bottom Water spreading from the western side of the ridge, mainly in its central part (Fig. 3.32).

Figure 3.31 shows quasi-isolated anticyclonic circulation in the northern part of the Scotia Sea, which includes the entire Circumpolar Bottom Water layer and even

Weddell Sea Deep Water unlike the circulation in the western part of the Yagan Basin. This type of circulation precisely appears due to the following reasons:

1. The depth of the upper boundary surface of Circumpolar Bottom Water in the northern part of the sea reaches the values that exceed the threshold depths characteristic of the Shag Rocks Passage (2900 m) and those in the eastern part of the Scotia Sea (3200 m) (Fig. 3.31). Thus, parts of the volume of Circumpolar Bottom Water and underlying Weddell Sea Deep Water are blocked for the eastward propagation with the ACC.
2. Very large differences in the density ranges of bottom waters between the Yagan Basin and the northern part of the Scotia Sea suggest that the circulations in these regions of the Scotia Sea are relatively isolated from each other.
3. Measurements on the continental slope of South America in the northern part of the Scotia Sea, including mooring measurements (Whitworth et al. 1982), climatic distributions of some characteristics at the bottom (for example, salinity), and distributions of salinity and temperature over the sections at the foot of the continental slope of South America (see Sect. 3.3.3.3) indicate the western direction of the slope current, which forms the northern branch of anticyclonic circulations in the Yagan Basin and in the northern Scotia Sea.

However, we cannot exclude that the idea of mean circulation in the northern part of the Scotia Sea, as well as in the Yagan Basin may be wrong because the current field here can be represented by intense abyssal eddies of different signs.

Mini-cell of the Global Conveyor Belt in the Western Part of the Scotia Sea Volumes of Circumpolar Bottom Water and Weddell Sea Deep Water blocked for the easterly transport in the entire western part of the Scotia Sea (region 2 in Fig. 3.30) are estimated at the same value of 30×10^4 km^3, which is approximately two fifths of the total Circumpolar Bottom Water volume and two thirds of the total Weddell Sea Deep Water volume in this region (Table 3.6). Diapycnal mixing with the overlying waters, isopycnal transport to the southern periphery of the Antarctic Circumpolar Current due to the formation of anticyclonic eddies with their further dissipation, or formation of a westerly flow to the Pacific over the northern part of the Shackleton Ridge can be considered as possible mechanisms of this transport. According to the results of numerical analysis of circulation (Naveira Garabato et al. 2003), intense diapycnal mixing occurs in the Scotia Sea between Weddell Sea Deep Water and overlying layers of circumpolar waters. The authors of the paper cited above relate this fact to the interaction with rough bottom topography. A total of 1–3 Sv is transferred from Weddell Sea Deep Water to Circumpolar Bottom Water. Along with mixing, the authors of the model note intense isopycnal ventilation of Lower Circumpolar Deep Water due to the influence of waters from the Weddell Sea and the formation of a local meridional circulation cell with an intensity of 8 Sv. Its lower limb, which includes almost the entire Circumpolar Bottom Water layer, forms heat and salt flux directed from the northern periphery of the Antarctic Circumpolar Current to the polar region. The upper circulation limb directed to the north includes Antarctic Intermediate Water and the upper layers of Upper Circumpolar Deep Water. According to (Naveira Garabato et al. 2003), eddy formation in the South Polar Frontal Zone is one of the possible mechanisms of the appearance of this cell. The existence of this cell plays an important role in the formation of heat and salt balance not only in the Scotia Sea, but also in the Weddell Gyre. We note that the latest estimates, for example, Klatt et al. (2005) show that about two thirds of the entire heat flux from the ocean to the atmosphere in the Weddell Gyre should be provided by the transport of warm circumpolar waters into the gyre within the zone between the northern end of the Antarctic Peninsula and Prime meridian.

3.3.3.4.5 Circumpolar Bottom Water North of the Scotia Sea

The structure of currents in Circumpolar Bottom Water layer near the Falkland Plateau, Georgia Basin, and North Scotia Ridge is based mainly on (Nowlin and Zenk 1988; Orsi et al. 1993; Orsi et al. 1999; Arhan et al. 1999, 2002; Naveira Garabato et al. 2002a) combined with the distributions of fN^2 shown in Fig. 3.33 and distributions of some bottom properties from Sect. 3.3.3.3 (Fig. 3.23). The lower layer of Circumpolar Bottom Water overflows the North Scotia Ridge only through the Shag Rocks Passage, and then, following along the Falkland Trough, enters the Georgia Basin. The upper layer crosses the North Scotia Ridge also through the Black Rock Passage. Water exchange occurs also in the upper layer of Circumpolar Bottom Water across the Falkland Plateau between the Falkland Trough and the

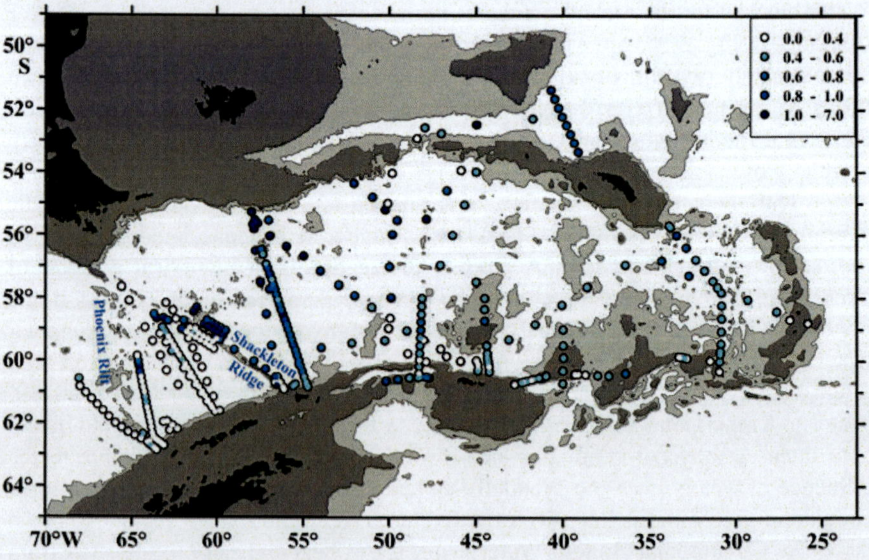

Fig. 3.33 Distribution of mean values of fN^2 (10^{-11} s^{-3}) in the upper $28.16 < \gamma^n < 28.23$ (**a**) and lower $28.23 < \gamma^n < 28.26$ (**b**) layers of Circumpolar Bottom Water in the Scotia Sea and Drake Passage. (Reproduced from Tarakanov 2010, with permission from Pleiades Publishers)

Argentine Basin. According to (Arhan et al. 1999), the southward flow in this layer across the plateau is an anticyclonic recirculation around the Maurice Ewing Bank. The abyssal circulation in the Georgia Basin is anticyclonic.

3.3.3.4.6 Quantitative Characteristics of Circumpolar Bottom Water

According to the scheme of Circumpolar Bottom Water transformation in the Scotia Sea and Drake Passage, the data presented in Table 3.6 demonstrate gradual freshening and oxygenation of Circumpolar Bottom Water from the western part of the study region to its eastern part due to mixing with the waters of the Weddell Sea. Already in the western part of the region (region 1 in Fig. 3.30), Circumpolar Bottom Water appears fresher, cooler, and more enriched with oxygen than in the Antarctic Circumpolar Current zone in the eastern part of the Pacific Ocean. According to (Koshlyakov and Tarakanov 2003b), mean values of potential temperature, salinity, neutral density γ^n, and dissolved oxygen were 0.57 °C, 34.710 PSU, 28.210 kg/m^3, and 4.70 ml/l, respectively. A slight temperature increase in region 2 (Fig. 3.30) is related to the inclusion of a vast northern periphery of the Antarctic Circumpolar Current in the Scotia Sea into this sector.

3.3.3.5 Warm Deep Water in the Scotia Sea

It was previously mentioned that Warm Deep Water and Lower Circumpolar Deep Water are characterized by a salinity maximum. However, unlike Lower Circumpolar Deep Water, this maximum in the northwestern periphery of the Weddell Gyre appears displaced to the density range of Circumpolar Bottom Water, which is responsible for the main difference of Warm Deep Water from Lower Circumpolar Deep Water and Circumpolar Bottom Water. In the Scotia Sea and at the northern tip of the Antarctic Peninsula, Warm Deep Water is located under the colder and fresher waters transported from the Bransfield Strait. This is why the upper boundary surface of Warm Deep Water is located more than 500 m deeper in this region (Fig. 3.31). As the waters spread eastward to the southern and southeastern parts of the Scotia Sea, the upper boundary surface of Warm Deep Water ascends to a depth of approximately 250 m (Fig. 3.31), where Antarctic Surface Water is located above Warm Deep Water. As Warm Deep Water propagates further, it becomes more saline and warmer (Table 3.6) owing both to mixing with circumpolar waters and water exchange with the warmer and more saline Warm Deep Water from the central part of the Weddell Gyre. We note that, despite notable differences in temperature, salinity, and density between the waters from the Bransfield Strait and Antarctic Surface Water of the open ocean, these waters are included into Antarctic Surface Water in Fig. 3.20, because waters of the Antarctic shelf (with $\theta > -1.7$ °C) are modifications of Antarctic Surface Water.

3.4 Antarctic Bottom Water in the Argentine Basin

Bottom water structure in the Argentine Basin is likely the most complicated and less studied structure among all Atlantic regions considered here. This is caused by hydrological and dynamic fronts, branching of the pathways for the propagation of Antarctic waters from the south, and drastic change in the propagation direction of these waters (branching from the Antarctic Circumpolar Current, which covers the southern part of the Argentine Basin (Peterson and Whitworth 1989)) and waters of the North Atlantic origin propagating to the Argentine Basin from the north. The latter waters change the direction of their propagation in this region (approximately at 30–40° S).

According to (Reid et al. 1977), the relatively cold, fresh, and oxygen-rich Weddell Sea Deep Water is located in the Argentine Basin at the bottom. Using the circumpolar classification by (Gordon 1967; Orsi et al. 1999) this water is equivalent to Antarctic Bottom Water. A thick layer above Weddell Sea Deep Water is occupied by the warmer oxygen-depleted Circumpolar Deep Water that propagates from the Antarctic Circumpolar Current region. The more saline North Atlantic Deep Water wedges into this layer from the north (Reid et al. 1977), dividing Circumpolar Deep Water into Upper and Lower Circumpolar Water (according to terminology in (Reid et al. 1977)). In general, lower oxygen concentration in Circumpolar Deep Water (with respect to the overlying and underlying waters in the Antarctic Circumpolar Current and to North Atlantic Deep Water) makes it possible to use the local oxygen minima as the main distinguishing indicators of Upper and Lower Circumpolar Water. Note that Upper Circumpolar Water (according to classification by (Reid et al. 1977)) and Upper Circumpolar Deep Water (according to the three-layer circumpolar classification by (Gordon 1967; Orsi et al. 1999)) is actually the same water, whereas Lower Circumpolar Water includes Circumpolar Bottom Water and partly Lower Circumpolar Deep Water. We also note that Lower Circumpolar Water and Weddell Sea Deep Water represent Antarctic Bottom Water according to the classification by Wüst (1936).

Waters of Antarctic origin propagate to the Argentine Basin along several pathways. It is considered that the Falkland Gap in the Falkland Ridge (Whitworth et al. 1991) is the main (and, probably, the only one) passage for Weddell Sea Deep Water to the Argentine Basin. However, trajectory of this water flow from the sources in the Weddell Sea to the Argentine Basin is divided into several branches (Fig. 3.34). It is also possible that Weddell Sea Deep Water propagates to the Argentine Basin east of the South Orkney Rise (Whitworth et al. 1991). However, the existence of this pathway has not been confirmed yet. Circumpolar waters propagate to the Argentine Basin together with the branches of the Antarctic Circumpolar Current related to the South Polar Front and Subantarctic Front mainly over the Falkland Plateau and east of the Maurice Ewing Bank over the Falkland Ridge (Fig. 3.34). The complex pattern and its branching indicate that the same waters propagating along different pathways can acquire their specific properties. Thus, there is a possibility to distinguish the waters, which passed along different

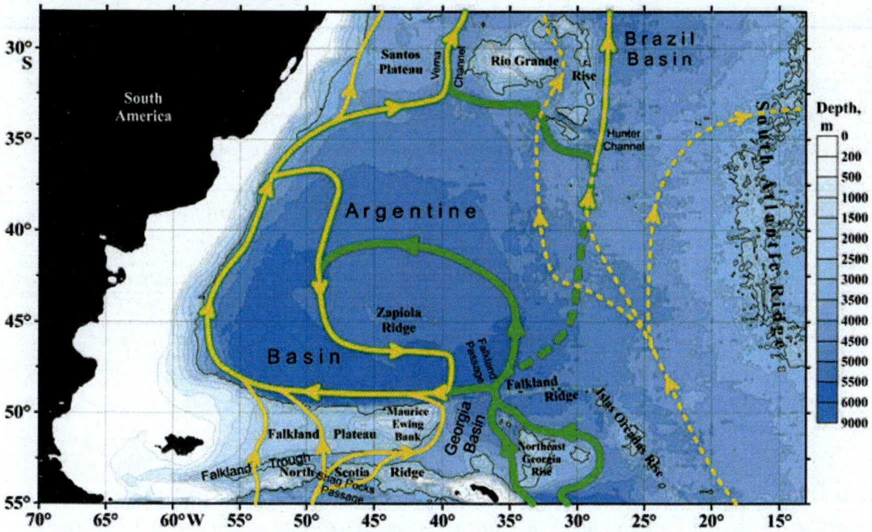

Fig. 3.34 Deep and bottom water circulation in the Argentine Basin. The thick green solid line shows Weddell Sea Deep Water pathways (Arhan et al. 1999; Smythe-Wright and Boswell 1998; Coles et al. 1996). The thin yellow line shows Lower Circumpolar Water/Lower Circumpolar Deep Water pathways (Arhan et al. 1999; Coles et al. 1996; Valla et al. 2019). The thick gray line shows WSDW, LCPW/LCDW pathways. Dashed lines represent circulation in the same layers based on (Larque et al. 1997)

pathways, at the site where they merge again. Arhan et al. (1999) distinguished cores of five types of water masses at the northern side of the Falkland Plateau in the Argentine Basin: two cores each in Lower Circumpolar Deep Water (LCDW-1 and LCDW-2) and Weddell Sea Deep Water (Upper Weddell Sea Deep Water and Lower Weddell Sea Deep Water) and the Southeast Pacific Deep Water layer between them. All these water masses are separated by maxima of stability. Southeast Pacific Deep Water is a special type of Circumpolar Bottom Water in the Drake Passage. Division of Weddell Sea Deep Water into two layers is related to the existence of a trajectory for the upper layer of Weddell Sea Deep Water that passes through the Scotia Sea (see the previous section). We also note that, precisely, Upper Weddell Sea Deep Water can be traced up to the equator (Reid 1989). Division of Lower Circumpolar Deep Water into the denser LCDW-2 and lighter LCDW-1 is related to the appearance of the upper layers of Lower Circumpolar Deep Water in the same density range with the more saline North Atlantic Deep Water (Arhan et al. 1999).

In general, the structure of the circulation of bottom waters in the Argentine Basin is complicated. It has a form of a cyclonic closed "C"-shaped ring in the deepest southwestern and western parts of the basin (Coles et al. 1996; Memery et al. 2000; Valla et al. 2019). After propagating to the Argentine Basin, the bottom waters turn to the west along the slope of the Falkland Plateau, flow to the north along the slope of South America, and then propagate to the Brazil Basin (Reid et al. 1977; Georgi

1981; Mantyla and Reid 1983; Whitworth et al. 1991). Along with this transport scheme in the central part of the Argentine Basin, anticyclonic circulation of bottom waters exists around the Zapiola Ridge with a center approximately at 45° S, 45° W. This fact was emphasized in (Smythe-Wright and Boswell 1998). We also note that, according to (Coles et al. 1996), the closed cyclonic "C"-shaped circulation in the northern part of the Argentine Basin is accompanied by a water inflow that propagates from the Mid-Atlantic Ridge; part of this inflow propagates to the north through the Hunter Channel. However, Coles et al. (1996) did not analyze the origin of this water. Figure 3.34 also shows elements of the circulation scheme presented in (Larque et al. 1997), which is based on the Optimum Multiparameter Analyses (Tomczak 1981) and seems the simplest one. In addition to the flow of Lower Circumpolar Deep Water and Weddell Sea Deep Water along the continental slope of South America, one can see quasi-meridional flows from the region of South Georgia to the Hunter Channel. Besides, Lower Circumpolar Deep Water branches off from this region and flows through the Mid-Atlantic Ridge and Walvis Ridge to the Cape Basin, which seems to be questionable. We think that conclusions in (Larque et al. 1997) are caused by confusion of thermohaline indices corresponding to Weddell Sea Deep Water, Lower Circumpolar Deep Water, and North Atlantic Deep Water resulting in overestimation of the role of Lower Circumpolar Deep Water in the circulation of bottom waters in the Southwest Atlantic.

It is difficult to compare Antarctic water transports in the Argentine Basin owing to the complex vertical structure of water masses, resulting in different locations of their upper boundaries. Available data from the high-precision zonal hydrographic sections in the Argentine Basin are not as numerous as, for example, in the Brazil Basin. That is why there are not so many estimates of transports in literature. The western transport of Antarctic Bottom Water (Weddell Sea Deep Water) with potential temperature $\theta < 0.2$ °C on the northern slope of the Falkland Ridge (41° W) was estimated at 8.2 Sv in (Whitworth et al. 1991). The authors indicate that the major part of this transport is related to Weddell Sea Deep Water recirculation in the Argentine Basin and also to the existence of a Weddell Sea Deep Water flow trapped by the Subantarctic Front propagating from the Falkland Gap at 36° W to the east. The westerly transport of relatively new water in the basin with potential temperature cooler than $\theta = -0.2$ °C is as high as 2.5 Sv (Whitworth et al. 1991). Further to the west at the foot of the Falkland Rise at 55° W, the transport of Weddell Sea Deep Water with $\sigma_4 < 46.04$ is estimated at 4.0 Sv (Arhan et al. 1999). Transports of other water components of the bottom structure in this region are as follows: Southeast Pacific Deep Water ($45.98 < \sigma_4 < 46.04$) 1.0 Sv; Lower Circumpolar Deep Water ($45.78 < \sigma_4 < 45.98$) 6.0 Sv (Arhan et al. 1999). The total transport of water with density $\gamma^n > 28.30$ (lower part of Weddell Sea Deep Water) across 47° S is 1.1 ± 0.5 Sv (Sloyan and Rintoul 2001b). At 45° S, the available estimates of Antarctic waters are close to each other: 5 ± 2 Sv (Ganachaud 2003) ($\gamma^n > 28.11$) and (Vanicek and Siedler 2002) ($\gamma^n > 28.12$) (Table 3.7).

Table 3.7 Transport of Antarctic Bottom Water in the Argentine Basin from literature data

Authors	AABW boundary	Transport, Sv	Comments
Georgi (1981)	$\theta = 0.0$	5	Transport in the northern part of the South Sandwich Trench; Geostrophic calculations with reference level at $\theta = 0.0$ °C
Whitworth et al. (1991)	$\theta = 0.2$	8 ± 2	Moorings at Falkland Escarpment
Saunders and King (1995)	$\sigma_4 = 45.95$	5.7–7.7	Shipborne ADCP+geostrophic calculations+ inverse model at 45°S
Arhan et al. (1999)	$\sigma_4 = 45.78$	11	Integral transport of bottom waters from the south
	$\sigma_4 = 46.04$	4	Transport of WSDW with the deep current near the Falkland Plateau over A17 section. Combined geostrophic calculations and SADCP
	$45.98 < \sigma_4 < 46.04$	1	Transport of Southeast Pacific Deep Water with the deep current near the Falkland Plateau over A17 section.
	$45.78 < \sigma_4 < 45.98$	6	Transport of LCDW with the deep current near the Falkland Plateau over A17 section.
Sloyan and Rintoul (2001a)	$\gamma^n = 28.30$	1.1 ± 0.5	Hydrographic sections and inverse model (47° S)
Vanicek and Siedler (2002)	$\gamma^n = 28.12$	5 ± 2	Inverse model. Transport of LCPW and WSDW across A11 section along 45° S
Ganachaud (2003)	$\gamma^n = 28.11$	5 ± 2	Inverse model (45° S)
McDonagh and King (2005)	$\sigma_3 = 41.58$	5.5	Inverse model (45° S)
Naveira Garabato et al. (2014)	$\gamma^n = 28.18$	4.8	Inverse model (45° S)
Manta et al. (2021)	$\gamma^n = 28.27$	2.4 ± 1.6	Estimate of Abyssal Overturning Cell at 34.5° S

References

Abrahamsen EP, Meijers AJS, Polzin KL, Naveira Garabato AC, King BA, Firing YL et al (2019) Stabilization of dense Antarctic water supply to the Atlantic Ocean overturning circulation. Nat Clim Chang 9(10):742–746. https://doi.org/10.1038/s41558-019-0561-2

Arhan M, Heywood KJ, King BA (1999) The deep waters from the Southern Ocean at the entry to the Argentine Basin. Deep-Sea Res II 46:475–499

Arhan M, Naveira Garabato AC, Heywood KJ, Stevens DP (2002) The Antarctic Circumpolar Current between the Falkland Islands and South Georgia. J Phys Oceanogr 32(6):1914–1931

Arhan M, Mercier H, Park Y-H (2003) On the deep water circulation of the eastern South Atlantic Ocean. Deep-Sea Res I 50:889–916

Baines PG, Condie S (1998) Observation and modelling of Antarctic downslope flows: a review, ocean, ice, and atmosphere: interactions at the Antarctic Continental Margin, Antarctic Res. Ser., vol 75. AGU, Washington, DC, pp 29–49. https://doi.org/10.1029/AR075p0029

Barre N, Provost C, Sennechael N, Lee JH (2008) Circulation in the Ona Basin, southern Drake Passage. J Geophys Res Oceans 113:C04033. https://doi.org/10.1029/2007JC004549

Broecker WS, Peacock SL, Walker S, Weiss R, Fahrbach E, Schröder M, Mikolajewicz U, Heinze C, Key R, Peng T-H, Rubin S (1998) How much deep water is formed in the Southern Ocean? J Geophys Res Oceans 103(C8):15833–15843

Carmack EC (1977) Water characteristics of the Southern Ocean south of the Polar Front. In: Deacon G, Angel M (eds) A voyage of discovery, 70th anniversary volume. Pergamon Press, Oxford, pp 15–41

Carmack EC, Foster TD (1975) On the flow of water out of the Weddell Sea. Deep-Sea Res I 22:711–724

Chereskin TK, Donohue KA, Watts DR, Tracy KL, Firing YL, Cutting AL (2009) Strong bottom currents and cyclogenesis in Drake Passage. Geophys Res Lett 36:L23602. https://doi.org/10.1029/2009GL040940

Coles VJ, McCartney MS, Olson BD, Smethie WJ Jr (1996) Changes in Antarctic Bottom Water properties in the western South Atlantic in the late 1980s. J Geophys Res Oceans 101(C4): 8957–8970

Connary SD, Ewing M (1974) Penetration of Antarctic Bottom Water from the Cape Basin into the Angola Basin. J Geophys Res Oceans 79:463–469

Deacon G (1979) The Weddell Gyre. Deep-Sea Res I 26:981–995

Egbert GD, Erofeeva SY (2002) Efficient inverse modeling of barotropic ocean tides. J Atmos Ocean Tech 19:183–204. https://doi.org/10.1175/1520-0426(2002)019<0183:EIMOBO>2.0.CO;2

Fahrbach E, Rohardt G, Schröder J, Strass V (1994) Transport and structure of the Weddell Gyre. Ann Geophys 12:840–855

Fahrbach E, Rohardt G, Scheele N, Schröder M, Strass V, Wisotzki A (1995) Formation and discharge of deep and bottom water in the Northwestern Weddell Sea. J Mar Res 53:515–538

Fahrbach E, Harms S, Rohardt G, Schröder M, Woodgate R (2001) Flow of bottom water in the northwestern Weddell Sea. J Geophys Res Oceans 106:2761–2778

Fahrbach E, Hoppema M, Rohardt G, Schröder M, Wisotzki A (2004) Decadal-scale variations of water mass properties in the deep Weddell Sea. Ocean Dyn 54:77–91. https://doi.org/10.1007/s10236-003-0082-3

Foldvik A, Gammelsrød T, Tørresen T (1985) Circulation and water masses on the southern Weddell Sea shelf. In: Jacobs SS (ed) Oceanology of the Antarctic Continental Shelf, Antarctic Res. Ser., vol 43. AGU, Washington, DC, pp 5–20

Foldvik A, Gammelsrød T, Østerhus S, Fahrbach E, Rohardt G, Schröder M, Nicholls KW, Padman L, Woodgate RA (2004) Ice shelf water overflow and bottom water formation in the southern Weddell Sea. J Geophys Res Oceans 109:C02015. https://doi.org/10.1029/2003JC002008

Foster TD, Carmack EC (1976) Frontal zone mixing and Antarctic Bottom Water formation in the southern Weddell Sea. Deep-Sea Res I 23:301–317. https://doi.org/10.1016/0011-7471(76)90872-X

Franco BC, Mata MM, Piola AR, Garcia CAE (2007) Northwestern Weddell Sea deep outflow into the Scotia Sea during the austral summers of 2000 and 2001 estimated by inverse methods. Deep-Sea Res I 55:1815–1840

Ganachaud A (2003) Large-scale mass transports, water mass formation, and diffusivities estimated from World Ocean Circulation Experiment (WOCE) hydrographic data. J Geophys Res Oceans 108(C7):3213. https://doi.org/10.1029/2002JC001565

Georgi DT (1981) Circulation of bottom waters in the southwestern South Atlantic. Deep-Sea Res 28:959–979

Gill A (1973) Circulation and bottom water production in the Weddell Sea. Deep-Sea Res I 20:111–140

Gill AE (1982) Atmosphere-ocean dynamics, International geophysics series, vol 30. Academic, p 662

Gladyshev SV, Koshlyakov MN, Tarakanov RY (2008) Currents in the Drake Passage based on observations in 2007. Oceanology 48(6):759–770

Gordon AL (1966) Potential temperature, oxygen and circulation of bottom water in the Southern Ocean. Deep-Sea Res I 13:1125–1138

Gordon AL (1967) Structure of Antarctic waters between 20° W and 170° W. In: Bushell VC (ed) Antarctic map folio series (folio 6). American Geographic Society, New York

Gordon AL (1971) Oceanography of Antarctic waters. In: Reid JL (ed) Antarctic oceanography, Antarctic research series, vol 15. American Geophysical Union, Washington, DC, pp 169–203

Gordon AL (1972) Spreading of Antarctic Bottom Waters II. In: Studies in physical oceanography – a tribute of George Wüst on his 80th birthday, vol 2. Gordon and Breach, New York, pp 1–17

Gordon AL (1998) Western Weddell Sea thermohaline stratification. In: Jacobs SS, Weiss RF (eds) Ocean, ice, and atmosphere: interactions at the Antarctic Continental Margin, Antarctic research series, vol 75, pp 215–240

Gordon AL, Huber BA, Hellmer H, Ffield A (1993) Deep and bottom water of the Weddell Sea's western rim. Science 262:95–97

Gordon AL, Mensch M, Dong Z, Smethie WM, de Bettencourt J (2000) Deep and bottom water of the Bransfield Strait eastern and central basins. J Geophys Res Oceans 105(C5):11337–11346

Gordon AL, Visbeck M, Huber B (2001) Export of Weddell Sea deep and bottom water. J Geophys Res Oceans 106(C5):9005–9017

Gordon AL, Huber BA, McKee D, Visbeck MH (2010) A seasonal cycle in the export of bottom water from the Weddell Sea. Nat Geosci 3(8):551–556. https://doi.org/10.1038/ngeo916

Gordon AL, Huber BA, Abrahamsen EP (2020) Interannual variability of the outflow of Weddell Sea Bottom Water. Geophys Res Lett 47:e2020GL087014. https://doi.org/10.1029/2020GL087014

Gouretski VV, Danilov AI (1993) Weddell Gyre: structure of the eastern boundary. Deep-Sea Res I 40(3):561–582

Gouretski V, Koltermann KP (2004) WOCE global hydrographic climatology. Ber. Bundesamtes Seeschiffahrt Hydrogr 35:1–52

Haine TWN, Watson AJ, Liddicoat MI, Dickson RR (1998) The flow of Antarctic Bottom Water to the Southwest Indian Ocean estimated using CFCs. J Geophys Res Oceans 103(C12): 27637–27653

Harms S, Fahrbach E, Strass VH (2001) Sea ice transports in the Weddell Sea. J Geophys Res Oceans 106(C5):9057–9073

Hellmer HH, Beckmann A (2001) The Southern Ocean: a ventilation contributor with multiple sources. Geophys Res Lett 28(15):2927–2930

Hollister CD, Heezen BC (1967) Chapter 17: The floor of the Bellingshausen Sea. In: Hersey JB (ed) Deep-sea photography, The John Hopkins Studies, No. 3. The John Hopkins Press, Baltimore, pp 177–189

Hoppema M, Klatt O, Roether W, Fahrbach E, Bulsiewicz K, Rodehacke C, Rohardt G (2001) Prominent renewal of Weddell Sea Deep Water from a remote source. J Mar Res 59:257–279

Huhn O, Hellmer HH, Rhein M, Rodehacke C, Roether W, Schodlok MP, Schröder M (2008) Evidence of deep- and bottom-water formation in the western Weddell Sea. Deep Sea Res II 55:1098–1116

Johnson JC (1998) Deep water properties, velocities, and dynamics over ocean trenches. J Mar Res 56:329–347

Jullion L, Naveira Garabato AC, Bacon S, Meredith MP, Brown PJ, Torres-Valdes S, Speer KG, Holland PR, Dong J, Bakker D, Hoppema M, Loose B, Venables HJ, Jenkins WJ, Messias M-J, Fahrbach E (2014) The contribution of the Weddell Gyre to the lower limb of the Global

Overturning Circulation. J Geophys Res Oceans 119:3357–3377. https://doi.org/10.1002/2013JC009725

Kerr R, Heywood KJ, Mata MM, Garcia CAE (2012) On the outflow of dense water from the Weddell and Ross Seas in OCCAM model. Ocean Sci 8(3):369–388. https://doi.org/10.5194/os-8-369-2012

Klatt O, Roether W, Hoppema M, Bulsiewicz K, Fleischmann U, Rodehacke C, Fahrbach E, Weiss RF, Bullister JL (2002) Repeated CFC sections at the Greenwich Meridian in the Weddell Sea. J Geophys Res Oceans 107(C4):3030. https://doi.org/10.1029/2000JC000731

Klatt O, Fahrbach E, Hoppema M, Rohardt G (2005) The transport of the Weddell Gyre across the Prime Meridian. Deep-Sea Res II 52:513–528

Koshlyakov MN, Sazhina TG (1995) Meridional water and heat transport by large-scale geostrophic currents in the Pacific sector of Antarctica. Oceanology 35(6):842–853

Koshlyakov MN, Tarakanov RY (1999) Water masses of the Pacific Antarctic. Oceanology 39(1):5–15

Koshlyakov MN, Tarakanov RY (2003a) Antarctic Bottom Water in the Pacific sector of the Southern Ocean. Oceanology 43(1):5–20

Koshlyakov MN, Tarakanov RY (2003b) Antarctic Circumpolar Water in the southern part of the Pacific Ocean. Oceanology 43(5):607–621

Koshlyakov MN, Tarakanov RY (2004) Pacific Deep Water in the Southern Ocean. Oceanology 44(3):325–340

Koshlyakov MN, Tarakanov RY (2011) Water transport across the Subantarctic Front and the Global Ocean Conveyer Belt. Oceanology 51(5):721–735

Koshlyakov MN, Lisina II, Morozov EG, Tarakanov RY (2007) Absolute geostrophic currents in the Drake Passage based on observations in 2003 and 2005. Oceanology 47(4):451–463

Koshlyakov MN, Gladyshev SV, Ryu T, Ryzhikov NI (2010) Deep currents in the central part of the Drake Passage based on the data of the 2008 Hydrographic Survey. Oceanology 50(6):821–828

Koshlyakov MN, Gladyshev SV, Ryu T, Fedorov DA (2013) Currents in the Drake Passage by observations in October–November of 2011. Oceanology 53(1):1–12

Larque L, Maamaatuaiahutapu K, Garcon VC (1997) On the intermediate and deep-water flow in the South Atlantic Ocean. J Geophys Res Oceans 102(C6):12425–12440

Locarnini RA, Whitworth T III, Nowlin WD Jr (1993) The importance of the Scotia Sea on the outflow of Weddell See Deep Water. J Mar Res 51:135–153

Manta G, Speich S, Karstensen J, Hummels R, Kersalé M, Laxenaire R et al (2021) The South Atlantic meridional overturning circulation and mesoscale eddies in the first GO-SHIP section at 34.5°S. J Geoph Res Oceans 126:e2020JC016962. https://doi.org/10.1029/2020JC016962

Mantyla AW, Reid JL (1983) Abyssal characteristics of the World Ocean waters. Deep-Sea Res I 30(8):805–833. https://doi.org/10.1016/0198-0149(83)90002-X

McCartney MS, Curry RA (1993) Trans-equatorial flow of Antarctic Bottom Water in the western Atlantic Ocean: abyssal geostrophy at the equator. J Phys Oceanogr 23:1264–1276

McDonagh EL, King BA (2005) Oceanic fluxes in the South Atlantic. J Phys Oceanogr 35:109–122. https://doi.org/10.1175/JPO-2666.1

Memery L, Arhan M, Alvarez-Salgado XA, Messias M-J, Mercier H, Castro CG, Rios AF (2000) The water masses along the western boundary of the south and Equatorial Atlantic. Prog Oceanogr 47:69–98

Mensch M, Bayer R, Bullister JL, Schlosser P, Weiss RF (1996) The distribution of tritiumand CFCs in the Weddell Sea during the mid-1980s. Prog Oceanogr 38:377–414

Mensch M, Simon A, Bayer R (1998) Tritium and CFC functions for the Weddell Sea. J Geophys Res Oceans 103(C8):15923–15937

Meredith MP, Locarnini RA, van Scoy KA, Watson AJ, Heywood KJ, King BA (2000) On the sources of Weddell Gyre Antarctic Bottom Water. J Geophys Res Oceans 105(1):1093–1104

Meredith MP, Naveira Garabato AC, Stevens DP, Heywood KJ, Sanders RJ (2001a) Deep and bottom waters in the Eastern Scotia Sea: rapid changes in properties and circulation. J Phys Oceanogr 31(8):2157–2168

Meredith MP, Watson A, Scoy KV (2001b) Chlorofluorocarbon-derived formation rates of the deep and bottom waters of the Weddell Sea. J Geophys Res Oceans 106(C2):2899–2919

Meredith MP, Hughes CW, Foden PR (2003) Downslope convection north of Elephant Island, Antarctica: influence on deep waters and dependence on ENSO. Geophys Res Lett 30(9):1462. https://doi.org/10.1029/2003GL017074

Meredith MP, Naveira Garabato AC, Gordon AL, Johnson GC (2008) Evolution of the deep and bottom waters of the Scotia Sea, Southern Ocean, during 1995–2005. J Clim 21:3327–3343

Morozov EG, Frey DI, Tarakanov RY (2020) Antarctic Bottom Water flow through the eastern part of the Philip Passage in the Weddell Sea. Oceanology 60(5):589–592

Muench RD, Gordon AL (1995) Circulation and transport of water along the western Weddell Sea margin. J Geophys Res Oceans 100:18503–18515

Naveira Garabato AC, Heywood KJ, Stevens DP (2002a) Modification and pathways of Southern Ocean deep waters in the Scotia Sea. Deep-Sea Res I 49:681–705

Naveira Garabato AC, McDonagh EL, Stevens DP, Heywood KJ, Sanders RJ (2002b) On the export of Antarctic Bottom Water from the Weddell Sea. Deep-Sea Res II 49:4715–4742

Naveira Garabato AC, Stevens DP, Heywood KJ (2003) Water mass conversion, fluxes, and mixing in the Scotia Sea diagnosed by an inverse model. J Phys Oceanogr 33(12):2565–2587

Naveira Garabato AC, Polzin K, King B, Heywood K, Visbeck M (2004) Widespread intense turbulent mixing in the Southern Ocean. Science 303:210–213

Naveira Garabato AC, Williams AP, Bacon S (2014) The three-dimensional overturning circulation of the Southern Ocean during the WOCE era. Prog Oceanogr 120:41–78

Naveira Garabato AC, Zika JD, Jullion L, Brown PJ, Holland PR, Meredith MP, Bacon S (2016) The thermodynamic balance of the Weddell Gyre. Geophys Res Lett 43(1):317–325. https://doi.org/10.1002/2015gl066658

Needham HD, Carré D, Sibuet J-C (1986) Carte Bathymétrique de la Ride de Walvis: Océan Atlantique Sud. Echelle 1/4. 382, 832. IFREMER, Direction de l'Environnement et des Recherches Océaniques, Département Géosciences Marines

Nicholls KW, Østerhus S, Makinson K, Gammelsrød T, Fahrbach E (2009) Ice-ocean processes over the continental shelf of the southern Weddell Sea, Antarctica: a review. Rev Geophys 47: RG3003. https://doi.org/10.1029/2007RG000250

Nikurashin M, Ferrari R (2010a) Radiation and dissipation of internal waves generated by geostrophic flows impinging on small-scale topography: theory. J Phys Oceanogr 40(5):1055–1074

Nikurashin M, Ferrari R (2010b) Radiation and dissipation of internal waves generated by geostrophic motions impinging on small-scale topography: application to the Southern Ocean. J Phys Oceanogr 40(9):2025–2042

Nowlin WD, Zenk W (1988) Westward bottom currents along the margin of the South Shetland island arc. Deep-Sea Res I 35:269–301

Orsi AH, Nowlin WD Jr, Whitworth T III (1993) On the circulation and stratification of the Weddell Gyre. Deep-Sea Res I 40:169–203

Orsi AH, Whitworth T III, Nowlin WD Jr (1995) On the meridional extent and fronts of the Antarctic Circumpolar Current. Deep-Sea Res I 42(5):641–673

Orsi AH, Johnsson GC, Bullister JL (1999) Circulation, mixing, and production of Antarctic Bottom Water. Prog Oceanogr 43:55–109. https://doi.org/10.1016/S0079-6611(99)00004-X

Orsi AH, Smethie WM, Bullister JL (2002) On the total input of Antarctic waters to the deep ocean: a preliminary estimate from chlorofluorocarbon measurements. J Geophys Res Oceans 107(C8):3122. https://doi.org/10.1029/2001JC000976

Palmer M, Gomis D, del Mar FM, Jorda G, Jullion L, Tsubouchi T, Naveira Garabato AC (2012) Water mass pathways and transports over the South Scotia Ridge west of 50°W. Deep-Sea Res I 59:8–24

Park Y-H, Charriaud E, Craneguy P, Kartavtseff A (2001) Fronts, transport, and Weddell Gyre at 30° E between Africa and Antarctica. J Geophys Res Oceans 106(C2):2857–2879
Patara L, Böning CW (2014) Abyssal Ocean warming around Antarctica strengthens the Atlantic overturning circulation. Geophys Res Lett 41:3972–3978
Peterson RG, Whitworth T III (1989) The Subantarctic and Polar fronts in relation to deep water masses through the Southwestern Atlantic. J Geophys Res Oceans 94(C8):10817–10838
Purkey SG, Johnson GC (2012) Global contraction of Antarctic Bottom Water between the 1980s and 2000s. J Clim 25:5830–5844
Purkey SG, Johnson GC (2013) Antarctic Bottom Water warming and freshening: contributions to sea level rise, ocean freshwater budgets, and global heat gain. J Clim 26:6105–6122
Reeve KA, Boebel O, Strass V, Kanzow T, Gerdes R (2019) Horizontal circulation and volume transports in the Weddell Gyre derived from Argo float data. Progr Oceanogr 175:263–283
Reid JL (1989) On thc total geostrophic circulation of the South Atlantic Ocean: flow pattern, tracers and transports. Prog Oceanogr 23:149–244
Reid JL, Nowlin WD, Patzert WC (1977) On the characteristics and circulation of the Southwestern Atlantic Ocean. J Phys Oceanogr 7(1):62–91
Rintoul SR (1998) On the origin and influence of Adelie Land Bottom Water, ocean, ice, and atmosphere: interactions at the Antarctic Continental Margin. Antarct Res Ser 75:151–171
Ryan S, Schröder M, Huhn O, Timmermann R (2016) On the warm inflow at the eastern boundary of the Weddell Gyre. Deep-Sea Res I 107:70–81
Sarukhanyan EI, Smirnov NP (1979) New data on the structure of the Antarctic Circumpolar Current at the exit from the Scotia Sea. Doklady AN SSSR 247(4):956–959
Saunders PM, King B (1995) Oceanic fluxes on the WOCE A11 section. J Phys Oceanogr 25(9):1942–1958
Schodlok MP, Hellmer HH, Beckmann A (2002) On the transport, variability and origin of dense water masses crossing the South Scotia Ridge. Deep-Sea Res II 49:4807–4825
Shannon LV, Chapman P (1991) Evidence of Antarctic Bottom Water in the Angola Basin at 32°S. Deep-Sea Res I 38(10):1299–1304
Shannon LV, van Rijswijck M (1969) Physical oceanography of the Walvis Ridge region. Investigational report no 70, Division of Sea Fisheries of RSA [Available from the library of the University of Cape Town]
Sheen KL, Brearley JA, Naveira Garabato AC, Smeed DA, Waterman S, Ledwell JR et al (2013) Rates and mechanisms of turbulent dissipation and mixing in the Southern Ocean: results from the Diapycnal and Isopycnal Mixing Experiment in the Southern Ocean (DIMES). J Geophys Res 118:2774–2762. https://doi.org/10.1002/jgrc.20217
Sievers HA, Nowlin WD Jr (1984) The stratification and water masses at Drake Passage. J Geophys Res Oceans 89(C6):10489–10514
Sloyan BM, Rintoul SR (2001a) Circulation, renewal, and modification of Antarctic Mode and Intermediate Water. J Phys Oceanogr 31(4):1005–1030
Sloyan BM, Rintoul SR (2001b) The Southern Ocean limb of the global deep overturning circulation. J Phys Oceanogr 31(1):143–173
Smith WHF, Sandwell DT (1997) Global Sea floor topography from satellite altimetry and ship depth soundings. Science 277:1956–1962. https://doi.org/10.1126/science.277.5334.1956
Smith IJ, Stevens DP, Heywood KJ, Meredith MP (2010) The flow of the Antarctic Circumpolar Current over the North Scotia Ridge. Deep-Sea Res I 57(1):14–28
Smythe-Wright D, Boswell S (1998) Abyssal circulation in the Argentine Basin. J Geophys Res Oceans 103(C8):15845–15851
Sokolov S, Rintoul SR (2009a) The circumpolar structure and distribution of the Antarctic Circumpolar Current fronts. Part A: Mean circumpolar paths. J Geophys Res 114(C11018). https://doi.org/10.1029/2008JC005108
Sokolov S, Rintoul SR (2009b) The circumpolar structure and distribution of the Antarctic Circumpolar Current fronts. Part B: Variability and relationship to sea surface height. J Geophys Res 114(C11019). https://doi.org/10.1029/2008JC005248

Tarakanov RY (2009) Antarctic Bottom Water in the Scotia Sea and the Drake Passage. Oceanology 49(5):607–621

Tarakanov RY (2010) Circumpolar Bottom Water in the Scotia Sea and Drake Passage. Oceanology 50(1):5–22

Tarakanov RY (2012) The Drake Passage and the Scotia Sea as an orographic barrier for the Antarctic Circumpolar Current. Oceanology 52:157–170

Tomczak M (1981) A multi-parameter extension of temperature/salinity diagram techniques for the analysis of nonisopycnal mixing. Prog Oceanogr 10:147–171

Tucholke BE, Embley RW (1984) Cenozoic regional erosion of the abyssal sea floor off South Africa. In: Schlee JS (ed) Interregional unconformities and hydrocarbon accumulation. AAPG Memoir 36, Tulsa, pp 145–164

Valla D, Piola AR, Meinen CS, Campos E (2019) Abyssal transport variations in the southwest South Atlantic: first insights from a long-term observation array at 34.5°S. Geophys Res Lett 46:6699–6705. https://doi.org/10.1029/2019GL082740

van Caspel MR, Absy JM, Wang Q, Hellmer HH, Schröder M (2015a) The flow of dense water plumes in the western Weddell Sea simulated with the Finite Element Ocean Model (FEOM). In: Lohmann G, Meggers H, Unnithan V, Wolf-Gladrow D, Notholt J, Bracher A (eds) Towards an interdisciplinary approach in earth system science, Springer earth system sciences. Springer, Cham, pp 125–129. https://doi.org/10.1007/978-3-319-13865-7_14

van Caspel M, Schröder M, Huhn O, Hellmer H (2015b) Precursors of Antarctic Bottom Water formed on the continental shelf off Larsen Ice Shelf. Deep-Sea Res I 99:1–9

Vanicek M, Siedler G (2002) Zonal fluxes in the deep water layers of the western South Atlantic Ocean. J Phys Oceanogr 32(8):2205–2235

Vernet M, Geibert W, Hoppema M, Brown PJ, Haas C, Hellmer HH et al (2019) The Weddell Gyre, Southern Ocean: present knowledge and future challenges. Rev Geophys 57:623–708. https://doi.org/10.1029/2018RG000604

Visbeck M (2002) Deep velocity profiling using Lowered Acoustic Doppler Current Profiler: bottom track and inverse solution. J Atmos Ocean Technol 19(5):794–807. https://doi.org/10.1175/1520-0426(2002)019<0794:DVPULA>2.0.CO;27

von Gyldenfeldt AB, Fahrbach E, García MA, Schröder M (2002) Flow variability at the tip of the Antarctic Peninsula. Deep-Sea Res II 49:4743–4766

Walkden GJ, Heywood KJ, Stevens DP (2008) Eddy heat fluxes from direct current measurements of the Antarctic, Polar Front in Shag Rocks Passage. Geophys Res Lett 35:L06602

Wang Q, Danilov S, Schröter J (2009) Bottom water formation in the southern Weddell Sea and the influence of submarine ridges: idealized numerical simulations. Ocean Modell 28(1–3):50–59. https://doi.org/10.1016/j.ocemod.2008.08.003

Warren BA, Speer KG (1991) Deep circulation in the eastern South Atlantic Ocean. Deep-Sea Res I 38(Suppl. 1):281–322

Weiss RF, Östlund HG, Craig H (1979) Geochemical studies of the Weddell Sea. Deep-Sea Res I 26:1093–1120

Well R, Roether W, Stevens DP (2003) An additional deep-water mass in Drake Passage as revealed by 3He data. Deep-Sea Res I 50:1079–1098

Weppernig R, Schlosser P, Khatiwala S, Fairbanks RG (1996) Isotope data from Ice Station Weddell: implications for deep water formation in the Weddell Sea. J Geophys Res Oceans 101(C9):25723–25739

Whitehead JA (1989a) Surges of Antarctic bottom water into the North Atlantic. J Phys Oceanogr 19(6):853–861

Whitehead JA (1989b) Giant Ocean cataracts. Sci Am 260:50–57. https://doi.org/10.1038/scientificamerican0289-50

Whitworth T, Nowlin WD, Worley SJ (1982) The net transport of the Antarctic Circumpolar Current through Drake Passage. J Phys Oceanogr 12(9):960–971

Whitworth T, Nowlin WD, Pillsbury RD, Moore MI, Weiss RF (1991) Observations of the Antarctic Circumpolar Current and Deep Boundary Current in the Southwest Atlantic. J Geophys Res Oceans 96(15):105–118

Whitworth T, Nowlin WD, Orsi AH, Locarnini RA, Smith SG (1994) Weddell Sea Shelf Water in the Bransfield Strait and Weddell-Scotia Confluence. Deep-Sea Res I 41:629–641

Whitworth T, Orsi AH, Kim S-J, Nowlin WD, Locarnini RA (1998) Water masses and mixing near the Antarctic Slope Front. In: Jacobs S, Weiss R (eds) Ocean, ice, and atmosphere: interactions at the Antarctic Continental Margin, Antarctic Res Ser, vol 75, pp 1–27

Wittstock R-R, Zenk W (1983) Some current observations and surface T/S distribution from the Scotia Sea and the Bransfield Strait during early austral summer 1980/81. Meteor Forschung-Ergebnisse, Reihe A/B. No 24. pp 77–86

WOD18 (2018) World Ocean Database 2018. https://www.nodc.noaa.gov/OC5/WOD/pr_wod.html

Wüst G (1936) Schichtung und Zirkulation des Atlantischen Ozeans, Das Bodenwasser und die Stratosphäre. In: Defant A (ed) Wissenschaftliche Ergebnisse, Deutsche Atlantische Expedition auf dem Forschungs – und Vermessungsschiff "Meteor" 1925–1927, 6(1). Walter de Gruyter & Co, Berlin. 411pg

Yaremchuk M, Nechaev D, Schroter J, Fahrbach E (1998) A dynamically consistent analysis of circulation and transports in the southwestern Weddell Sea. Ann Geophys 16:1024–1038

Zhou M, Zhu Y, Dorland R, Measures CI (2010) Dynamics of the current system in the southern Drake Passage. Deep-Sea Res I 57:1039–1048

Chapter 4
Exchange Between the Argentine and Brazil Basins; Abyssal Pathways and Bottom Flow Channels (for Waters of the Antarctic Origin)

Abstract The bottom waters of Antarctic origin flow to the Brazil Basin along three pathways: through the Vema Channel, through the Hunter Channel, and over the Santos Plateau. The depths of the Vema Channel exceed 4600 m, while the background depths of the Santos Plateau and Hunter Channel are approximately 4000 m. The Vema Channel is a narrow passage between two terraces with a minimum width slightly greater than 15 km. Investigations in these channels provide evidence that the Vema Channel plays a dominating role in the transport of Antarctic waters compared to the other passages. The propagation of bottom waters in the deep-water Vema Channel occurs in the form of a well-mixed jet. The coldest part of the flow is displaced to the eastern slope of the deep-water channel due to the influence of the bottom Ekman friction. The second core of the flow is observed in the upper part of the channel over its western slope. Part of the Antarctic waters propagating to the Brazil Basin through these passages continues propagation mainly along the western slope of the basin to the north. In the northern part of the Brazil Basin, the bottom flow separates into the flows directed to the Equatorial Channel and to the eastern basin through the Romanche and Chain fracture zones.

Keywords Vema Channel · Topography · Water transport · Structure of bottom flow · Warming trend · Moorings and LADCP

4.1 General Description

The zonally aligned Rio Grande Rise separates the Argentine Basin in the south from the Brazil Basin in the north. It is a high topographic obstacle for bottom water propagation to the north. Two meridional gaps intersect the Rio Grande Rise at ~39° W and ~ 28°W (Vema and Hunter channels, respectively). The depth in the Vema Channel exceeds 4600 m as compared to the background depths of the Rio Grande

With a contribution by Walter Zenk, GEOMAR Helmholtz Centre for Ocean Research Kiel, Germany

E. G. Morozov et al., *Bottom Gravity Currents and Overflows in Deep Channels of the Atlantic Ocean*, https://doi.org/10.1007/978-3-030-83074-8_4

Rise (~4000 m). The Hunter Channel is much shallower and the greatest depth does not exceed 4000 m. To make the discussion clearer we give a chart of the Vema Channel region in Fig. 4.1. We also show distribution of potential temperatures at the bottom in Fig. 4.2. This distribution shows that the coldest waters are found only in the channel and after the flow exits the channel.

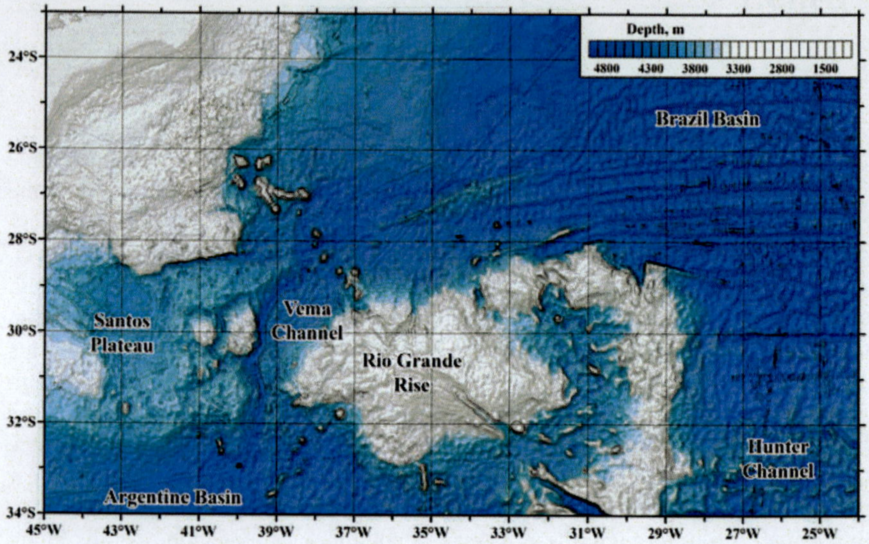

Fig. 4.1 Chart of the Vema Channel region, Santos Plateau, and Hunter Channel. The bathymetry source is the GEBCO 2019 database

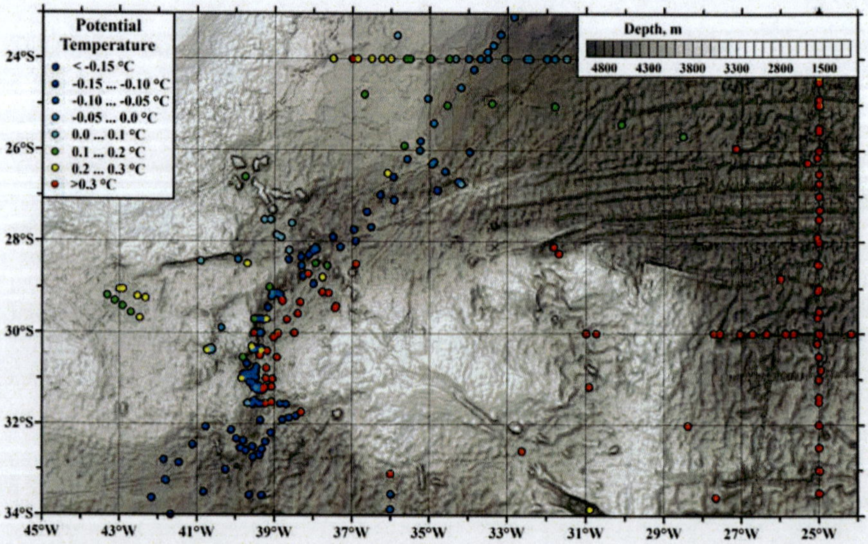

Fig. 4.2 Bottom topography based on GEBCO 2019 data (m) and distribution of potential temperature (°C) at the bottom in the Vema Channel region. Potential temperatures below 3500 m based on WOD18 are indicated at the points of measurements with colors

Repeated observations from hydrographic stations and current-meter arrays deployed in the Vema and Hunter channels unambiguously confirm the prominent role of the Vema Channel for the transport of Antarctic Bottom Water with respect to the Hunter Channel in the east and the Santos Plateau in the west (see progressive vector diagrams in Fig. 9, curves 2, 3, 4, 5, 6, and 1). In this section, we use the definition of Antarctic Bottom Water by Wüst (1936) (potential temperature of water $\theta < 2.0$ °C) and divide this water into Weddell Sea Deep Water and Lower Circumpolar Water (Reid et al. 1977). Based on moored current-meter observations in combination with geostrophic velocity computations from hydrographic stations, the total Antarctic Bottom Water transport across the Rio Grande Rise and Santos Plateau was estimated at 6.9 Sv (Hogg et al. 1999). On the average, about half of this volume passes through the Vema Channel. Earlier estimates in (Speer and Zenk 1993) inferred from hydrography alone yielded a northward net flow of 6.6 Sv distributed at a ratio of 30: 59: 11 on the western boundary current system, and the Vema and Hunter channels, respectively. However, our estimates make us think that this value is overestimated. Recent publications based on direct measurements of currents and our research show that the Vema Channel plays an even more important role providing the northward transport of the dominating part of Antarctic Bottom Water rather than the flows over the Santos Plateau and in the Hunter Channel (Morozov et al. 2010).

Some authors also distinguish the western slope of the Rio Grande Rise, as a structure similar to the Santos Plateau but less extended. It is considered as an independent pathway for bottom water penetration to the Brazil Basin, which is separated from the deep Vema Channel (McDonagh et al. 2002). The majority of publications consider this slope together with the Vema Channel.

Table 4.1 shows classic estimates of Antarctic Bottom Water transport over pathways in the region around the Rio Grande Rise based on different publications. As seen from the table, most researchers estimate the integral transport of Antarctic waters along these pathways at 2–4 Sv. The Antarctic Bottom Water transport through the Vema Channel is estimated based on measurements at 1.3–4.0 Sv (McDonagh et al. 2002; Hogg et al. 1982) and 3.1–3.7 Sv from model simulations (Cai and Greatbatch 1995; Holfort and Siedler 2001. Our estimates based on six LADCP sections at 31°12′ S are within 2.1–5.4 Sv. However, we have to distinguish the transport through the repeated standard section at 31°12′ S and the transport from the Vema Channel approximately at 26°40′ S in the region of the outlet of the bottom water into the Brazil Basin, which is notably smaller than that at the standard section.

Estimates of transports through the other channels and pathways are much smaller. Transport estimates through the Hunter Channel are as follows: 0.7 Sv (Speer et al. 1992), 2.7 ± 1.3 (Lenz 1997), and 2.92 ± 1.24 (Zenk et al. 1999). The flow over the Santos Plateau ranges from −1.5 to 2 Sv (McDonagh et al. 2002; Speer and Zenk 1993), indicating that water exchange over the Santos Plateau is possible in both directions. Flows of opposite directions were also found in the Hunter Channel (Zenk et al. 1999). According to (McDonagh et al. 2002), the transport over the Santos Plateau is a branch of the upper part of a deep flow in the Vema Channel at 27° S, which makes a large loop and eventually turns to the north near 30°–31° S

Table 4.1 Transport of Antarctic Bottom Water over pathways in the region of the Rio Grande Rise: Vema Channel, Hunter Channel, and Santos Plateau

Authors	AABW boundary	Transport, Sv	Comments
Hogg et al. 1982	$45.87 < \sigma_4 < 45.93$	4.15 ± 1.27 (Vema)	Mean geostrophic transport over six sections, reference surface: $\theta = 2°$
Reid 1989		1.8 (Hunter)	
Zemba 1991	$\sigma_4 = 45.85$	4.1 (Vema)	Geostrophic transport at 31° S, reference surface: $\sigma_4 = 45.85$
		0.2 (Santos)	
Speer and Zenk 1993	θ =2 °C	3.9[a] (Vema)	Geostrophic transport (*Meteor* 15, reference surface: $\theta = 2°$
		0.7 (Hunter)	
		2 (Santos)	
Holfort and Siedler 2001	θ =2 °C	5.1–5.3[a] (Vema)	Current meter data
Zenk and Hogg 1996	θ =2 °C	2[a] (Vema)	Geostrophic transport over A10 section, reference surface: $\theta = 2°$
		1.8 (Santos)	
Lenz 1997	θ =2 °C	2.7±1.3 (Hunter)	Current meter data
Hogg et al. 1999	θ =2 °C	4.0 (Vema)	Current meter data
		2[a] (Vema)	Geostrophic transport over A10 section. First estimate for Santos: *Meteor* 15
		−0.8-0.0 (Santos)	
Wienders et al. 2000		3.5[a] (Vema)	Geostrophic transport over A17 section, reference surface is based on inverse method
Cai and Greatbatch 1995		3.7 entire ocean	Model
Zenk et al. 1999	θ =2 °C	2.92 Hunter	Hunter Channel
McDonaugh et al. (2002)	θ =2 °C	3.7 ± 0.7[a] (Vema)	A17 section
	θ =2 °C	0.1 ± 2.5 (Santos)	A10 section
		3.7 ± 0.7[a] (Vema)	
	θ =2 °C	0.1 ± 1.5 (Santos)	A23 section
		2.7 ± 0.7[a] (Vema)	
		1.3 (Vema)	
Zenk (2008a)	<0.2 °C	2.07 ± 0.73	WSDW at Vema Sill only two current meters, direct observations
Morozov et al. (2010)	θ =2 °C	2.5–3.7 (Vema)	LADCP

[a]Without inclusion or with partial inclusion of transport at the western slope of the Rio Grande Rise

region and flows over the Santos Plateau. A southerly flow exists over the western slope of the Rio Grande Rise with a transport ranging from 0.4 to 2.4 Sv (McDonagh et al. 2002).

Other authors (Hogg et al. 1982; Zemba 1991; Speer and Zenk 1993) integrate this flow with the flow in the deep Vema Channel. Barnier et al. (1998) and Marchesiello et al. (1998) modeled the seasonal variability of bottom water transport at 30° S. According to their calculations, water transport varied only slightly from 1 to 2 Sv. A seasonal cycle was also found in the record of moored current meters in the Vema Channel deployed by IFM-GEOMAR (Kiel) in 2003–2007.

Analysis of the division of Antarctic Bottom Water into components shows that most publications distinguish Weddell Sea Deep Water and Lower Circumpolar Water. Weddell Sea Deep Water flows to the Brazil Basin only through the Vema Channel (Reid et al. 1977; Sandoval and Weatherly 2001). At the same time, Lower Circumpolar Water flows through both Vema and Hunter channels. A different scheme of circulation in the region is presented in (Larque et al. 1997): Weddell Sea Deep Water propagates to the Brazil Basin through both channels. Speer and Zenk (1993) conclude that Weddell Sea Deep Water (bounded by $\sigma_4 > 46.05$) accounts for 40% of waters of the bottom layer flowing to the Brazil Basin. This agrees with the estimates in (Reid 1989), according to which Circumpolar Deep Water accounts for 60%.

4.2 Vema Channel

4.2.1 Topography and General Description

For the first time this region was systematically investigated during the *Deutsche Atlantische Expedition* of the R/V *Meteor* in 1925–1927 (Spiess 1932; Emery and Zenk 2019). The application of the newly invented echo sounding equipment allowed German researchers to reveal many objects of the bottom topography on a quasi-continuous track (Maurer and Stocks 1933). The *Meteor* found a narrow valley with a depth of 4920 m at 29°47.1′ S, 39°27.5′ W. Decades later Cruise 22 of the R/V *Vema* (for more historical remarks see Sect. 4.2.2) was especially dedicated to the study of bathymetry of this region and the stratification of deep and bottom waters enabling water transport from the Argentine Basin to the Brazil Basin.

The bottom topography around the Vema Channel is shown in Fig. 4.3. The Vema Channel is the deepest one among the passages existing for Antarctic Bottom Water (hereinafter we use the definition by (Wüst 1936); see the previous section). Therefore, the coldest water (Weddell Sea Deep Water) can exit the Argentine Basin in the equatorward direction only through this channel (Zenk et al. 1993). The Vema Channel is a narrow conduit between two terraces located to the east and west of this canyon (Johnson et al. 1976). Its width slightly exceeds 15 km in the narrowest part of the channel. The Vema Channel has been eroded for many millions of years by the abyssal currents of Antarctic Bottom Water (Gamboa et al. 1983).

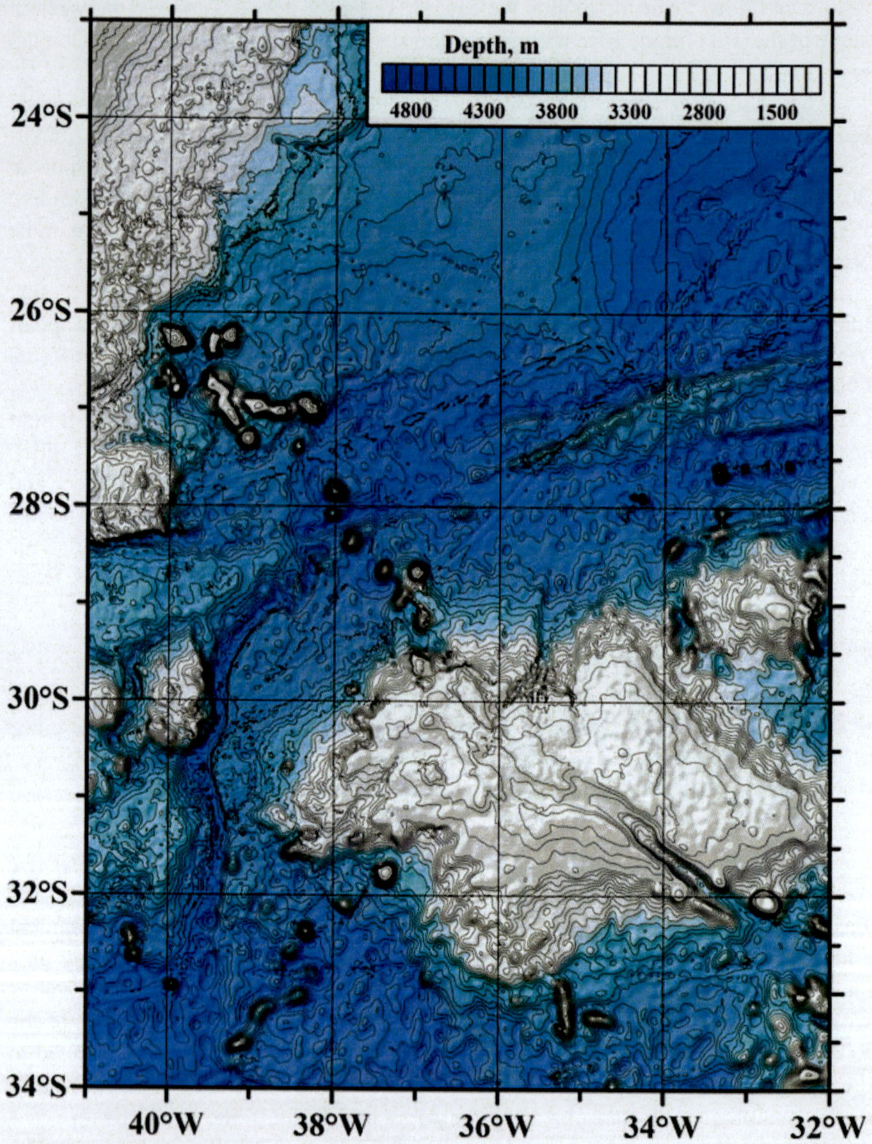

Fig. 4.3 Bottom topography around the Vema Channel

A virtual "pipeline" for the bottom waters of Antarctic origin with the highest density in the whole water column crosses the Rio Grande Rise. North of its entrance on the Argentine side, the Vema Channel depth contours follow an almost meridional direction. Depth in the Vema Channel exceeds 4600 m against the background depth of 4200 m. A small constraining sill (Vema Sill) located at 31°12′ S, 39°24′ W crosses the channel. At its deepest spot on the crest of the Vema Sill the depth is approximately 4614 m. The width of the channel here is roughly 15 km between isobaths 4300 m (Zenk et al. 1993).

The presence of the Vema Channel explains the existing structure of deep waters in the Atlantic Ocean. Conditions of water formation in the Arctic and Antarctic regions do not differ significantly with respect to temperature at the surface. The Arctic waters flow into the Atlantic over the shallow thresholds between Greenland and Iceland and between Iceland and Scotland; the depths of the thresholds are 650 and 850 m, respectively. Therefore, only the intermediate waters of lower density from the Arctic Basin actually overflow the thresholds. On the other hand, the depth of the sill in the Vema Channel allows the densest waters of the Antarctic origin to overflow this sill and occupy the bottom layers in the Atlantic Ocean.

The northern part of the Vema Channel is wider. The name Vema Extension was suggested by Zenk and Morozov (2007) for this region at about 27° S, 34° W. In the region of 27–29° S, the channel turns northeast, filling a buffering reservoir for bottom water at 27° S and broadens slightly farther north of 29° S. The local topography facilitates the further easterly flow of the bottom waters after outflowing from the channel. Two branches of the bottom waters are distinguished in the Brazil Basin (Sandoval and Weatherly 2001). The upper branch turns to the northwest immediately after outflowing from the channel. Possible cyclonic or anticyclonic rotation around the Rio Grande Rise is suggested in (Hogg et al. 1982; Hogg et al. 1996).

Seen from the bottom, the canyon-like walls on both sides of the channel near the Vema Sill at 31°12′ S are about 400 m high (Zenk et al. 1993). Relatively flat areas located on either side of the channel were called the eastern and western terraces (Johnson et al. 1976). Since 1991, the majority of revisits with CTD casts were concentrated on a narrow strip at that latitude. A detailed chart of the Vema Sill region is shown in Fig. 4.4. We have compiled an inventory of a total of 110 CTD stations from 30 visits to the Vema Sill (Table 4.2). The first 11 successful visits to the sill region were already evaluated in the earlier work of Hogg and Zenk (1997).

4.2.2 *History of Research and Datasets of Long-Term Observations*

Initially, the Vema Channel was known as the Rio Grande Rinne or Rio Grande Gap (for more details see (Zenk et al. 1993)). The width of this valley was estimated less than 30 km. In recognition of the intense work done by the former research vessel *Vema* the name Vema Channel was first suggested in (Le Pichon et al. 1971) for this deep canyon of the Rio Grande Rise. The *Vema* (Fig. 4.5) made an outstanding contribution to physical oceanography and marine geology. Originally this 202-ft. pleasure yacht was built in Copenhagen in 1923 as a luxury yacht *Hussar*. Later, the ship was renamed by the owner. Its name was derived from the name **Ma**ud **Ve**tlesen, the wife of the first owner. Displacement of the ship is 750 tons. In

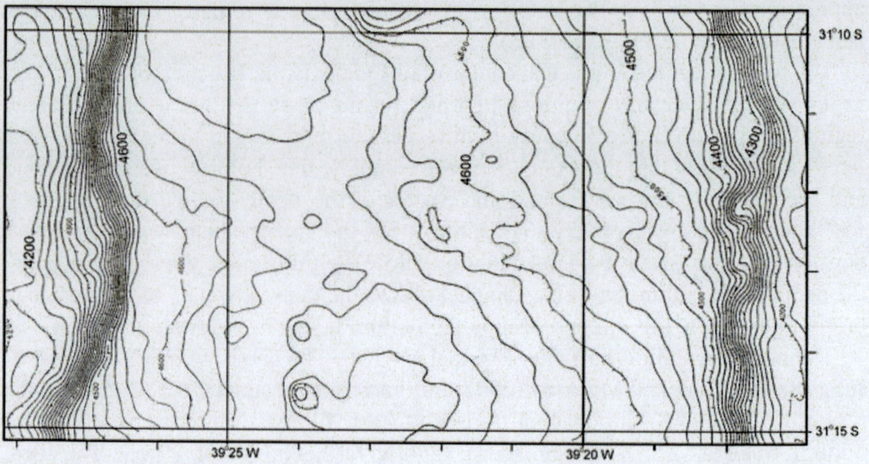

Fig. 4.4 Detailed bathymetry near the Vema Sill based on a multibeam survey (Zenk et al. 1993). Depths are in meters

1953–1981, it belonged to the former Lamont-Doherty Geological Observatory (now Lamont-Doherty Earth Observatory) in Palisades, New York. The vessel covered 1,250,000 miles with research works in all parts of the world ocean. Information in the Internet reads that the ship still serves as the cruising yacht *Mandalay* in the Caribbean belonging to Sailwindjammer company based on the information in 2019 (https://sailwindjammer.com/). Tours were sold for cruises in 2020. The ship is in operation.

Many prominent seafloor features were discovered from the *Vema* and named after this famous research vessel. They are the following:

The Vema Fracture Zone, crosses the Mid-Atlantic Ridge at 10°44′ N, 42°20′ W, was discovered in 1956.

The Vema Trench in the Indian Ocean crosses the Central Indian Ridge at 8°59′ S, 67°34′ E was discovered in 1958.

The Vema Seamount rises from a depth of almost 5000 m to 25 m below sea level in the South Atlantic 31°38′ S, 8°20′ E was discovered in 1959

The Vema Knoll at 23°40′ N, 66°58′ W and Vema Gap at 23°30′ N, 67°00′ W are located adjacent to each other, 500 km north of Puerto Rico.

The Vema Deep is the deepest place in the Romanche Fracture Zone (7856 m).

The Vema Channel is a deep channel in the Rio Grande Rise of the South Atlantic at 31°12′ S, 39°24′ W was investigated in detail in 1962.

The first deep oceanographic measurements in the Vema Channel region were made from the famous R/V *Meteor* in 1925. In the following years, this region was studied within the expedition of the International Geophysical Year, the "GEOSECS" program, and a number of other expeditions. Thirty (mainly German, Russian, and British) expeditions with deep water measurements were carried out in

Table 4.2 Historical measurements near the Vema Sill

Vema Sill Visit No	Cruise/Expedition	Country of ship	Date mmm/yyyy	Number of stations
1	Cato, leg 6	US	Dec/1979	1
	GEOSECS			
2	Atlantis II-107, leg 2	US	Nov/1979	2
3	Atlantis II-107, leg 8	US	Jun/1980	4
4	Thomas Washington	US	Dec/1984	4
5	Meteor 15, leg 1	DE	Jan/1991	6
6	Meteor 22, leg 4	DE	Dec/1992	6
7	Polarstern ANT 12/1	DE	Nov/1994	1
8	Meteor 34, leg 3	DE	Mar/1996	3
9	Meteor 41, leg 3	DE	Apr/1998	4
10	Meteor 46	DE	Apr/2000	2
11	Ak. Ioffe 11	RU	Nov/2002	7
12	Ak. S. Vavilov 17	RU	Nov/2003	6
13	RSS discovery 276	UK/DE	Dec/2003	1
				Mooring deployment
14	Ak. Ioffe 16	RU	Nov/2004	5
15	Ak. Ioffe 17	RU	Mar/2005	5
16	Polarstern ANT 22/5	DE	May/2005	5
				Mooring recovery
				Mooring deployment
17	Ak. Ioffe 19	RU	Oct/2005	5
18	Ak. Ioffe 22	RU	Oct/2006	5
19	James Clark Ross 160	UK/DE	May/2007	1
				Mooring recovery
20	Polarstern ANT 24/4	DE	Apr/2008	1
21	Ak. Ioffe 27	RU	Apr/2009	5
22	Polarstern	DE	Apr/2009	2
23	Polarstern	DE	Apr/2010	2
24	Ak. S. Vavilov 44	RU	Apr/2017	11
25	Polarstern	DE	May/2018	1
26	Ak. S. Vavilov 46	RU	Oct/2018	7
27	Antares	BR	Feb 2019	Mooring deployment
28	Ak. M. Keldysh 79	RU	Jan/2020	1
29	Ak. M. Keldysh 79	RU	April/2020	9
30	Merchant ship Macaé	BR	Aug 2020	Mooring recovery
	Total CTD stations			110

Fig. 4.5 Photo of the *Vema* research ship that made an enormous contribution to the research of bottom topography (top) and a modern photo of cruising yacht *Mandalay* (bottom)

this region since 1979. The data used in this study were either collected by the authors of this work or downloaded from the World Ocean Database (WOD18 2018; Boyer et al. 2018) dataset. During the World Ocean Circulation Experiment

(WOCE) significant progress was achieved in observation of abyssal variability in the South Atlantic. The Deep Basin Experiment, a core project of WOCE, was focused on the key passages that control the equatorward interbasin exchange of bottom waters (Hogg 2001).

After the WOCE completion, further surveys related to this program were mainly concentrated on the Vema Channel. Investigators from many countries, including Germany and Russia, were involved in the research. Revisits of the site in this post-WOCE phase were inspired by a warming trend in bottom water at the Vema Sill, which was first noted in (Zenk and Hogg 1996), and long-term variations in other properties of water and transport in the entire channel. Recently, the slowly growing time series of the coldest water measurements in the Vema Sill (Zenk et al. 2003) are known as CLIVAR research topics of national programs in Germany ("Marin-2", terminated in 2005) and Russia ("Meridian") (Morozov et al. 2003) and continuations of these programs. A compilation of international activities near the Vema Sill (31°12′ S) is given in Table 4.2.

Recent studies in the channel continue the earlier measurements. Numerous expeditions to this region have confirmed its utmost importance for the entire water dynamics in the Atlantic Ocean. In our research, we divide datasets from the region into three parts: Vema Sill, Santos Plateau, and Vema Extension.

In the beginning of the time series (1972–1990), satellite-supported navigation with today's high accuracy was not permanently available on research vessels. The local topography was less known. Hence, CTD casts were taken in a wider range south and north of the exact location of the sill. Since 1991, the lowest temperatures were encountered consistently on the eastern part of the sections across the Vema Sill on all repeated sections (Hogg and Zenk 1997). This phenomenon reflects the frictionally induced thick bottom boundary layer on the right side of the throughflow jet in the Southern Hemisphere. The concurrent secondary or helical circulation sharpens the interface above the mixed bottom layer. It causes the well-known isotherm pinching in the vertical temperature distribution (Jungclaus and Vanicek 1999).

Data accuracy and stability are important issues for all long-term observations. A steady technical improvement of CTD-profilers was undertaken in the last decades. The data we analyzed had been obtained either with Neil Brown instruments or Sea-Bird instruments (in the last decade). In the search for systematic trends in abyssal bottom temperatures in the Weddell Sea, Robertson et al. (2002) also faced uncertainties in different cruises similar to our investigation. According to the WOCE standards, the accuracy of the high-quality CTD data is leveled nowadays at $\theta = 0.002$–0.003 °C for potential temperature and approximately 0.003 PSU for salinity. Absolute pressure uncertainties for deep-sea sensors are considered to be below ±3 dbar. Most expeditions to the Vema area were carried out with pre- and post-cruise calibration checks of the operated CTD probes. Instrumental stability has improved significantly over the last three decades.

After the Deep Basin Experiment has been completed as part of the World Ocean Circulation Experiment, the study of Antarctic Bottom Water transport was concentrated mainly in the Vema Channel. In the late 1990s and beyond, a detailed study of the abyssal flow was carried out during several cruises of German research vessels *Meteor* and *Polarstern* to the Vema Channel. Besides long-term observations of the stratification and its conjectured changes, the studies were concentrated on the high-speed flow structure and the quantification of mixing between near-bottom water masses (Zenk 2008a).

In 2002, Russian scientists joined the international research program in the region. In November 2002, the first Russian section across the Vema Channel was occupied within the Russian National Program "Meridian". These studies were planned together with the German colleagues from the *Institut für Meereskunde* in Kiel (now GEOMAR). The zonal section crossed the Vema Channel along 31°14′ S. Seven stations were occupied using the Neil Brown CTD instrument. In November 2003, three sections across the channel were occupied in the northern, central, and southern parts of the channel. The sections were made across three narrow parts of the channel. In addition, during the cruise onboard the R/V *Akademik Sergey Vavilov* in November 2003, a CTD section along the Vema Channel was occupied with a spacing of 30–40 miles. Stations along the 600-km-long sections were located at the eastern wall of the channel.

In November 2004, the section at the Vema Sill was repeated again. The section at the Vema Sill was displaced 2 miles northward to 31°12′ S to avoid collision with the German moorings at 31°15′ S. Two more stations were occupied in the northern (Vema Extension) and central parts of the channel. In March and October 2005, the section across the Vema Channel at 31°14′ S was repeated twice. In October 2006 and March 2009, the section was repeated again. Measurements continued in 2017, 2018, and 2020.

During the period from 2003 to 2020, the Vema Channel was alternatively revisited by the German and Russian scientists. In 2003, German scientists performed measurements in this region from the R/V *Discovery* shortly after the Russian visit to the channel. This made it possible to intercalibrate the field measurements made by the German and Russian scientists. Differences in the measurements of temperature and salinity at one point in the coldest jet near the Vema Sill were equal to 0.001 °C and 0.001 PSU, respectively. Similar insignificant observational differences were recorded in the spring of 2009. Potential temperatures and salinity values measured from the R/V *Akademik Ioffe* and *Polarstern* only 3 weeks apart differ by 0.001 °C and 0.003 PSU, respectively. In 2009, the section at 31°12′ S was repeated. In 2017, the section was extended to the west. In 2018, and 2020, the section was repeated again. Many stations were made north of the Vema Channel and on the Santos Plateau. Brazilian scientists deployed a mooring near the standard section in February 2019. The mooring was recovered in August 29, 2020. More measurements are planned.

The deepest part of the southern entrance to the channel is located in its southern part at 33°33.6′ S. A seamount with its summit at 3900 m forms the eastern slope of the channel here. A sedimentary ridge (Emery Drift) with the depth of the crest at 4500 m against the background depth of 4650 m is located at the western side of the channel. In October 2005, a section was occupied here from the deepest point of the seamount slope located at a depth of 4900 m in the western direction to the slope of the Emery Drift. Contour currents usually develop around such bottom topography structures (Zenk 2008b).

Since 2005, the CTD measurements were supplemented by lowered ADCP profilers. All Russian measurements were conducted with the participation of the authors of this work. Visits to the Vema Sill are schematically shown in Fig. 4.6 as sections of potential temperature across the channel.

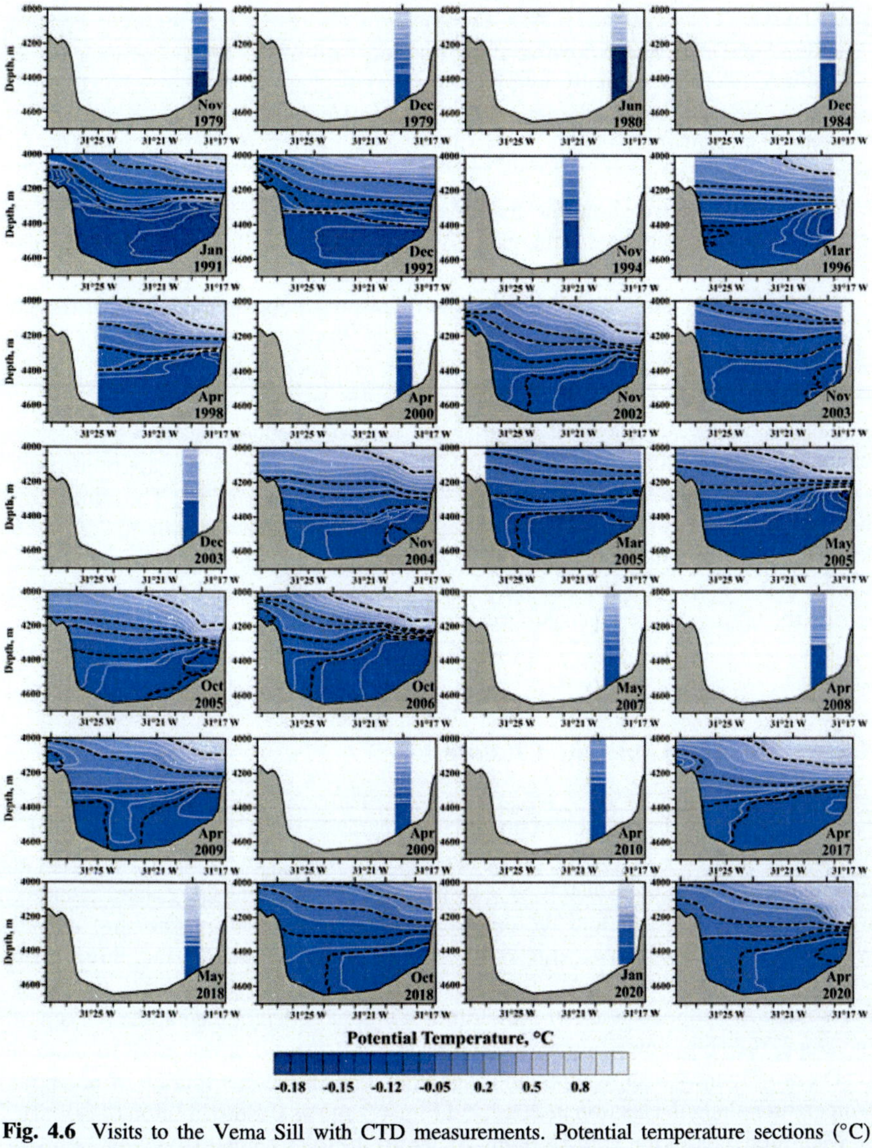

Fig. 4.6 Visits to the Vema Sill with CTD measurements. Potential temperature sections (°C) across the channel are shown if several stations were occupied in the channel. A narrow stripe shows individual stations in the Vema Sill region occupied when research vessels passed the region

4.2.3 Deep and Bottom Waters

The Vema Channel is interesting first of all as the natural pipeline for Weddell Sea Deep Water, the densest and coldest component of Antarctic Bottom Water. Its flow through the channel from the Argentine Basin to the north is controlled by the depths of sills along the channel. This choke point of bottom water transport to the north is an excellent region for studying the mechanisms of the northward bottom water propagation. The CTD and LADCP measurements reveal that the cold core of Weddell Sea Deep Water flowing along the deep bed of the channel propagates as a well-mixed jet (occasionally split into two jets) displaced to the eastern wall of the channel. The thickness of the jet is 150–300 m. Within the sections of the jet, the variation range of its potential temperature and salinity is 0.1 °C and 0.005 PSU, respectively.

In this section, we show the thermohaline properties of bottom waters in the channel and waters above the channel. Figure 4.7 shows the θ,*S*-curves for deep and abyssal waters. Names of water masses are indicated.

Researchers of Antarctic Bottom Water draw its upper boundary at different levels varying from $\sigma_4 = 45.87$ (Memery et al. 2000) to $\sigma_4 = 45.92$ (Fu 1981; Roemmich 1983). Thus, the difference between locations of the boundary isopycnal surface in the Vema Channel can be as large as 150 m.

Long-term measurements show that the core of Weddell Sea Deep Water is characterized by minimum potential temperature θ equal to approximately −0.110 °C (in situ temperature +0.215 °C) and salinity 34.67 PSU. The concentration of silicates is as high as 118 μmol/kg and that of phosphates is up to 2.3 μmol/kg. The local minimum of oxygen concentration (220 μmol/kg) is located at a distance of 500 m from the bottom, which allowed Vanicek and Siedler (2002) to divide the bottom water layer into two water masses.

4.2.4 Section Along the Channel

Let us consider the distribution of potential temperature in the bottom layer along the Vema Channel. In November 2003, the stations were set up along the line of maximum depths approximately at a distance of 700 m from the eastern slope of the channel. In order to find the desired location of the station, the ship moved westward along the slope with echo sounder measurements using three echo sounders: narrow beam ELAC LAZ 4700, multibeam ECHOS XDM, and Atlas Parasound. After the location of each future station was determined at the bottom of the slope, the ship slowed down, turned back, and stopped at the point to make a CTD cast. Figure 4.8 shows the latitudinal variation in the distribution of potential temperature along the section in 2003 and Fig. 4.9 shows the distribution of salinity.

Measurements in the cold jet along the entire Vema Channel show that temperature in the jet increases due to mixing with overlying waters during the bottom

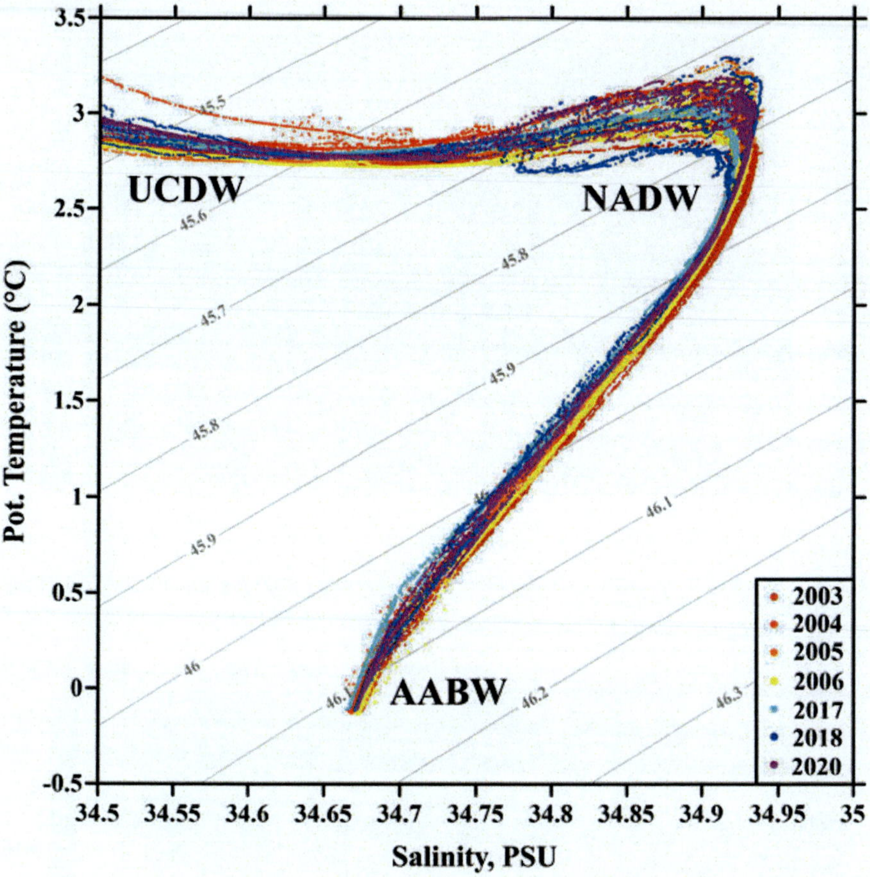

Fig. 4.7 Curves of θ,*S*-distributions for deep and abyssal waters in the Vema Sill region based on the results of CTD profiling in different years over the section at 31°12′ S. Contour lines of potential density σ_2 are shown. The waters occupying the corresponding layers are indicated

water propagation from south to north. The isotherms and isohalines deepen from south to north. Thus, the layer of coldest waters becomes thinner as the flow propagates along the channel. Latitudinal variation in temperature in the coldest part of the jet is shown in Fig. 4.10. Approximately the same linear trend of temperature was recorded in 2003, 2004, 2005, 2009, and 2020. Fewer stations along the channel were made in 2004, 2005, 2009, and 2020. However, the measurements in 2009 and 2018 revealed a stronger warming of bottom waters in the northern part of the channel, but we could miss the coldest part of the jet.

The channel widens in the Vema Extension between 27° S and 28° S. It is noteworthy that the cold jet was not found in this widening near the eastern slope of the channel. Therefore, the measured temperatures are not plotted on the graph.

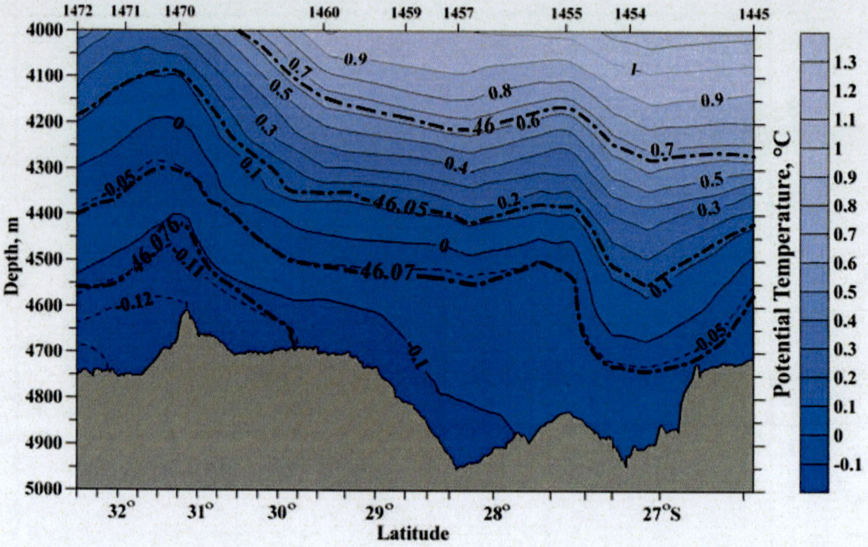

Fig. 4.8 Potential temperature (°C) along the Vema Channel in 2003 (measured at the foot of the eastern slope)

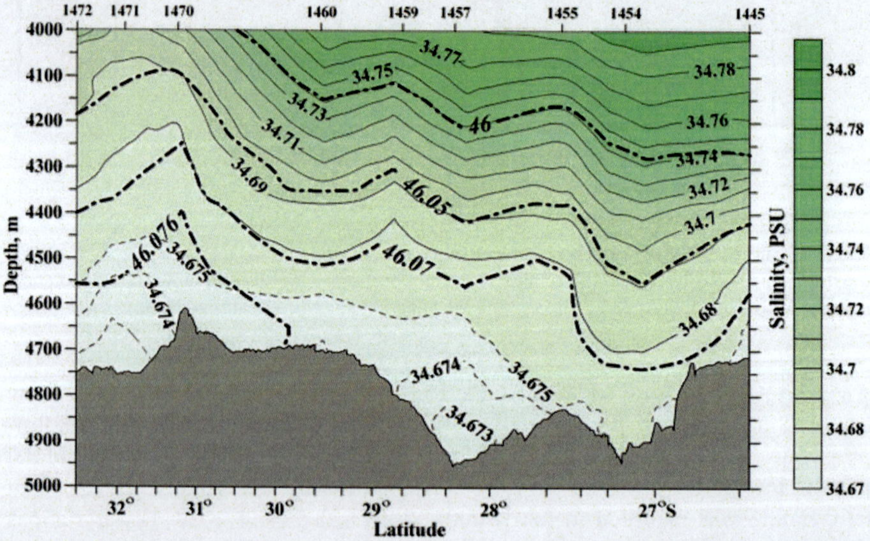

Fig. 4.9 Salinity (PSU) along the Vema Channel in 2003 (measured at the foot of the eastern slope)

Due to widening of the channel here, the Ekman friction that displaces the jet to the eastern slope has less strength.

The figures show a clear increase in the values of these characteristics. Data in 2003 show that potential temperature increases from south to north from −0.135° to

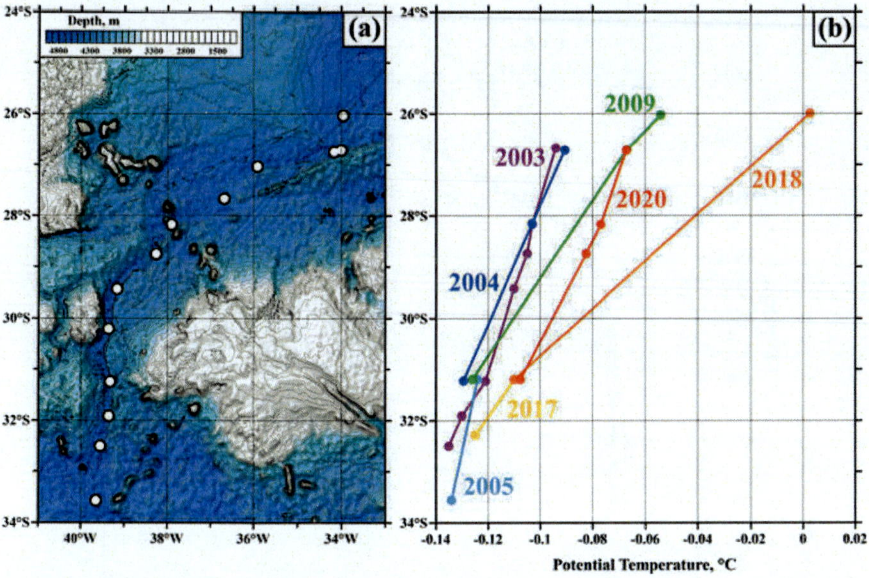

Fig. 4.10 Latitudinal variation in potential temperature (°C) along the Vema Channel in 2003, 2004, 2005, and 2009 (right panel). White dots in the left panel show the stations at the bottom of the eastern slope in 2003. Latitude scales are the same in the left and right panels of the graph

−0.094 °C, while salinity increases only from 34.67 to more than 34.68 PSU at the extreme southern and northern points of the section. The maximum latitudinal gradients are confined to the narrowest part of the channel. The coolest and densest part of Weddell Sea Deep Water cannot overflow the sills in the channel. Moreover, mixing with the overlying waters increases the temperature and salinity of the flow.

Four temperature profiles along the channel below 4000 dbar (in the northern, central, and southern parts) based on the measurements in 2003 are shown in Fig. 4.11. The greatest temperature increase is observed between 4000 and 4500 dbar at the boundary with North Atlantic Deep Water. This boundary displaces to the greater depths as the bottom water flows to the north. The corresponding velocity shear due to oppositely directed flows is the greatest here. The minimum temperature at the bottom also increases to the north. A temperature increase is not accompanied by strong changes in θ,S-characteristics.

Linear regression of the sequence of temperature minima of Weddell Sea Deep Water along the Vema Channel suggests a horizontal temperature gradient of ~$5.7 \cdot 10^{-8}$ °C m^{-1}. Assuming pathway lengths of 220 and 760 km between the sill position and the downstream sites, one expects a spatially conditioned offset in potential temperature of ~0.013 °C and ~ 0.043 °C, respectively. The difference between two salinity curves comprises the latitudinal increase of near-bottom salinity equal to ~2.10^{-9} PSU m^{-1}. The salinity signal, however, is noisier than the temperature signal. Nevertheless, we note a density-compensating horizontal

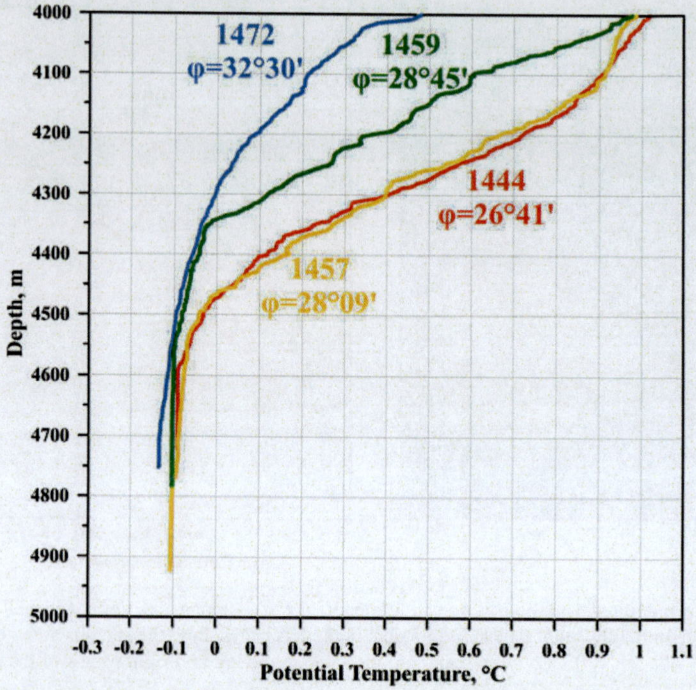

Fig. 4.11 Changes in the vertical profile of potential temperature (°C) along the Vema Channel in 2003

correlation between potential temperature and salinity along the flow path of Weddell Sea Deep Water.

4.2.5 *Structure of the Flow. Sections Across the Channel*

As a rule and traditionally, CTD sections across the Vema Channel are occupied near the Vema Sill at 31°12′ S. Since 1991, 15 sections have been occupied. These sections returned much data to analyze the structure of the flow of Antarctic Bottom Water and variation of this flow in time. An inventory of the data is given in Table 4.3.

Let us first consider sections across the channel occupied not along 31° S but in different locations along the channel. In November 2003, three CTD sections were made across the channel in the northern, central, and southern (standard section along 31° S) parts of the channel. Two sections of potential temperature and salinity across the channel occupied in 2003 in the northern and central parts of the channel are shown in Figs. 4.12 and 4.13, respectively. In March 2020, we also made three sections across the channel approximately at the same locations as in 2003. The

Table 4.3 Historical sections in the Vema Channel near the Vema Sill

Time	Ship	Country
January–February 1991	RV *Meteor*	Germany
December 1992	RV *Meteor*	Germany
March 1996	RV *Meteor*	Germany
April 1998	RV *Meteor*	Germany
November 2002	RV *Akademik Ioffe*	Russia
November 2003	RV *Akademik Sergey Vavilov*	Russia
November 2004	RV *Akademik Ioffe*	Russia
March 2005	RV *Akademik Ioffe*	Russia
May 2005	RV *Polarstern*	Germany
October 2005	RV *Akademik Ioffe*	Russia
November 2006	RV *Akademik Ioffe*	Russia
April 2009	RV *Akademik Ioffe*	Russia
April 2017	RV *Akademik Sergey Vavilov*	Russia
October 2018	RV *Akademik Sergey Vavilov*	Russia
April 2020	RV *Akademik Mstislav Keldysh*	Russia

section in the middle part occupied in 2020 is shown in Fig. 4.14 together with a chart of the stations in 2003 and 2020. The third section at 28°09–28°30′ S is shown in Fig. 4.15. The flow in this part of the Vema Channel splits into two parts. The flow along the main deep channel, which is located to the northwest of the smaller channel, is stronger. We made stations in the smaller channel twice: in 2018 and in 2020. In 2018, the velocities in the shallower channel were directed to the northeast. In 2020 they were almost zero and even a slow reversal flow to the south was noticed. All sections were made in the narrow parts of the channel. In the deeper channel, the cold jet on all sections was displaced to the eastern slope of the channel owing to the Ekman bottom friction. Correspondingly, the temperature in the coldest parts of the jet was increasing from southeast to northwest.

Distributions of temperature and salinity over sections across the channel near the Vema Sill along 31°12′ S or 31°14′ S in 1991, 1992, 1996, 1998, 2002, 2003, 2004, 2005 (March, May, and November), 2006, 2009, 2017, (2018), 2018 and March 2020 are shown in Figs. 4.16 and 4.17. Contour lines of density are also shown in the figures. Measurements made in many expeditions showed that the cold core of Antarctic Bottom Water is usually displaced to the eastern slope of the channel. The core is a jet of well-mixed water. The thickness of the core is approximately 150–300 m. The flow is strongly ageostrophic.

Usually, the width of the cold core is approximately one fourth of the channel. The upper boundary of the core migrates between 4300 and 4400 m. High location of the boundary results in a sharp thermocline over the core as, for example, in 2006 (Fig. 4.16k). In 2003 and 2004, it was much higher over the bottom than in other years. Displacement of the jet from the easternmost position to a position closer to the middle of the channel is likely caused by meandering of the jet. The bottom Ekman friction force induces secondary helical circulation relative to the general

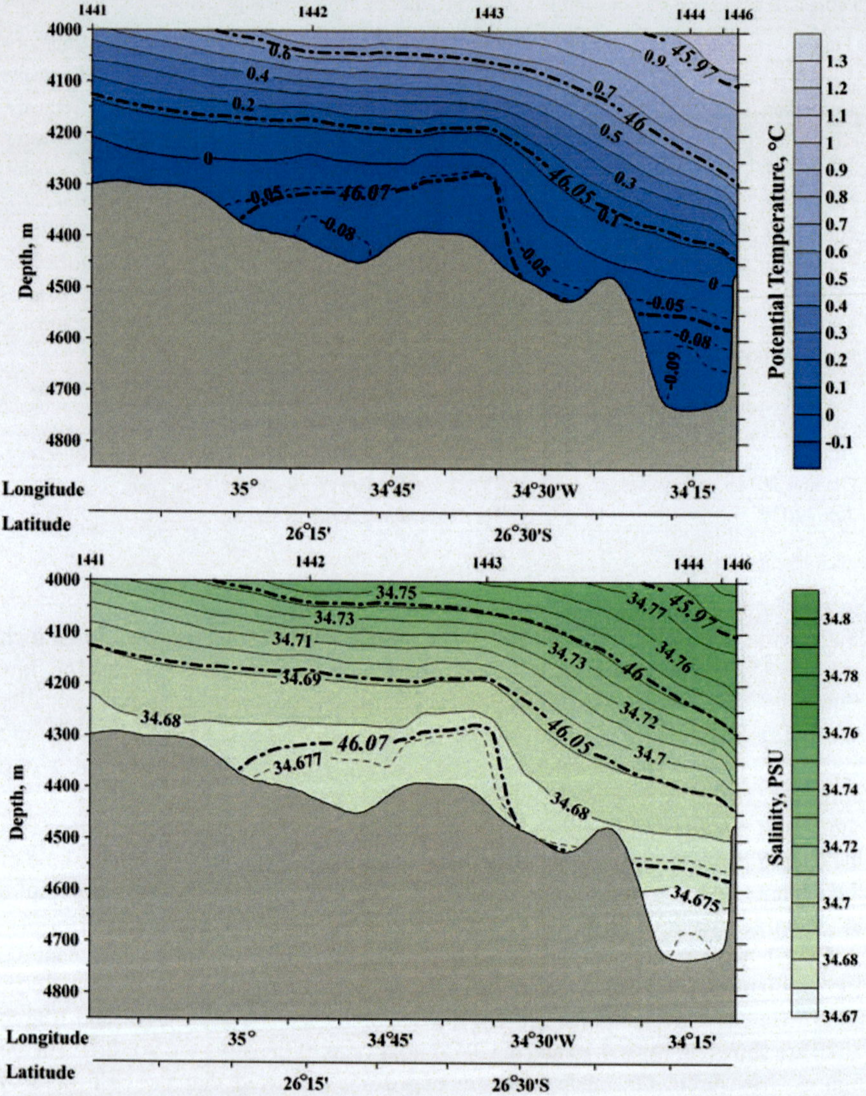

Fig. 4.12 Sections of potential temperature (°C) (top) and salinity (PSU) (bottom) across the Vema Channel in 2003 in the northern part of the Vema Channel at 26°43′ S (Vema Extension)

current, which displaces the cold-water core to the right (to the east if the current is directed to the north) in the Southern Hemisphere (Johnson et al. 1976). In the Southern Hemisphere, the bottom Ekman friction is directed to the right of the flow, which agrees with the modeling results of Jungclaus and Vanicek (1999). In March 2005, the core occupied the most distant position from the slope. This is the only observed case of such displacement. Possibility of meandering of the flow even in a

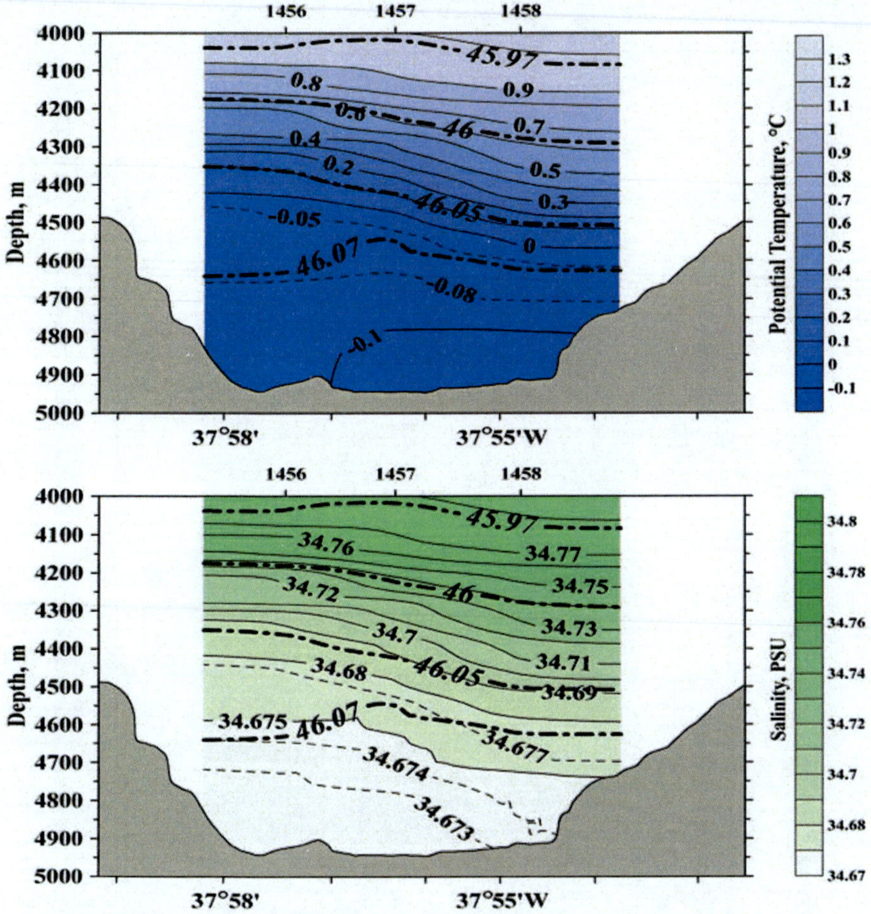

Fig. 4.13 Sections of potential temperature (°C) (top) and salinity (PSU) (bottom) across the Vema Channel in 2003 in the middle part of the Vema Channel at 28°09′ S

very narrow channel is suggested by the fact that the jet can notably shift to the east and up the slope and simultaneously occupy a position closer to the middle of the channel. Mixing with the overlying waters is related to lateral and bottom friction.

In the upper part of the flow above the channel, the cold low saline core of the flow is displaced to the west. This core has been usually found over the western wall of the channel. This phenomenon was observed in 1991, 1992, 2002, 2006, 2009, 2017, 2018, and 2020. In 2002, the low values of temperature and salinity observed over the western slope of the channel were as follows: potential temperature was less than −0.04 °C and salinity was less than 34.675 PSU. At the eastern slope, such values of temperature and salinity are observed at depths greater by 150–200 m. Only in these years, the measurements were made not only in the deepest part of the channel but also over the eastern and western slopes of the channel. We suppose that

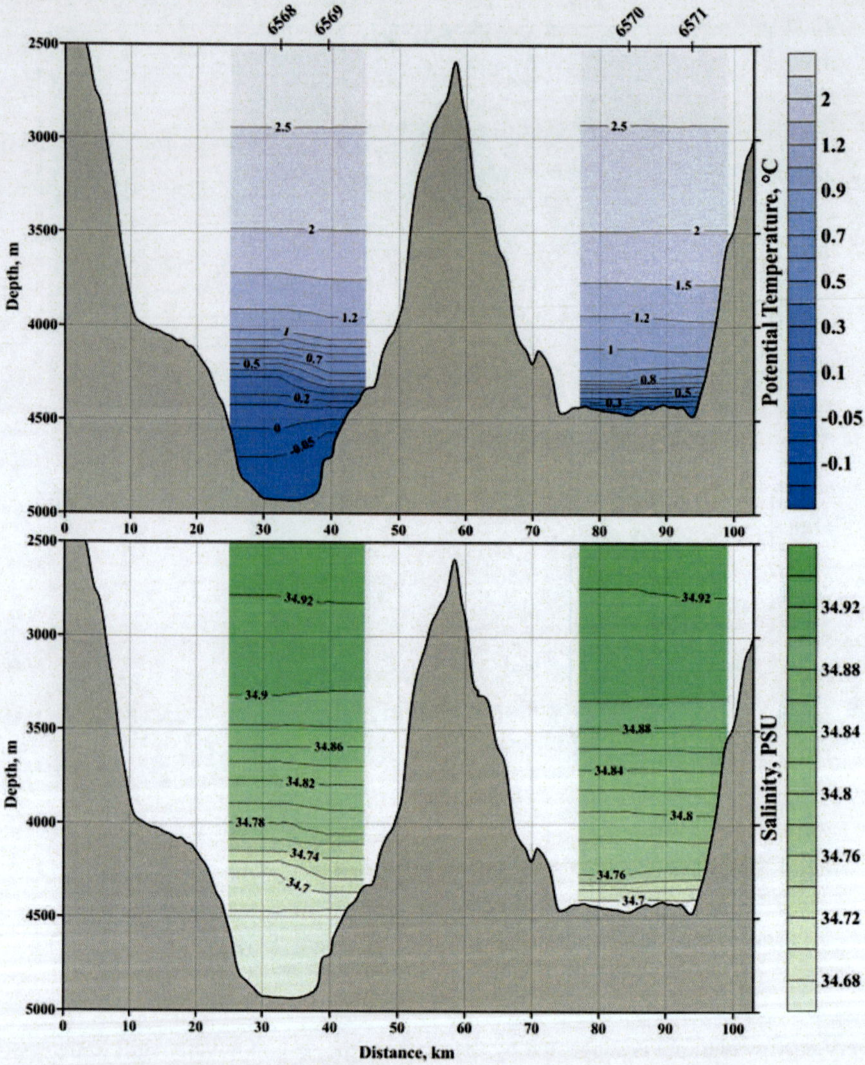

Fig. 4.14 Sections of potential temperature (°C) (top), salinity (PSU) (bottom) across the Vema Channel in 2019 in the middle part of the Vema Channel at 28°09′ S and 28°30′ S

this upper core over the western wall exists always. However, observations at the upper part of the western slope are not available for each section, because several times the western part of the section did not reach the western slope. This jet exists due to the displacement of the flow by the Coriolis force.

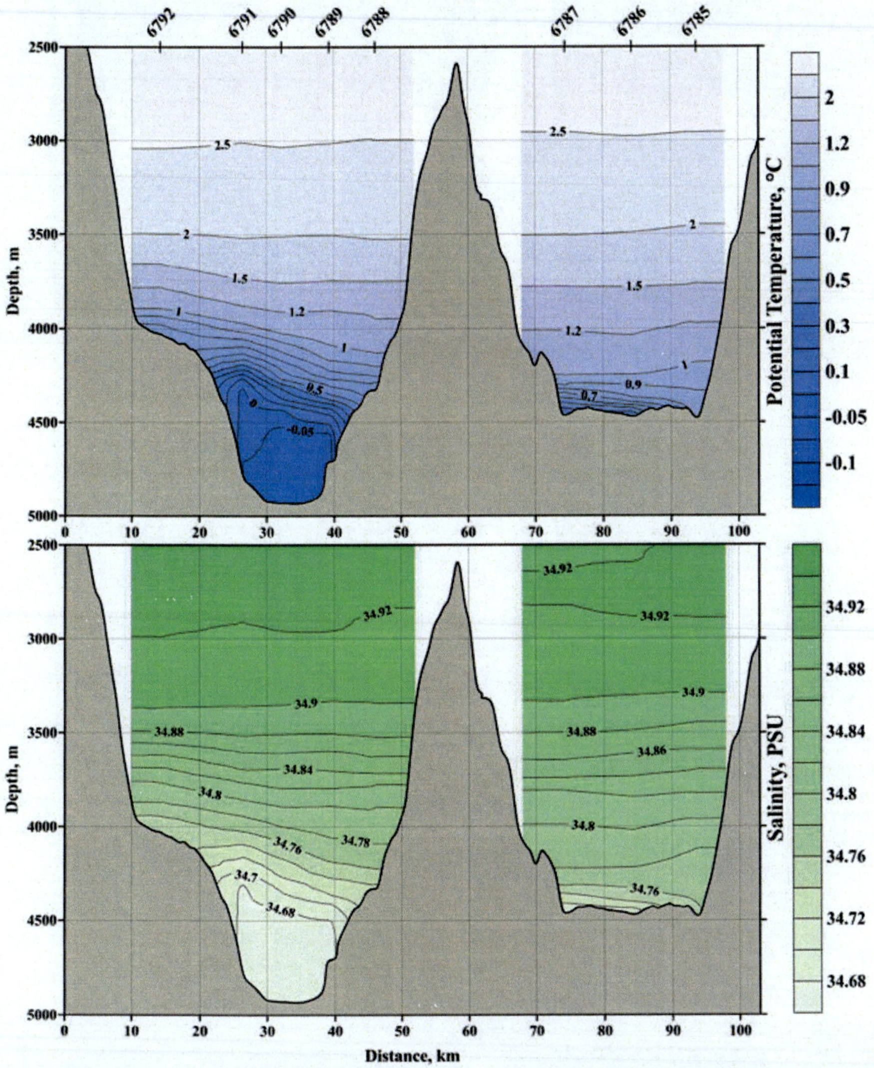

Fig. 4.15 Sections of potential temperature (°C) (top), salinity (PSU) (bottom) across the Vema Channel in 2020 in the middle part of the Vema Channel at 28°09′ S and 28°30′ S

Observations in different years show that the structure details of the flow change every time as we make measurements. However, the general feature associated with the displacement of the jet to the eastern slope at the bottom is almost the same.

In 1991, we noted that the values of potential temperature and salinity were the lowest ($\theta < -0.17$ °C, $S < 34.672$ PSU). In 2003, the distributions of temperature and salinity were more homogeneous compared to the sections in other years, indicating the fact of stronger mixing in the bottom layer.

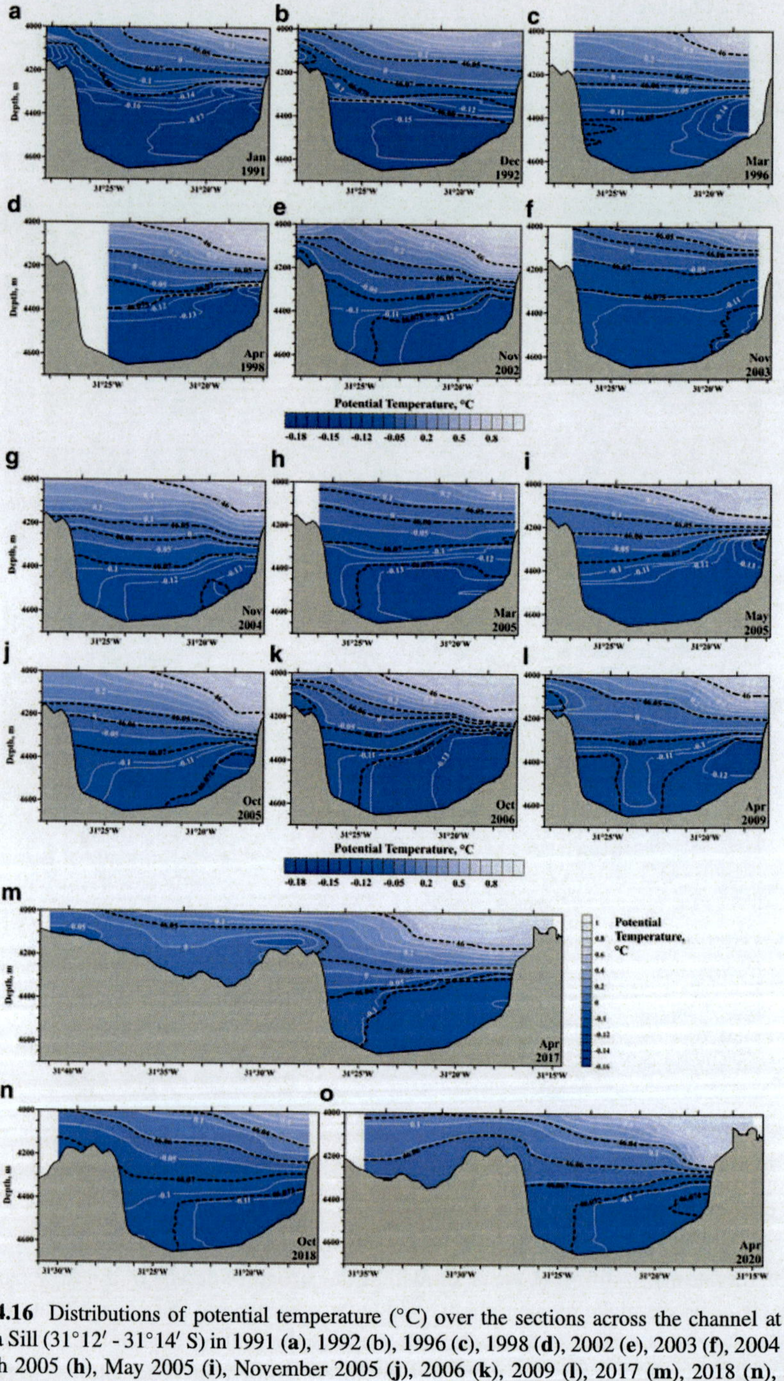

Fig. 4.16 Distributions of potential temperature (°C) over the sections across the channel at the Vema Sill (31°12′ - 31°14′ S) in 1991 (**a**), 1992 (**b**), 1996 (**c**), 1998 (**d**), 2002 (**e**), 2003 (**f**), 2004 (**g**), March 2005 (**h**), May 2005 (**i**), November 2005 (**j**), 2006 (**k**), 2009 (**l**), 2017 (**m**), 2018 (**n**), and 2020 (**o**). Axes *y* and *x* show depth and longitude, respectively. The year and month of measurements are indicated at each section. Contour lines of density are shown with thick dash-dotted lines

The lowest values of temperature and salinity observed in November 2004 and March 2005 as well as greater elevation of isotherms and isohalines over the bottom were related to greater amounts of Antarctic Bottom Water in the bottom layer of the channel. In 2004, the strongest horizontal stratification of the flow was observed compared to all other years of observation. Thermohaline properties of the flow in the eastern and western parts of the channel are slightly different. The western part of the flow is usually more saline and warmer.

Year to year observations show changes in the jet structure. The cold core is mixed well in the vertical direction, but horizontal stratification exists always. The changes in stratification, location of the jet, and its structure over a period from 1991 to 2020 can be seen from Figs. 4.16 and 4.17.

We determined the position of the Antarctic Bottom Water core using the classic principles: local minima of temperature and salinity (Wüst 1936). In many cases, they do not coincide on the sections across the channel. The most typical spatial structure of the flow is the location of waters with minimum temperature and salinity and maximum density in the lower part of the flow at the eastern slope. Sometimes, the distribution was even more homogeneous. In 1991 and 2009, density increased from the center to the eastern and western slopes. In some years of measurements, a strong temperature difference was observed between the eastern and western parts of the channel (>0.04 °C).

When the cold jet is strongly displaced to the eastern slope and elevated, a strong temperature gradient is observed at the upper boundary of the slope, as was the case in 2006 (Fig. 4.18). This happens because the lower isotherms are elevated by the core of the flow, while the upper remain at their previous depths. The vertical gradient is 0.02 °C/m. In the previous years, such a strong gradient was not found. Strong layering is also characteristic of the gradient zone.

4.2.6 Trends in Potential Temperature and Salinity of the Coldest Bottom Water Observed Since 1972

While analyzing the variations in the flow over a long period we found variation of the position of isotherms and isohalines by depth, which can naturally be explained by different intensities of Antarctic Bottom Water transport. Variations in the depth of the 0 °C and 0.2 °C potential temperature isotherms can be as high as ~150 m between the uppermost and lowermost positions. The latter isotherm marks the upper boundary of Weddell Sea Deep Water. A general trend to deepening of both isotherms over decades is observed. Of course, our measurements are very rare. Nevertheless, we found that when large volumes of bottom water pass through the Vema Channel over the sill at 31°12′ S, the isotherms ascend by 100–150 m. The

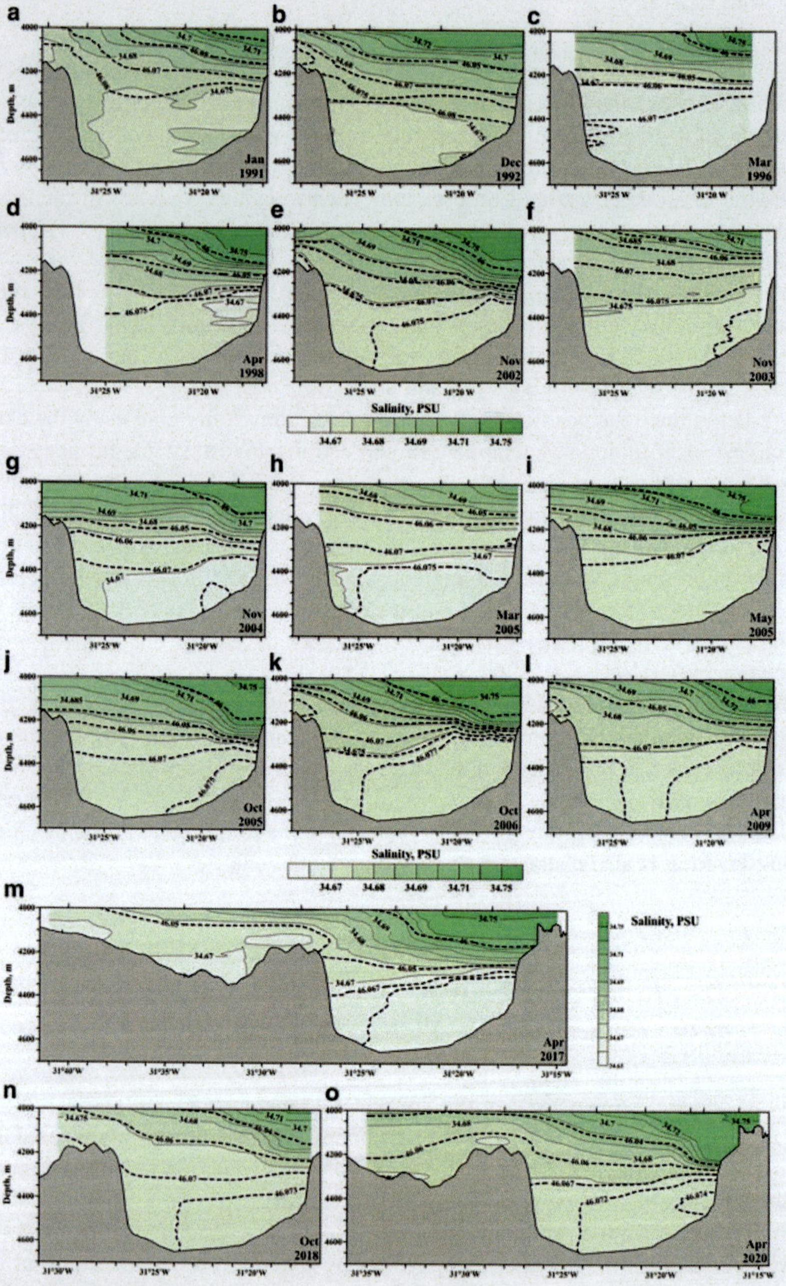

Fig. 4.17 Distribution of salinity (PSU) over the sections across the channel at the Vema Sill (31°12′ - 31°14′ S) in 1991 (**a**), 1992 (**b**), 1996 (**c**), 1998 (**d**), 2002 (**e**), 2003 (**f**), 2004 (**g**), March 2005 (**h**), May 2005 (**i**), November 2005 (**j**), 2006 (**k**), 2009 (**l**), 2017 (**m**), 2018 (**n**), and 2020 (**o**). Axes *y* and *x* show depth and longitude, respectively. The year and month of measurements are indicated at each section. Contour lines of density are shown with thick dash-dotted lines

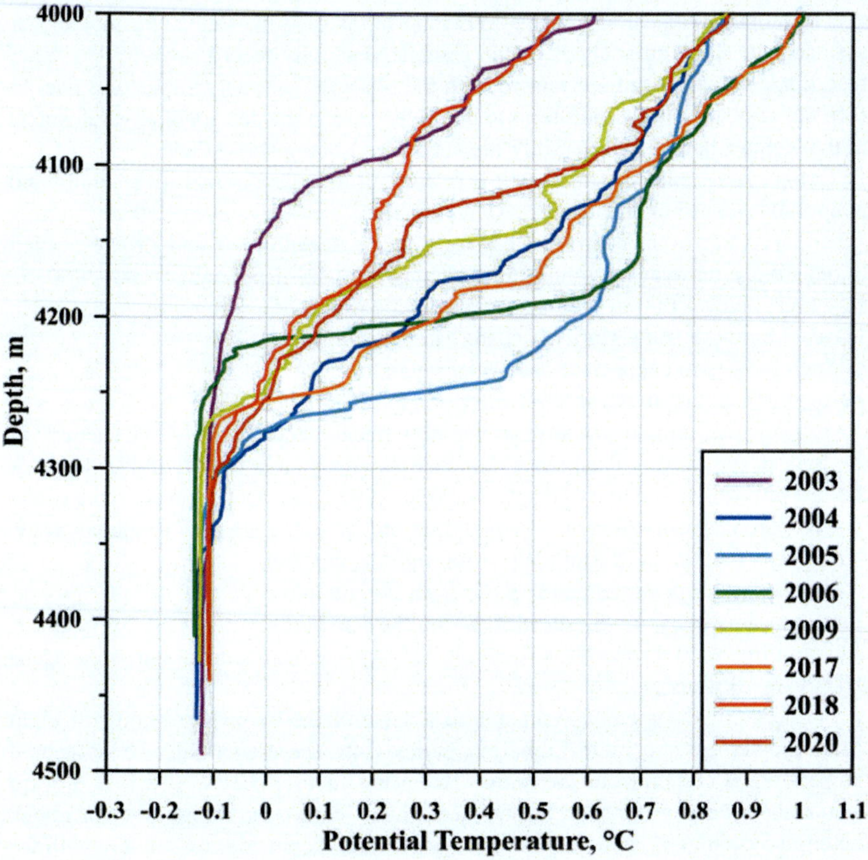

Fig. 4.18 Vertical profiles of potential temperature (°C) at the eastern slope of the Vema Channel. In 2006, a transition zone with sharp vertical temperature gradients was found above the mixed jet of Antarctic Bottom Water

highest position of the 0.2 °C and 0.0 °C isotherms was recorded in November 2004, March 2005, and January 2020 (Fig. 4.19).

Antarctic Bottom Water represents the coldest and deepest layer of the western South Atlantic. Significance of abyssal pathways for global thermohaline circulation and ventilation and the role of Antarctic Bottom Water in interhemispheric exchange have turned the Vema Channel into an object for continued observations beyond the end of the WOCE decade. The ongoing motivation for numerous revisits (by now 30) is maintained by a general decadal warming trend in Antarctic Bottom Water that was first noted in (Zenk and Hogg 1996). Due to the small magnitude of the expected signals, detection of significant changes in bottom water properties requires equipment with the highest accuracy and stability.

The length and accuracy of the CTD time series from the Vema Channel appear sufficient to allow an analysis of the associated salinity record as well. The latter record appears to be anti-correlated with the observed decadal temperature rise. In fact, this result implies a long-term trend of decrease in the density of abyssal waters in the Vema Channel and possibly the entire deep Southwest Atlantic.

Comparative analysis shows that it is difficult to isolate a trend in spatially and temporally aliased shorter series. As an example, Fukasawa et al. (2004) found a general warming of the deep Pacific along 47° N between 140° E and 120° W, which is negatively correlated with salinity changes in the depth range 500–3000 m. However, below 5000 m they also reported a basin-wide warming by 0.005 °C in 14 years but with "no significant change in the salinity field".

Such long-term changes of deep-water mass characteristics are needed as observational constraints in ocean circulation models. The objective of this section is to describe the evolution of the abyssal warming phenomenon in the Vema Channel in the light of the historic database and recent site visits more than two decades after the last progress reported by (Hogg and Zenk 1997). Thanks to annual visits of the Russian and German research vessels to this site, new data are now available on the variability of Antarctic Bottom Water through the channel.

Time series of potential temperature from the site in the Vema Channel (approximate coordinates 31°12′ S, 39°18.3′ W) on the standard section at the Vema Sill is shown in Fig. 4.20. Linear fit is shown in red. Approximate temperature increase is 0.02 °C in 10 years.

Furthermore, the potential temperature series from the Vema Extension with eight visits supplement the more numerous series from the Vema Sill. The potential temperature θ curves from the Vema Extension in Fig. 4.21 are shifted upward against the reference curve (from the Vema Sill). Within the statistical error range, the implicit spatial increase in potential temperature θ coincides reasonably with the horizontal temperature structure from a quasi-meridional section along the Vema Channel in November 2003.

The section across the Vema Channel used to compare temperature measurements in different years is located at 31°12′-31°14′ S. Long-term observations reveal a trend of increase in potential temperature in the core of Antarctic Bottom Water with time. Before 1993, potential temperature at 31°14′ S was −0.18 °C, while the deviations did not exceed 0.01 °C. From 1972, potential temperature has been increasing (Fig. 4.20). During the period from 1993 to 1998, the temperature increased to −0.135 °C. This was the period of significant temperature rise reported in (Zenk and Hogg 1996; Hogg and Zenk 1997). Actually, we can speak about the temperature increase that started after 1991. From this period, observations were made over the standard section located within 31°12′-31°14′ N, 39°17′ - 39°19′ W. In November 2002, the measurements recorded a temperature of −0.123 °C. In 2003, the temperature increased to −0.120 °C. In 2004 and 2005, the temperature remained at the level of 2003. During the following years, temperature fluctuations of the cold jet were observed with an amplitude of approximately 0.01–0.02 °C with a slight warming. This warming has been continuing until recently.

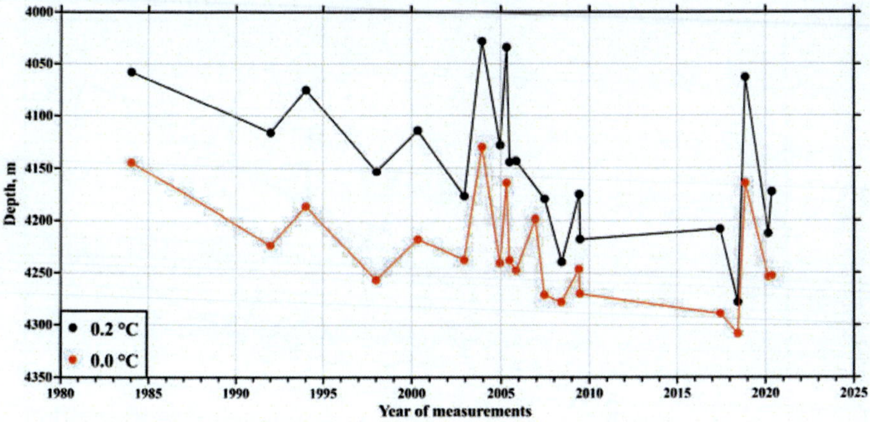

Fig. 4.19 Variations in the depths of 0.2 °C (black) and 0.0 °C (red) isotherms from 1984 to 2020

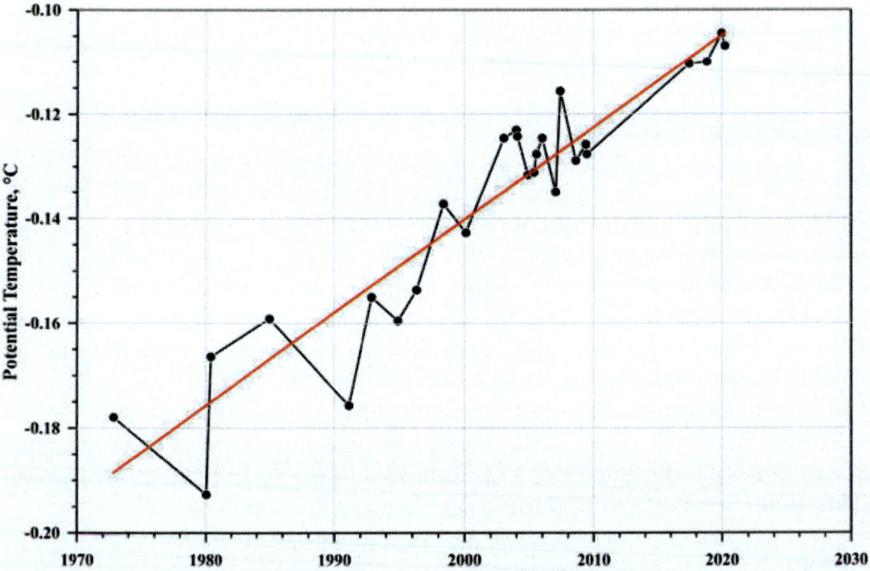

Fig. 4.20 Time evolution of potential temperature (°C) of the coldest Antarctic Bottom Water in the Vema Channel (point at standard section at 31°12′ S, 39°19′ W). Linear fit is shown in red

The graph shown in Fig. 4.22a illustrates the time evolution of the vertical distribution of potential temperature on the standard section across the channel at 31°12′-31°14′ S, while Fig. 4.22b shows the time evolution of the vertical distribution of salinity. It seems that the warming trend observed before 2003 was related to

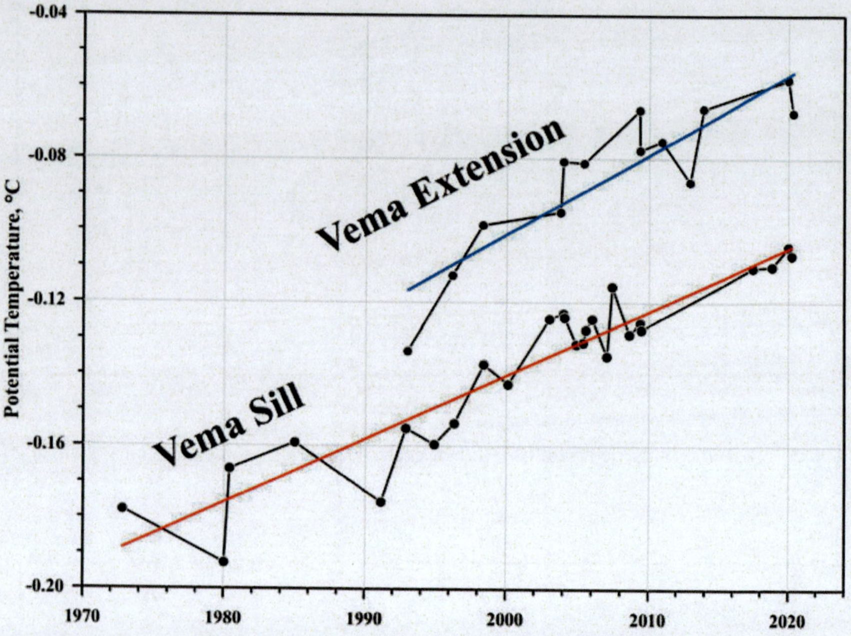

Fig. 4.21 Time evolution of potential temperature (°C) of the coldest Antarctic Bottom Water in the Vema Channel and Vema Extension

the propagation of a large mass of cold fresh water, which changed later to fluctuations of smaller amounts of water.

Next, we investigate time evolution of the coldest Antarctic Bottom Water (Weddell Sea Deep Water) along the Vema Channel in more detail. The first part of the record has already been analyzed in (Hogg and Zenk 1997). They discussed in detail at least 14 visits to the Vema Channel until 1996 with high-quality hydrographic measurements. Eleven datasets of the total 14 measurements were judged successful. Zenk and Hogg (1996) found little variation of the coldest Antarctic Bottom Water passing through different pathways across the Rio Grande Rise before 1990. In the Vema Channel, fluctuations were smaller than 0.005 °C. After such stagnant temperatures over 20 years, a change of +0.030 °C in the interval 1991/1992 was rated as "quite significant."

Our measurements show that the annual temperature increase exists near the Vema Sill. Usually, time gradient of temperature is in the range of 0.002–0.003 °C year^{-1} (though with occasional fluctuations of O(15) 10^{-3} °C year^{-1} on shorter scales (Zenk and Hogg 1996)).

A most relevant observation was reported in (Robertson et al. 2002). Based on the data from 1912 to 2000, they found indications for a slight temperature rise at depths of 1500–3500 m in the eastern Weddell Sea, i.e., at the Weddell Sea Deep Water

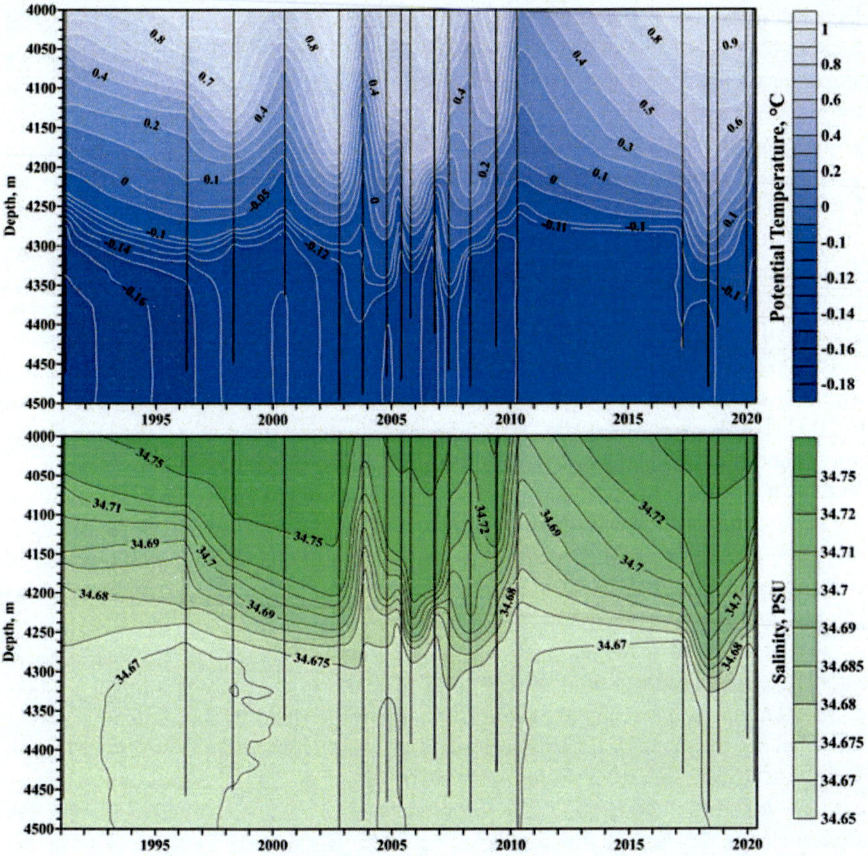

Fig. 4.22 Time evolution of potential temperature (°C) (top) and salinity (PSU) (bottom) at a point on the standard section at the Vema Sill at 31°12′ S, 39°18′ W. Vertical lines indicate times of measurements

source. The weak temperature signal seemed to be uncompensated or insignificantly compensated by salinity. The issue of how far the warming is related to the changes in the seasonal cooling or to the advection modulated by the Weddell Front remains open in this study.

Bottom water properties in the Southern Ocean are subjected to interannual and long-term variations (Fahrbach et al. 2004; Rintoul 2007). An increase in potential temperature near the Weddell Sea bottom is one of the possible explanations for the temperature trend observed in the Vema Channel. It is likely that this trend is related to the general warming in the Southern Hemisphere in the twentieth century; mean temperature of air, sea surface, and land in the Southern Hemisphere increased approximately by half a degree (New et al. 2000). Recent temperature variation over the globe is being discussed among scientists in relation to the global warming.

Depending on the region of the Earth, the temperature increase in the last 100 years is as high as 1 °C. Analysis of satellite data from 1979 to 1998 demonstrates that this warming was accompanied with an increase in Weddell Sea surface temperature by 0.01–0.02 °C/year (Comiso 2000).

A correlation between the variability of conditions at the Southern Ocean surface and flux of bottom water was reported in a series of publications (Comiso and Gordon 1998; Fahrbach et al. 1995; Jacobs and Giulivi 1998). As to the deep and bottom waters in the Weddell Sea, Weddell Sea Deep Water was becoming warmer near the source of its formation at a rate of +0.003 °C/year (Fahrbach et al. 2004). From 1970 to 1990, the warming trend of Weddell Sea Deep Water was not statistically significant (Robertson et al. 2002). According to (Robertson et al. 2002), Warm Deep Water in the Weddell Sea has a significant statistical warming trend of +0.012 (±0.007)°C/year from 1970 to 1990 and a negative trend from 1998 to 2002 (Fahrbach et al. 2004). The temperature increase trend in the Weddell Sea is not monotonic: this fact suggests that the explanation of temperature increase in the Vema Channel and its correlation with the bottom temperature in the Weddell Sea is not straightforward. Continuous warming in the South Atlantic in recent years and its influence on the meridional overturning circulation in the Atlantic was reported in (Johnson and Doney 2006; Zhu and Liu 2020; Gronholz et al. 2020; Johnson and Lyman 2020).

The increase in water temperature in the Vema Channel can be considered a remote response (with a time shift of tens of years) to the aforementioned temperature increase in the Antarctic region. Anomalies formed in the Antarctic region spread along the slope of South America to the Southwest Atlantic (Whitworth et al. 1991; Locarnini et al. 1993; Coles et al. 1996; Orsi et al. 1999). According to (Coles et al. 1996), these anomalies were recorded near 18° S approximately 5–11 years after their appearance in the Weddell Sea. Analysis of hydrographic data in the Argentine Basin in 1988–1989 and their comparison with the results of measurements 8–10 years before this period showed a decrease in the amount of coldest waters with potential temperature θ less than −0.2 °C (Coles et al. 1996). On the other hand, Smythe-Wright and Boswell (1998) reported that the age of bottom waters in the Argentine Basin is approximately 30 years. Thus, the mean velocity of Weddell Sea Deep Water spreading in the basin can be estimated at ~0.7 cm s^{-1}, which seems reasonable to the order of magnitude. A warming in the bottom waters of approximately +0.04 °C was found in 1989–1995 and 2005–2003 in the Brazil, Argentine, and South Georgia basins (Johnson and Doney 2006).

The temperature of Weddell Sea Deep Water near the Argentine Basin bottom increased significantly from 1994 to 2003. It is especially well manifested in the southern part of the A17 WOCE section. In 1994, the mean potential temperature in the zone between 37° and 48° S in the layer deeper than 4800 dbar was −0.130 °C, which is by 0.024 °C lower than the mean potential temperature of water (−0.106 °C) in the same layer, calculated from the measurements in 2003. Figure 4.23 presents the difference between section A17 in 1994 and 2003 in the Argentine Basin. A temperature increase in the bottom layer is clearly seen. Coles

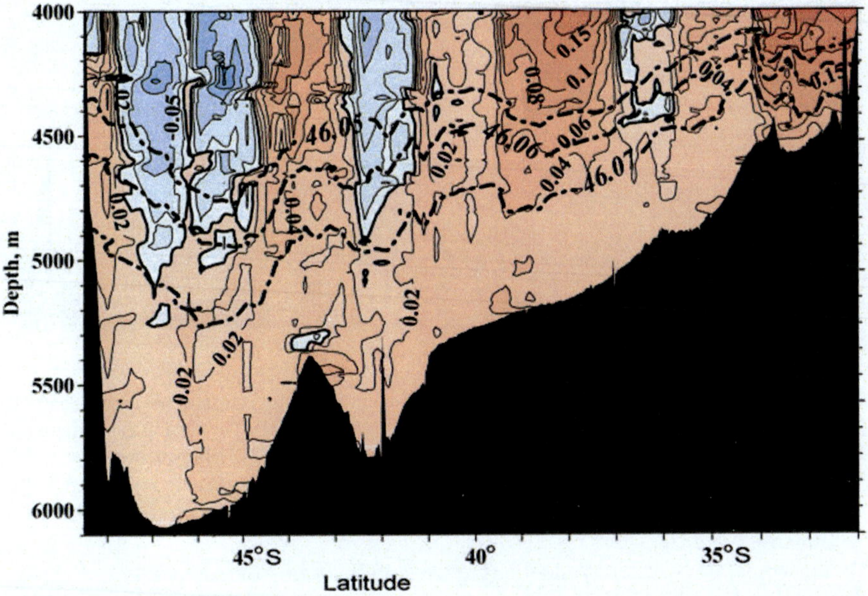

Fig. 4.23 Variation in potential temperature (°C) in 2003 compared to 1994 in the southern part of section A17. Contour lines of σ_4 are also shown

et al. (1996) reported a temperature increase in the bottom waters in the 1980s and concluded that this climatic signal has already reached the Brazil Basin.

One can also see in this figure that potential temperature variations in the Argentine Basin at a level of Vema Channel depths are ±0.02 °C. The long-term record of temperature fluctuations in the cold jet of the channel also demonstrates alternating temperature fluctuations within ±0.02 °C. This fact is interpreted as a consequence of temperature increase in the bottom layer of the Weddell Sea (Fahrbach et al. 1998) that was responsible for the temperature increase of the Weddell Sea Deep Water flow in the Vema Channel.

Here, we describe another interpretation of temperature variation in the Argentine Basin that can correlate with the temperature fluctuations in the Vema Channel, because bottom waters in the Argentine Basin are sources of the flow through the Vema Channel. The idea put forward in (Meredith et al. 2008) based on the idea in (Coles et al. 1996) seems promising. Changes in atmospheric cyclonic forcing over the Weddell Gyre influence the Weddell Sea Deep Water export across the South Scotia Ridge. Isopycnal surfaces in the Weddell Sea become either steeper or flatter in response to spinup or spindown of the gyre induced by winds, resulting in the changes in the density of the deepest Weddell Sea Deep Water exported across the South Scotia Ridge to the Scotia Sea. The strong Weddell Gyre during the period of strong winds entraps the coldest water south of the South Scotia Ridge and produces the less dense (warmer) Weddell Sea Deep Water that crosses the South Scotia Ridge and enters the Scotia Sea through the Georgia Passage. During the period of not very

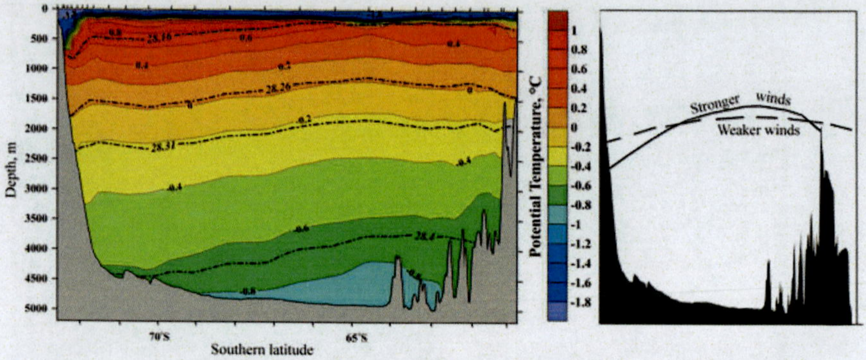

Fig. 4.24 Section of potential temperature (°C) and neutral density across the Weddell Sea (left panel) and scheme of isopycnals within the Weddell Sea (right panel). The section is based on the A23 section data (southern part) collected in March and April 1995. Isopycnals on the graph approximately correspond to the boundaries of water masses (from bottom to top): WSBW/LWSDW, LWSDW/UWSDW, UWSDW/CBW, and CBW/LCDW

strong winds, the weaker Weddell Gyre leads to flattening of isopycnals and allows denser (colder) Weddell Sea Deep Water cross the South Scotia Ridge. Two pathways for Weddell Sea Deep Water from the Weddell Sea exist: one through the South Sandwich Trench, the other through the Scotia Sea. The pathway through the Scotia Sea is longer and water is subjected to mixing with saline circumpolar water. Later both pathways merge in the Georgia Basin and then propagate to the Argentine Basin. Waters that passed different pathways attain different properties and this process produces an additional modulation of bottom water properties. Then, volumes of colder or warmer water that passed through the Scotia Sea and South Sandwich Trench circulate in the Argentine Basin and change temperature of the flow in the Vema Channel, depending on the temperature of water that appears at the southern entrance to the channel. A scheme of isopycnals within the Weddell Sea and a section of potential temperature and density across the Weddell Sea are shown in Fig. 4.24. Two types of isopycnals are shown: during "normal" deep water formation under moderate wind forcing (dashed lines) and during strong forcing leading to gyral spinup and midgyre convection (solid lines).

4.2.7 *Salinity Variations*

During the initial studies of salinity in the region of the Vema Channel it was not possible to identify time variations in salinity. Coles et al. (1996) reported a freshening of bottom water upstream in the northwestern part of the Argentine Basin; Hogg and Zenk (1997) were unable to find a significant change in salinity within the initial part of the available time series. After 1997, the number of stations

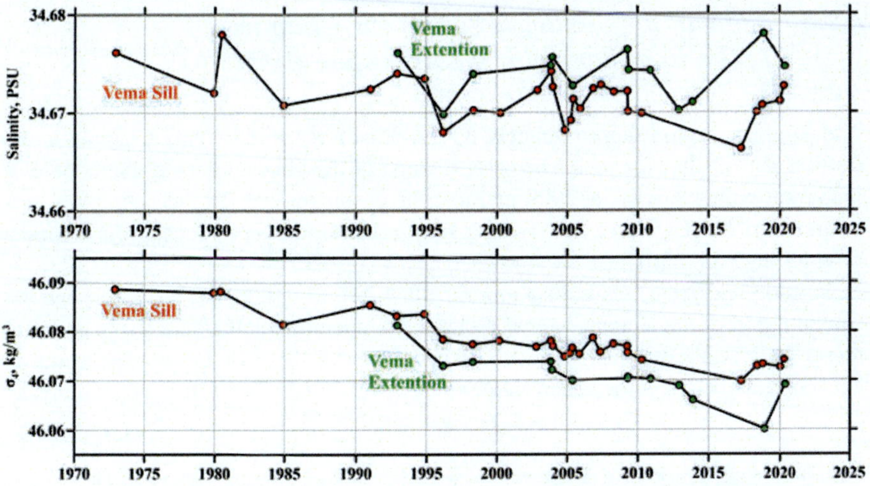

Fig. 4.25 Evolution of salinity (top panel) and density (bottom panel) at the reference points of potential temperature minimum in the Vema Channel (Vema Sill and Vema Extension)

in the Vema Channel increased allowing us to perform an analysis of salinity variations in time.

We consider salinity variations based on the measurements on the standard section at 31°12′ - 31°14′ S since 1980. Salinity variations in the bottom layer of the Vema Channel and instrumental errors are shown in Fig. 4.25. Salinity variations are very close to the instrumental accuracy, especially those measured in the 1980s–1990s. There is no significant salinity trend, although salinity variations by 0.002–0.005 PSU were recorded in 1994–1998 and 2003–2006. We note that minimum salinities were measured in 1996 and then in 2005. It is possible that many variations can be caused by small deviations of the instrumental calibration from the standards. However, no salinity variations were found in the layer above the depths of the Vema Channel.

The graphs in Figs. 4.25 show time evolution of salinity and density at the reference points of potential temperature minimum in three major regions of observations from the 1970s to 2020. Variations in salinity over the Vema Sill are shown in Fig. 4.22b. Potential temperature θ increases systematically over three decades, but the overall series in salinity begin to show a negative tendency ranging from -0.3 to $-1.2{\cdot}10^{-4}$ PSU year^{-1}. The associated density (σ_4) decrease measures about -0.4 g m^{-3} year^{-1}.

Our new data on increase in the minimum temperatures in the Vema Channel are substantially more robust and they confirm the previously reported temperature rising trend. At the present state and in view of intrinsic instrumental errors, we have to admit that the observed salinity changes are not compelling.

4.2.8 *Flow in the Southern Part of the Channel at the Boundary with the Argentine Basin*

The southern region (approximately 33°10–33°40′ S) of the Vema Channel is the deepest part of this channel. The eastern slope of the channel here is the slope of a seamount with a summit at 3900 m under the ocean surface. The western slope is a sedimentary ridge (Emery Drift) with a crest at 4480 m over the background depths of 4650 m (Fig. 4.26).

In 2003 and 2005, several stations (1708, 1709, Fig. 4.26) were occupied in the southern part of the Vema Channel. Potential temperature at the bottom measured at these stations shown in Fig. 4.28 indicates that the flow of the coldest water is concentrated between the Emery Drift and the previously mentioned seamount. The

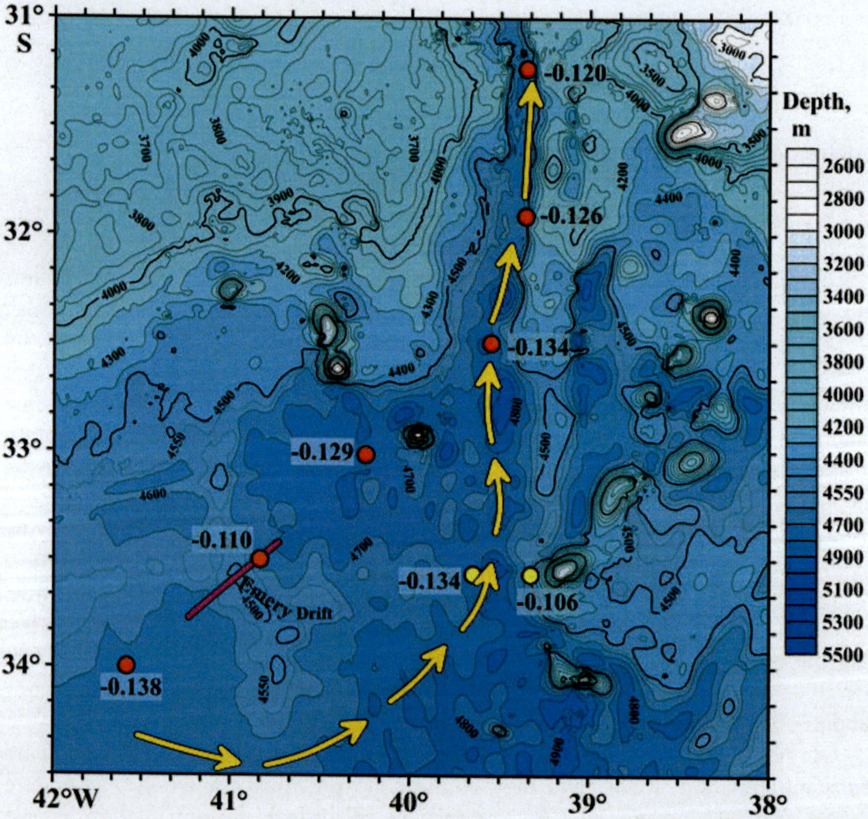

Fig. 4.26 Potential temperatures at the bottom (°C) at characteristic points of measurements near the entrance to the Vema Channel. Two stations (1708, 1709) in 2005 are highlighted with yellow color in the center of circles. Stations in 2003 are indicated with red color. Arrows show the most likely AABW flow. Potential temperatures at the bottom at key points of measurements are indicated. The violet line shows the section in 2003 across the Emery Drift (see Fig. 4.27)

Fig. 4.27 Sedimentary ridge (Emery Drift) south of the Vema Channel (a record by Atlas parasound echo sounder)

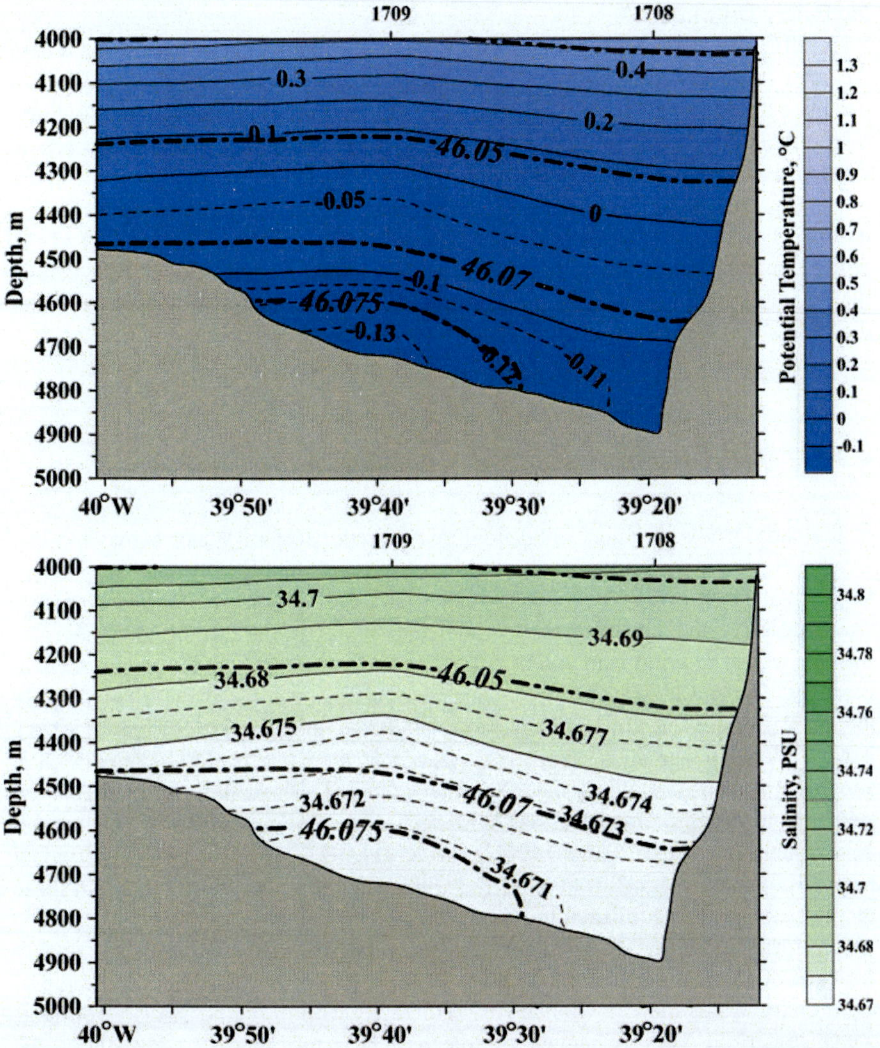

Fig. 4.28 Distribution of potential temperature (°C) (top) and salinity (PSU) (bottom) over the section across the deep passage (33°34′ S) at the entrance to the Vema Channel in 2005

data of two stations in 2005 at 33°34′ S, (Fig. 4.26), show that the flow of cold and less saline bottom water is displaced from the maximum depth of the channel to the western slope. The temperature at the deepest point was −0.106 °C, which is warmer than at the slope (−0.134 °C). The densest water is also located not in the deepest part of the trench, but over the slope. Such structure of the flow indicates the dominating role of the Coriolis force in displacing the core of the flow across the channel here, unlike Ekman friction near the Vema Sill in a narrow and long channel. Most likely, the coldest water does not overflow the Emery Drift, because the temperature north of the ridge is higher by almost 0.03 °C than south of the ridge (−0.138 °C and −0.110 °C, respectively). Warmer temperatures were recorded over the Emery Drift in 2003.

In 2017, we occupied a CTD/LADCP section in the southern part of the Vema Channel in the region of Antarctic Bottom Water inflow. Three stations were occupied across the channel along 32°17′ S. The sections of temperature and salinity distributions are shown in Figs 4.29a, b. The core of cold and less saline water is displaced to the eastern wall of the channel, which is characteristic of the flow in the Vema Channel. In 2003, we made a CTD cast at a close location to the eastern station in 2017. Potential temperature in 2003 was −0.134 °C; in 2017, it was −0.125 °C, which confirms the idea of warming of the flow. The velocities were lower than at the standard section at 31°12′ S. The velocities were within 10–14 cm s^{-1} and directed to the north.

4.2.9 *Moored Observations of Velocities in the Channel*

In the early 1990s, the horizontal flow of Antarctic Bottom Water across the Rio Grande Rise of the South Atlantic was intensely studied during the Deep Basin Experiment (DBE) of WOCE (Hogg et al. 1996). A core objective of this campaign was to describe and quantify the circulation and mixing of the sub-thermocline water masses while entering and exiting the Brazil Basin through its major abyssal passages. Starting in January 1991, an array of 13 moorings was deployed roughly along 29° S between the continental rise off Brazil and the eastern terrace of the narrow Vema Channel at about 39° W. After recovery in December 1992, a subgroup of instruments was moved farther eastward, where they were redeployed primarily in five mooring locations across the wider Hunter Channel. This second deployment lasted until early 1996 (Zenk et al. 1998). In most cases, individual records covered a period of at least 18 months. Comprehensive results from these moored arrays were published in (Hogg et al. 1999).

The basin-scale nature of the Deep Basin Experiment enabled a reasonably resolved quantification of the total horizontal transport, its variability, and temperature fluxes across the Rio Grande Rise. However, with only one available current-meter set in the Vema Channel, the database was too sparse to investigate the

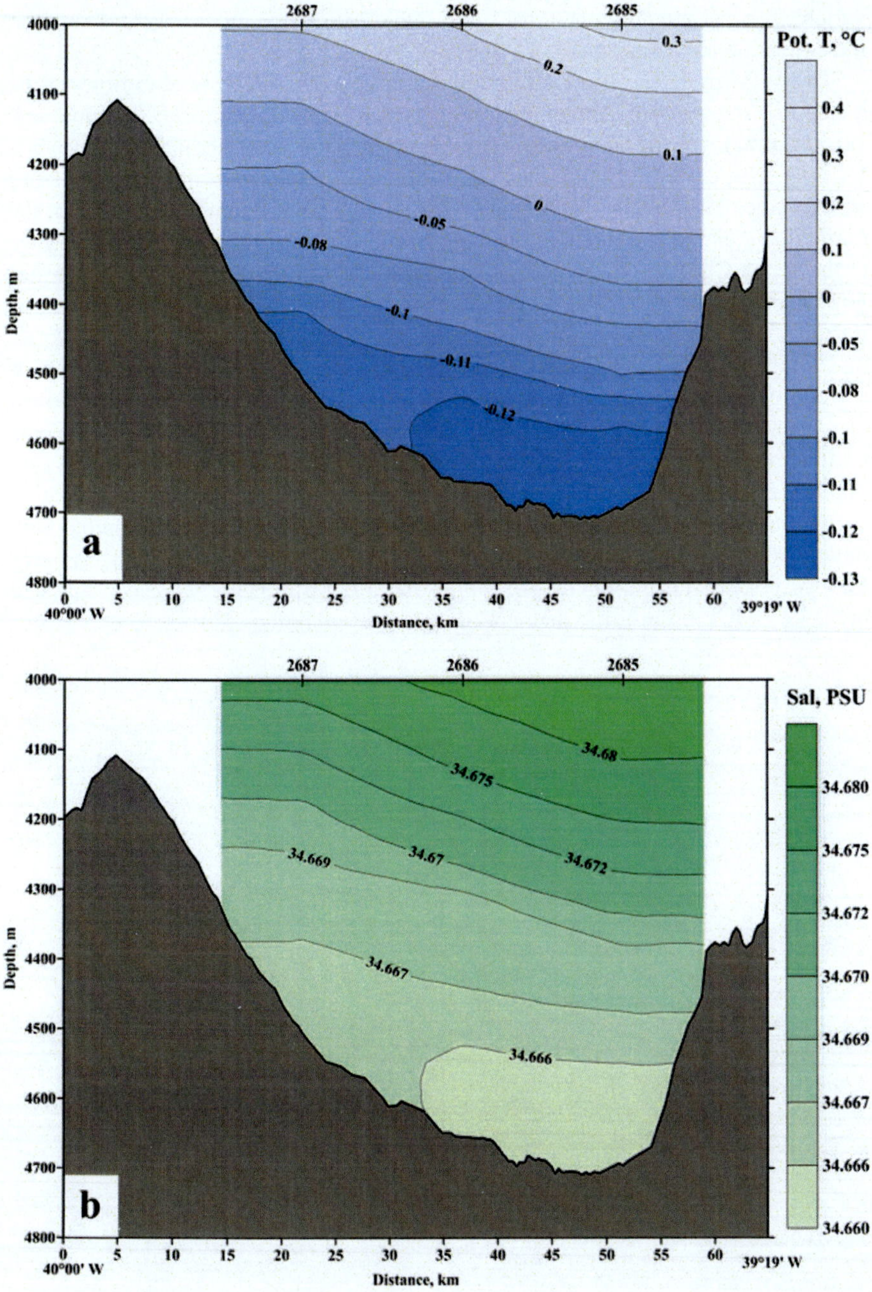

Fig. 4.29 Distribution of potential temperature (°C) (**a**), salinity (PSU) (**b**) over the section across the Vema Channel along 32°17′ S in April 2017. Isopycnals referenced to 4000 m (σ_4 in kg/m^3) are shown

variable vertical and horizontal structures of the abyssal flow in this passage between the Argentine and Brazil basins.

Twelve years later, a new effort was started by IFM-GEOMAR to observe the Vema Sill overflow of Antarctic Bottom Water in a 3-D approach (longitudinal, vertical, time). Part of the rationale for the new array was to establish an ocean site for long-term current observations based on the representative numbers of earlier records since 1979. After a significant change in the abyssal temperature field of the northern Argentine Basin (Coles et al. 1996) and analysis of the flow at the entrance to the Vema Channel (Hogg et al. 1999; Zenk and Morozov 2007) the problem of how a change in mass and temperature of the flow could manifest themselves remained open.

During three visits between December 2003 and May 2007, a twin array of current meters and temperature recorders was installed and completely redeployed on opposite flanks of the Vema Sill, the narrowest section of the whole Vema Channel (Zenk et al. 1993). The mooring sites were zonally separated by 14 km. Figure 4.30 composed by W. Zenk displays the potential temperature stratification during the cruise of the R/V *Polarstern* in May 2005. Positions of the initial twin moorings are overlaid. Bottom contours are inferred from nocturnal Hydrosweep surveys on the site during mooring preparations. Instruments were concentrated in the deepest 1500-m-thick segment of the water column.

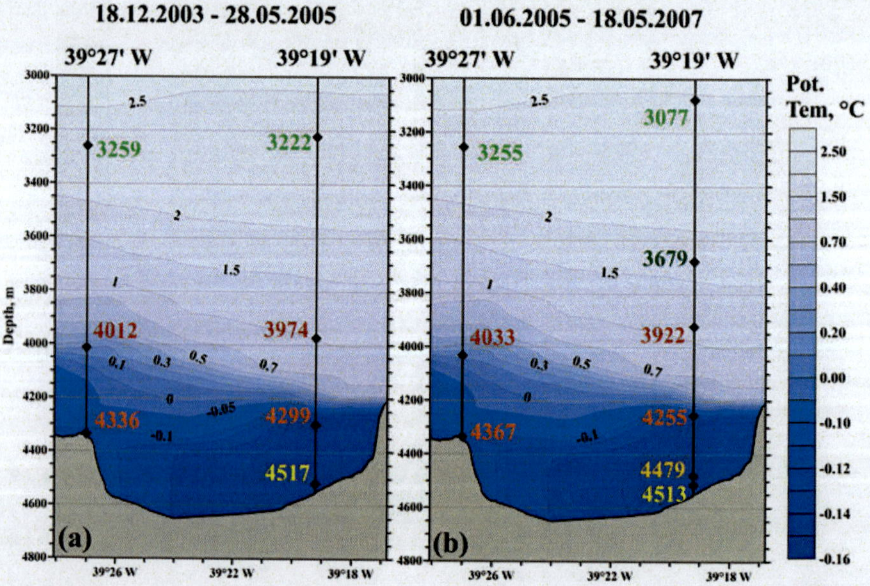

Fig. 4.30 Zonal deep distribution of potential temperature (°C) across the Vema Sill. The data were collected in May 2005 from the R/V *Polarstern*. Station locations are depicted at the top axis. Positions of the moored self-recording current meters are indicated by diamonds on both sides of the channel

After the turn-around of equipment, seven current records distributed on the western (3 records) and eastern (4 records) moorings were collected. Time series on the west side were continued as planned. For a better resolution in the vertical, the east side array was even enhanced by two additional instruments. During the second deployment, at least 16 simultaneous records were collected, most of them covering by then in total a period of 3.5 years. Table 4.4 shows the data inventory.

A qualitative analysis of velocity time series shows that below 4000 m the topographic influence starts to dominate the flow direction. The structure of the

Table 4.4 Mooring inventory Vema Sill 1991–2020 IFM-GEOMAR Kiel and Sao Paulo University, Brazil

Latitude	Longitude	Date	Date	
South	West	Start	End	Instrument depths, m
31°08.2′ S	39°26.0′ W	Jan 12, 1991	Dec 5, 1992	2900
				3850
				4150
				4425
				4625
31°14.3′ S	39°20.0′ W	Apr 21, 1998	Mar 6, 2000	4130
				4513
				4530
31°15.3′ S	39°19.0′ W	Dec 18, 2003	May 28, 2005	3222
				3974
				4299
				4517
31°15.8′ S	39°26.9′ W	Dec 18, 2003	May 28, 2005	3259
				4012
				4336
31°15.6′ S	39°27.2′ W	May 30, 2005	May 18, 2007	3255
				4033
				4367
31°15.3′ S	39°19.0′ W	May 31, 2005	May 18, 2007	3077
				3679
				3922
				4255
				4479
				4513
31°15.6′ S	39°27.0′ W	Dec. 18, 2003	May. 27, 2005	3259
				4012
				4336
31°15.6′ S	39°19.2′ W	May 30, 2005	May 18, 2007	3255
				4033
				4367
31°14.0′ S	39°23.0′ W	Feb 1, 2019	Aug 28, 2020	4450

along-channel (meridional) current component shows a zero line at ~3550 m separating the southward flow of North Atlantic Deep Water from the equatorward flow of Antarctic Bottom Water. While searching for a reference level of no motion for geostrophic analyses, we find the 3600-m level to be a good choice for the whole Vema Sill based on the 3.5 years average of direct current observations.

Continuing the program of moorings in the Vema Channel, scientists from the San Paolo University (Brazil) deployed a mooring at 31°14′ S, 39°23′ W from February 1, 2019 to August 28, 2020 with velocity and temperature meter near the bottom.

4.2.10 *Measurements with the Lowered ADCP*

In 2005, 2006, 2009, 2017, 2018, and 2020, the CTD measurements in the Vema Channel were accompanied with velocity measurements using a Lowered Acoustic Doppler Current Profiler (LADCP). These measurements allowed us to distinguish the spatial structure of the flow in the channel. The structure of the current field changes in time. The minimum measured (LADCP) velocities were 29 cm s^{-1} in 2005, while in 2009 they increased to 35 cm s^{-1}. The maximum velocities equal to 55 cm s^{-1} were measured in 2017. In 2018, the maximum velocities did not exceed 30 cm s^{-1}. Figure 4.31a shows the velocity section as measured by the LADCP instruments in October 2005. The distribution of velocities was almost uniform with a slight increase at the eastern (up to 29 cm s^{-1}) and western (up to 26 cm s^{-1}) slopes of the channel. In 2006, the maximum velocities were observed in the upper western part of the channel (Fig. 4.31b). In 2009, two regions of maximum velocities were detected at the eastern and western slopes of the channel. In 2009, a strong current (up to 45 cm s^{-1}) was recorded over the upper part of the western slope, while a southerly countercurrent was found over the eastern slope of the channel (Fig. 4.31c). Measurements with lowered ADCP were continued in 2017, 2018, and 2020, which generally revealed two jets of the current: one in the middle of the channel and the other above its western wall.

In 2006, the northerly flow was intensified in the central part of the channel. Normal velocities to the section across the channel (northern component) reached 40 cm s^{-1}, which is almost 1.5 times greater than those measured in October 2005. During measurements in 2006, the current at the western slope appeared stronger than at the eastern slope. This effect correlates with the moored velocity measurements.

Transport of Antarctic Bottom Water in the Vema Channel was estimated from the LADCP data. Table 4.5 presents the northward transport in each year of measurements based on the measurements in the layer bounded by the $\theta = 0.2°$ and $\theta = 2.0$ °C isotherms of potential temperature. The transports were calculated for two versions of lateral boundaries. One version is limited by the stations of LADCP profiling. The second version is limited by the walls of the channel. The latter version needed extrapolation in some cases, but it allows one to compare the

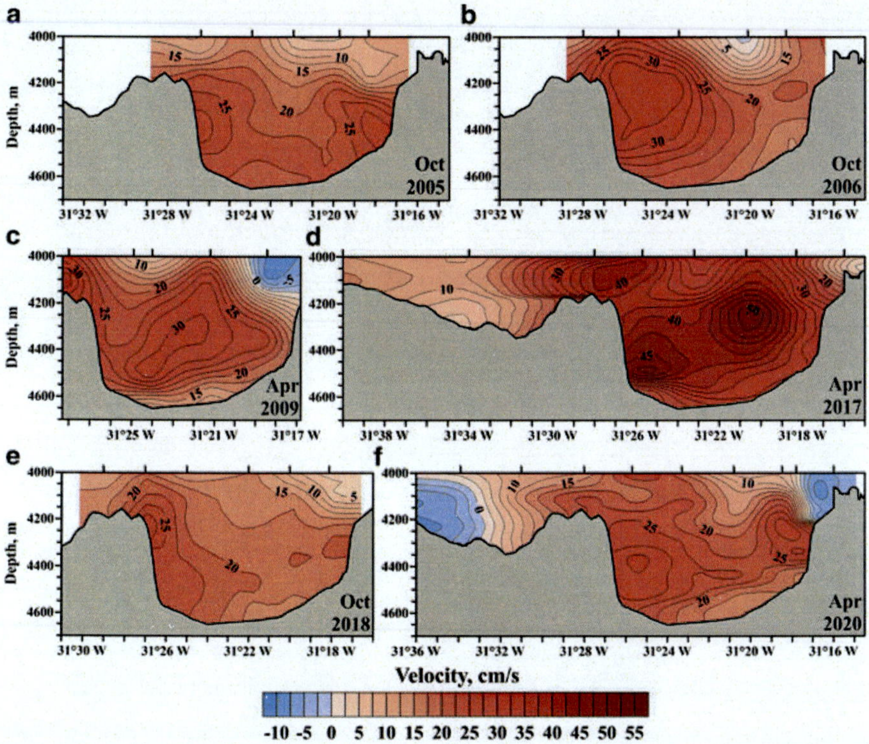

Fig. 4.31 Distribution of velocities (cm s^{-1}) over the standard section across the Vema Channel at 31°12′ S based on the LADCP data in October 2005 (**a**); in October 2006 (**b**); in April 2009 (**c**); in April 2017 (**d**), October 2018 (**e**), and April 2020 (**f**)

transports in different years. The former version (between the stations of LADCP profiling depends on the span of the stations. In 2017, the section was very wide; thus, it resulted in high transport of AABW in addition to the high transport compared to other years of measurements also seen in the estimates between the walls of the channel.

A wide range of Antarctic Bottom Water transport in the Vema Channel is found in the literature. Sometimes, the publications include the region between the channel and Rio Grande Rise, thus including the southerly transport. According to (McDonagh et al. 2002), this transport is estimated at 1.4 ± 1.0 Sv. All estimates range between 2 and 4.0 Sv (Hogg et al. 1999; McDonagh et al. 2002; Wienders et al. 2000; Zenk and Hogg 1996; Speer and Zenk 1993). Our estimates based on LADCP data with limits located at the stations of profiling show a range of transport below 2 °C potential temperature within 1.6–2.7 Sv, which is closer to the lower boundary of this interval and close to the range in (Hogg et al. 1999; McDonagh et al. 2002). If we extrapolate the transport to the limits of the channel walls, the range increases to 2.4–3.7 Sv.

The data of LADCP measurements reveal a strong southerly countercurrent in the depth interval 4000–4100 m. This countercurrent was recorded twice in 2009 and

Table 4.5 Transport of Antarctic Bottom Water (Sv) in the Vema Channel at the standard section measured by LADCP in different years

Years of measurements	Limits of LADCP profiles		Western limit 39°28' W; Eastern limit 39°16' W		
	Water transport below	Water transport below	Integrated transport (northward + southward)	Pure northward transport	Integrated transport (northward + southward)
	θ < 0.2 °C	θ < 2.0 °C	θ < 0.2 °C	θ < 2.0 °C	θ < 2.0 °C
2005	1.31 ± 0.32	1.94 ± 1.04	1.66 ± 0.33	2.81 ± 1.11	2.77 ± 1.13
2006	2.02 ± 0.55	2.59 ± 2.33	2.04 ± 0.55	3.13 ± 1.72	2.93 ± 2.44
2009	2.02 ± 0.30	1.79 ± 1.18	2.06 ± 0.30	2.51 ± 0.55	1.82 ± 1.22
2017	3.55 ± 0.15	7.10 ± 0.89	2.64 ± 0.11	5.47 ± 0.59	5.37 ± 0.63
2018	1.86 ± 0.12	1.87 ± 0.31	1.90 ± 0.12	2.17 ± 0.20	1.79 ± 0.32
2020	2.02 ± 0.13	3.17 ± 0.49	1.69 ± 0.12	2.52 ± 0.43	2.36 ± 0.45

Note: Section in 2017 was extended to the west; hence transports between LADCP casts were increased due to the longer section

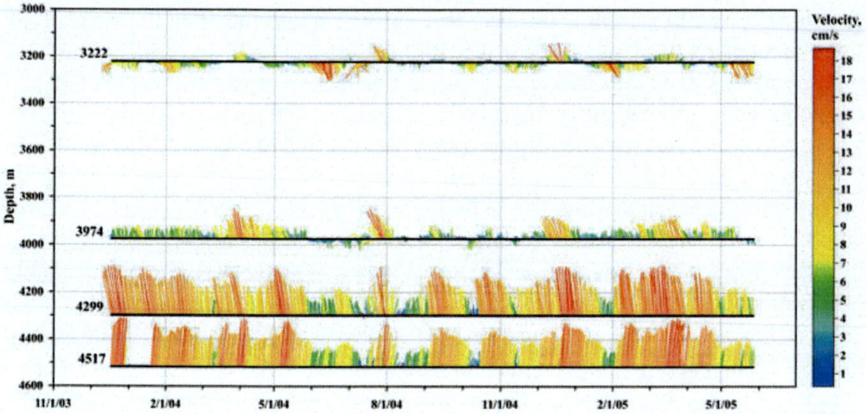

Fig. 4.32 Meridional component of current (cm s^{-1}) at the mooring in the Vema Channel operating in 2004–2005. Measurements at levels of 3222, 3074, 4299, and 4517 m are shown. Maximum velocities were recorded at 4299 m

2020. The velocities of this southward current reach 20 cm s^{-1}. A similar countercurrent was recorded by the moored instrument at 3974 m (Fig. 4.32). The mooring records indicate that this countercurrent appears when the velocities of the bottom flow in the channel are not very high. This southerly current can be generated by the inflow of water through the Vema Channel and elevation of water due to this inflow. The countercurrent can feed the circulation over the part of the Santos Plateau adjacent to the Vema Channel. In 2020, the countercurrent was recorded again.

Current measurements using an LADCP instrument in the northern part of the Vema Channel revealed low velocities close to 10 cm s^{-1}. This fact can be interpreted as splitting of the flow to the branch that flows down to the abyssal depths in the Brazil Basin and retroflection of the flow in the layer above the channel, which was recorded at the standard section.

4.2.11 Extreme Transport Velocities of Antarctic Bottom Water in the Deep-Water Vema Channel

Here we estimate the maximum and minimum velocities and transports in the bottom layer of Antarctic Bottom Water based on our data and other available measurements in the Vema Channel. We compared the data of the Shirshov Institute of Oceanology with the results of German scientists from the GEOMAR Institute, which in 1991–1992, 2003–2005, and in 2005–2007 deployed long-term moorings with current meters in the Vema Channel at latitudes of 31°08′ S and 31°15′ S (Hogg et al. 1999; Zenk and Visbeck 2013). Their current meters were located in the bottom layer in the channel at a distance of 15–50 m from the bottom. Usually, the mean velocities here were in the range of 20–35 cm s^{-1}. At the same time, the maximum speeds of the northward flow reached 56.1 cm s^{-1}. It was found on the basis of the large amount of the German data that in some cases the instruments recorded a minimum velocity in the AABW layer, up to a change in the sign of its direction and the occurrence of transport to the south at a speed of 2.8 cm s^{-1}. We note that the southward speed was recorded within a very short period of time only of a few hours (Morozov et al. 2010).

The traditional standard section in the Vema Channel, which data can be used for comparing the measurements of the hydrophysical characteristics of AABW in different years, was occupied at 31°12′-31°14′ S. On the basis of all available observational results, it became clear that the thermodynamic structure of the AABW flux in the channel is variable (Morozov et al. 2010). We found a trend towards an increase in the potential temperature in the AABW core: in the last two decades, it was as high as a few hundredths of a degree (Zenk and Morozov 2007). There are significant fluctuations in the flow velocity and in the meridional transport of AABW to the north. The most comprehensive review of all long-term mooring measurements in the Vema Channel is given in (Zenk and Visbeck 2013).

In April 2017, during cruise 44 of the R/V *Akademik Sergey Vavilov*, we carried out repeated measurements over the standard section across the Vema Channel at a latitude of 31°12′ S. The section, consisting of eight stations, was extended to the west of the Vema Channel in order to obtain a more detailed pattern of the distribution of current velocity over the entire cross-section of the channel and further to the Santos Plateau. Figure 4.33a shows the distribution of velocity in the Vema Channel based on our measurements in 2017 using the LADCP profiler. As before, the core of cold water with low salinity was displaced to the eastern wall of the channel.

The main characteristic feature of the distribution of the northward meridional component of the flow velocity in the section is that the main flow of the bottom water and maximum current velocities were confined to the western and central parts of the channel, which is a usual feature of the current here (Morozov et al. 2010; Jungclaus and Vanicek 1999). The magnitude of the velocity component normal to the section shown in Fig. 4.33 was close to the absolute value of its vector and reached 55 cm s^{-1}, which is more than 1.5 times greater than the average values measured in different years (30–35 cm s^{-1}). Thus, in 2017 we were able to record the absolute positive extremum of the AABW velocity in the Vema Channel over the entire period of our observations in the region.

At the same time, the observations showed that over the eastern slope of the channel the flow velocity was on average almost by 10 cm s^{-1} lower than that over the western slope. In accordance with the characteristics of the velocity field, an estimate was made of the northward volume transport in the bottom water of the channel through its section.

First, we calculated the water transport in the layer below the potential temperature isotherm $\theta = 2\,°C$. The northward transport of the flow associated with the total mass of AABW between the walls of the channel in 2017 was as high as 5.47 Sv (1 Sv = 10^6 m^3 s^{-1}) (Table 4.5). This was the maximum transport ever measured over the standard section using the LADCP profiling. Then, in order to compare the

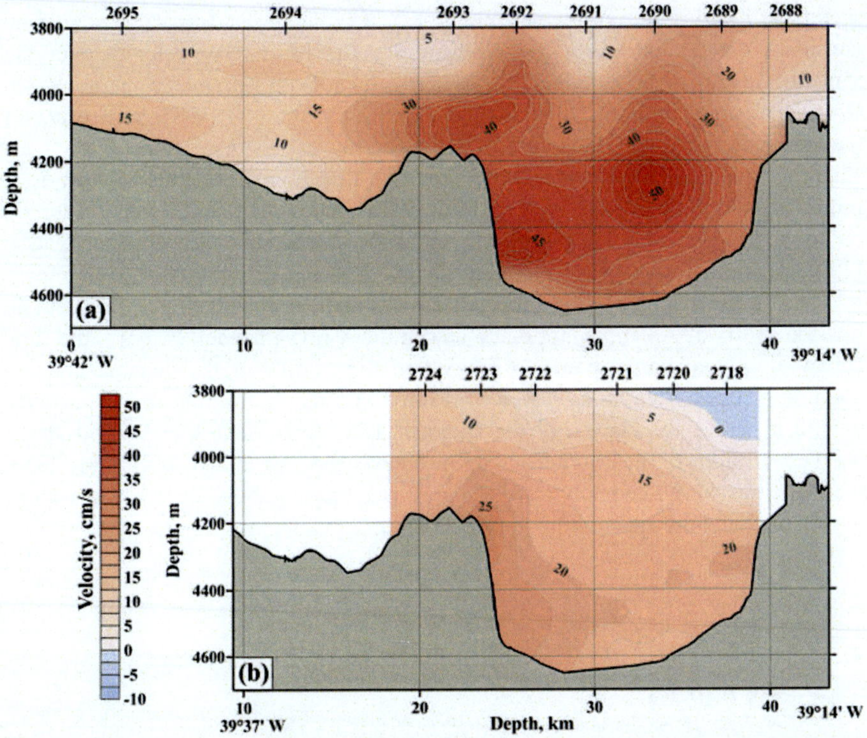

Fig. 4.33 Distributions of the meridional velocity component over the section at 31°12′ S in April 2017 (**a**) and in October 2018 (**b**). Numbers of stations where measurements were made are indicated at the top of the section. (Reproduced from Morozov et al. (2019) with permission from Pleiades Publishers)

order of magnitude, the calculation of transport was performed in a layer with a temperature less than $\theta = 0.2$ °C, between the vertical liquid boundaries at 39°41′ W and 39°14′ W. The transport of the coldest part of AABW calculated within such a northward flow between the limits of the channel walls was 2.6 Sv.

In October 2018, during cruise 46 of the R/V *Akademik Sergey Vavilov,* we occupied six stations on a standard section across the Vema Channel at 31°12′ S. This time, the section from east to west was made only up to a longitude of 39°29′ W, since in the previous year only very weak currents were observed west of this meridian. Although the velocities at the section in 2018 also turned out to be low, they exceeded 10 cm s^{-1} everywhere in the bottom 100-meter layer. The core of cold water of lower salinity was displaced to the eastern wall of the channel as was recorded in the previous observations. A section of the meridional component of velocity, constructed from the observational data of 2018, is shown in Fig. 4.33b.

The figure shows that in 2018, the maximum velocities of the northward flow of bottom water in the Vema Channel were recorded near the western wall of the channel. However, the maximum value of the velocity component normal to the section in its core did not exceed 30 cm s^{-1}. It is likely that this value should be considered as an extremely low maximum value of AABW velocity over the standard section ever recorded by our instrumental measurements in the Vema Channel. Correspondingly, the calculated mass transport of the bottom water in the layer below the 2 °C isotherm turned out to be 2.17 Sv, i.e., almost three times smaller compared to the value of a similar estimate in the previous year. It is important that the transport of water in the layer below the 0.2 °C isotherm turned out to be 2.64 Sv in 2017 and 1.90 in 2018, within the accuracy limits of calculation (Table 4.5).

Thus, in April 2017 and in October 2018, we found extremely high and extremely low transports of Antarctic Bottom Water in the Vema Channel. In 2017, the flow consisted of two cores: one was located in the middle of the channel and the maximum speeds (55 cm s^{-1}) were measured at a depth of 4300 m, at a distance of 300 m above the bottom of the channel. The second core was located above the western slope of the channel and its speed exceeded 40 cm s^{-1}. In October 2018, the velocities were extremely low. The core was located above the western wall of the channel and the speed did not exceed 30 cm s^{-1}.

4.2.12 *Flow of Antarctic Bottom Water from the Vema Channel*

In this section we will consider bottom flows in the region north of the Vema Channel. A map of the Vema Channel region is shown in Fig. 4.34.

Repeated measurements of temperature, salinity, and velocity profiles at hydrographic stations confirm the prominent role of the Vema Channel for the transport of Antarctic Bottom Water with respect to the Hunter Channel in the east and the Santos Plateau in the west. The water transport in the Vema Channel based on moored current-meter observations and LADCP sections across the Vema Channel at 31°12′ S ranges between 2.17 and 5.47 Sv (Morozov et al. 2010, 2018) (Table 4.5).

For more than 20 years, repeated studies of the bottom water flow though the Vema Channel have been carried out on the standard section across the channel at 31°12′ S (Morozov et al. 2008, 2010). This section is located approximately in the middle of the channel. Antarctic Bottom Water flows out from the Vema Channel to the Brazil Basin in the region of Vema Extension at 26°40′ S. Comparison of our measurements over the standard section with the section in the northern part of the Vema Channel shows that not all the water that flows through the standard section at 31° S reaches the northern part of the channel.

Two sections of velocities through the Vema Channel at 31°12′ S and 26°40′ S are shown in Fig. 4.35.

Let us compare the water transport across the standard section in the Vema Channel (31°12′ S) and across the section in the northern part of the channel. In

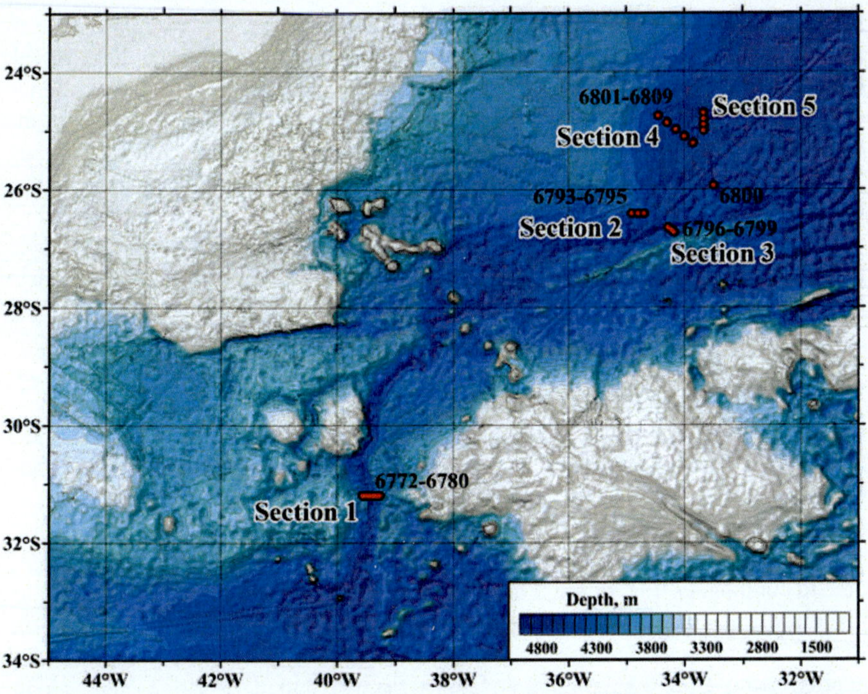

Fig. 4.34 Bottom topography in the region of the Rio Grande Plateau (the Vema Channel and its continuation). The numbers of stations and CTD/LADCP sections in 2020 are indicated. (Reproduced from Morozov et al. 2020)

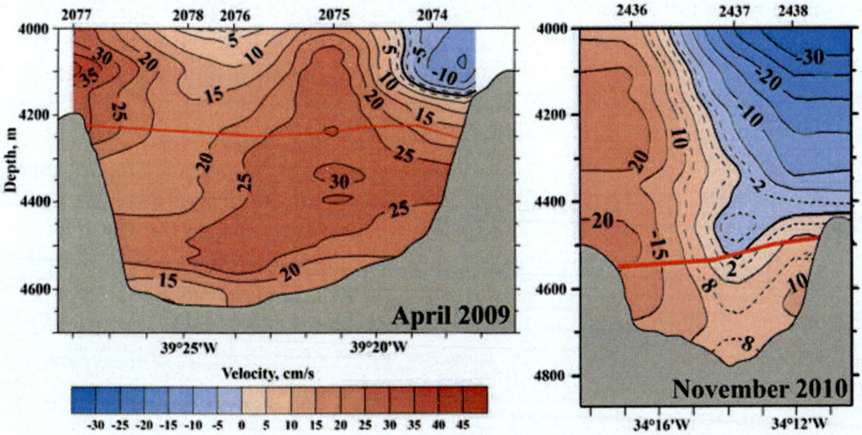

Fig. 4.35 Sections of velocities across the Vema Channel. Positive velocities (red tones) point north, blue tones are directed to the south. Left panel shows currents over the standard section at 31°12′ S in 2009, right panel shows currents over the section in the northern part of the channel at 26°40′ S in 2010. The red line shows the 0 °C isotherm of potential temperature

the northern part, the channel becomes deeper and narrower. However, isotherms of lower potential temperatures than 0 °C are located between the walls of the channel. On both sections we see two cold jets: one over the deep bed of the channel and the second above the western wall. Based on the available data, we can only compare the flows of cold water with a temperature of $\theta < 0$ °C. The mean velocities of the water with such a temperature on the standard section at 31°12′ S are 23 cm s^{-1}, while on the northern section they decrease to 11 cm s^{-1}. A jet of Antarctic bottom water with a higher temperature flows above the western slope of the channel at speeds of more than 20 cm s^{-1}, however, in this expedition, we were unable to extend the section to the west for a more accurate calculation of water transport there.

The section of the channel on the standard section for water with a temperature colder than 0 °C has an area of $6 \cdot 10^6$ m^2, and on the northern section the same area is more than four times smaller ($1.4 \cdot 10^6$ m^2). A large amount of the coldest water does not reach the northern section, remaining behind orographic obstacles along the channel and mixing with the overlying layers. The flow of water with a temperature of less than 0 °C through the standard section is 1.4 Sv, and through the northern section is 0.16 Sv, which is almost 10 times smaller.

The bottom friction prevents free flow of water, and the bottom water, not being able to completely flow through the channel, finds other pathways. The conclusion that not all the current flowing in the middle of the Vema Channel flows into the Brazil Basin was reported even earlier (McDonagh et al. 2002). They suggested that approximately at 27°30′ S the stream turns to the west and makes a large anticyclonic loop, which transports about 2 Sv of the upper warmer part of the flow.

Intense research in the region was performed in 2020. Analysis of measurements in the region north of the Vema Channel (Tarakanov and Morozov 2015; Morozov et al. 2019, 2020; Tarakanov et al. 2020) revealed that the flow of Antarctic Bottom Water forms at least two cold cores (Fig. 4.36). The core in the deep channel that has been studied by many authors is a discharge of cold, dense water accumulated in the abyssal depths of the Argentine Basin. This is a hydraulically controlled flow driven by the pressure gradient along the Vema Channel. The second core, which has been repeatedly recorded at a distance of 300–400 m above the gentle western slope of the Vema Channel, is a continuation of the quasi-geostrophic Deep Western Boundary Current transporting Antarctic Bottom Water along the continental slope of South America in the Argentine and Brazil basins (Morozov et al. 2019). The research made in (Tarakanov et al., 2020) is related to the study of hydrodynamic properties of the bottom flow and underwater spillways. Another study is related to numerical simulations of the flow in the Vema Channel including the region north of the channel (Frey et al., 2019a, 2019b); however, direct measurements of the velocities and thermohaline structure of these bottom flows remain rare.

Locations of our and historical stations in the region are shown in Fig. 4.37. Potential temperatures indicated in the figure give an idea about the pathways of the cold jets of Antarctic Bottom Water north of the Vema Channel.

In this section, we study the pathways of propagation of the coldest and densest Antarctic Bottom Water ($\theta < 0.0$ °C) into the Brazil Basin through the Vema Channel based on the data of our measurements in 2003, 2009, 2010, 2012, 2013,

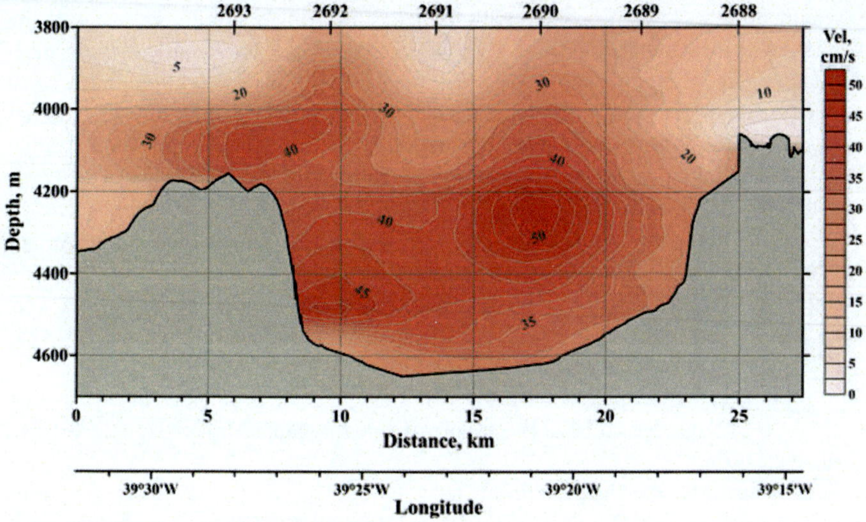

Fig. 4.36 Section of the meridional velocity component along 31°12′ S in April 2017. The numbers of stations are indicated at the top of the section

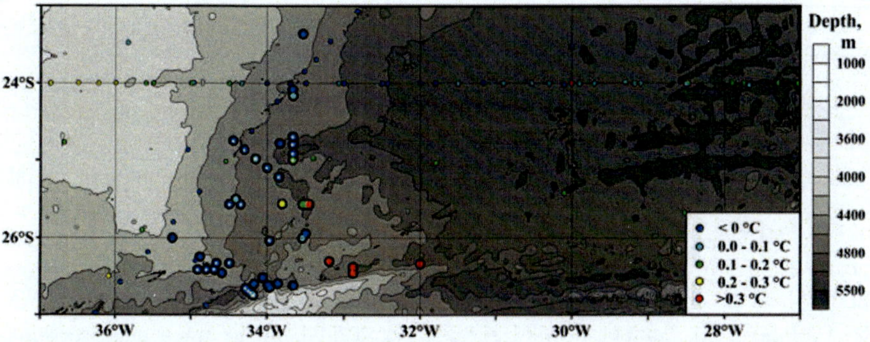

Fig. 4.37 Bathymetry of the study region north of the Vema Channel based on (GEBCO 2019). Bathymetry was corrected based on our echo-sounder measurements in 2009–2020. Locations of our CTD/LADCP casts in 2003–2020 (large circles) and historical stations (small circles) are shown. The corresponding potential temperatures at the bottom (5–20 m above the bottom) are shown in color. Stations in December 2019 and March–April 2020 are indicated with a white dot in the middle. Reproduced from (Morozov et al. 2020)

2018, 2019, and 2020. The main goal is to locate these pathways geographically. The analysis is focused on the outflow of AABW from the Vema Channel. The research is based on our data of measurements in the region as well as on the data of zonal section A09 occupied in 2009 and 2018 along 24° S (http://cchdo.ucsd.edu), the data of the WOCE A17 section in 1994 (WHP 2002), and A10 section in 2011 along 30° S. We also used the data of several stations from the WOD18 database. In 2009, 2010, 2012, 2013, 2018, 2019, and 2020, the CTD-casts from the Russian research ships were accompanied with the measurements of current velocities using the LADCP instrument (LADCP, RDI WHS 300 kHz). The details of these measurements are given in Table 4.6. We traced the flow by finding the coldest temperatures at the bottom and by measuring velocities in the cold-water flows using LADCP. We also took into consideration that the bottom flow should either follow the isobaths or descend. Locations of the stations are shown in Fig. 4.37.

The digital 15″ array of the predicted bottom relief (GEBCO 2019) was used as the basis of this work. We corrected the depths over the route of the vessels in our expeditions using ELAC LAZ 4700 single beam echo sounder in 2003–2013 and Kongsberg EA600 in 2018, 2019, and 2020. The echo-sounder measurements were corrected for the sound speed deviations from 1500 m s^{-1} based on the CTD-data.

We made our CTD/LADCP-casts as deep as 5 m above the bottom. The barotropic tide was removed from the resulting velocity profiles using the NASA 0.25° × 0.25° database (TPXO9 model) of Oregon State University (Egbert and Erofeeva 2002).

Let us consider the deeper current near the outflow of AABW from the Vema Channel based on our CTD/LADCP casts in 2010–2020. The locations of the casts are in the region 26°30–26°50 S, 33°30–35°00 W (Fig. 4.37). The CTD/LADCP sections across the Vema Channel and especially in its northern part occupied in different years reveal the existence of two types of currents of the coldest AABW ($\theta < 0.2$ °C), which are slightly different by their thermohaline properties. The deeper flow that has been studied in many publications is confined to the deep channel bed of the Vema Channel. It is seen in Fig. 4.36 that the deepest core of the current in the middle of the Vema Channel at 31°12′ S is located at depths of 4200–4500 m. In the region of AABW outflow from the channel, the ocean bottom is deeper, and the core occupies the depths of 4600–4800 m. The minimum potential temperature of this type of AABW in the northern region of the channel (26°43.2′ S, 34°12.1′ W) was recorded in 2003 ($\theta = -0.094$ °C). It is noteworthy that there is a positive trend in the bottom potential temperature with time (Zenk and Morozov 2007). The potential temperature at the bottom at the same point in 2020 was $\theta = -0.067$ °C. Another flow of AABW was found on the transversal sections across the Vema Channel. The core of this flow was detected over the northwestern slope of the channel. The core of this current with the minimum potential temperature equal to $\theta = -0.084$ °C is located at a depth of approximately 4100–4200 m (150 m above the bottom) and approximately 600 m shallower than the core of AABW that flows at the bottom of the channel (Fig. 4.36).

Both cores are seen in Fig. 4.38 over sections #2 and #3 (shown on the map in Fig. 4.34) (26°30′- 27°00′ S) in 2020. The western section #2 does not fully show the upper jet because we had to stop the casts in stormy weather. Section #2 was made along 26°25.0′ S on April 3, 2020. The maximum velocity was recorded at the eastern station 6795 at 25°25.0′ S, 34°41.9′ W (12 cm s^{-1}). On December 31, 2019 we made two stations along 26°20.0′ S at 34°30.0′ W and 34°40.0′ W. The velocity in the bottom layer at the western station was directed to the northeast with a maximum of 9 cm s^{-1} at 130 m above the bottom. Velocity at the bottom at the eastern station was lower, 6–7 cm s^{-1}. Thus, the maximum velocities at 25°20′ S are

Table 4.6 Russian CTD/LADCP stations in the southwestern part of the Brazil Basin in 2013–2020

Stations	Date	Coordinates of the instrument at the bottom	Lower Measurement/Ocean Depth
		R/V *Akademik Ioffe*	
1441	02.11.2003	26°00.0′ S, 35°15.0′ W	4229/4237
1442	02.11.2003	26°14.9′ S, 34°53.0′ W	4361/4370
1443	02.11.2003	26°27.5′ S, 34°35.8′ W	4435/4442
1444	03.11.2003	26°40.9′ S, 34°15.4′ W	4723/4732
1445	03.11.2003	26°43.2′ S, 34°12.1′ W	4698/4715
1446	03.11.2003	26°44.4′ S, 34°10.8′ W	4453/4479
1454	09.11.2003	27°05.5′ S, 35°54.7′ W	4906/4920
2079	18.04.2009	26°42.9′ S, 34°12.2′ W	4699/4710
2080	18.04.2009	26°01.9′ S, 33°58.0′ W	4596/4609
2437	06.11.2010	26°39.9′ S, 34°16.8′ W	4687/4696
2438	06.11.2010	26°42.2′ S, 34°13.8′ W	4774/4783
2439	06.11.2010	26°43.4′ S, 34°12.1′ W	4667/4676
		R/V *Akademik Sergey Vavilov*	
2494	05.11.2012	26°36.8′ S, 33°59.3′ W	4729/4737
2495	05.11.2012	26°35.9′ S, 34°10.2′ W	4652/4659
2496	06.11.2012	26°31.1′ S, 34°03.3′ W	4724/4731
2497	06.11.2012	26°35.8′ S, 33°51.7′ W	4698/4706
2521	18.10.2013	26°20.5′ S, 32°00.1′ W	4859/4864
2522	18.10.2013	26°27.7′ S, 32°53.1′ W	4756/4762
2523	18.10.2013	26°23.0′ S, 32°53.1′ W	4768/4773
2525	19.10.2013	26°18.6′ S, 33°11.5′ W	4809/4816
2526	20.10.2013	26°35.7′ S, 33°51.6′ W	4696/4704
2527	20.10.2013	26°37.3′ S, 33°39.6′ W	4554/4559
2532	06.04.2014	31°37.2′ S, 43°02.6′ W	4002/4008
2534	07.04.2014	31°25.9′ S, 42°26.5′ W	3968/3974
2535	08.04.2014	28°55.3′ S, 38°00.5′ W	4419/4426
2559	18.10.2014	27°05.5′ S, 39°36.0′ W	4196/4201
2560	19.10.2014	28°39.8′ S, 41°54.0′ W	4199/4204
2561	19.10.2014	28°39.2′ S, 41°54.9′ W	4192/4197
2562	19.10.2014	29°06.0′ S, 41°48.2′ W	4150/4155
2563	20.10.2014	29°51.0′ S, 41°17.2′ W	3946/3949
2564	20.10.2014	30°35.0′ S, 41°09.6′ W	3979/3984
2565	20.10.2014	30°40.0′ S, 41°07.8′ W	3982/3987
2566	21.10.2014	31°21.3′ S, 43°07.7′ W	3926/3931
2567	21.10.2014	31°28.6′ S, 43°25.6′ W	3926/3931
2579	*05.04.2016*	*31°20.0′ S, 43°22.0′ W*	*3853/3861*
2580	*06.04.2016*	*31°00.0′ S, 41°08.2′ W*	*4080/4089*
2581	*06.04.2016*	*30°57.0′ S, 41°07.2′ W*	*40,854,094*
2582	*06.04.2016*	*30°54.0′ S, 41°06.9′ W*	*4025/4037*
2583	*07.04.2016*	*30°51.0′ S, 41°06.8′ W*	*4025/4037*

(continued)

Table 4.6 (continued)

Stations	Date	Coordinates of the instrument at the bottom	Lower Measurement/Ocean Depth
2584	*07.04.2016*	*30°20.6′ S, 40°57.1′ W*	*3992/4001*
2585	*07.04.2016*	*30°24.4′ S, 41°20.1′ W*	*3881/3890*
2586	*08.04.2016*	*30°17.0′ S, 41°19.3′ W*	*4021/4030*
2587	*08.04.2016*	*28°41.0′ S, 41°52.2′ W*	*4125/4133*
2588	*08.04.2016*	*28°40.3′ S, 41°52.2′ W*	*4200/4208*
2589	*09.04.2016*	*28°39.8′ S, 41°54.1′ W*	*4200/4206*
2590	*09.04.2016*	*28°39.2′ S, 41°55.0′ W*	*4190/4195*
2591	*09.04.2016*	*28°38.6′ S, 41°55.9′ W*	*3987/3992*
2592	*09.04.2016*	*28°37.8′ S, 41°57.2′ W*	*3919/3923*
2593	*09.04.2016*	*28°37.2′ S, 41°58.2′ W*	*3773/3781*
2676	04.11.2016	29°20.0′ S, 41°49.0′ W	4148/4153
2677	04.11.2016	29°31.5′ S, 41°54.3′ W	4016/4021
2678	05.11.2016	30°35.1′ S, 43°23.9′ W	3904/3908
2679	05.11.2016	30°35.0′ S, 43°29.9′ W	3797/3803
2680	06.11.2016	31°20.0′ S, 43°22.0′ W	3853/3857
2681	06.11.2016	31°28.6′ S, 43°25.6′ W	3926/3931
2710	22.10.2018	25°34.1′ S, 33°29.5′ W	4904/4906
2711	22.10.2018	25°34.0′ S, 33°31.1′ W	4844/4847
2712	22.10.2018	25°34.0′ S, 33°27.4′ W	4770/4772
2713	22.10.2018	25°34.0′ S, 33°32.6′ W	4807/4808
2714	23.10.2018	25°33.6′ S, 33°48.3′ W	4805/4810
2716	23.10.2018	26°00.0′ S, 33°32.6′ W	4813/4817
		R/V *Akademik Mstislav Keldysh*	
6563	30.12.2019	24°10.1′ S, 33°37.9′ W	4681/4686
6564	30.12.2019	25°34.0′ S, 34°21.2′ W	4615/4629
6565	31.12.2019	25°33.9′ S, 34°30.0′ W	4560/4564
6566	31.12.2019	26°20.0′ S, 34°29.9′ W	4521/4529
6567	31.12.2019	26°20.0′ S, 34°40.0′ W	4442/4450
6568	31.12.2019	28°07.4′ S, 37°52.3′ W	4925/4930
6569	31.12.2020	28°10.9′ S, 37°50.7′ W	4705/4710
6570	01.01.2020	28°29.8′ S, 37°35.0′ W	4483/4489
6571	02.01.2020	28°33.0′ S, 37°30.6′ W	4491/4496
6572	03.01.2020	31°11.9′ S, 39°17.9′ W	4382/4387
6573	04.01.2020	32°47.4′ S, 45°12.2′ W	4122/4127
6574	05.01.2020	33°15.6′ S, 47°00.0′ W	3693/3698
6771	27.03.2020	31°20.0′ S, 43°22.0′ W	3875/3881
6772	28.03.2020	31°12.0′ S, 39°16.0′ W	4081/4084
6773	28.03.2020	31°11.9′ S, 39°17.8′ W	4201/4208
6774	28.03.2020	31°12.0′ S, 39°18.5′ W	4453/4458
6775	28.03.2020	31°12.0′ S, 39°20.7′ W	4570/4574
6776	29.03.2020	31°12.0′ S, 39°23.3′ W	4618/4623
6777	29.03.2020	31°12.0′ S, 39°25.7′ W	4591/4597

(continued)

Table 4.6 (continued)

Stations	Date	Coordinates of the instrument at the bottom	Lower Measurement/Ocean Depth
6778	29.03.2020	31°12.0′ S, 39°28.0′ W	4161/4168
6779	29.03.2020	31°12.0′ S, 38°31.0′ W	4322/4328
6780	29.03.2020	31°12.0′ S, 39°33.9′ W	4325/4330
6781	30.03.2020	30°39.8′ S, 39°56.5′ W	4110/4114
6782	30.03.2020	30°02.4′ S, 40°25.8′ W	4165/4170
6783	30.03.2020	29°18.0′ S, 39°57.8′ W	4240/4244
6784	31.03.2020	28°45.2′ S, 38°16.1′ W	4790/4794
6785	31.03.2020	28°33.0′ S, 37°30.4′ W	4494/4500
6786	01.04.2020	28°30.0′ S, 37°35.0′ W	4485/4490
6787	01.04.2020	28°26.5′ S, 37°39.7′ W	4478/4482
6788	01.04.2020	28°14.2′ S, 37°49.1′ W	4331/4336
6789	01.04.2020	28°10.9′ S, 37°50.8′ W	4710/4714
6790	01.04.2020	28°07.4′ S, 37°52.4′ W	4920/4924
6791	01.04.2020	28°04.4′ S, 37°53.7′ W	4775/4782
6792	02.04.2020	27°58.3′ S, 37°56.4′ W	4028/4032
6793	03.04.2020	26°25.0′ S, 34°54.6′ W	4365/4371
6794	03.04.2020	26°25.0′ S, 34°48.0′ W	4401/4406
6795	03.04.2020	26°25.0′ S, 34°41.9′ W	4420/4425
6796	06.04.2020	26°39.2′ S, 34°16.8′ W	4658/4662
6797	06.04.2020	26°41.4′ S, 34°14.2′ W	4772/4779
6798	06.04.2020	26°43.1′ S, 34°12.1′ W	4711/4718
6799	06.04.2020	26°44.4′ S, 34°10.8′ W	4490/4495
6800	07.04.2020	25°56.4′ S, 33°30.0′ W	4812/4817
6801	07.04.2020	25°13.0′ S, 33°51.0′ W	4850/4854
6802	07.04.2020	24°00.0′ S, 33°40.0′ W	4750/4755
6803	07.04.2020	24°55.1′ S, 33°40.0′ W	4920/4924
6804	07.04.2020	24°48.0′ S, 33°40.0′ W	4835/4841
6805	08.04.2020	24°42.0′ S, 33°40.0′ W	4810/4815
6806	08.04.2020	25°06.0′ S, 34°00.0′ W	4762/4768
6807	08.04.2020	24°59.0′ S, 34°09.0′ W	4705/4710
6808	08.04.2020	24°52.0′ S, 34°17.9′ W	4648/4653
6809	08.04.2020	24°45.0′ S, 34°27.0′ W	4580/4585

close to 34°32′ W longitude. This jet can be traced based on the coldest water at the bottom and maximum velocities generally directed to the north and northeast on several sections in the study region (Figs. 4.38, 4.39, 4.40).

These two types of relatively isolated cores of the cold AABW current are separated by a few hundred meters of depth. They exist because the lower flow is formed mainly from the AABW that fills the vast abyssal region in the Argentine Basin, whereas the upper type is a continuation of the Deep Western Boundary Current along the continental slope of South America. The cold water of the upper

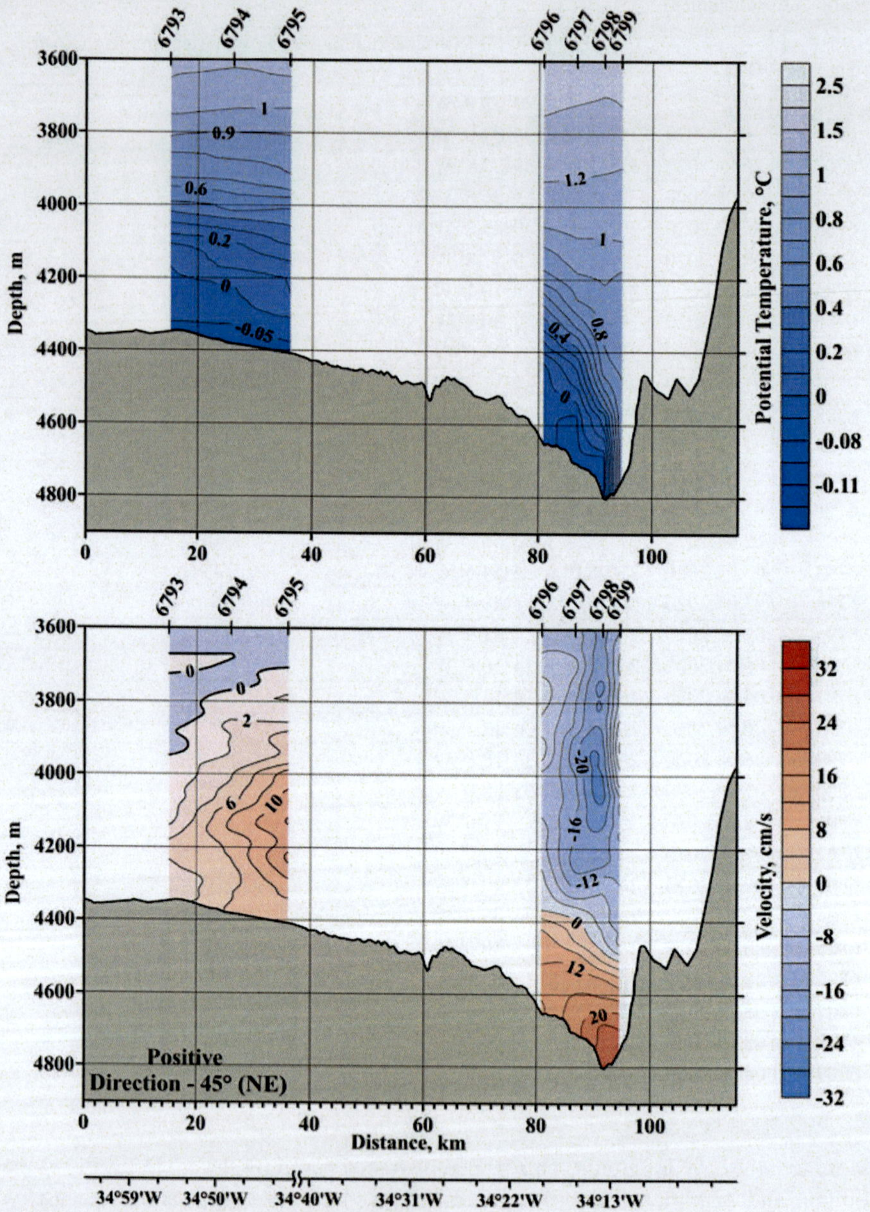

Fig. 4.38 Sections #2 (western) and #3 (eastern) (see Fig. 4.34) of potential temperature (top panel) and velocity (bottom panel) across the northern part of the Vema Channel along 26°30′ - 27°00′ S in April 2020. Numbers of stations are shown along the top axis. Reproduced from (Morozov et al. 2020)

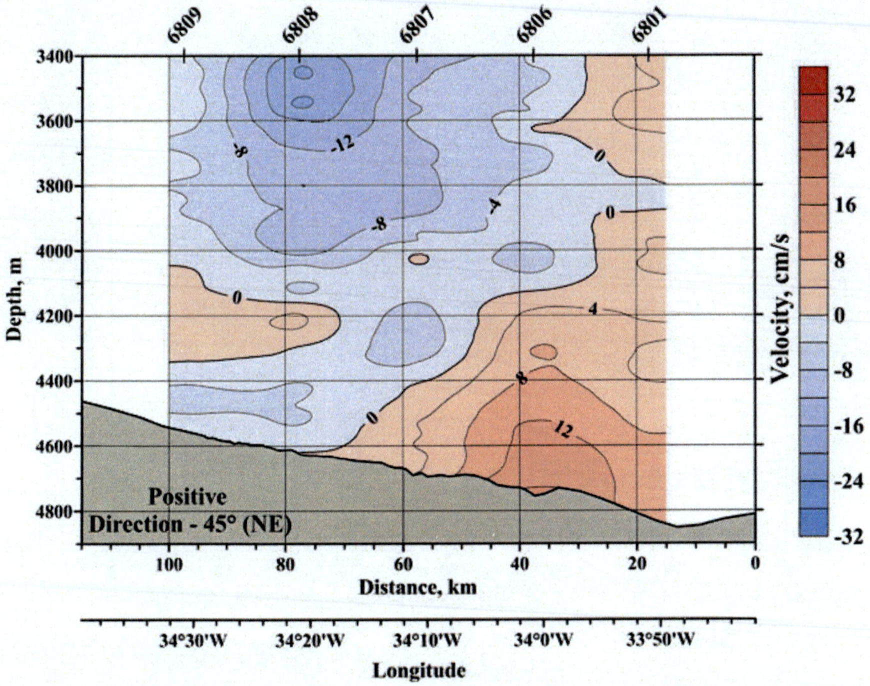

Fig. 4.39 Section #4 (see Fig. 4.34 for location) of velocity north of the Vema Channel. (Reproduced from Morozov et al. 2020)

flow propagates from the Argentine Basin over a different pathway than along the bed of the main Vema Channel. Since the potential temperature of this bottom water is very low, its propagation over the lower Santos Plateau is not possible. The characteristic values of potential temperature at the bottom over the Santos Plateau based on the WOD18 database and our measurements exceed +0.09 °C almost everywhere.

Considering the two jets of AABW from the Vema Channel, let us first analyze a region in close vicinity to the outflow of AABW from the channel, which is called the Vema Extension (Zenk and Morozov 2007). Intense measurements in this region were carried out in 2010–2013 and 2020. This region close to the exit from the Vema Channel has been investigated to study the continuation of the flow in the Vema Channel bed. Vectors of bottom currents based on the LADCP casts north of the Vema Extension are shown in Fig. 4.41. This figure shows the deepest measured vectors of velocity at all stations.

The CTD/LADCP casts in the region north of the Vema Channel showed that the continuation of lower Antarctic Bottom Water flow over the deep bed of the channel is no longer tied to a deep channel, but it spreads as a wider fan-shaped stream. The measurements showed that the continuation of the coldest AABW flow splits into two bottom jets. One of the jets of cold water ($\theta = -0.080$ °C) is directed initially to

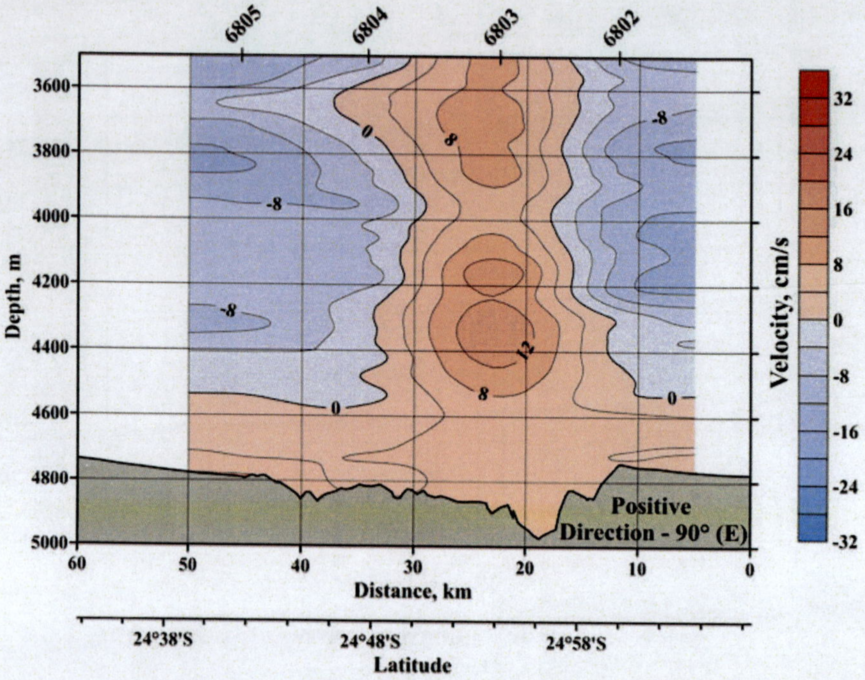

Fig. 4.40 Section #5 (see Fig. 4.34 for location) of velocity north of the Vema Channel across the canyon. (Reproduced from Morozov et al. 2020)

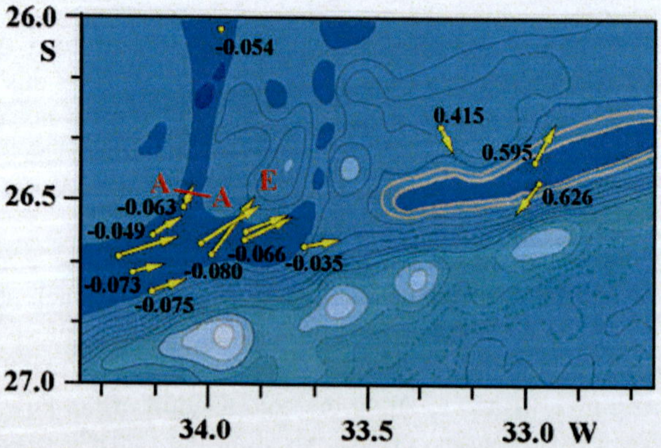

Fig. 4.41 Bathymetry of the study region based on (Smith and Sandwell 1997) and corrected with our echo-sounder measurements in 2010–2020. The stations of CTD/LADCP casts are shown with yellow dots. The yellow arrows show the velocity vectors of bottom currents. Black numbers at the dots of stations indicate potential temperature at the bottom. The A-A section shows the line of the bottom profile discussed below (Fig. 4.42). Letter "E" indicates an elevation discussed in the text. (Reproduced from Morozov et al. 2018)

the northeast. The second jet with warmer water (θ = −0.063 °C) is directed almost to the north (Morozov and Tarakanov 2014; Tarakanov and Morozov 2015). It was found on the basis of echo sounder surveys that the continuation flows of AABW in the northern part of the Vema Channel exist not as flows in clearly pronounced channels, but the pathways are located between underwater elevations and mountains so that bottom water finds its way to the north along depressions in the bottom topography. The flow is confined to the feet and slopes of higher mountains, while on the outer side of the flow there is a flatter elevation, and the channel is ~200–250 m deeper than the elevation. Usually, the channel width is ~10 km. A typical profile of the bottom topography transversal to the channel is shown in Fig. 4.42.

We found that the continuation of the bottom flow in the bed of the channel splits into two branches. One of them is the eastward continuation of the AABW flow in the Vema Channel, the second branch of this stream is directed to the north. During an echo sounder survey on the western side of the hill with the center at 26°30′ S, 33°48′ W, the vessel three times crossed a channel oriented approximately from south to north. This hill (elevation) is indicated with letter "E" in Fig. 4.41. The channel is about 4 km wide and 70 m deeper than the relatively flat plain in the west. The slopes of the walls of this small channel range within 7–11%. Measurements at stations in such small channels usually show a current with a velocity of about 5 cm s^{-1}. The thermohaline properties of water in this current are similar to the water properties in the Vema Channel. Later in 2013 and 2018, we found continuations of these jets farther to the north; both of the channels are directed to the north.

In 2012, we occupied a station north of the Vema Channel at 26°35.8′ S, 33°51.7′ W. This was a station in a small channel. The potential temperature of water at the bottom was θ = −0.087 °C; the current was directed to the east-northeast. In 2013, this station (2497) was repeated almost at the same coordinates 26°35.7′ S, 33°51.7′ W. The potential temperature at this station in 2013 was θ = −0.066 °C (unlike the

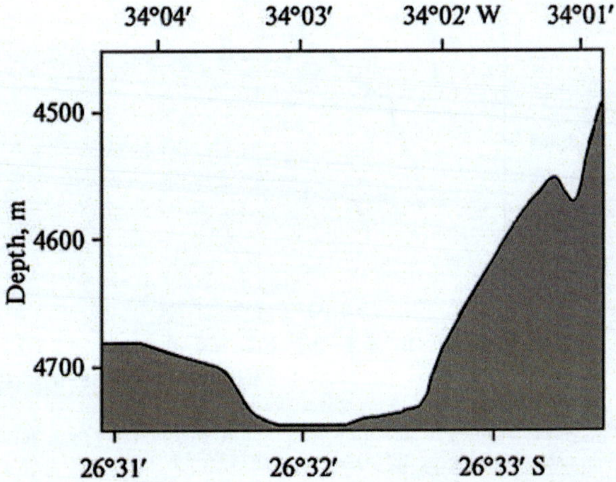

Fig. 4.42 Bottom topography profile along the line of the echo sounding across the quasi-meridional branching from the main channel bed of the Vema Channel. The line of the survey (interval A–A′) is highlighted in Fig. 4.41

colder water in 2012, θ = −0.080 °C). We think that this happened because the instrument did not get exactly into the stream of the coldest water, or maybe the water became warmer by 0.02 °C. The direction of the current did not change because both stations were made in a small channel. An echo-sounder survey around this station showed that the channel bed becomes shallower (up to 4550 m) and in the region of longitude 33°39′ W it turns from the eastern direction to the northeast because shallower topographic elevations are located further to the east and do not allow the flow of the bottom water propagate farther to the east. Station 2527 was occupied at 26°37.2′ S, 33°39.6′ W at a depth of 4554 m east of the stations (2497 and 2526) at 26°35.7′ S, 33°51.7′ W (Figs. 4.43 and 4.44). The current at this station was directed to the northeast (12 cm s^{-1}), while the bottom temperature increased to θ = −0.035 °C.

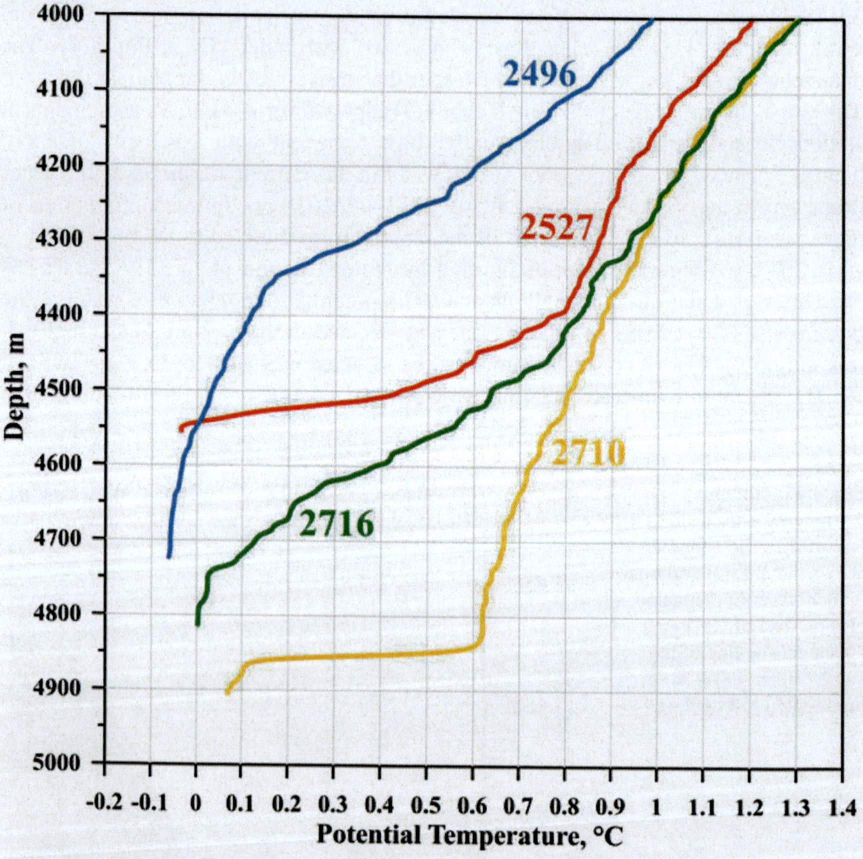

Fig. 4.43 Profiles of potential temperature below 4200 m at stations 2496 (26°31.1′ S, 34°03.3′ W), 2710 (25°34.0′ S, 33°29.5′ W), 2527 (26°37.2′ S, 33°39.6′ W), and 2716 (26°00.0′ S, 33°32.5′ W) in the depressions (see Fig. 4.44)

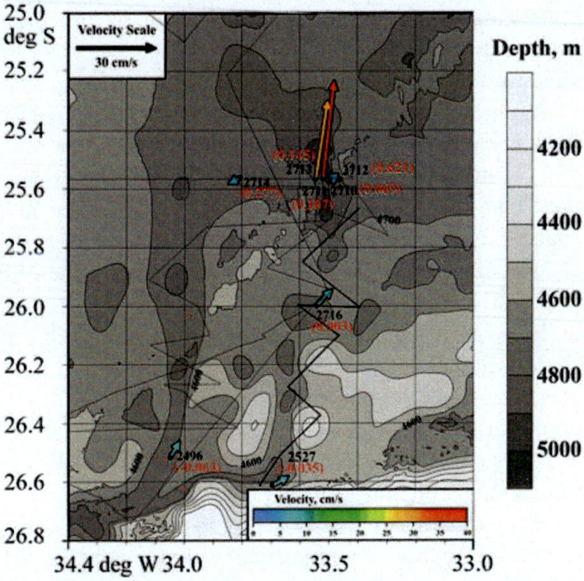

Fig. 4.44 Locations of stations in two deep channels north of the Vema Channel. Vectors of the LADCP velocities averaged over the 50-m bottom layer, and potential temperatures measured at the bottom (numbers in brackets) are shown. Thin lines show the tracks of echo-sounder measurements

The bottom flow in the small channel several times overflows the thresholds with a depth of 4550–4560 m. The bottom current of Weddell Sea Deep Water occupies a thin bottom layer approximately 40 m thick. The vertical temperature profile at station 2527 at 26°37.2′ S, 33°39.6′ W illustrates this (Fig. 4.43). The depth of the channel at this station is 4554 m. A layer of a very sharp temperature gradient was observed over this layer 40 m thick. The velocity of the bottom flow was 15 cm s^{-1}.

We searched for the continuation of the coldest AABW flow to the northeast of the channel. The stations were made in a quasi-zonal valley nearly at 26°30′ S east of 33°30′ W, which seemed to be a topographic continuation of the Vema Channel (Figs. 4.37; 4.41). We did not find potential temperatures below zero at the bottom there, despite the fact that the sill depth at 33°30′ W is only a few tens of meters shallower than 4500 m. The vertical temperature gradient at the bottom (0.46°/50 m) west of this sill is very high so that potential temperature at station 2527 (26°37.2′ S, 33°39.6′ W) at a distance of 50 m above the bottom is $\theta \sim 0.4$ °C unlike $\theta = -0.035$ °C at the bottom (4554 m). The flow with potential temperatures below 0.2 °C is confined to a bottom layer 40 m thick. The coldest continuation of the flow turns to the north because of the existence of this barrier. The CTD measurements at the stations occupied to the east of not very high topographic barrier showed that such topographic blocking is sufficient to stop the cold-water current so that extremely cold water cannot pass to the valley located east of 33°30′ W (Fig. 4.41).

Thus, the continuation of the bottom water current to the east is closed east of 33°30′ W due to the existence of a barrier and the flow turns to the north steered by topography. The AABW that is found east of this longitude with a temperature of $\theta = +0.4$ °C was transported from the east or was formed by the water overflowing underwater barriers. A summary of velocity measurements and reconstruction of bathymetry in the region is shown in Fig. 4.41. The lowest potential temperatures measured by CTD-probes near the bottom are indicated.

Along with the eastward continuation of the AABW flow in the Vema Channel, there is also a branch of this stream directed to the north (approximately along 34° W). Let us analyze the measurements in the second branch of the AABW flow from the Vema Channel, which flows to the north after outflowing from the bed of the channel. In 2012, the CTD/LADCP measurements at 26°31.1′ S, 34°03.3′ W in this channel revealed a northerly current with a velocity of approximately 5 cm s^{-1} (Fig. 4.41). The thermohaline properties of water in this current are similar to those in the Vema Channel. The potential temperature of bottom water in this flow was slightly higher (−0.063 °C in 2012) compared to the water (−0.080 °C in 2012) measured in the eastward channel. Even earlier in 2009, we occupied a station to the north of the Vema Channel (station 2080 at 26°01.9′ S, 33°58.0′ W) approximately at a distance of 80 km from the channel. This flow is a continuation of the flow from the channel. Potential temperature at the bottom was $\theta = -0.054$ °C. The northward current was slow, approximately 4 cm s^{-1} (Fig. 4.41).

A zonal section of four stations was occupied in 2018 along 33°30′ W nearly at 25°34′ S where a small channel was found based on the digital database (GEBCO 2019). The position of the section was chosen based on the data of multi-beam survey of bottom topography. The channel is approximately 150–200 m deep relative to the surrounding plateau (4700 m) (Fig. 4.44). This channel is apparently the continuation of the bottom flow that turned to the north at 33°30′ W. The width of the channel here is about 7 km at a level of the 4800 m isobath, the maximum depth is 4950 m. Echo-sounder measurements confirmed the existence of this small channel. In addition, two more CTD/LADCP stations were made: at the same latitude 25 km to the west, and south at 26° S near the eastern wall of the channel (Fig. 4.44).

The measurements of currents, temperature, and salinity over a section across this channel approximately at 25°36′ S (Figs. 4.45 and 4.46) show the presence of a high-speed (more than 35 cm s^{-1}) core of the flow confined to the bottom. The flow is directed to the north. This flow is approximately 100–150 m thick. Potential temperature of the bottom water in this channel was higher than zero. The velocity core of the flow near the bottom is displaced to the western slope of the channel. This layer with high velocities is located approximately below the $\theta = 0.62$ °C isotherm. The core of cold fresher water is displaced to the eastern slope (i.e., to the right of the flow direction) (Fig. 4.45). A sharp deep thermocline exists over the flow. Such temperature distribution corresponds to the effect of bottom friction in the channel in the Southern Hemisphere (Johnson et al. 1976; Jungclaus and Vanicek 1999) displacing the cold less saline core of the jet to the right side of the flow similar to the flow in the Vema Channel. This displacement is also seen in Fig. 4.38 over the section in the northern part of the Vema Channel. The displacement is seen in the channel, but there is no displacement over the flat western terrace.

We used the procedure described in (Tarakanov et al. 2018) and estimated the transport in the channel in the bottom layer below $\theta < 0.62$ ° C, which was equal to

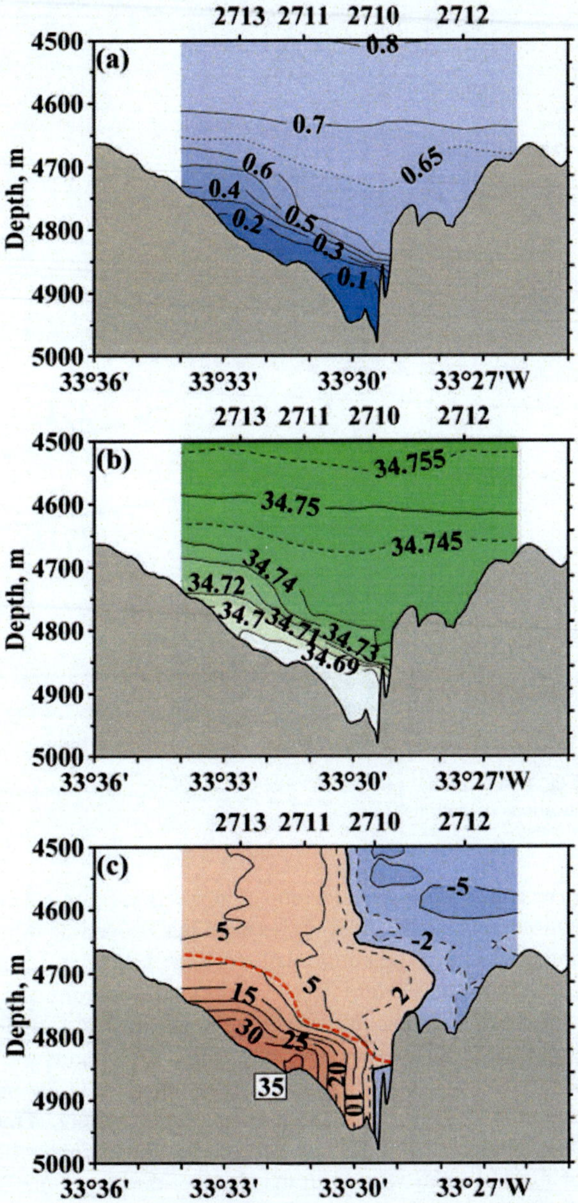

Fig. 4.45 Distributions of temperature, salinity, and meridional velocity in the bottom part of the section across the northward channel in the southwestern part of the Brazil Basin: potential temperature, °C (**a**); salinity, PSU (**b**); velocity component normal to the section, cm s^{-1} (**c**). The red dashed line in Fig. 4.44c shows the 0.6 °C potential temperature isotherm. Reproduced from (Tarakanov et al. 2020)

0.150 ± 0.007 Sv. This transport of bottom water is more than one order of magnitude smaller than the estimates of the total AABW flow through the Vema Channel across the standard section at 31°12′ S.

The difference in the temperature profiles along the meridional channel and at a station in the beginning of the channel (station 2496 at 26°31.1′ S, 34°03.3′ W), and

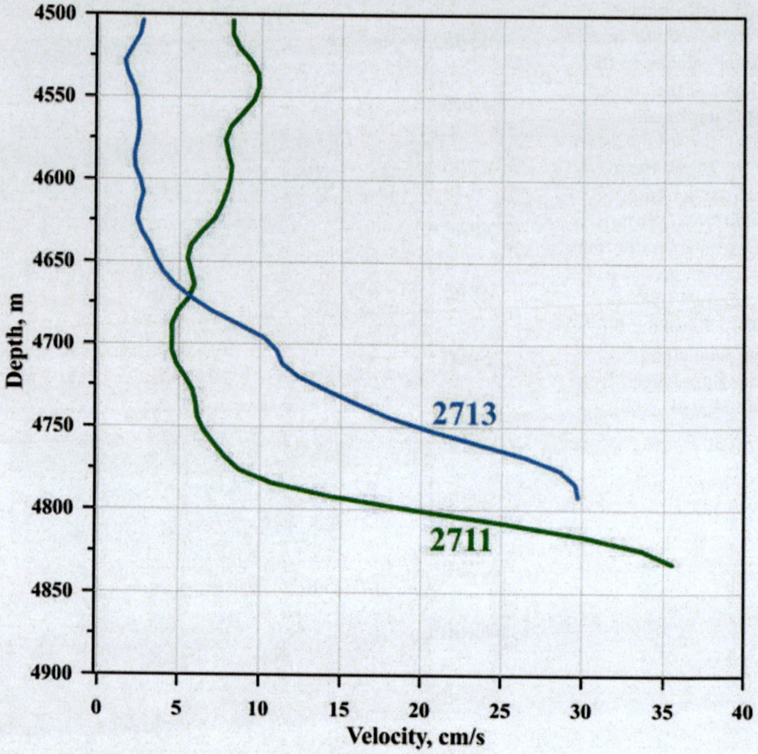

Fig. 4.46 Vertical profiles of meridional velocity at stations 2711 and 2713 (see Fig. 4.44 for locations of stations)

in its continuation in the region of the section (station 2710 at 25°34.0′ S, 33°29.5′ W and stations 2713, 2716) shows a gradual deepening of the isotherms (up to ~400 m) and thinning of the bottom layer (Fig. 4.43). This is especially notable in the layer colder than 0.60–0.65 °C. Such a slope of isotherms of a gravitational flow (the flow of AABW) corresponds to a horizontal pressure gradient increasing with depth. Since the only source of cold water in the section is the Vema Channel, this means that the motion of the abyssal flow to the north in the meridional channel is largely determined by the longitudinal pressure gradient. Thus, this small channel is a continuation of the deepest bed of the Vema Channel as a path for the discharge of AABW accumulating in the abyss of the Argentine Basin.

As the bottom flow propagates to the north its temperature increases and the thickness of the layer decreases, which leads to a decrease in the transport of the coldest water. It is likely that this flow eventually disappears. Potential temperature of the flow at 25°30′ S becomes higher than zero.

Let us now consider the upper current of Antarctic Bottom Water. This is a pathway of the upper cold AABW flow over the western terrace of the Vema Channel beginning from the Argentine Basin, which can be traced by cold temperatures at the bottom (Fig. 4.37). Such an isolated core over the western terrace of the Vema Channel found on the standard section (31°12′ S, Fig. 4.36) was observed in the temperature distributions in 1984 (−0.15 °C), 1991 (−0.16 °C), 1992 (−0.13 °C), 2006 (−0.10 °C), 2009 (−0.10 °C), 2017 (−0.06 °C), 2018 (−0.08 °C), and 2020 (−0.06 °C). Velocity measurements performed since 2006 confirm the existence of this core; they range within 30–36 cm s^{-1} (Morozov et al. 2010).

In 2009, a CTD-section along 24° S has been occupied by British scientists. The section was repeated in 2018 (https://cchdo.ucsd.edu). In 2009, two individual cores of low bottom temperatures were found on this section with potential temperatures in the first core ranging between $\theta = -0.04$ °C and $\theta = -0.05$ °C at 33°50′ W (approximately at a depth of 4650 m) and in the second core at 31°25′ W (in the depth range 5000–5200 m) the potential temperatures were between $\theta = -0.03$ °C and $\theta = -004$ °C. These cores slightly displaced in 2018. The potential temperature sections along this line in 2009 and in 2018 (24° S) are shown in Fig. 4.47.

It is clear that the previously considered two deeper jets could not have such a cold continuation at 24° S because the upper core at 24°S was located at shallower depths (~4600 m) than the two deeper continuation flows (~4700 m) from the deep Vema Channel bed. The potential temperature of the cold water found in 2009 at 24°00′ S, 33°42′ W at a depth of 4600 m was −0.049 °C. In 2018, the potential temperature increased to $\theta = -0.029$ °C. The lower core at 24° S was much colder ($\theta = -0.03$ °C) than the two deeper jets of the continuation of AABW flow from the deep channel bed. Potential temperature at 25°30′ S ranged within 0.069°–0.145 °C above zero (Fig. 4.44).

It is likely that these two regions of relatively cold bottom temperature (at 24°S) could be associated with the continuation of the upper current of AABW from the Vema Channel that was initially spreading above the western wall of the channel. There is no other source of such cold water because the potential temperature of bottom water in the two channels continuing the Vema Channel bed is not as cold as recorded on the 24° S section. The cold water found at this latitude (24° S) is located at shallower depths (4600 m) than the two continuation flows (4700–4800 m) from the deep Vema Channel bed. The potential temperature of the two continuations of the flow gradually increases and is close to zero already at latitude 25°30′ S. The temperature of the cold water found at 24° S, 33°42′ W at a depth of 4626 m is $\theta = -0.049$ °C. The measurements in 2009 and 2018 were made at a distance of 300 km north of the channel. The inclination of the continental slope in this region is rather small: approximately 1: 100.

A jet of cold water in the motion can maintain its position above the slope balanced by the gravity and Coriolis forces (Zhmur and Nazarenko 1994). Shallow depressions on the continental slope (~50–100 m) were found exactly in the region of the location of cold-water cores, which could be eroded by the bottom water flow (Morozov and Tarakanov 2014). We emphasize that there are no sources of such cold water other than the Vema Channel. Hence, this water arrives here with the upper jet of AABW. In 2018, 2019, and 2020, we made stations to locate the upper jet north of the Vema Channel based on the existence of cold temperatures and

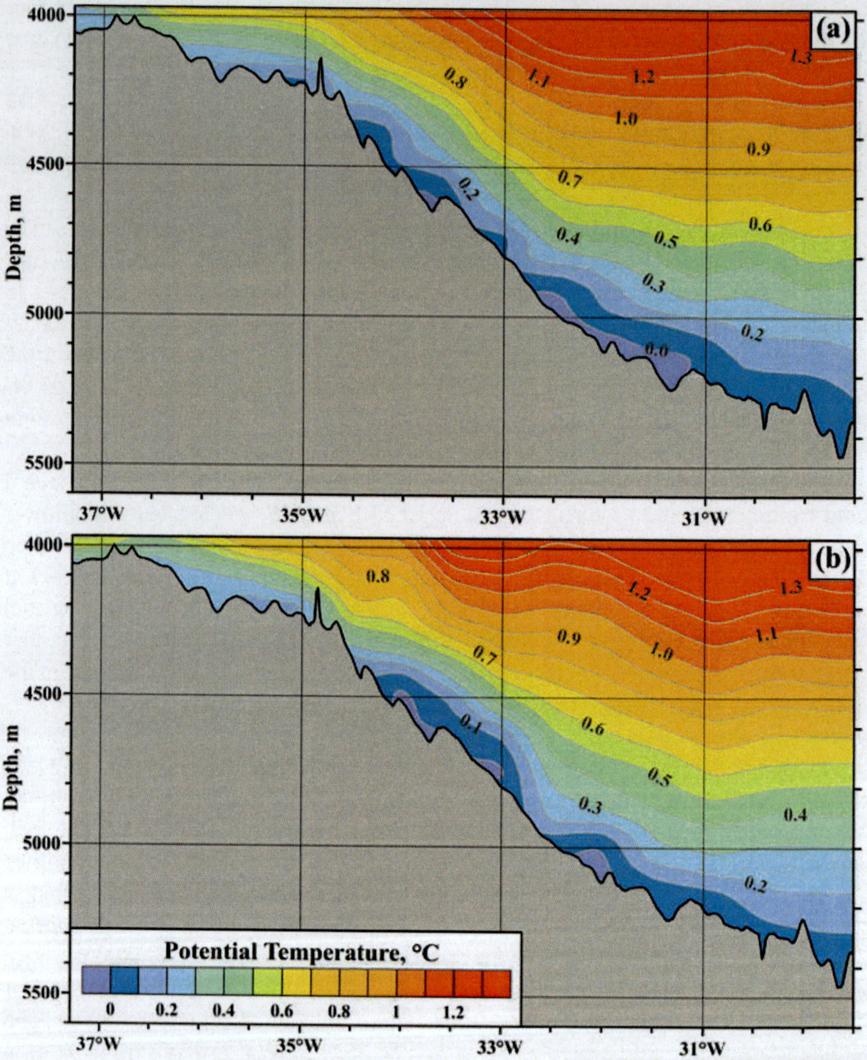

Fig. 4.47 Distribution of potential temperature θ over the bottom part of the A09 section in 2009 (**a**) and 2018 (**b**) over the continental slope of South America along 24° S. Potential temperatures in the cold cores of the flows at the bottom are indicated

direction of the flow with a northerly component. We confirmed its location at latitude 25°34′ S and on sections #4 and #5 around 25° S (Figs. 4.34, 4.39 and 4.40). We also found this jet in the southern region closer to the Vema Channel at 26°20′ S, 34°30′ W.

The jet that follows over the western slope of the Vema Channel was traced based on the low potential temperatures and vectors of velocity at the bottom shown with yellow arrows in Fig. 4.41. In 2020, we made two CTD/LADCP sections (Figs. 4.34,

4.39 and 4.40) to show that this bottom flow is directed into the canyon, whose location is seen in Fig. 4.37 approximately at 24°50′ S.

An important form of bottom topography is located on the pathway of the upper AABW flow. It is a canyon in the continental slope. Figures 4.37 and 4.48 show that a zonal canyon is located at 24°45′ S; the canyon ends with an expansion at about 32°30′ W, which is called 'Vema Mouth' in (Tarakanov et al. 2020). As seen from the distribution of temperature at the bottom from the historical data set (Fig. 4.37), this is exactly the canyon that is associated with the widening of the cold AABW

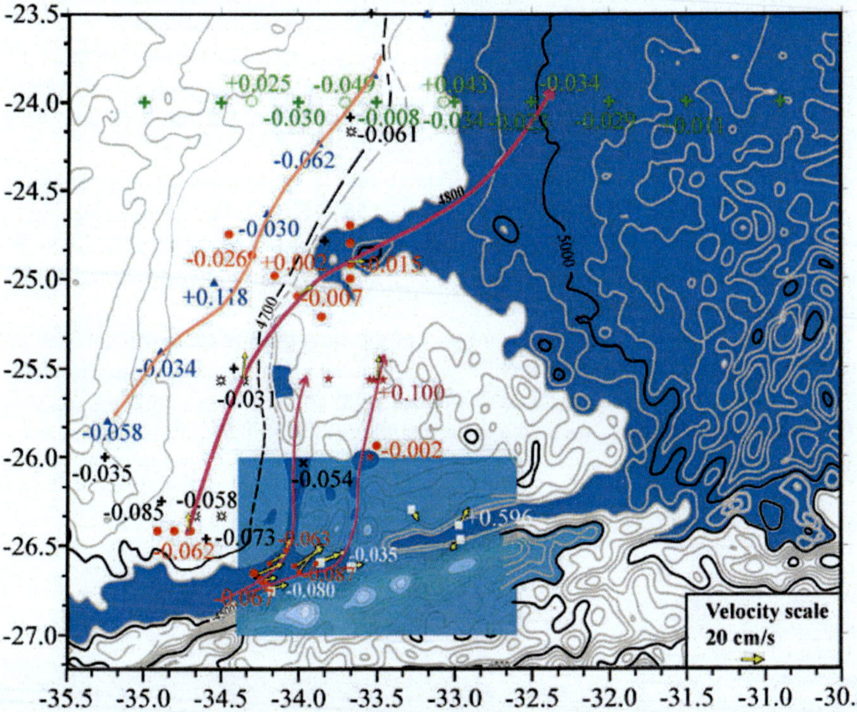

Fig. 4.48 Bathymetry of the study region based on (GEBCO 2019) and corrected with our echo-sounder measurements in 2010–2020. Locations of CTD/LADCP casts are shown. Blue triangles (▲) show the stations of the WOCE A17 section in 1994. Black crosses (+) show our stations in 2003. Black cross (x) shows the station in 2009. Green circles (◯) show stations of the A09 (24°S) section in 2009 and green crosses (✚) show the A09 stations in 2018. Small red squares (■) are our stations in 2012. Larger white squares are our stations in 2013. Violet stars (★) show our stations in 2018. Circles with rays (B&W) (☼) show stations in 2019. Red circles (●) show stations in 2020. Yellow arrows show the directions of currents and their velocity (vectors). Numbers in color at the dots of stations (the same color) indicate potential temperature at the bottom. Blue colors show deep-water regions, especially the canyon at 24°40′ S. The blue rectangle shows a region, which was studied best of all in 2010–2020 (see Fig. 4.41). The magenta lines show a scheme of the pathways of AABW. The beige line shows the pathway of a shallow jet based on CTD data in 1994 and 2003. Dashed lines show isobaths 4700 and 4750 m based on our echo-sounder measurements. (Reproduced from Morozov et al. 2020)

flow on the continental slope of South America. This canyon can influence the flow of cold water in the upper quasi-geostrophic core of the AABW flow and divert part of the flow of the coldest AABW down the slope. In such a flow, the pressure gradient directed normal to the flow direction increases because of the steeper slope in the canyon. If we estimate the geometric parameters of the canyon and its orientation across the geostrophic flow a conclusion follows that the geostrophic balance fails below the edges of the canyon. In this case, a geostrophically uncompensated gravity current would appear down the canyon in the direction of the pressure gradient. Thus, the canyon operates as a channel that separates and diverts part of the bottom water from the upper, quasi-geostrophic stream of AABW to the abyssal depths of the Brazil Basin. As an example of the operation of such a mechanism, we note that it is key for descending of dense cold water "cascading" (which forms over the Antarctic slope and flows along this slope in a quasi-geostrophic flow) into the abysses of the basins adjacent to Antarctica (Baines and Condie 1998).

We could detect this jet in 2020 by the maximum velocities and low temperatures at 26°25′ S, 26°20′ S, 25°34′ S, 25°06′ S, and 24°55′ S (Fig. 4.48). At the latter point, the jet descends down the canyon. The θ,*S*-curves based on the data in 2020 at several points of the pathway are shown in Fig. 4.49. The curves shown in the figure are very close.

It is not likely that this jet splits and part of the flow goes to the north because in this case the flow should go upslope. We analyzed the bottom topography from the digital databases GEBCO and ETOPO and our echo-sounder data in the region. The 4700 m isobath is actually located approximately 50 km to the southeast relative to

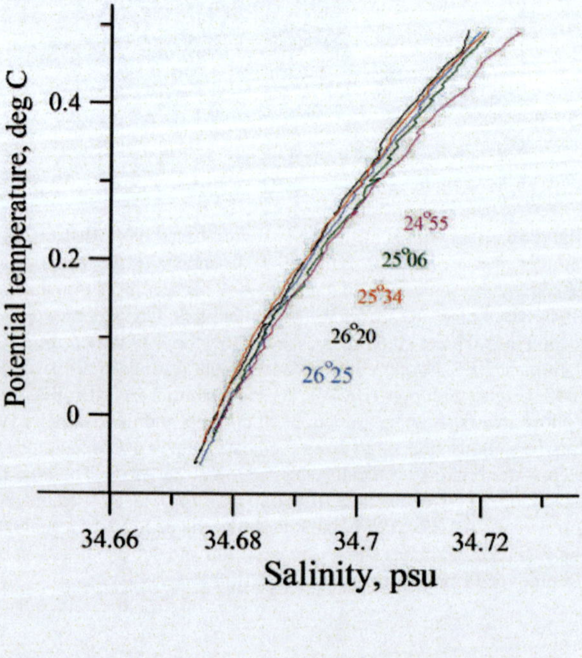

Fig. 4.49 Curves of θ,*S*-distributions along the pathway of the upper jet. Latitudes of the points are indicated on the graph with the same color as the curves. (Reproduced from Morozov et al. 2020)

its position that can be plotted from the digital data. Based on our echo-sounder data we plotted isobaths 4700 m and 4750 m with dashed lines in Fig. 4.48.

The currently available data allow us to state that two flows of cold water (θ < 0 °C) are found along the line of the A09 section in 2009 and 2018 (Fig. 4.47) over the continental slope in the depth range 4650–5150 m.

We conclude that the coldest type of AABW (θ < 0.0 °C) flows north from the Vema Channel as two jets, one of which is located in the deep channel bed and the other approximately 200 m above the western slope of the channel. The deep channel bed of the Vema Channel does not end in the region 26°40′ S, 34°00′ W but continues as a widening to the north in the Brazil Basin. This flow generally diverges into two bottom flows along smaller channels. The potential temperature of the flow increases in the course of water spreading. Approximately at 25°30′ S, the potential temperature in these flows becomes warmer than 0 °C.

It is likely that a third jet exists at shallower depths. We could trace it based on the CTD casts in 1994 and 2003, which revealed potential temperatures below zero; however, there were no LADCP measurements (Fig. 4.48). We plotted this jet with a beige line. This line is drawn close to the points of stations with cold potential temperatures. The station with a potential temperature of +0.118 °C is located over an elevation and we have drawn the line deeper. This jet continues to the north and can be seen in Fig. 4.47 as the upper cold core at 24° S, 33°-34° W. It is not reasonable to compare the θ,*S*-curves from the data of different years (1994, 2003, 2009, 2018) because the properties of AABW fluctuate as shown in our book (Morozov et al. 2010).

The potential temperature of the upper jet remains below 0 °C even at 24° S. Measurements north of this latitude also revealed negative potential temperatures of the flow. This can be a continuation of the AABW jet described in (Sandoval and Weatherly 2001). Thus, we can construct a scheme for the continuation of the AABW flow from the Vema Channel based on our and historical measurements north and northeast of the Vema Channel (Fig. 4.48).

A chart of bottom topography with corrections based on our echo sounder surveys is also shown in Fig. 4.48. Potential temperatures measured at the bottom at stations in different years are indicated, which gives an idea of how jets of cold water propagated after AABW outflows from the Vema Channel. The most important result of our field works in the region of AABW outflow from the Vema Channel is that there are two outflowing jets. The existence of an even shallower third jet can be seen only from the CTD data.

The flow in the deep channel bed splits into two parts, one of which turns to the north and the second continues in the eastern direction but also turns to the north because it cannot overflow a barrier at 33°30′ W. The thickness of both flows decreases and the temperature increases so than they disappear approximately at 25°30′ S.

The upper jet that initially flows above the western slope of the channel spreads further to the north than the bottom flows in the small channels. Hence, this is exactly the jet that transports AABW farther to the north and fills the deep basins in the North Atlantic.

Thus, the coldest type of AABW (θ < 0.0 °C) flows to the north in the Vema Channel as two jets, one of which is located in the deep channel bed, and the other approximately 300 m above over the western slope of the channel. The deep channel bed of the Vema Channel does not end in the region 26°40′ S, 34°00′ W as a widening to the north in the Brazil Basin as it follows from the digital bottom topography database (GEBCO 2019). The core of the coldest AABW that flows along the channel bed splits into two parts, one of which turns to the north and the second continues the motion in the eastern direction.

4.3 Modeling of Antarctic Bottom Water Flow in the South Atlantic

The Mid-Atlantic Ridge divides the Atlantic Ocean into two parts and causes significant differences between the deep and bottom waters of the West and East Atlantic. In this section we compare temperature and salinity structure of the abyssal waters in the Southeast and Southwest Atlantic. Both the results of high spatial resolution modeling and hydrographic measurements were used in this study. We also simulated the velocity field in the bottom layer of the South Atlantic.

We used a three-dimensional ocean circulation model with high spatial resolution in bottom layers for a better description of the AABW structure in the Southeast and Southwest Atlantic. Two hydrographic sections were used for verification of the model and for studying the evolution of bottom water properties along the pathways of AABW propagation. Some integral characteristics of the waters in deep-water basins were also calculated.

Sparse direct observations in the abyssal ocean layer make investigations of bottom waters structure difficult. Generally, simulations with modern numerical models focus on the upper part of the ocean; usually, vertical resolution of these models in the deep layers is not sufficient for describing abyssal processes. In this work, we combined two approaches: first of all, we simulated the deep-water circulation using a regional model with high resolution in the deep layers; in addition, we used direct hydrographic measurements for verification of the model and for detailed analysis of AABW evolution along the pathways of its spreading.

The Institute of Numerical Mathematics Ocean Model (INMOM) (Zalesny et al. 2010) was used for the simulations. The INMOM is a σ-coordinate numerical model based on the primitive equations of hydrothermodynamics with the Boussinesq and hydrostatic approximations (Diansky et al. 2002; Zalesny et al. 2012; Frey et al. 2017a, 2017b).

This model is based on the full system of primitive thermo-hydrodynamic equations in spherical coordinates with the hydrostatic and Boussinesq approximations. The model has been well tested in different parts of the Atlantic Ocean including the Coordinated Ocean-ice Reference Experiments (CORE) international program (Danabasoglu et al. 2014). The INMOM uses a vertical coordinate $\sigma = \sigma(x, y,t)$ scaled by the depth of the ocean at the given point. It is specified by relation:

$$\sigma = \frac{z - \zeta(x, y, t)}{H(x, y) - \zeta(x, y, t)}, \qquad (4.1)$$

where, z is the ordinary vertical coordinate in meters measured down from the undisturbed sea surface, H(x,y) is the ocean depth, $\zeta(x,y,t)$ is the sea surface height, (x,y) are the longitude and latitude, respectively. The σ-coordinates, which are also called isobathic, follow the topography more exactly and allow simulation of bottom circulation in a wide range of depths (Jungclaus and Vanicek 1999; Griffies et al. 2005).

The model domain covered the part of the Atlantic Ocean from the equator to 50° S and from 60° W to 20° E (the domain is shown with thick black lines in Fig. 4.50). The horizontal resolution was 0.05° by latitude and longitude, which gives 1600*1000 model nodes in the horizontal plane. The domain covers the deep basins

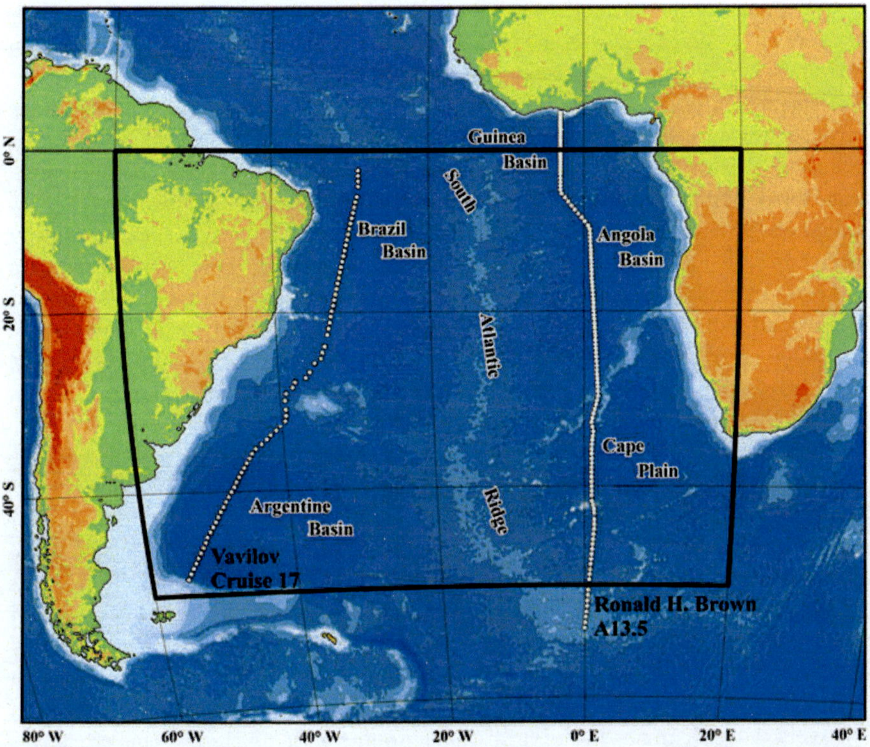

Fig. 4.50 Bottom topography of the South Atlantic based on the GEBCO 2014 data. The model domain is shown with thick black lines. White dots show the two CTD sections occupied in cruise 17 of the R/V *Akademik Sergey Vavilov* (Morozov 2005) and A13.5 section in the expedition of the R/V *Ronald H. Brown* (Bullister et al. 2010). (Reproduced from Frey et al. 2019b)

of the West and East Atlantic: Angola, Cape, Argentine, and Brazil basins are the regions, which we want to investigate and compare in this study.

The bottom topography was interpolated to the model domain from the GEBCO 2014 digital database. The climatological data from the World Ocean Atlas 2013 (Locarnini et al. 2010; Zweng et al. 2013) were used as initial temperature and salinity fields. These data were also specified in a buffer zone with a width of 10 grid points at the liquid boundaries (Klinck 1995). The atmosphere–ocean interaction was prescribed by specifying the fluxes of heat, freshwater, and momentum at the boundary. The atmospheric characteristics included air temperature, humidity, and wind velocity at a height of 10 m, atmospheric pressure, longwave and shortwave irradiance, and monthly mean precipitation were taken from the CORE database (Large and Yeager 2009). The numerical experiment was performed in two stages: at first, the model was subjected to the spin-up procedure using «frozen» temperature and salinity fields; then, the simulation included adjustment of thermohaline properties. This method was suggested in (Sarkisyan 1991) for computing the velocity field based on known temperature and salinity data.

Prognostic variables of the model are three-dimensional fields of potential temperature, salinity, and components of horizontal velocity. Here, we analyzed temperature and salinity fields on the last day of the calculation. We used the data from two quasi-meridional hydrographic sections (the locations of stations over these sections are indicated with white dots in Fig. 4.50). We chose these two sections because they follow the deep part of the basins in the West and East Atlantic and represent transformation of AABW properties. The first section across the Argentine and Brazil basins was occupied during cruise 17 of the R/V *Akademik Sergey Vavilov* in 2003 (Morozov 2005). The route of this cruise covered the main pathway of Antarctic Bottom Water spreading. The data acquisition system consisted of a General Oceanics Rosette multi-bottle array system and Neil Brown CTD profiler. The second section across the eastern basins was carried out during the cruise of the R/V *Ronald H. Brown* along line A13.5 (Bullister et al. 2010). The data were collected using Sea-Bird Electronics (SBE) 9plus CTD profiler mounted in an SBE 32 aluminum frame.

The results of numerical modeling were compared with direct measurements over two sections in the West and East Atlantic (Fig. 4.50) because bottom water structures in these two regions differ significantly. Quasi-meridional direction of these sections allows us to verify how the model reproduces the evolution of AABW properties. Moreover, the locations of stations in cruise 17 of the R/V *Akademik Sergey Vavilov* were especially chosen along the main pathway of AABW spreading. These stations include the Vema Channel with the most intense flow of AABW in the Atlantic. In the channel, CTD casts were performed near its eastern wall (Morozov 2005), because the Ekman friction displaces the coldest waters to the right relatively to the main northward flow (in the Southern Hemisphere). This effect was investigated in the Vema Channel numerically (Jungclaus and Vanicek 1999) and observed many times in different expeditions (Morozov et al. 2008, 2010).

The data used for verification are shown in Fig. 4.51. We compared two properties of AABW: the depth of its upper boundary (Fig. 4.51a) and potential

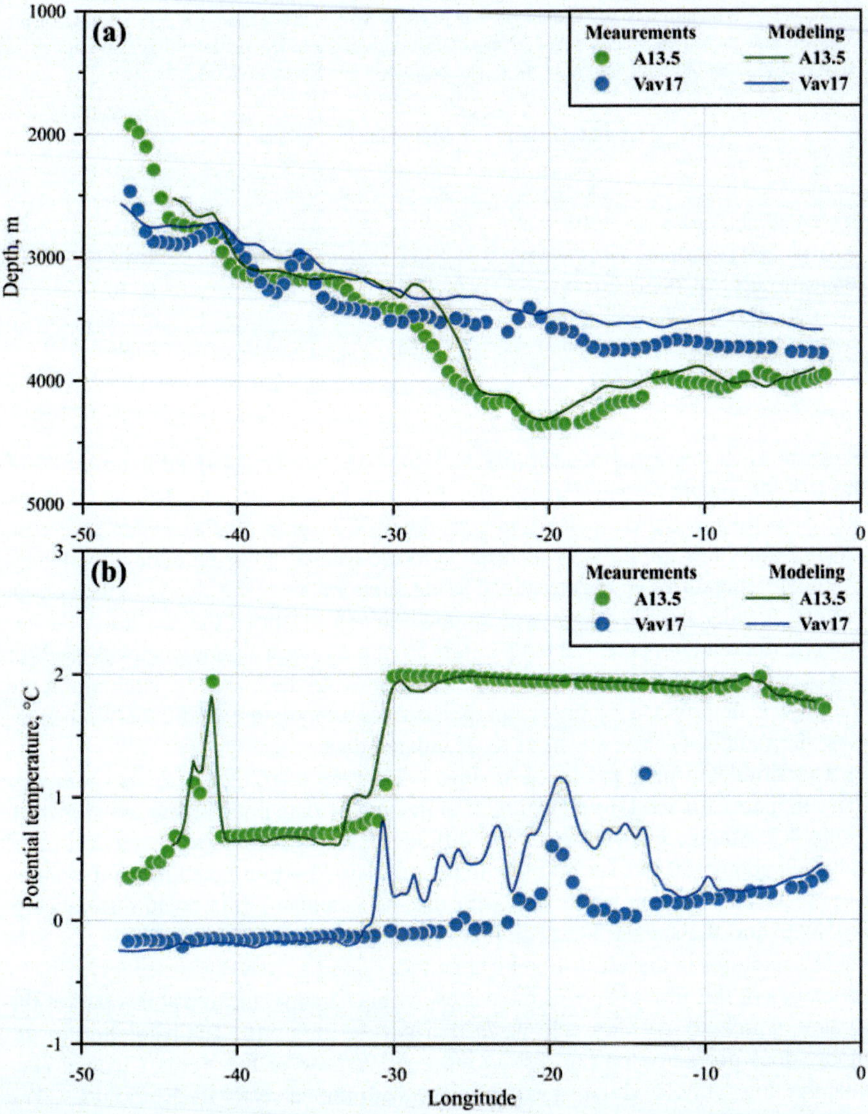

Fig. 4.51 The data used for verification of the model: depth of $\theta = 2$ °C potential temperature isotherm (**a**) and potential temperature at the ocean bottom (**b**) along the hydrographic sections in the West Atlantic (shown with blue colors) and East Atlantic (green colors) Atlantic. Circles indicate the bottom potential temperatures at the CTD-stations; solid lines are the data of numerical modeling interpolated to the trajectory of the sections. Some statistical parameters of these data are given in Table 4.7. (Reproduced from Frey et al. 2019b)

Table 4.7 Verification of the model on the basis of two CTD-sections. X_{model} and X_{meas} are modeled and measured values, respectively. These values include calculated upper boundary of AABW and potential temperature at the ocean bottom in the East and West Atlantic

	Maximum	Minimum	Mean	Standard Deviation
	(X_{model} – X_{meas})	(X_{model} – X_{meas})	(X_{model} – X_{meas})	(X_{model} – X_{meas})
AABW level (East), m	103	−402	−117	109
AABW level (West), m	217	−303	−159	101
Bottom Pot. Temp. (East), °C	0.55	−0.20	0.018	0.110
Bottom Pot. Temp. (West), °C	0.66	−0.85	0.092	0.401

temperature at the ocean bottom (Fig. 4.51b). The stations occupied in cruise 17 of the R/V *Akademik Sergey Vavilov* are shown in blue color, and stations from the A13.5 section occupied in the cruise of the R/V *Ronald H. Brown* are shown in green. Solid lines in Fig. 4.51 correspond to the results of our numerical modeling. Bottom temperatures were determined as the temperature of the closest to the bottom sample; usually the CTD cast was stopped at 3–5 m above the sea floor. As for numerical modeling, potential temperature from the lowest sigma-level was used for comparison with the measured value. The location of the $\theta = 2$ °C isotherm at the station was determined by linear interpolation of temperature profile between samples; the same approach was used for simulated data.

The simulated data presented in Fig. 4.51 were interpolated to the points of measurements for model verification. Parameters of data comparison are shown in Table 4.7. Standard deviation of the difference between simulated and measured values is about 100 m over both sections. Simulated bottom potential temperatures over the A13.5 section agree well with the measurements (the mean difference is 0.018 °C and the standard deviation is 0.11 °C). As can be seen from Fig. 4.51b, the temperatures in the West Atlantic at latitudes of 32°-15° S differ strongly. One of the causes of this discrepancy is the narrow Vema Channel located at this part of the section, which allows very cold-water transport to the north. The resolution of the model does not allow one to restore the AABW flow in the Vema Channel adequately; thus, simulated values do not agree with the correctly measured ones. The correct simulation is presented in (Frey et al. 2017a) using the same model with much higher horizontal resolution.

The results of numerical modeling allow us to study the spatial structure of bottom waters in the South Atlantic. For this analysis, we interpolated three-dimensional temperature and salinity fields from vertical sigma-levels to regular z-levels with a high resolution. Using interpolated data, we calculated horizontal velocities in the AABW layer in the entire South Atlantic (Fig. 4.52). Here we present the velocities at a depth of 4300 m, which corresponds to a depth of the main flow in the Vema Channel. Generally, these velocities do not exceed 5 cm s^{-1}; more intense flows are observed in the western part of the Argentine Basin and in the

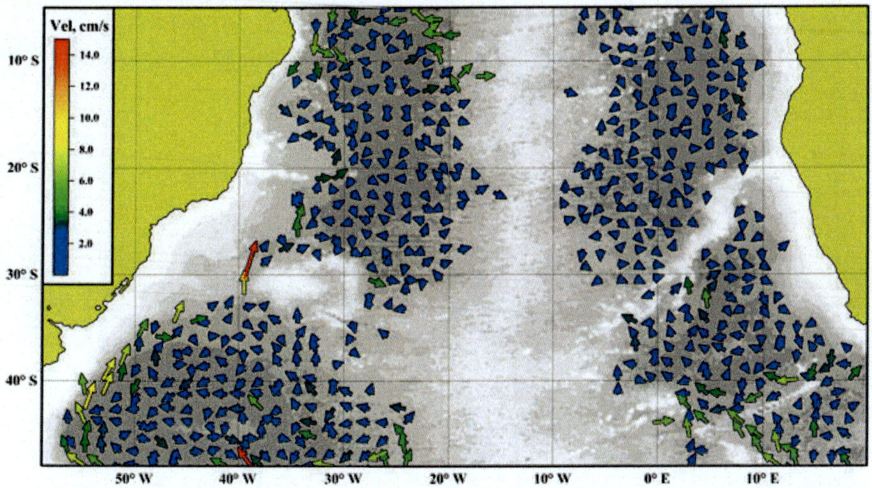

Fig. 4.52 Vectors of horizontal velocities at a depth of 4300 m in the South Atlantic based on the results of numerical modeling. (Reproduced from Frey et al. 2019b)

Vema Channel. Correct modeling of these flows requires higher resolution. The main feature of the bottom circulation in the western part of the South Atlantic is the bottom flow of the Deep Western Boundary Current directed to the north. This current is clearly pronounced south of the Rio Grande Rise (south of 30° S). Strong intensification of the bottom current occurs in the Vema Channel (27°-33° S). Based on the numerical modeling, the northerly current becomes less intense north of 27° S and almost disappears in the northern part of the Brazil Basin. However, direct measurements in the ocean reveal that the northerly bottom flow continues further to the equator and north of the equator (Sandoval and Weatherly 2001).

In the East Atlantic, the inflow of AABW to the Angola Basin based on numerical simulations is revealed in the southern part of the Angola Basin. This conclusion was also made in (Morozov et al. 2010) based on the distribution of potential temperature at the bottom. The circulation in the bottom layer of the Angola Basin is generally very weak.

Some aspects of deep-water circulation in this region are discussed in (Frey et al. 2017b). Distribution of bottom potential temperature (Fig. 4.53) is helpful for determining the main pathways of AABW spreading. The values of potential temperature were taken from the bottom σ-level of our model.

We calculated some integral parameters of AABW in the main basins of the South Atlantic. The parameters include minimum potential temperature and salinity, maximum potential density referenced to 4000 dbar (σ_4), and total volume of Antarctic waters below the 2 °C potential temperature isotherm (Table 4.8). The calculation was made for the Argentine, Brazil, Cape, and Angola basins separately. The minimum temperatures were determined from the deepest sigma-level of the model. Usually, the waters with the minimum temperatures are located in the deep part of the basins near the channel with inflow of bottom waters. The minimum

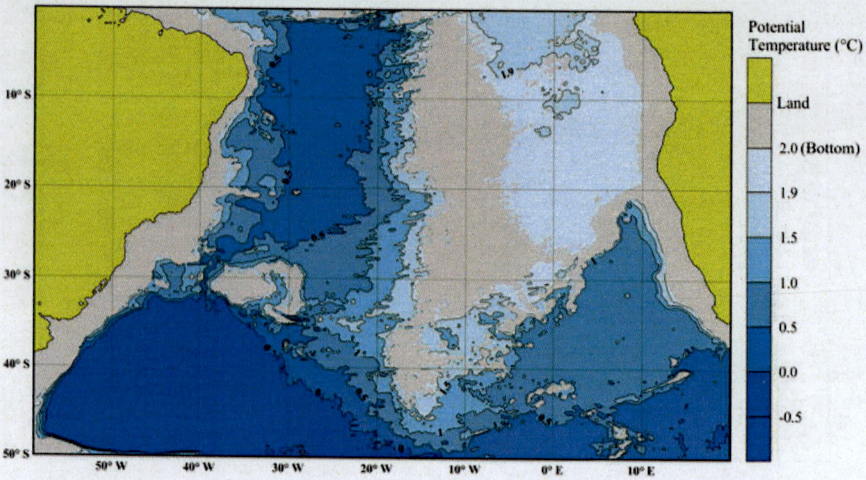

Fig. 4.53 Bottom potential temperature in the bottom layer of the Atlantic. The data were taken from the deepest σ-level of the model. The ocean basins with bottom potential temperature greater than 2 °C are shown in gray color. (Reproduced from Frey et al. 2019b)

Table 4.8 AABW properties in the deep basins of the South Atlantic based on the numerical simulations

Basin	Minimum Pot. Temperature, °C	Minimum Salinity, psu	Maximum Density σ^4, kg/m^3	Total AABW volume, 10^6 km^3
Argentine	−0.25	34.67	46.09	10.8
Brazil	0.15	34.69	46.05	9.2
Cape	0.6	34.71	45.99	5.6
Angola	1.9	34.88	45.88	3.5

salinity in Table 4.8 was determined as salinity at the points with minimum temperature. Then, maximum potential density referenced to 4000 dbar was derived using the EOS80 equations (Lewis and Perkin 1981). Three-dimensional temperature field interpolated to z-levels was used for the calculation of the AABW volume. Horizontal resolution of the field was 0.05° (the same as the model resolution); the vertical resolution was 20 m. Then, we made the following check for each point of the grid: if potential temperature was lower than $\theta = 2$ °C and location of the node was above the bottom, then this volume was added to the corresponding ocean basin. We can see that the AABW volume in the West Atlantic is two times larger than in the East Atlantic; in addition, the waters west of the Mid-Atlantic Ridge are colder and less saline, which results in the higher density (Table 4.8).

Thus, a significant increase in bottom temperature is observed over underwater ridges on the pathway of AABW propagation. In addition, its upper boundary becomes much deeper as it spreads to the north. For example, after crossing the ridge at 30° S in the Southeast Atlantic, the bottom potential temperature increases from 0.7 °C up to 1.9 °C, while the 2 °C isotherm descends almost by 1 km. Strong differences are observed in the abyssal waters in the Southwest and Southeast Atlantic due to the local peculiarities of the bottom topography. Thus, the total volume of AABW in the West Atlantic is two times greater than in the East Atlantic; the bottom potential temperature variations between them reach 2 °C over the same zonal section. The flow of AABW to the north is generally concentrated in the abyssal Deep Western Boundary Current. It is well pronounced in the Argentine Basin but becomes weaker in the Brazil Basin.

4.4 Modeling of Antarctic Bottom Water Flow Through the Vema Channel

Since all the measurements of velocities in the channel were carried out at single-point locations they do not make possible presenting an entire pattern of currents along the length of the channel. The full dynamics can be calculated only using numerical modeling; however, the spatial resolution of modern global models of the ocean is not sufficient to simulate the current in a narrow abyssal channel. Only one publication was dedicated to the simulation of the flow in the Vema Channel (Jungclaus and Vanicek 1999). This paper is related to the influence of the bottom friction on the current over the section at the sill. The numerical simulation was performed using a two-dimensional model and does not describe the flow in the longitudinal direction.

Here, we report about numerical simulation of the flow in the abyssal Vema Channel using the Institute of Numerical Mathematics Ocean Model (INMOM), which is a modern three-dimensional model of the ocean dynamics (Diansky 2013).

The INMOM model is based on the full system of the so-called primitive thermo-hydrodynamic equations written in spherical coordinates with the hydrostatic and Boussinesq approximations (Diansky 2013). It is a σ-model and it is worth noting that these models of the ocean describe bottom currents more adequately than the z-models, because the lower layers in the σ-coordinates, which are also called isobathic, follow the bottom topography more exactly (Jungclaus and Vanicek 1999; Griffies et al. 2005).

We selected a rectilinear domain in the Southwest Atlantic for modeling between 37.0° and 22.8° S and between 45.2° and 29.0° W. This domain completely covers the Vema Channel, the Santos Plateau, and the Rio-Grande Rise as well as parts of the Brazil and Argentine basins. The bottom topography data with an initial resolution of 30′ were taken from the electronic atlas (Smith and Sandwell: http://topex.ucsd.edu/cgi-bin/get_data.cgi); the depths in the channel were corrected on the basis of the multi-beam echo-sounder survey (Zenk et al. 1993) and later measurements. The topographic data were interpolated to the modeling grid and smoothed several times using the Tukey filter. The temperature and salinity fields from the World Ocean Atlas 2013 (WOA2013 2013; Locarnini et al. 2010; Antonov et al. 2010) were specified as the initial conditions. We specified the zero initial velocity field.

The WOA2013 data were also used to specify the annual evolution of temperature and salinity at liquid boundaries. The temperature and salinity values from the database were specified in a buffer zone with a width of 10 boundary nodes to the simulation grid at the liquid boundaries (the "sponge" layer condition (Klinck 1995)). In these buffer zones, the model fits the ocean level and currents to the specified density of the ocean water practically according to the geostrophic relations. The atmosphere-ocean interaction was prescribed by specifying the fluxes of heat, freshwater, and momentum at the boundary. The turbulent fluxes were described on the basis of the bulk-formulas (Diansky 2013). All atmospheric surface synoptic characteristics were taken from the CORE database (Large and Yeager 2009). The atmospheric characteristics include air temperature, humidity, and wind velocity at a height of 10 m, atmospheric pressure, longwave and shortwave irradiance, and monthly mean precipitation. The CORE data of the so-called normalized year were used for the simulations.

The prognostic parameters of the model were: fields of the horizontal velocity, potential temperature, and salinity at sigma-levels, sea surface height deviation from the undisturbed surface. The horizontal resolution was 0.02° by latitude and longitude, which gives 814*764 points in the horizontal plane and approximately 10 calculation points in the transversal section of the Vema Channel. The vertical sigma-levels were specified not uniformly; the best resolution was at the bottom. The total number of sigma-levels was 33. The maximum depth was 4850 m; the 10 deepest levels were specified in the bottom layers from 4400 to 4850 m with a step of 50 m. The numerical experiment was carried out in the regime of diagnosis and adjustment suggested by A.S. Sarkisyan (1991). This method allows us to restore the ocean circulation on the basis of numerical modeling using the known data of temperature and salinity measurements. The total duration of the simulations was 45 days. In the first stage, the model was spun up during 30 days in the diagnosis regime, while the temperature and salinity fields were "frozen". In the second stage, the numerical experiment has been continued for 15 days in the adjustment regime. The last day of simulation was used for the analysis.

The use of sigma-levels as the vertical coordinate allows us to describe the bottom ocean dynamics in arbitrary depth interval. A good resolution at the bottom is obtained both in the Vema Channel and in the deeper Brazil and Argentine basins. Taking into account that the depth of the sill of the channel is 4614 m and a sufficient part of the channel is located at depths of 4650–4700 m, the simulated velocity data were interpolated and presented in Fig. 4.54 at a depth of 4600 m.

Simulated velocities from the lowest σ-level are presented in Fig. 4.55a. The model shows insignificant bottom currents up to 5–10 cm s^{-1} in the Argentine and Brazil basins. On the contrary, the velocities in the Vema Channel are relatively high; they varied between 20 and 30 cm s^{-1}. The maximum velocities reaching 35 cm s^{-1} were observed in the narrowest part of the channel. High velocities are observed along the entire length of the channel from 33° to 26° S. A slight decrease in velocities is revealed in the wide middle part of the channel (Vema Extension) between 28° and 29° S; the maximum velocities here reached 20 cm s^{-1}. The total transport was directed from south to north. It was observed over the entire length of

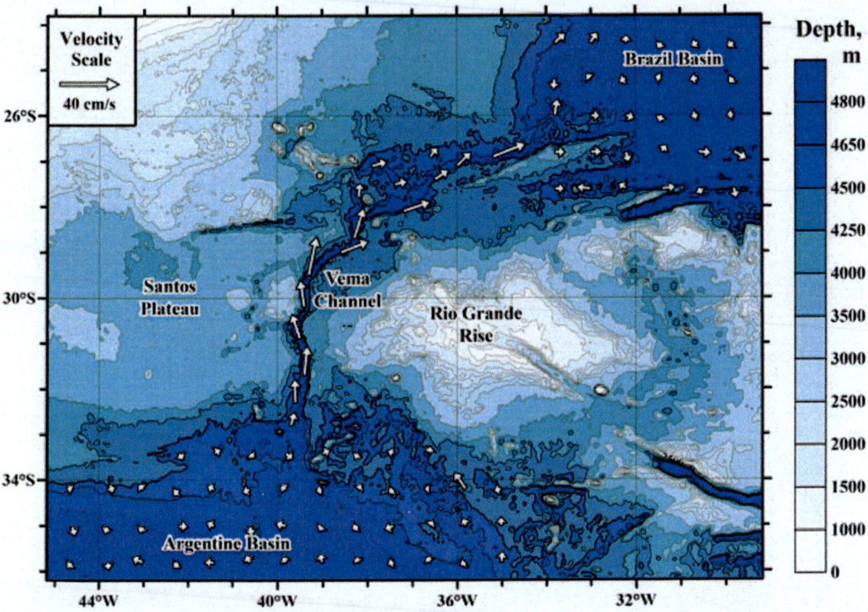

Fig. 4.54 Bathymetry of the study site and vectors of current velocities at a depth of 4600 m based on numerical simulation. (Reproduced from Frey et al. (2017a) with permission from Pleiades Publishers)

the channel. We used our direct measurements of currents in the Vema Channel as the reference data to verify the results of the numerical experiments.

We performed LADCP velocity measurements at three sections across the channel at 32°17′, 31°12′ (Fig. 4.56a), and 26°23′ S (Fig. 4.56b). The bottom velocities based on both experimental observations and numerical modeling are 10–15 cm s^{-1} at 32°17′ S and 25–30 cm s^{-1} at 31°12′ S; the currents south of 32°30′ S are insignificant. As for the outflow of AABW from the Vema Channel at 26° S, the speed of the bottom current is up to 30 cm s^{-1}, which is the same as in the southern part. However, unlike the inflow of bottom waters from the Argentine Basin, high velocities are observed in the Brazil Basin far away from the end of the channel. The estimates of the measured and simulated velocities are presented in Table 4.9.

A detailed map of the currents in the bottom layer is shown in Fig. 4.56. The best studied part of the channel is the Vema Sill at 31°12′ S (Fig. 4.56a). More than 20 CTD sections across the channel were made in this area since the 1970s. Long-term current measurements performed near the bottom at the sill show that the speed of the current varies from 0 to 60 cm s^{-1} with a mean value of 30 cm s^{-1} (Hogg et al. 1999). The LADCP measurements performed at 31°12′ S show the northward current along the entire section. Two branches of the current are observed in the middle part of the channel (Fig. 4.56b). The main jet is located in the deepest part of the channel with depths of approximately 4700 m; it is the main passage for the coldest waters. The additional southern jet with velocities up to 20 cm s^{-1} is

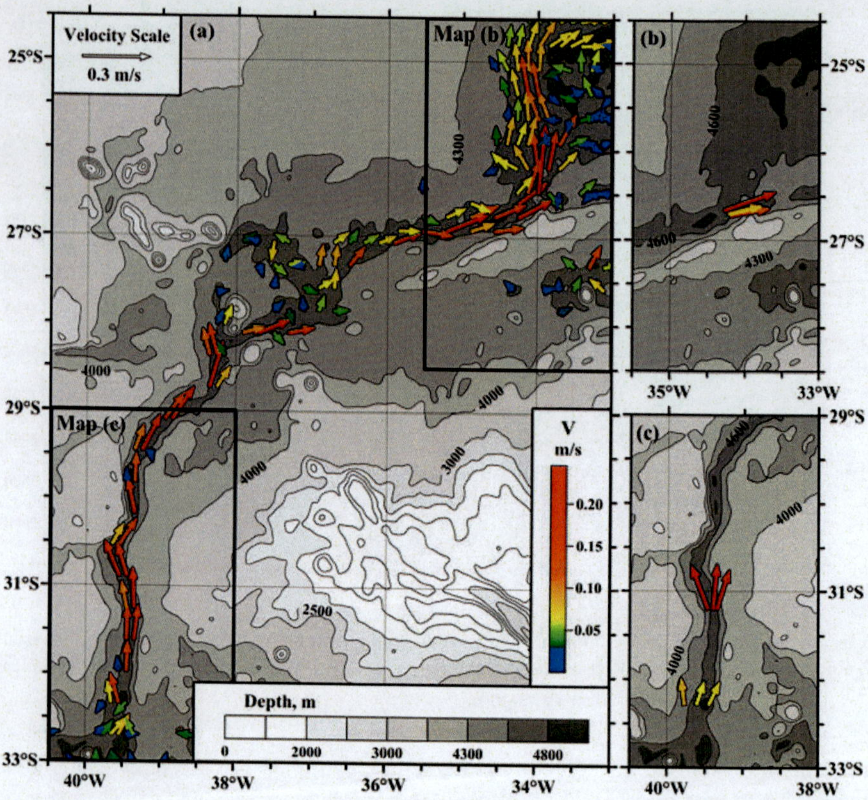

Fig. 4.55 Velocities in the bottom layer in the Vema Channel based on numerical modeling (**a**) and corresponding direct measurements at a few locations in the channel (**b**, **c**). The velocities are presented from the data of the deepest σ-level (approximately 50 meters above the seafloor) for the depths greater than 4550 m. The rectangles in Fig. 4.55a show the regions in Fig. 4.55b and c. (Reproduced from Frey et al. 2019a)

observed at depths of 4300–4400 m on the southern (right) slope of the channel. Based on the model simulations the lengths of the jets (from splitting into two jets to merging) are 200 km and the maximum distance between them is 50 km.

The distribution of potential temperature in the channel based on different data sets is shown in Fig. 4.57. Direct CTD measurements (Fig. 4.57c) reveal the physical phenomenon that the jet of the coldest waters displaces to the eastern slope. The difference in the bottom potential temperatures near the western and eastern walls of the channel is about 0.02° – 0.03 °C. The eastern displacement of the coldest core is explained by the Ekman flux due to the bottom friction (Jungclaus and Vanicek 1999). This effect is not observed in the WOA data (Fig. 4.57a) due to low resolution. However, the simulation (Fig. 4.57b) performed on the basis of the WOA data revealed the displacement of the cold waters to the eastern part of the channel similar to our field measurements (Fig. 4.57c). Both the model and

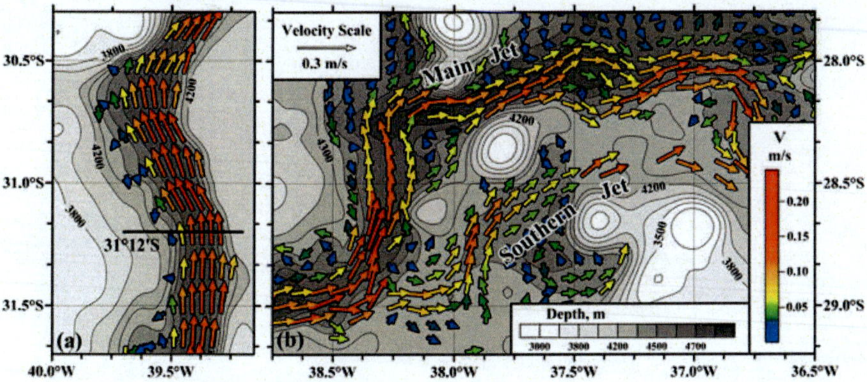

Fig. 4.56 Simulated velocities in the bottom layer near the Vema Sill (31°12′ S) (**a**) and in the wide middle part of the channel (**b**). Thick line indicates the experimentally well studied repeated standard section of the Vema Channel at 31°12′ S (Morozov et al. 2010). Two branches of the current are observed in the middle part. (Reproduced from Frey et al. 2019a)

Table 4.9 Maximum velocities and AABW transports in four parts of the Vema Channel

	Section of the channel	Measurements	Modeling
Maximum velocity, cm/s	32°17′ S	12	14
	31°12′ S	0–65	26
	26°23′ S	30	31
AABW transport, Sv	32°17′ S	–	2.5
	31°12′ S	1.6–2.7	2.0
	26°23′ S	–	2.7

Reproduced from Frey et al. (2019a)

measurements show inclination of the isotherms in the upper part of the current and a thermocline at a depth of 4300–4400 m near the eastern part of the channel. We examined this effect using two different versions of the model, with zero bottom friction (Fig. 4.57d), and zero Coriolis force (Fig. 4.57e). In the model without friction, the core of the coldest waters displaces to the west due to the Coriolis force only, which confirms that the Ekman flux is the key cause of the observed temperature shift. As can be seen from Fig. 4.57e, the inclination of the isotherms is explained by the influence of the Coriolis force. Our model shows that this inclination and the cross-current Ekman flux appears in the bottom layer at any section of the Vema Channel; the simulated difference of bottom temperatures is 0.02 °C.

Our simulation shows the existence of a southerly countercurrent at smaller depths beginning approximately from 4000 m. In the widest part of the channel in the region of 28–27° S, the core of the current as a relatively narrow jet is confined to the southeastern slopes. The transversal horizontal velocity gradient is found also in the narrow part of the channel where the flow is confined to the right wall, which also corresponds to the data of observations.

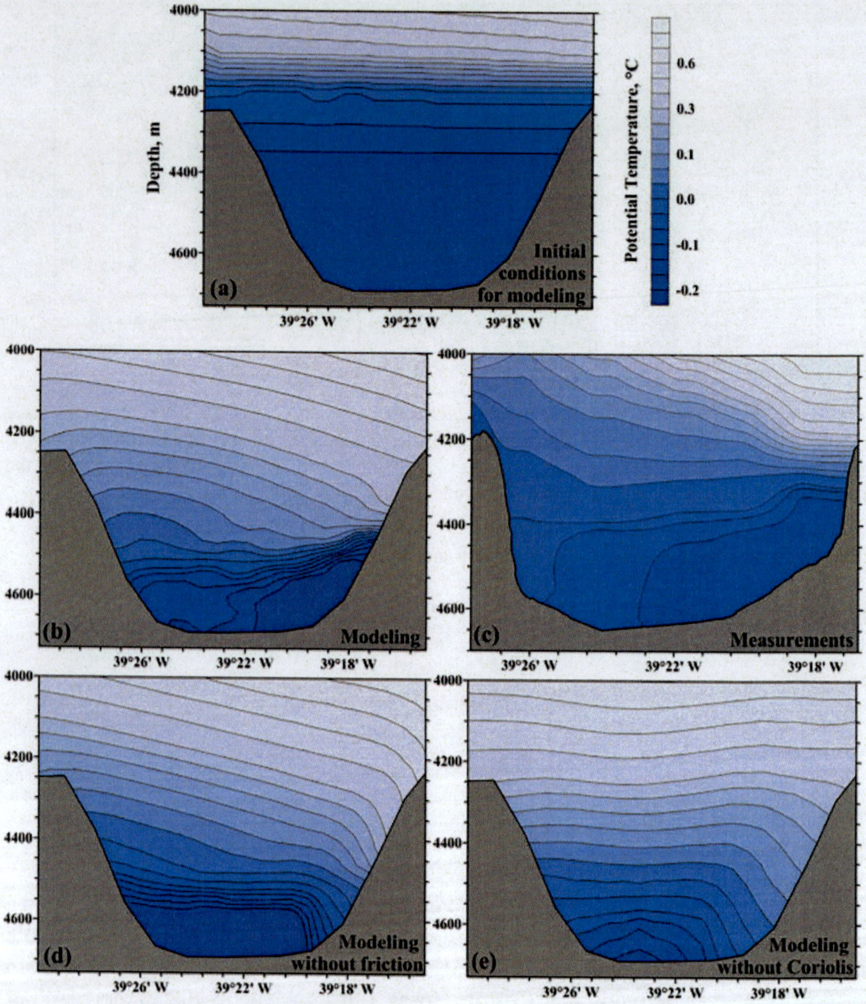

Fig. 4.57 Potential temperature sections across the Vema Channel at 31°12′ S (location of the section is shown in Fig. 4.56a with a thick black line) based on different data: interpolated WOA data were specified as the initial field for simulations (**a**); numerical modeling (**b**); direct CTD measurements (**c**); simulations without friction (**d**); and without Coriolis force (**e**). (Reproduced from Frey et al. 2019a)

Due to the limited power of computers, the existing global models of the ocean have insufficient spatial resolution for the simulation of velocity characteristics in narrow abyssal channels. In addition, such models are oriented on more detailed modeling of the upper layer. Their resolution in the deeper part does not allow us to qualitatively simulate the dynamics of the bottom layer. The powerful deep-water flows of the densest and coldest waters that influence the general heat balance of the ocean are important for the simulation of the global ocean circulation and modeling of the climate. Therefore, creation and development of three-dimensional models of oceanic circulation with high horizontal resolution and high vertical resolution in the bottom layers are needed for the investigation of deep circulation in the regions with abyssal channels (Vema, Hunter, and other channels).

We simulated the water transport in the Vema Channel using a model with a high spatial resolution in the regional model for the Southwest Atlantic. The numerical simulation showed the existence of an intense flow of Antarctic Bottom Water from the Argentine to the Brazil Basin whose major part follows the abyssal Vema Channel. The current velocities over the entire length of the channel were approximately 20–30 cm s^{-1}. The simulated velocities agree with the data of the instrumental measurements at several points in three different parts of the channel. The model simulation adequately presents the kinematic and spatial structure of Antarctic Bottom Water flow over the entire length of the abyssal Vema Channel and allows us to analyze the transversal structure of the current and its evolution along the channel.

4.5 Santos Plateau

The Santos Plateau is located between the continental slope of South America and the Vema Channel. The major part of the plateau is located at a depth of 4000 m, which is well below the depth of the 2 °C potential temperature isotherm. The AABW layer is 600–700 m thick; hence, it is likely that the bottom currents can provide the transport of at least the upper portion of Antarctic Bottom Water (Speer and Zenk 1993).

Geostrophic transport calculations based on the hydrographic section in 1991 (Speer and Zenk 1993), relative to the surface of zero motion coinciding with the 2 °C potential temperature isotherm give a total northward transport below the $\theta = 2$ °C isotherm of approximately 3 Sv and a total southward transport of 1 Sv, resulting in the northward net flow equal to 2 Sv. The flow can merge with the upper part of the flow through the Vema Channel and flow further to the north.

Much of our knowledge about water transport over the Santos Plateau comes from (Hogg et al. 1999). The data include time series of currents measured at an array of moorings deployed across the southern boundary of the Brazil Basin in early 1991. The length of the time series exceeds 1.5 year. The array spanned from the lower continental slope of South America across the Santos Plateau, Vema Channel, and Hunter Channel. A quasi-zonal hydrographic section was occupied from the continental slope to the Rio Grande Rise between latitudes 35° and 30° S. The measurements were made by cooperative efforts of scientists from the *Institut für Meereskunde* and the *Woods Hole Oceanographic Institution* (Moorings 1077, 1082, 1084, and 1088). Information about the moorings is given in Table 4.10.

In 2003, Russian scientists deployed two moorings on the Santos Plateau with almost 1 year time series of measurements 20 m and 30 m above the bottom (moorings 1066 and 1069). Progressive diagrams for the near-bottom instrument at the six moorings are shown in Fig. 4.58. The diagrams cover a period of 3 months. The particle travel distances correspond to the scale of the figure.

Table 4.10 Moorings on the Santos Plateau

Mooring number	Latitude	Longitude	Country, time	Deepest instrument level	Ocean depth
1077	28°28′ S	44°28′ W	US, Ger, 1991–1992	3532 m	3632 m
1082	29°03′ S	43°30′ W	US, Ger, 1991–1992	3850 m	3950 m
1084	29°32′ S	42°42′ W	US, Ger, 1991–1992	3918 m	4019 m
1088	30°05′ S	41°44′ W	US, Ger, 1991–1992	3714 m	3814 m
1066	31°12′ S	41°51′ W	Rus, 2002–2003	3750 m	3770 m
1069	30°39′ S	45°01′ W	Rus, 2002–2003	3460 m	3490 m

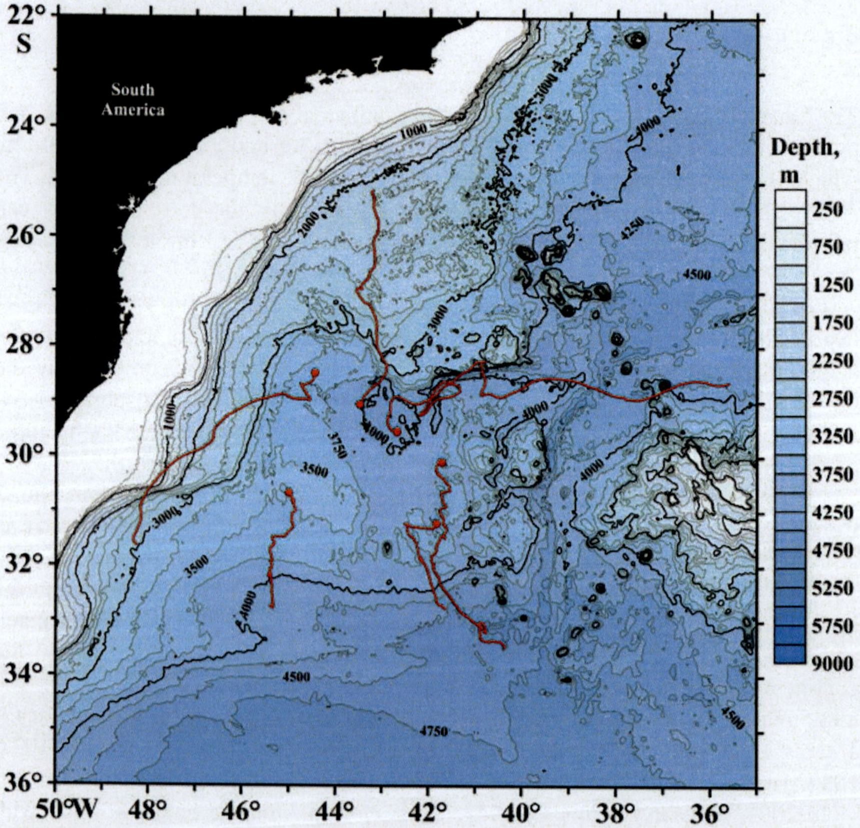

Fig. 4.58 Bottom topography of the Santos Plateau (meters) and three-months-long progressive diagrams for the near-bottom current measurements on the Santos Plateau

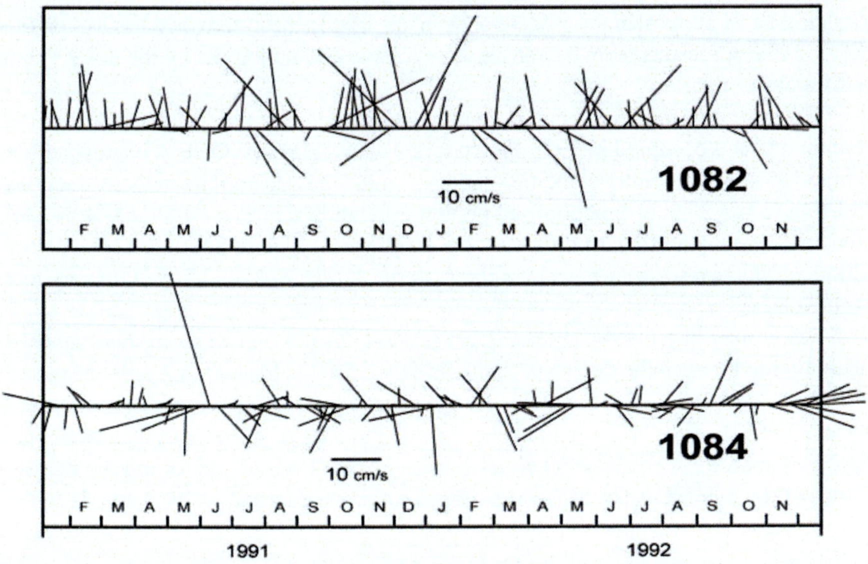

Fig. 4.59 Vectors of horizontal velocities as a function of time at moorings 1082 (29°03′ S, 43°30′ W; instrument at 3850 m) and 1084 (29°32′ S, 42°42′ W; instrument at 3918 m). Currents were low-pass filtered to remove frequencies higher than 1/48 cph. Vectors are shown with a week interval. Letters indicate months of the year

Progressive diagrams both in 1991 and 2003 show that there is no net northward transport of bottom waters over the Santos Plateau. Similarly, the vectors of horizontal velocities demonstrate alternative currents to the north and to the south. Figure 4.59 shows the vectors based on the data of two moorings in 1991 and 1992 deployed on the Santos Plateau. The moored array data provided direct current measurements to estimate the bottom water flux from the Argentine Basin to the Brazil Basin. Although the current-meter moorings were deployed sparsely, it is possible to combine the direct measurements with the computations of geostrophic velocities. While the geostrophic computations give a uniform northerly flow, current-meter data reveal two regions of the southward flow.

Hogg et al. (1999) separated geostrophic and barotropic components of the flow on the basis of moored measurements between two CTD stations. The combination of moored measurements and geostrophic calculations from cruises 15 (1991) and 22 (1992) of the R/V *Meteor* and from the cruise of the R/V *Thomas Washington* (1984) under the "Marathon" program resulted in the barotropic and geostrophic transport curves as functions of longitude (see Fig. 6 in Hogg et al. 1999). Thus, the total transport of Antarctic Bottom Water is the sum of the barotropic and baroclinic contributions. The authors conclude that the northward geostrophic transport in the range 1.3–1.8 Sv calculated from the data of cruise 22 of the R/V *Meteor* is nullified by a comparable southward transport. Upon emerging from the Vema Channel, the upper (shallow) components of Antarctic Bottom Water recirculate over the Santos Plateau and there is very little if any net northward flow over the plateau. This conclusion is supported in (McDonagh et al. 2002): near-zero net flow over the Santos Plateau results from a near-closed cyclonic circulation fed by the deep Vema Channel throughflow.

Geostrophic calculations based on the data of hydrographic sections across the Santos Plateau occupied in 1993 (WOCE A10), and 2003 (WOCE A10) resulted in compilation of baroclinic velocities and transport. The geostrophic velocity sections relative to the 2 °C of potential temperature are shown in Figs. 4.60 and 4.61. All sections reveal a northward flow over the continental slope of South America below 3000 m. The two sections reveal similar patterns of the northward flow with very low velocities.

Two cruises of the R/V *Akademik Sergey Vavilov* (41 and 43) were held in 2016 to study the flows in the deep-sea channels on the Santos Plateau in April – May and October – November, 2016. One of the goals was to study the bottom flows in small channels west of the Vema Channel. The general idea of these field measurements was to search for the deepest channels on the Santos Plateau and to find the bottom flows of the coldest water. Therefore, the efforts were focused on the analysis of the

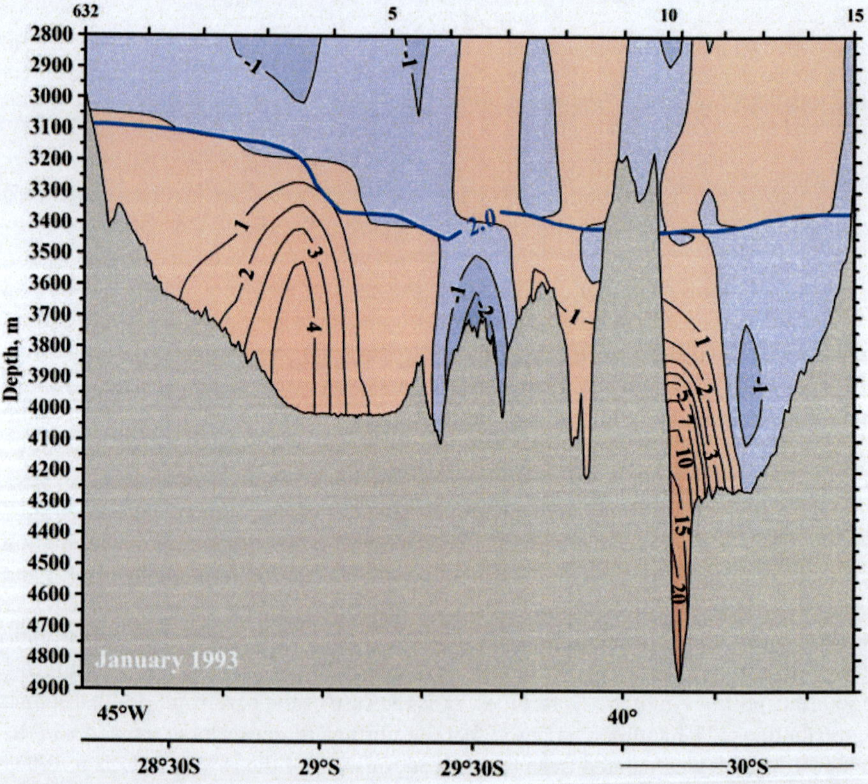

Fig. 4.60 Geostrophic velocities (cm s^{-1}) over WOCE section A10 in 1993. The light brown color shows the northward flow and blue color shows the southward flow. Stations of measurements are shown at the top axis

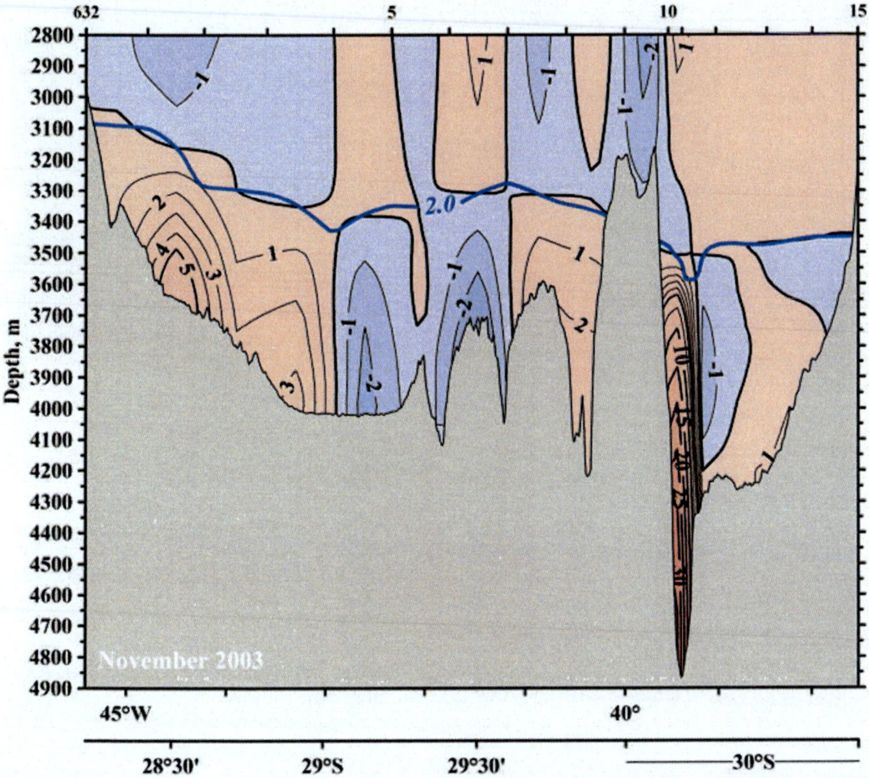

Fig. 4.61 Geostrophic velocities (cm s^{-1}) over WOCE section A10 in 2003. The light brown color shows the northward flow and blue color shows the southward flow. Stations of measurements are shown at the top axis

available bottom topography of the seabed and the implementation of additional echo-sounding measurements along the vessel. The ship routes passed over the ridges in the northeastern, southeastern, and southwestern parts of the plateau (Fig. 4.62), while the depressions in these ridges were studied from the point of view of searching for intense bottom currents.

A total of 15 stations in the spring and six stations in the fall of 2016 were occupied in this area (Fig. 4.62). Measurements at one of the points in the southwestern part were made twice both in spring and autumn (stations 2579 and 2680); strong flow variability was detected. In spring, in the southeastern part, four stations were occupied along the AABW flow (2580–2583). In the northeastern part of the plateau, a section of seven stations was occupied across the zonally-oriented channel connecting the Santos Plateau with the northern part of the Vema Channel (stations 2587–2593). The continuation of this channel towards the Santos Plateau was investigated in the fall of 2016 primarily from the point of view of topography. It was found that the channel continues in the meridional direction to the south and

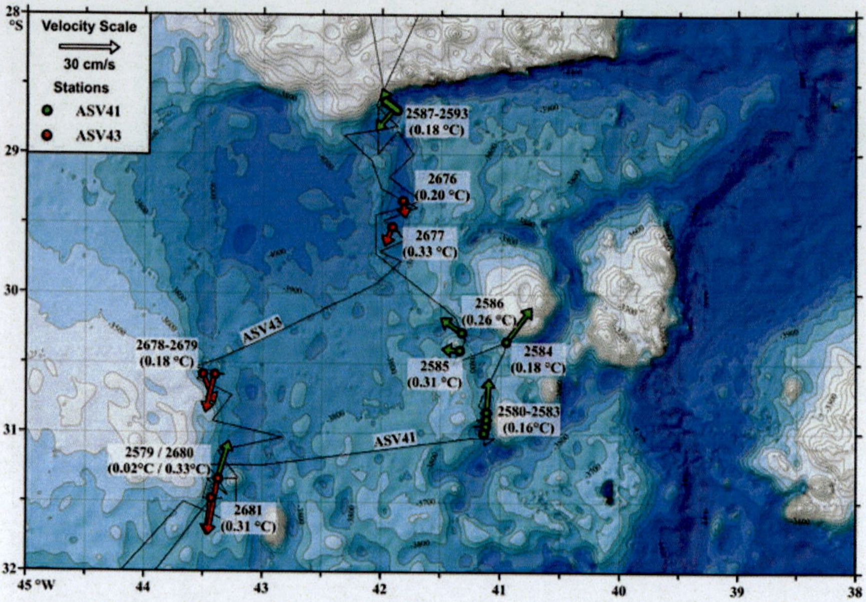

Fig. 4.62 Bottom topography, ship routes, and deep-water stations in the Santos Plateau region west of the Vema Channel. Velocity vectors at the bottom from LADCP measurements in ASV cruises ASV41 in the spring of 2016 and ASV43 in the autumn are shown with green and red arrows

there are no deep-sea passages in the ridge to the west of it. We occupied stations (2676 and 2677) in this channel.

Analysis of the measurements carried out in the spring and autumn of 2016 shows a strong variability of bottom flows over the Santos Plateau. Not only the current velocities vary strongly, but also the thermohaline properties of the waters are subjected to strong changes. Below are the results of our field observations in different parts of the plateau, performed at different times, and their comparison.

A relatively long meridionally oriented channel exists in the bottom topography in the southwestern part of the plateau, in which our measurements were concentrated in 2016. In the fall, a section of two stations 2678–2679 was occupied here, which showed an intense southward flow in the lower layer 1 km thick; the current velocities reached 15 cm s^{-1} (Fig. 4.63). A potential isotherm of $\theta = 2$ °C is located at a depth of 3200–3250 m. Two more stations were made in the southern part of this channel. The profiles of characteristics measured at station 2681 are shown in Fig. 4.64. Here, one can also see the continuation of the flow to the south. The velocities were within 10–15 cm s^{-1}, the currents were observed in the entire lower layer at least 1 km thick. A potential temperature isotherm of 2 °C was located at a depth of 3320 m (not at 3200–3250 m) already, which indicates the gradual mixing of colder bottom waters with warmer deep waters. A layer of well-mixed water with a thickness of 150 m and a potential temperature of 0.311 °C and a salinity of 34.697 PSU is distinguished in the lower part.

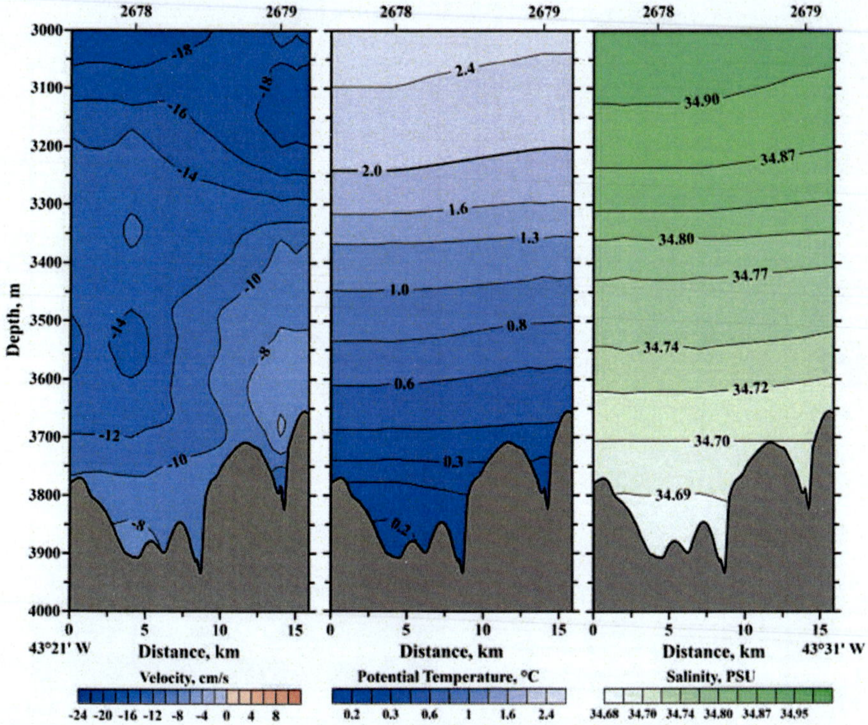

Fig. 4.63 Section of the meridional component of velocity (left), potential temperature (middle), and salinity (right) across the channel in the southwestern part of the plateau. The section line goes along latitude 30°35′ S. Negative values of the velocity (blue colors) correspond to the southern direction of the current

The measurements were performed twice, in the spring and autumn of 2016 at another point of this channel with coordinates 31°20′ S, 43°22.1′ W. The water structure and current velocities in these measurements are strongly different (Fig. 4.65). The velocities in the bottom layer change their direction from the northern (in April 2016) to the southern (in October 2016). The absolute values of velocity in both cases are of the order of 10–15 cm s^{-1}. The boundary between the AABW and NADW layers is along 3320 m at both stations, the velocities and temperatures in the NADW layer coincide. The changes concern only the lower 700-meter layer of AABW, the potential temperature here in autumn of 2016 was by 0.3 °C higher than in spring. Thus, during the flow of water from the south, cold Antarctic water flows into the plateau, then it mixes on the plateau and flows south with a higher temperature in autumn.

According to the measurements in the southeastern part of the plateau in the spring of 2016, the AABW inflows to the plateau from the east, from the side of the Vema Channel. The measurements at stations 2585 and 2586 indicate that the flow is directed westward, with characteristic velocities up to 10 cm s^{-1} (Fig. 4.66). The

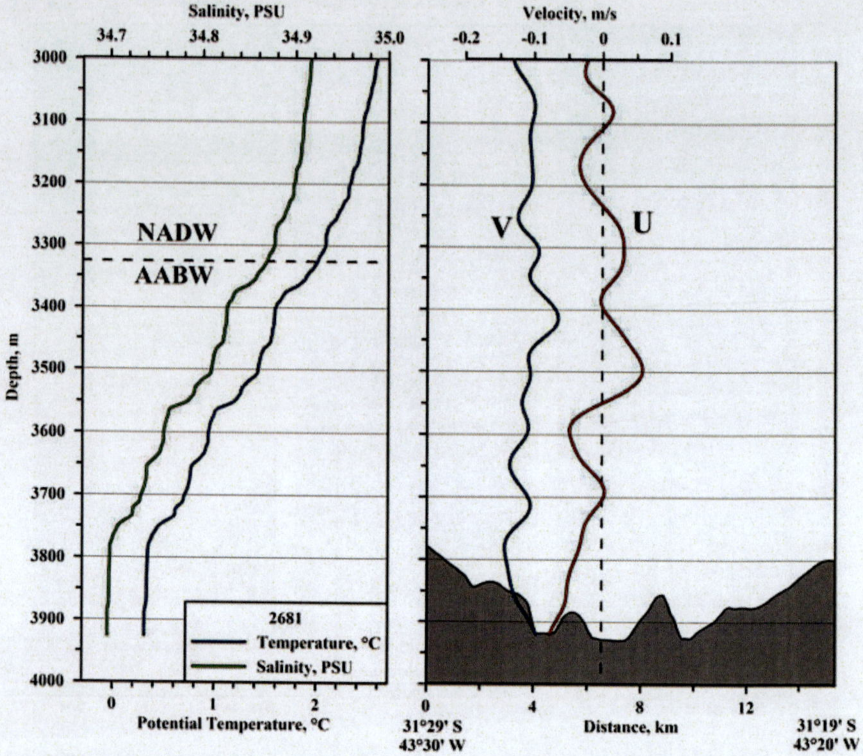

Fig. 4.64 Profiles of potential temperature, salinity, and velocities components in the southern part of the Santos Plateau. Notations *U* and *V* in the right panel correspond to the zonal and meridional components of the velocities, positive values are to the east and north, respectively

thickness of the AABW is the same as in the southwestern part: 700 m. The thermohaline structure of water in this part of the plateau is similar at different stations, the depth of the potential temperature isotherm 2 °C is 3320 m. There is a homogeneous layer in the lower 200 m above the bottom. The bottom potential temperature at station 2586 is 0.263 °C, while the salinity is 34.691 PSU. At a distance of 30 miles south of stations 2584–2586, four stations were occupied along the stream (2580–2583) with a step of 3 miles between neighboring stations. The flow here is directed to the north, the characteristic velocities are 10 cm s^{-1}, no monotonic increase in temperature along the channel has been recorded, the potential temperature ranges from 0.16° to 0.20 °C.

In the northeastern part of the plateau, the water exchange between the Santos Plateau and the Brazil Basin passes through a zonally-oriented channel to the northeast of the plateau. In the spring of 2016, a section of seven stations was made across the channel (Fig. 4.67). The velocity of the main stream was as high as 20 cm s^{-1}; it was directed to the southwest towards the plateau, while a countercurrent with maximum speeds of 6–8 cm s^{-1} was found closer to the northern wall of

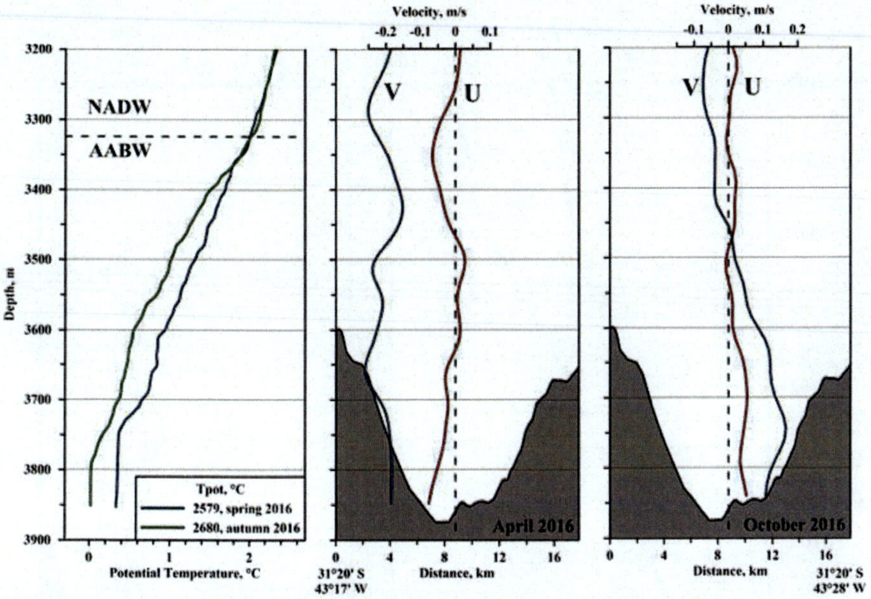

Fig. 4.65 Comparison of measurements in the spring and autumn of 2016 at 31°20′ S, 43°22.1′ W in the southwestern part of the Santos Plateau. The graphs show the profiles of potential temperature (left) and zonal and meridional components of velocities in April 2016 (station 2579) (middle) and October 2016 (station 2680) (right). Positive velocities correspond to the flow to the east (profile *U*) and north (*V*). The bottom is shown with gray color

the channel. The maximum depth of the channel is 4200 m, the depth of the 2 °C isotherm was about 3130 m, i.e., the thickness of the AABW layer here exceeds 1 km. To construct the velocity sections, barotropic tide velocities were calculated, based on the altimetry data using the model of Oregon State University (Egbert and Erofeeva 2002). The components of the tidal wave velocities reach 3 cm s^{-1}. Some stations were occupied in the maximum tidal phase (for example, 2589, 2591, 2593), which makes accounting for tidal velocities especially important.

In the fall of 2016, the measurements in this part were supplemented by two stations (2676 and 2677) further downstream in a meridionally oriented channel south of the section. At both stations, a southward flow was detected with speeds up to 10 cm s^{-1}, the potential temperature successively increased: −0.18 °C on the section, 0.20 °C at station 2676, and 0.33 °C at station 2677 (Fig. 4.68).

We performed numerical modeling of currents over the Santos Plateau (Frey et al. 2018). Due to the close location of the Santos Plateau to the Vema Channel, the simulation of the velocity field in these areas is possible only within one joint numerical experiment. Simulation at the sigma levels provides sufficient resolution at the bottom both in the Vema Channel with a depth of 4650 m and over the Santos Plateau with depths of 3900–4000 m. In both cases, we specified 10 bottom levels with a step of about 50 m (in terms of the z-levels). Numerical modeling was carried out in a rectangular region in the range 45.2°–29.0° W, 37.0°–22.8° S, the horizontal

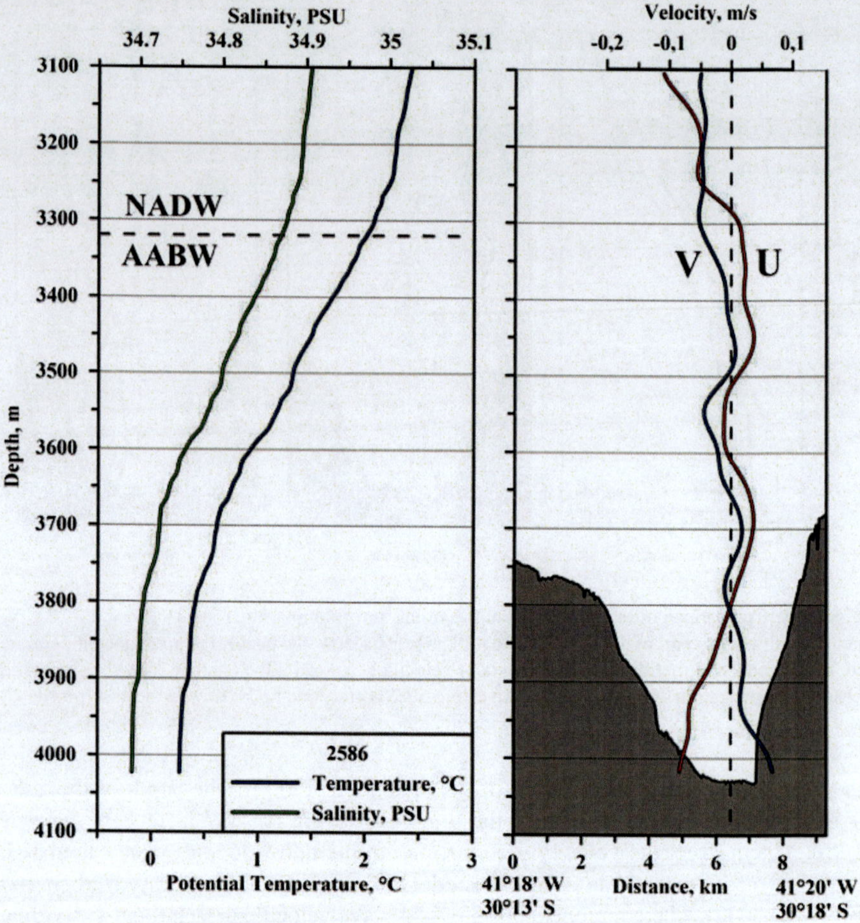

Fig. 4.66 Profiles of potential temperature, zonal (U), and meridional (V) velocity components at station 2586 in the southeastern part of the Santos Plateau

resolution was 0.02° by longitude and latitude. The results of velocity simulations were interpolated to a depth of 3800 m (the right panel in Fig. 4.69), which corresponds to a height of 100–200 m above the bottom in the region of the Santos Plateau. The left panel in Fig. 4.69 presents the minimum potential temperature and current velocities in the bottom layer.

According to the simulation results, the current velocities in the middle of the plateau do not exceed 5–10 cm s^{-1}; intense currents are found at the edges of the plateau. Water exchange occurs mainly in the southwestern and northeastern parts of the plateau through the shallow bottom channels. A strong current is observed along the western edge of the Santos Plateau; the velocities are directed to the south and reach 30 cm s^{-1}. After the waters approach the Argentine Basin, this stream turns

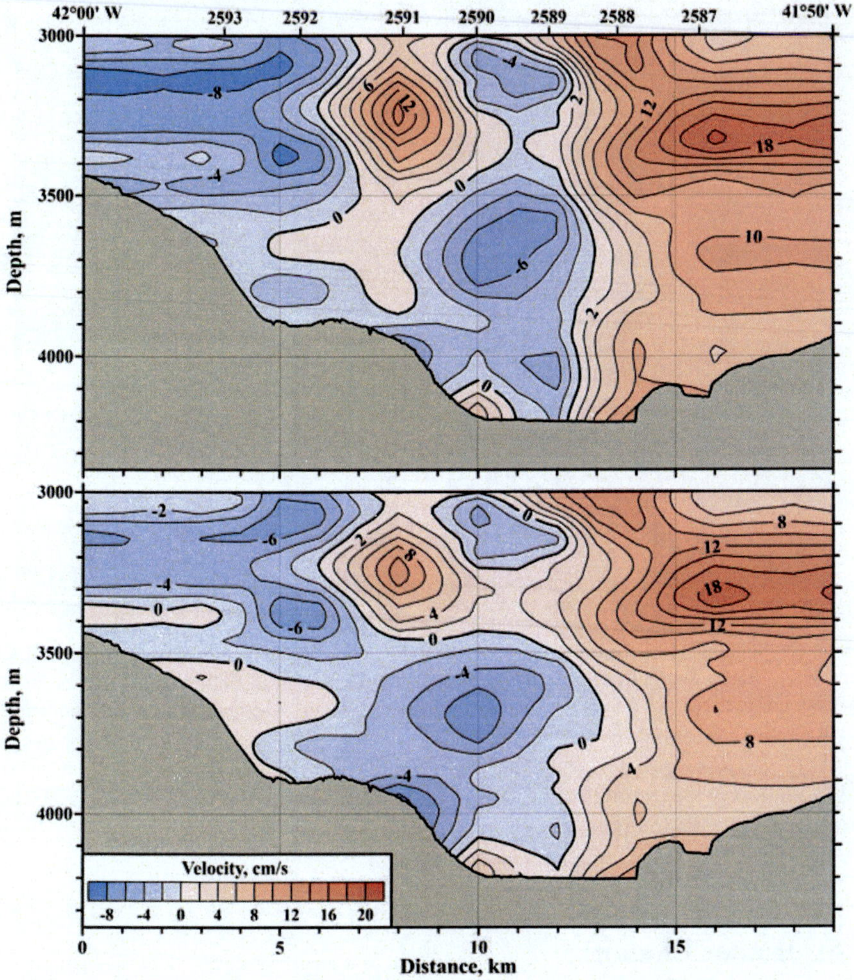

Fig. 4.67 Velocity section across a zonally-oriented channel in the northeastern part of the Santos Plateau, without taking into account (top) and taking into account (bottom) the tide. Positive speeds correspond to the southwestern direction of the current

east towards the Vema Chanel. Similar currents were revealed from the measurements in the fall of 2016, when we found a southern direction in this region with the maximum speeds of 20–30 cm s^{-1}. A more detailed comparison of the velocity pattern in this region is presented in Fig. 4.69.

Despite a good agreement between the numerical calculation in the southwestern part of the plateau and the experimental observations in the fall of 2016, the simulation does not reproduce the variability of the flow at this point. The physical mechanism of a complete change in the direction of the flow from north to south also remains unknown. The streams in the other parts of the plateau require further

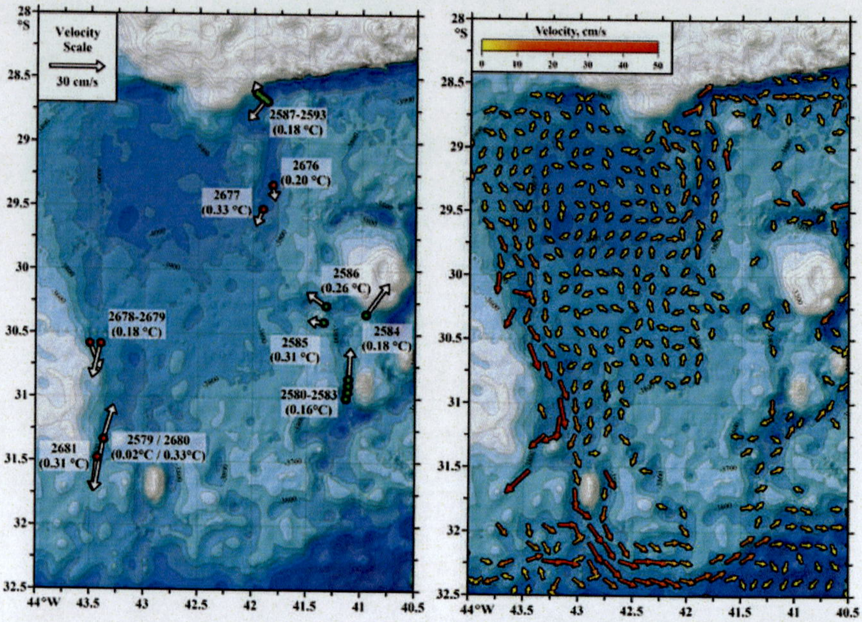

Fig. 4.68 Stations in the Santos Plateau region, bottom potential temperature, and current velocities are indicated. Stations occupied in April are highlighted in green, and those in October 2016 are highlighted in red (left panel). Velocities at a depth of 3800 m based on the numerical simulation (right panel)

investigation, while numerical modeling and experimental observations can provide additional information here.

4.6 Hunter Channel

The Hunter Channel is a more than 200 km wide valley located between two ridges at 32–35° S. The maximum depth in the channel is approximately 4300 m, which is only about 300 m shallower than the Vema Channel. The main sill is located at about 35°12′ S, 27°47′ W, i.e., approximately 70 km upstream of the CTD sections occupied in different years. Special researches aimed at the study of the flow in the channel were carried out in 1991, 1992, 1994, and 2002. Locations of hydrographic stations and moorings in the Hunter Channel are shown in Fig. 4.70.

The first standard hydrographic measurements were carried out in the Hunter Channel in 1959 using Nansen bottles and reversing thermometers. A total of seven stations were occupied along 32° S between the Rio Grande and Mid-Atlantic ridges (Fuglister 1960). During the WOCE field program, ten CTD stations were made across the narrow part of the Hunter Channel during cruise 15 of the R/V *Meteor* in January 1991. Later, the region was visited in December 1992, during cruise 22 of

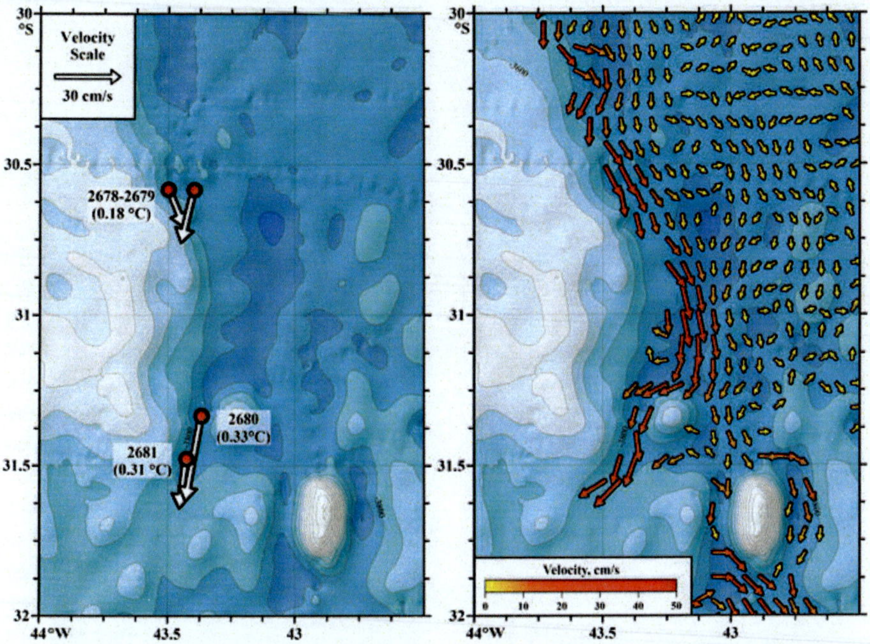

Fig. 4.69 Velocities in the western part of the Santos Plateau based on the results of the survey in autumn 2016 (left) and numerical simulation (right). The velocities are shown at a depth of 3800 m. In the southern part, the flow ascends to shallower depths following the bottom topography

the R/V *Meteor* (four CTD stations) and then in May 1994; another section (five CTD stations) was occupied during cruise 28 of the R/V *Meteor*. In November 2002, a section (five CTD stations) was occupied during cruise 11 of the R/V *Akademik Ioffe*. The most complete dataset was provided during cruise 15 of the R/V *Meteor* when ten CTD stations were occupied in 1991 (Zenk et al. 1999).

During the measurements in 1991, the coldest bottom water with a potential temperature of about 0.2 °C was found on the eastern side of the gap, which is similar but more prominent in the Vema Channel. At shallower depths in the water layer with potential temperatures between 0.5 °C and 2.0 °C, the coolest water was found at the western boundary of the section. The isotherms slope downward to the east, indicating the northward flow of bottom water relative to the warmer surface. The salinity field varies mainly in the vertical direction and correlates with temperatures such that isopycnals are parallel to isotherms below 3000 m.

The horizontal and vertical resolution of measurements in 1959 was not sufficient to distinguish a weak flow near the bottom. Reid (1989) was the first to estimate the deep transport through the Hunter Channel at 1.8 Sv. Two high-resolution geostrophic transport estimates of bottom-water overflow became available based on the sections in 1991 and 1994.

Using the *Meteor* (cruise15) data, Speer et al. (1992) calculated the geostrophic currents in the channel. Taking the flow to be zero at the 2 °C surface, the bottom

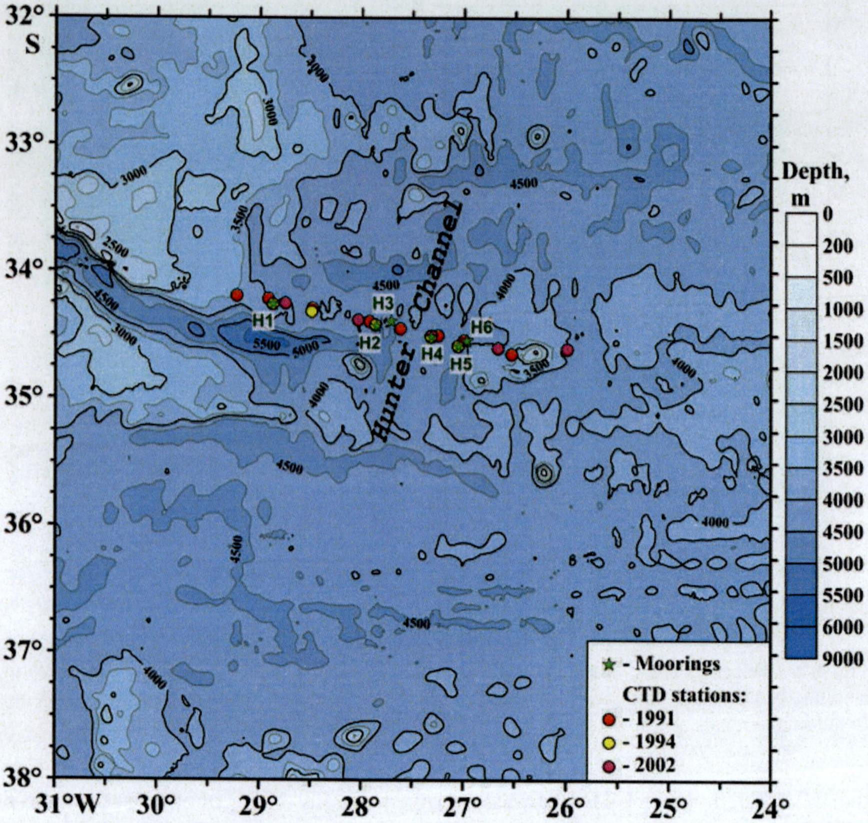

Fig. 4.70 Bottom topography (meters) of the Hunter Channel and locations of moorings and CTD-stations occupied in different years

(geostrophic) northward velocities were typically 1 to 2 cm s^{-1}, and the total transport below 2 °C was 0.7 Sv. The accuracy was estimated within ±50%. Geostrophic velocities were calculated relative to other surfaces of zero velocity. The results ranged within small variations near this value. It is considered that the above transport estimate, relative to the zero surface at 2 °C, is accurate to better than 50%. The calculation was based on the assumption that the velocity below the deepest common level of station pairs is representative of the whole bottom triangle set up by these stations. Speer et al. (1992) conjectured that a higher horizontal station resolution revealing some bottom intensified flow would increase the transport closer to 1 Sv. Northward bottom velocities were of approximately 1 cm s^{-1}. A weak countercurrent of about −0.5 cm s^{-1} was found at 27°30′ W.

The second CTD section in the Hunter Channel was occupied 3 years later during cruise 28 of the R/V *Meteor* (May 1994). Again, the 2 °C isotherm was chosen as an upper limit of bottom water and a reference level of zero velocity for geostrophic calculations. For the bottom triangles, a linear decrease of the current was assumed at the lowest data pair to zero over a distance of 1000 m. The structure of the geostrophic current distribution appears as an eddy with the rotational velocity exceeding 0.5 cm s^{-1} in its core. The bottom water export to the Brazil Basin had almost ceased (<0.1 Sv).

The difference in the depth of the Vema and Hunter channels eliminates water colder than $\theta = 0$ °C from the Hunter Channel. Thus, the average potential temperature of the bottom current in the Hunter Channel is higher by 1 °C than in the Vema Channel. The warmer Hunter Channel flow supplies water to the warmer southeastern part of the Brazil Basin.

From December 1992 to May 1994, deep flow through the Hunter Channel was directly observed by an array of moored current meters and thermistor chains. Direct measurements of the Hunter Channel throughflow reveal a persistent northward transport of bottom waters. Current reversals exist in all records, with time scales of 1–2 weeks.

Highest speeds were observed on the eastern side of the Hunter Channel 15 m above the bottom. The typical values of both current components were above 4 cm s^{-1}, resulting in an average speed of 7.1 cm s^{-1} and a maximum speed of 17.6 cm s^{-1}. The strong eastward flow at this site is caused by the local orientation of a deep northeast-oriented valley. Mean meridional components of about 0.9 cm s^{-1} are characteristic for the central and western sides of the Hunter Channel.

The first long-term moored current-meter observations from the Hunter Channel were reported in (Pätzold et al. 1996). Six moorings were deployed in the Hunter Channel. The sampling rate was 2 h. The mean northerly transport in the channel was estimated from moored current meters at 2.92 Sv, which is significantly higher than earlier estimates based on the geostrophic calculations. The maximum transport exceeding 11 Sv was recorded in June 1993. Its impact on the mean transport was compensated only 6 weeks later by the minimum transport of −5 Sv during a current reversal event of only a few days duration.

A tendency toward increased bottom-water temperatures was observed during the WOCE observational period. The lowest potential temperature in 1991 was 0.223 °C. In 1994, the minimum bottom temperatures barely reached below 0.301 °C and the lowest potential temperature measured in 2002 was 0.219 °C. The lowest salinity observed in 1991 was 34.693 PSU. In 1994, they were replaced by the values above 34.695 PSU. In 2002, the lowest salinities at the bottom were 34.689 PSU.

Sections of potential temperature, salinity, and geostrophic velocity measured in 1991, 1994 are shown in Figs. 4.71a, b; 4.72a, b; and 4.73a, b (Zenk et al. 1999). Those measured in 2002 are shown in Fig. 4.71c, 4.72c, and 4.73c. All graphs were calculated and plotted using the original data and the same programming code and graphic program. The graphs are similar, but lower temperatures were not found in 2002 at the eastern wall of the channel. However, only four stations were located in the gap in 2002. The eastern station was not occupied in the deepest place of the eastern part of the gap.

Distributions of salinities do not show strong differences among three measurements. The geostrophic velocity section relative to the depth of the 2 °C potential

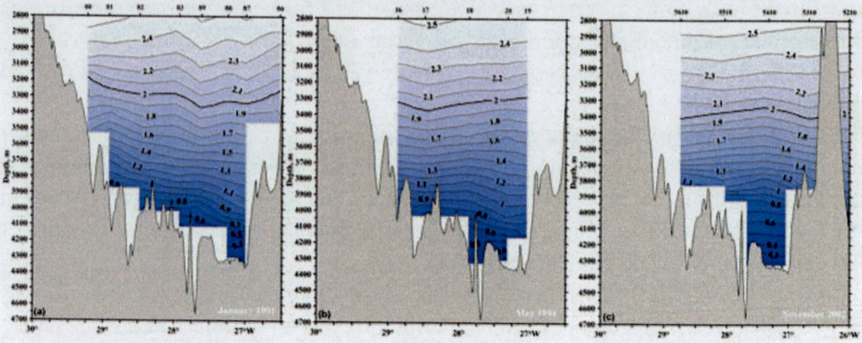

Fig. 4.71 Potential temperature (°C) sections across the Hunter Channel in January 1991 (**a**), May 1994 (**b**), and November 2002 (**c**). The bottom topography was taken from (Smith and Sandwell 1997)

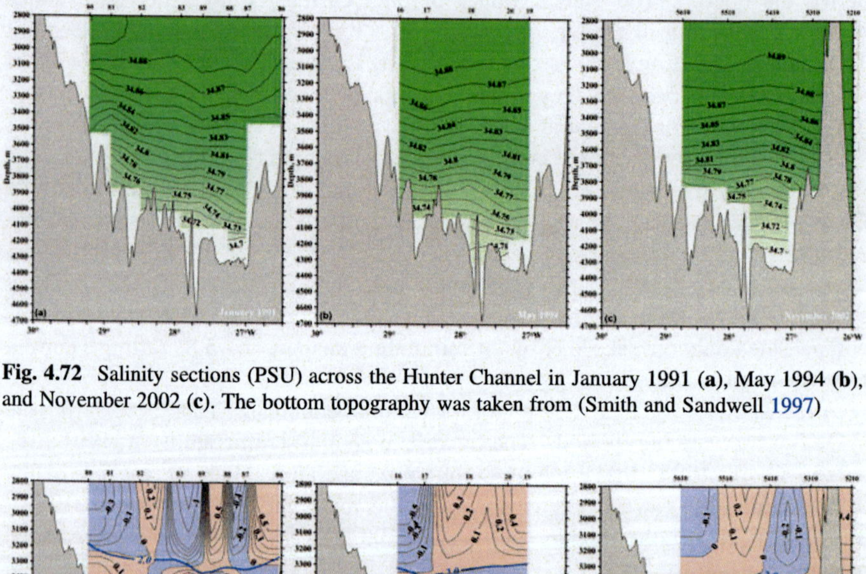

Fig. 4.72 Salinity sections (PSU) across the Hunter Channel in January 1991 (**a**), May 1994 (**b**), and November 2002 (**c**). The bottom topography was taken from (Smith and Sandwell 1997)

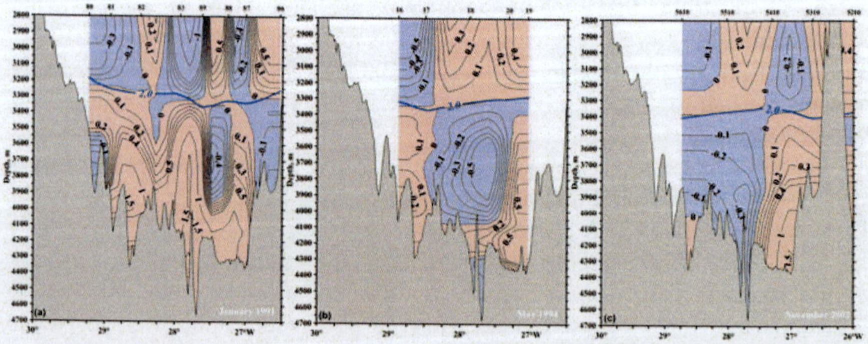

Fig. 4.73 Distribution of geostrophic velocity (cm s^{-1}) across the Hunter Channel in January 1991 (**a**), May 1994 (**b**), and November 2002 (**c**) relative to the 2 °C isotherm depth in Figs. 4.71. The light brown color indicates the northward transport. The bottom topography was taken from (Smith and Sandwell 1997)

temperature isotherm is shown in Fig. 4.73. They result in lower velocities than those measured at moored buoys in 1994. The data of three sections (1991, 1994, and 2002) were processed using the same approach. Note a good coincidence between the velocity distributions in 1991 and 1994 published in (Zenk et al. 1999). Geostrophic transports of water layer below the 2 °C isotherm in 1991, 1994, and 2002 are 0.75 Sv to the north, 0.10 Sv to the south, and 0.13 Sv to the north, respectively. The latter two transports are close to zero with a strong reverse southward flow in a part of the channel. Mean velocities and water transport measured in 1992–1994 by current meters on buoys (2.92 $\pm$ 1.24 Sv), which include the barotropic component, are notably greater.

References

Antonov JI, Seidov D, Boyer TP et al (2010) World Ocean atlas 2009, Vol. 2: salinity, Ed. by S. Levitus. U.S. Government Print. Office, Washington, DC, 184 p

Baines PG, Condie S (1998) Observation and modelling of Antarctic downslope flows: a review, ocean, ice, and atmosphere: interactions at the Antarctic continental margin; Antarctic Res. Ser., Vol. 75, pp. 29–49. AGU, Washington, DC. https://doi.org/10.1029/AR075p0029

Barnier B, Marchesiello P, De Miranda AP, Molines JM, Coulibaly M (1998) A sigma-coordinate primitive equation model for studying the circulation in the South Atlantic. Pt I: model configuration with error estimates. Deep-Sea Res I 45:543–572

Boyer TP, Baranova OK, Coleman C, Garcia HE, Grodsky A, Locarnini RA, Mishonov AV, Paver CR, Reagan JR, Seidov D, Smolyar IV, Weathers K, Zweng MM (2018) World Ocean database 2018. AV Mishonov, technical Ed., NOAA Atlas NESDIS 87. https://www.ncei.noaa.gov/sites/default/files/2020-04/wod_intro_0.pdf

Bullister JL, Feely RA, Wanninkhoff R, Dickson AG, Hansell DA, Key RM (2010) Carbon dioxide, hydrographic, and chemical data obtained during the R/V Ronald H. Brown Cruise in the Atlantic Ocean on CLIVAR repeat hydrography section A13.5 (Mar. 08 - Apr. 17, 2010). Oak Ridge: Carbon Dioxide Information Analysis Center, Oak Ridge National Laboratory, U.S. Department of Energy

Cai W, Greatbatch R (1995) Compensation for the NADW outflow in a global ocean general circulation model. J Phys Oceanogr 25(2):226–241

Coles VJ, McCartney MS, Olson BD, Smethie WJ Jr (1996) Changes in Antarctic bottom water properties in the western South Atlantic in the late 1980s. J Geophys Res Oceans 101 (C4):8957–8970

Comiso JC (2000) Variability and trends in Antarctic surface temperatures from in situ and satellite inferred measurements. J Clim 13:1674–1696

Comiso JC, Gordon AL (1998) Interannual variability in summer sea ice minimum, coastal polynyas and bottom water formation in the Weddell Sea. In: Jeffries MO (ed) Antarctic Sea ice: physical processes, interactions and variability, Antarctic res. Ser., Vol. 74. AGU, Washington, DC, pp 293–315

Danabasoglu G, Yeager SG, Bailey D et al (2014) Ocean Modell 73:76–107

Diansky NA, Bagno AV, Zalesny VB (2002) Sigma model of global ocean circulation and its sensitivity to variations in wind stress. Izvestia Atmosph Oceanic Phys 38(4):537–556

Diansky NA (2013) Simulation of ocean circulation and investigation of its reaction on short-term and long-term atmospheric impacts (Fizmatlit, Moscow, 2013) [in Russian]

Egbert GD, Erofeeva SY (2002) Efficient inverse modeling of barotropic ocean tides. J Atmos Ocean Tech 19:183–204. https://doi.org/10.1175/1520-0426(2002)019<0183:EIMOBO>2.0.CO;2

Emery WJ, Zenk W (2019) Captain F. Spiess and the German Meteor expedition of 1925-27. Brown Walker Press, Irvine, CA, 275 pp

Fahrbach E, Meyer R, Rohardt G, Schröder M, Woodgate RA (1998) Gradual warming of the Weddell Sea deep and bottom water. In: Oerter H (ed) Filchner-Ronne Ice Shelf Programme (FRISP), Report No. 12. Alfred-Wegener-Institut, Bremerhaven, pp 24–34

Fahrbach E, Hoppema M, Rohardt G, Schröder M, Wisotzki A (2004) Decadal-scale variations of water mass properties in the deep Weddell Sea. Ocean Dyn 54:77–91. https://doi.org/10.1007/s10236-003-0082-3

Fahrbach E, Rohardt G, Scheele N, Schröder M, Strass V, Wisotzki A (1995) Formation and discharge of deep and bottom water in the Northwestern Weddell Sea. J Mar Res 53:515–538

Frey DI, Fomin VV, Diansky NA, Morozov EG, Neiman VG (2017a) New model and field data on estimates of Antarctic bottom water flow through the deep Vema Channel. Dokl Earth Sci 474 (1):561–564

Frey DI, Novigatsky AN, Kravchishina MD, Morozov EG (2017b) Water structure and currents in the Bear Island trough in July–August 2017. Russ J Earth Sci 17:ES3003. https://doi.org/10.2205/2017ES000602

Frey DI, Fomin VV, Tarakanov RY, Diansky NA, Makarenko NI (2018) Bottom water flows in the Vema channel and over the Santos plateau based on the field and numerical experiments. In: Velarde MG, Tarakanov RY, Marchenko AV (eds) The ocean in motion: circulation, waves, polar oceanography. Springer, Cham, pp 475–485. https://doi.org/10.1007/978-3-319-71934-4_29

Frey DI, Morozov EG, Fomin VV, Diansky NA, Tarakanov RY (2019a) Regional modeling of Antarctic bottom water flows in the key passages of the Atlantic. J Geophys Res Oceans 124:8414–8428. https://doi.org/10.1029/2019JC015315

Frey DI, Morozov EG, Ansorge I, Fomin VV, Diansky NA, Tarakanov RY (2019b) Thermohaline structure of Antarctic bottom water in the abyssal basins of the South Atlantic. Russ J Earth Sciences 19:ES5005. https://doi.org/10.2205/2019ES000679

Fu LL (1981) The general circulation and meridional heat transport of the Subtropical South Atlantic determined by inverse methods. J Phys Oceanogr 11(9):1171–1193

Fuglister FC (1960) Atlantic Ocean atlas of temperature and salinity profiles and data from the international geophysical year of 1957–1958 (WHOI atlas series, 1), 209 pp, Woods Hole, MA

Fukasawa M, Freeland H, Perkin R, Watanabe T, Uchida H, Nishina A (2004) Bottom water warming in the North Pacific Ocean. Nature 427:825–827

Gamboa LAP, Buffler RT, Barker PF (1983) Seismic stratigraphy and geologic history of the Rio Grande Gap and Southern Brazil Basin. Init. Reports of the Deep Sea Drilling Project Vol. 72, US Government Printing Office, 481–498. https://doi.org/10.2973/dsdp.proc.72.119.1983

GEBCO Bathymetric Compilation Group 2019 (2019) The GEBCO_2019 Grid - a continuous terrain model of the global oceans and land. British Oceanographic Data Centre, National Oceanography Centre, NERC, UK. https://doi.org/10.5285/836f016a-33be-6ddc-e053-6c86abc0788e

Griffies S, Gnanadesikan A, Dixon KW, Dunne J, Gerdes R, Harrison MJ, Rosati A, Russell JL, Samuels BL, Spelman MJ, Winton M, Zhang R (2005) Formulation of an ocean model for global climate simulations. Ocean Sci 1:45–79

Gronholz A, Dong S, Lopez H, Lee S-K, Goni G, Baringer M (2020) Interannual variability of the South Atlantic Ocean heat content in a high-resolution versus a low-resolution general circulation model. Geophys Res Lett 47(23):e2020GL089908. https://doi.org/10.1029/2020GL089908

Hogg N, Biscaye P, Gardener W, Schmitz WJ (1982) On the transport and modification of Antarctic bottom water in the Vema Channel. J Mar Res 40(suppl):231–263

Hogg NG, Owens WB, Siedler G, Zenk W (1996) Circulation in the deep Brazil Basin. In: Wefer G, Berger WH, Siedler G, Webb DJ (eds) The South Atlantic: present and past circulation. Springer Verlag, Berlin, Heidelberg, pp 249–260

Hogg NG, Zenk W (1997) Long-period changes in the bottom water flowing through Vema Channel. J Geophys Res Oceans 102(C7):15639–15646

Hogg N, Siedler G, Zenk W (1999) Circulation and variability at the southern boundary of the Brazil Basin. J Phys Oceanogr 29:145–157

Hogg NG (2001) Quantification of the deep circulation. In: Siedler G, Church J, Gould J (eds) Ocean circulation and climate. Academic, San Diego, pp 259–270

Holfort J, Siedler G (2001) The meridional oceanic transport of heat and nutrients in the South Atlantic. J Phys Oceanogr 31(1):5–28

Jacobs SS, Giulivi CF (1998) Interannual Ocean and sea ice variability in the Ross Sea. In: Jacobs SS, Weiss RF (eds) Ocean, ice, and atmosphere: interactions at the Antarctic continental margin, Antarctic res. Ser., Vol. 75. AGU, Washington DC, pp 135–150

Johnson DA, McDowell SE, Sullivan LG, Biscaye PE (1976) Abyssal topography, nephelometry, currents, and benthic boundary layer structure in the Vema Channel. J Geophys Res Oceans 81:5771–5786. https://doi.org/10.1029/JC081i033p05771

Johnson GC, Doney SC (2006) Recent western South Atlantic bottom water warming. Geophys Res Lett 33:L14614. https://doi.org/10.1029/2006GL026769

Johnson GC, Lyman JM (2020) Warming trends increasingly dominate global ocean. Nat Clim Chang 10:757–761. https://doi.org/10.1038/s41558-020-0822-0

Jungclaus J, Vanicek M (1999) Frictionally modified flow in a deep ocean channel: application to the Vema Channel. J Geophys Res Oceans 104(C9):21123–21136. https://doi.org/10.1029/1998JC900055

Klinck JM (1995) Thermohaline structure of an eddy-resolving North-Atlantic model: the influence of boundary conditions. J Phys Oceanogr 25:1174–1196

Large WG, Yeager S (2009) The global climatology of an interannually varying air-sea flux data set. Clim Dyn 33:341–364

Larque L, Maamaatuaiahutapu K, Garcon VC (1997) On the intermediate and deep-water flow in the South Atlantic Ocean. J Geophys Res Oceans 102(C6):12425–12440

Lenz B (1997) Bodenwassertransporte über die Rio-Grande-Schwelle. Diplom thesis, Kiel University, 68 pp. [Available from library, Christian-Albrechts-Universität, D-24098 Kiel, Germany

Le Pichon X, Ewing M, Truchan M (1971) Sediment transport and distribution in the argentine basin, 2, Antarctic bottom current passage into the Brazil Basin. In: Ahrens LH, Press F, Runcorn SK, Urey HC (eds) Physics and chemistry of the earth, vol 8. Pergamon Press, Oxford, pp 29–48

Lewis EL, Perkin RG (1981) The practical salinity scale 1978: conversion of existing data. Deep-Sea Res I 28(4):307–328

Locarnini RA, Whitworth T III, Nowlin WD Jr (1993) The importance of the Scotia Sea on the outflow of Weddell see deep water. J Mar Res 51:135–153

Locarnini RA, Mishonov AV, Antonov JI et al (2010) In: Levitus S (ed) World Ocean atlas 2009, Vol. 1: temperature. U.S. Government Print. Office, Washington, DC

Marchesiello P, Barnier B, de Miranda AP (1998) A sigma-coordinate primitive equation model for studying the circulation in the South Atlantic. Pt II: meridional transports and seasonal cycle. Deep-Sea Res I 45:573–608

Maurer H, Stocks T (1933) Die Echolotungen des "METEOR". In: Defant A (ed) Wissenschaftliche Ergebnisse, Deutsche Atlantische Expedition auf dem Forschungs und Vermessungsschiff "METEOR" 1925–1927, vol 2. Walter de Gruyter & Co, Berlin

McDonaugh EL, Arhan M, Heywood KJ (2002) On the circulation of bottom water in the region of the Vema Channel. Deep-Sea Res I 49:1119–1139

Memery L, Arhan M, Alvarez-Salgado XA, Messias M-J, Mercier H, Castro CG, Rios AF (2000) The water masses along the western boundary of the south and equatorial Atlantic. Prog Oceanogr 47:69–98

Meredith MP, Naveira Garabato AC, Gordon AL, Johnson GC (2008) Evolution of the deep and bottom waters of the Scotia Sea, Southern Ocean, during 1995–2005. J Clim 21:3327–3343

Morozov EG, Demidova TA, Lappo SS, Pisarev SV, Sokov AV (2003) Spreading of Antarctic bottom water in the Vema Channel. Dokl Earth Sci 390:593–596

Morozov EG (2005) Cruise 17 of R/V *Akademik Sergei Vavilov* in the Atlantic Ocean. Oceanology 45(3):441–443

Morozov EG, Demidov AN, Tarakanov RY (2008) Transport of Antarctic waters in the deep channels of the Atlantic Ocean. Dokl Earth Sci 423(8):1286–1289. https://doi.org/10.1134/S1028334X08080230

Morozov EG, Demidov AN, Tarakanov RY, Zenk W (2010) Abyssal channels in the Atlantic Ocean: water structure and flows. Springer, Dordrecht, 266 p. https://doi.org/10.1007/978-90-481-9358-5

Morozov EG, Tarakanov RY (2014) The flow of Antarctic bottom water from the Vema Channel to the Brazil Basin. Dokl Earth Sci 456:598–601. https://doi.org/10.1134/S1028334X14050274

Morozov EG, Tarakanov RY, Frey DI, Borisov DG (2018) Currents and water structure north of the Vema Channel. Russ J Earth Sci 18:ES5006. https://doi.org/10.2205/2018ES000630

Morozov EG, Frey DI, Neiman VG, Makarenko NI, Tarakanov RY (2019) Extreme transport velocities of Antarctic bottom water flow through the deep Vema Channel. Dokl Earth Sci 486:659–662

Morozov EG, Frey DI, Tarakanov RY (2020) Flow of Antarctic bottom water from the Vema Channel. Geosci Lett 7:Article number 16. https://doi.org/10.1186/s40562-020-00166-4

New M, Hulme M, Jones P (2000) Representing twentieth-century space-time variability, Pt. II: development of 1901-1996 monthly grids of terrestrial surface climate. J Clim 13:2217–2238

Orsi AH, Johnsson GC, Bullister JL (1999) Circulation, mixing, and production of Antarctic bottom water. Prog Oceanogr 43:55–109. https://doi.org/10.1016/S0079-6611(99)00004-X

Pätzold J, Heidland K, Zenk W, Siedler G (1996) On bathymetry of the Hunter Channel. In: Wefer G, Berger WH, Siedler G, Webb DJ (eds) The South Atlantic: present and past circulation. Springer, pp 355–361

Reid JL, Nowlin WD, Patzert WC (1977) On the characteristics and circulation of the Southwestern Atlantic Ocean. J Phys Oceanogr 7(1):62–91

Reid JL (1989) On the total geostrophic circulation of the South Atlantic Ocean: flow pattern, tracers and transports. Prog Oceanogr 23:149–244

Rintoul SR (2007) Rapid freshening of Antarctic bottom water formed in the Indian and Pacific oceans. Geophys Res Lett 34:L06606. https://doi.org/10.1029/2006GL028550

Robertson R, Visbeck M, Gordon A, Fahrbach E (2002) Long-term temperature trends in the deep waters of the Weddell Sea. Deep-Sea Res II 49:4791–4806

Roemmich D (1983) The balance of geostrophic and Ekman transports in the tropical Atlantic Ocean. J Phys Oceanogr 13(8):1534–1539

Sandoval FJ, Weatherly GL (2001) Evolution of the deep western boundary current of Antarctic bottom water in the Brazil Basin. J Phys Oceanogr 31(6):1440–1460. https://doi.org/10.1175/1520-0485(2001)031<1440:EOTDWB>2.0.CO;2

Sarkisyan AS (1991) Simulation of oceanic dynamics. Gidrometeoizdat, St. Petersburg. [in Russian]

Smith WHF, Sandwell DT (1997) Global Sea floor topography from satellite altimetry and ship depth soundings. Science 277:1956–1962. http://topex.ucsd.edu/cgi-bin/get_data.cgi. https://doi.org/10.1126/science.277.5334.1956

Smythe-Wright D, Boswell S (1998) Abyssal circulation in the argentine basin. J Geophys Res Oceans 103(C8):15845–15851

Speer KG, Zenk W, Siedler G, Pätzold J, Heidland C (1992) First resolution of flow through the Hunter Channel in the South Atlantic. Earth Planet Sci Lett 113:287–292

Speer KG, Zenk W (1993) The flow of Antarctic bottom water into the Brazil Basin. J Phys Oceanogr 23:2667–2682

Spiess F (1932) Das Forschungsschiff und seine Reise. Dt. Atl. Exp. "Meteor" 1925-1927, Bd 1. 442 pp. [in German]

Tarakanov RY, Morozov EG (2015) Flow of Antarctic bottom water at the output of the Vema Channel. Oceanology 55(2):153–161. https://doi.org/10.1134/S0001437015010166

Tarakanov RY, Morozov EG, van Haren H, Makarenko NI, Demidova TA (2018) Structure of the deep spillway in the western part of the Romanche fracture zone. J Geophys Res Oceans 123:8508–8531. https://doi.org/10.1029/2018JC013961

Tarakanov RY, Morozov EG, Frey DI (2020) Hydraulic continuation of the abyssal flow from the Vema Channel in the southwestern part of the Brazil Basin. J Geophys Res Oceans 125: e2020JC016232. https://doi.org/10.1029/2020JC016232

Vanicek M, Siedler G (2002) Zonal fluxes in the deep water layers of the western South Atlantic Ocean. J Phys Oceanogr 32(8):2205–2235

Whitworth T, Nowlin WD, Pillsbury RD, Moore MI, Weiss RF (1991) Observations of the Antarctic circumpolar current and deep boundary current in the Southwest Atlantic. J Geophys Res Oceans 96(15):105–118

WHP (2002) WOCE hydrographic Programme, WHP (2002): physical oceanography during Maurice Ewing cruise 3230CITHER2_1 on section A17C. PANGAEA, https://doi.org/10.1594/PANGAEA.293955

Wienders N, Arhan M, Mercier H (2000) Circulation at the western boundary of the south and equatorial Atlantic: exchanges with the ocean interior. J Mar Res 58:1007–1039

WOA2013 (2013) World Ocean Atlas https://www.nodc.noaa.gov/OC5/woa13/woa13data.html

WOD18 (2018) World Ocean Database 2018 https://www.nodc.noaa.gov/OC5/WOD/pr_wod.html

Wüst G (1936) Schichtung und Zirkulation des Atlantischen Ozeans, Das Bodenwasser und die Stratosphäre. In: A. Defant (Ed.), Wissenschaftliche Ergebnisse, Deutsche Atlantische Expedition auf dem Forschungs - und Vermessungsschiff „Meteor“ 1925-1927, 6(1), Walter de Gruyter & Co, Berlin, 411 pg

Zalesny VB, Marchuk GI, Agoshkov VI, Bagno AV, Gusev AV, Diansky NA, Moshonkin SN, Tamsalu R, Volodin EM (2010) Numerical simulation of large-scale ocean circulation based on the multicomponent splitting method. Russ J Num Anal Math Model 25(6):581–609. https://doi.org/10.1515/rjnamm.2010.036

Zalesny VB, Diansky NA, Fomin VV, Moshonkin SN, Demyshev SG (2012) Numerical model of the circulation of the Black Sea and the sea of Azov. Russ J Num Anal Math Model 27 (1):95–111. https://doi.org/10.1515/rnam-2012-0006

Zemba JC (1991) The structure and transport of the Brazil current between 27° and 36°S. Ph.D. dissertation, WHOI-91-37, 160 pp. [Available from Institut für Meereskunde, Düstenbrooker Weg 20, 24105 Kiel, Germany]

Zenk W, Speer KG, Hogg NG (1993) Bathymetry at the Vema sill. Deep-Sea Res I 40:1925–1933

Zenk W, Hogg NG (1996) Warming trend in Antarctic bottom water flowing into the Brazil Basin. Deep-Sea Res I 43(9):1461–1473

Zenk W, Becker S, Jungclaus J, Link R (1998) Physical oceanography. In: Schulz HD, Devey CW, Pätzold J, Fischer G (eds) GEO Bremen/GPI Kiel South Atlantic, cruise no 41. Universität Hamburg, METEOR-Berichte, pp 99, 341 pp–93

Zenk W, Siedler G, Lenz B, Hogg NG (1999) Antarctic bottom water flow through the Hunter Channel. J Phys Oceanogr 29(11):2785–2801

Zenk W, Morozov E, Sokov A, Müller TJ (2003) Vema Channel: Antarctic bottom water temperatures continue to rise. CLIVAR Exchanges 8(1):24–26

Zenk W, Morozov EG (2007) Decadal warming of the coldest Antarctic bottom water flow through the Vema Channel. Geophys Res Lett 34:L14607. https://doi.org/10.1029/2007GL030340

Zenk W (2008a) Temperature fluctuations and current shear in Antarctic bottom water at the Vema Sill. Prog Oceanogr 77:276–284

Zenk W (2008b) Abyssal and contour currents. In: Rebesco M, Camerlenghi A (eds) Developments in sedimentology, Contourites, vol 60. Elsevier, Amsterdam, pp 37–57

Zenk W, Visbeck M (2013) Structure and evolution of the Abyssal Jet in the Vema channel of the South Atlantic. Deep-Sea Res II 85:244–260. https://doi.org/10.1016/j.dsr2.2012.07.033

Zhmur VV, Nazarenko DV (1994) Dynamics of dense fluid in an ocean boundary layer near a sloping bottom. Oceanology 34(2):193–200

Zhu C, Liu Z (2020) Weakening Atlantic overturning circulation causes South Atlantic salinity pile-up. Nat Clim Chang 10:998–1003. https://doi.org/10.1038/s41558-020-0897-7

Zweng MM, Reagan JR, Antonov JI, Locarnini RA, Mishonov AV, Boyer TP, Garcia HE, Baranova OK, Johnson DR, Seidov D, Biddle MM (2013) World Ocean Atlas. Vol. 2, Salinity.

Chapter 5
Further Propagation of Antarctic Bottom Water from the Brazil Basin

Abstract We analyze the flow of Antarctic Bottom Water in the Brazil Basin and the flow of bottom water from the Guiana Basin through the westward equatorial channels into the North American Basin. Much consideration is related to the eastward flow in the equatorial fracture zones in the Mid-Atlantic Ridge: the Romanche and Chain fracture zones. A special region for analysis is the inflow of Antarctic Bottom Water into the western part of the Romanche Fracture Zone at 22°30′ W where a deep spillway is formed. This region is the beginning of the bottom water flow in the Romanche Fracture Zone. The bottom water flows into the fracture over two sills: one with a depth of 4430 m in the narrow main channel of the fracture (3.5 km wide along the 4000 m isobath) and another with a depth of 4570 m in the gap of the Southern Wall of the fracture (8.8 km wide along the 4000 m isobath). We also analyze overflows in the region of the main sills of the equatorial channels and outflow of Antarctic Bottom Water to the Guinea Basin. Analysis is based on the CTD casts combined with LADCP profiles, and a mooring deployed in the region. We assume Wüst's (1936) definition of Antarctic Bottom Water propagation. In other words, this is the bottom water of Antarctic rather than North Atlantic origin. However, there is no generally accepted isotherm or value of any characteristics in literature for the upper boundary of Antarctic Bottom Water in the Equatorial and North Atlantic. Therefore, it is essential that quantitative estimates of Antarctic Bottom Water properties and transport depend on the choice of this boundary.

Keywords Brazil Basin · Equatorial Channel · Romanche and chain fracture zones · Water transport

5.1 Brazil Basin

The Brazil Basin is a major transit region for the propagation of Antarctic waters in the abyssal depths. After the outflow from the Vema Channel, intensity of the bottom flow and vertical mixing decreases strongly in this large and deep basin. In the southern part, the basin is separated from the eastern basin of the Atlantic by the

E. G. Morozov et al., *Bottom Gravity Currents and Overflows in Deep Channels of the Atlantic Ocean*, https://doi.org/10.1007/978-3-030-83074-8_5

relatively high Mid-Atlantic Ridge without deep fractures that are crucial for the bottom water transport to the east. In the northern part, the ridge is crosscut by the Romanche and Chain fracture zones.

Not all publications support the division of Antarctic Bottom Water in the Brazil Basin into several components. Many authors consider that Weddell Sea Deep Water disappears in the northern equatorial part of the basin (Reid 1989; Sandoval and Weatherly 2001). However, Weddell Sea Deep Water is distinguished at 5° S based on the measurements of CFC concentration in (Fischer et al. 1996; Andrie et al. 1998). The distinguishing indicators of Lower Circumpolar Water, the minimum of oxygen concentration (Reid et al. 1977) and the local maximum of silicates disappear in the Brazil Basin because relatively high oxygen concentration in the bottom layer of the Argentine Basin changes to a bottom minimum in the Brazil Basin (Mantyla and Reid 1983).

Part of Antarctic Bottom Water propagating from the Argentine Basin through the passages to the Brazil Basin continues its flow mainly along the western slope of the basin (Stommel and Arons 1960; Rhein et al. 1995; Sandoval and Weatherly 2001). A number of elevations located east of the continental slope of South America form obstacles for the propagation of these waters, and the flow turns around these topographic features. However, according to the scheme in (Hogg and Owens 1999), these obstacles are possibly not very significant. The bottom waters propagate along the western slope of the Brazil Basin approximately to 5° S. According to (De Madron and Weatherly 1994), a branch of the northerly Antarctic Bottom Water flow turns to the south in this region. This flow reaches the southern boundary of the Brazil Basin, then again turns to the north, and flows along the Mid-Atlantic Ridge. These authors consider that a cyclonic eddy is formed in the 4–11° S region near the ridge. This eddy entrains the branch propagating along the Mid-Atlantic Ridge and the branch flowing here directly from the western boundary region. Stephens and Marshall (2000) consider that waters from this eddy propagate further through the Romanche and Chain fractures into the eastern basin. Sandoval and Weatherly (2001) suggest another scheme of Antarctic Bottom Water propagation in the bottom layer of the Brazil Basin based on the results of geostrophic calculations over different sections. They distinguish two main flows of these waters, which they traced from the Vema Channel. The flows are located at a distance of 150–200 km from each other at depths of 4700 and 4250 m. They flow almost parallel to each other in the 16° S region and turn together around the Vitoria-Trindade Ridge of seamounts. The lower flow propagates around the Bahia Mountains and then turns to the east to the Romanche and Chain fractures in the 4° S region. The upper flow propagates along the western coast of the ocean and further to the Equatorial Channel. Thus, the flows are separated into two branches (one flowing to the Equatorial Channel and the other flowing to the eastern basin) immediately after the Vema Channel. However, direct velocity measurements over the continental slope of South America reveal, on average, a return southward current in the

Antarctic Bottom Water layer in the upper branch zone, which is almost equal by intensity to the deep branch (Weatherly et al. 2000). According to (Larque et al. 1997), Lower Circumpolar Water (in terms of (Reid et al. 1977)) and Weddell Sea Deep Water in the Brazil Basin make up several quasi-meridional jets flowing from the Vema and Hunter channels almost directly to the north.

Many earlier estimates of integral bottom water transport in the Brazil Basin are available in the literature and partially summarized in Tables 5.1, 5.2 and 5.3 (Onken 1995; Evans et al. 2017; De Madron and Weatherly 1994, Hogg et al. 1999, and others). Transport of Antarctic Bottom Water decreases strongly from 5–7 to 3 Sv due to vertical mixing as it flows in the Brazil Basin. According to (McCartney and Curry 1993), a decrease in water transport can be caused by weak upwelling flow of water due to diapycnal mixing. The estimates of the transport of Antarctic Bottom Water in the bottom layer presented by researchers differ strongly not only owing to the difference in the definition of the upper boundary of these waters. The choice of different reference surfaces is another cause. Below, we present the available estimates of the position of reference surfaces in the northern part of the Brazil Basin: $\sigma_{1.5} = 32.15$ (Stramma 1991); $\sigma_4 = 45.90$ (Rhein et al. 1998); 2000–3000 m (Friedrichs and Hall 1993); 2000 and 4000 dbar (Fu 1981). A summary of bottom water transport in the Brazil Basin is given in Tables 5.1, 5.2, and 5.3. A scheme of circulation is shown in Fig. 5.1.

The major part of estimates given in these tables was obtained using the data of inverse models (or box-models) for large-scale basins of the ocean limited by quasi-latitudinal sections. Although the authors used almost the same sections, the transport estimates of Antarctic Bottom Water sometimes differ even in sign. The transport can be even directed to the south, for example, in (Fu 1981). This is likely explained by the assumption about zero total transport and the small proportion of Antarctic Bottom Water in the general mass transport. It is also noteworthy that the differences between the most recent publications became much smaller (1–2 Sv), which is within the errors presented in papers (Vanicek and Siedler 2002; Ganachaud 2003). Calculations of geostrophic velocities in (Roemmich 1983) estimate the transport of Antarctic Bottom Water below $\sigma_4 = 45.92$ as 0.2 Sv at 24°S and 0.6 Sv at 8°S.

The geostrophic calculations based on the WOCE data with different reference surfaces yielded the following values of Antarctic Bottom Water transport to the north (isopycnal surfaces $\sigma_2 = 37.10$–37.11 were assumed as the upper boundary of Antarctic Bottom Water layer depending on latitude): 1.5–5.7 Sv at 30° S, 1.9–5.6 at 19° S, 1.6–3.3 Sv at 11° S, 0.5–3.0 at 4° S, and 0.3–0.9 Sv at 8° S. The relative square of the Antarctic Bottom Water layer section as part of the entire ocean zonal section decreases from 16% at 30° S to 4% at 8° S. The flow of Antarctic Bottom Water in the Brazil Basin decreases from south to north due to mixing with overlying waters. Mixing coefficients within the Antarctic Bottom Water layer are in the range $k \sim (1\text{–}5) \cdot 10^{-4}\ \mathrm{m^2\ s^{-1}}$ (Morris et al. 2001).

Table 5.1 Transport in the Southern part of the Brazil Basin (28°-32° S, section A10 along 30° S)

Authors	AABW boundary	Transport, Sv	Comments
Sverdrup et al. (1942)		3.0	Transport across 30° S; geostrophic calculation; reference surface at the boundary between AAIW and deep waters; net northward transport of bottom waters
Wright (1970)	$\theta = 2.0$ °C	5.1	Geostrophic transport at 32° S; reference surface $\theta = 2.0$ °C
Rintoul (1991)	$\sigma_3 = 41.66$	4.0	Inverse model; transport at 32° S
Macdonald (1993)	$\sigma_3 = 41.54$	7.5	Inverse model
Matano and Philander (1993)	z = 3783 m	2.5	Numerical model; transport at 32° S
Speer and Zenk (1993)	$\theta = 2.0$ °C	6.6	Geostrophic transport, meteor 15, reference surface $\theta = 2.0$ °C; total transport over Rio Grande Rise
Saunders and Thompson (1993)	3900 m	1.5	FRAM model at 32° S
De Madron and Weatherly (1994)	$\sigma_4 = 45.87$	6.2	Geostrophic transport at SAVE, 25° S; reference surface $\sigma_4 = 45.87$; AABW transport
	$\sigma_4 = 46.01$	1.7	Geostrophic transport at SAVE, 25° S; reference surface $\sigma_4 = 45.87$; WSDW transport
Cai and Greatbatch (1995)	3700 m	0.3	OGCM model at 32° S
Schlitzer (1996)	$\sigma_4 = 45.9$	4.1–4.2	Box model
Marchesiello et al. (1998)	$\sigma_3 = 41.7$	2.6	Numerical model, annual cycle
Hogg et al. (1999)	$\theta = 2.0$	6.9	Geostrophic transport + moorings over Rio Grande Rise
Holfort and Siedler (2001)	$\sigma_4 = 45.93$	6.8	Inverse model
Vanicek and Siedler (2002)	$\gamma^n = 28.12$	7.0 ± 2.0	Inverse model; transport of LCDW and AABW across A10 section
Ganachaud (2003)	$\gamma^n = 28.11$	6.0 ± 1.3	Inverse model
Demidov (2003)	$\sigma_2 = 37.1$	1.5–2.4	Geostrophic transport over A10 section, reference surfaces $\sigma_2 = 37.11$ and 37.095
McDonagh and King (2005)	$\sigma_3 = 41.58$	5.5	Inverse model (30° S)
Naveira Garabato et al. (2014)	$\gamma^n = 28.18$	5.0	Inverse model (30° S)
Hernández-Guerra et al. (2019)	$\gamma^n = 28.15$	3.6–8.5	Inverse model (30° S)

Table 5.2 Transport in the Central part of the Brazil Basin (18–24° S)

Authors	AABW boundary	Transport, Sv	Comments
Sections along 18–19° S			
Wright (1970)	$\theta = 2.0$ °C	6.4	Geostrophic transport at 24° S; reference surface $\theta = 2.0$ °C
Roemmich (1983)	$\sigma_4 = 45.92$	0.2	Geostrophic transport + Ekman's transport at 24° S.
McCartney and Curry (1993)	$\theta = 1.9$ °C	6.7	Geostrophic transport at 23° S; reference surface $\theta = 1.9$ °C; *Oceanus* 1983
Speer and Zenk (1993)	$\theta = 2.0$ °C	5.0	Geostrophic transport, *Meteor* 15, reference surface $\theta = 2.0$ °C; total transport at 24° S
Macdonald (1998)	$\theta = 1.7$ °C	6 ± 1	Inverse model based on IGY section at 23° S, *Oceanus* 1983
		5 ± 1	Inverse model based on IGY section at 27° S, *Knorr* 1988 (SAVE). Transport 0 ± 1 east of the Mid Atlantic Ridge
Vanicek and Siedler (2002)	$\gamma^n = 28.12$	7 ± 2	Inverse model. Transport of LCDW and AABW across sections SAVE 3 and 4 at 25° S
		7 ± 2	Inverse model. Transport of LCDW and AABW across sections at 23° S; *Oceanus*
15–21° S, section A09 along 19°S			
Wright 1970	$\theta = 2.0$ °C	2.3	Geostrophic transport at 16° S; reference surface $\theta = 2.0$
Speer and Zenk 1993	$\theta = 2.0$ °C	4.5	Geostrophic transport, *Meteor* 15, reference surface $\theta = 2.0$ °C; total transport at 19° S
De Madron and Weatherly (1994)	$\sigma_4 = 45.87$	4.3	Geostrophic transport at SAVE, 18° S; reference surface $\sigma_4 = 45.87$; AABW transport
	$\sigma_4 = 46.01$	1.6	Geostrophic transport at SAVE, 18° S; reference surface $\sigma_4 = 45.87$; WSDW transport
Vanicek and Siedler (2002)	$\gamma^n = 28.12$	7 ± 2	Inverse model. Transport of LCDW and AABW across A09 section
Ganachaud (2003)	$\gamma^n = 28.11$	3.4 ± 1.1	Inverse model.
Demidov (2003)	$\sigma_2 = 37.1$	1.9–2.2	Geostrophic transport across A09 section; reference surfaces $\sigma_2 = 37.11$ and 37.10
Weatherly et al. (2000)	$\sigma_4 = 45.87$	4	DWBC of AABW; moorings at 18°S; seasonal variation ~1 Sv
Evans et al. (2017)	$\gamma^n = 28.27$	2.6 ± 0.5	Inverse model at 24°S

5.2 Flow in the Guiana Basin and Westward Equatorial Channels

Intensity of Antarctic Bottom Water transport in the abyssal ocean decreases strongly north of the Brazil Basin. These waters are found here only in strongly transformed form. A large portion is transported to the eastern basin mainly through

Table 5.3 Transport in the Northern part of the Brazil Basin

Authors	AABW boundary	Transport, Sv	Comments
Section A08 along 11° S			
Roemmich (1983)	$\sigma_4 = 45.92$	0.6	Geostrophic transport + Ekman's transport at 8° S
Macdonald (1993)	$\sigma_3 = 41.54$	8.1	Inverse model
McCartney and Curry (1993)	$\theta = 1.9$ °C	5.5	Geostrophic transport; reference surface θ =1.9 °C at 11° S; *Oceanus* 1983
Speer and Zenk (1993)	$\theta = 2.0$ °C	3.1	Geostrophic transport, *Meteor* 15, reference surface θ =2.0 °C; total transport at 11° S.
Cai and Greatbach (1995)	3700 m	3.7	OGCM, 11° S
Macdonald (1998)	$\theta = 1.8$ °C	4 ± 1	Inverse model based on section at 11° S; *Oceanus* 1983
Sloyan and Rintoul (2001)	$\gamma^n = 28.30$	0	Inverse model
Vanicek and Siedler (2002)	$\gamma^n = 28.12$	4.0 ± 3.0	Inverse model. Transport of LCDW and AABW across section at A08
		5.0 ± 2.0	Inverse model. Transport of LCDW and AABW across section at 11° S; *Oceanus*
Ganachaud (2003)	$\gamma^n = 28.11$	3.1 ± 1.5	Inverse model
Demidov (2003)	$\sigma_2 = 37.10$	1.5–1.6	Geostrophic transport at section A08; reference surfaces $\sigma_2 = 37.11$ and 37.10
Section A07 at 4.5°S			
Wright (1970)	$\theta = 2.0$ °C	2.8	Geostrophic transport at 8° S; reference surface θ =2.0 °C
De Madron and Weatherly (1994)	$\sigma_4 = 45.87$	0.7	Geostrophic transport at SAVE, 10°–0°S; reference surface $\sigma_4 = 45.87$; AABW transport
	$\sigma_4 = 46.01$	2.8	Geostrophic transport at SAVE, 10°–0°S; reference surface $\sigma_4 = 45.87$; WSDW transport
Vanicek and Siedler (2002)	$\gamma^n = 28.12$	3.0 ± 2.0	Inverse model. Transport of LCDW and AABW across section A08
Ganachaud (2003)	$\gamma^n = 28.11$	3.4 ± 1.2	Inverse model
Demidov (2003)	$\sigma_2 = 37.10$	3.0	Geostrophic transport at section A07; reference surfaces $\sigma_2 = 37.1$

the Romanche, Chain, and Vema fracture zones as well as through other fractures in the northern part of the Mid-Atlantic Ridge (Fig. 5.2).

Antarctic Bottom Water propagates to the western part of the North Atlantic through the Equatorial Channel located north of the Brazil Basin, which is almost zonally oriented. The width of this channel is approximately 300 km, and the maximum depth is 4500 m. According to (Hall et al. 1997), the Equatorial Channel is located between the Parnaiba Ridge (2° S) at the continental slope and the ridge of

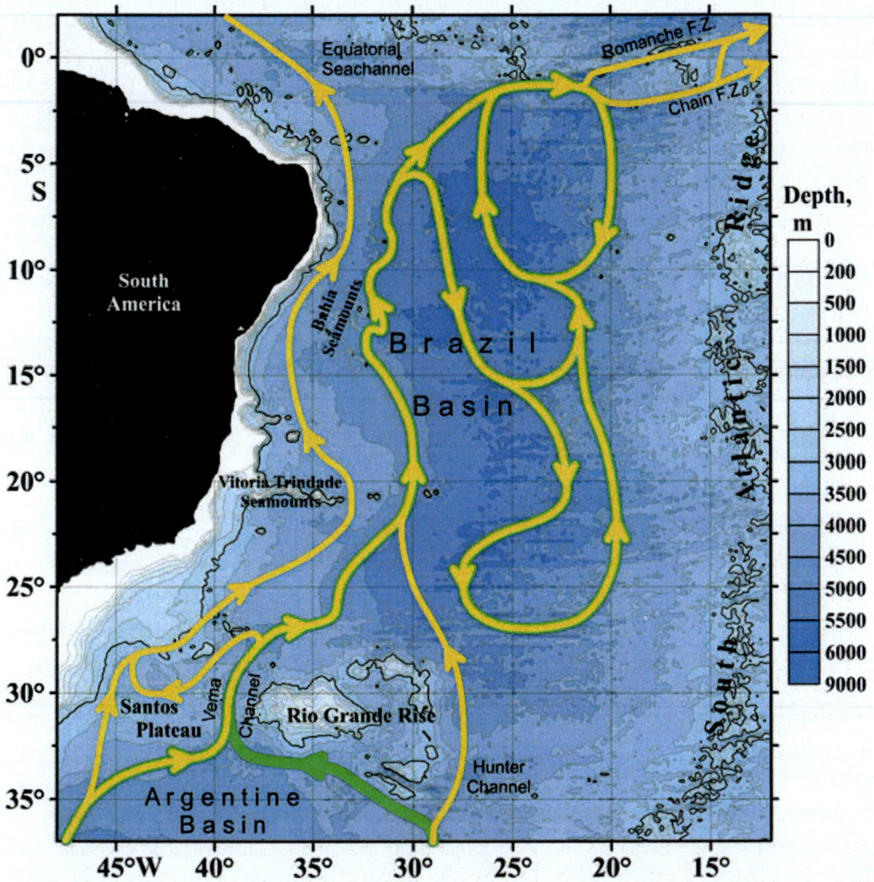

Fig. 5.1 Bottom circulation in the Brazil Basin (De Madron and Weatherly 1994; Sandoval and Weatherly 2001; Morozov et al. 2008). Thick green lines show circulation of Weddell Sea Deep Water and Lower Circumpolar Water. Thin yellow lines show circulation only of Lower Circumpolar Water

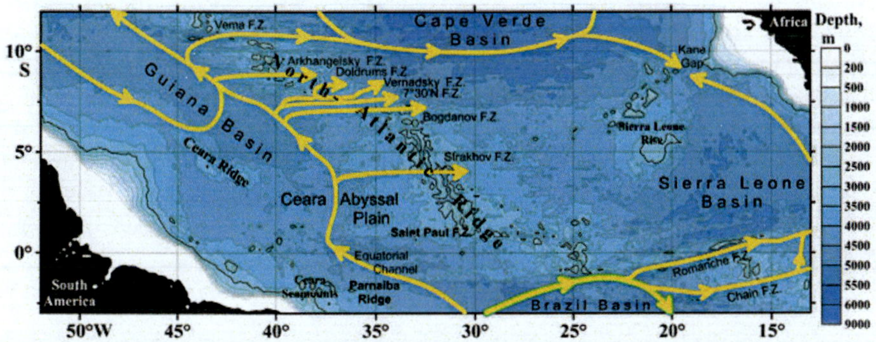

Fig. 5.2 Circulation of Antarctic Bottom Water in the Equatorial Channel and Guiana Basin and eastward flows through the Mid-Atlantic Ridge (Schmitz 1996; Friedrichs and Hall 1993; McCartney et al. 1991). Thick lines show the circulation of Weddell Sea Deep Water and Lower Circumpolar Water. Thin lines show circulation only of Lower Circumpolar Water

the Saint Paul Fracture Zone (1.5° N). The southern part of the channel, formed by the extension of one of the fractures of the Mid-Atlantic Ridge with a sill depth of about 4500 m, topographically blocks the northward propagation of waters colder than 0.6 °C (McCartney and Curry 1993). On the northern side, the same threshold depths are noted above the Ceara Abyssal Plain, which retains the water colder than 1 °C at the bottom (Whitehead and Worthington 1982). These two thresholds represent giant deep cataracts along the path of Antarctic waters spreading into the abyssal of the Atlantic.

According to (Mantyla and Reid 1983), the coolest and least saline Antarctic water (among the bottom waters of all oceans) propagates in the western part of the Atlantic Ocean. The properties of this water change after it passes the equator. The formerly coolest and least saline water becomes warmer and more saline than in the other oceans. Mantyla and Reid (1983) conclude on the basis of radiocarbon concentration that Antarctic waters in the North Atlantic are younger compared to similar waters in the North Pacific. All these properties provide evidence that Atlantic Antarctic waters are strongly transformed under the influence of North Atlantic Deep Water.

Since the Equatorial Channel is located at the equator, the estimates of Antarctic Bottom Water transport in the channel are mainly the data of direct instrumental measurements of currents (rather than geostrophic estimates). Publications dedicated to the interpretation of these data provide evidence that a strong seasonal variability exists from 0.2 Sv in February and March to 3.1 Sv in September and October (Hall et al. 1997) (Table 5.4). The authors of this paper also note that the deep thermocline between the Antarctic and North Atlantic waters ascends and descends synchronously with the intensity of the flow of Antarctic waters. All available estimates of transports from Table 5.4 fall practically within the range of seasonal variation presented above.

According to the scheme of circulation, Antarctic Bottom Water propagates in the Equatorial Channel and further in the Guiana Basin along the western slope of the northern Mid-Atlantic Ridge (Friedrichs and Hall 1993) (Fig. 5.2). A quasi-stationary cyclonic eddy existing north of the Ceara Ridge extends up to 17° N. This internal transport of this eddy is approximately 3 Sv of bottom waters (Friedrichs and Hall 1993; Friedrichs et al. 1994). The flow of Antarctic Bottom Water splits in the Vema Fracture Zone region (11° N): part of this water flows through the Vema Fracture Zone to the east, while the other part flows to the north of the Guiana Basin. The estimate in (Rhein et al. 1998) indicates that 1.1 Sv of the total 1.4 Sv of bottom waters transported through the Equatorial Channel flow to the eastern part of the Atlantic and only 0.3 Sv remain in the western basin. However, other estimates presented in Table 5.5 indicate that the remaining part is much greater.

Table 5.4 Antarctic Bottom Water transport across the Equator and in the Equatorial Channel

Authors	AABW boundary	Transport, Sv	Comments
Sverdrup et al. (1942)		1	Indirect estimates; net northward transport of bottom waters across the equator
Whitehead and Worthington (1982)	θ = 1.9 °C	0.8	Current meters at 4° N north of Ceara abyssal plain
		2.0	Geostrophic transport at 4° N north of Ceara abyssal plain; reference surface θ =1.9 °C
McCartney and Curry (1993)	θ = 1.9 °C	4.3	Equatorial geostrophic relation for zonal velocity with reference surface 4000 dbar; westward transport over the section across 37° W
		4.0	Geostrophic transport at 4° N north of the Ceara abyssal plain; reference surface θ =1.9 °C;
Luyten et al. (1993)	θ = 1.9 °C	5.1	Geostrophic transport at 4° N north of the Ceara abyssal plain; reference surface θ =3.7
		4.6	Reference surface θ =2.0 °C
		4.4	Reference surface θ =1.9 °C
Hall et al. (1997)	θ = 1.9 °C	1.8 ± 1.6 (from −2.3 to 8.6)	Line of current meters across the Equatorial Channel at 36°W
	θ = 1.8 °C	2.0 ± 1.6 (from −2.2 to 8.3)	
		0.2–3.1 annual cycle of transport (February–September)	
	θ = 1.7 °C	2.1 ± 1.5 (from −2.5 to 8.9)	
Rhein et al. (1995)	$\sigma_4 = 45.90$	2.6 ± 2.0	LADCP measurements in the Equatorial Channel at 35°W
Schlitzer (1996)	$\sigma_4 = 45.90$	3.2–3.7	Box model; transport across the equator
Rhein et al. (1998)	$\sigma_4 = 45.90$	1.8	LADCP measurements combined with Pegasus profiling system in the Equatorial Channel at 35°W
		2.3	Geostrophic calculations at 4°30′ N north of the Ceara abyssal plain and 44° W; reference surfaces $\sigma_4 = 45.90$
Schott et al. (2003)	$\sigma_4 = 45.90$	1.7	LADCP measurements in the Equatorial Channel at 35° W
Limeburner et al. (2005)		2.0	Line of current meters across the Equatorial Channel at 36° W

Table 5.5 Antarctic Bottom Water transport in the Guiana Basin (8–13° N, section A06)

Authors	AABW boundary	Transport, Sv	Comments
Wright (1970)	$\theta = 2.0$ °C	2.7	Geostrophic transport at 8° N; reference surface $\theta = 2.0$ °C
		1.4	Geostrophic transport at 16° N;
Roemmich (1983)	$\sigma_4 = 45.92$	0.5	Geostrophic transport + Ekman's transport at 8° N
Molinari et al. (1992)	$\theta = 1.8$ °C	2.2 ± 0.9	Average geostrophic transport over several sections between 4° N and 14°30′ N; reference surface $\theta = 4.7$ °C
Friedrichs and Hall (1993)	$\theta = 1.8$ °C	2.1 ± 1.0	Geostrophic transport at 11° N; variable reference level based on the a priori constraints and optimization procedure
Klein et al. (1995)	$\sigma_4 = 45.90$	0.69	Geostrophic transport with inverse correction; transport at 8° N.
		0.31	Transport at 14°30′ N
Macdonald (1998)	$\theta = 1.8$ °C	0.0 ± 1.0	Inverse model; section along 11° N
Ganachaud (2003)	$\gamma^n = 28.11$	3.6 ± 1.2	Inverse model; section along 7°30′ N
Demidov (2003)	$\sigma_2 = 37.11$	0.3–0.9	Geostrophic transport across A06 section; reference surface $\sigma_2 = 37.11$ and 37.10
Hernández-Guerra et al. (2014)	$\gamma^n = 28.1295$	1.6 ± 0.9	Inverse model; two sections at 7°30′ N
		1.5 ± 0.8	
Fu et al. (2018)	$\gamma^n = 28.141$	2.2 ± 2.3	Inverse model; 2 sections along 14°30′ N
		1.4 ± 2.4	

5.3 North American Basin

After passing the Equatorial Channel and Guiana Basin, Antarctic Bottom Water flows to the North American Basin. Many researchers (Wunsch 1984; Stephens and Marshall 2000) consider that Antarctic Bottom Water propagates mainly near the western slope of the Mid-Atlantic Ridge and, generally, the circulation in the basin is cyclonic (Stephens and Marshall 2000; Weatherly and Kelley 1982; Lavin et al. 2003) (Fig. 5.3).

The authors of many publications consider that the Newfoundland Bank region (42°–53° N) is the limit for the propagation of waters of Antarctic origin (Mantyla and Reid 1983; Mamayev 1992; Dobrovolsky and Zalogin 1992). The $\theta = 2.0$ °C isotherm is located at the bottom in this region. According to the classification in (Wüst 1936), this isotherm is the northern boundary of Antarctic Bottom Water propagation. It is the opinion of Schmitz (1996) that only 1 Sv of this water of the

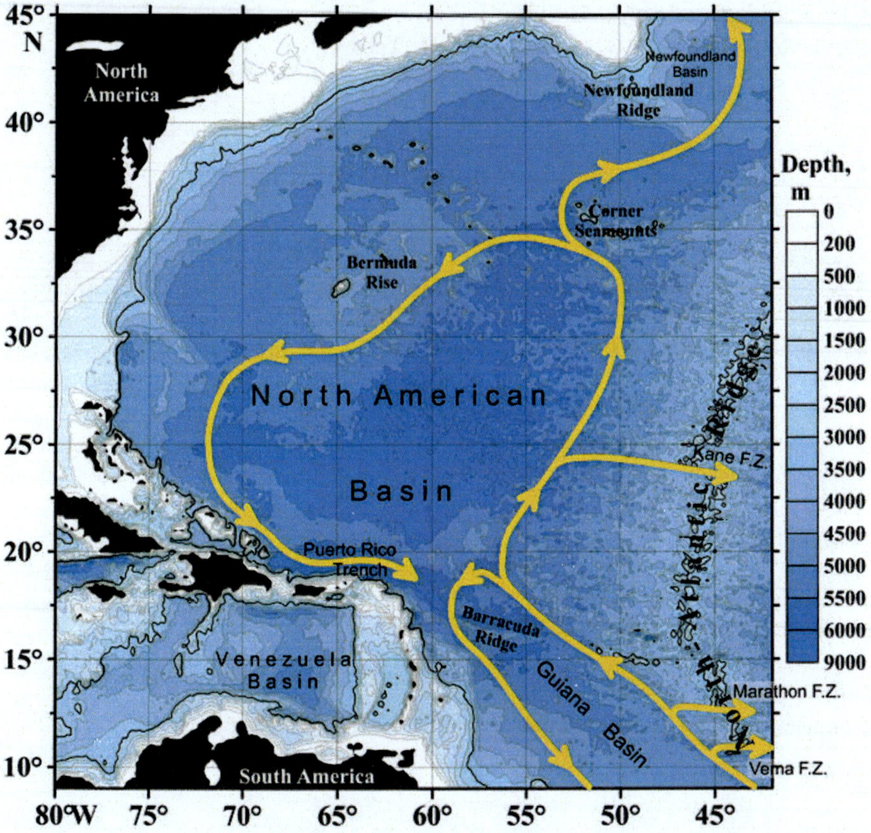

Fig. 5.3 Circulation of Antarctic Bottom Water (Lower Circumpolar Water) in the Northwest Atlantic (Schmitz 1996; Friedrichs and Hall 1993)

total 2 Sv, which passes the 25° N region, reaches the Newfoundland Bank. Traces of Antarctic Bottom Water were also found in the bottom layer at 50° N (Clarke and Gascard 1983). Antarctic Bottom Water returns to the south with the Deep Western Boundary Current (Tucholke et al. 1973; Weatherly and Kelley 1982; Schmitz 1996; Friedrichs and Hall 1993).

Other authors consider that only one-half of Antarctic Bottom Water from the Guiana Basin flows to the Vema Fracture Zone, and its transport is estimated at 2.1–2.3 Sv (McCartney et al. 1991). The calculation of geostrophic velocities made in (Molinari et al. 1992) provides evidence that 2.2 ± 0.9 Sv of Antarctic Bottom Water is transported to the Guiana Basin. Other authors (Klein et al. 1995) estimate that the transport of Antarctic Bottom Waters north of the equator is much smaller:

0.69 Sv at 8° N and only 0.31 Sv at 14.5° N. A scheme of bottom water circulation in the Northwest Atlantic is shown in Fig. 5.3 (Table 5.6).

5.4 Eastward Equatorial Channels. The Romanche and Chain Fracture Zones

The Equatorial Romanche and Chain fractures zones are pathways for Antarctic Bottom Water spreading from the abyssal depths of the West Atlantic to the equatorial part of the East Atlantic (Fig. 5.4). The Chain Fracture Zone is located south of the equator, 200 km south of the Romanche Fracture Zone. Both fracture zones are corridors for water flow between the Brazil and Guinea basins. The $\theta = 1.9$ °C isotherm is traditionally accepted as the upper boundary of the Antarctic Bottom Water in the equatorial fractures. From the Guinea Basin, these waters then spread north and south and fill the abyssal of the East Atlantic from the Angola Basin in the south to the Western European Basin in the north. Detailed description of field research, bottom topography, measurements of currents in the eastern parts of the fractures, temperature distribution along fractures, long-term variations in temperature and salinity are given in the subsequent sections. Analyses of the bottom flows in the deep-sea spillway in the western part of the Romanche Fracture Zone and spillways on the Main Sill of the Chain Fracture Zone and Nameless Sill of the Romanche Fracture Zone are presented.

5.4.1 Research History

The authors of many papers demonstrated that bottom topography of the Atlantic Ocean allows propagation of Antarctic Bottom Water to the northern latitudes of the East Atlantic only through the Romanche and Chain fracture zones (at the equator) and the Vema Fracture Zone (11° N) (Wüst 1936; Leontieva 1985; McCartney et al. 1991; Messias et al. 1999). Other small and shallower passages are less significant.

The Romanche Fracture Zone is a deep passage in the Mid-Atlantic Ridge 800 km long and 10–40 km wide. Together with the Chain Fracture Zone, they form an equatorial pathway for Antarctic Bottom Water to the East Atlantic. The main sill across the Romanche Fracture Zone is located at a depth of 4359 m (13°40′ W). The Vema Deep (7856 m) is the deepest place in the Romanche Fracture Zone. The Chain Fracture Zone is located south of the equator, 200 km south of the Romanche Fracture Zone. The main sill of the Chain Fracture Zone is located at a depth of 4050 m (12°22′ W). Both fracture zones are corridors for water flow between the Brazil and Guinea basins.

The history of hydrographic research in the Romanche Fracture Zone is longer than a century. The first measurements in this region were made in 1901. Later, the

Table 5.6 Antarctic Bottom Water transport in the North American Basin

Authors	AABW boundary	Transport, Sv	Comments
24°-30° N section A05			
Tucholke et al. (1973)	$\theta = 1.84°$	−1.6	Current meters and geostrophic calculation only in the DWBC at 24° N; reference surface 4700 m
Roemmich (1983)	$\sigma_4 = 45.92$	0.4	Geostrophic transport + Ekman's transport at 24° N
Schlitzer (1996)	$\sigma_4 = 45.90$	0.7–1.1	Box model; 30° N;
Macdonald (1998)	$\theta = 1.90°$	−4.0 ± 2.0	Inverse model; section along 24° N in 1981
Lavin et al. (2003)	4900 m	0.9	Geostrophic transport at 24° N; variable reference level based on the a priori constraints; −1.1 Sv in the DWBC and 2.0 Sv in the rest part of the Western Basin
Ganachaud (2003)	$\gamma^n = 28.11$	1.0 ± 1.0	Inverse model; section along 24° N
Marsh et al. (2005)	$\theta = 2.0°$	3.0	Numerical model; section A05 along 26° N
Frajka-Williams et al. (2019)	p = 4100 db	2.8 ± 0.6	Average geostrophic transport over 6 sections at 24° N; reference surface p = 4100db
	$\theta = 1.80°$	1.9–4.0	Geostrophic transport over 6 sections at 24° N; reference surface $\theta = 1.8°$
Atkinson et al. (2012)	Z = 5000 m	1.4–2.8	Geostrophic+Ekman's transports at 6 sections adjusted by rapid-WATCH data at 26° N
		2.0 ± 0.5 (0.5 ± 0.4)	Rapid-WATCH timeseries at 26° N; mean ± STD (season cycle±STD)
	$\theta = 1.80°$	1.4–3.7	Geostrophic+Ekman's transports at 6 sections adjusted by rapid-WATCH data at 26° N
Hernández-Guerra et al. (2014)	$\gamma^n = 28.1295$	1.4 ± 1.8 1.1 ± 1.8	Inverse model; 2 sections at 24° N
Fu et al. (2018)	$\gamma^n = 28.141$	2.1 ± 2.5 1.2 ± 2.6	Inverse model; 2 sections along 24°30' N
36° - 40° N; section A03			
Weatherly and Kelley (1982)	$\theta = 1.82$	−0.5	Current meters only in the DWBC at 40° N
Macdonald (1998)	$\theta = 1.90$	−7.0 ± 1.0	Inverse model; section along 36° N
McDonagh et al. (2010)	$\theta = 1.8$	4.7	Inverse model; section along 36° N
48° N, section A02			
Macdonald (1998)	$\theta = 2$	−10.0 ± 1.0	Inverse model; section along 48° N
Ganachaud (2003)	$\gamma^n = 28.11$	0	Inverse model; section along 48° N

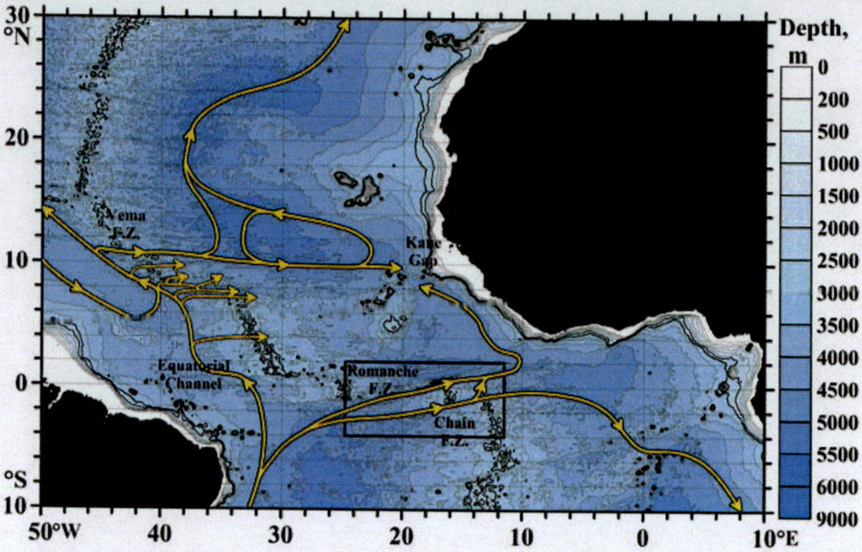

Fig. 5.4 Scheme of Antarctic Bottom Water spreading in the equatorial and East Atlantic adapted from (McCartney et al. 1991) and (Mercier et al. 1994). The Romanche and Chain region in the rectangular frame is shown on a larger scale in Fig. 5.5b

fracture zone was studied during the International Geophysical Year (1958) and during Russian expeditions in the 1960s and 1970s, in particular, in 1967 onboard the R/V *Akademik Kurchatov*. The eastern parts of the Romanche and Chain fracture zones were intensely studied in 1991–1994 during the French expeditions within the Romanche-I, Romanche-II, and Romanche-III projects (Fig. 5.5a), which were parts of the International WOCE program. Many CTD/LADCP stations were occupied in the region and moorings were deployed for long-term operation. These measurements with modern instruments were the most precise ones. A CTD-section across the Romanche Fracture Zone at 16°03′ W was occupied by Russian scientists in 2005 and repeated in 2009 (cruises 19 and 29 of the R/V *Akademik Ioffe*) (Demidov et al. 2006).

In 2011–2014, during the expeditions of the R/V *Akademik Ioffe* and *Akademik Sergey Vavilov* in the western part of the Romanche Fracture Zone (near 22°30′ W), 24 CTD stations were occupied (Fig. 5.5b). All CTD casts in 2005, 2009, and 2011–2014 were accompanied with LADCP profiling. In 2013–2014, a mooring was deployed in the same region, which operated for 6 months.

The first hydrographic measurements in the region of the Chain Fracture Zone were carried out in the western part of the channel in 1903 and 1927. Unfortunately, they were not made in the axial part of the fracture. The high quality of

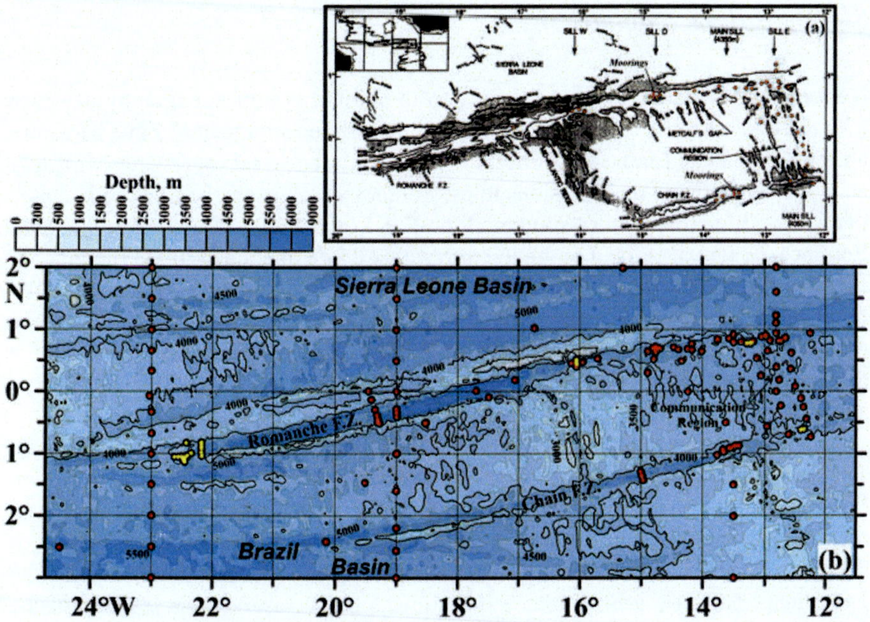

Fig. 5.5 Bottom topography in the region of the Romanche and Chain fracture zones based on the data of multibeam echo sounding from (Mercier et al. 1994) (**a**), and the database (Smith and Sandwell, 1997, version V19.1) (**b**). Red circles show the position of CTD stations from the WOD18 database (WOD18 2018; Boyer et al. 2018), yellow dots show stations with CTD/LADCP casts from the ships of the Shirshov Institute of Oceanology *Akademik Ioffe* and *Akademik Sergei Vavilov* in 2005, 2009, and 2011–2014

measurements during the *Meteor* expedition in 1927 is notable. Their measurements are close to the data collected in 1991. Much work in the region was conducted in 1961. A number of transform faults including the Chain transform fault were found during cruise 17 of the R/V *Chain* in 1961. They are described in (Metcalf et al. 1964; Heezen et al. 1964). It is likely that the first station in this cruise was occupied in the axial part of the fracture zone at 13°10′ W (depth 4823 m). The bottom potential temperature was equal to 0.806 °C. The Chain Fracture Zone was studied in 1991–1994 during the French expeditions together with the Romanche Fracture Zone. In 2009, a Russian expedition during cruise 27 of the R/V *Akademik Ioffe* occupied two CTD stations near the main sill of the Chain Fracture Zone. All CTD casts in 2005 and 2009 were accompanied with LADCP profiling.

Figure 5.5 shows the position of all CTD stations with measurements to the bottom available from the WOD18 database, stations of our measurements in 2005, 2009 and 2011–2014, as well as the position of moorings deployed within the ROMANCHE programs in the Romanche and Chain fracture zones.

5.4.2 Bottom Topography

Romanche Fracture Zone According to the digital bottom topography database (Smith and Sandwell 1997, version V19.1), the Romanche Fracture Zone begins as an abyssal channel for the eastward flow of the coldest abyssal waters approximately at 22°30′ W in the two depressions in the Southern Wall of the fracture with depths of about 4530 m (hereinafter Southern Entrance-1) and 4440 m (hereinafter Southern Entrance-2), and over a sill of the fracture with a depth of about 4200 m (hereinafter Western Sill-1) (Figs. 5.6 and 5.7). This region we designate as the entrance to the fracture channel. Another sill with a depth of about 4600 m is located in the same region between Southern Entrances 1 and 2 in the fracture channel (Fig. 5.7). In the northern part of the fracture there is another fracture valley with a threshold depth of about 3900 m (hereinafter Western Sill-2).

Among the five indicated threshold points, four are marked as derived from satellite altimetry data in the database (Smith and Sandwell 1997), and only Western Sill-1 as revealed from the echo sounder data. The depth of the latter can serve as a reliable estimate of the maximum possible threshold depth of the southern valley of the Romanche Fracture Zone (i.e., not deeper than 4200 m), provided that the one-minute database grid (Smith and Sandwell 1997) resolves the features of the seabed. The depth of the northern valley of the fracture (Western Sill-2), based on the depths at the points of the valley, marked as echo sounding, does not exceed 4300 m. Note that detailed studies of the bottom topography are required to establish the real pattern of the depths at the entrance to the fracture, which is characterized here by very strong roughness. Such measurements were carried out during our expeditions in 2011–2014. These measurements revealed significant differences in the measured depths in comparison with the digital array; the differences reached 700 m. The measurement results are described in more detail in Sect. 5.4.7.

The bottom topography in the eastern part of the fracture (east of 19°30’ W) was studied in detail using a multibeam echo sounder during the French projects ROMANCHE-I, -II, -III in 1991–1992 as part of the international WOCE program. The map constructed on the basis of these measurements is shown in Fig. 5.5a (Mercier and Morin 1997). According to these studies, the exit from the fracture channel, i.e., the region behind which the coldest part of the flow spreads along the Sierra Leone Basin abyssal is the region of the Romanche Fracture Zone at approximately 12°30′-13°00′ W near Sill E with a depth of about 4550 m and Northern Exit with a depth exceeding 4600 m, which is a depression in the Northern Wall of the fracture located west of Sill E (Figs. 5.5a and 5.6).

The deepest region of the Romanche Fracture Zone (7856 m) is located in the middle part of the fracture, the Vema Deep (Mercier and Morin 1997) (Fig. 5.6). The minimum sill depth in the channel is 4359 m at 13°45′ W (Main Sill) (Mercier and Morin 1997) (Fig. 5.5a and 5.6). Note that, according to the database (Smith and Sandwell 1997), the main sill of the fracture is located approximately 10 miles west of the actual Main Sill determined from the multibeam echo sounding (Figs. 5.5a and 5.6). According to the same database, the depth above the Southern Wall of the

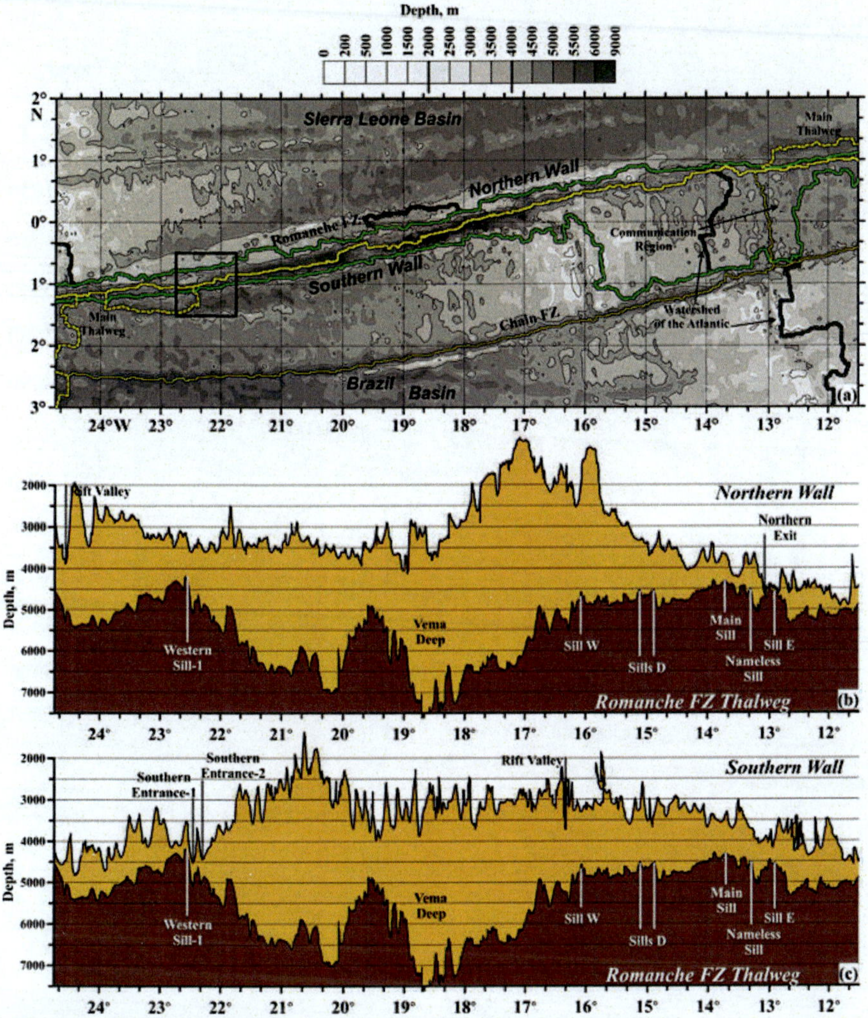

Fig. 5.6 Bottom topography in the region of the Romanche and Chain fracture zones according to the database (Smith and Sandwell 1997, version V19.1) (**a**); Bottom topography along the Northern Wall and thalweg (line of maximum depts) of the Romanche Fracture Zone according to the database (**b**); Bottom topography along the Southern Wall and thalweg of the Romanche Fracture Zone (**c**). Bold green lines in Fig. 5.6a show the ridges of the Southern and Northern walls of the fracture; east of 16° W the Southern Wall is drawn along the watershed line of the Romanche and Chain fractures; the yellow lines show the thalweg of the Romanche and Chain fractures. A sector of the Romanche Fracture Zone marked with a square in Fig. 5.6a is shown in Fig. 5.7

fracture between the fracture entrance and the Main Sill is nowhere greater than 4000 m (Fig. 5.6c). The greatest depths above the wall in this section are 3900–3950 m; they are found in the Vema Deep region (18°40′–19°40′ W) and closer to the entrance to the fracture (21°20′–21°50′ W) (Fig. 5.6c).

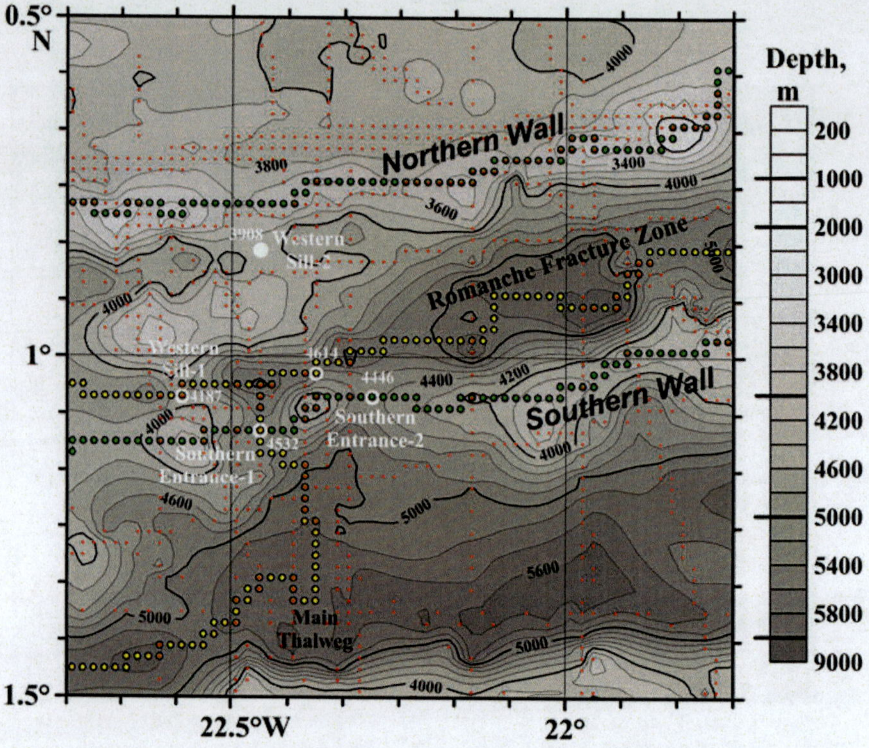

Fig. 5.7 Bottom topography in the region of the entrance to the Romanche Fracture Zone (at 22°30' W) according to the database (Smith and Sandwell 1997, version V19.1). Red dots denote grid nodes marked in this database as echo sounder measurements; yellow dots are grid points corresponding to the thalweg line, green dots correspond to the ridges of the Northern and Southern walls, red dots show grid points marked in the database (Smith and Sandwell 1997) as reconstructed from echo sounding data; large white circles and numbers are points and depths of sills according to the database (Smith and Sandwell 1997) at the entrance to the fracture

These regions require a more detailed study of the bottom topography to establish real depths. The Communication Region is located east of the Main Sill and south of the fracture channel (Figs. 5.5 and 5.6a), along which the abyssal waters flow from the Chain Fracture Zone to the Romanche Fracture Zone (Mercier et al. 1994). The threshold depth in this region is estimated at 4200 m according to the multibeam echo sounder data (Mercier and Morin 1997). The depth above the Northern Wall of the Romanche Fracture Zone in the entire area between the entrance and the Main Sill is shallower than 4000 m, which is confirmed by echo sounding data (Fig. 5.5a). The greatest depth above the Northern Wall (about 4000 m) is noted in depressions in the Vema Deep area (Fig. 5.5a and 5.6b). Slightly overestimated depths over the Northern Wall, according to the database (Smith and Sandwell 1997), apparently occur in the area between Sill D and the Main Sill (Fig. 5.6b).

Chain Fracture Zone According to the digital bottom topography database (Smith and Sandwell 1997, version V19.1), the western part of the Chain Fracture Zone is more than 5000 m deep with relatively shallow walls, especially the southern one, which is generally over 5000 m deep (Fig. 5.8). Starting at approximately 19°30′ W in the eastern direction, the depths above both walls are generally shallower than 4000 m. In the region of 17° W in the Southern Wall of the fracture there is a gap

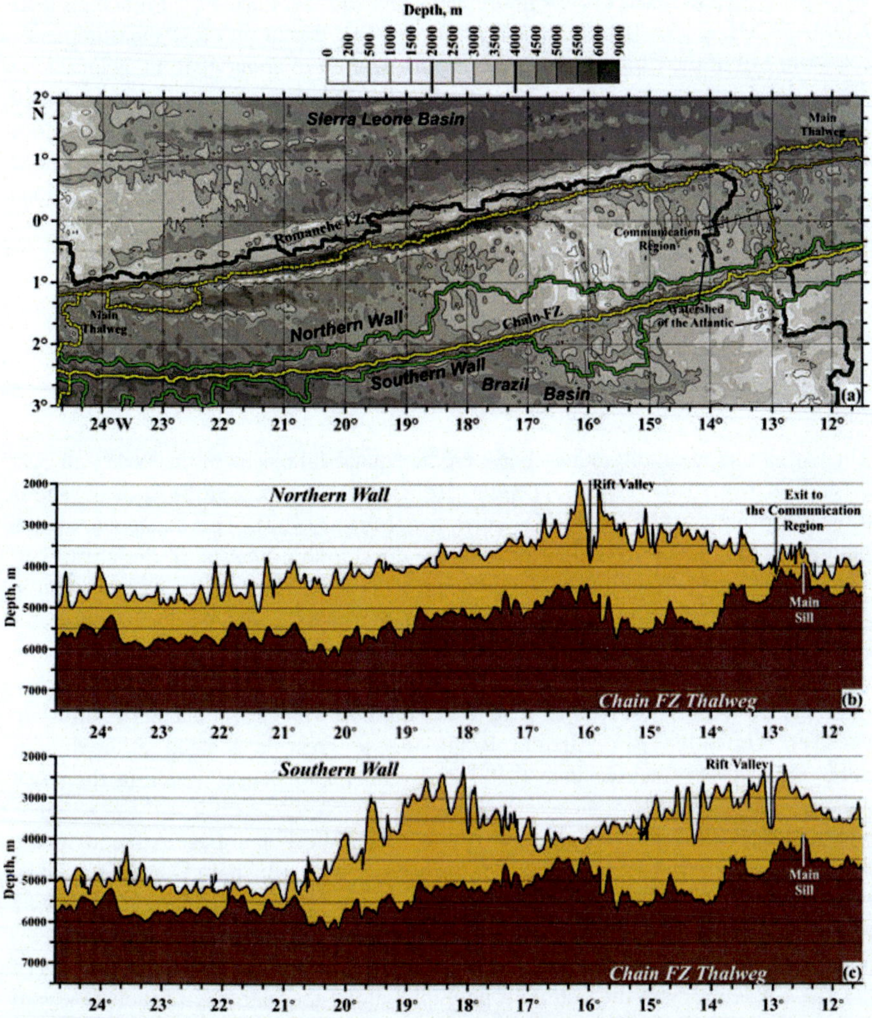

Fig. 5.8 Bottom topography in the region of the Romanche and Chain fractures according to the database (Smith and Sandwell 1997, version V19.1) (**a**); Bottom topography along the Northern Wall and thalweg (the line of greatest depths) of the Chain Fracture Zone according to the database (**b**); Bottom topography along the Southern Wall and the thalweg of the Chain Fault (**c**). Bold green lines in Fig. 5.8a show the ridges of the Southern and Northern walls of the Chain Fracture Zone; the yellow lines show the thalweg of the Romanche and Chain fractures

with a depth of about 4200 m. Starting from 15° W in the eastern direction, the depths above the Southern Wall of the fracture are everywhere shallower than 4000 m. In the section of the fracture from 17°20′ W to 16°00′ W and near 13°30′ W, there are several sills with depths of about 4500 m, and in the section 12°50′-12°20′ W, a few sills with depths of about 4000 m were found.

The bottom topography in the eastern part of the Chain Fracture Zone (east of 15° W) was studied in detail using a multibeam echo sounder during the French projects ROMANCHE-I, -II, -III in 1991–1992 (Fig. 5.5a). Based on these measurements, the Main Sill of the Chain Fracture Zone with a depth of about 4050 m, is located at 12°23’ W. In the database (Smith and Sandwell 1997, version V19.1), the Main Sill with a depth of about 3900 m is displaced to the east approximately by 5 miles relative to the actual sill based on the echo sounder survey. In the region of 13°15′-12°55′ W in the Northern Wall of the fracture there are several depressions with depths exceeding 4000 m, the deepest of which is about 4200 m.

5.4.3 *Hydrography of the Romanche and Chain Fracture Zones*

The baroclinic pressure gradient between the equatorial basins of the West and East Atlantic is formed at the abyssal depths (deeper than 3800 m near the Equator) and is directed generally to the east (Mercier and Morin 1997). The existence of this gradient is caused by the differences in the density stratification of the abysses in these basins. A thermocline (pycnocline) exists in the abyss of the West Atlantic marking the transition from Antarctic Bottom Water to North Atlantic Deep Water (Mercier and Morin 1997); but the abyssal waters in the East Atlantic are generally more homogeneous and less dense. The difference in the bottom temperature (density) between these basins can be as high as 1.5 °C (about 0.16 kg/m^3 in density). Thus, the Mid-Atlantic Ridge whose crest is located at depths of 2500–3500 m serves as a dam that keeps colder and denser water in the West Atlantic.

The baroclinic pressure gradient between the West and East Atlantic in the absence of the geostrophic balance in the Romanche and Chain fractures is compensated by the friction and acceleration of liquid particles (Monin 1988). As a result, the deep thermocline in the fractures becomes eroded and the bottom temperature increases due to the interaction of the thermocline waters from the abyss of the West Atlantic with the bottom topography along the pathway of their eastward motion (Mercier and Morin 1997). This process is accompanied by the release of kinetic energy from the available potential energy of the gravitational field (owing to the work of the buoyancy forces) associated with the horizontal density difference between the waters of the deep thermocline of the West Atlantic and more homogeneous abyssal waters of the East Atlantic. Such a release should be localized in the areas where the abyssal currents overflow the sills on their pathway and/or contract passing the narrows of the abyssal channels of the deep-water fractures, since acceleration of the abyssal flow can occur namely in these regions. It is clear that the dissipation of the released kinetic energy due to the bottom friction and turbulization of the flow leads to mixing of the abyssal waters. Note that overflows and flow contractions are different variants of hydraulically control process (Pratt and Whitehead 2007).

It was indicated in Tarakanov et al. (2018) that the long Romanche and Chain fractures are series of overflows and flow contractions that are combined in a deep cataract (great waterfall) between the West and East Atlantic. Term cataract was introduced by Whitehead (1989). The strongest transformation of the bottom flow properties in these series due to mixing occurs where the flow spills from the main sills of the fractures (near 13°40′ W in the Romanche Fracture Zone, and 12°22′ W in the Chain Fracture Zone) and entrains the warmer layers overlying the deep thermocline (Mercier and Morin 1997). Here the thermocline vanishes and the bottom temperature sharply increases (Mercier and Morin 1997). "Strait mixing" is essential in setting the stratification of bottom waters in the World Ocean when the near-bottom current overflows a sill subjected to some kind hydraulic control process and plunges down into the downstream basin (Bryden and Nurser 2003).

5.4.4 Currents in the Eastern Parts of the Romanche and Chain Fracture Zones

In 1992–1994, a series of measurements was carried out in the Romanche and Chain fracture zones by French oceanographers to obtain reliable estimates of the eastward bottom water transport and its variability (Mercier and Speer 1998). The program included moored velocity and temperature measurements over a period of 2 years from November 1992 to November 1994. Four moorings were deployed across valleys in each of the fracture zones. Mooring R2 in the Romanche Fracture Zone was lost. In both fracture zones, the sites of mooring deployment were located near the first sill where strong bottom water modification was observed at 14°46′ W in the Romanche Fracture Zone and 13°32′ W in the Chain Fracture Zone (Figs. 5.5a and 5.9). A total of 44 current and temperature meters were set in the bottom layer 500 m thick. The total return rate of the data was about 90%. Sections of mean along-channel velocities components (u) over the period of measurements (2 years) in both fracture zones are shown in Fig. 5.10. The figures are plotted on the basis of data in (Mercier and Speer 1998). The moorings in the Romanche Fracture Zone were located in the deep northern channel of the fracture (Fig. 5.9). A smaller portion of bottom water could propagate through the southern channel.

Direct measurements of velocities presented in (Mercier and Speer 1998) revealed the easterly transport of Antarctic Bottom Water at a rate of 1.22 Sv through both fracture zones (Table 5.7). The transport share of the two fracture zones is 0.66 Sv and 0.56 Sv in the Romanche and Chain fracture zones, respectively.

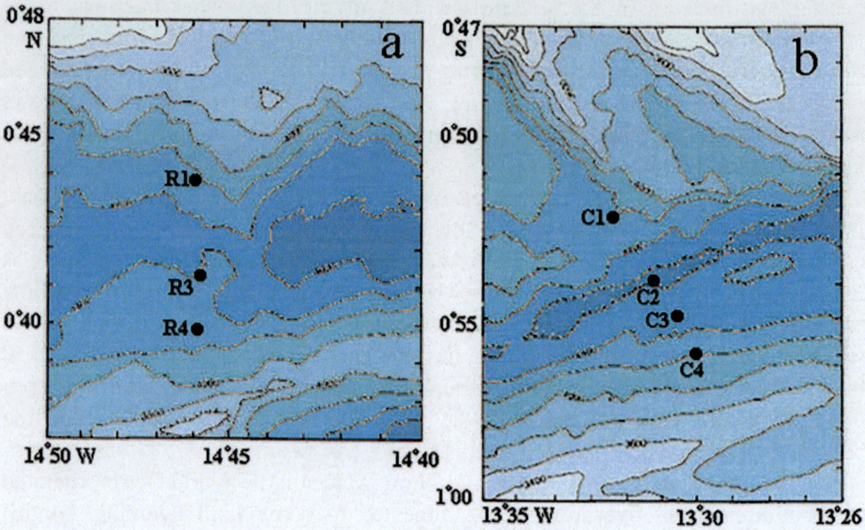

Fig. 5.9 Locations of moorings in the Romanche Fracture Zone (**a**) and Chain Fracture Zone (**b**). Bathymetry is based on multibeam echo sounder survey (Mercier et al. 1994). Isobaths are shown with a step of 200 m. R1, R3, R4 and C1, C2, C3, C4 indicate moorings in the Romanche and Chain fracture zones, respectively

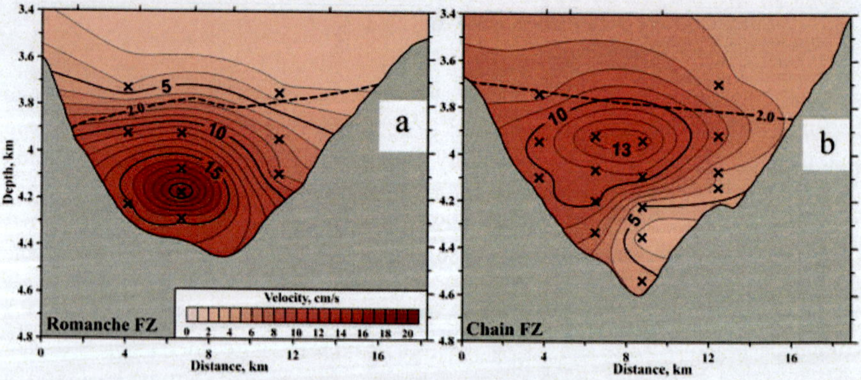

Fig. 5.10 Mean along-channel velocities (cm s^{-1}) in the Romanche (**a**) and Chain (**b**) fracture zones based on moored measurements during the French expeditions in 1992–1994. Potential temperature isotherm 2.0 °C based on CTD section in the same expedition is shown with a dashed line. Crosses show locations of current meters

Previous studies resulted in greater transports. According to the estimates in (Schlitzer 1987; Warren and Speer 1991; Polzin et al. 1996), the transport in the zones deeper than 4000 m is equal to 2 Sv (Table 5.7). The maximum transport of Antarctic Bottom Water over a two-year-long period is as large as 0.9·Sv in the Romanche Fracture Zone and 0.8·Sv in the Chain Fracture Zone (Mercier and Speer 1998). The minimum transport of Antarctic Bottom Water, which was observed in

Table 5.7 Propagation of Antarctic Bottom Water to the eastern basin of the Atlantic Ocean: the Romanche and Chain fracture zones

Authors	AABW boundary	Transport, Sv	Comments
Schlitzer (1987)	θ = 2.07 °C	2.6–5.1	Box model
Warren and Speer (1991)	4000 m	2.0	
Polzin et al. (1996)		1.4	Profiling
Mercier and Speer (1998)	θ = 1.90 °C	1.22 ± 0.25	Current meter. Total transport in Romanche and Chain FZ
		0.3	Minimal transport through the Romanche FZ over 2 years based on current meter data
		0.3	Minimal transport through the Chain FZ over 2 years based on current meter data
		0.66 ± 0.14	Mean transport through the Romanche FZ over 2 years based on current meter data
		0.56 ± 0.17	Mean transport through the Chain FZ over 2 years based on current meter data
Stephens and Marshall (2000)	θ =1.80 °C	~1.0	
Demidov et al. (2006)	θ =1.80–2.00 °C	0.15–0.78	LADCP

November, is 0.3·Sv in both fracture zones. The apparent variability period is approximately 1 year. This variability is observed mainly in the warmer (1.38–2.18 °C) layers.

The maximum velocities averaged over the two-year-long period in the Antarctic Bottom Water layer (θ < 2.0 °C) exceeded 21 cm s^{-1} in the Romanche Fracture Zone and 14 cm s^{-1} in the Chain Fracture Zone (Fig. 5.10a, b). Based on the temperature data of moorings the lowest mean potential temperature was recorded in the Chain Fracture Zone (0.83 °C) compared to 0.89 °C in the Romanche Fracture Zone. In both fracture zones, the velocities are highest in the middle of the section of channels and decay toward the walls and bottom.

The cores of maximum velocities based on moored current measurements were found at 4200 m in the Romanche Fracture Zone and 4000 m in the Chain Fracture Zone. Both cores were located approximately 200–300 m above the bottom. In the Chain Fracture Zone, the potential temperature (θ =1.88 °C) in the core of the bottom water current was warmer than in the Romanche Fracture Zone (1.38 °C), although the Chain Fracture Zone is located south of the Romanche Fracture Zone. This is caused by stronger mixing in the Chain Fracture Zone possibly because of rougher topography.

In 2005 and 2009, Russian CTD casts at 16°03′ W were accompanied with LADCP measurements (Fig. 5.5b and 5.11). The sections of along-channel velocity based on the LADCP data in 2005 and 2009 are shown in Fig. 5.12. Tidal currents were excluded from the LADCP data using the TPXO9 model (Egbert and Erofeeva

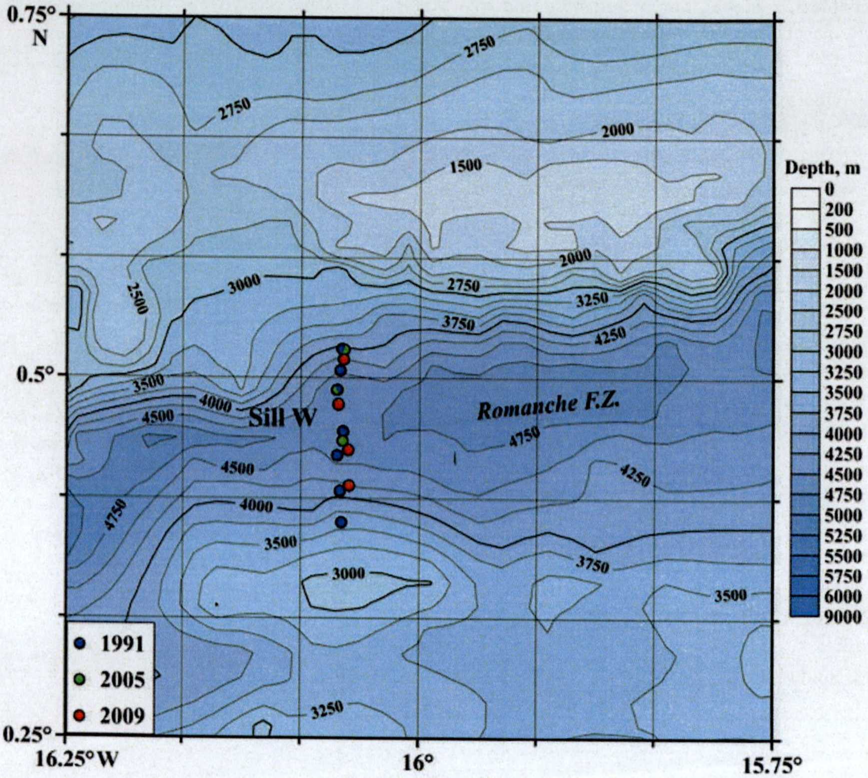

Fig. 5.11 Bottom topography near Sill W in the Romanche Fracture Zone (16°03′ W). The CTD stations in 1991, 2005, and 2009 are shown

2002). The 2005 section was not completed due to technical problems and therefore did not reach the Southern Wall of the fracture. According to the LADCP measurements in 2005 and 2009 (at 16°03′ W), the core of the Antarctic Bottom Water current in the Romanche Fracture Zone was located at a depth of 3900–4000 m. The maximum velocities of the flow in 2009 exceeded 25 cm s^{-1}, whereas in 2005 they were lower than 15 cm s^{-1}. The potential temperature isotherm $\theta = 2$ °C, which is considered the upper boundary of the bottom water flow, was located approximately at 3800 m. The transport estimates of Antarctic Bottom Water flow are 0.35 Sv in 2005 and 0.88 Sv in 2009. Transports of bottom water in the Romanche Fracture Zone are given in Table 5.8.

The Russian expedition in 2009 measured velocities and temperatures at two stations in the region of the Main Sill of the Chain Fracture Zone using CTD/LADCP casts (Fig. 5.13). Figure 5.14 shows profiles of potential temperature and the along-channel velocity component in the fracture zone measured at the station (2082) over the Main Sill. Velocities up to 20 cm s^{-1} were recorded at this station. Velocities up to 10 cm s^{-1} were recorded at the deeper station downstream, which occurs because

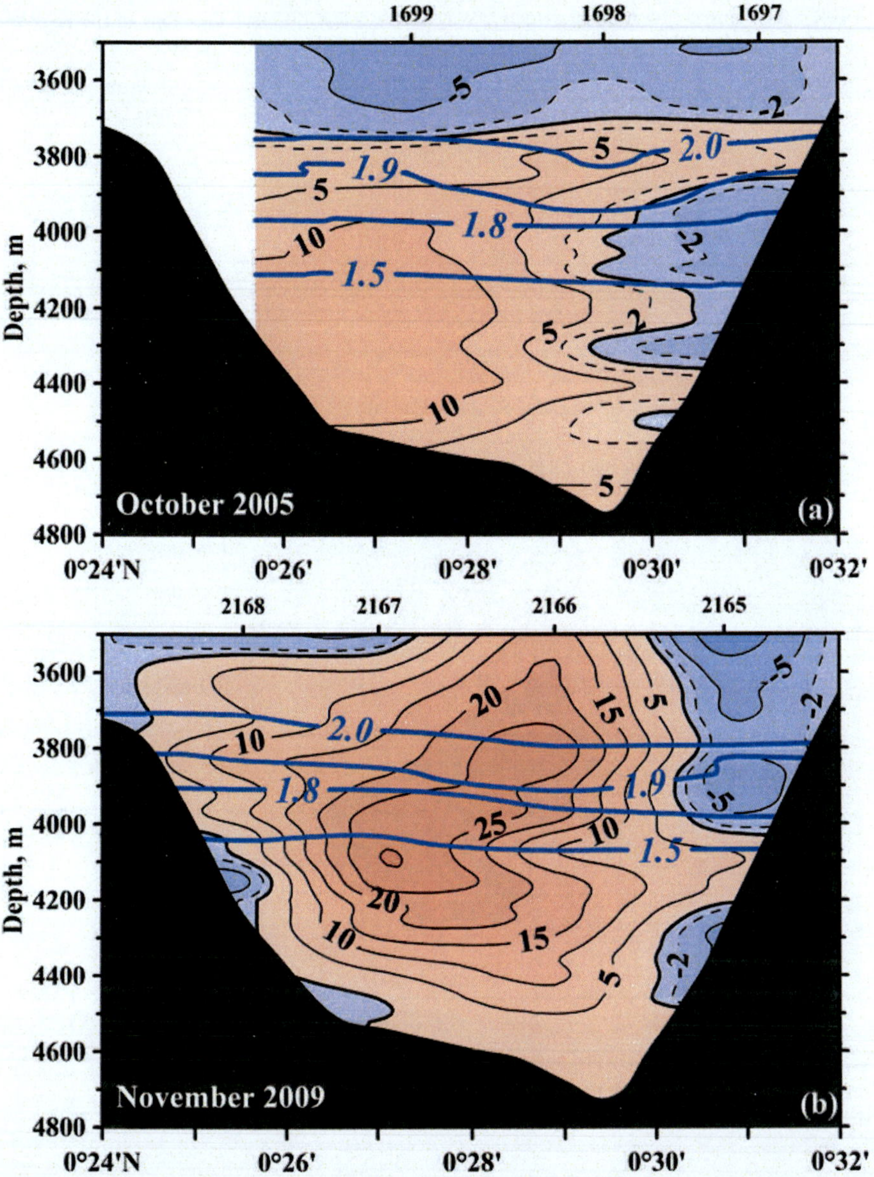

Fig. 5.12 Sections of along-channel velocity (cm s^{-1}) based on LADCP data in October 2005 (**a**) and November 2009 (**b**). Stations are indicated along the top axis. The lines of isotherms used to calculate Table 5.8 are shown

the fracture widens almost twice. It is seen from Fig. 5.14 that the zones of maximum velocities are confined to increased temperature and density gradients. We used the topography from the database in (Smith and Sandwell 1997) to construct the bottom

Table 5.8 Transport of Antarctic Bottom Water (Sv) below different selections of the AABW upper boundary in the Romanche Fracture Zone; measurements in 2005 and 2009

Water mass	Upper boundary, θ, °C	Transport in 2005	Transport in 2009
AABW	Below θ = 2.0 °C	0.45	0.88
	Below θ = 1.9 °C	0.42	0.69
	Below θ = 1.8 °C	0.37	0.62
	Below θ = 1.7 °C	0.33	0.55
	Below θ = 1.5 °C	0.29	0.43

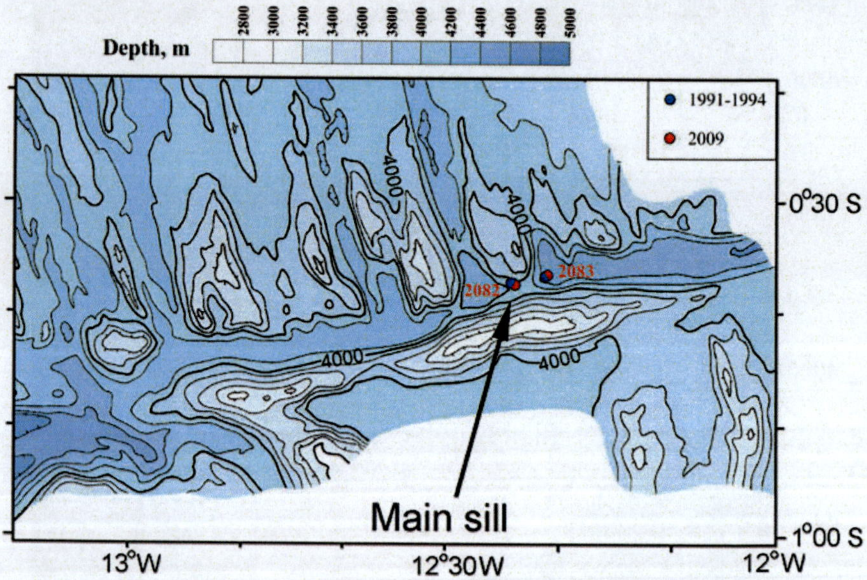

Fig. 5.13 Bathymetry of the Main Sill of the Chain Fracture Zone from a multibeam echo sounder survey (Mercier et al. 1994). Locations of French CTD-stations in 1991 and 1994 and Russian stations in April 2009 are shown

section normal to the flow (Fig. 5.14) and to calculate the transport of bottom waters and Lower North Atlantic Deep Water in the channel. If we accept the upper boundary of bottom waters at θ = 1.9 °C and the boundary of Lower North Atlantic Deep Water at θ = 2.1 °C as in (Mercier and Speer 1998), the transport of the bottom water and Lower North Atlantic Deep Water would be equal to 0.11 and 0.26 Sv, respectively. The estimate of Lower North Atlantic Deep Water transport equal to 0.26 Sv is close to 0.22 Sv estimated in (Mercier and Speer 1998). If we take

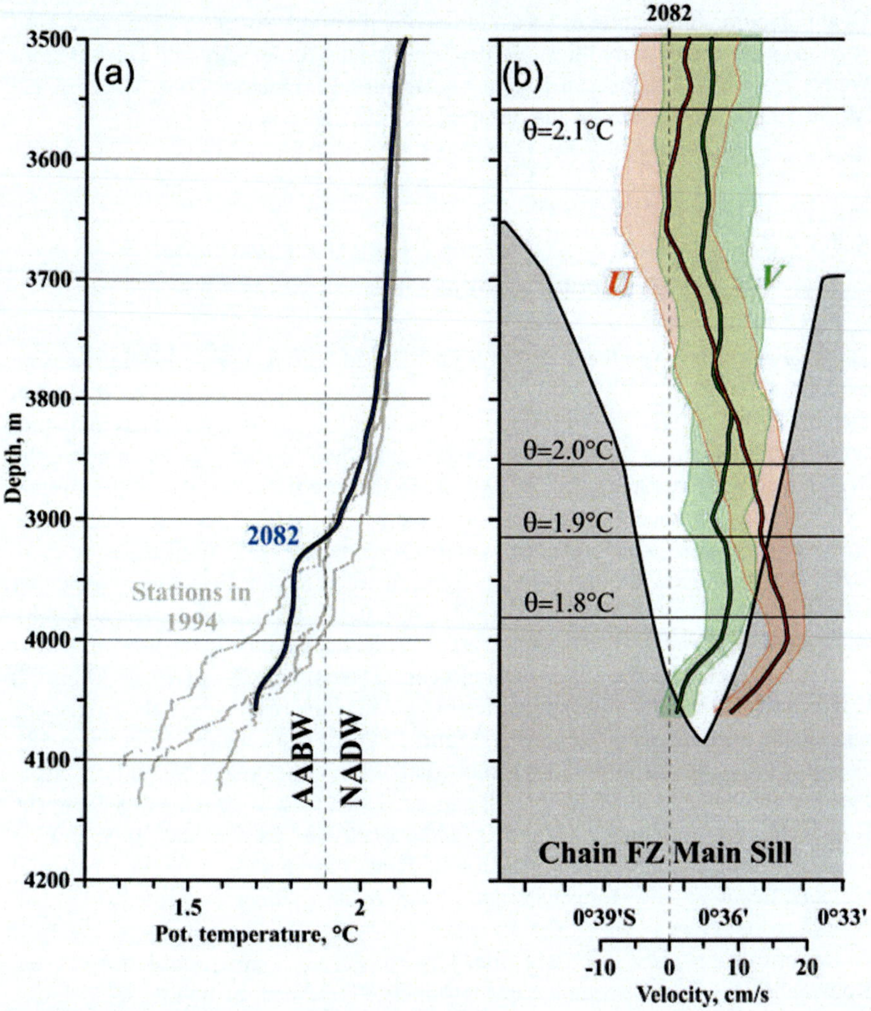

Fig. 5.14 Potential temperature profiles near the Main Sill of the Chain Fracture Zone; stations in 2009 (station 2082, blue) and 1994 (gray) (**a**); Vertical profiles and errors of zonal (U) and meridional (V) components of the measured current velocity in 2009 at station 2082 (**b**). Boundaries ($\theta = 2.1$ °C, 2.0 °C, 1.9 °C, and 1.8 °C) for the calculation of transport from LADCP measurements in the Chain Fracture Zone at 12°23.9′ W are indicated in Fig. 5.14b

isotherm $\theta = 2$ °C as the upper boundary of bottom waters, the transport would increase from 0.11 Sv up to 0.17 Sv.

Thus, bottom water transport in the main sill region of the Chain Fracture Zone can be estimated at 0.11–0.17 Sv, which is less than 0.3 Sv estimated in (Mercier and Speer 1998). However, the French moorings in 1991–1992 (13°30′ W) in the Chain Fracture Zone were deployed west of the Russian LADCP measurements in 2009 (12°20′ W). Between these regions, near 12°30′–13°00′ W, the northern slope of the

Chain Fracture Zone is broken by a number of valleys (the so-called Communication Region) with a maximum threshold depth of about 4200 m, through which bottom waters from this fracture can flow to the Romanche Fracture Zone (Mercier and Morin 1997). This can cause the difference in the transports.

5.4.5 *Temperature Distributions Along the Romanche and Chain Fracture Zones*

Our analysis is based on the results of French expeditions in 1991–1994 and Russian expeditions in 2005, 2009, 2013, and 2014. We also used the data of World Ocean Database 2018 (WOD18 2018). Figure 5.15 shows longitudinal variations in bottom potential temperatures along the Romanche and Chain fracture zones based on the WOD18 and our measurements in 2009–2014 in these fractures. The coldest waters with a potential temperature of 0.51 °C were found by CTD measurements at the Entrance Region of the Romanche Fracture Zone near 22°30′ W (Tarakanov et al. 2013, 2018) (region of the Western Sill-1 in Fig. 5.15a). The lowest potential temperature at the mooring in this region was 0.479 °C (Tarakanov et al. 2018). According to the WOD18 data the lowest potential temperature in the western part of the Chain Fracture Zone was measured near 23°W (0.21 °C) and 19°W (0.27 °C) (Fig. 5.15b). The potential temperatures measured at the main sills of the Romanche and Chain fracture zones were 0.92 °C and 1.32 °C, respectively (WOD18 2018). Potential temperatures in the easternmost parts of the fractures at the outflows from the fractures already in the Guinea Basin are 1.66 °C at 1°30′ N, 8°30′ W in the Romanche Fracture Zone, 1.71 °C at 0°00′ N, 10°W, and 1.62 °C at 0°40′ S, 7° W in the Chain Fracture Zone (WOD18 2018). Thus, the increase in bottom potential temperature along the Romanche and Chain fracture zones exceeds 1.0 °C and 1.4 °C, respectively.

Bottom waters are intensely transformed during their propagation in the Romanche and Chain fracture zones. Horizontal gradients of bottom temperature along the fracture zones are as high as 0.15 °C/100 km and 0.6 °C/100 km, respectively. One can see from Fig. 5.15a that the increase in the bottom potential temperature along the Romance Fracture Zone does not occur gradually, but is tied to individual regions of the fracture. In the Western Sill-1 (22°30′ W) and Sill E regions (13° W), this increase is clearly pronounced step-wise and tied to the regions of 10–15 miles long just beyond the sill that the current overflows. Near Western Sill-1 (22°30′ W), the increase is approximately 0.14 ° C, and near Sill E (13° W), it is close to 0.18 °C. The longitudinal interval from Sill D to Sill E has a number of sills with comparable depths of 4350–4600 m. A sharp potential temperature increase is observed here, which is not divided into separate stages by the available set of measurements. The total temperature jump in this region is about 0.5 °C.

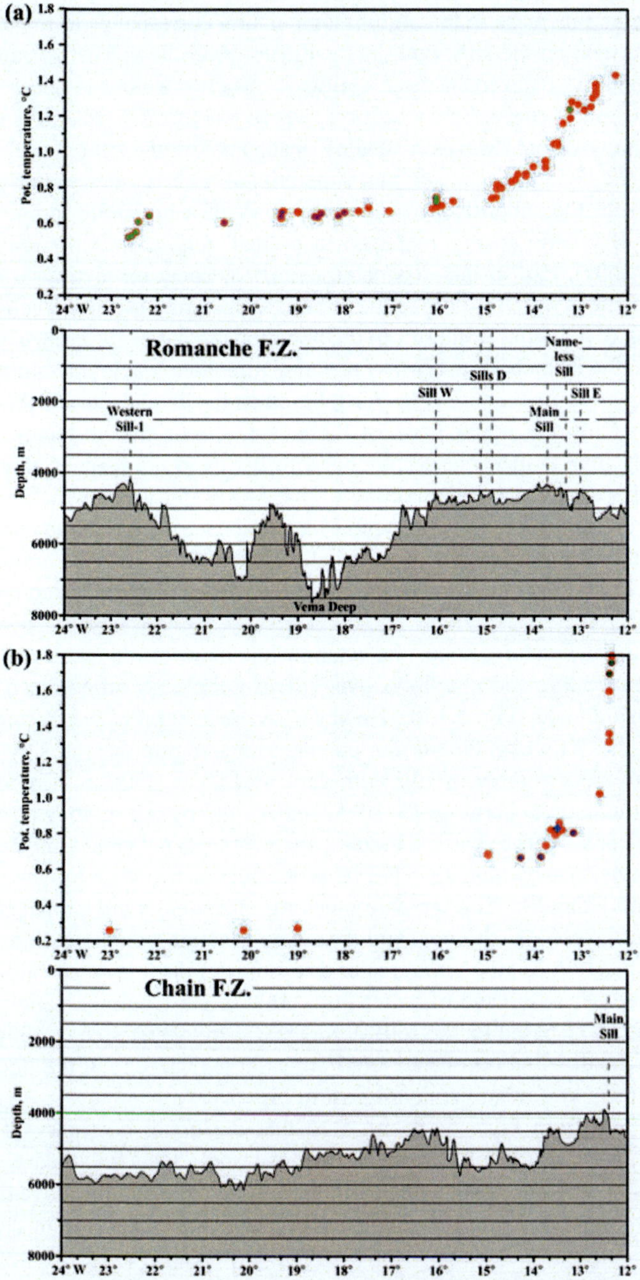

Fig. 5.15 Bottom potential temperature (°C) vs. longitude in the Romanche Fracture Zone (a) and Chain Fracture Zone (b) based on measurements in different years. Red dots correspond to CTD-measurements from 1983 to 1994, green dots are CTD-data of the Shirshov Institute in 2009–2014, purple dots are measurements with reversible thermometers in 1961–1971. Blue diamonds in Fig. 5.15b correspond to the mean potential temperatures measured at the deepest level on the mooring in the Chain Fracture Zone. Ocean bottom along the thalweg line is shown in gray in the bottom panels

The step-like increase in the bottom temperature corresponds to the existence of spillways formed on the sills along the pathway of the flow. The hydraulic control overflow mode is realized in these spillways, when the flow is accelerated from subcritical (Froude number: $Fr = u/c < 1$, where u is the flow velocity, and c is the velocity of gravitational waves in shallow water) before the sill (critical section) to the critical ($Fr = 1$) over the sill, and supercritical ($Fr > 1$) mode after the sill and returns back to the subcritical mode in the form of a hydraulic jump, i.e., shock dissipation of kinetic energy localized in a small region after the sill (Pratt and Whitehead 2007). Part of this dissipating energy is spent on mixing of the bottom flow, as a result of which a jump in the bottom temperature is formed behind the sill. Outcropping of potential isotherms to the bottom behind the sills is clearly seen over other sills in the Romanche Fracture Zone along the potential temperature section in the eastern part of the fracture (Fig. 5.16) based on the data of French expeditions in 1991–1994. Only the upper parts of these waters pass the obstacles, which are subsequently transformed due to mixing with the overlying Lower North Atlantic Deep Water. A more detailed structure of one of the spillways, which is formed in the Entrance Region of the Romanche Fracture Zone, is considered in Sect. 5.4.7 based on publications (Tarakanov et al. 2013, 2018). Strong mixing in the Romanche Fracture Zone was reported in (Ferron et al. 1998; Tarakanov et al. 2013; van Haren et al. 2014).

In the Chain Fracture Zone, the relationship between the increase in bottom potential temperature and overflows over sills is not clearly pronounced due to the lack of measurements (Fig. 5.15b). However, in the interval of the fracture between 19° W and 15° W, where the bottom potential temperature increases by more than 0.4 °C, there are a number of sills with depths of about 4500 m. The temperature increase in the interval between 13°10′ W and 12°40′ by 0.2 °C may also be due to the existence of sills. The sharpest increase in the bottom potential temperature in the Chain Fracture Zone by more than 0.3 °C is observed in the Main Sill area before the Main Sill not behind it. Topographic blocking of the eastward propagation of the coldest water layers, which flow from the Chain Fracture Zone to the north through the Communication Region with greater threshold depths, may also cause this temperature increase. A scheme of water flow through the Communication Region is shown in Fig. 5.17 based on the scheme suggested in (Mercier et al. 1994) with the account for the distribution of measured potential temperature at the bottom in the region between two fracture zones shown in Fig. 5.17.

One can see from Figs. 5.14 and 5.16 that the upper boundary of Antarctic Bottom Water ($\theta = 1.9$ ° C) in the Romanche and Chain fractures is located at depths of 3800–3900 m. If we compare this depth with the depths above the walls of the Romance Fracture Zone (Fig. 5.6) we conclude that in addition to the Entrance Region the inflow of Antarctic Bottom Water probably exists through several depressions in the Southern Wall of the fracture between 19°30′ W and 18°30′ W,

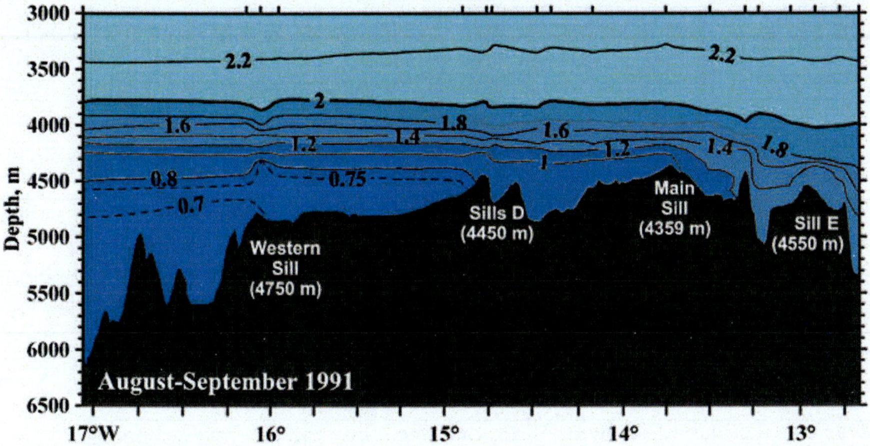

Fig. 5.16 Potential temperature section (°C) along the Romanche Fracture Zone based on the data in 1991. Tics on the top axis indicate locations of CTD-casts

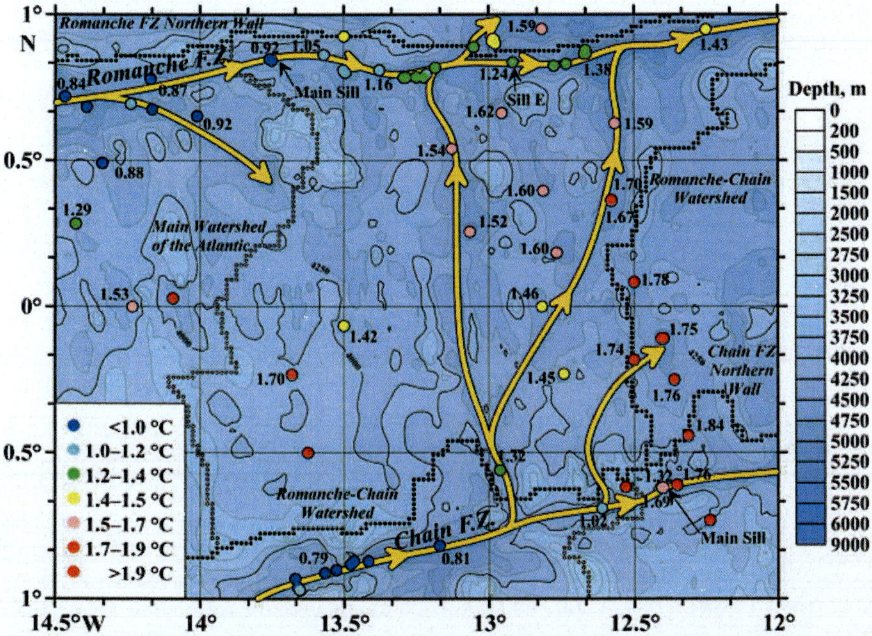

Fig. 5.17 Distribution of bottom potential temperatures (°C) in the Communication Region between the Romanche and Chain fracture zones. Bottom topography is based on (Smith and Sandwell 1997, version V19.1). Arrows indicate possible pathways of the flows between the fracture zones based on the decrease of potential temperatures along the flow and conclusions reported in (Mercier et al. 1994). Watersheds between different regions are shown with lines of black and white dots

the depths above which are approximately 4000 m according to the database (Smith and Sandwell, 1997). In the Chain Fracture Zone, the inflow of this water can freely occur west of 19°30′ W as well as through several depressions in the Northern Wall of the fracture in the interval between 19°30′–18°30′ W, and through the Southern Wall in the interval between 19°30′–15°00′ W with threshold depths of about 4000 m (Fig. 5.8).

Since the geothermal flux in the Romanche and Chain fracture zones is rather high, it could influence the bottom water temperature. We shall consider this problem using the data along the Chain Fracture Zone. Variations in the bottom temperature/salinity ratio as function of longitude in the Chain Fracture Zone are shown in Fig. 5.18. The θ,*S*-curve for a station in the western part of the Chain Fracture Zone is shown with a line of small dots. The deepest θ,*S*-distributions at the bottom at other stations along the axis of the fracture are shown with large red dots. One can see that all of them fall on one line. Only the minimum temperature and minimum salinity at each station increase with longitude. This demonstrates the negligible effect of the geothermal flux on warming of the bottom water layer. If the warming caused by the geothermal flux had been strong, the dots related to the station along the fracture would be elevated (temperature increase at fixed salinities) relative to the θ,*S*-curve of the western station. Variations in temperature and salinity as functions of longitude are similar, suggesting that these variations are caused only by mixing or because the coldest waters do not overflow numerous sills in the fracture.

5.4.6 Long-Term Variations in Temperature and Salinity

Long-term time variations in the bottom flow in the Romanche Fracture Zone are clearly seen from comparison of cross-sections along the same track at 16°03′ W within the ROMANCHE-I program and those repeated by the Russian expeditions in 2005 and 2009. The sections crossed the local Sill W of the fracture along 16°03′ W (the shallowest depth is 4750 m). Positions of the stations of the sections are shown in Fig. 5.11. Potential temperature distributions over the 1991 and 2009 sections are shown in Fig. 5.19, and salinity sections are shown in Fig. 5.20. The curves of θ, S-distributions in the deep-waters layers based on the data of the deepest stations in the expeditions of 1991, 2005, and 2009 are shown in Fig. 5.21.

Fourteen and eighteen years passed between the French measurements in 1991 and Russian expeditions in 2005 and 2009. Comparison of measurements shows that no principal changes occurred in the structure of the flow in 2005 and 2009 since the first measurements in 1991. Equatorial location of the fracture zone leads to symmetrical distribution of temperature and salinity with respect to the axis of the

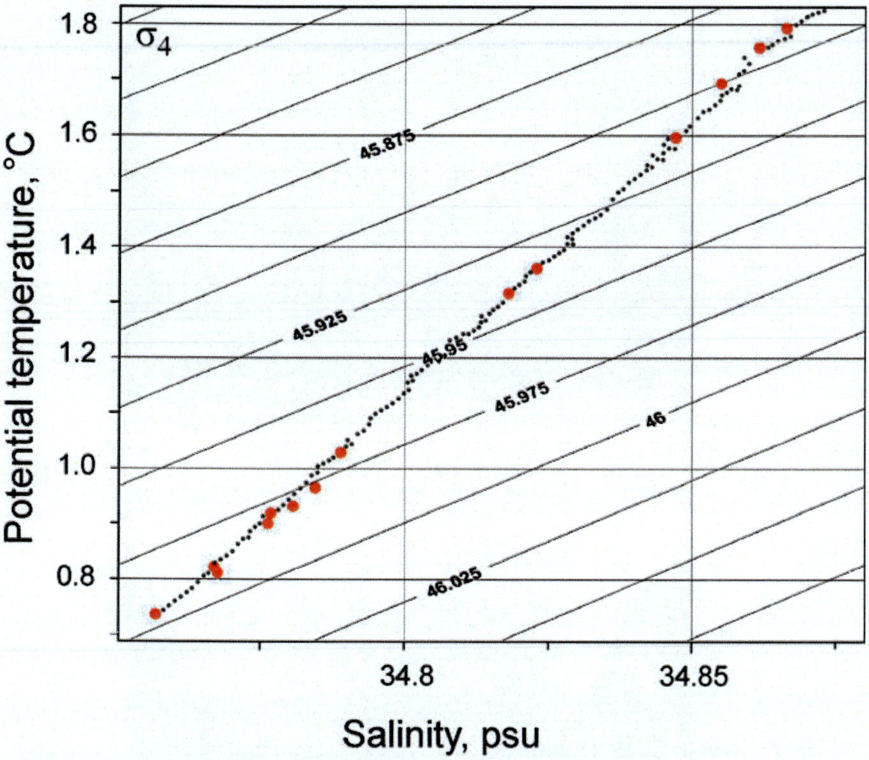

Fig. 5.18 Curves of θ,*S*-distribution for bottom measurements along the Chain Fracture Zone demonstrating that geothermal warming of bottom waters in the fracture zone is negligible. The line of small dots represents the θ,*S*-relation for a station in the extreme western part of the fracture zone. Other larger red dots falling on the same line are related to the bottom measurements at points located along the fracture axis. Isopycnals referenced to 4000 dbar (density contour lines σ_4) are shown

channel. Notable displacements of isothermal and isohaline surfaces during the years between the expeditions occurred only in the lower 400–500 m as deepening of isotherms, isohalines, and isopycnals (Figs. 5.19 and 5.20). Deepening of isopycnals is most clearly seen. This is associated with cooling and a decrease in salinity on isopycnic surfaces in the abyssal flow. It is also seen in the θ, S-diagram (Fig. 5.21) as well as from comparing the depths of isotherms and isopycnals (Fig. 5.19).

The minimum potential temperature at the bottom in 2009 (0.752 °C) was higher by 0.057 °C than in 1991 (0.695 °C) and by 0.022 °C higher than in 2005 (0.730 °C). Salinity minimum at the bottom almost did not change.

An increase in the bottom temperature at the bottom in the Romanche Fracture Zone is also seen in Fig. 5.15a. Later measurements appear a few hundredths of a

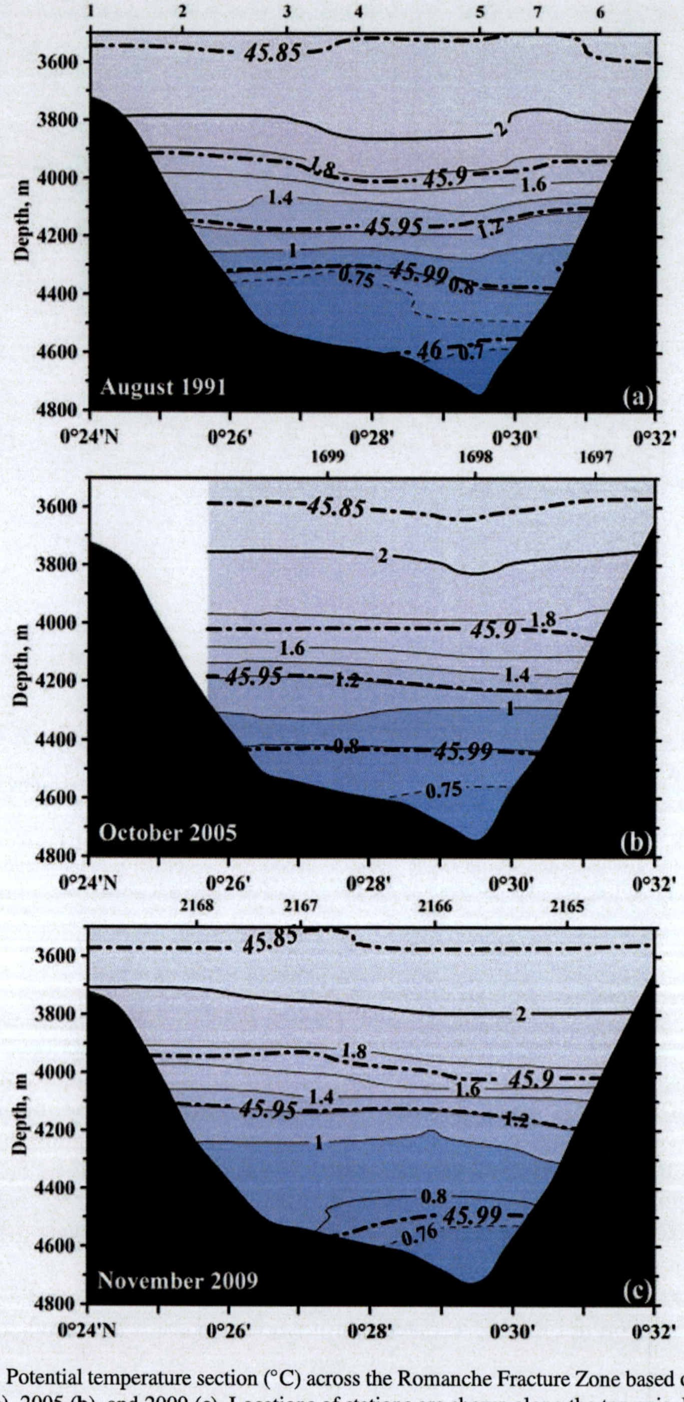

Fig. 5.19 Potential temperature section (°C) across the Romanche Fracture Zone based on the data in 1991 (a), 2005 (b), and 2009 (c). Locations of stations are shown along the top axis. Isopycnals referenced to 4000 dbar (density contour lines σ_4) are shown

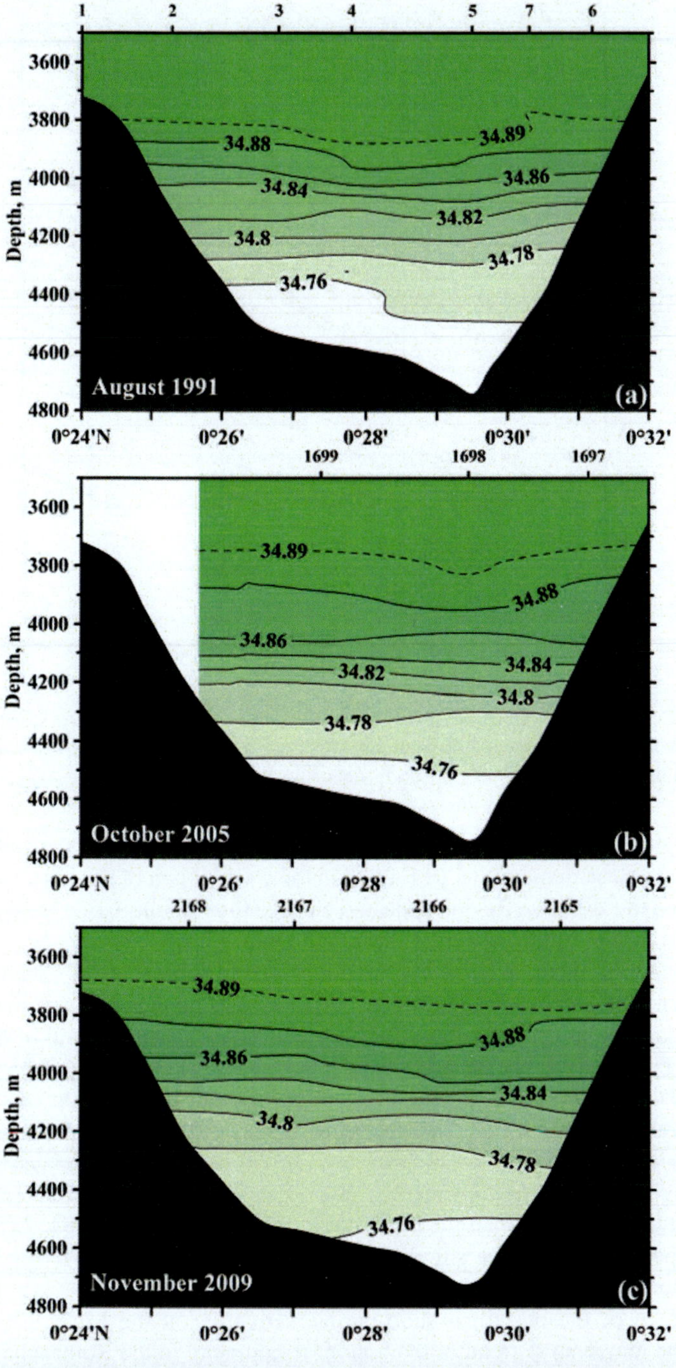

Fig. 5.20 Salinity section (PSU) across the Romanche Fracture Zone based on the data in 1991 (**a**), 2005 (**b**), and 2009 (**c**). Locations of stations are shown along the top axis

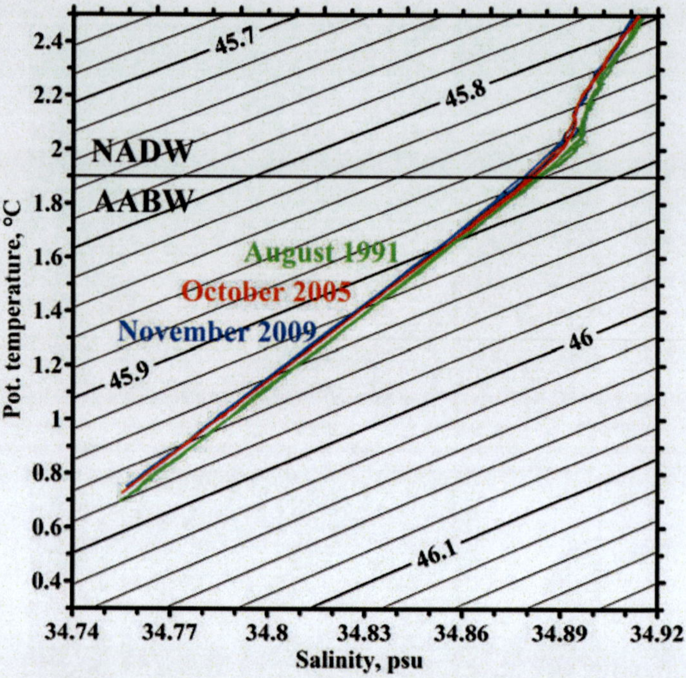

Fig. 5.21 Curves of θ,S-distributions for deep waters based on the data of the deepest stations in the expeditions of 1991, 2005, and 2009. Names of water masses and the boundary between them are indicated. Contour lines of potential density σ_4 are shown

degree higher than earlier measurements at the same points. Potential temperature based on the measurements in 2011–2014 in the western part of the fracture after the flow passed the Entrance Region appeared close to those in the earlier measurements carried out further to the east up to the Vema Deep.

The long-term increase in the bottom potential temperature in the Chain Fracture Zone is not as evident as in the Romanche Fracture Zone, possibly due to a lack of measurements. Comparison is possible only at two points near the sill of the Chain Fracture Zone, where the stations were made in 1991–1994 and then in 2009 (Fig. 5.13). In 1994, three CTD-casts were carried out at one point on the Main Sill, one after the other within 6 hours. Deeper than 3850 m, the temperature profiles significantly differed from each other (Fig. 5.14), and the bottom temperatures differed by 0.28 °C (!). In 2009, at the same point, the potential bottom temperature was about 0.1 °C higher than the highest bottom temperature in 1994. At the same time, at the station after the sill, which repeated the point of 1991 (Fig. 5.13), a temperature drop of 0.04 °C was recorded.

The range of the described short-period variability at the Main Sill strongly exceeds the temperature trend of the bottom temperature in the Romanche Fracture

Zone. Such variability may be due to fluctuations in the flow intensity, including that caused by the tidal motions by means of the so-called Bernoulli aspiration (Kinder and Bryden 1990). The effect occurs because in a hydraulic controlled overflow, water located below the sill is pumped unto the sill from the basin upstream. Moreover, the higher the flow velocity, the deeper and, accordingly, colder layers of water are transported. The large range of fluctuations in the bottom temperature can also be caused by the fact that the depth of the sill corresponds to a layer of an increased vertical temperature gradient (Fig. 5.14). This strong variation also provides evidence that the flow is not stationary and can be perturbed by tidal forcing that causes the oscillatory transport of bottom water along the channel axis near the main sill.

Considering the time variations in temperature near the sills of the Romanche Fracture Zone, it would be helpful to analyze additional data of earlier observations. Although these data had a lower accuracy as compared with the modern data, they can still be applied to determine the dominating tendency of water temperature change at abyssal depths. Therefore, we present Table 5.9 with a time series of instrumental measurements of potential temperature in the Vema Deep, the deepest part of the Romanche Fracture Zone within the layers of North Atlantic Deep Water (4000 m) and Antarctic Bottom Water (7000 m) based on the data of several known expeditions (Leontieva 1985). By presenting this Table we also give credit to the oceanographers of earlier years that performed measurements in the deep ocean layer in this region.

The data presented in the table demonstrate significant fluctuations of temperature in the North Atlantic Deep Water layer, but no gradual secular climatic trend in North Atlantic Deep Water can be seen. At the same time, warming of Antarctic Bottom Water by 0.04 °C has undoubtedly been observed starting approximately from the second half of the twentieth century.

5.4.7 *Inflow of Antarctic Bottom Water and Deep Spillway in the Western Part of the Romanche Fracture Zone*

The goal of this research is based on the idea that we wanted to find the entrance where Antarctic Bottom Water flows into the Romanche Fracture Zone and then the bottom current continues along the deep channel of the fracture to the east eventually discharging into the Guinea Basin. We thought that it would be interesting to study the hydrodynamic properties of the bottom flow, which converges from the slow motion in the Brazil Basin to a strong jet of high velocity in the narrow fracture zone. Analysis of the digital maps of bottom topography revealed a narrow gap in the Southern Wall of the fracture approximately at 1°05’ S, 22°28′ W. Our first measurements in 2011 revealed that we found exactly the place we wanted to find (Tarakanov et al. 2013).

Table 5.9 Potential temperature (°C) of deep and bottom waters in the Romanche Fracture Zone (historical data)

Level	Expedition and year						
	Albatross, 1948	Crawford, 1958	Chain, 1965	Meteor, 1965	Akademik Kurchatov, 1967	Akademik Vernadsky, 1969	Mikhail Lomonosov, 1972
4000 m	1.75	1.66	1.61	1.64	1.77	1.57	1.54
7000 m	0.63	0.64	0.62	0.60	0.65	0.66	0.67

High velocity currents of bottom water in the equatorial fractures are maintained by the pressure gradient in the abyssal depths (deeper than 3800 m). This gradient, which is generally directed to the east, exists due to the differences in stratification in the abyssal depths (Mercier and Morin 1997). The deep waters in the East Atlantic are generally less stratified and less dense than the waters of the West Atlantic. The Mid-Atlantic Ridge whose crest is located at depths of 2500–3500 m is an obstacle that prevents spreading of colder and denser water of the West Atlantic to the east.

In this section, we present the details of the measurements in the western part of the Romanche Fracture Zone. A total of 25 stations (Table 5.10 and Fig. 5.22) to the bottom were made in 2011–2014 in the Romanche Fracture Zone entrance region and a mooring was deployed there in addition to four surveys. The locations of stations and other information about the stations are given in Figs. 5.22, 5.24 and 5.25 and Table 5.10. In addition, a multi-beam echo-sounder survey was made in 2016 from the R/V *Akademik Nikolay Strakhov* (Fig. 5.24) (Kapustina et al. 2021).

5.4.7.1 Description of Data and Data Processing

At all stations in the region where AABW enters the Romanche Fracture Zone, we made CTD-casts using SBE-19 plus profiler and velocity measurements with the lowered profiler LADCP RDI WHS, 300 kHz. Immediately before the deployment of the mooring (station 2513 in 2013) and immediately after its recovery (station 2537 in 2014) "yo-yo" CTD profiles were made at the point of the mooring with three upcasts and downcasts of the instrument by 300 m above the bottom (up to 4320 m). The difference between the locations of the ship when the "yo-yo" casts reached the bottom was within 10 m controlled by keeping the ship at the point. A difference of 10 m is not true for the location of the instrument on the other end of the cable because the instrument could be displaced by the strong bottom currents while it approached the bottom. The difference between the time moments when the instrument reached the bottom during the "yo-yo" profiles (from 3 to 7 m above the bottom) was 15–20 min. In 2013, three stations (2509, 2513, and 2511) were occupied along the narrow gap to study the cascading. The distance between the neighboring stations was one mile. In addition to our measurements, we also used the historical database WOD18.

The CTD measurements were processed by the standard processing package SBE Data Processing V. 7.23.2. and the data of lowered ADCP were processed using the LADCP Processing, version IX.8, whose principles are described in (Visbeck 2002). LADCP processing outputs the profiles of zonal and meridional velocity components with a vertical step of 8 m. Such a step approximately corresponds to the digitization of the hardware averaging of the velocity profiles measured by the LADCP and provides an estimate of the velocity recovery error at each level close to the lowest possible in the processing algorithm. This velocity error is also a result of processing the data using the specified software. In the bottom layer up to 150 m

Table 5.10 CTD/LADCP stations and mooring in the western part of the Romanche Fracture Zone

Time GMT (at the bottom)	CTD stations and mooring	Coordinates of the instrument at the bottom	Lower measurement/ ocean depth, m
		Cruise 34 ASV	
23.10.2011 01:52	2466	01°02.0′ S, 22°10.3′ W	4097/4107
23.10.2011 06:27	2467	00°59.4′ S, 22°20.5′ W	4915/4925
23.10.2011 11:12	2468	00°58.4′ S, 22°10.5′ W	5571/5581
23.10.2011 15:27	2469	00°54.9′ S, 22°10.5′ W	4907/4917
23.10.2011 19:18	2470	00°51.9′ S, 22°10.5′ W	5245/5255
23.10.2011 23:24	2471	00°48.9′ S, 22°10.5′ W	4325/4335
24.10.2011 05:25	2472	01°06.5′ S, 22°27.4′ W	4613/4623
		Cruise 36 ASV	
27.10.2012 09:57	2480	01°02.9′ S, 22°33.4′ W	4430/4437
27.10.2012 15:18	2481	01°08.3′ S, 22°27.4′ W	4668/4674
27.10.2012 19:00	2482	01°05.8′ S, 22°27.3′ W	4565/4571
28.10.2012 00:04	2483	00°49.3′ S, 22°25.5′ W	4368/4374
28.10.2012 01:25	2484	00°49.2′ S, 22°25.4′ W	4000/4420
28.10.2012 08:30	2485	01°02.9′ S, 22°30.1′ W	4728/4735
28.10.2012 12:20	2486	01°03.5′ S, 22°27.4′ W	4923/4930
28.10.2012 16:14	2487	01°04.8′ S, 22°27.4′ W	4697/4704
		Cruise 37 ASV	
07.10.2013 10:38	2509	01°06.1′ S, 22°27.5′ W	4588/4598
07.10.2013 14:14	2510	01°06.0′ S, 22°28.6′ W	4391/4400
07.10.2013 17:55	2511	01°04.0′ S, 22°28.0′ W	4987/4997
07.10.2013 22:07	2512	01°04.0′ S, 22°29.0′ W	4847/4857
08.10.2013 12:31	2513	01°05.1′ S, 22°27.7′ W	4698/4705
08.10.2013 13:45 (time of deployment)	2514	01°05.034′ S,	4629
	Mooring deployment	22°27.652′ W	
08.10.2013 16:28	2515	01°03.0′ S, 22°37.5′ W	4580/4590
09.10.2013 10:47	2516	01°15.5′ S, 23°54.0′ W	4981/4991
09.10.2013 20:00	2517	01°25.8′ S, 24°00.6′ W	4899/4908
		Cruise 38 ASV	
16.04.2014 08:20 (time of deployment)	2536	01°05.227′ S,	4629
	Mooring recovery	22°27.637′ W	
16.04.2014 17:10	2537	01°05.0′ S, 22°27.7′ W	4703/4706
16.04.2014 21:10	2538	01°01.8′ S, 22°24.7′ W	4866/4872

Reproduced from Tarakanov et al. (2018)

from the bottom, where the RDI BottomTrack operates, the error of the baroclinic velocity component is 1–2 cm s^{-1}, and at a greater distance from the bottom it is 7–10 cm s^{-1}. We interpreted it for the transport calculations as a continuous depth function in a discrete depth representation.

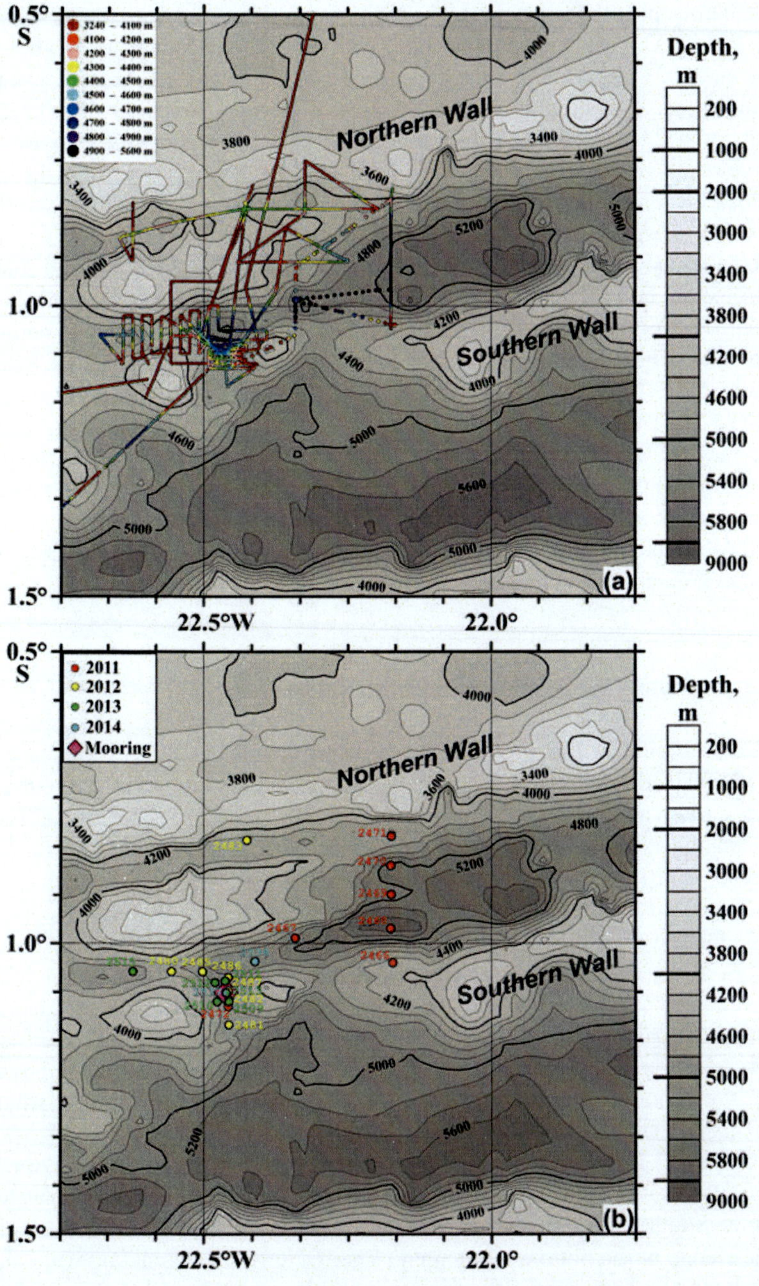

Fig. 5.22 Bottom topography in the region of the entrance to the Romanche Fracture Zone based on the data of the recent version of database (Smith and Sandwell 1997, version V18.1) (**a**) and corrected using our echo soundings in 2011–2014 (**b**). Color lines in Fig. 5.22a show the lines and depths of echo-sounder measurements (see the upper left legend). Color circles and numerals in Fig. 5.22b indicate the points and numbers of stations and mooring in 2011–2014 (see the bottom left legend). The color of the dots indicates year of measurements. (Reproduced from Tarakanov et al. 2018)

The barotropic tidal velocities in the region were calculated from the satellite data using the TPXO 9.1 model available at Oregon State University (Egbert and Erofeeva 2002). In addition, the total tidal velocities were estimated from the mooring measurements. Based on the results of the comparison of these methods of estimating tidal velocities, we subtracted the barotropic tidal velocities from the measurements of LADCP data. The results of processing show that the mean velocities are one order of magnitude greater than the velocity of the barotropic tide, which is estimated to be a few cm s^{-1}.

We deployed a mooring in Southern Entrance - 1 with two Aquadopp (1.5 MHz) current meters at 4421 and 4611 m and 99 temperature sensors with 2 m spacing between sensors (at 01°05.0′ S, 22°27.7′ W; location of the mooring is shown in Fig. 5.18b). The sampling interval was 10 min. The mooring operated during 6 months from October 8, 2013 to April 16, 2014. Details of its location are given in Table 5.10. The temperature measurements at 99 sensors were synchronized with an accuracy to 1 s. Their accuracy was 0.0005 °C (van Haren et al. 2014). The *in situ* temperatures of these sensors were recalculated into the potential temperatures using the linear relation between temperature and salinity in the bottom layer from the CTD data at stations near the mooring: stations 2513 and 2537 in the layer from bottom to a depth of 300 m above the bottom (Fig. 5.23).

5.4.7.2 Bottom Topography in the Deep Spillway Region

The data of digital bottom topography were downloaded from the 1-minute digital array (Smith and Sandwell 1997, version V19.1). It was surprising but our echo-sounder survey in 2011 revealed that that the difference between the digital bottom array and the actual depths can be as high as 700 m (Tarakanov et al. 2013). Therefore, in 2012, we carried out a detailed survey of the region with parallel tracks separated by 0.5–1.0 miles using the ELAC LAZ 4700 echo-sounder. The results of these measurements were corrected considering that the actual sound velocity estimated from the CTD-measurements is different from a sound velocity of 1500 $m{\cdot}s^{-1}$, to which the echo-sounder was initially tuned. Figure 5.22a shows the bottom topography based on (Smith and Sandwell 1997) and corrected according to our echo-sounder measurements. The route of the ship is also shown in Fig. 5.22a. A multi-beam echo-sounder survey was performed in 2016 from the R/V *Akademik Nikolay Strakhov* (Fig. 5.24).

Description of the spillway structure in the region of the inflow of Antarctic Bottom Water to the zonal channel of the Romanche Fracture Zone through a narrow gap in the Southern Wall of the fracture in the western part of the fracture requires the knowledge of the bottom topography. The fracture is a channel oriented almost from west to east between the Southern and Northern walls (ridges) (Figs. 5.22 and 5.24). The characteristic depths above the crests of the Southern Wall around the

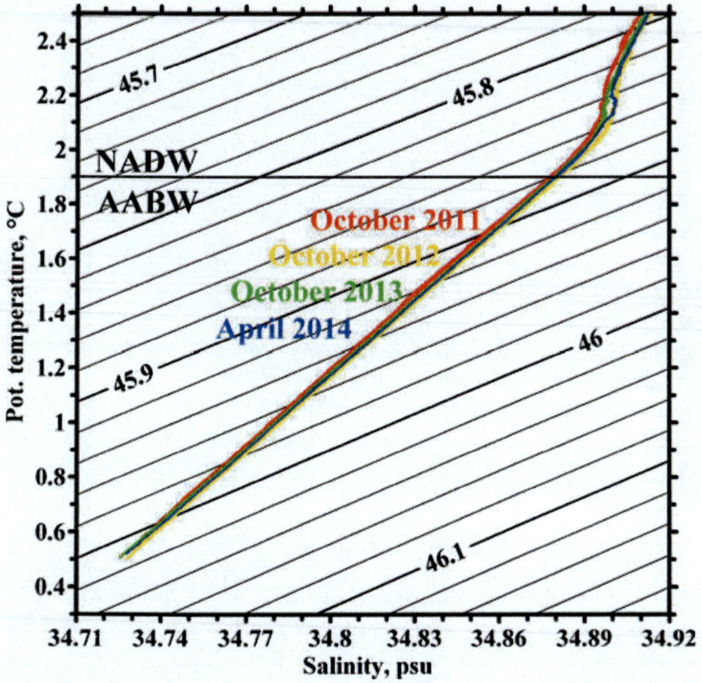

Fig. 5.23 Curves of θ,*S*-distributions for the abyssal layers in the Romanche Fracture Zone. The red color indicates the stations in 2011, yellow in 2012, green in 2013, and blue in 2014. Solid black lines indicate density anomaly referenced to 4000 dbar. The horizontal line along θ = 1.9 °C shows the boundary between NADW and AABW. Reproduced from (Tarakanov et al. 2018)

spillway are 3500–4100 m, while the depths at the Northern Wall are 3200–3600 m. The Median Ridge extends along the channel approximately in the middle and divides the channel into the Southern and Northern valleys (channels) (Fig. 5.22). The crest of this ridge rises to a depth of 3400–4000 m approximately at 22°20′–22°40′ W and rapidly descends to the east to almost 5000 m already at 22°10′ W. The bottom of the Northern Valley is almost flat. A local deep depression in the Northern Valley, which could be a hydraulic bath, was not found. The width of the valley at a level of the 4150 m isobath is 3–5 miles (Fig. 5.22b); the choice of this particular isobath for characterizing the width is related to the dynamic structure of the spillway in the western part of the Romanche Fracture Zone. This isobath approximately corresponds to the 1.2 °C isotherm, which is the upper boundary of the spillway layer (Tarakanov et al. 2013).

The depth of the sill in the Northern Valley does not exceed 4300 m. In the eastern direction, the valley becomes deeper than 5200 m already at 22°10′ W. The Southern Valley at 22°33.5′ W has a sill with a depth of about 4430 m (Western Sill-1 according to Tarakanov et al. 2013) (Fig. 5.22). This sill also coincides with the

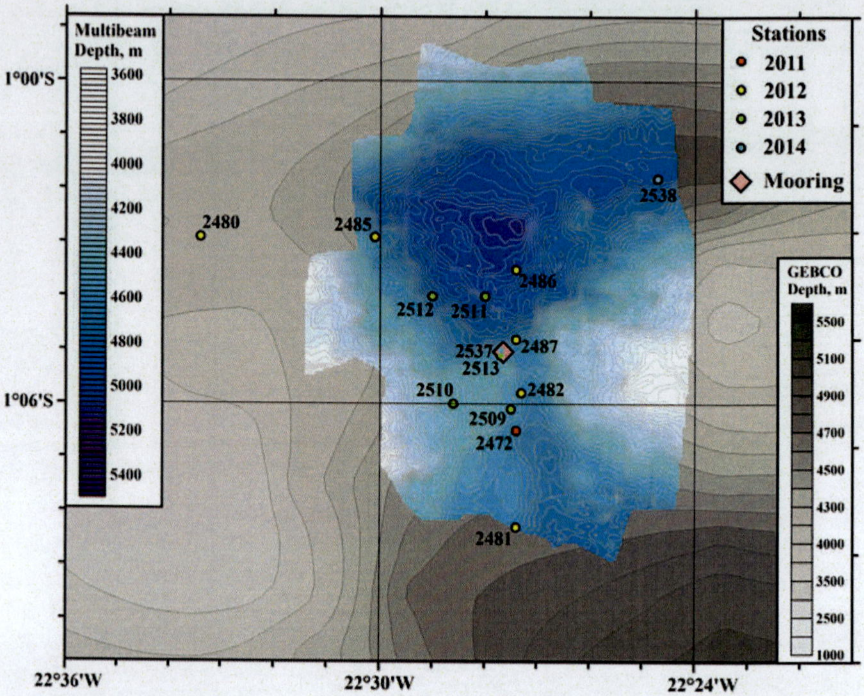

Fig. 5.24 Bottom topography in the region of the entrance of AABW flow to the Romanche Fracture Zone based on the data of the recent multi-beam survey from the R/V *Akademik Nikolay Strakhov* in 2016 (Kapustina et al. 2021)

extreme narrowness of the valley. Here, its width at the 4150 m isobath slightly exceeds one mile. Further to the east, the valley widens and deepens reaching 5000 m at 22°28′ W (hereinafter, the Southern Valley). To the east of this depression, the Southern Valley becomes shallower (4870 m) and narrows to three miles at 22°24.5′ W (hereinafter, the Intermediate Sill), and again deepens to 5600 m at 22°10′ W. A gap (passage) with a sill depth of about 4560 m exists in the Southern Wall of the fracture at the longitude of the Southern Depression (Southern Entrance-1). The width of this passage, which connects the Romanche Fracture Zone with the fracture to the south, is about 4 miles at a depth of the 4150 m isobath.

5.4.7.3 CTD/LADCP Measurements in the Deep Spillway Region in the Western Part of the Romanche Fracture Zone

Indications of a deep spillway were initially found at the entrance of Antarctic Bottom Water flow to the Romanche Fracture Zone at 22°30′ W. One prominent indication is the increase in temperature at the bottom over a very short distance

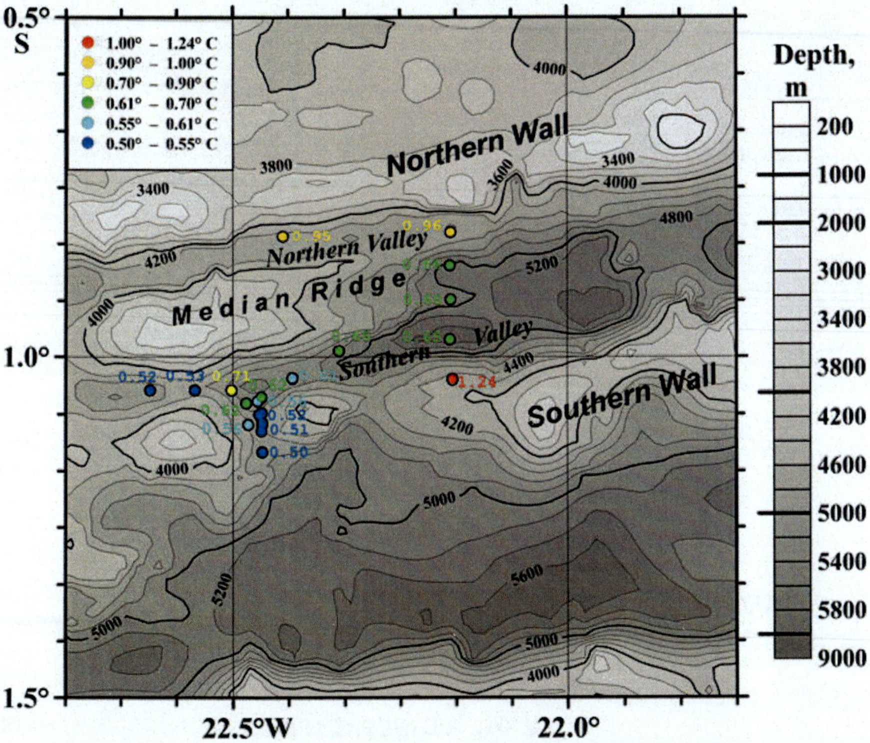

Fig. 5.25 Map of bottom (~5–10 m above the ocean bottom) potential temperatures at the network of stations in the small-scale study site. Numerals and colors show potential temperatures at the lowest CTD point (~10 m) above the ocean bottom based on the CTD-measurements at stations in 2011–2014. (Reproduced from Tarakanov et al. 2018)

along the abyssal flow. Another indication is the erosion of the lower part of the deep thermocline (Tarakanov et al. 2013). A strong increase in the bottom potential temperature as the cold-water flows into the fracture zone can be seen in Fig. 5.25. Another indication is the erosion of the lower part of the deep thermocline (Tarakanov et al. 2013) that can be seen from profiles of potential temperature at all stations in 2011–2014 (Fig. 5.26). Profiles of potential temperature at stations in 2011, 2012, 2013, and 2014 are shown in Figs. 5.27, 5.28, 5.29, and 5.30, respectively. Since such indications exist and were observed, we made detailed investigations to find the deep spillway in this region (Tarakanov et al. 2018).

Analysis of CTD/LADCP data in 2011–2014 revealed the following features of the spillway in the western part of the Romanche Fracture Zone:

(1) Figure 5.26 shows that all profiles in Western Sill-1 and Southern Entrance-1 in all years of observation form one group similar in shape, the profiles within the Southern Valley and further to the east along the fracture form the other, and the profiles above the fracture slope opposite to Southern Entrance-1 form the third.

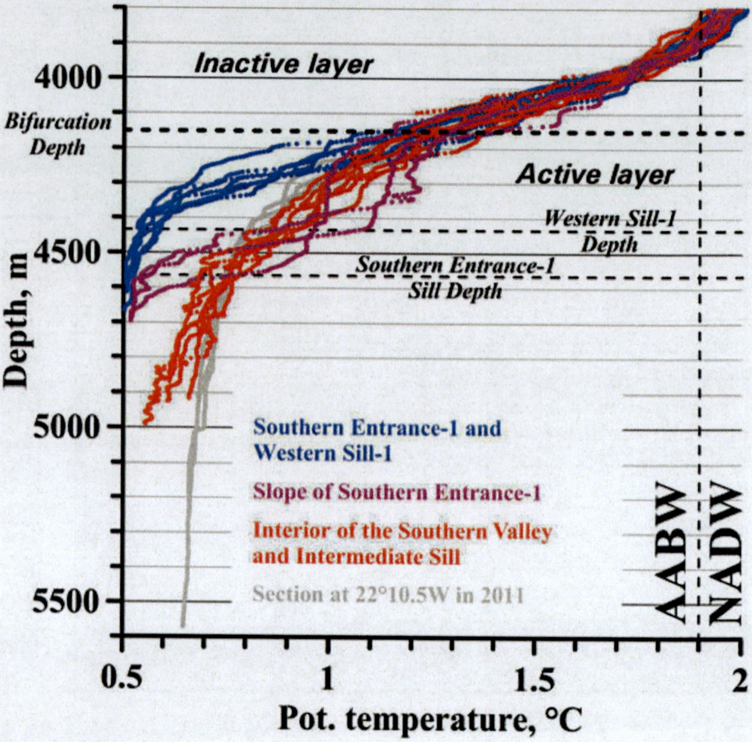

Fig. 5.26 Potential temperature curves at stations near Southern Entrance-1 and Western Sill-1 (blue), at the slope of Southern Entrance-1 (violet), in the interior of the Southern Valley and Intermediate Sill (red), and section at 22°10.5′ W in 2011 (gray). The bifurcation depth (4150 m) (in the region between the sills and the interior of the Southern Valley), below which temperature profiles split, is shown. This level corresponds to a boundary between inactive (motionless) and active layer in 1.5 layer model of the hydraulically controlled flow through the channel with a rectangular section between two basins (Whitehead et al. 1974). The sill depths of Southern Entrance-1 and Western Sill-1 are shown

Thus, the spatial variability of the temperatures between these regions during the measurement periods is much greater than the time variability between surveys. Profiles in the group merge above a level of 4150 m and diverge below this depth, which corresponds to a temperature of $\theta = 1.2$°C. Following Whitehead (1998) we will call this level a bifurcation depth.

(2) Diffusion (erosion) of the lower part of the deep thermocline (below $\theta = 1.2$ °C) in the region between the sills and in the Southern Depression indicates the existence of a hydrostatic pressure gradient directed into the Southern Depression. The gradient is increasing with depth (as a sequence of the hydrostatic equation) with nearly the same values based on the measurements in different years. This pressure difference drives the abyssal flow, which is similar (both qualitatively and quantitatively) in each survey of our measurements.

(3) We found that Southern Entrance-1 is not the only source of cold abyssal waters with $\theta < 1.2$ °C for the eastward flow in the Romanche Fracture Zone, as we

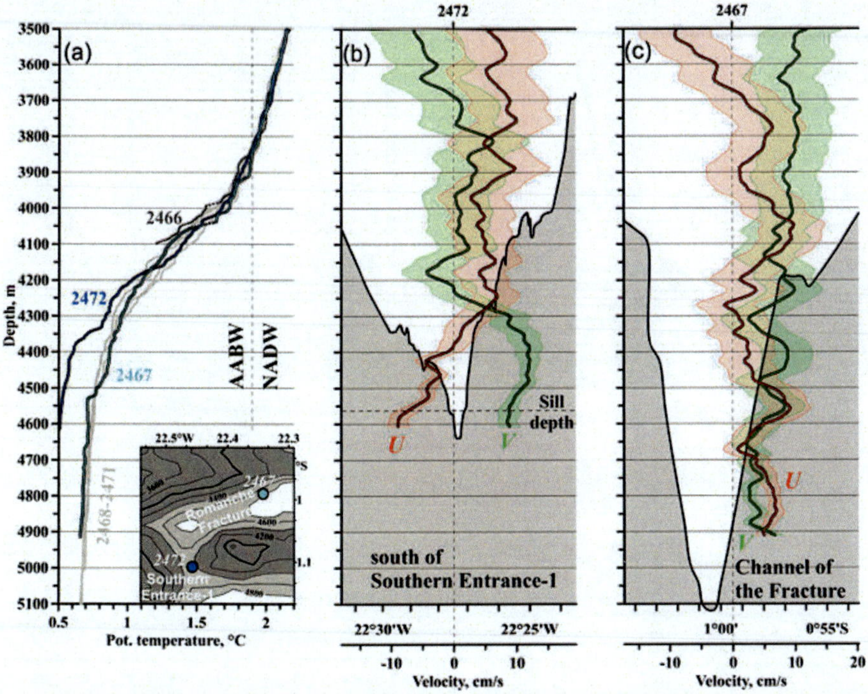

Fig. 5.27 Profiles of potential temperature in the deep part of the study region in 2011 (**a**); zonal (U) and meridional (V) velocity components and errors based on LADCP measurements at station 2472 south of Southern Entrance-1 (**b**); and at solitary station 2467 (see Fig. 5.22) in the southern canyon of the fracture (c). The bottom topography shown in Fig. 5.27b and c in the section of Southern Entrance-1 and channel of the fracture is based on the data of echo soundings in 2011. A scheme of locations of stations 2467 and 2472 in the fracture is shown in the inset of Fig. 5.27a. (Reproduced from Tarakanov et al. 2018)

assumed in (Tarakanov et al. 2013), because another source of cold abyssal waters exists in the western part of the fracture through Western Sill-1. The bottom potential temperature at Western Sill-1 in the Southern Valley of the fracture was 0.534 °C in 2012, and 0.524 °C in 2013 almost at the point of mooring before its deployment and after its recovery (Figs. 5.24, 5.25). These temperatures are close to those measured at the sill of Southern Entrance-1 (0.513 °C in 2012 and 0.506 °C in 2013, Figs. 5.25, 5.28, and 5.29). Thus, a deep spillway is formed from two flows of water in almost the same temperature range ($\theta < 1.2$ °C) that merge after descending from these sills into the Southern Depression.

The minimum potential CTD temperature measured at the bottom in 2011–2014 was 0.501 °C at CTD station 2481 at Southern Entrance-1 (Figs. 5.24 and 5.25). The lower instrument on the mooring at a location close to this point recorded even a lower temperature of 0.479 °C on December 26, 2013, while the mean temperature over the period of measurements was 0.501 °C exactly as the minimum temperature recorded by CTD.

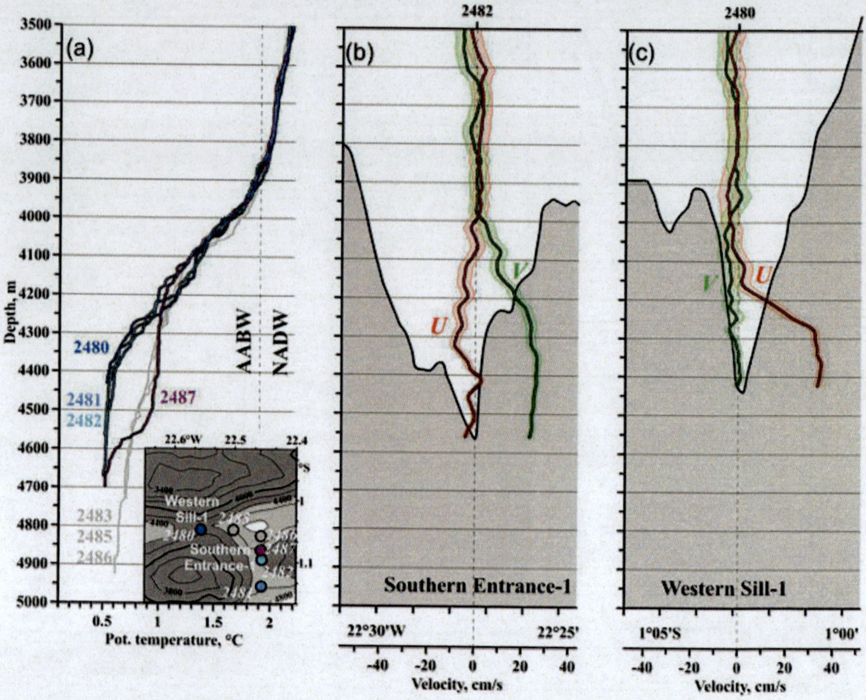

Fig. 5.28 Profiles of potential temperature in the deep part of the study region in 2012 (**a**), zonal (U) and meridional (V) velocity components and errors based on LADCP measurements at station 2482 in Southern Entrance-1 (**b**) and at station 2480 at Western Sill-1 in the southern canyon of the fracture (c). The bottom topography shown in Fig. 5.28b and c in the section of Southern Entrance-1 and Western Sill-1 is based on the data of echo soundings in 2012. The inset in Fig. 5.28a shows a scheme of the location of stations in 2012. (Reproduced from Tarakanov et al. 2018)

(4) Bottom potential temperatures generally increase from both sills to the Southern Depression, where the temperature increase is the greatest; then, the potential temperatures increase to the east over the Intermediate Sill to the deep part of the Romanche Fracture Zone. This temperature increase is consistent with the condition for the conservation of heat during mixing in the abyssal flow. Its high intensity in this region has been shown in (Tarakanov et al. 2013; van Haren et al. 2014). The presence of a local temperature maximum in the western part of the Southern Valley (Southern Depression) (yellow circle in Fig. 5.25, $\theta = 0.71$ °C) is because the corresponding station on the slope of Western Sill-1 was not located exactly in the coldest flow flowing down from this sill. The local maximum (0.96 °C) immediately after the Intermediate Sill ($\theta = 0.66$ °C, green circle in Fig. 5.25) appears because this station was located on the northern slope of the Southern Valley.

(5) The bottom potential temperature at station 2483 in the Northern Valley in 2012 was 0.949 °C (Figs. 5.19 and 5.25). The deep thermocline at this station in the layer with $\theta < 1.2$ °C is not sharp as one can see from the vertical profile

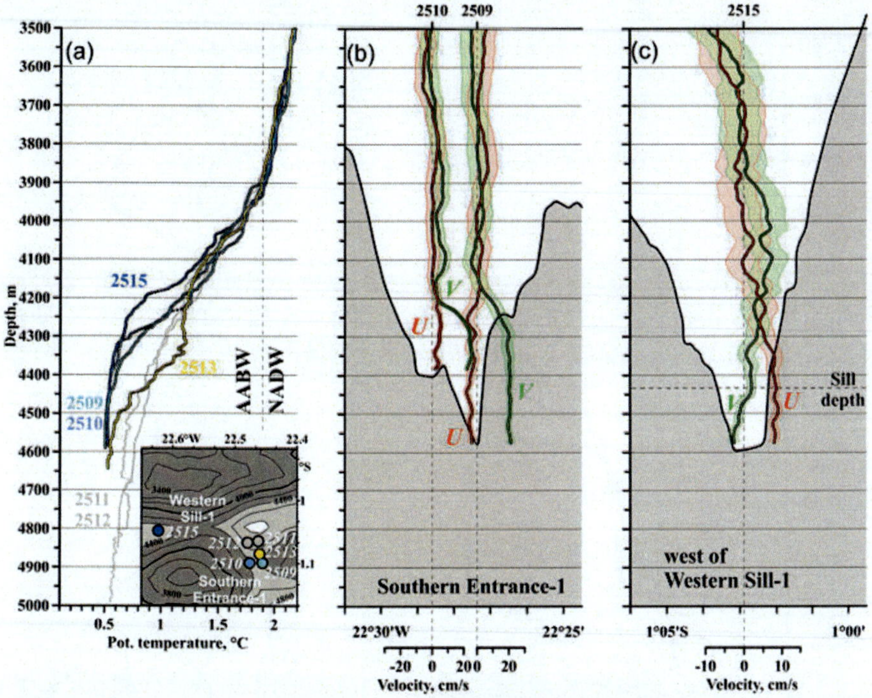

Fig. 5.29 Profiles of potential temperature in the deep part of the study region in 2013 (**a**), two profiles of the zonal (*U*) and meridional (*V*) velocity components and errors based on LADCP measurements across Southern Entrance-1, stations 2509, 2510 (**b**), and zonal (*U*) and meridional (*V*) velocity components at station 2515 west of Western Sill-1 in the southern canyon of the fracture (**c**). The bottom topography shown in Fig. 5.29b and c in the section of Southern Entrance-1 and Western Sill-1 is based on the data of echo soundings in 2013. The inset in Fig 5.29a shows a scheme of the location of stations in 2013. (Reproduced from Tarakanov et al. 2018)

(Fig. 5.28); hence, this station was located downstream the flow in a region where this thermocline erosion takes place. We did not find a local depression, in which a hydraulic jump should exist in the deep Northern Valley where this thermocline erosion could take place. Therefore, the part of the valley, in which our station was located, is most likely filled with the water through the Southern Valley region rather than from the western channel along the valley.

In the southern channel of the Romanche Fracture Zone near Western Sill-1, we measured a slightly warmer temperature 0.523 °C at station 2515 (Fig. 5.29), which is almost by 0.022 °C lower than the minimum temperatures ever measured in the channel of the fracture (WOD18 database, WOD18 2018). The latter fact indicates that the Romanche Fracture Zone as a channel for the cold AABW begins to the west of our study region. A temperature increase up to 0.554 °C was found at station 2511 in the aforesaid depression of the channel where two flows merge (Fig. 5.29). This station was occupied north of Southern Entrance-1 and east of Western Sill-1 along

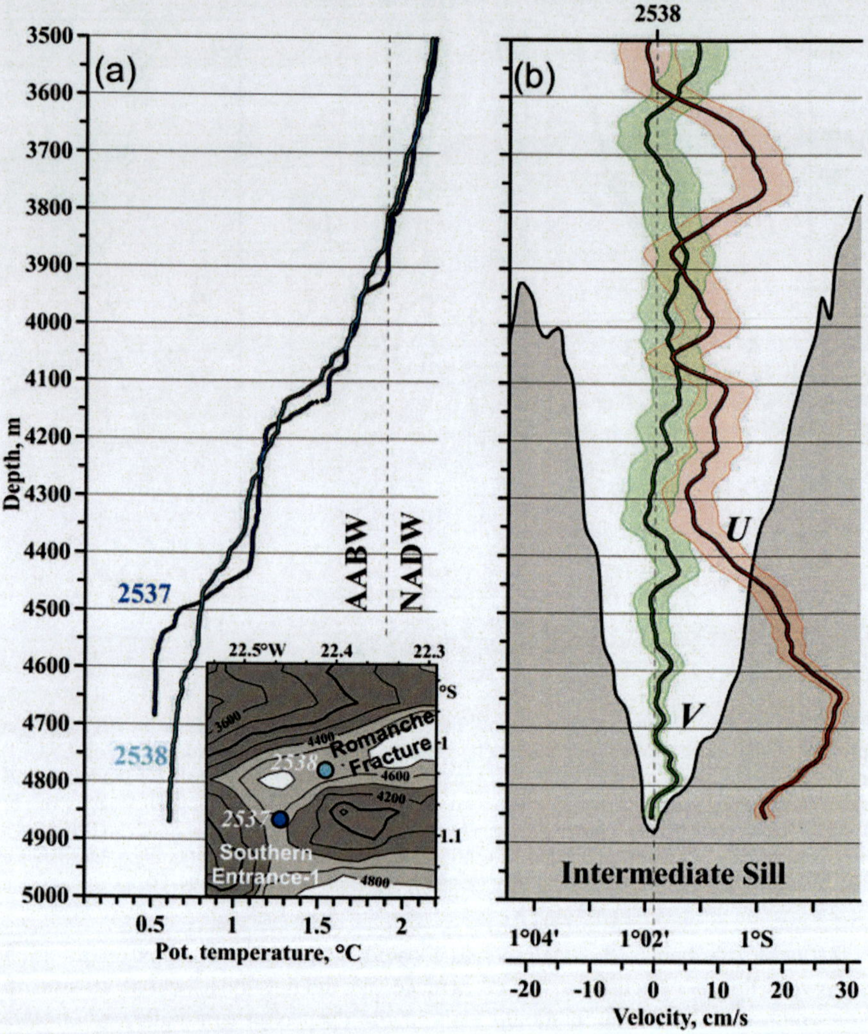

Fig. 5.30 Profiles of potential temperature in the deep part of the study region in 2014 (**a**), zonal (U) and meridional (V) velocity components and errors based on LADCP measurements at station 2538 in the southern canyon of the fracture (**b**). The bottom topography shown in Fig. 5.30b is based on the data of echo soundings in 2014. The inset in Fig 5.30a shows a scheme of the location of stations in 2014. (Reproduced from Tarakanov et al. 2018)

the channel. Higher bottom temperatures also observed in this depression (0.707 °C at station 2485, 0.620 °C at station 2512, and 0.615 °C at station 2486) were measured in different locations of the depression. They indicate that the two jets that flow into the depression do not fill it completely with the cold water but propagate further and the temperature at station 2538 downstream appeared lower (0.613 °C) than at the western stations. The minimum bottom temperature in the northern channel of the Romanche Fracture Zone (Western Sill –2) is 0.953 °C that shows that this channel is not a pathway for the coldest water.

5.4.7.4 Deep Spillway at the Entrance to the Romanche Fracture Zone, Structure of Bottom Currents

All velocity profiles based on the LADCP measurements over the sills show increasing intensity of the abyssal flow deeper than 4150 m: Southern Entrance-1 (stations 2472, 2481, 2482, 2509, 2510), Western Sill - 1 (stations 2480, 2515). The flow at Southern Entrance-1 is directed to the north, while the flow at Western Sill-1 is directed to the east. Above this level (4150 m), the velocities are usually low or the currents are even oppositely directed.

The following figures illustrate the spatial structure of currents in the region of the spillway in the western part of the Romanche Fracture Zone. Figure 5.31 is a chart of

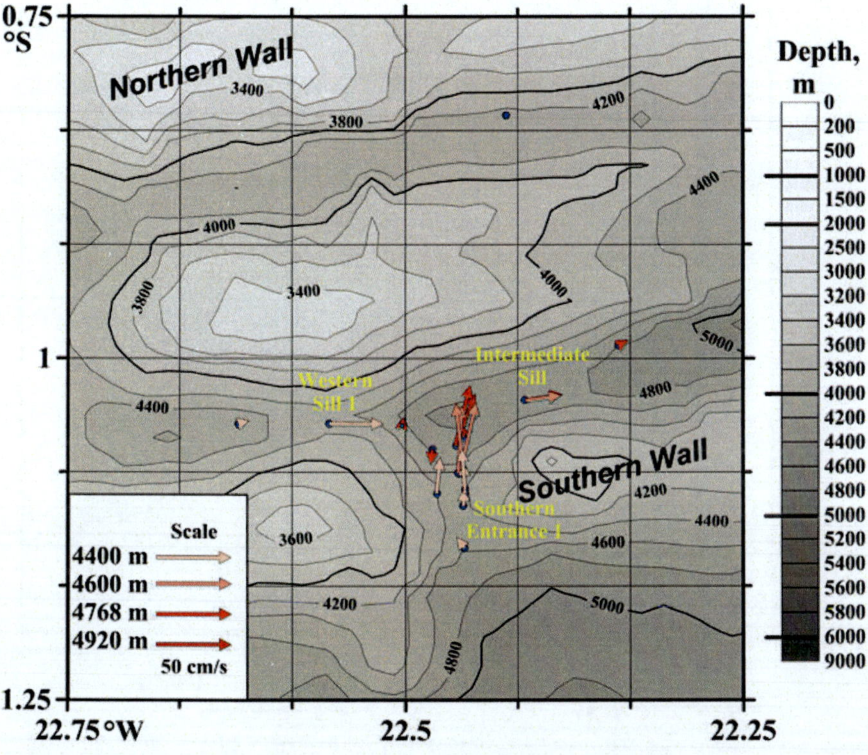

Fig. 5.31 Velocity vectors of the abyssal flow at the entrance to the Romanche Fracture Zone based on the data of LADCP measurements at stations (blue dots) in 2011–2014. The vectors are shown at the depths approximately corresponding to the maximum velocity. Velocity vector at the station in the Northern Valley is not shown because the maximum depth at this station is shallower than 4400 m. (Reproduced from Tarakanov et al. 2018)

velocity vectors in the bottom layer based on the LADCP measurements in all years of observations. Fig. 5.31 gives us an idea about the structure of the velocity field in the dynamically most active part of the abyssal flow (i.e., deeper than 4150 m) in the region of the entrance of the AABW flow to the Romanche Fracture Zone. The figure is plotted from the LADCP data at all stations in 2011–2014 near the entrance to the fracture.

Figures 5.27, 5.28, 5.29 and 5.30 show velocity profiles (along with the accuracy range) in the bottom depth layers at stations near the sills in 2011–2014. Figures 5.32 and 5.33 show the sections of potential temperature across the sill of Southern Entrance-1 and profiles of velocity in 2012 and in 2013 along the flow descending from this sill.

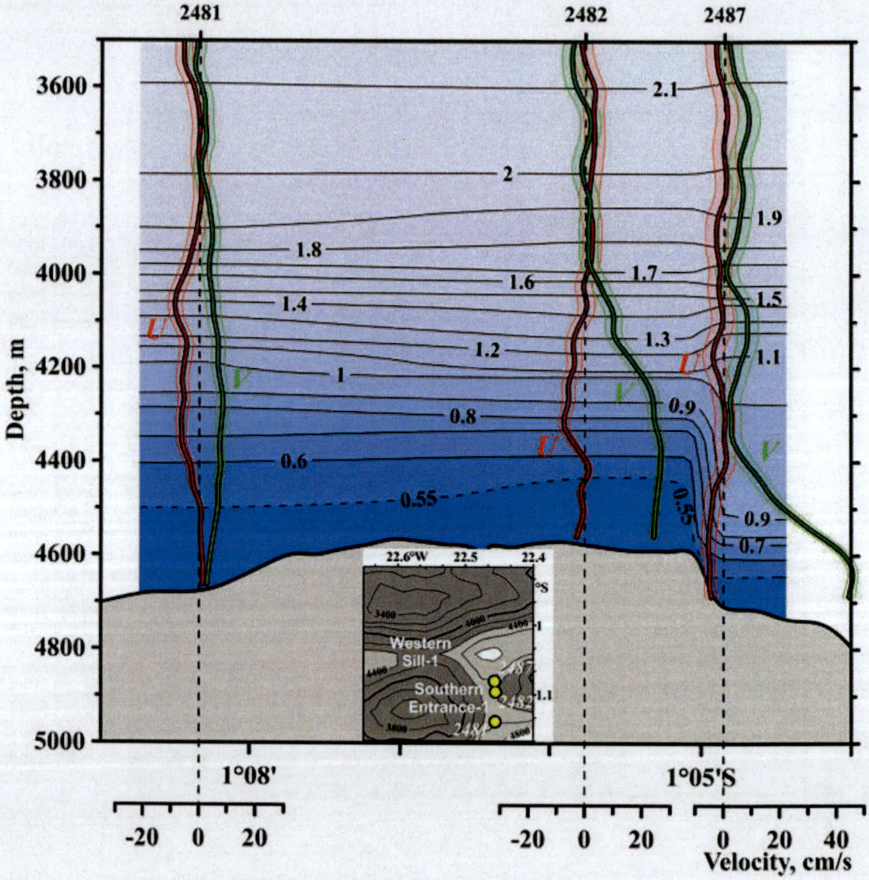

Fig. 5.32 Distribution of potential temperature (°C) and profiles of the zonal (U) and meridional (V) current components and errors in the bottom layer over the meridional section across Southern Entrance-1 to the abyssal depths of the Southern Valley in 2012. A scheme of the stations over the section is shown in the inset. (Reproduced from Tarakanov et al. 2018)

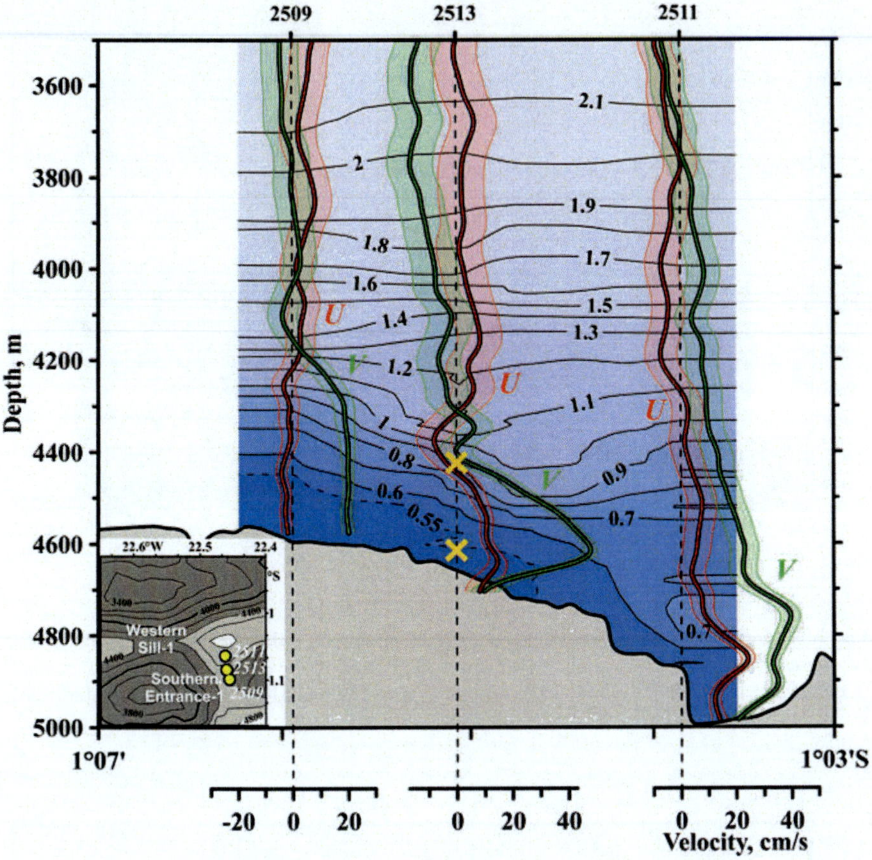

Fig. 5.33 Distribution of potential temperature (°C) and profiles of the zonal (U) and meridional (V) current components and errors in the bottom layer over the meridional section across Southern Entrance-1 in 2013. A scheme of the stations over the section is shown in the inset. Yellow crosses indicate the levels of two Aquadopp current meters at 4421 and 4611 m. (Reproduced from Tarakanov et al. 2018)

Figure 5.34 is based on the data of the «yo-yo»-casts (station 2513 in 2013 and station 2537 in 2014) over the southern slope of the Southern Valley (Southern depression) on the sill of Southern Entrance-1. Figure 5.35 shows the section across the Romance Fracture Zone occupied in 2011 at 22°10.5′ W. The direction of currents in the bottom layer in Fig. 5.31 corresponds to the direction of the temperature (and pressure) gradient from the sills into the Southern Valley: the abyssal waters flow through Southern Entrance-1 and Western Sill-1 to the north and to the east, respectively, and descend into the Southern Valley. Then, they merge into the flow that ascends eastward to the Intermediate Sill and continues to flow further east along the fracture channel. The measured velocities of the abyssal flow were 5–10 cm s^{-1} before overflowing the sill of Southern Entrance-1 and Western Sill-

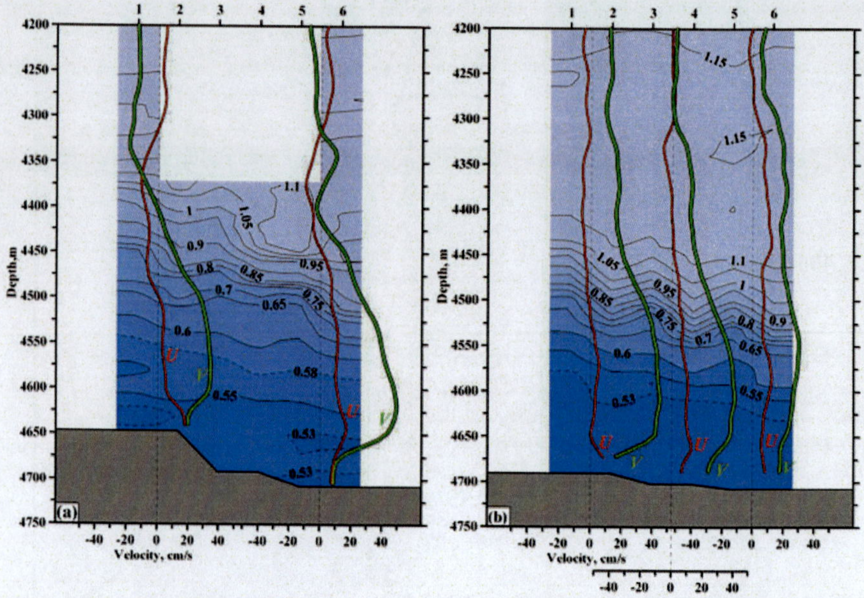

Fig. 5.34 Distribution of potential temperature (°C) and profiles of the zonal (U) and meridional (V) current component in the bottom layer based on the data of yo-yo casts over the northern slope of the fracture Southern Wall in Southern Entrance – 1 at station 2513 in 2013 (**a**) and station 2537 in 2014 (**b**) at the location of the mooring before its deployment and after its recovery, respectively

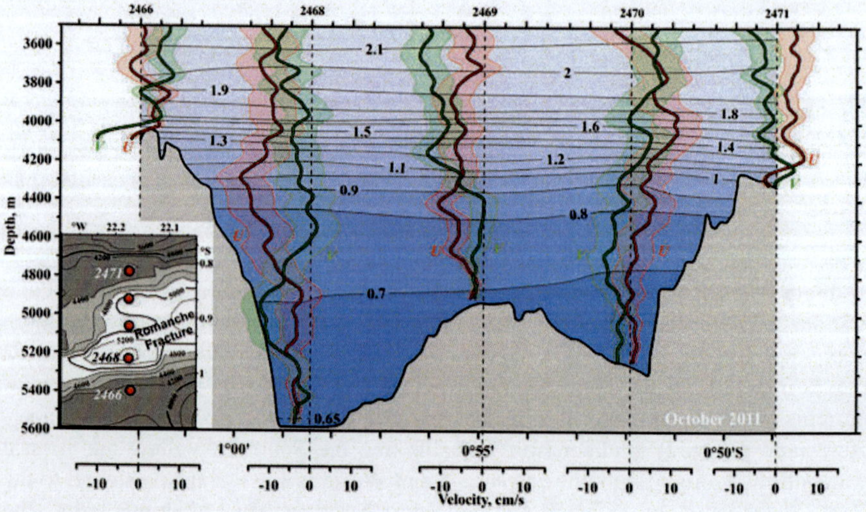

Fig. 5.35 Distribution of potential temperature (°C) and profiles of the zonal (U) and meridional (V) current components and errors in the bottom layer over the meridional section across Romanche Fracture Zone in 2011 at 22°10.5′ W

1, 20–25 cm s^{-1} at Southern Entrance-1, and 35 cm s^{-1} at Western Sill-1 (Figs. 5.27, 5.28, 5.29 and 5.31). The maximum velocity (up to 48 cm s^{-1}) was measured on the slope of the Romanche FZ at Southern Entrance-1. Lower velocities up to 28 cm s^{-1} were fixed at the Intermediate Sill (Fig. 5.31).

The velocity field shown in Figs. 5.31 is typical for an overflow with hydraulic control. The flow accelerates in the subcritical regime (Froude number $Fr < 1$) before the control point (sill or narrows), reaches the critical regime ($Fr = 1$) at this point, and then the regime changes to the supercritical ($Fr > 1$) after passing the critical point to return back to the subcritical regime downstream as a hydraulic jump. The Froude number is defined as $Fr = u/\sqrt{g'H}$; where u is the flow velocity, $\sqrt{g'H}$ is the velocity of gravity waves in shallow water in the flow; H is the thickness of the flow; $g' = g\Delta\rho/\rho$ is reduced acceleration due to gravity; ρ is the *in situ* density; Δρ, which is small, is the difference between the density of the flow and surrounding water. The data available from our measurements provide more indirect evidences of a hydraulically controlled overflow across the sills of Western Sill-1 and Southern Entrance-1, which are as follows.

(1) The so-called predicted transport over the sills between two basins based on an inviscid 1½-layer model of hydraulically controlled overflow through a channel with a rectangular cross-section has been calculated as in (Whitehead et al. 1974). The flow transport in this steady model Q is derived from the Bernoulli's law for the lower active layer (Fig. 5.26) using the potential vorticity conservation and geostrophic balance. As far as the Romance Fracture Zone is located in the equatorial zone, transport Q depends only on the width W of the channel, upstream thickness of the active layer h_u above the bottom, and reduced gravity g':

$$Q = \left(\frac{2}{3}\right)^{\frac{3}{2}} W h_u \sqrt{g' h_u}. \tag{5.1}$$

Following Whitehead (1998) we estimated g' from the upstream-downstream density difference Δρ. There were only a few distant upstream stations from the sill. Therefore, we used the density difference between the stations at the sills and mean density in the Southern Valley over all measurements, which is Δρ = 0.027 kg/m^3 (at a depth of 4400 m). The estimated reduced gravity is $g' = 2.7 \times 10^{-4}$ m·s^{-2}. Thickness h_u is estimated as the depth difference between the bifurcation depth and the channel bottom at the sill. This thickness is equal to 280 m over Western Sill-1, and 410 m over Southern Entrance-1. Since the cross-sectional square of the passages is much smaller than the value of the rectangular model, we used the mean values of W over the depths (below the bifurcation depth), which were equal to 1.4 km and 3.2 km, respectively. Using these parameters, we estimate the transport as 0.06 Sv over Western Sill-1, and 0.24 Sv over Southern Entrance-1. These estimates agree well with the transport estimates based on the direct measurements.

(2) We estimated the horizontal gravity wave speed as $\sqrt{g'H}$ using the same 1½-layer model. Since there is difference between the real cross-section over the sills and the section in the rectangular model, we used not the h_u, but the average thickness of the layer below bifurcation depth, which was 160 m at Western Sill-1 and 185 m at Southern Entrance-1. The gravity wave speeds based on these parameters were 21, and 22 cm·s^{-1}, respectively. These values are close to the measured velocities at Southern Entrance-1 (20–25 cm·s^{-1}) and are significantly smaller than at Western Sill-1 (35 cm·s^{-1}). Thus, the ratio of the gravity wave speed to the measured velocity at Western Sill-1 corresponds to the supercritical flow.

(3) Another indication of the hydraulic control of the flow over the sill is an uplift of the deep upstream water, which is termed Bernoulli aspiration in (Kinder and Bryden 1990). This uplift is manifested in the inclination of isopycnals (isotherms) almost parallel to the bottom slope. In addition, acceleration of this deep upstream flow is observed. We observed this uplift over the section in 2012 across the sill of Southern Entrance-1 (Fig. 5.32). The elevation of isotherms from the station in front of the sill was observed in the layer approximately below $\theta < 0.75$ °C. The velocities of the current directed toward the sills in the layer below the sill depth were measured at another two stations displaced upstream from the sills: stations 2472 south of the sill of Southern Entrance-1 (Fig. 5.27), and 2515 west of Western Sill-1 (Fig. 5.29). Following (Clément et al. 2017) we estimated the thickness of the layer below the sills to find the magnitude of the uplift. This is u/N where u is the velocity at the sill, N is the characteristic Brunt-Väisälä frequency in the upstream basin. Assuming that $N = 0.7 \times 10^{-3}$ s^{-1} (in the bottom layer at station 2481 south of Southern Entrance-1), $u = 25$ cm·s^{-1} at Southern Entrance-1, and 35 cm·s^{-1} at Western Sill-1, we get that the uplifts are 350 m and 500 m respectively. Thus, the estimates of the bottom layer thickness in the fracture south of the Romanche Fracture Zone at depths of 5200–5700 m are typical; hence, the bottom layers can be blocked from the inflow into the Romanche Fracture Zone.

(4) The properties of the internal hydraulic jump are seen on the section of three stations occupied in 2013 from Southern Entrance-1 sill to the abyssal depths of the Southern Valley (Fig. 5.33). As the abyssal flow descends, its bottom core accelerates and the inclination of isotherms in this core becomes almost parallel to the slope of bottom topography of the fracture. A similar pattern of isotherms is shown in Figs. 5.32 and 5.34. The latter figure is plotted from the data of casts at stations 2513 and 2537 over the southern slope of the Southern Valley. During the cast when the instrument was close to the bottom it was lowered and raised three times over 300 m distance at the bottom. Each time the instrument was lowered to the bottom we found that strong currents displaced it down the slope and we had to increase the wire to get closer to the bottom. It is seen in Fig. 5.33 that as soon as the velocity core reaches the bottom layer of the Southern Valley it slows down and becomes thicker.

For example, the core thickness on the sill over the 18 cm·s^{-1} isotach (which is the boundary of the flow core with quasi-constant velocity) was 300 m over the sill. The thickness decreased to 150 m on the slope and then increased to 470 m in the deep part of the Southern Valley. The elevation of the isotherms from the depth in the layer $\theta > 0.8$ °C is consistent with this thickening (Fig. 5.33). The slowdown and thickening of the flow correspond to the shock discharge (dissipation) of the kinetic energy in the zone of the hydraulic jump. It is possible that station 2511 in the Southern Valley was located immediately in the zone of the hydraulic jump with very strong vertical turbulent mixing, as indicated by the records of several temperature inversions in the layer from the bottom to about 4400 m (Fig. 5.33).

Since the spatial variations in the temperature profiles in each of the surveys are similar, we conclude that the overflows through Southern Entrance-1 and Western Sill-1 have been hydraulically controlled during both surveys. Similarity between the spatial distributions of velocity in the abyssal flow in Fig. 5.26 with the distribution characterizing the hydraulically controlled overflow such that an internal hydraulic jump is formed indicates that the variations in velocity between surveys as well as the variations in temperature did not introduce qualitative distortions into this pattern of the current.

The existence of low velocities, and even reverse (southward) currents at several points in the Southern Valley is an interesting property of the spillway (Fig. 5.31). The explanation of their existence, given the presence of a flow that descends from the sills measured in each survey, is that these stations were located not in the narrow jet of this flow. The presence of jets in the Southern Valley can also be shown by considering the transport balance along the flow descending from the sill. An increase in the flow velocity when it descended requires a reduction of its cross section due to the condition of the transport conservation. We estimated the widths of the flow descending from the Southern Entrance-1 based on this balance using the measurement in 2013 (Fig. 5.33). The calculation was made assuming the horizontal homogeneity of the velocity field in the flow separately for the triangular and rectangular cross-sections of the flow. The width of the passage at the sill along the bifurcation depth is 7.1 km. In the case of a triangular cross-section, the width of the flow on the slope was about 7.1 km at the 4400 m isobath (zero velocity above the velocity core of the flow) and about 2.7 km at 4150 m in the deep part of the Southern Valley. The flow widths of a rectangular section were 2.1 and 1.0 km, respectively. Just to compare the widths we indicate that the characteristic size of the Southern Valley at the 4400 and 4150 m isobaths (Figs. 5.24) is 8–11 km. These estimates show that the abyssal flows descending from the sills in the Southern Valley form narrow jets. Their horizontal transverse dimension may be substantially smaller than the size of this basin.

The lower velocities as the currents flow up the Intermediate sill (up to 28 cm s^{-1} at station 2538, Fig. 5.30) may be caused by the time variability of the intensity of this flow (at least, the transverse inhomogeneity of the velocity field) and by a decrease in the kinetic energy of the flow as it ascends from the depth of the Southern Valley. The real causes are not clear because the measurements in the Intermediate Sill and Southern Valley were performed in different years. The overestimated transport compared with the estimated transport sum of the cross-sections of Western Sill-1 and Southern Entrance-1 and small velocities (3–8 cm·s^{-1}) at station 2467 over the northern slope of the Southern Valley (Fig. 5.27) are consistent with the idea of the existence of a jet at the Intermediate Sill and in the region of downstream in the Romanche Fracture Zone. The jet structure of the flow even further downstream is unambiguously confirmed by the LADCP measurements over the section in 2011 at 22°10.5′ W (Fig. 5.35). One can see that abyssal currents of alternative direction with velocities up to 15 cm·s^{-1} were observed here.

The total transport of the abyssal flow appeared negative (i.e., westward) (Tarakanov et al. 2013). Hence, the existence of the flow down from the upstream sills in 2011 means that the flows directed to the east slipped through the 3-mile grid of stations of this section. Northeasterly bottom velocities of 3–5 cm·s^{-1} were also recorded at station 2483 in the Northern Valley of the fracture north of the spillway zone (Fig. 5.28). Small values of velocity in this region agree with the indirect indications of the absence of an intense abyssal flow in this valley.

The estimates of vertical velocities in the flow spilling from the sills were based on the measured horizontal velocity and average inclination of the bottom slope at Southern Entrance-1 equal to 1/10. The velocities measured at the stations along the bottom flow were 25–40 cm·s^{-1}. Thus, the estimates of the average vertical component are 3–4 cm·s^{-1}. The average inclination of the bottom of the Southern Valley of the fracture after the flow passes Western Sill-1 is 1/15. Assuming that the velocity of the current does not decrease below the velocities measured at the sill (35 cm·s^{-1}), the average vertical component can be estimated here as not less than 2.3 cm·s^{-1}. All estimates indicate that very high vertical velocities exist in the spillway.

5.4.7.5 Mooring Observations

Moored measurements of velocities and temperature allow us to get direct estimates of time variations in velocities and temperature at the point of mooring deployment. First of all, we shall analyze the average and statistical characteristics of the mooring time series of currents and temperature. After we found extremely high velocities of the Antarctic Bottom Water flow into the Romanche Fracture Zone based on LADCP data (Tarakanov et al. 2013) we deployed a mooring for 6 months in the narrow gap at 01°05.034′ S, 22°27.652′ W (Southern Entrance-1). This point is located at the northern slope of the Southern Wall of the fracture. Before the deployment and after the recovery, we made CTD-casts at this point (at a distance not exceeding 400 m from the mooring point). These were stations 2513 and 2537 in 2013 and 2014, respectively. The mooring was equipped with two Aquadopp (1.5 MHz) current meters at 4421 and 4611 m (sampling interval 10 min). We collected the mooring data from October 8, 2013 to April 16, 2014.

Mean and Statistical Parameters of Velocity and Temperature The distribution of the intensity of currents with respect to directions on the basis of the data of two current meters at the mooring station is shown in Fig. 5.36. We calculated the

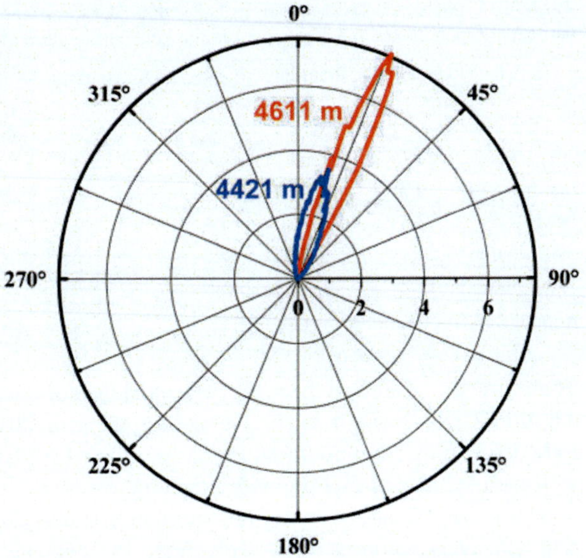

Fig. 5.36 Distribution of the intensity of currents by directions (rose diagram) (percent per one degree angle) over a period of 6 months based on the data of two moored current meters located at 4421 m (blue line) and 4611 m (red contour). Reproduced from (Tarakanov et al. 2018)

transport roses for the two current meters (similar to the wind rose graph), which are shown in Fig. 5.36. The full circle was divided into 360 sectors. The sum of the currents in each of the sectors was divided by the number of time moments when the current was directed to this sector. Thus, we found the mean velocity in the sector. In order to calculate the transport, this sum was divided by the entire period of observations, and then we calculated the full sum over 360 sectors, which was assumed as 100%. In this case, the transport is mean weighted absolute value in the given direction normalized by the total absolute value over 360 sectors. In our case, the transport rose almost repeats the rose of currents, i.e., the distribution of currents by directions.

The directions of the mean flow at the depths of the lower and upper current meters were 23° and 13°, respectively. We estimate that 95% of the measured directions appear within the range of 11–32° at the lower meter and 336–45° at the upper meter. The rotation of the mean flow direction of currents from 23° at the lower to 13° at the upper instrument is obviously caused by the influence of the bottom topography steering.

Table 5.11 shows the statistical characteristics of the time series of velocity component in the mean flow direction and vertical velocity component (W) at the upper and lower current meters of the mooring and of the highest and lowest thermistors.

Figure 5.37a shows the time variation in the daily average velocity flow. This figure together with the data in Table 5.11 show that the velocities measured by the lower current meter are high and quasi-permanent (~40 cm·s^{-1}); their fluctuations are almost one order of magnitude smaller than the mean velocities. These values are in a good agreement with the LADCP-measured velocities (40–45 cm·s^{-1}) at three stations (2487 in 2012, 2513 in 2013, and 2537 in 2014) near the mooring point.

Table 5.11 Statistical characteristics of the moored measurements of temperature and velocity: vertical component and the component in the general direction of the flow ($V_{mean\ flow}$)

	Upper ADCP current meter		Lower ADCP current meter		Upper thermistor, θ, °C	Lower thermistor, θ, °C
	$V_{mean\ flow}$, cm·s^{-1}	W, cm·s^{-1}	$V_{mean\ flow}$, cm·s^{-1}	W, cm·s^{-1}		
Mean	19.2	−4.9	40.7	−6.6	0.781	0.527
RMS	13.4	4.1	7.5	2.1	0.078	0.015
Maximum	56.1	10.3	66.2	2.4	1.173	0.756
Minimum	−13.2	−21.3	5.0	−15.2	0.534	0.479

Reproduced from Tarakanov et al. (2018)

Their direction is almost the same. The existence of the near-bottom jet current here was previously shown based on the measurements in 2013 over the section along the flow descending from Southern Entrance-1. The LADCP measured velocities agree well with the data of mooring velocities in the near-bottom jet. This correspondence between LADCP and mooring measurements shows that the near-bottom jet in the Southern Depression is quasi-permanent. The average velocity at the level of the upper current meter is 19 cm·s^{-1}; the amplitudes of fluctuations are of the same order of magnitude as the mean velocity (Fig. 5.37a and Table 5.11). The vertical velocities at the deep spillway are much higher than those expected in the open ocean (Fig. 5.37b and Table 5.11). The average measured vertical velocities of the descending flow at the lower and upper current meters were 6.6 cm·s^{-1} and 4.9 cm·s^{-1} respectively; the highest instantaneous values of vertical velocities were 15.2 and 21.3 cm·s^{-1}, respectively. These measured mean velocities agree with the estimates based on the LADCP measurements but are slightly greater. This fact is related to the steeper local bottom slope (1/6) at the location of the mooring, which was estimated from the ratio of the vertical and horizontal velocity components at the lower current meter. The potential temperatures (Table 5.11) near the bottom at the CTD stations near the sills of Southern Entrance-1 and Western Sill-1 fall within the RMS limits of the average value at the lower thermistor (0.527° ± 0.015 °C) or they are slightly (by 0.003 °C) cooler than the lower limit due to the upstream location of these stations relative to the mooring. The smallest instantaneous measured potential temperature near the bottom at the mooring was 0.473 °C. Thus, even colder water than that measured during CTD-casts may sporadically flow into the Romanche Fracture in the spillway region. The lowest and highest temperatures may be caused by intense turbulent overturning due to the shear instability observed at the point of the mooring (van Haren et al. 2014).

The vertical velocities are extremely high (Fig. 5.38). We assume that the flow follows the bathymetric slope. Since the slope of the underwater spillway is approximately 1/10 and the mean horizontal velocities measured by two instruments are 21 and 41 cm s^{-1} we find that the mean vertical velocities are 2 and 4 cm s^{-1} at the upper and lower levels of instruments, respectively. The estimated maximum vertical velocity can be as high as 7 cm s^{-1}. Indeed, the average vertical velocities measured by the current meters are 5 and 6 cm s^{-1}, while the maximum velocities are 10 and 15 cm s^{-1} at the upper and lower levels, respectively.

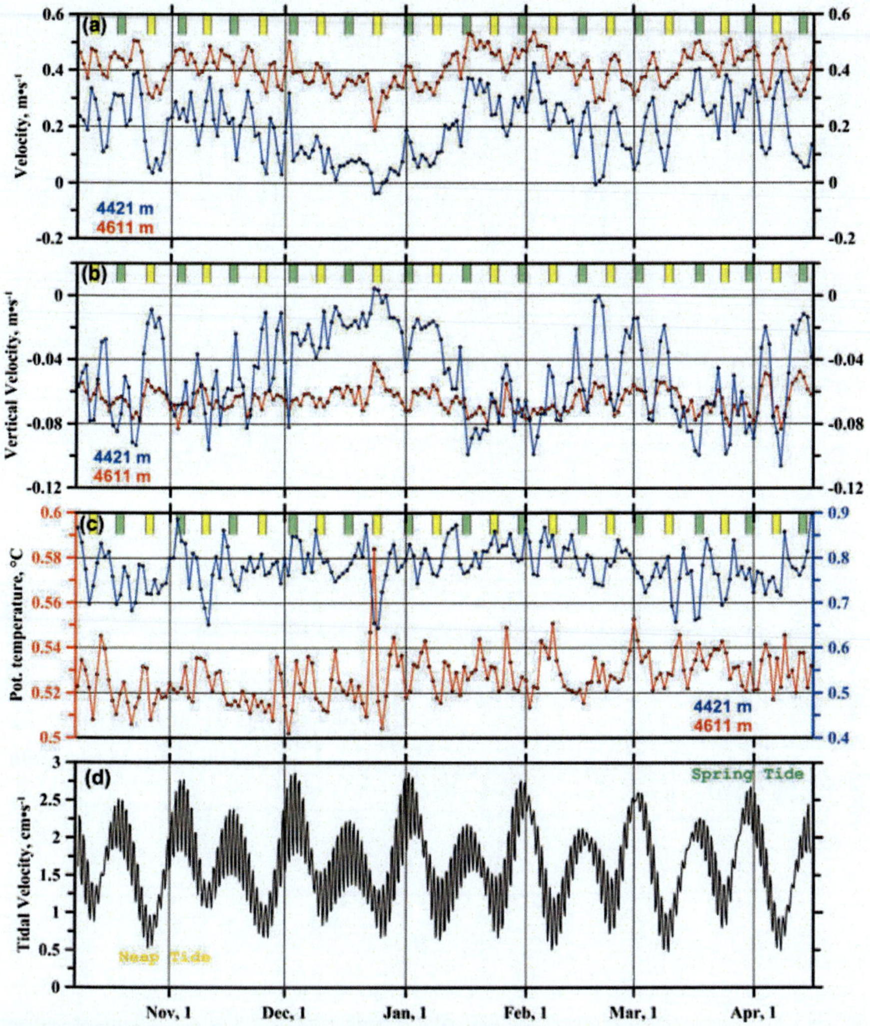

Fig. 5.37 Daily mean current velocity component in the mean flow direction (which is almost northward) at the upper (blue line) and lower (red line) Aquadopp current meters (**a**), vertical velocity components (**b**), and potential temperature at the upper and lowest temperature sensors (**c**), total barotropic tide on the basis of the TPXO9 model (**d**); positive values in Fig. 5.37b correspond to the upward direction. Yellow (green) marks at the upper axes correspond to the neap (spring) tide. (Reproduced from Tarakanov et al. 2018)

Time Variability of Temperature and Currents Including Tidal Variations The graphs of daily mean horizontal velocity components (in the direction of the mean flow current) measured by the upper and lower current meters (Fig. 5.37a and 5.38) show high correlation at sub-inertial frequencies over the vertical distance between the instruments. It is obvious that there is a long-term oscillation, which we

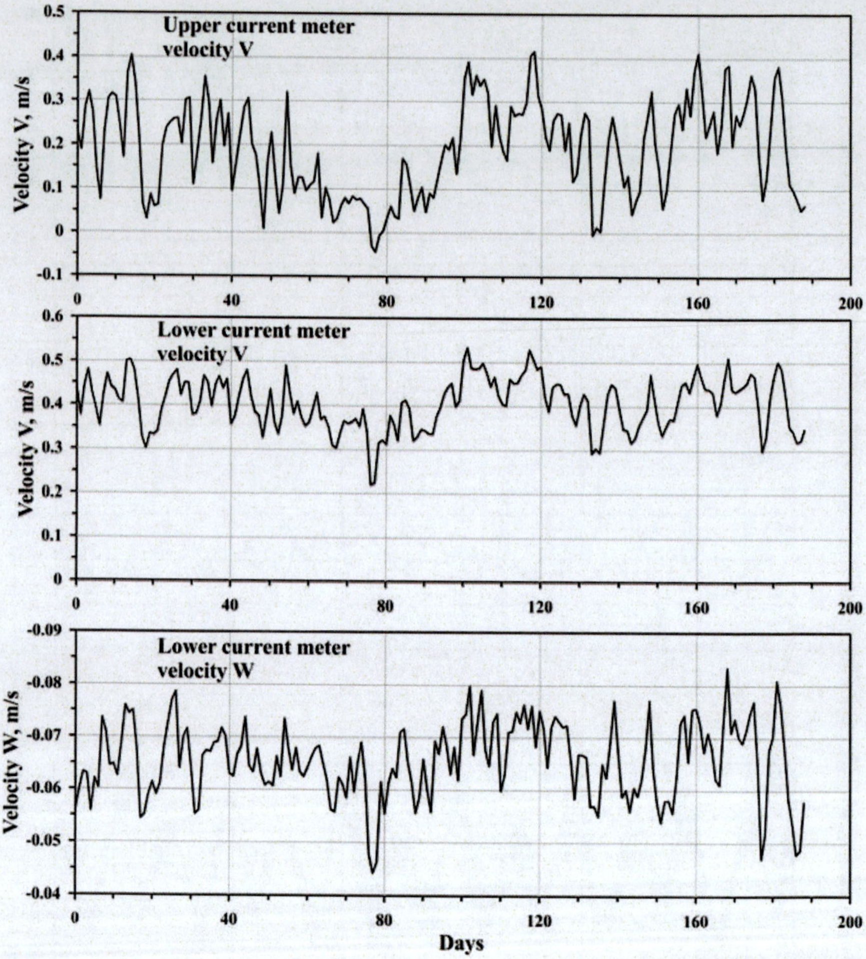

Fig. 5.38 Daily average currents: meridional current component, upper current meter (top panel), meridional current component, lower current meter (middle panel), vertical current, lower current meter (bottom panel)

estimated as 90-day periodicity over the 189 days of velocity measurements, but the time series is too short to get a statistically reliable spectral estimate. Both current meters recorded a minimum flow on the last days of December 2013 and a maximum in the beginning of February 2014. The vertical velocities (Fig. 5.37b) are in the negative correlation with the horizontal velocities since the velocity vector is directed along the sloping bottom. The indicated near-90-day period is not distinguished in the temperature time series (Fig. 5.37c).

The existence of oscillations with a period of several days (hereinafter low-frequency oscillations) is clearly seen on the graphs of the horizontal and vertical velocity components and on the graph of the lower temperature sensor

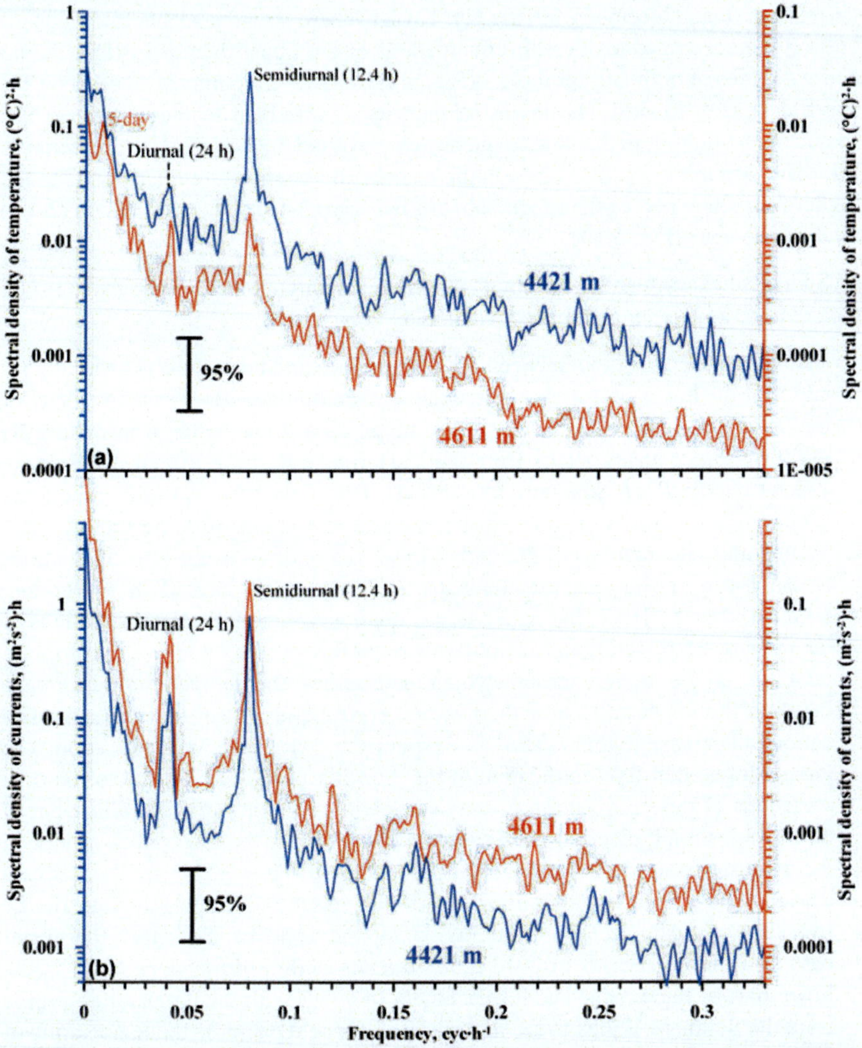

Fig. 5.39 Spectra of potential temperature (**a**), and components of velocities in the mean flow direction of currents (**b**) at the upper (blue line) and lower (red line) thermistors. 95% confidence interval is shown. (Reproduced from Tarakanov et al. 2018)

(9 m above the bottom) (Fig. 5.37). These oscillations correlate well. An increase in the horizontal velocities corresponds to an increase in the vertical downward velocities and a decrease in the temperature. The spectral analysis of the temperature distribution at the lower thermistor shows approximately a 5-day peak of the spectral density (but it is statistically insignificant) (Fig. 5.39). The temperature distribution at the upper thermistor demonstrates the segments of the positive correlation with the temperature at the lower thermistor (e.g., October, 2013) and of the negative

correlation (e.g., December, 2013) (Fig. 5.37a). There are no peaks at low frequencies on the spectra of temperature fluctuations at the upper thermistor (Fig. 5.39a) and velocities at both current meters (Fig. 5.39b). Only a significant increase in the spectral density towards the lower frequencies is observed in these spectra. The frequency resolution in this low frequency range based on the 189-day time series is low. The dominating periodicities in the current and temperature fluctuations in the higher-frequency part of the spectra are diurnal (near 24 h) and semidiurnal (12.4 h) tidal frequencies (Fig. 5.39).

Statistical characteristics of temperature and velocity fluctuations of the mooring data require additional comments (Tarakanov et al. 2018):

1. The negative correlation between the temperature and horizontal velocity fluctuations at the lower sensor shows that the temperature decreases as the velocity of the overflow increases. Hence, there is an increase in the depth, from which the uplift of colder water occurs from the fracture south of the Romanche Fracture Zone to the sill of Southern Entrance-1. Thus, as was assumed earlier, the Southern Wall blocks the uplift of the deepest cold water from the south.
2. Temperature fluctuations of the upper sensor many times exceed the fluctuations at the lower sensor because the latter is likely to be located in the bottom boundary layer. Hence, the upper sensor qualitatively reflects the fluctuations of the density stratification in the bottom layer at the mooring point. A temperature increase at the upper sensor, which determines the temperature difference between the upper and lower thermistors, corresponds to an increase in stratification; a decrease in temperature corresponds to a decrease in stratification. The variability of density stratification at the mooring point may be due to the time variations of the stratification of the overflowing flow or to the establishment of the stratification typical for the Southern Valley during the low-velocity phase at the mooring point due to the displacement of these weakly stratified waters towards the sill or even termination of the hydraulic control regime of the overflow. As can be seen from Fig. 5.33, the level of the upper thermistor based on the CTD casts in 2013 along the descending flow was exactly in the lower part of this range. The displacement of water with the temperature profile from the Southern Valley to the mooring point would decrease the temperature of the upper thermistor. The most pronounced manifestation of this situation is the period in the end of December 2013 when the minimum daily and instantaneous velocities were recorded as well as the maximum temperature at the lower level, and the maximum daily and instantaneous negative velocities and one of the lowest temperatures at the upper level were found (Fig. 5.37 and Table 5.11). If the depth range corresponding to the elevation of the isotherms in the direction of the temperature jump is above the level of the upper thermistor, then displacement of weakly stratified waters from the Southern Valley toward the sill or termination of the hydraulic control overflow would cause the temperature increase at the upper thermistor. This situation of positive correlation of the lower and upper thermistors was observed, for example, in the first half of October 2013 (Fig. 5.37).

3. The influence of tidal spring-neap modulation (14-day cycle) on turbulent mixing and overflow regime in the canyons and narrows of the fractures in the South Atlantic Ridge was shown in (Clément et al. 2017). We found similar variations in the Romanche Fracture Zone. Figure 5.37d shows the distribution of the absolute values of the tidal velocities based on the TPXO9 model (Egbert and Erofeeva 2002). Comparison of the velocities of the tidal currents with the mean currents in Fig. 5.37 (after subtracting tidal motions from the velocity measurements) shows that the episode of the end of December 2013 described above with extremely low velocities of the flow corresponds to a neap tide. Such a correlation can be revealed only from the analysis of time series but it is not detected by spectral analysis. At the same time, there is no general correlation between the episodes of high and low velocities and spring/neap variations.

The time series of velocities were band-filtered using the Tukey filter to find the dominating directions. The filter was tuned to the central frequency of the band 1/11.9–1/12.9 cph to isolate the semidiurnal tide. The directional distribution of this semidiurnal range of currents at the upper and lower meters is shown in Fig. 5.40. One can see from the figure that the dominating directions of the semidiurnal tidal currents at the lower and upper current meters do not coincide. At the lower current meter, they fluctuate around opposing directions 45° and 225°, while the corresponding directions at the upper instrument are 22° and 202°. Directions of fluctuations of the diurnal tide (filter cutoff at 1/23.2 to 1/24.8 cph) are almost the same as the directions of the semidiurnal tidal fluctuations measured by both current meters. The estimated fluctuations of directions in the low frequency range also fall into the tidal direction range. Figure 5.40 shows the mean distribution of intensities of the barotropic tide over directions based on the TPXO9 model over 6 months. The

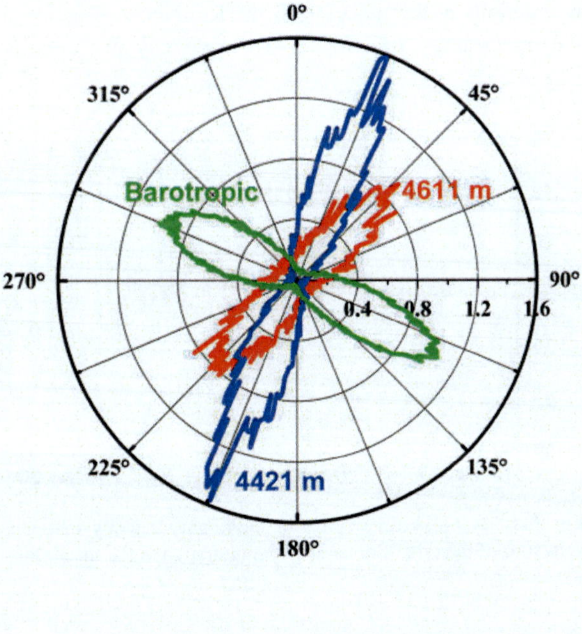

Fig. 5.40 Distribution of the intensity of semidiurnal tidal currents by directions (percent per one-degree angle) over a period of 6 months based on the data of two moored current meters located at 4421 m (blue line) and 4611 m (red contour), and full barotropic tide (TPXO9 model) (green). (Reproduced from Tarakanov et al. 2018)

directions of the large axes of the barotropic tide ellipses are almost normal (115°–295°) to the direction of the tide at the upper current meter. Note that the fluctuations of the model barotropic tide are related to the entire water column. These tidal currents induce internal motions in the channel that allow propagation of AABW into the fracture channel (Southern Entrance-1).

The intensity of fluctuations in different frequency bands is well illustrated in Fig. 5.42, which shows the cumulative (starting from zero frequency) distributions of the specific kinetic energy of fluctuations at two current meters. The total specific energy of the oscillations at the upper and lower current meters is estimated at 89 cm^2s^{-2} and 28 cm^2s^{-2}, which is equivalent to the root-mean-square values of current velocities of 13.4 $cm{\cdot}s^{-1}$ and 7.5 $cm{\cdot}s^{-1}$ (Table 5.10), respectively. The major part of the energy of oscillations (approximately 70%) both at the lower and upper meters is related to the low frequency. We estimated that 38–45% of this energy corresponds to the periods from 2 to 12 days. The high-low frequency cutoffs at 1/11 and 1/14 cph were used to isolate the semidiurnal tide (central frequency is 1/12.4 cph) (Lozovatsky et al. 2003), and cutoffs at 1/22 and 1/28 cph were used for the diurnal tide. Thus, about 10–15% and 4–5% of energy is concentrated at the semidiurnal and diurnal tidal frequencies, respectively. The superposition and proportion between the amplitudes of the tidal velocities, short period fluctuations, and the mean flow are well illustrated in Fig. 5.41.

The frequency distributions of energies and velocities are shown in Fig. 5.42. This figure also shows the energy level of the barotropic tide, which is approximately 30 times lower than the total energy of oscillations at the lower current meter, and 100 times lower compared with the upper one.

Thus, we can conclude that the available potential energy of the deep thermocline, transforms to the kinetic energy in the zone of the overflow. This energy, but not the energy of the barotropic tide, seems to be the main source of the energy of fluctuations at the mooring in the Romanche Fracture Zone. However, we observe the tidal energy transfer to low-frequency fluctuations related to the spring-neap

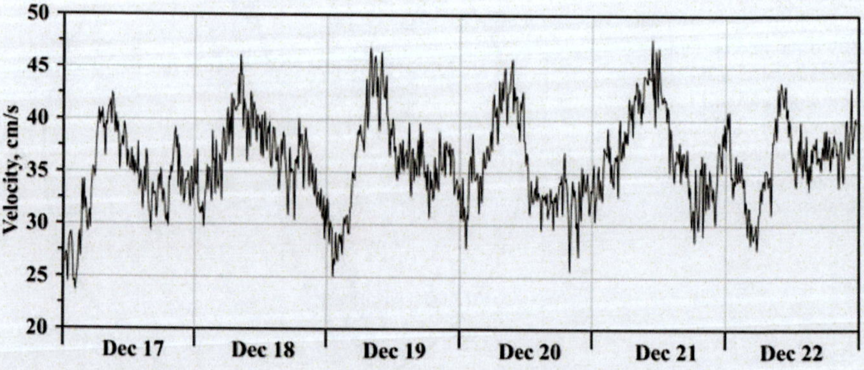

Fig. 5.41 A 6-day time series of meridional velocity that illustrates the relation between the velocities of the tide (diurnal and semidiurnal) and the mean flow

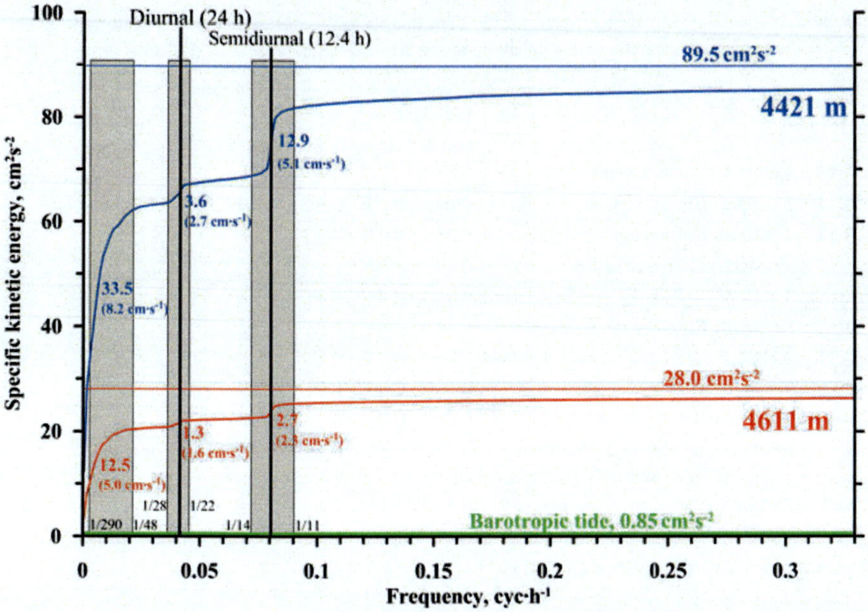

Fig. 5.42 Cumulative specific kinetic energy of oscillations as function of frequency at 4421 m (blue curve) and 4611 m (red curve). The levels of the total energy of the fluctuations and the energy of the barotropic tide are shown. Energies and RMS-velocities corresponding to the frequency bands identified by the gray rectangles are also given. (Reproduced from Tarakanov et al. 2018)

variability. The other sources of energy can be, for example, seiches or fluctuations in the available potential energy of the deep thermocline. Note that recently Clément et al. (2017) observed small fluctuations of the kinetic energy many times exceeding the energy of the barotropic tide based on the data of moorings in the canyon and narrows of the fracture at 21° S in the Brazil Basin.

5.4.7.6 Transport of Abyssal Waters in the Deep Spillway Region

The abyssal flow transport of AABW flowing to the Romanche Fracture Zone was estimated from formula:

$$Q = \sum_{i=1,n} \left(\sum_{Z_{1i}}^{Z_{2i}} U_i(z) w_i(z) \Delta z \right) \pm \sqrt{\sum_{i=1,n} (\delta Q_i)^2}, \qquad (5.2)$$

where n is the total number of stations along the section across the passage, $U_i(z)$ is the velocity component normal to the section of the passage, $w_i(z)$ is the width of the section segment at depth z corresponding to the i-th station (or the width of the passage if only one station was made), δQ_i is the transport error calculated on the

Table 5.12 AABW transports based on the data of LADCP-measurements and bottom potential temperature in different parts of the spillway in the Romanche FZ (see Fig. 5.22b)

Year	Place of measurements	Transport in the layer θ < 1.2 °C, Sv	Transport in the layer 1.2 < θ < 1.7 °C, Sv	Bottom potential temperature, °C
2011	Southern Entrance-1 (st. 2472)	0.08 ± 0.05	0.00 ± 0.11	0.508
2012	Southern Entrance-1 (st. 2482)	0.28 ± 0.02	0.11 ± 0.05	0.513
	Western Sill-1 (st. 2480)	0.07 ± 0.01	0.00 ± 0.03	0.534
	Total	0.35 ± 0.02	0.11 ± 0.06	
2013	Southern Entrance-1 (st. 2509, 2510)	0.19 ± 0.02	0.04 ± 0.04	0.506
	Western Sill-1 (st. 2515)	0.10 ± 0.04	0.01 ± 0.07	0.525
	Total	0.29 ± 0.04	0.05 ± 0.08	
2014	Intermediate sill in the Southern Valley of Romanche FZ (st. 2538)	0.44 ± 0.08	0.14 ± 0.12	0.613
2011	Southern Valley of Romanche FZ (st. 2467)	0.23 ± 0.19	0.22 ± 0.18	0.660

Reproduced from Tarakanov et al. (2018)

basis of the formal velocity error $\delta U_i(z)$ evaluated using the LADCP software package:

$$\delta Q_i = \sqrt{\sum_{Z_1}^{Z_2} (\delta U_i(z))^2 w_i(z) \Delta z \sum_{Z_1}^{Z_2} w_i(z) \Delta z}. \tag{5.3}$$

The estimates of abyssal flow transports through several sections in the Romanche Fracture Zone based on the LADCP-measurements in 2011 (Tarakanov et al. 2013) and in 2012–2014 are given in Table 5.12. Using the above formulas, we assumed that the flow structure is horizontally homogeneous normal to the flow direction, i.e., that the flow velocity is constant over the section. We excluded from the calculations all the stations in the Southern Valley due to the obvious horizontal inhomogeneity of currents as well as station 2481 located south of Southern Entrance-1 sill on the slope of the adjacent fracture where the bottom topography measurements were not performed.

The most reliable estimates of the transport across all individual sections in the layer θ < 1.2 °C are at the sill of Southern Entrance-1 in 2012 and 2013. We can state this because:

(1) A section of two stations in 2013 across the gap of Southern Entrance-1 at the sill was made to test the horizontal homogeneity of the flow. These measurements showed that the maximal horizontal velocity below the bifurcation depth only slightly varies across the gap (from 20 to 25 cm s^{-1}, Fig. 5.29a and b).
(2) We have shown before that the transport estimates and the measured velocities agree well with the predicted transport and the velocity of the gravity waves near

this sill, thereby confirming that this section is close to the hydraulically control point. We found that the mean conditions between surveys (stratification, currents) are similar. The velocities are close to the mean velocities, and the measured velocities in two surveys are also close to each other (20 and 25 cm·s^{-1}). The transports measured in 2012 and 2013 seem real. Based on the same considerations the flow through the section south of Southern Entrance-1 in 2011 is likely to be underestimated due to the horizontal inhomogeneity of the flow in the wider section in front of this sill.

Although we do not know whether the flow at Western Sill-1 in 2012 and west of it in 2013 was horizontally homogeneous the transports in the layer $\theta < 1.2$ °C seem real because they correspond to the predicted values, which are many times smaller than the transport through Southern Entrance-1. The estimates of the total transport at Southern Entrance-1 and Western Sill-1 are based on two velocity profiles in 2012 and three in 2013. Thus, we conclude that these LADCP measurements are good estimates of the average transports in the spillway in the layer $\theta < 1.20$ °C (Table 5.12). This transport is equal to 0.29–0.35 Sv, which is about 50–60% of the total eastward AABW transport (0.6 Sv) estimated in (Mercier and Speer 1998; Morozov et al. 2010) in the middle part of the fracture (16° W). The estimates of the total transport are supplemented by the measurements at two more sections on Intermediate Sill in 2014 and in 2011 to the east of it (Table 5.12).

The transport estimates in the water layer $1.20 < \theta < 1.70$ °C are given in Table 5.12. The upper limit is based on the analysis of bottom topography along the Romanche Fracture Zone (Tarakanov et al. 2013). This analysis showed that downstream of the study site (~22° W) there are several gaps in the Northern and Southern Walls of the fracture (at 18°00′–19°30′ W) reaching the upper boundary of this layer ($\theta < 1.70$ °C ~ 4000 m) before the watershed line between the West and East Atlantic. Water exchange with the surrounding basins can occur through these gaps. Additional water exchange in the water layer $1.20 < \theta < 1.70$ °C between the downstream part of the flow in the Romanche Fracture Zone and the West Atlantic seems more limited. The water of this temperature range can pass through the Northern Valley of the fracture in addition to Southern Entrance-1 and Western Sill-1. Water exchange through the passages at 18°00′–19°30′ W can occur only if there are narrow passages in the walls, which are missing in the database (Smith and Sandwell 1997). As can be seen from Figs. 5.29 and 5.34, oppositely directed currents are observed over the 1.20 °C isotherm even above the sill of Southern Entrance-1, i.e., horizontal inhomogeneity of the flows exists in this layer. Therefore, the estimates of transport in this layer are not reliable. This fact is confirmed by the estimates of transport accuracy in this layer, which is equal and even greater than the transport estimates (Table 5.12). Nevertheless, we note that the total water transport in the layer $\theta < 1.70$ °C, on the one hand, is higher than in the layer with $\theta < 1.20$ °C, which was expected, and on the other hand it does not exceed the previously mentioned total AABW transport in the middle part of the fracture (16°W).

5.4.7.7 Mixing at the Entrance to the Romanche Fracture Zone

During the oceanographic studies in 2011–2014 at the entrance of Antarctic Bottom Water flow to the Romanche Fracture Zone a significant increase in the bottom temperature was found due to mixing over short distances as the cold abyssal water flows to the east. For example, on the basis of measurements in 2011, the increase in the minimum bottom temperature was as high as 0.15 °C over a distance of only 15 miles, i.e., from the sill at Southern Entrance-1 (0.50 °C) to the section along 22°10.5′ W (0.65 °C) (Figs. 5.22, 5.25, Table 5.10). The intensity of mixing can be roughly estimated on the basis of temperature and velocity measurements in the study region. The calculation of the vertical mixing intensity was performed using two methods:

(1) The sections across Southern Entrance-1 and Western Sill-1 of the fracture and any isothermal surface with $\theta < 1.00$ °C outcropping to the bottom of the fracture as the water flows along the fracture form a closed box. Assuming stationarity, the balance should be satisfied between the ingoing advective heat flux that crosses the section of the channel and the outgoing advective flux and diffuse heat flux across this isothermal surface. This balance results in the following formula to estimate the coefficient of cross-isothermal mixing (adopted from (Ferron et al. 1998), which we modified):

$$K_T = \frac{Q^{net}}{A\partial_z\theta}, \tag{5.4}$$

where, Q^{net} is a value proportional to the pure advective heat flux ingoing to the box; $\partial_z\theta$ is the mean vertical temperature gradient over isothermal surface with area A and potential temperature θ_s; $Q^{net} = Q_\theta - Q\theta_s$, where Q and Q_θ are the ingoing water transport and potential temperature flux, respectively in the layer $\theta < \theta_s$.

The water transports and heat fluxes were calculated in two boxes shown in Fig. 5.43 with red and green colors to estimate the coefficients of cross-isothermal mixing at the entrance to the Romanche Fracture Zone. The liquid walls of one box were formed by the vertical sections of Southern Entrance-1 and Western Sill-1; the upper cover of the first box was formed by isothermal surface $\theta_s = 0.61$ °C that outcrops to the bottom immediately before the location of station 2538 in 2014 in the southern channel of the fracture. The vertical wall of the second box was formed by the section of the southern channel of the fracture at this point, while the upper boundary of the second box was formed by isothermal surface $\theta_s = 0.64$ °C that outcrops to the bottom at the line of the section in 2011 at 22°10.5′ W. The calculations for box 1 were performed using the data in 2011, 2012, and 2013 assuming stationarity and constant bottom temperature at the location of station 2538. The calculation for box 2 was performed similarly based on the data in 2014. The areas of isothermal surfaces $\theta_s = 0.61$ °C and $\theta_s = 0.64$ °C were determined from the bottom topography of the recent version of (Smith and Sandwell 1997)

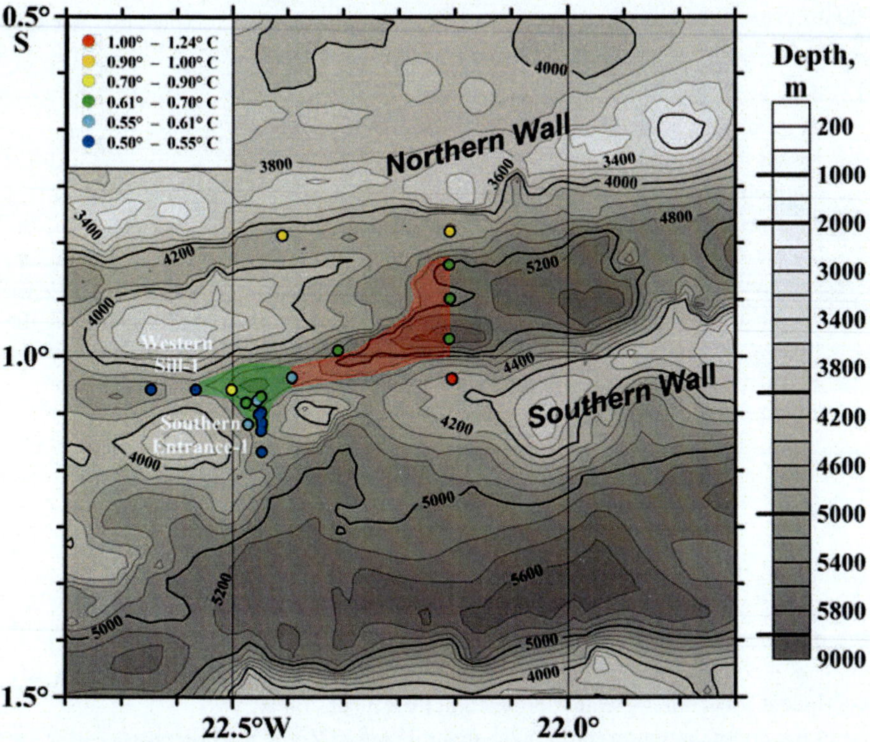

Fig. 5.43 Detailed map of bottom (~5–10 m above the ocean bottom) potential temperatures at the network of stations in the small-scale study site. Colors show potential temperatures at the lowest CTD point (~10 m) above the ocean bottom based on the CTD-measurements at stations in 2011–2014 (See Fig. 5.25). Green and red colors indicate boxes 1 and 2 for the calculations in Table 5.13

corrected using our echo soundings (Fig. 5.43). The procedure was as follows. In the first box, the area was determined as half-sum of the sectors limited by isobaths 4870 and 4370 m. The 4870 m isobath corresponds to the depth at station 2538, and the 4370 m isobath corresponds to the mean depth of isotherm 0.61 °C measured at six stations at Southern Entrance-1 and Western Sill-1. In the second box, the area was calculated from the line from station 2538 to station 2467 and from station 2467 to the line of the section in 2011 as the areas of the sectors limited by isobaths 4870 and 4915 m, respectively. The depth of the latter isobath is the maximum depth at station 2467. Since this station was occupied not at the deepest point of the fracture section, and the minimum temperature at this point was 0.66 °C, we assume that isothermal surface $\theta = 0.64$ °C can be traced further to the east downstream the flow at depths deeper than 4915 m. The temperature gradient in box 1 was calculated as the mean value from six stations at Southern Entrance-1 and Western Sill-1. The scatter of the values ranges from $-1.1{\cdot}10^{-3}$ to $-2.2{\cdot}10^{-3}$ °C/m. The temperature gradient for box 2 was taken from station 2538. The estimates of the areas, temperature gradients, and fluxes, and eventually the coefficients of cross-isothermal mixing are given in Table 5.13. One can see that K_T ranges within (55–630) 10^{-4} m^2 s^{-1}.

(2) The coefficient of the vertical turbulent mixing K_z can be calculated on the basis of its physical sense and considerations of dimensions. Coefficient K_z characterizes the turbulent interaction between two layers. In a two-dimensional qualitative model, the mass and momentum exchange between them occurs through the interface owing to the turbulence. Coefficient K_z characterizes the degree of interaction. It is calculated as the water transport by the turbulent mechanism through the area of this surface that spreads to a depth, which we denote as ΔH. It is clear that the smaller the depth of turbulent flux spreading the smaller the coefficient. In our case, we substitute the turbulent flux by the advective flux assuming that the previously mentioned surface is the depth of the threshold. Coefficient K_z can be written as.

$$K_Z = b\frac{Q}{S}\Delta H, \tag{5.5}$$

where Q is the water transport through horizontal surface with area S. The transport is directed to a layer with thickness ΔH; the decreasing coefficient b is determined by the distribution of transport Q over this layer (see Fig. 5.44). If the transport is concentrated closer to the bottom (transition of curve q_1 to curve q_3 in Fig. 5.44), coefficient b becomes greater and eventually tends to unity.

In the entrance region to the Romanche Fracture Zone, the horizontal surface at 4576 m (the sill depth of Southern Entrance-1) and the vertical cross-section of the fracture below this depth at the Intermediate Sill at station 2538 form a closed box, for which we can perform calculation of coefficient K_z using formula (5.5) given above. Transport Q is calculated from the ADCP data at this station. The value of decreasing coefficient b can be roughly estimated equal to 4.0 based on two considerations: the first is that transport Q is distributed over layer ΔH, but not fully displaced to this depth, and the second is that the cross-section of the fracture at station 2538 deeper than the 4576 m isobath has a quasi-triangle but not rectangular form (Fig. 5.24). In our case: $Q = 0.192$ Sv, $S = 58.8$ km^2, $\Delta H = 296$ m. Then, the estimate of K_z is $2.4 \cdot 10^{-1}$ m^2 s^{-1}.

Comparing the values of K_z and K_T we note that although the physical sense of these coefficients is different, they do not differ strongly in the open ocean (far beyond the region of abyssal channels) (much less than by one order of magnitude). We note that the values of coefficient K_T from Table 5.13 are most likely underestimated (by a few times) because the parameters for the calculation of this coefficient in formula (3) were knowingly selected not to overestimate it.

We think that the value of $\partial_z\theta$ is most likely overestimated because it is calculated from the temperature profiles over the sills. Before, we mentioned that when the current flows down the slope the lower part of the abyssal thermocline breaks in the layer $\theta < 1.20$ °C (Figs. 5.27a, 5.28a and 5.29a). It is likely that the areas of isothermal surfaces $\theta = 0.61$ °C and $\theta = 0.64$ °C are also overestimated. It is seen

Table 5.13 Parameters to calculate the coefficients of cross-isothermal mixing K_T

Box number	Isothermal surface, θ_s, °C	Area of isothermal surface, $10^8 m^2$	Temperature gradient, $\partial_z\theta$, 10^{-3} °C/m	Year and place of measurements	Transport Q in layer with $\theta < \theta_s$, 10^3 m³/s		Potential temperature flux Q_θ in the layer with $\theta < \theta_s$, 10^3 °C·m³/s		Q^{net}, 10^3 °C·m³/s	Coefficient of cross-isothermal mixing, K_T, 10^{-4} m²/s
Box-1 (green in Fig. 5.43)	0.61	0.6	−1.6	2011, southern Entrance-1	40		22.4		−2.0	210
				2012, southern Entrance-1	51	66	28.6	37.3	−3.0	310
				2012, Western Sill-1	15		8.7			
				2013, southern Entrance-1	60	116	34.1	64.7	−6.1	630
				2013, Western Sill-1	56		30.6			
Box-2 (red in Fig. 5.43)	0.64	1.1	−0.6	2014, southern channel of the fracture	44		27.8		−0.4	55

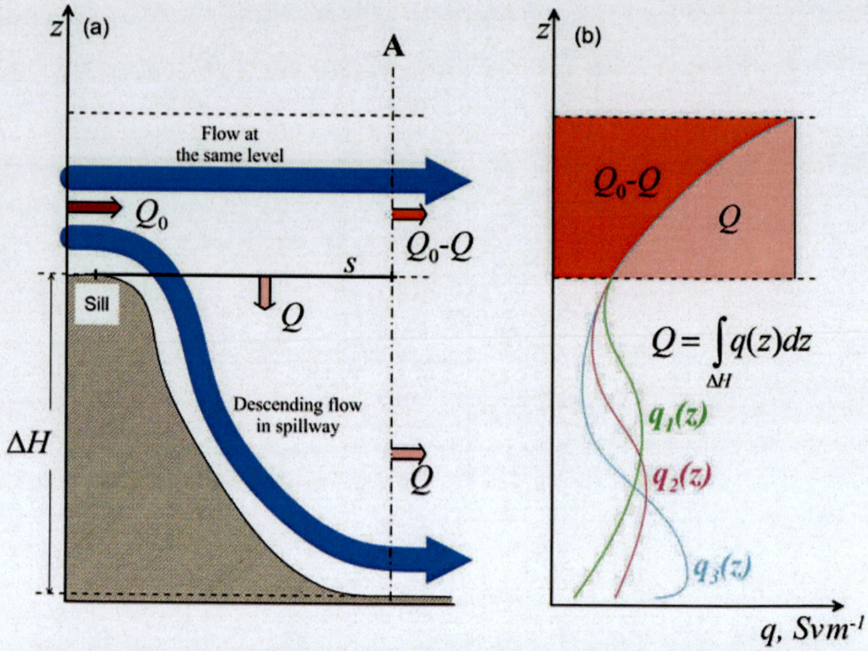

Fig. 5.44 Estimate of the vertical diffusivity k_z: scheme of the descending flow from the sill; Q_0 is the flow transport over the sill, Q is the transport of the descending part of the flow through horizontal section s at the depth of the sill and through section A lower than the sill downstream (**a**); vertical distribution of transport $q(z)$ over section A**;** $q_1(z)$, $q_2(z)$, $q_3(z)$ are different versions of the depth distribution of the transport (**b**)

in box 1 because the temperatures within this box are higher than $\theta = 0.61$ °C at the bottom of depression in the southern channel of the fracture, into which the flows descend from the two sills (Fig. 5.44). On the other hand, the value of K_z using formula (5.5) can be slightly overestimated due to overestimation of the transport based on the data of station 2538, which was clearly located in the core of the abyssal flow. However, the reality of such high values of K_z in this region can be confirmed by the values of $K_z > 2000 \cdot 10^{-4}$ m^2 s^{-1}, estimated using high-resolution moored temperature data at Southern Entrance-1 (see Fig. 3b in (van Haren et al. 2014)). Comparison of our estimates of K_z and K_T with the estimates of K_T at the parts of the fracture immediately before the Main Sill and after it (where this value based on the observations reaches the maximum values of $120 \cdot 10^{-4}$ m^2/C and $370 \cdot 10^{-4}$ m^2/C, respectively (Ferron et al. 1998)) indicates that the region of the entrance to the fracture is one of the fracture parts with extremely intense mixing.

Individual profiles of local potential density relative to 4500 dbar at several stations occupied in 2011–2014 (Fig. 5.45) also show high intensity of mixing. The stations were occupied in the deep parts of the fracture channel: in the depression on the southern channel (stations 2486 and 2511), at solitary station in 2011 in the same channel (station 2467), and at one of the stations of the section in 2011

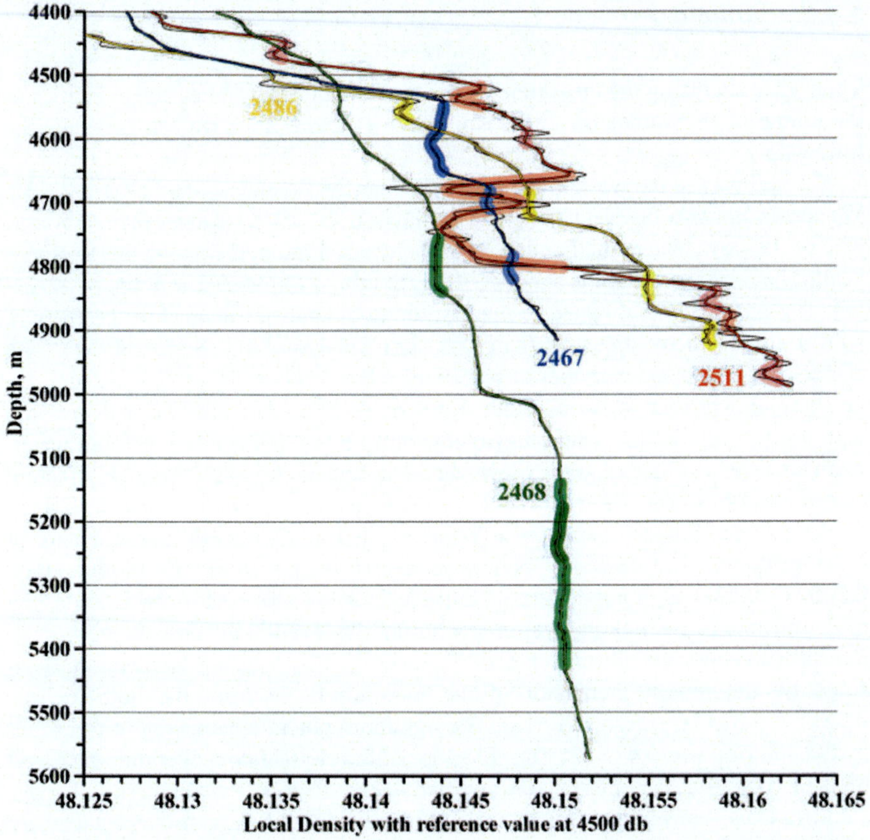

Fig. 5.45 Profiles of the local potential density relative to 4500 dbar at four stations in the channel of the Romanche Fracture Zone. Two stations (2486 and 2511) are located in the depression of the southern canyon opposite to Southern Entrance – 1; one station (2467) is located to the east of the deep sill and east of the depression; and one more station (2468) is at the line of section in 2011 in the southern canyon of the fracture. The profiles were smoothed by running Tukey filter with 12 dbar cutoff (thin black lines) and 40 dbar cutoff (thick grey lines). The intervals with hydrostatic instability are made thicker on the latter lines

(station 2468). The local potential density is an integral of hydrostatic stability with respect to depth. The reference level is 4500 dbar; a constant correction was introduced equal to the density in situ at the same depth. Thus, a decrease of this value with depth corresponds to the local hydrostatic instability. One can see from Fig. 5.45 that the vertical scale of such instabilities reached 100–200 m, and the maximum amplitude of density fluctuations was observed in the depression in the southern channel, which indicates that mixing here is most intense. The same is seen in Table 5.13: the estimates of cross-isothermal mixing coefficient here are almost one order of magnitude greater than downstream the abyssal flow.

5.4.8 Summary

Thus, after analyzing the measurements of thermohaline properties and velocities of the inflow of abyssal waters to the Romanche Fracture Zone we can conclude the following:

The spillway in the western part of the Romanche Fracture Zone (22°27′–22°32′ W) consists of two flows and exists in the layer $\theta < 1.20$ °C (deeper than 4150 m). One of the flows enters the fracture from the south through the gap in the Southern Wall. The depth of the sill is about 4570 m (Southern Entrance-1 at 1°05′ S, 22°27′ W). The other flow is directed to the east along the narrow channel of the fracture with a sill depth of about 4430 m (Western Sill-1 at 1°03′ S, 22°34′ W). The characteristic depths over the surrounding ridges and walls of the fracture are much shallower than 4100 m. Both flows descend to the same local depression with a depth of 5000 m. There they merge and fill the depression with cold water. Then the flow continues and overflows the Intermediate Sill (4870 m) of the fracture east of the depression. The entire spillway exists in the layer cooler than $\theta < 1.20$ °C.

We measured the temperatures at Western Sill-1 and Southern Entrance-1 using CTD profilers. The recorded bottom potential temperatures are in the range 0.506 °C–0.534 °C. Temperature measurements were also performed using a line of thermistors at the bottom layer on a mooring deployed in the descending flow in the spillway from the sill of Southern Entrance-1. The potential temperature measured by the lowest thermistor of the mooring (9 m from the bottom) was 0.527° ± 0.015 °C (mean ± RMS). The minimum potential temperature measured by the mooring was 0.479 °C. The velocities of the abyssal flow over the sills based on the LADCP-measurements were approximately 20 $cm{\cdot}s^{-1}$ at Southern Entrance-1 and 40 $cm{\cdot}s^{-1}$ at Western Sill-1. As the flows descend the slope from the sills, they accelerate and form narrow jets near the bottom. The mean horizontal velocities based on the data of the mooring were 40 $cm{\cdot}s^{-1}$, while the maximum recorded horizontal velocities were 65 $cm{\cdot}s^{-1}$. The average vertical downward velocity component near the bottom based on the mooring measurements was 6.6 $cm{\cdot}s^{-1}$, the maximum velocity was 15 $cm{\cdot}s^{-1}$. The bottom inclination at the mooring location based on the ratio of the vertical and horizontal velocity components is 1/6. The mean slope of the bottom along this flow in the spillway based on the echo-sounder measurements is estimated at 1/10. Velocities of about 40 $cm{\cdot}s^{-1}$ were detected in the bottom layer in the Southern Valley of the Romanche Fracture Zone. As the flow ascends to the Intermediate Sill in the eastern part of the Southern Valley the velocity decreases to 30 $cm{\cdot}s^{-1}$.

The energy of velocity fluctuations based on the 189-day time series of mooring measurements appeared very high. The energy density of fluctuations is estimated at the mooring location as $2.8{\cdot}10^{-3}$ m^2s^{-2} at a distance of 200 m from the bottom and as $9.9{\cdot}10^{-3}$ m^2s^{-2} near the bottom, which is equivalent to the root-mean-square deviation of the velocities equal to 13.4 $cm{\cdot}s^{-1}$ and 7.5 $cm{\cdot}s^{-1}$, respectively. Approximately 10–15% and 4–5% of this energy is related to the semidiurnal (~12.4 h) and diurnal (~24 h) frequency ranges, respectively. About 70% of the total fluctuation energy corresponds to the low frequencies (the periods exceeding 2-days), 40% of this energy corresponds to the periods of 2–12-days. These energies exceed many times the energy of the barotropic tide, which is estimated at 0.85 $cm^2 s^{-2}$ based on the TPXO9 model. Thus, the only energy source of the fluctuations measured by the mooring is the available potential energy of the deep thermocline of the West Atlantic that transforms to the kinetic energy of the deep spillway. The direction of all semidiurnal velocity fluctuations at the location of the mooring in a narrow gap in the fracture wall is caused by the configuration of the bottom topography and tidal forcing. It is strongly different from the direction of the barotropic tide and roughly corresponds to the direction of the mean current.

We estimate the total transport of bottom water below the $\theta = 1.2$ °C isotherm (i.e., in the layer of the deep spillway) on the basis of LADCP measurements as 0.29–0.35 Sv. This transport is two times smaller than the estimate of the total eastward transport of Antarctic Bottom Water (0.6 Sv in the layer colder than 1.9 °C) through the Romanche Fracture Zone in its middle part at 16° W. The total transport between $\theta = 1.2$ °C and $\theta = 1.7$ °C is 0.05–0.11 Sv. The sum of the estimates shows that the other sources of AABW inflow of the water warmer than $\theta = 1.2$ °C into the fracture can exist west of the investigated region. These sources can provide a possible eastward flow through the Northern Valley of the fracture. At the same time, the inflow of water in the layer $\theta < 1.7$ °C east of the spillway region, as shown in (Tarakanov et al. 2013), most likely does not exist.

An extremely high intensity of mixing has been revealed, occurring in the spillway zone at the entrance to the Romanche Fracture Zone at 22°30' W. In particular, the coefficient of cross-isothermal mixing here reaches $600 \times 10^{-4}\ m^2 s^{-1}$, which makes this region one of the "hot spots" of mixing the World Ocean. From the modern point of view the existence of such mixing "hot spots", is necessary to maintain the density stratification actually observed in the World Ocean.

5.4.9 Abyssal Spillway at the Main Sill in the Chain Fracture Zone

Two CTD/LADCP stations (2082, 2083) were occupied during cruise of the R/V *Akademik Ioffe* in 2009 in the region of the Main Sill in the Chain Fracture Zone. These stations repeated the measurements in 1991 and 1994 (Fig. 5.13). One station was located at the main sill, the other one three miles to the east downstream from the bottom flow at a depth greater by 500 m than at the first station. The stations were located at 00°37.4′ S, 12°23.9′ W (4067 m) and 00°36.5′ S, 12°21.0′ W (4573 m). This difference in depths gives a slope of 10°, however the mean slope from the digital database of bottom topography gives a slope of 5°. Figure 5.46 shows potential temperature distributions over a section drawn based on the data of these two stations and profiles of the zonal and meridional velocities. Velocities of the barotropic tide were subtracted based on the TPXO9.1 model (Egbert and Erofeeva

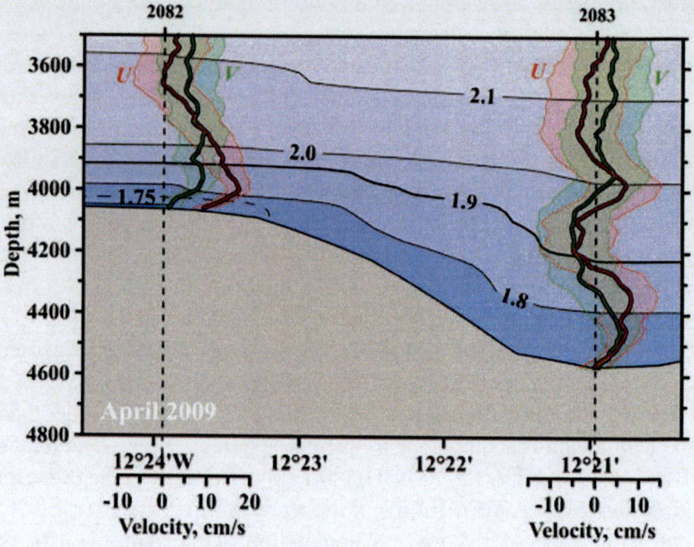

Fig. 5.46 Section of potential temperature in the abyssal layer in the region of the Main Sill of the Chain Fracture Zone and profiles of the zonal (U) and meridional (V) velocity components and errors (cm s^{-1}) based on the CTD/LADCP data at two stations. Vertical lines show the location of profiling stations. The bottom is shown with light-gray color

2002). The tidal velocities over the Main Sill were less than 1 cm s^{-1}, and deeper in the Depression they was as high as 3.5 cm s^{-1} in the eastern direction during the station.

The flow of Antarctic Bottom Water descends from the Main Sill into the Depression (Fig. 5.46). Such descending water flow is a spillway. The properties of the flow indirectly indicate that a spillway is formed here with a hydraulic control overflow over the sill. In particular:

1. A pressure gradient between the Main Sill and Depression exists in the abyssal increasing with depth below the divergence of temperature profiles. This gradient is directed towards the Depression. We have estimated this depth, called Bifurcation Depth, at 3800 m. As shown in Fig. 5.47 the differences in the temperature profiles at the Sill and Depression were also recorded in 1991–1994, which indicates the quasi-constant regime of the hydraulic control overflow mode.
2. We estimated the predicted transport using the 1½-layer model (Whitehead et al. 1974) according to formula (5.1) from the previous subsection. We used the density difference between the station at the Sill and in the Depression, which is $\Delta\rho = 0.028$ kg/m^3 (at a depth of 4057 m). The resultant reduced gravity is $g' = 2.8 \times 10^{-4}$ m s^{-2}. Thickness h_u is estimated at 357 m as the depth difference between the bifurcation depth and the channel bottom at the sill. The mean width of the fracture W at depths greater than 3800 m based on the digital topography (Smith and Sandwell 1997) is 6 km. Thus, a predicted transport estimate of 0.23 Sv has been estimated, which is in good agreement with the transport of

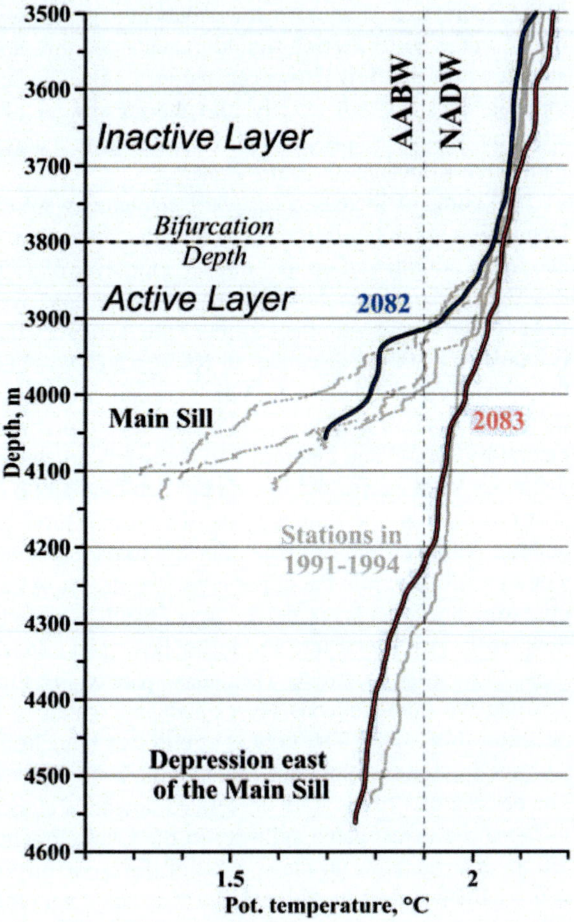

Fig. 5.47 Potential temperature profiles (°C) at stations near the Main Sill of the Chain Fracture Zone and Depression east of the sill. Bifurcation Depth is indicated, above which is Inactive Layer, below is Active Layer of the spillway. Numbers 2082 and 2083 are station numbers

the so-called active layer of 0.20 ± 0.06 Sv, obtained on the basis of measurements of the flow velocity at the sill from formula (5.2).

3. We estimated the horizontal speed of gravitational waves as $\sqrt{g'H}$ using the same 1½-layer model. We used the average thickness of the layer below bifurcation depth, which was 162 m at the Main Sill of the Chain Fracture Zone. The gravitational wave speed based on these parameters was estimated at 21 cm·s^{-1}. This value is close to the measured velocities at the Sill (18 cm·s^{-1}).

Resolution of the spillway with just two stations along the stream does not allow one to unambiguously determine its configuration. When the flow descends down from the sill in the hydraulic control overflow mode, an acceleration of the flow should appear compared to the velocity at the sill, and then deceleration of the flow again to values smaller than those over the sill with a simultaneous thickening of the flow. Velocities of the flow of about 7 cm s^{-1} to the east were recorded at the station in the Depression. They were lower than over the sill (up to 18 cm s^{-1} with a rotation of 30° to the north), which should indicate that this station is already in the zone of slowdown of the flow. However, reduced velocities are also possible if the station location does not fall exactly into the narrow jet of the accelerating flow. This situation was described in the previous subsection about the spillway in the western part of the Romanche Fracture Zone.

Two cores of increased eastward velocities were observed at the station in the Depression in the Chain Fracture Zone: one up to 8 cm s^{-1} in a layer of about 300 m above the bottom and up to 7 cm s^{-1} in a layer of 3900–4050 m. The second core exactly corresponds to the depths of the stream core over the sill. This flow pattern is in good agreement with the results of the laboratory experiment by V. Liapidevsky (Liapidevsky 2004; Zatsepin et al. 2005), which was intended to simulate the mixing that occurs when a denser fluid descends down a slope in a stratified fluid environment. He made experiments with two layers of fluid of different densities in a laboratory setup. The laboratory experiments conducted at the Institute of Hydrodynamics were carried out in a tank 5.2 m long and 0.6 m high. A two-layer stably stratified fluid at rest (imitating density stratification) filled a channel with sloping bottom. A tinted fluid of greater density in the tank was poured from the left down along the slope, while the upper layer was at rest. As a result of the entrainment of fluid from the upper layer, the density of the flow became non-uniform by depth. The flow partly penetrated into the lower layer and continued its spreading along the slope as a density current. The mixed part of the flow remained at the interface between the homogeneous layers and propagated to the right in the form of an intrusion (Fig. 5.48). The flow split into two branches. Part of the flow that mixed with the fluid at rest continued its motion in the horizontal direction due to inertia. The densest part of the fluid descended down the slope due to the gravity force.

Since the investigated spillway in the Chain Fracture Zone forms on the Main Sill, its interpretation in the context of the laboratory experiment described above seems quite justified. In this case, the layers of water typical for the depths of the East Atlantic act as a stratified medium, into which a denser fluid penetrates with the flow over the Main Sill. Note that in this case, momentum is transferred directly in the spillway zone not along the bottom surface, as in the model of immiscible fluids, but horizontally, i.e., across isopycnic (isothermal) surfaces.

5.4.10 *Abyssal Spillway at the Nameless Sill of the Romanche Fracture Zone*

The Romanche Fracture zone located between 22°30′ W and 12°00′ W almost at the equator is approximately 800 km long. The width of the fracture varies from 10 to 40 km. Several topographic sills are located in the fracture along the pathways of the easterly abyssal water flow. The main sill of the Romanche Fracture Zone is located at 13°42′ W, its depth is 4350 m (Fig. 5.9). A section of potential temperature along

Fig. 5.48 Scheme of laboratory experiments by Liapidevsky at the Institute of Hydrodynamics (Siberian Branch of the Russian Academy of Sciences). Splitting of water flow is observed as it descends down a sloping bottom

the Romanche Fracture Zone based on the data of the French expedition in 1991–1992 shown in Fig. 5.16 demonstrates that each of the sills prevents the overflow of the cold bottom water and only overlying warmer water overflows the sills. Under specific conditions, the water flow down from the sill can form an abyssal spillway. In 2009 during cruise 27 of the R/V *Akademik Ioffe*, such a spillway was found at the main sill in the eastern part of the Chain Fracture Zone. The bottom topography of the Romanche Fracture Zone similar to that in the Chain

Fracture Zone suggests that such spillways can also exist in the Romanche Fracture Zone.

According to the data of the previous expeditions, a sharp increase in the bottom temperature and salinity was found in the region of the main sill of the Romanche Fracture Zone (Fig. 5.15). Such an increase is obviously related to the blocking of the coldest water propagation by the underwater barrier. Strong mixing exactly in the sill region can be caused by the tidal currents, possible spillways, and generation of tidal internal waves.

The finding of an abyssal spillway in the Chain Fracture Zone that was described in the previous subsection initiated similar investigations and search for a spillway in the Romanche Fracture Zone, in which the geological structure of the sills is very similar to the Chain Fracture Zone. Before, abyssal spillways were found in the Strait of Denmark, where cold arctic water overflows the threshold between Iceland and Greenland and flows down along the slope. A similar spillway is known in the Strait of Gibraltar, in which the Mediterranean Water flows to the Atlantic (Whitehead 1989).

A region of sharply increasing depth along the axis of the Romanche Fracture Zone between longitudes 13°18′-13°13′ W near a Nameless Sill with a depth of 4350 m was selected as the study site in 2010 in cruise 32 of the R/V *Akademik Ioffe*. The bathymetry of this region strongly resembles the bathymetry of the region in the Chain Fracture Zone, where in 2009 we found a powerful abyssal spillway with a water transport of approximately 0.5 Sv (the estimate is given for the layer of Antarctic Bottom Water with potential temperature less than 1.9 °C). The depth of the sill is approximately 4500 m. A depression deeper than 5000 m is located to the east of the sill. The mean bottom slope along the fracture zone in this region is estimated at 5°. A sharp increase in the fracture depth exists near this sill of the Romanche Fracture Zone approximately from 4350 m to more than 5000 m. The Nameless Sill is located east of the Main Sill of the Romanche Fracture Zone, i.e., downstream the abyssal stream. During the cruise, three CTD/LADCP casts (stations 2428, 2429, 2430) were occupied in the Nameless Sill region that formed a zonal section about three miles long (Fig. 5.49). Before the measurements, we performed an echo-sounder survey along the pathway of the section and used these data to select the coordinates of three stations. The location of the deepest station was selected so that it was located in the lower part of a sharp cliff with a depth jump from 4800 to 4970 m at a distance of 200–300 m from the cliff. The stations were located at 00°47.0′ N, 13°17.0′ W (4265 m), 00°47.0′ N, 13°14.9′ W (4674 m), and 00°47.0′ N, 13°13.4′ W (4972 m).

Figure 5.50 shows the depth dependence of potential temperature along the spillway in the Romanche Fracture Zone aligned with the zonal velocity component measured with lowered ADCP at three stations. Velocities of the barotropic tide were subtracted based on the TPXO9.1 model (Egbert and Erofeeva 2002). At the station on the Sill, the tidal velocities were lower than 1 cm s^{-1}; at the middle station, the velocities were 3.8 cm s^{-1} to the east, at the extreme eastern station they were equal to 2.1 cm s^{-1} to the west. The stations on the Sill and in the Depression almost repeated the points of French stations in 1992 and 1991, respectively (Fig. 5.49).

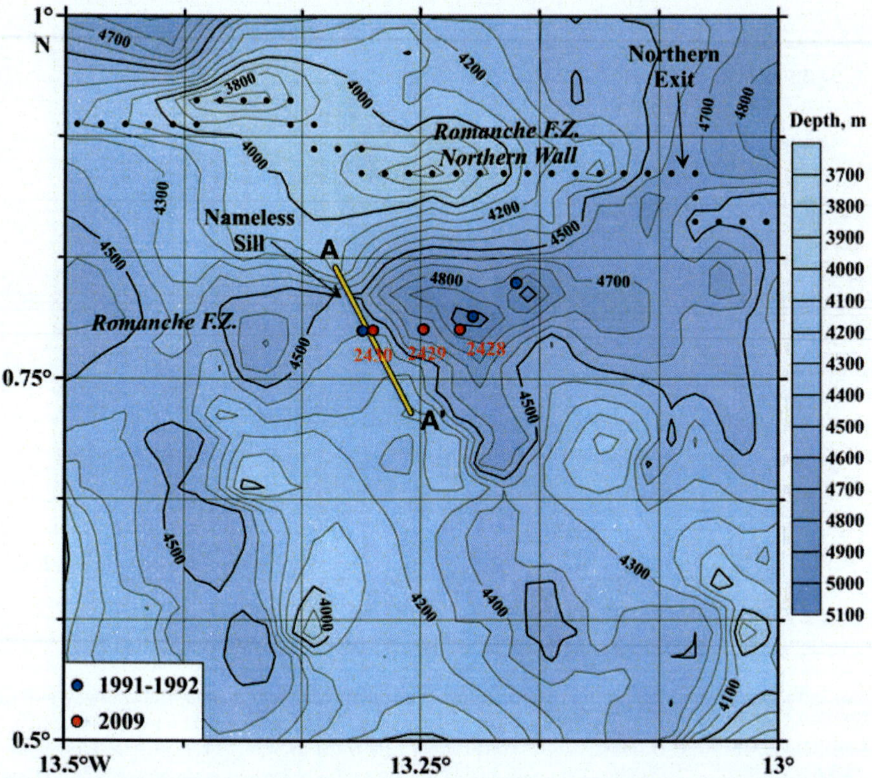

Fig. 5.49 Bottom topography near the Nameless Sill in the Romanche Fracture Zone (13°18′ W). The position of CTD stations in 1991–1992 and 2010 is shown. The black dots correspond to the Northern Wall of the Romanche Fracture Zone. The yellow line shows the section A-A‘, for which the gravity wave speed is calculated

Figure 5.51 shows the profiles of the potential temperature of three French stations in 1991–1992 and three stations of our measurements in 2010. The figure illustrates the flow structure variation as the flow descends from the sill. At the station occupied immediately over the sill, the vertical temperature gradient exceeded 1 °C per 200 m, and the bottom velocities were as high as 32 cm s^{-1}. It is noteworthy that such a high temperature gradient is not characteristic of the bottom layers but of the subsurface thermocline layer.

The distributions of temperature and velocity in Fig. 5.50 show that the stream descends from the Sill to the Depression forming a spillway. The properties of the flow indirectly indicate that a spillway is formed here with a hydraulic control overflow the sill. In particular:

1. A pressure gradient is formed in the abyssal between the Sill and Depression, which increases with depth below the depth of divergence of temperature profiles. The gradient is directed towards the Depression. Measurements in 1991–1992

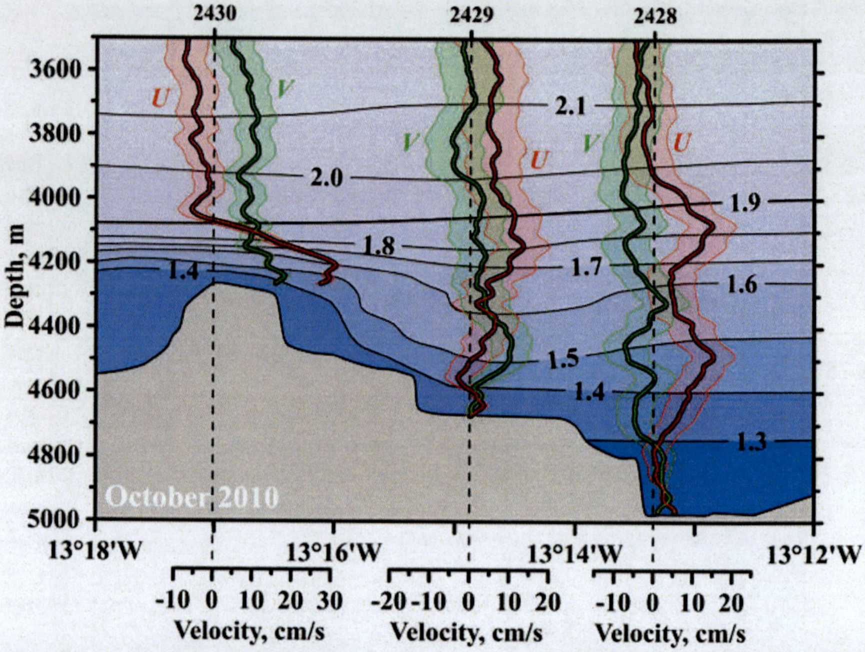

Fig. 5.50 Latitudinal section in October 2010 of potential temperature along the spillway at the Nameless Sill of the Romanche Fracture Zone and zonal (U) and meridional (V) velocity components (cm/s) and errors at three stations. Vertical lines show the location of CTD/LADCP profiling stations. The numbers of stations are indicated at the top axis. The bottom is shown with light-gray color

(Fig. 5.51) show the same restructuring of the temperature profiles between the Sill and Depression, which indicates that a quasi-constant hydraulic control overflow exists. The depth of divergence of temperature profiles, called Bifurcation Depth, is located approximately at a depth of 4125 m. Unfortunately, due to the very large width of the fracture along the isobath corresponding to this depth (Fig. 5.49), it is impossible to make an adequate comparison of predicted transport based on the model (Whitehead et al. 1974) and the transport in the active layer based on the field measurements. The abyssal flow most likely spreads as a narrow jet, the width of which at this depth is many times smaller than the width of the fracture.

2. We estimated the horizontal speed of gravitational waves $\sqrt{g'H}$ using the 1½-layer model (Whitehead et al. 1974). The mean depth of the active layer near the Nameless Sill of the Romanche Fracture Zone was calculated for the A – A‘ section shown in Fig. 5.50. The northern edge of the section extends to the depths shallower than 4125 m, and the southern edge is over an elevation with a depth of about 4180 m based on the data in (Smith and Sandwell 1997). The section length is 13.5 km, the maximum depth is 4502 m, the average depth is 4341 m. Thus, the average depth of the active layer is estimated at 216 m. We

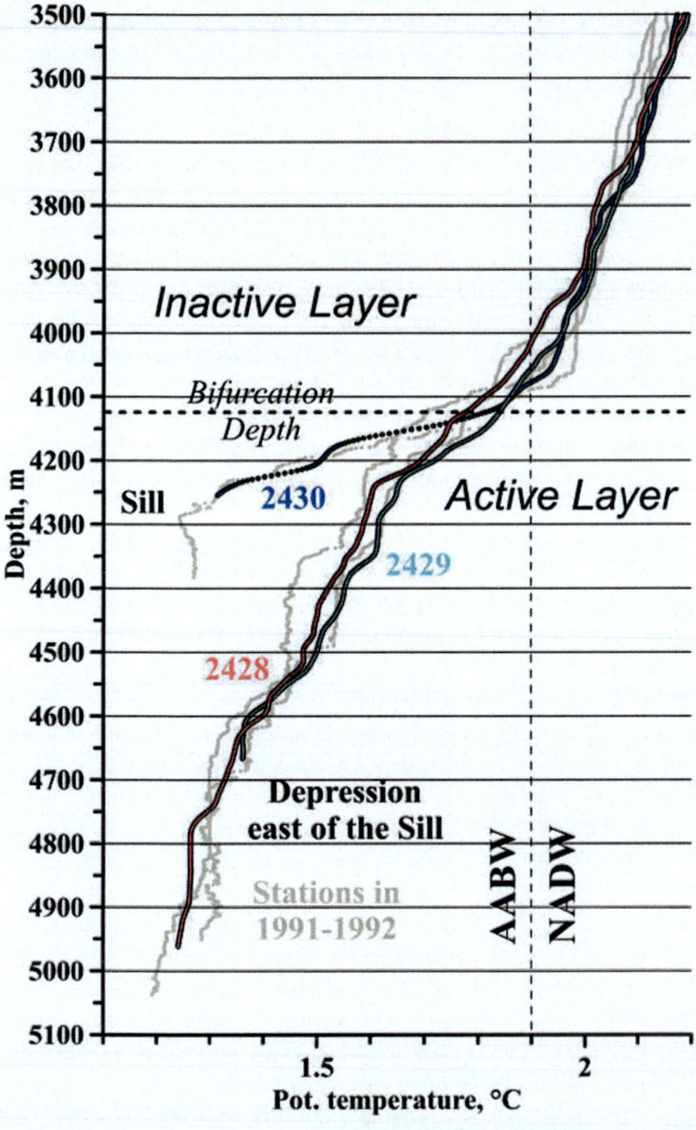

Fig. 5.51 Potential temperature profiles (°C) at stations at the Nameless Sill (13°17′ W) of the Romanche Fracture Zone and Depression east of the Sill. Bifurcation Depth is also noted, above which is Inactive Layer, below is Active Layer of the spillway. Numbers 2428, 2429, 2430 are stations

used the density difference between the station at the Sill and in the Depression, which is $\Delta\rho = 0.034$ kg/m^3 (at a depth of 4255 m). The estimated reduced gravity is $g' = 3.4 \times 10–4$ m s^{-2}. The gravity wave speed based on these parameters was 27 cm s^{-1}. This value is close to the measured velocities at the Sill (31 cm s^{-1})

3. The temperature distribution along the section (Fig. 5.50) shows signs of a hydraulic jump in the form of deepening of 1.5 °C and 1.6 °C potential isotherms at the middle station with their ascent at the eastern station.

The lowest potential temperature (1.24 °C) among the three stations was recorded at the Depression station. This is most likely because the section line was located 1.5–2.0 miles south of the fracture thalweg (deepest line) (Fig. 5.49), along which the coldest water spreads. It also cannot be ruled out that the coldest water flow into the Depression occurs during periodical bursts with strong currents when the bottom water overflows the Main Sill and the Nameless Sill at 13°18′ W. Our French colleagues found a similar phenomenon in 1991 (Messias et al. 1999). The bottom flow velocity at the sill reached 33 cm s^{-1}, and the flow was oriented to the east with a deviation to the north by almost 30° (Fig. 5.50). This also indicates that the other two stations were not located in the core of the jet descending from the sill. This most likely also explains that the velocity values at these stations are lower than those on the sill (up to 13 cm s^{-1} at the middle station and up to 16 cm s^{-1} at the eastern) and become almost zero at the bottom.

Two velocity cores of the stream were observed on the velocity profiles at the middle and eastern stations. At the middle station, the lower core (up to 10 cm s^{-1}) was located at a depth of 4350–4600 m and was oriented almost to the north, while the upper (up to 13 cm s^{-1}) was oriented eastward at the same depths as the stream core at the sill. At the eastern station, the depths of both cores were almost the same as the depths of the cores at the middle station. Both cores (the lower up to 16 cm s^{-1} and the upper up to 15 cm s^{-1}) were oriented to the east with some deviation to the south. As in the case of the spillway at the Main Sill in the Chain Fracture Zone, this velocity distribution over depth is in good agreement with the results of laboratory modeling of mixing during the descent of a density flow in a stratified fluid (Zatsepin et al. 2005; Liapidevsky 2004). As in the case of the Chain Fracture Zone, the interpretation of this structure in the context of this model seems justified because, first, the Nameless Sill is located east of the Main Sill of the Romanche Fracture Zone, i.e., downstream of the pathway of the abyssal stream; second, the next sill to the east along the pathway of the stream (Northern Exit, Fig. 5.49), where it exits into the open part of the East Atlantic, has a greater depth (about 4700 m). Thus, the deep waters of the East Atlantic act as a stratified medium into which a denser fluid flows over the Nameless Sill of the Romanche Fracture Zone.

After overflowing the sill, the flow splits into three layers. The part of the flow with the smallest density continues its motion at the depths of the sill (4200–4300 m) as an intrusion. The water of the greatest density slides down the slope. The vertical velocity component estimated from the geometrical viewpoint is 1.0–2.5 cm s^{-1}. As a result of mixing, a supercritical flow of intermediate density is formed in the layer between the bottom jet and the surrounding fluid that gains energy as it deepens owing to a decrease in the potential energy of the flow. This process is an evolution of the mixed layer over the lee slope (Bacmeister and Pierrehumbert 1988). At the final stage of the development of the descending gravity current, the entrainment of the surrounding fluid and further turbulent mixing balance the downwelling force and the velocity of the flow stabilizes. The negative buoyancy of the mixed layer relative to the stratified fluid at rest gradually decreases, and the flow transforms into intrusion when it reaches the depth with denser fluid. The supercritical current jet formed during this process transforms to a subcritical current as an internal hydraulic jump at a specific distance from the spillway foot. A steep ascent of the isopycnal and isothermal surfaces is observed at this place (with a depth difference of the order of 100 m), which is a characteristic indicator of the hydraulic jump of the cold Antarctic water.

An opposing current above 3900 m is an interesting peculiarity of the observed current structure. Actually, at this level, the current velocity turns to zero, and below in the core of the powerful flow of Antarctic water the velocity gradients are very high. Such high velocity gradients indicate a possible existence of developed wave structure in the current beyond the sill formed by intense internal waves. The structures of this type are well known in the stratified atmospheric currents over the terrain obstacles. They are called lee waves (Bacmeister and Pierrehumbert 1988; Nappo 2002).

Under favorable conditions in the atmosphere even if the obstacles are relatively not high (700–800 m) the generated lee waves can spread upwards to a height of 7000 m. However, if opposing currents exist, the upward propagation of these perturbations in the layers is blocked in the intermediate layer, where the velocity of the flow turns to zero. Such a critical layer plays the role of a fluid boundary preventing propagation of waves from the region where they are generated to the adjacent regions of the current. Therefore, we can suppose that the critical layer at a depth of 3900 m limits from above the zone of high velocity gradients in the vicinity of the spillway. Similar localized structures in stratified currents with a hydraulic jump downstream the obstacle and a critical layer above it were observed, for example in the laboratory experiments (Liapidevsky 2004).

A velocity decrease almost to zero near the bottom due to friction is a characteristic peculiarity of the velocity distribution at the shallowest and intermediate stations. The flow slides down over this boundary friction layer. At the same time, the flow at the deepest station does not penetrate into a deep depression (underwater lake of cold water) but flows along the 45.91 isopycnal (isotherm 1.6 °C) and even ascends. It is likely that a rise in the bottom topography east of our stations at another sill of the fracture at an approximate longitude of 13° W (Sill E in Fig. 5.9) facilitates the ascent of the flow.

The water temperature in the deep depression (east of 13°35′ W) is cooler than 1.25 °C. We did not find such cold water at the sill and at the intermediate station. The water with this temperature was found only west of 13°17′ W. Our French colleagues found a similar phenomenon in 1991 (Messias et al. 1999). This coldest water can fill the deep depression during periodical bursts with strong currents when the bottom water overflows the main sill and the sill at 13°17′ W. It is also possible that a narrow gap exists in the bottom of the fracture along which the coldest water permanently flows down, but our station was not located in this gap.

5.4.11 Modeling of the Flow in the Romanche Fracture Zone

We simulated velocities in the Romanche and Chain fracture zones with a high resolution in bottom layers using the INMOM numerical model. The domain completely covered the fracture zones: from 28° to 8° W and from 4° S to 5° N. The horizontal resolution was equal to 0.02° or approximately 2 km. Using simulated three-dimensional velocity fields, we analyzed the AABW flow in a few narrow places, in which the local bottom currents are intense.

The inflow of AABW occurs in the western part of the Romanche Fracture Zone (Fig. 5.52). The general flow into the fracture channel crosses the first zonally oriented sill approximately at 22°26′ W, 1°06′ S (Southern Entrance-1), then turns to the east and propagates through the second sill located at 22°20′ W, 1°00′ S. The maximum velocities are 25 cm s^{-1} at the first and 28 cm s^{-1} at the second sill. An additional inflow is located at 22°35′ W, 1°03′ S west of the main southern entrance; the maximum velocities here do not exceed 20 cm s^{-1}. Our direct LADCP velocity measurements were performed at a few locations in different parts of the region (Fig. 5.52b); characteristic velocities vary from 20 to 40 cm s^{-1} (Tarakanov et al. 2018). It should be noted that the jet displaces to the northern wall of the fracture after crossing all the sills. This effect is pronounced both in the simulated and measured (meridional section at 22°10′ W) data: the velocity at station 2470 is the highest among five stations at this section (Fig. 5.53b).

High zonal velocities are observed at the sills in the eastern part of the Romanche Fracture Zone. We analyzed sections of zonal velocity over two sills at 16°03′ W and 15°08′ W; the simulated velocities reach 45 cm s^{-1} and 35 cm s^{-1}, respectively (Fig. 5.53). In 2005 and 2009, we made CTD/LADCP measurements across the Romanche Fracture Zone at 16°03′ W. The along-channel velocity distribution was

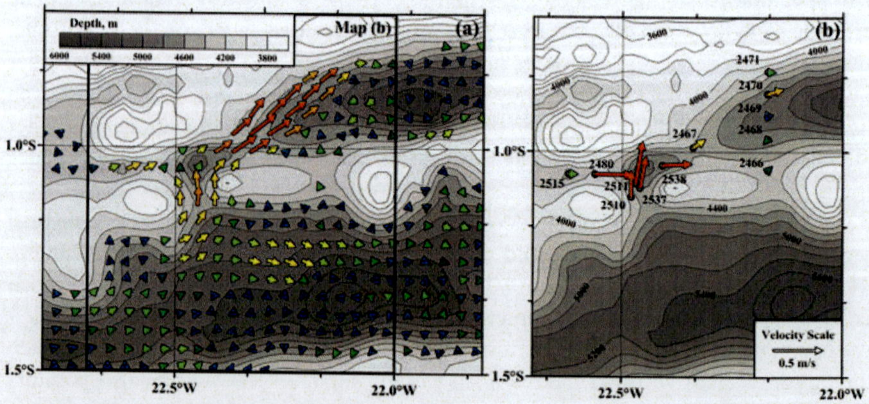

Fig. 5.52 The inflow of AABW into the Romanche Fracture Zone in its western part based on numerical modeling (left panel, a) and direct measurements (right panel, b). The velocities are presented at a height of 100 m above the seafloor; the ocean depths are greater than 4300 m. (Reproduced from Frey et al. 2019)

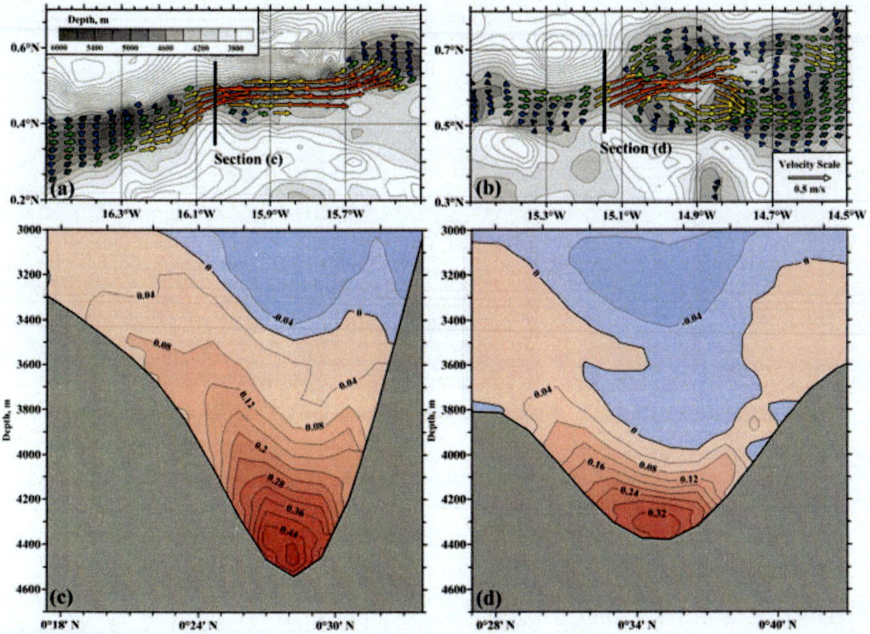

Fig. 5.53 Simulated velocity distributions across the Romanche Fracture Zone at 16°05′ W (**a, c**) and 15°08′ W (**b, d**). The bottom is shown with gray color. The sections of measurements are indicated with vertical black lines in Figs. (a) and (b). (Reproduced from Frey et al. 2019)

Table 5.14 Maximum velocities and AABW transports at a few sections across the Romanche Fracture Zone. Locations of sections are shown with thick lines in Fig. 5.53

Section №	Location of section	Maximum velocities, cm s^{-1}		Modeled AABW transports, Sv (θ < 2 °C)
		Measurements	Modeling	
1	01°05′ S	25	19	1.55
2	22°20′ W	15	42	1.11
3	22°06′ W	16	10	0.96
4	16°03′ W	27	41	–
5	15°08′ W	–	35	–

quite similar to the one shown in Fig. 5.10a with the maximum velocities equal to 27 cm s^{-1}. The velocity estimates for all studied sections are presented in Table 5.14. We also calculated transports of AABW through the fracture at different sections; their positions are shown and indicated in Fig. 5.53a, b. The total transport of waters with potential temperature below 2 °C reaches 2 Sv.

We compare potential temperature distributions across the studied channels at locations with intense bottom currents in the Vema Channel, Vema and Romanche fracture zones (Fig. 5.54). It is clear that the displacement of the coldest waters in the bottom layer across the current depends on the position of the channel relative to the equator. Isotherms of potential temperature are almost horizontal in the Romanche

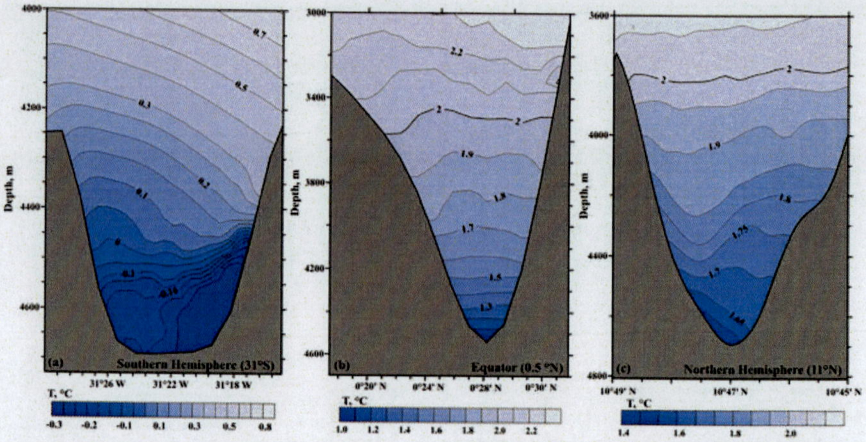

Fig. 5.54 Comparison of modeled potential temperature distributions across the Vema Channel in the Southern Hemisphere at 31° S (**a**), the Romanche Fracture Zone near the Equator at 0.5° N (**b**), and the Vema Fracture Zone in the Northern Hemisphere at 11° N (**c**). (Reproduced from Frey et al. 2019)

Fracture Zone (0.5°N) while in the Southern and Northern hemispheres we observe the displacement of the cold jet across the channel to the right and left walls, respectively.

Thus, the bottom current in the Romanche Fracture Zone changes strongly along the channel. Velocities up to 30–50 cm s^{-1} are observed at a few narrow sills; the current slows down up to zero velocities in the wide parts of the channel. We estimated transports of bottom waters with different potential temperatures (which were selected as the upper boundary of the AABW current) through the fracture; the total transport of AABW with potential temperature $\theta < 2$ °C exceeds 1 Sv.

References

Andrie C, Ternon JF, Messias MJ, Memery L, Bourlès B (1998) Chlorofluormethane distribution in the deep equatorial Atlantic during January-March 1993. Deep-Sea Res I 45:903–930

Atkinson CP, Bryden HL, Cunningham SA, King BA (2012) Atlantic transport variability at 25° N in six hydrographic sections. Ocean Sci 8:497–523. https://doi.org/10.5194/os-8-497-2012

Bacmeister JT, Pierrehumbert RT (1988) On high drag states of nonlinear stratified flow over an obstacle. J Atmos Sci 45:63–80

Boyer TP, Baranova OK, Coleman C, Garcia HE, Grodsky A, Locarnini RA, Mishonov AV, Paver CR, Reagan JR, Seidov D, Smolyar IV, Weathers K, Zweng MM (2018) World Ocean Database 2018. AV Mishonov, Technical Ed., NOAA Atlas NESDIS 87. https://www.ncei.noaa.gov/sites/default/files/2020-04/wod_intro_0.pdf

Bryden HL, Nurser AJG (2003) Effect of strait mixing on ocean stratification. J Phys Oceanogr 33(8):1870–1872. https://doi.org/10.1175/1520-0485(2003)033<1870:EOSMOO>2.0.CO;2

Cai W, Greatbatch R (1995) Compensation for the NADW outflow in a global ocean general circulation model. J Phys Oceanogr 25(2):226–241

Clarke RA, Gascard JC (1983) The formation of Labrador Sea water, part I: large-scale processes. J Phys Oceanogr 13(10):1764–1778

Clément L, Thurnherr AM, St Laurent LC (2017) Turbulent mixing in a deep fracture zone on the mid-Atlantic ridge. J Phys Oceanogr 47(8):1873–1896. https://doi.org/10.1175/JPO-D-16-0264.1

De Madron XD, Weatherly G (1994) Circulation, transport and bottom boundary layers of the deep currents in the Brazil Basin. J Mar Res 52:583–638

Demidov AN (2003) Distinguishing the intermediate and deep water masses in the South Atlantic. Oceanology 43:153–163

Demidov AN, Morozov EG, Neiman VG (2006) Structure and variability of deep waters in the Romanche fracture zone. Doklady Earth Sci 410:1136–1140. https://doi.org/10.1134/S1028334X06070300

Dobrovolsky AD, Zalogin BS (1992) Regional oceanography. MGU Publishers, Moscow. [in Russian]

Egbert GD, Erofeeva SY (2002) Efficient inverse modeling of barotropic ocean tides. J Atmos Ocean Tech 19:183–204. https://doi.org/10.1175/1520-0426(2002)019<0183:EIMOBO>2.0.CO;2

Evans GR, McDonagh EL, King BA, Bryden HL et al (2017) South Atlantic interbasin exchanges of mass, heat, salt and anthropogenic carbon. Prog Oceanogr 151:62–82. https://doi.org/10.1016/j.pocean.2016.11.005

Ferron B, Mercier H, Speer K, Gargett A, Polzin K (1998) Mixing in the Romanche fracture zone. J Phys Oceanogr 28:1929–1945. https://doi.org/10.1175/1520-0485(1998)028<1929:MITRFZ>2.0.CO;2

Fischer J, Rhein M, Schott F, Stramma L (1996) Deep water masses and transports in the Vema fracture zone. Deep-Sea Res I 43(7):1067–1074

Frajka-Williams E, Ansorge IG, Baehr J, Bryden HL, Paz Chidichimo M, Cunningham SA et al (2019) Atlantic meridional overturning circulation: observed transport and variability. Frontiers Mar Science 6:260. https://doi.org/10.3389/fmars.2019.00260

Frey DI, Morozov EG, Fomin VV, Diansky NA, Tarakanov RY (2019) Regional modeling of Antarctic bottom water flows in the key passages of the Atlantic. J Geophys Res Oceans 124(11):8414. https://doi.org/10.1029/2019JC015315

Friedrichs MA, Hall MM (1993) Deep circulation in the tropical North Atlantic. J Mar Res 51(4):697–736

Friedrichs MA, McCartney MS, Hall MM (1994) Hemispheric asymmetry of deep water transport modes in the western Atlantic. J Geophys Res Oceans 99(C12):25165–25179

Fu LL (1981) The general circulation and meridional heat transport of the subtropical South Atlantic determined by inverse methods. J Phys Oceanogr 11(9):1171–1193

Fu Y, Karstensen J, Brandt P (2018) Atlantic meridional overturning circulation at 14.5° N in 1989 and 2013 and 24.5° N in 1992 and 2015: volume, heat, and freshwater transports. Ocean Sci 14:589–616

Ganachaud A (2003) Large-scale mass transports, water mass formation, and diffusivities estimated from World Ocean circulation experiment (WOCE) hydrographic data. J Geophys Res Oceans 108(C7):3213. https://doi.org/10.1029/2002JC001565

Hall MM, McCartney MS, Whitehead JA (1997) Antarctic bottom water flux in the equatorial Western Atlantic. J Phys Oceanogr 27(9):1903–1926. https://doi.org/10.1175/1520-0485(1997)027<1903:ABWFIT>2.0.CO;2

Heezen B, Bunce E, Hersey J, Tharp M (1964) Chain and Romanche fracture zones. Deep Sea Res I 11:30–33. https://doi.org/10.1016/0011-7471(64)91079-4

Hernández-Guerra A, Pelegrí JL, Fraile-Nuez E, Benítez-Barrios V, Emelianov M, Pérez-Hernández MD, Vélez-Belchí P (2014) Meridional overturning transports at 7.5N and 24.5N in the Atlantic Ocean during 1992–93 and 2010–11. Prog Oceanogr 128:98–114. https://doi.org/10.1016/j.pocean.2014.08.016

Hernández-Guerra A, Talley LD, Pelegríc JL, Vélez-Belchíd P, Baringer MO, Macdonald AM, McDonagh EL (2019) The upper, deep, abyssal and overturning circulation in the Atlantic Ocean at 30°S in 2003 and 2011. Progr Oceanogr 176:102136

Hogg NG, Owens WB (1999) Direct measurement of the deep circulation within the Brazil Basin. Deep-Sea Res II 46:335–353

Hogg N, Siedler G, Zenk W (1999) Circulation and variability at the southern boundary of the Brazil Basin. J Phys Oceanogr 29:145–157

Holfort J, Siedler G (2001) The meridional oceanic transport of heat and nutrients in the South Atlantic. J Phys Oceanogr 31(1):5–28

Kapustina MV, Dorokhov DV, Sivkov VV (2021) Multibeam bathymetry data of the western part of the Romanche trench (equatorial Atlantic). Data in Brief (in press)

Kinder TH, Bryden HL (1990) The aspiration of deep waters through straits. In: Pratt LJ (ed) The physical oceanography of Sea Straits. Kluwer Academic Publ, Netherlands, pp 295–319

Klein B, Molinari RL, Muller TJ, Seidler G (1995) A transatlantic section at 14.5°N: meridional volume and heat fluxes. J Mar Res 53:929–957

Larque L, Maamaatuaiahutapu K, Garcon VC (1997) On the intermediate and deep-water flow in the South Atlantic Ocean. J Geophys Res Oceans 102(C6):12425–12440

Lavin AM, Bryden HL, Parrilla G (2003) Mechanisms of heat, freshwater, oxygen and nutrient transports and budgets at 24.5° N in the subtropical North Atlantic. Deep-Sea Res I 50:1099–1128

Leontieva VV (1985) Hydrology of trenches in the World Ocean. Nauka, Moscow. [in Russian]

Liapidevsky VY (2004) Mixed layer at the lee side of an obstacle. J Appl Mechanics Technical Physics 45(2):62–67

Limeburner R, Whitehead JA, Cenedese C (2005) Variability of Antarctic bottom water flow into the North Atlantic. Deep Sea Res II 52:495–512. https://doi.org/10.1016/j.dsr2.2004.12.012

Lozovatsky ID, Morozov EG, Fernando HJS (2003) Spatial decay of energy density of tidal internal waves. J Geophys Res Oceans 108(6):3201–3216. https://doi.org/10.1029/2001JC001169

Luyten J, McCartney MS, Stommel H, Dickson R, Gmitrowicz E (1993) On the sources of North Atlantic deep water. J Phys Oceanogr 23:1885–1892

Macdonald A (1993) Property fluxes at 30° S and their implications for the Pacific-Indian throughflow and the global heat budget. J Geophys Res Oceans 98(4):6851–6868

Macdonald A (1998) The global ocean circulation a hydrographic estimate and regional analysis. Prog Oceanogr 41:281–382

Marchesiello P, Barnier B, de Miranda AP (1998) A sigma-coordinate primitive equation model for studying the circulation in the South Atlantic. Pt II: meridional transports and seasonal cycle. Deep-Sea Res I 45:573–608

Mamayev OI (1992) Abyssal waters of the World Ocean. IRO Publishers, Moscow. [in Russian]

Mantyla AW, Reid JL (1983) Abyssal characteristics of the World Ocean waters. Deep-Sea Res I 30(8):805–833. https://doi.org/10.1016/0198-0149(83)90002-X

Marsh R, de Cuevas BA, Coward AC, Nurser AJG, Josey SA (2005) Water mass transformation in the North Atlantic over 1985-2002 simulated in an eddy-permitting model. Ocean Sci 1:127–144

Matano RP, Philander SGH (1993) Heat and mass balances of the South Atlantic Ocean calculated from a numerical model. J Geophys Res 98:977–984

McCartney MS, Bennet SL, Woodgate-Jones ME (1991) Eastward flow through the mid-Atlantic ridge at 11° N and its influence on the abyss of the eastern basin. J Phys Oceanogr 21(8):1089–1121. https://doi.org/10.1175/1520-0485(1991)021<1089:EFTTMA>2.0.CO;2

McCartney MS, Curry RA (1993) Trans-equatorial flow of Antarctic bottom water in the western Atlantic Ocean: abyssal geostrophy at the equator. J Phys Oceanogr 23:1264–1276

McDonagh EL, King BA (2005) Oceanic fluxes in the South Atlantic. J Phys Oceanogr 35:109–122. https://doi.org/10.1175/JPO-2666.1

McDonagh ML, McLeod P, King BA, Bryden HL, Torres Valdéz S (2010) Circulation, heat, and freshwater transport at 36°N in the Atlantic. J Phys Oceanogr 40:2661–2678

Mercier H, Morin P (1997) Hydrography of the Romanche and chain fracture zones. J Geophys Res Oceans 102(C5):10,373–10,389. https://doi.org/10.1029/97JC00229

Mercier H, Speer KG (1998) Transport of bottom water in the Romanche fracture zone and the chain fracture zone. J Phys Oceanogr 28(5):779–790. https://doi.org/10.1175/1520-0485(1998)028<0779:TOBWIT>2.0.CO;2

Mercier H, Speer KS, Honnorez J (1994) Romanche flow pathways of bottom water through the Romanche and Chain Fracture Zones. Deep-Sea Res 41(10):1457–1477. https://doi.org/10.1016/0967-0637(94)90055-8

Messias M-J, Andrie C, Memery L, Mercier H (1999) Tracing the North Atlantic deep water through the Romanche and chain fracture zones with chlorofluoromethanes. Deep-Sea Res I 46:1247–1278. https://doi.org/10.1016/S0967-0637(99)00005-9

Metcalf WG, Heezen BC, Stalcup MC (1964) The sill depth of the mid-Atlantic ridge in the equatorial region. Deep-Sea Res I 11:1–10

Molinari RL, Fine RA, Johns E (1992) The deep Western boundary current in the tropical North Atlantic. Deep-Sea Res I 39:1967–1984

Monin AS (1988) Theoretical principles of geophysical hydrodynamics. Gidrometeoizdat, Leningrad. [in Russian]

Morozov EG, Demidov AN, Tarakanov RY (2008) Transport of Antarctic waters in the deep channels of the Atlantic Ocean. Dokl Earth Sci 423(8):1286–1289. https://doi.org/10.1134/S1028334X08080230

Morozov EG, Demidov AN, Tarakanov RY, Zenk W (2010) Abyssal channels in the Atlantic Ocean: water structure and flows. Springer, Dordrecht, 266 p. https://doi.org/10.1007/978-90-481-9358-5

Morris MY, Hall MM, St. Laurent LC, Hogg NG (2001) Abyssal mixing in the Brazil Basin. J Phys Oceanogr 31:3331–3348

Nappo CJ (2002) An introduction to atmospheric gravity waves. Academic, San Diego, 247 p

Naveira Garabato AC, Williams AP, Bacon S (2014) The three-dimensional overturning circulation of the Southern Ocean during the WOCE era. Prog Oceanogr 120:41–78

Onken R (1995) The spreading of lower circumpolar water in the Atlantic Ocean. J Phys Oceanogr 25(12):3051–3063

Polzin KL, Speer KG, Toole JM, Schmitt RW (1996) Intense mixing of Antarctic bottom water in the equatorial Atlantic Ocean. Nature 380:54–57. https://doi.org/10.1038/380054a0

Pratt LJ, Whitehead JA (2007) Rotating hydraulics: nonlinear topographic effects in the ocean and atmosphere. Springer, New York, 550 p. https://doi.org/10.1007/978-0-387-49572-9

Reid JL, Nowlin WD, Patzert WC (1977) On the characteristics and circulation of the southwestern Atlantic Ocean. J Phys Oceanogr 7(1):62–91

Reid JL (1989) On the total geostrophic circulation of the South Atlantic Ocean: flow pattern, tracers and transports. Prog Oceanogr 23:149–244

Rhein M, Stramma L, Send U (1995) The Atlantic deep western boundary current: water masses and transports near the equator. J Geophys Res Oceans 100:2441–2457

Rhein M, Stramma L, Krahmann G (1998) The spreading of Antarctic bottom water in the tropical Atlantic. Deep-Sea Res I 45:507–527

Rintoul SR (1991) South Atlantic interbasin exchange. J Geophys Res Oceans 96:2675–2692

Roemmich D (1983) The balance of geostrophic and Ekman transports in the tropical Atlantic Ocean. J Phys Oceanogr 13(8):1534–1539

Sandoval FJ, Weatherly GL (2001) Evolution of the deep western boundary current of Antarctic bottom water in the Brazil Basin. J Phys Oceanogr 31(6):1440–1460. https://doi.org/10.1175/1520-0485(2001)031<1440:EOTDWB>2.0.CO;2

Saunders PM, Thompson SR (1993) Transport, heat, and freshwater fluxes within a diagnostic numerical model (FRAM). J Phys Oceanogr 23:452–464

Schlitzer R (1987) Renewal rates of East Atlantic deep water estimated by inversion of 14C data. J Geophys Res Oceans 92:2953–2980

Schlitzer R (1996) Mass and heat transports in the South Atlantic derived from historical hydrographic data. In: Wefer G, Berger WH, Siedler G, Webb DJ (eds) The South Atlantic: present and past circulation. Springer-Verlag, pp 83–104

Schmitz WJ (1996) On the World Ocean circulation: Volume I, Some global features. North Atlantic circulation (Technical report WHOI–96–03). Woods hole: Woods Hole Oceanographic Institution 142 p

Schott FA, Dengler M, Brandt P, Affler K et al (2003) The zonal currents and transports at 35° W in the tropical Atlantic. Geophys Res Lett 30(7):1349. https://doi.org/10.1029/2002GL016849

Sloyan BM, Rintoul SR (2001) Circulation, renewal, and modification of Antarctic mode and intermediate water. J Phys Oceanogr 31(4):1005–1030

Smith WHF, Sandwell DT (1997) Global Sea floor topography from satellite altimetry and ship depth soundings. Science 277:1956–1962. http://topex.ucsd.edu/cgi-bin/get_data.cgi. https://doi.org/10.1126/science.277.5334.1956

Speer KG, Zenk W (1993) The flow of Antarctic bottom water into the Brazil Basin. J Phys Oceanogr 23:2667–2682

Stephens JC, Marshall DP (2000) Dynamical pathways of Antarctic bottom water in the Atlantic. J Phys Oceanogr 30(3):622–640

Stommel H, Arons AB (1960) On the abyssal circulation of the World Ocean. II. An idealized model of the circulation pattern and amplitude in oceanic basins. Deep-Sea Res (1953) 6(3):217–233

Stramma L (1991) Geostrophic transports of the south equatorial current in the Atlantic. J Mar Res 49:281–294

Sverdrup H, Johnson M, Fleming R (1942) The oceans. Their physics, chemistry and general biology. Prentice Hall Inc, New York

Tarakanov RY, Makarenko NI, Morozov EG (2013) Antarctic bottom water flow in the western part of the Romanche fracture zone based on the measurements in October of 2011. Oceanology 53(6):655–667. https://doi.org/10.1134/S0001437013050147

Tarakanov RY, Morozov EG, van Haren H, Makarenko NI, Demidova TA (2018) Structure of the deep spillway in the western part of the Romanche fracture zone. J Geophys Res Oceans 123:8508–8531. https://doi.org/10.1029/2018JC013961

Tucholke BE, Wright WR, Hollister CD (1973) Abyssal circulation over the greater Antilles outer ridge. Deep Sea Res I 20:973–995

van Haren H, Gostiaux L, Morozov E, Tarakanov R (2014) Extremely long kelvin–Helmholtz billow trains in the Romanche fracture zone. Geophys Res Lett 41(23):8445–8451. https://doi.org/10.1002/2014GL062421

Vanicek M, Siedler G (2002) Zonal fluxes in the deep water layers of the western South Atlantic Ocean. J Phys Oceanogr 32(8):2205–2235

Visbeck M (2002) Deep velocity profiling using lowered acoustic doppler current profiler: bottom track and inverse solution. J Atmosph Oceanic Technol 19(5):794–807. https://doi.org/10.1175/1520-0426(2002)019<0794:DVPULA>2.0.CO;27

Warren BA, Speer KG (1991) Deep circulation in the eastern South Atlantic Ocean. Deep-Sea Res I 38(Suppl. 1):281–322

Weatherly GL, Kelley EA (1982) 'Too cold' bottom layers at the base of the Scotian rise. J Mar Res 40(4):985–1012

Weatherly GL, Kim YY, Kontar EA (2000) Eulerian measurements of the North Atlantic deep water deep Western boundary current at 18°S. J Phys Oceanogr 30:971–986

Whitehead JA, Leetma A, Knox RA (1974) Rotating hydraulics of strait and sill flows. Geophys Fluid Dyn 6(2):101–125. https://doi.org/10.1080/03091927409365790

Whitehead JA, Worthington LV (1982) The flux and mixing rates of Antarctic bottom water within the North Atlantic. J Geophys Res Oceans 87(C10):7903–7924. https://doi.org/10.1029/jc087ic10p07903

Whitehead JA (1989) Giant ocean cataracts. Sci Am 260:50–57. https://doi.org/10.1038/scientificamerican0289-50

Whitehead JA (1998) Topographic control of oceanic flows in deep passages and straits. Rev Geophys 36(3):423–440. https://doi.org/10.1029/98RG01014

WOD18 (2018) World Ocean Database 2018. https://www.nodc.noaa.gov/OC5/WOD/pr_wod.html

Wright WR (1970) Northward transport of Antarctic bottom water in the Western Atlantic Ocean. Deep-Sea Res I 17:367–371

Wunsch C (1984) An eclectic Atlantic Ocean circulation model. Part I: the meridional flux of heat. J Phys Oceanogr 14(11):1712–1733

Wüst G (1936) Schichtung und Zirkulation des Atlantischen Ozeans, Das Bodenwasser und die Stratosphäre. In: A. Defant (Ed.), Wissenschaftliche Ergebnisse, Deutsche Atlantische Expedition auf dem Forschungs - und Vermessungsschiff „Meteor“ 1925-1927, 6(1), Walter de Gruyter & Co, Berlin, 411 pg

Zatsepin AG, Gritsenko VA, Kremenetskii VV, Poyarkov SG, Stroganov OY (2005) Laboratory and numerical study of gravity currents over a sloping bottom. Oceanology 45(1):5–15

Chapter 6
Fractures in the Mid-Atlantic Ridge of the North Atlantic

Abstract Flows of Antarctic Bottom Water in the Vema Fracture Zone and other fractures in the northern part of the Mid-Atlantic Ridge are analyzed. We study the flows of Antarctic Bottom Water through the fracture zones in the northern part of the Mid-Atlantic Ridge based on the CTD and LADCP observations in 2006, 2014, 2015, and 2016. We measured the thermohaline properties and velocities and analyzed the flows of bottom water in the Strakhov, Bogdanov, Nameless (07°28′ N), Vernadsky, Doldrums, Arkhangelsky, Ten Degree, Vema, Marathon, Fifteen Twenty, and Kane fracture zones. These abyssal channels connect the deep basins of the East and West Atlantic. In addition to the known fact that the main portion of water (approximately 1 Sv) propagates through the Vema Fracture Zone (11° N) we revealed that approximately half of this volume propagates through the other fractures.

Keywords Vema fracture zone · Smaller fractures in the northern mid-Atlantic ridge · Water transport

6.1 Vema Fracture Zone

6.1.1 Bottom Topography

During the German expedition in 1927–1929 on the R/V *Meteor*, Wüst (1936) found cold abyssal waters with potential temperature θ = 1.74 °C at the bottom of the Gambia Abyssal Plain. He suggested that these waters were of Antarctic origin. He attributed their existence to the penetration through the Romanche Fracture Zone (known already by that time) and a fracture in the Mid-Atlantic Ridge presumptively located near 7–13° N. Only in 1956, deep sounding from the R/V *Vema* revealed a deep fracture near 11°N, which was later called the Vema Fracture Zone after the name of the ship (Heezen et al. 1964). Later, the morphology of the Vema Fracture Zone was studied in detail in (Vangriesheim 1980). The fracture is located between 43.5° and 41.0° W and connects the Demerara and Gambia abyssal plains. The width of the fracture zone is 8–10 km and the maximum depth is approximately 5200 m.

E. G. Morozov et al., *Bottom Gravity Currents and Overflows in Deep Channels of the Atlantic Ocean*, https://doi.org/10.1007/978-3-030-83074-8_6

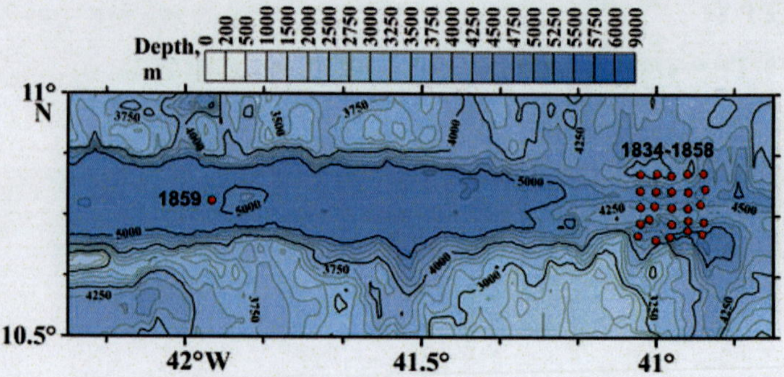

Fig. 6.1 Bathymetric chart of the Vema Fracture Zone and locations of CTD stations occupied in 2006 (red circles). Depth is given in meters. The remote western station at 10°47′ N, 41°56′ W in 1994 and 2006 is shown

Vangriesheim (1980) indicated the existence of three main sills of the fracture at 41°02′, 40°55′, and 40°53′ W, respectively. According to her data, sills are located at depths of 4690, 4650, and 4710 m. During the geophysical expedition in 1998 onboard the R/V *Akademik Nikolay Strakhov*, the existence of these sills was confirmed, but the measured depths were reported as 4690, 4660, and 4660 m, respectively. Thus, these sills seem to be located at the same depth. Based on the data in (Smith and Sandwell 1997), the main sill of the fracture zone is located at 41°01′ W at a depth of 4571 m. Charts of the Vema Fracture Zone bottom topography based on (Smith and Sandwell 1997) and GEBCO are shown in Figs. 6.1, 6.2, and 6.4.

6.1.2 Measurements

Among all fractures of the Northern Mid-Atlantic Ridge only the Vema Fracture Zone has been previously investigated as a pathway for AABW. However, even in this abyssal channel the direct measurements of currents are very scarce. According to the WOD18 database (WOD18 2018; Boyer et al. 2018), the first temperature and salinity measurements in the Vema Fracture Zone were made in 1959. In 1977, a series of CTD-stations was occupied along the axial line (10°48′ N) of the fracture mainly in its eastern part (40°03′–40°55′ W). In the 1970s–1980s, the fracture zone was studied intensely. Stations in the fracture zone were made within the TTO program from the vessels *Knorr*, *Robert Conrad*, *Oceanus*, and others. In 1979, a section along the fracture was occupied at a distance from the fracture axis. In 1994, a series of three stations across the fracture near its western end was occupied from the R/V *Meteor* within the WOCE program.

The transport of AABW through the Vema Fracture Zone was estimated in several publications in the range between 1.0 and 2.4 Sv (1 Sv $= 10^6$ m^3 s^{-1}) (Fischer et al. 1996; Rhein et al. 1998; McCartney et al. 1991; Morozov et al. 2010). The estimates in these publications do not provide ranges of the interannual and

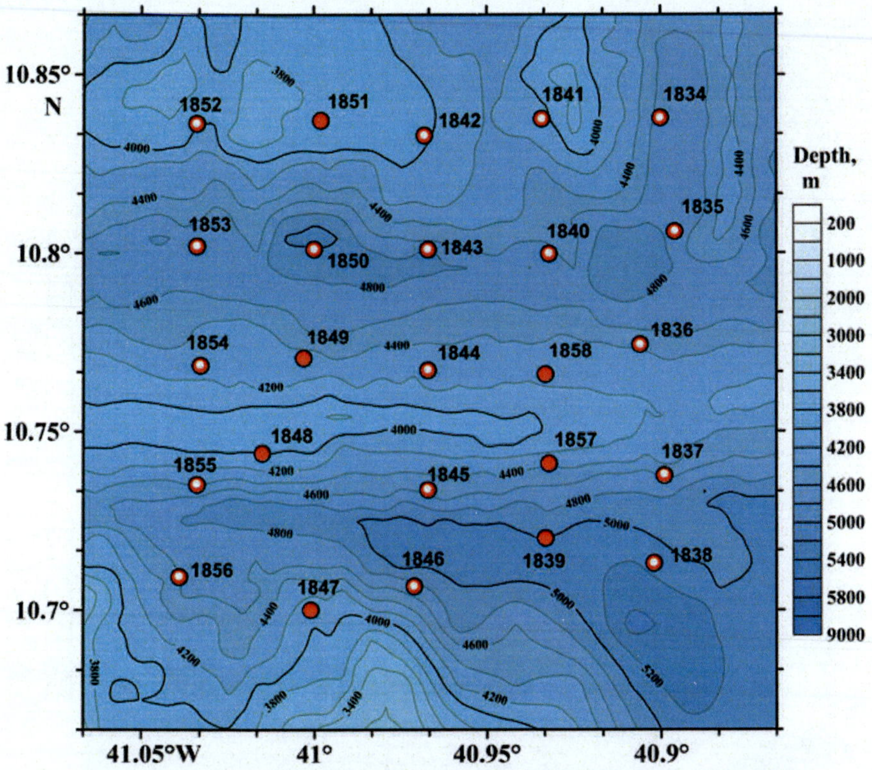

Fig. 6.2 Bathymetric chart (m) of the study region of the Vema Fracture Zone (main sill) based on the measurements in cruise 22 of the R/V *Akademik Ioffe* in 2006. Red dots indicate the locations of stations when the instrument was at the bottom. Dots indicating points with LADCP measurements are highlighted by the white color in the middle of the dots

mesoscale variability, which requires long-term data from moorings that are currently not available. It was shown in Whitehead (1998) that flows in the channels between neighboring deep-ocean basins are unidirectional depending on the source of bottom water. The interface between the bottom water and the overlying water layer over the sills in the channel adjusts itself to permit the bottom water flow from the basin with the source water. Before our studies in the region, the measurements of the AABW flows in the other fractures of the North Mid-Atlantic Ridge either do not exist or they are not available. The A05 and A06 lines of the World Ocean Circulation Experiment (WOCE) pass through the Kane (24° N) and Nameless (07°28′ N) fracture zones, respectively (Fig. 6.3). However, direct velocity measurements on these lines are not available. The conclusion about the leading role of the Vema Fracture Zone in the flow of AABW to the East Atlantic was based on the spatial distribution of potential temperature and salinity east of the Mid-Atlantic Ridge in the Northern Hemisphere (Mantyla and Reid 1983; McCartney et al. 1991).

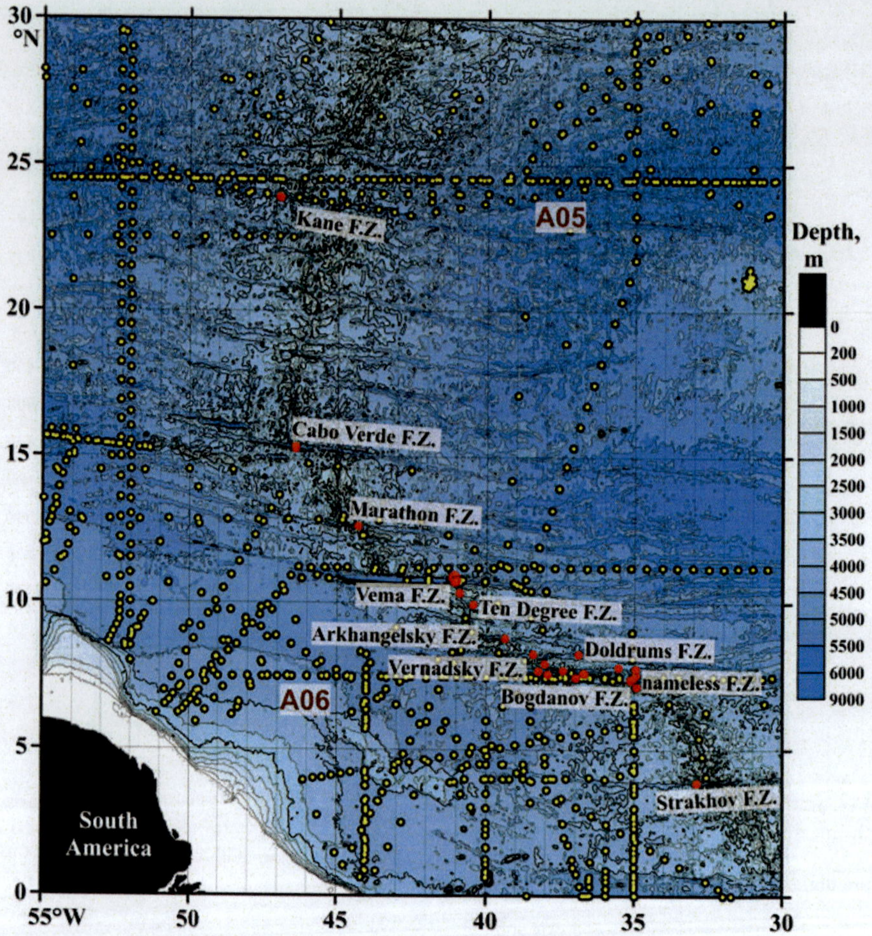

Fig. 6.3 Scheme of the transport of Antarctic Bottom Water in the North Atlantic. The most important abyssal channels (fracture zones) connecting the deep basins are indicated: Kane, Cabo Verde (Fifteen Twenty), Marathon, Vema, Ten Degree, Arkhangelsky, Doldrums, Vernadsky, Nameless, Bogdanov, and Strakhov. The bottom topography is based on (Smith and Sandwell 1997). The yellow dots indicate historical stations based on the WOD18 database. Sections A05 and A06 are the WOCE CTD sections. Our stations in the fractures in the southern part of the Northern Mid-Atlantic Ridge are shown with the red dots

In 2006, an expedition with CTD and LADCP measurements onboard the R/V *Akademik Ioffe* visited the region of the main sills. The depth measurements using an ELAC LAZ 470 echo sounder and an Atlas Parasound parametric echo sounder during cruise 22 of the R/V *Akademik Ioffe* in 2006 introduced more details in the bottom topography data in the region by combining the previous and new bottom soundings. The resulting bathymetric chart is shown in Fig. 6.1 and the study region is shown in more detail in Fig. 6.2.

During the expedition onboard the R/V *Akademik Ioffe* in 2006, a survey of 25 CTD-stations (SBE-25 Sea-logger) was carried out in the study region with a size of 16 × 16 km over the main sill of the Vema Fracture Zone at 40°30′–41°00′ W and a station west of this region (at 41°56′) was occupied repeating the *Meteor* station in 1994. A survey of 5 × 5 stations with a distance of ~4 km between stations was located in the main channel of the fracture and a depression separated by a sill from the main channel of the fracture. Currents were measured using an LADCP profiler (RDI Workhorse Sentinel) at 18 stations including 5 stations along the main channel (Fig. 6.2). Thus, in 2006, the total number of stations in the fracture zone increased from 21 to 47.

We continued our research in the Vema Fracture Zone and visited this region in 2014, 2015, and in April and October of 2016. The information about our measurements is given in Table 6.1.

6.1.3 Structure of Bottom Flow Based on the Measurements in 2006

The deepest part of the Vema Fracture Zone is occupied by Antarctic Bottom Water. In many places, the Mid-Atlantic Ridge serves as an obstacle to the inter-basin exchange of Antarctic Bottom Water and lower part of North Atlantic Deep Water, which can propagate only through some of the transform fractures, of which the Vema Fracture Zone is the deepest. The lower boundary of Lower North Atlantic Deep Water is located here within the layer approximately from 3700 to 4100 dbar (Morozov et al. 2010).

According to (Eittreim et al. 1983), Antarctic Bottom Water propagates in the Vema Fracture Zone as a homogeneous layer, which can be as thick as 700 m. The flow is well mixed due to the bottom and lateral friction. The authors distinguish two gradient zones ('benthic thermocline' and 'transition zone') at the Antarctic Bottom Water boundary with Lower North Atlantic Deep Water. These zones differ by the vertical temperature gradients equal to 0.0012 °C/m and 0.0004 °C/m, respectively. The maximum thickness of the transition zone can reach 300 m in the eastern part of the fracture zone. The benthic (bottom) thermocline is located approximately 1000 m above the bottom. The potential temperature jump in the benthic thermocline is between 1.5 °C and 2.0 °C. In (Eittreim et al. 1983), its origin is related to the waters of two-degree discontinuity distinguished in (Broecker et al. 1976), which are located at a significant distance south of the study region. Following the concept described in (Broecker et al. 1976), Eittreim et al. (1983) considered that these waters are formed in the Rio Grande Rise region due to the diffuse interaction between Antarctic Bottom Water and North Atlantic Deep Water, alternative intensification, or weakening of the inflow of North Atlantic Deep Water (or Antarctic Bottom Water) was revealed in (Broecker et al. 1976). Researchers have no common opinion whether the transition zone and benthic thermocline are included in the

Table 6.1 Coordinates of CTD/LADCP stations in 2006, 2014, 2015, and 2016 in the Vema Fracture Zone

Date	Station	Coordinates	Lower CTD depth/ ocean depth, m
26.10.2006	1834	10°50.0′ N, 40°54.0′ W	4551/4580
27.10.2006	1835	10°48.0′ N, 40°54.0′ W	4741/4752
27.10.2006	1836	10°45.9′ N, 40°54.0′ W	4389/4402
27.10.2006	1837	10°44.0′ N, 40°53.8′ W	4690/4695
27.10.2006	1838	10°42.0′ N, 40°54.0′ W	5272/5275
27.10.2006	1839	10°42.0′ N, 40°55.9′ W	5010/5014
28.10.2006	1840	10°48.0′ N, 40°56.0′ W	4623/4633
28.10.2006	1841	10°50.0′ N, 40°56.0′ W	4021/4030
28.10.2006	1842	10°50.0′ N, 40°57.7′ W	3942/3953
28.10.2006	1843	10°48.0′ N, 40°57.8′ W	4764/4774
28.10.2006	1844	10°45.9′ N, 40°58.0′ W	4214/4225
28.10.2006	1845	10°43.9′ N, 40°58.0′ W	4517/4521
28.10.2006	1846	10°42.0′ N, 40°58.0′ W	4557/4568
29.10.2006	1847	10°42.0′ N, 41°00.0′ W	4022/4032
29.10.2006	1848	10°44.0′ N, 41°00.2′ W	4061/4072
29.10.2006	1849	10°46.0′ N, 41°00.0′ W	4326/4330
29.10.2006	1850	10°48.0′ N, 41°00.0′ W	4931/4942
29.10.2006	1851	10°49.9′ N, 41°00.0′ W	3838/3849
29.10.2006	1852	10°50.0′ N, 41°02.0′ W	3931/3943
30.10.2006	1853	10°48.0′ N, 41°02.0′ W	4797/4809
30.10.2006	1854	10°46.0′ N, 41°02.0′ W	4309/4321
30.10.2006	1855	10°43.9′ N, 41°02.0′ W	4349/4360
30.10.2006	1856	10°42.0′ N, 41°02.0′ W	4608/4618
30.10.2006	1857	10°44.0′ N, 40°56.0′ W	4377/4389
30.10.2006	1858	10°46.0′ N, 40°56.0′ W	4301/4312
31.10.2006	1859	10°46.8′ N, 41°56.6′ W	5155/5165
08.10.2014	2550	10°52.6′ N, 41°07.3′ W	4309/4315
08.10.2014	2551	10°47.4′ N, 41°01.0′ W	4683/4688
08.10.2014	2552	10°48.3′ N, 41°01.0′ W	4597/4602
09.10.2014	2553	10°53.0′ N, 41°08.5′ W	4342/4347
09.10.2014	2554	10°55.1′ N, 41°09.5′ W	4309/4314
09.10.2014	2555	10°43.5′ N, 41°05.5′ W	4504/4510
21.09.2015	2569	10°48.3′ N, 41°01.0′ W	4420/4500
21.09.2015	2570	10°47.3′ N, 41°01.0′ W	4642/4648
21.09.2015	2571	10°43.6′ N, 41°05.4′ W	4491/4499
22.09.2015	2572	10°43.2′ N, 41°02.5′ W	4715/4722
22.04.2016	2606	10°43.5′ N, 41°05.1′ W	4536/4541
23.04.2016	2607	10°43.4′ N, 41°03.0′ W	4714/4719
23.04.2016	2608	10°46.3′ N, 41°01.0′ W	4246/4251
23.04.2016	2609	10°47.3′ N, 41°01.0′ W	4658/4663
23.04.2016	2610	10°48.3′ N, 41°01.0′ W	4507/4511

(continued)

Table 6.1 (continued)

Date	Station	Coordinates	Lower CTD depth/ ocean depth, m
23.04.2016	2611	10°51.3′ N, 41°09.0′ W	4451/4456
24.04.2016	2612	10°51.0′ N, 41°08.0′ W	4314/4319
21.10.2016	2642	10°58.5′ N, 41°04.1′ W	4521/4527
21.10.2016	2643	10°54.5′ N, 41°02.0′ W	4221/4225
21.10.2016	2644	10°48.3′ N, 41°01.0′ W	4486/4490
21.10.2016	2646	10°47.3′ N, 41°01.0′ W	4675/4680
22.10.2016	2648	10°46.3′ N, 41°01.0′ W	4248/4253
22.10.2016	2650	10°43.5′ N, 41°05.5′ W	4465/4470
22.10.2016	2651	10°21.0′ N, 40°54.5′ W	3843/3846

Antarctic Bottom Water layer. In rare cases, the transition zone is divided into several parts based on different estimates of the location of boundaries, which are within the range of potential temperature $\theta = 1.5$–2.0 °C and density $\sigma_4 = 45.895$–45.920.

During the cruise of the R/V *Akademik Ioffe* in 2006, a structure similar to the one described in (Eittreim et al. 1983) was found. A quasi-homogeneous layer of bottom water with a thickness of 400–500 m, potential temperature $\theta = 1.34$ °C, and salinity 34.829 PSU at the bottom was recorded at the distant western station (10°47′ N, 41°56′ W). In 2006, this station was repeated (station 1859) (Figs. 6.1 and 6.4). While propagating to the east, the flow reaches the main sill region at station 1853 (Fig. 6.2) after it passes the western sill and the region with relatively smooth bottom topography. At station 1853, the flow is approximately 100 m thick and its potential temperature is $\theta = 1.36$ °C and salinity is lower than 34.835 PSU (Fig. 6.5).

By the time the water in the main channel flows from station 1853 to station 1850 (only 2 miles, but down the sill), potential temperature θ increases from 1.360 °C to 1.395 °C, which probably indicates strong mixing in the bottom layer. Based on the bottom potential temperature (1.51 °C) at a station in the depression (station 1838) located south of the southern wall of the main channel (Fig. 6.2), we can conclude that the part of the Antarctic Bottom Water jet, which is characterized by the lowest temperatures, does not get into the depression basin or it becomes warmer after flowing down the slope. The temperature at the bottom of the depression is close to the temperature in the upper part of the transition zone in the main channel. This temperature corresponds to the water that overflows the deepest crest over the elevation between the main channel and depression. Twelve years later we made an echo-sounder survey along 10°44′ N over the median ridge dividing the central and southern channels. A narrow gap was found with a depth of 4462 m that could provide water exchange between channels so that the coldest water from the central channel could fill the bottom of the depression in the southern channel (Fig. 6.6).

Strong spatial variability of water mass characteristics was found in the study region over a scale of a few miles, which is seen well in Fig. 6.5. Bottom water becomes warmer and warmer as it overflows the sill in the main channel. However,

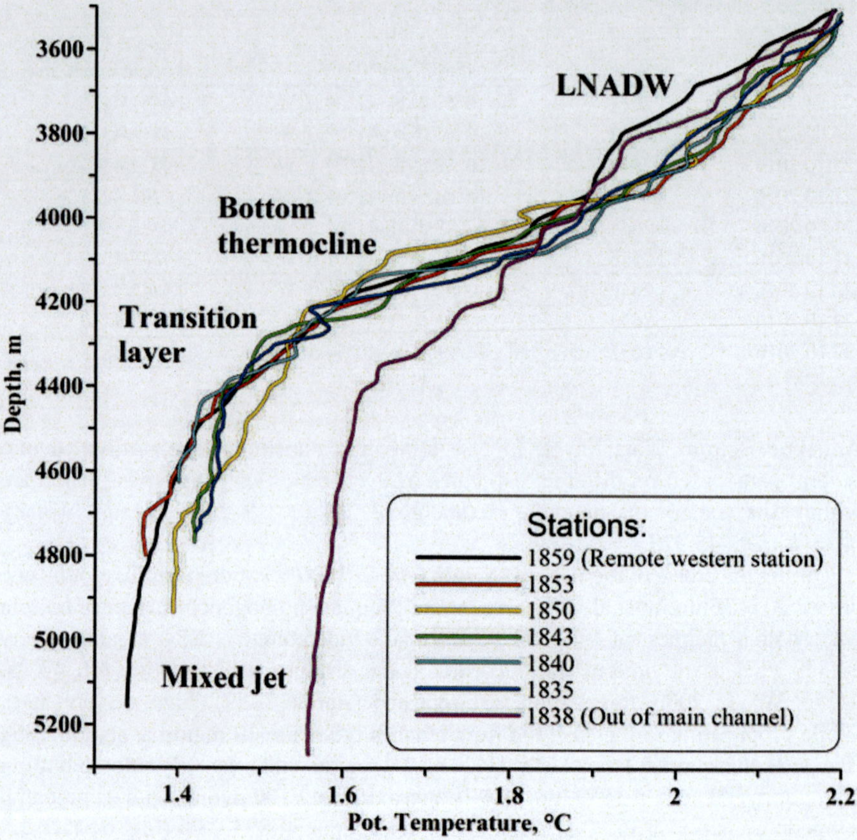

Fig. 6.4 Depth variation of potential temperature (°C) at several stations in the Vema Fracture Zone. Numbers of stations according to Fig. 6.2 are indicated. The curve related to station 1838 shows a profile in the depression south of the main channel. (See also Fig. 6.2)

the temperature increase is not monotonic probably because we could not follow the main cold jet of the flow. The θ,*S*-structure of water is similar to the structure found by (Eittreim et al. 1983), which was described above (Fig. 6.7).

As to the time variability of water properties at the stations located in the region of the sills, no notable long-term temperature variations were found while comparing the measurements in 1977, 2006, and 2016. Maximum temperature variations exceeding 0.06 °C were found near the 4000-m isobath. Potential temperatures at the bottom measured in 1977 were within 1.406–1.440 °C (up to 1.441 °C in the mixed jet), which coincides with the range of values at the bottom at four nearest stations in 2006 (1.399–1.440 °C).

An increase in potential temperature θ by 0.027 °C and salinity by 0.009 PSU in the Antarctic Bottom Water core was found by comparing the measurements at the remote western station in 2006 and the measurements of the R/V *Meteor* in 1994, which were made at the same place (10°47′ N, 41°56′ W). The depth of the station in

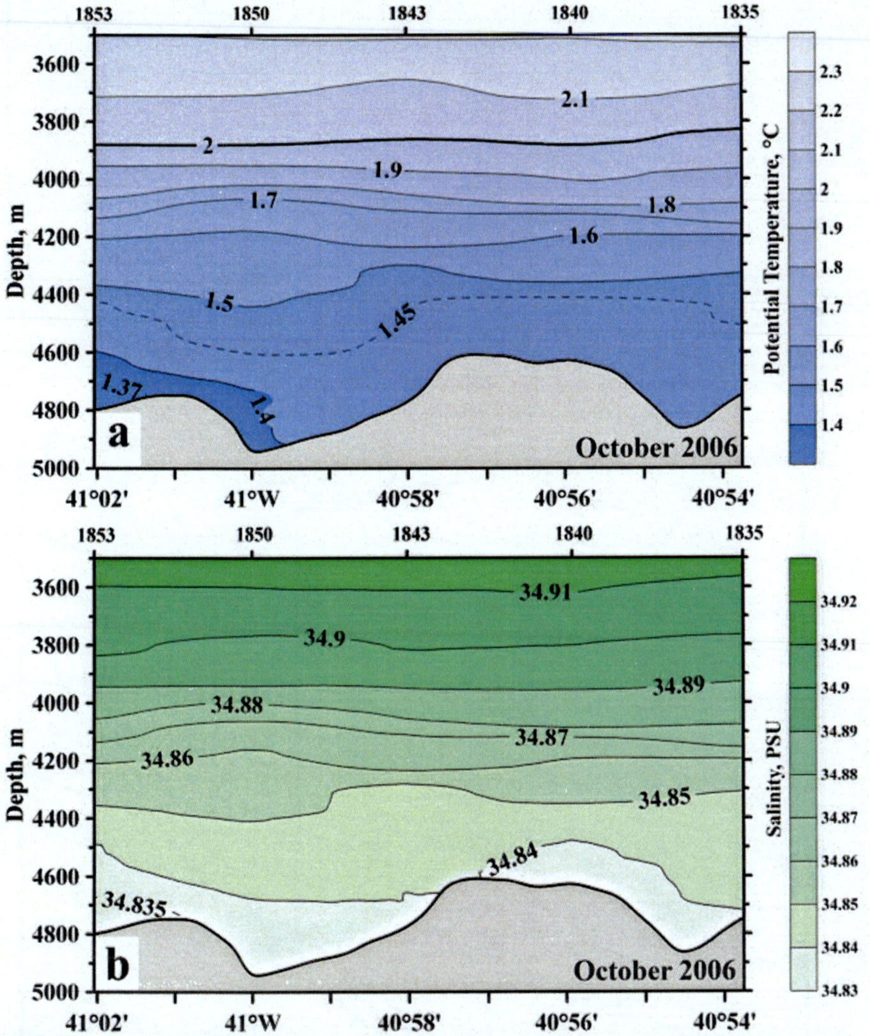

Fig. 6.5 West to east variation in potential temperature (°C) (**a**) and salinity (PSU) (**b**) sections along the main channel in the study region (Vema Fracture Zone). Stations are indicated at the top axis

2006 appeared 100 m greater than in 1994. We explain this fact as different positions of the instrument relative to the position of the vessel in different expeditions and steep slopes of the fracture walls. Coordinates of the station are assumed as the position of the ship; however, the coordinates of the instrument at the bottom can be different.

Figure 6.8 shows the distribution of potential temperature at the bottom as a function of the distance from the main sill of the Vema Fracture Zone to the east. The

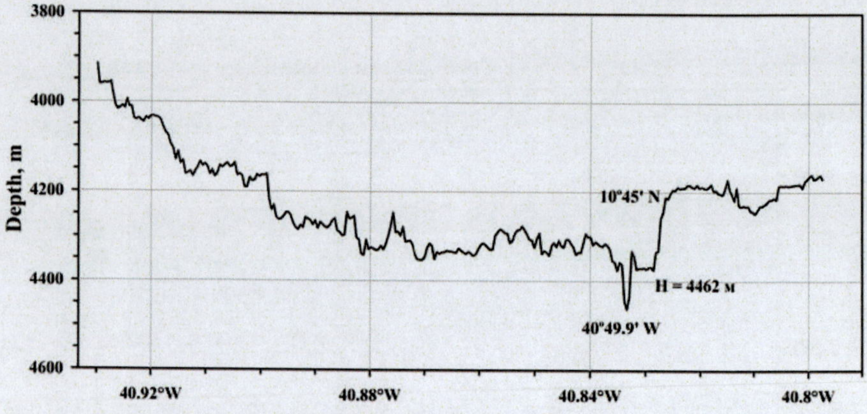

Fig. 6.6 Bottom topography along the median ridge at 10°44′ N

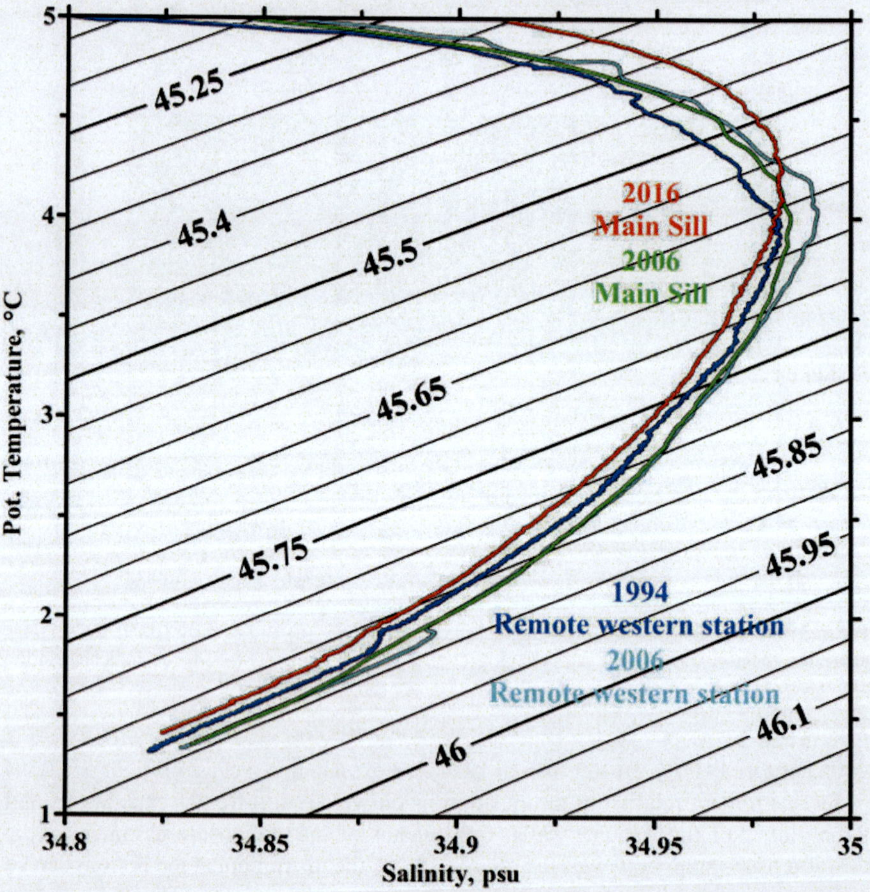

Fig. 6.7 Curves of θ,*S*-distributions for the bottom layer waters in the Vema Fracture Zone based on the data of 1994 (remote western station, blue color) with the nearest station in 2006 (light blue) and over the main sill in 2006 (green) and 2016 (red). Isopycnals referenced to 4000 m (σ_4 in kg/m^3) ($\sigma_4 = 45.3$–46.1) are shown

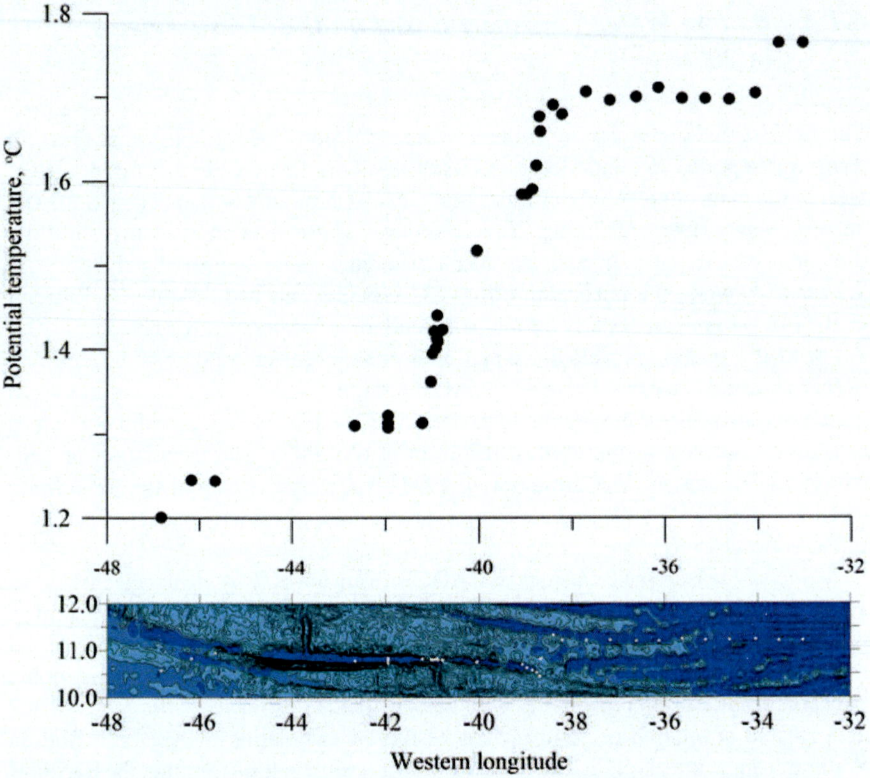

Fig. 6.8 Potential temperature (°C) at the bottom as function of the distance from the main sill of the Vema Fracture Zone (top panel). (Combined data from (Arhan et al. 1998) and the data of new expeditions are presented. White dots on the chart in the bottom panel show locations of stations)

figure presents combined data from (Arhan et al. 1998), WOD18 database, and the data of the new expeditions. In addition to the differences with the measurements in 1977 and 1994, transformation of properties of Antarctic Bottom Water in the study region caused by the elevation of bottom topography is also shown in the figure. Only the upper part of Antarctic Bottom Water can overflow this elevation. The greatest longitudinal gradients of potential temperature are found in the region of two sills, where the study site was located. This is caused by the peculiarities of the bottom water overflow (the densest waters can overflow only occasionally) and possible intensification of mixing in this region due to steep slopes of the fracture walls. This figure also shows that the increase in the water temperature and mixing is strongest near the slopes of the Mid-Atlantic Ridge and in the Vema Fracture Zone, mainly near the sills. After this, temperature changes only slightly during the further motion in the deep basin.

6.1.4 Bottom Water Transport Based on the Measurements in 2006

The first estimate of Antarctic Bottom Water transport through the Vema Fracture Zone was reported in (Vangriesheim 1980) based on the data of two current meters west of the main sill (40°54′ W) approximately at a depth of 5040 m (Table 6.2). The current meters operated during 26 and 9 days. These data showed that the flow direction varied strongly and the mean velocities were approximately 2.9 and 3.7 cm s^{-1}, while the maximum observed velocities reached 33 cm s^{-1}. Eittreim et al. (1983) found an easterly current at a depth of 4790 m (approximately at 41°35′ W) with the maximum velocity equal to 3 cm s^{-1} (mean velocity 0.07 cm s^{-1}) observed during 11 days. The authors measured the easterly flow not only in the sill region but also east of the region. However, underwater photography did not record any strong currents at the bottom (Eittreim et al. 1983). The calculation of geostrophic velocities in (McCartney et al. 1991) based on the data of the R/V *Knorr* (section along 35° W) allowed them to get an estimate of velocity equal to 20 cm s^{-1} in the bottom layer.

Velocity measurements using an LADCP from the R/V *Meteor* (Fischer et al. 1996) recorded the maximum velocities equal to 20 cm s^{-1} at a depth of 4200 m in the middle of the channel, which is close to the geostrophic estimates. The total transport was calculated by combining the direct measurements and the geostrophic transport from the reference level with zero velocities estimated from the LADCP data located at a depth of 3640 m. The results of calculation of Antarctic Bottom Water transport are given in Table 6.2. It was previously discussed that the transport estimates depend strongly on the choice of the upper boundary of Antarctic Bottom Water. For example, in the calculation of the geostrophic transport made in (McCartney et al. 1991), the total transport increased more than four times (!) after they changed the upper boundary of Antarctic Bottom Water from $\theta = 1.5$ °C to $\theta = 2$ °C. (Table 6.2).

In 2006, we repeated the station made in 1994 that allowed us to make a comparison of currents (at 10°47′ N, 41°56′ W). Comparison of the velocity distribution by depth obtained during direct measurements of currents in 1994 and 2006 showed a decrease in the transport in the bottom part of the channel in 2006 (Fig. 6.9). However, the vertical distribution of velocities is quite similar, and even more this distribution is close to the distribution of geostrophic velocities based on the data in 1994.

The measurements conducted in 2006 showed that the core of bottom water flow in the channel of the Vema Fracture Zone was observed not at the bottom but in the layer between 3700 and 4000 m (maximum LADCP velocities), which is close to the upper boundary of Antarctic Bottom Water (Fig. 6.9). This is a similar property of bottom flows in many abyssal channels (see Fig. 10 in the Preface of the book).

While estimating the transport of Antarctic Bottom Water we took into account the following:

Table 6.2 Transport of Antarctic Bottom Water to the eastern basin of the Atlantic Ocean through the Vema Fracture Zone based on literature data

Authors	AABW boundary	Transport, Sv	Comments
Vangriesheim 1980	θ = 1.5 °C	0.05–0.46	Moored current meter
Schlitzer 1987		0.0–0.7	Box model
McCartney et al. 1991	θ =2.0 °C	2.08–2.24	Geostrophic transport, reference surface, θ = 2.17–2.43 °C
	θ =1.5 °C	0.46	
Fischer et al. 1996	θ =2.0 °C	1.8–2.0	Combination of geostrophic transport and LADCP data at 42° W
	σ_4 = 45.9	2.1–2.4	
Rhein et al. 1998		1.1	Inclusion of part of LNADW into the AABW layer
Stephens and Marshall 2000	θ =1.8 °C	2.0	
Demidov et al. 2007	θ =1.8°-2.0 °C	0.12–0.64	LADCP measurements

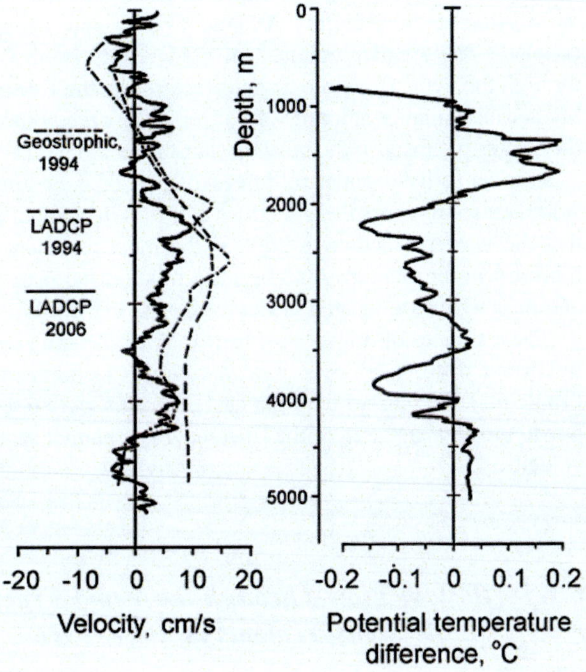

Fig. 6.9 Comparison of the vertical distribution of velocities (cm s^{-1}) and potential temperatures θ (°C) at a western distant station from the study region in 2006 and German station in 1994 both at 10°47′ N, 41°56′ W (left panel). Difference in potential temperatures between 1994 and 2006 are shown in the right panel

1. The estimates of transport differ significantly depending on the methods of determination of the section square and extrapolation of the current velocities to the section triangles, where the bottom waters exist but the measurements are lacking. Calculations taking into account the transport beyond the main channel within the study region, where there is no direct flow of bottom water, were carried out separately.
2. It is crucial that current velocity measured using an LADCP also includes the influence of tides and other periodical motions. The transports over the meridional sections differ by a factor greater than 1.5, which required the presentation of the final estimates only in the form of intervals.
3. Determination of the upper boundary of Antarctic Bottom Water is ambiguous and the total transport strongly depends on its choice. Therefore, in this work, we used different versions of the location of this boundary coinciding with isotherms of potential temperature $\theta = 1.5$, 1.8, 1.9, and 2.0 °C.

Transports of bottom water in the Vema Fracture Zone are given in Table 6.3. In general, the estimates of transport appeared significantly smaller than those reported in the literature with the upper boundary of Antarctic Bottom Water assumed as an isotherm of potential temperature $\theta = 2$ °C. The maximum estimate of transport in 2006 reached only 0.64 Sv.

If we compare two major channels, through which bottom water flows from the West Atlantic to the East Atlantic (Vema and Romanche fracture zones), we conclude that smaller volumes of cold waters (with $\theta < 1.5$ °C) are transported through the Vema Fracture Zone than through the Romanche Fracture Zone, possibly because smaller amounts of bottom waters reach the Vema Fracture Zone along the longer pathway in the western part of the Atlantic.

In order to have statistical representation of our estimates, we calculated bottom water transport across three sections in the main channel (Table 6.3). It is our opinion that the western section in 2006 is the most representative for characterizing the transport in the fracture zone because it is located near the main sill of the fracture and its stations are located closer to the channel walls.

The estimates of transport of Lower North Atlantic Deep Water are characterized by strong differences over five meridional sections in the study region in 2006 (Table 6.3). To our opinion, the best estimates were obtained at the western section due to the smallest area of the fracture zone channel section for the extrapolation of velocities.

6.1.5 Bottom Flow Through the Vema Fracture Zone Based on the Measurements in 2014–2016

In 2014–2016, we continued our studies and made six stations in 2014, four stations in 2015 (Fig. 6.10, left panel), seven stations in April 2016 (Fig. 6.10, middle panel),

Table 6.3 Transport of bottom waters (Sv) in the layers of Lower North Atlantic Deep Water (upper line) and at different selections of the Antarctic Bottom Water upper boundary in the Vema Fracture Zone based on the data in 2006

Water mass	θ (°C)	Western section, stations 1852–1856	Central section, stations 1842–1846	Eastern section, stations 1834–1838
LNADW	From θ = 2.4 °C to 2.0 °C	0.38–0.66	0.16–0.21	0.33–0.50
AABW	Below θ =2.0 °C	0.15–0.64	0.13–0.21	0.23–0.46
	Below θ =1.9 °C	0.14–0.51	0.13–0.21	0.19–0.40
	Below θ =1.8 °C	0.12–0.37	0.11–0.18	0.13–0.32
	Below θ =1.5 °C	0.01	0.04	0.02

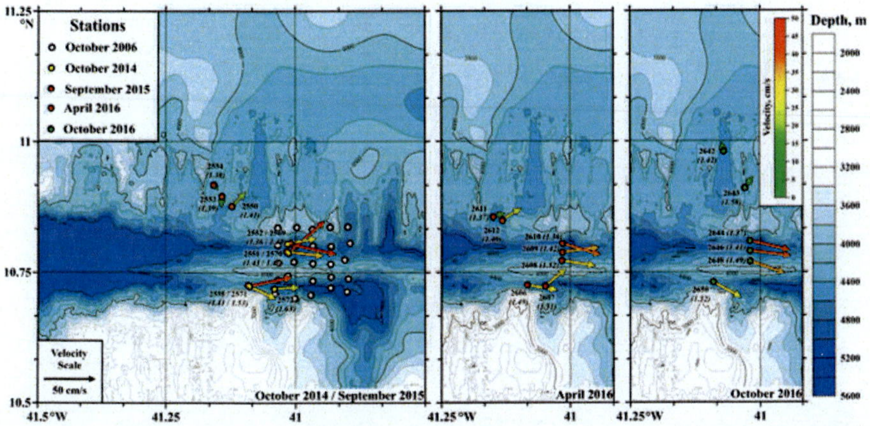

Fig. 6.10 Maps of the Vema Fracture Zone and vectors of velocity. The arrows show the vectors of current velocities at the bottom. LADCP velocities in 2006, 2014, and 2015. Gray circles indicate our stations in 2006. Yellow circles show stations in 2014 and orange circles show stations in 2015 (left panel). The length of the arrows of vectors of current velocities corresponds to the velocity color scale. LADCP velocities in April 2016. Red circles indicate our stations (middle panel); LADCP velocities in October 2016. Green circles indicate our stations (right panel). The numbers of stations according to Table 6.4 are indicated. The potential temperatures (°C) at each station at the bottom are written in parentheses after the station number. The bottom topography is based on (Smith and Sandwell 1997). (Modified from (Morozov et al. 2018))

and six stations in October 2016 (Fig. 6.10, right panel) near the main sills of the fracture at 10°48′ N, 41°01′ W. We found that in the Vema Fracture Zone, the flow splits into three jets.

Table 6.4 Coordinates of CTD/LADCP stations in 2014, 2015, 2016 in the fractures of the northern part of the Mid-Atlantic Ridge

Region	Date	Station	Coordinates, N and W	Lower CTD depth/ ocean depth, m
Kane	03.10.2014	2540	23°54.6′, 46°58.0′	4349/4354
Kane	03.10.2014	2541	23°53.0′, 46°55.9′	4864/4869
Fifteen Twenty	06.10.2014	2547	15°16.3′, 46°25.0′	4499/4505
Fifteen Twenty	06.10.2014	2548	15°20.5′, 46°25.0′	4166/4171
Marathon	07.10.2014	2549	12°37.4′, 44°18.0′	4215/4220
Vema	08.10.2014	2550	10°52.6′, 41°07.3′	4309/4315
Vema	08.10.2014	2551	10°47.4′, 41°01.0′	4683/4688
Vema	08.10.2014	2552	10°48.3′, 41°01.0′	4597/4602
Vema	09.10.2014	2553	10°53.0′, 41°08.5′	4342/4347
Vema	09.10.2014	2554	10°55.1′, 41°09.5′	4309/4314
Vema	09.10.2014	2555	10°43.5′, 41°05.5′	4504/4510
Rift valley of the Doldrums Fracture Zone	10.10.2014	2556	07°56.5′, 38°01.9′	4327/4332
Vernadsky	11.10.2014	2557	07°43.3′, 37°25.0′	4623/4628
Nameless	11.10.2014	2558	07°28.7′, 36°59.1′	4644/4648
Vema	21.09.2015	2569	10°48.3′, 41°01.0′	4420/4500
Vema	21.09.2015	2570	10°47.3′, 41°01.0′	4642/4648
Vema	21.09.2015	2571	10°43.6′, 41°05.4′	4491/4499
Vema	22.09.2015	2572	10°43.2′, 41°02.5′	4715/4722
Doldrums	23.09.2015	2573	08°15.8′, 36°53.4′	4870/4877
Vernadsky, southern wall	23.09.2015	2574	07°36.5′, 36°41.6′	4823/4830
Vernadsky	24.09.2015	2575	07°49.2′, 35°32.5′	4024/4031
Nameless	24.09.2015	2576	07°23.1′, 35°08.0′	4523/4529

(continued)

Table 6.4 (continued)

Region	Date	Station	Coordinates, N and W	Lower CTD depth/ ocean depth, m
Bogdanov	24.09.2015	2577	07°09.6′, 34°55.2′	3815/3820
Strakhov	18.04.2016	2595	03°52.5′, 32°53.4′	4017/4023
Bogdanov	19.04.2016	2596	07°09.7′, 34°55.1′	3934/3940
Nameless	19.04.2016	2597	07°23.0′, 35°08.0′	4525/4530
Nameless	19.04.2016	2598	07°24.5′, 35°08.0′	4423/4428
Nameless	19.04.2016	2599	07°28.1′, 35°01.5′	4348/4365
Nameless/Vernadsky	20.04.2016	2600	07°34.8′, 34°57.0′	4207/4211
Vernadsky	20.04.2016	2601	07°46.4′, 34°57.1′	4410/4415
Vernadsky	21.04.2016	2602	07°37.5′, 36°44.2′	4618/4621
Vernadsky	21.04.2016	2603	07°37.5′, 36°42.2′	4622/4627
Vernadsky	21.04.2016	2604	07°42.0′, 38°15.0′	4801/4806
Arkhangelsky	22.04.2016	2605	08°48.1′, 39°22.5′	3955/4001
Vema	22.04.2016	2606	10°43.5′, 41°05.1′	4536/4541
Vema	23.04.2016	2607	10°43.4′, 41°03.0′	4714/4719
Vema	23.04.2016	2608	10°46.3′, 41°01.0′	4246/4251
Vema	23.04.2016	2609	10°47.3′, 41°01.0′	4658/4663
Vema	23.04.2016	2610	10°48.3′, 41°01.0′	4507/4511
Vema	23.04.2016	2611	10°51.3′, 41°09.0′	4451/4456
Vema	24.04.2016	2612	10°51.0′, 41°08.0′	4314/4319
Vema	21.10.2016	2642	10°58.5′, 41°04.1′	4521/4527
Vema	21.10.2016	2643	10°54.5′, 41°02.0′	4221/4225
Vema	21.10.2016	2644	10°48.3′, 41°01.0′	4486/4490

(continued)

Table 6.4 (continued)

Region	Date	Station	Coordinates, N and W	Lower CTD depth/ ocean depth, m
Vema	21.10.2016	2646	10°47.3′, 41°01.0′	4675/4680
Vema	22.10.2016	2648	10°46.3′, 41°01.0′	4248/4253
Vema	22.10.2016	2650	10°43.5′, 41°05.5′	4465/4470
Vema rift valley	22.10.2016	2651	10°21.0′, 40°54.5′	3843/3846
Ten-degree	22.10.2016	2652	09°57.0′, 40°27.0′	3786/3790
Doldrums	23.10.2016	2655	08°16.1′, 38°25.2′	4034/4038
Vernadsky	24.10.2016	2658	07°35.0′, 37°55.7′	4081/4085

The wide northern channel directs the flow to the northeast. The central channel is the deepest channel of the fracture. Its direction is approximately zonal; it is divided by an underwater ridge into the central (main) and southern channels. The central channel is the direct pathway for AABW flowing to the east. The current in the southern Vema channel overflows the sill and fills a deep depression downstream (Morozov et al. 2010).

The jet flows to the northeast through a gap in the northern wall of the fracture, which is approximately 10 km wide along the 3950 m isobath (the depth of the 2 °C isotherm). The maximum depth of this passage is 4350 m. We made three CTD/LADCP casts across this gap in 2014 (stations 2550, 2553, and 2554; Figs. 6.10, left panel and 6.11). The minimum potential temperature was 1.409 °C; the maximum velocities reached 20.7 cm s^{-1}. The minimum salinity of AABW at the bottom was 34.823. The core of the maximum velocity and minimum potential temperature and salinity were displaced to the right of the flow direction. The AABW transport was estimated at 0.078 ± 0.008 Sv. In 2016, we made four stations north of the region studied in 2014.

Stations 2642 and 2643 were occupied here in October 2016. The minimum potential temperature at station 2642 was 1.419 °C at a depth of 4527 m. The velocities of the bottom flow directed to the northeast were 6–10 cm s^{-1}. The second station (2643) was at a shallower location (4225 m); the potential temperature at the bottom was 1.576 °C. The salinities at these stations at the bottom were relatively high: 34.899 and 34.949 PSU, respectively. The coldest water with minimum salinity cannot flow through these shallow passages. The walls of the fracture here do not keep AABW in the channel and it spreads above them as a wide flow.

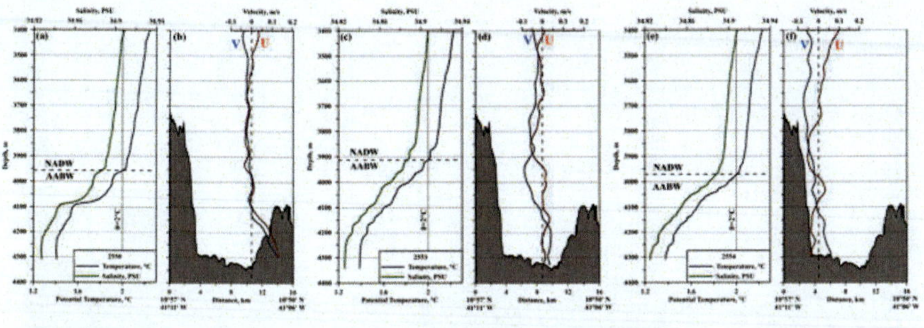

Fig. 6.11 Profiles of potential temperature, salinity, and velocity U (red) and V (blue) components in the northern channel of the Vema Fracture Zone. The vertical dashed line indicates the zero velocity and also the location of the station. The bottom is shown with gray color. The vertical lines at 2 °C (**a, c**) show the boundary of AABW with North Atlantic Deep Water (NADW): station 2550 (**a, b**), the boundary of AABW is at 3960 m; station 2553 (**c, d**), the boundary of AABW is at 3915 m; and station 2554 (**e, f**), the boundary of AABW is at 3975 m. (Modified from (Morozov et al. 2018))

It was found in 2006 (Demidov et al. 2007) that the strongest flow of AABW to the east through the Mid-Atlantic Ridge is located in the central channel of the fracture zone, which we consider the main channel. This conclusion was later confirmed by our measurements in 2014–2016.

After our first visit to the main channel of the Vema Fracture Zone in 2006, we made four visits more in 2014, 2015, and twice in 2016. The currents directed to the east flow along the main fracture, which is divided into the central and southern channels by a zonal ridge. The width of the central channel at a depth of 3850 m (the 2 °C isotherm) is 6.5 km; the maximum depth is 4690 m. Two stations (along 41°01.0′ W) were occupied across this channel in 2014 (2551, 2552) and repeated in 2015 (2569, 2570). In 2016, we twice occupied three stations across the channel (stations 2608, 2609, 2610 in April and stations 2644, 2646, 2648 in October). The distance between the stations across the channel was approximately 2 km.

Profiles of potential temperature, salinity, and velocity at station 2610 in April 2016 are shown in Fig. 6.12a, b. Sections of potential temperature across the fracture in different years are shown in Fig. 6.13a, d; salinity sections are shown in Fig. 6.14a, d; and zonal velocity sections are shown in Fig. 6.15a, d.

A section of two stations in 2014 (2551, 2552) showed that the minimum bottom temperatures and lowest salinities were found at station 2552 over thc northern slope. The bottom potential temperature at station 2552 was 1.362 °C. The salinity at this station was 34.823. The maximum velocity of the eastward flow was 33 cm s^{-1}. The maximum velocity core was displaced to the south, and the potential temperature and salinity minima were displaced to the north. The AABW transport in the central channel was 0.860 ± 0.080 Sv.

The distributions of potential temperature and salinity did not change strongly based on the measurements in 2014–2016 (Figs. 6.13 and 6.14). In 2015, the potential temperature at the deepest place of the bottom in the central channel (station 2570) was 1.458 °C, whereas the minimum potential temperature was found over the northern slope at station 2569 at a depth of 4416 m (1.403 °C). At

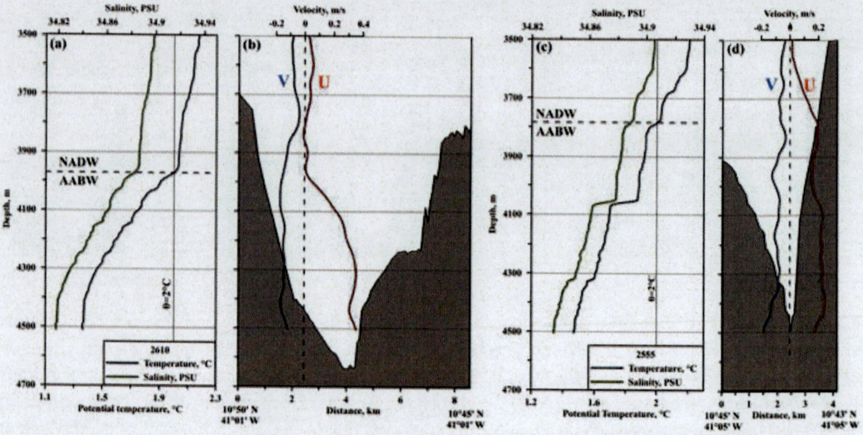

Fig. 6.12 Profiles of potential temperature, salinity, and velocity *U* (red) and *V* (blue) components in the central channel of the Vema Fracture Zone: station 2610, April 2016 (**a**, **b**), the upper boundary of AABW is at 3970 m; and in the southern channel of the Vema Fracture Zone: station 2555 in 2014 (**c**, **d**), the boundary of AABW is at 3780 m. (Modified from (Morozov et al. 2018))

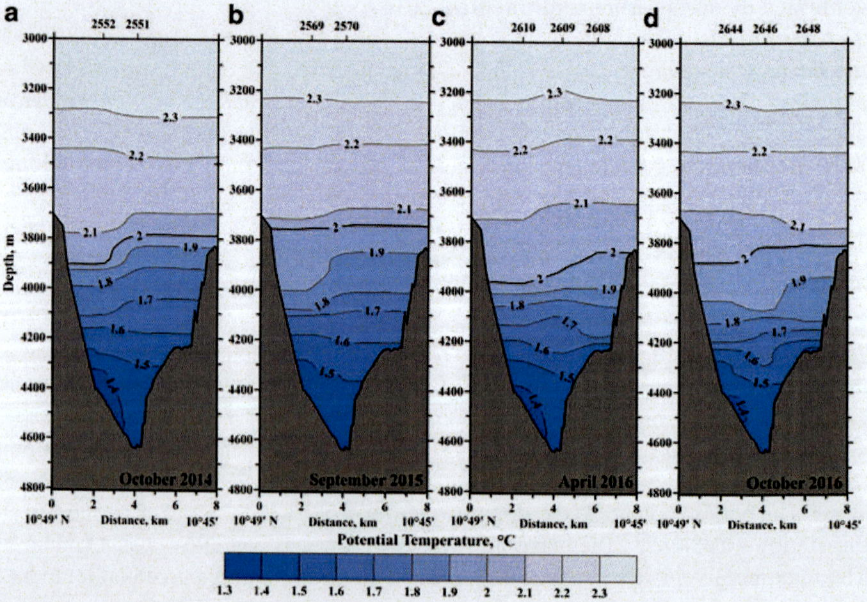

Fig. 6.13 Sections of potential temperature across the central channel of the Vema Fracture Zone in 2014 (stations 2551, 2552) (**a**); 2015 (stations 2569, 2570) (**b**); April 2016 (stations 2608–2610) (**c**); and October 2016 (stations 2644, 2646, 2648) (**d**). (Modified from (Morozov et al. 2018))

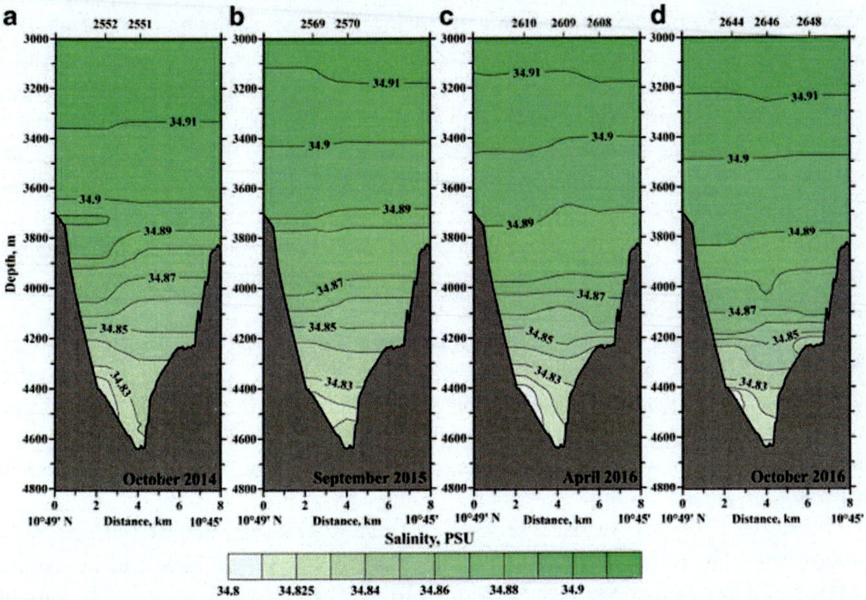

Fig. 6.14 Sections of salinity across the central channel of the Vema Fracture Zone in 2014 (**a**); 2015 (**b**); April 2016 (**c**); and October 2016 (**d**). (Modified from (Morozov et al. 2018))

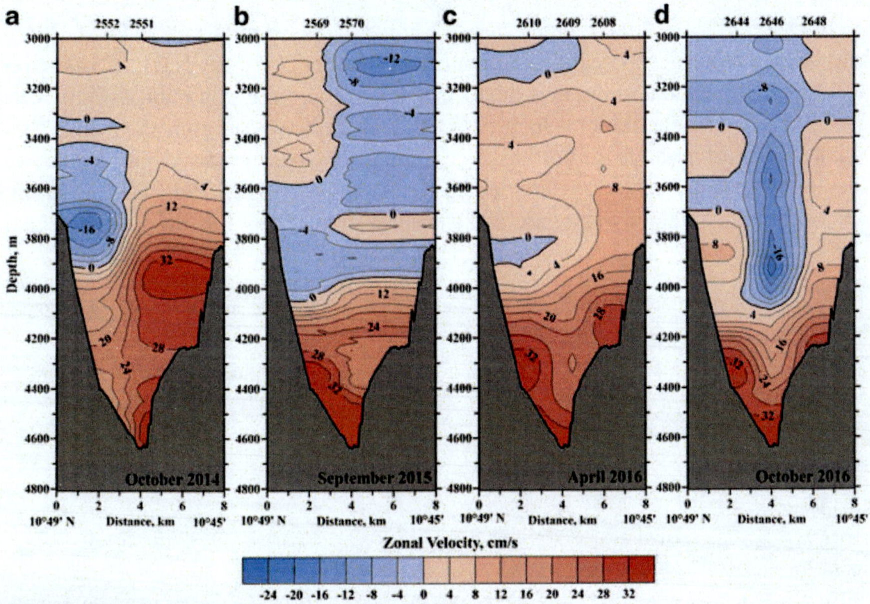

Fig. 6.15 Sections of the zonal velocity component across the central channel of the Vema Fracture Zone in 2014 (**a**); 2015 (**b**); April 2016 (**c**); and October 2016 (**d**). The position of stations is shown with tics at the upper axes. (Modified from (Morozov et al. 2018))

this station (2569), we did not lower the instrument close to the bottom because of the pinger signal failure. The minimum distance between the instrument and the bottom was 80 m. Maybe this was the cause why we did not find the cold low saline core over the northern slope (Figs. 6.13b and 6.14b). The core was deeper. However, the LADCP signal reached the bottom. The maximum velocity in 2015 up to 47 cm s^{-1} was found at the bottom where the fracture is narrow. In 2014, the maximum velocities were found in the core at a depth of 4000 m in a wider part of the fracture. In 2015, the coldest potential temperatures and low salinities were displaced to the northern wall, and the core of the maximum velocities was also displaced to the northern wall.

The transport in the central channel in 2015 was estimated at (0.45–0.58) ± 0.08 Sv, which is smaller than in 2014. A decrease in the transport despite increased maximum velocity can be explained by the fact that in 2014 the maximum velocity was found not immediately at the bottom but in the upper wider part of the fracture, whereas in 2015 the maximum velocities were at the bottom in a narrow part.

In April 2016, we made a section of three stations in the central channel: 2608–2610. The minimum potential temperature 1.358 °C was measured at station 2610 on the northern slope at a depth of 4510 m. The bottom salinity at this station was 34.818. These values of potential temperature and salinity were the lowest measured in all of the fractures in this region. At the deepest station 2609 (4659 m), the bottom potential temperature and salinity were higher than over the northern slope: 1.421 °C and 34.825 PSU.

In October 2016, three stations were occupied in the central channel: 2644, 2646, and 2648. The coldest potential temperature in October 2016 was 1.370 °C at station 2644 on the northern slope at a depth of 4486 m, and the minimum salinity was 34.820 PSU. The maximum velocities in October 2016 were as high as 40 cm s^{-1} at a distance of 50 m above the bottom. A countercurrent was found at depths shallower than 4100 m with westerly velocities up to 24 cm s^{-1} (Fig. 6.15d). Hence, strong variations in the velocity structure and transport through the fracture were revealed

Table 6.5 Minimum potential temperatures, and maximum velocities, and transport below potential isotherm 2.0 °C in the central channel of the Vema Fracture Zone

Time of measurements	Minimum θ, °C; salinity, PSU	Maximum velocity, cm/s	Transport, Sv
2014 October, st. 2552 over the northern slope	1.362; 34.823	24	0.86 ± 0.08
2014 October, st. 2551 over the southern slope	1.426; 34.830	34	
2015 October, st. 2569 over the northern slope	1.403; 34.820	46	0.50 ± 0.08
2016 April, st. 2610 over the northern slope	1.358; 34.818	37	0.70 ± 0.05
2016 October, st. 2644 over the northern slope	1.370; 34.820	38	0.52 ± 0.05

during three years of measurements from 2014 to 2016 (Fig. 6.15a and d). Table 6.5 presents the maximum velocities, minimum potential temperatures, and transports in the central channel of the Vema Fracture Zone. Generally, the eastward bottom current almost linearly increases from the depths of 3900–4100 m to the bottom.

One of the general features in the distribution of potential temperature and salinity in the fracture is the displacement of the potential temperature and salinity minima to the northern slope. We observed this phenomenon over four sections in 2014–2016 (Figs. 6.13 and 6.14). This effect was explained for the first time in (Johnson et al. 1976) in the Vema Channel located in the Southern Hemisphere, where the coldest water is displaced to the right side of a narrow channel despite the Coriolis forcing to the left. The presence of the bottom Ekman layer results in the secondary helical circulation relative to the general current, which displaces the cold water to the left in the Northern Hemisphere (Johnson et al. 1976).

The vertical gradients of potential temperature and salinity decrease in the layer close to a depth of 4400 m where the fracture sharply becomes narrower (Fig. 6.12). We attribute this to increased mixing in the bottom layer due to lateral friction in a narrow channel.

In 2014, one station (2555) was occupied in the shallower and narrower southern channel of the fracture at 41°05.5′ W. The width of the channel in this place at a depth of the 3780 m isobath (approximately the 2 °C isotherm) is 4 km; the depth of the fracture is 4500 m. The minimum potential temperature at the bottom was 1.469 °C, and the salinity was 34.835 PSU (Fig. 6.12c). The velocities of the eastern current were up to 26 cm s^{-1} (Fig. 6.12d). The AABW transport was estimated at 0.26 ± 0.03 Sv. Thus, the total AABW transport through the Vema Fracture Zone in 2014 was 1.20 ± 0.12 Sv. Such a transport estimate is almost two times greater than the one we estimated in 2006 (Morozov et al. 2010).

In 2015, we repeated the station in the southern channel (2571) and occupied one more station downstream to the east at a distance of 3.5 km believing that there was a spillway. In 2015, the minimum potential temperature at the sill (4552 m) of the southern channel (10°43.5′ N, 41°05.5′ W) was 1.534 °C, and the salinity was 34.838 PSU. The potential temperature and salinity at the station downstream (2572) at a depth of 4720 m were 1.627 °C and 34.848 PSU, respectively, which is higher than at the upper station on the sill.

In 2015, warming was observed in both channels compared to 2014: by 0.04 °C in the northern channel and by 0.15 °C in the southern channel. The velocities of the flows also increased compared with 2014: in the northern channel from 33 to 47 cm s^{-1} and from 29 to 40 cm s^{-1} in the southern channel. The transports in the southern channel in 2014 and 2015 were comparable: 0.20–0.30 $\pm$ 0.02 Sv.

An underwater spillway was found in the southern channel. The eastern station in 2015 (2572) in the lower part of the spillway was occupied in a narrow part of the channel. Over a distance of one kilometer, the bottom water rapidly descends from 4560 m to 4725 m (the slope is approximately 1/5). This flow in the spillway is concentrated in a narrow trough 1500 m wide. The velocity at the upper point at a depth of 4416 m was approximately 40 cm s^{-1}, but at the lower point the velocity decreased to 22 cm s^{-1}. It is worth noting that another velocity maximum (17 cm s^{-1}) was found at the deeper station at the level of the sill (4440 m). Sections

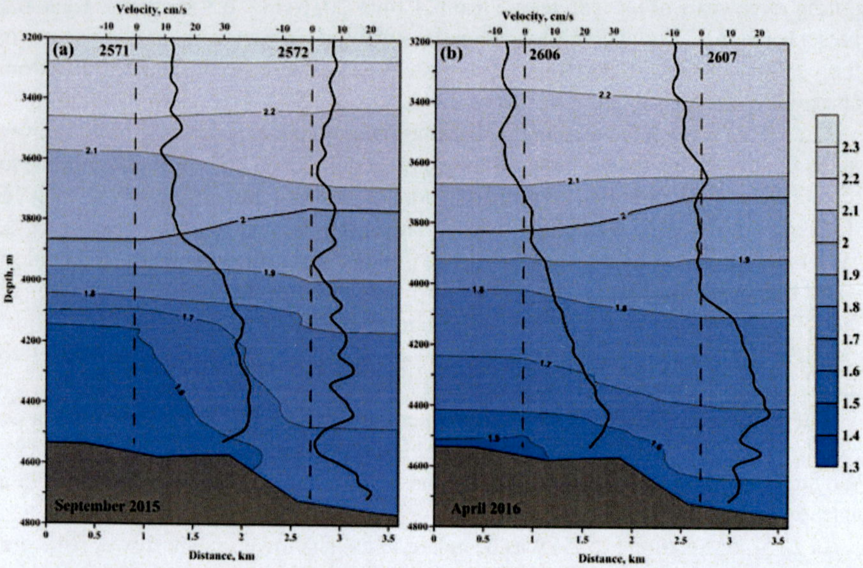

Fig. 6.16 Two sections of potential temperature and profiles of eastward velocity at the stations in the southern channel of the Vema Fracture Zone in the region of the underwater spillway: in 2015 (**a**), in 2016(**b**). Stations over the sill: 2571 in September 2015, and 2606 in April 2016. Stations downstream: 2572 and 2607. (Modified from (Morozov et al. 2018))

of potential temperature along the southern channel and profiles of current velocity based on the measurements in September 2015 and April 2016 are shown in Fig. 6.16a, b. Such vertical distribution of velocity with several maxima is characteristic of underwater spillways, which we found in the equatorial fracture zones and also in the laboratory experiments at the Institutes of Hydrodynamics and Oceanology, Russian Academy of Sciences (Morozov et al. 2012). The isotherms in the region of the spillway sharply deepen below the 1.8 °C potential isotherm.

The isotherms above the 1.9 °C remain at the same depth. A sharp potential temperature increase was observed in the course of the water flow over a distance of 4 km from 1.534 °C to 1.627 °C. Such an increase is observed when a gravity current of cold water descends down a steep slope with formation of intrusions and mixing. For example, a similar effect was found when AABW enters the Romanche Fracture Zone and descends after overflowing a sill (Morozov et al. 2010; Morozov et al. 2012; Tarakanov et al. 2013).

In April, 2016, two stations were occupied in the southern channel. One station (2606) was occupied over the sill at a depth of 4541 m, and the other (2607) downstream at a depth of 4719 m. The bottom potential temperature at station 2606 was 1.488 °C, the salinity was 34.834 PSU. The potential temperature at station 2607 downstream was warmer than at the sill (1.512 °C); the salinity was also slightly greater 34.836 PSU. The velocity at the upper station was 29 cm s^{-1}. Similarly to the observations in 2015, an underwater spillway was found

Table 6.6 Transport through the Vema Fracture Zone below $\theta = 2$ °C, Sv

	2006 (Morozov et al. 2010)	2014	2015	2016 April	2016 October
Vema northern channel		0.08 ± 0.01		0.20 ± 0.03	0.05 ± 0.01
Vema central channel	0.64 ± 0.20	0.86 ± 0.08	0.50 ± 0.05	0.70 ± 0.05	0.52 ± 0.04
Vema southern channel		0.26 ± 0.03	0.30 ± 0.03	0.10 ± 0.01	0.14 ± 0.02
Total	>0.64	1.20 ± 0.12	>0.80	1.00 ± 0.09	0.71 ± 0.08

downstream with three velocity maxima typical for a spillway: 21 cm s^{-1} at 4432 m; 19 cm s^{-1} at 4560 m, and 12 cm s^{-1} at 4648 m (Fig. 6.16).

In October 2016, only one station (2650) was occupied at the same point as station 2606 on the sill. The bottom potential temperature was 1.517 °C, the salinity was 34.838 PSU, and the velocity maximum at 4440 m was 23 cm s^{-1}. We observed not very strong fluctuations in potential temperature, salinity, and velocity each time of measurements. All measurements indicate that velocities in the spillway downstream are always lower than at the sill because the layer of the current becomes thicker.

We found that the structure of the bottom currents and AABW transport to the East Atlantic in the Vema Fracture Zone are strongly variable in time. Estimates of transports calculated on the basis of LADCP profiles in each series of measurements are given in Table 6.6.

The mean transport based on five series of measurements is close to 1 Sv. The major part of this transport occurs through the central channel, which is the deepest of the channels. The velocities of AABW flow in this channel sometimes exceed 40 cm s^{-1}. The transport of bottom water is not directly related to the maximum velocity. If the core of the maximum velocity is located close to the bottom in a narrow gorge of the fracture, the transport may become smaller than in the case when the maximum velocities are gained at a larger distance from the bottom where the fracture is wider.

Thus, the measurements of currents in the Vema Fracture Zone in 2006, 2014, 2015, and 2016 revealed a strong AABW flow with a transport of approximately 1.00 Sv (ranging from 0.7 to 1.2 Sv in different years), which exceeds the transports in any of the other fractures in the northern part of the Mid-Atlantic Ridge as was expected. A spillway with the descent of water by 170 m was found in the southern channel of the Vema Fracture Zone.

6.2 Other Fracture Zones of the Northern Mid-Atlantic Ridge

The goal of this section is to clarify the flow pattern and the volume transport of AABW through the other than Vema fracture zones in the northern part of the Mid-Atlantic Ridge based on the CTD and LADCP observations in 2014, 2015, and 2016. We started our research in the fractures of the Mid-Atlantic Ridge in 2006 in the Vema Fracture Zone (Morozov et al. 2010). In 2014–2016, we continued our measurements in the Vema Fracture Zone and widened the study region to the northern Kane and Marathon fractures and southern fractures: Doldrums, Vernadsky, Nameless (07°28′ N), Bogdanov, and Strakhov. The names of the fracture zones, in which we made measurements, were taken from (Hekinian 1982; Roest and Collette 1986; Müller and Roest 1992; Smoot 1989). The information about measurements is given in Table 6.4.

6.2.1 *Strakhov Fracture Zone (Four North Fracture Zone) (3°53′ N)*

The measurements in the Strakhov Fracture Zone were made in April 2016 at one CTD station (2595) at 03°52.5′ N, 32°53.4′ W (Fig. 6.17). The depth of the ocean at

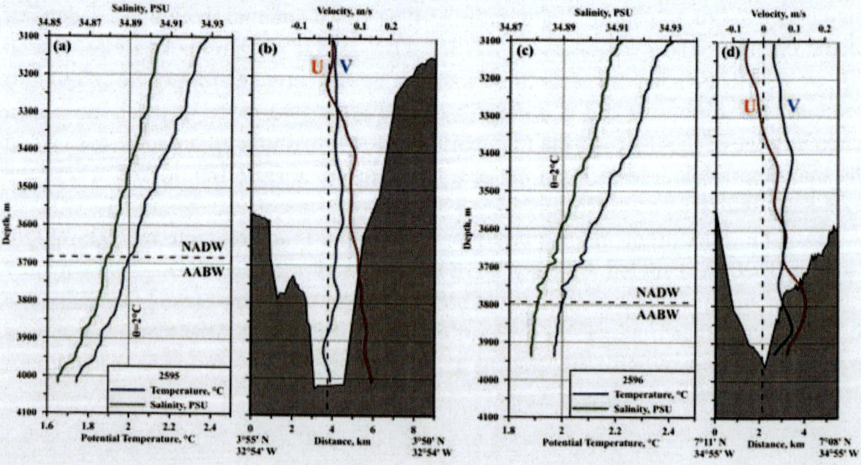

Fig. 6.17 Profiles of potential temperature, salinity, and velocity *U* (red) and *V* (blue) components in the Strakhov Fracture Zone (station 2595) (**a**, **b**) and Bogdanov Fracture Zone (station 2596) (**c**, **d**). The vertical lines at 2 °C (**a**, **c**) show the boundary of AABW with North Atlantic Deep Water (NADW) at 3680 m at station 2595 and at 3750 m at station 2596. The vertical dashed line in Fig. 6.17b, d indicates the zero velocity and also the location of the station. The bottom is shown with gray color. (Modified from (Morozov et al. 2018))

the place of measurements is 4023 m, the minimum measured potential temperature at the bottom was 1.742 °C, and the minimum salinity was 34.855 PSU (Fig. 6.17a).

It will be shown below that the minimum salinity of the water that flows through the Vema Fracture Zone is 34.818 PSU and the minimum potential temperature is 1.358 °C. Hence, the water that flows through the Strakhov Fracture Zone is neither the coldest nor the freshest. The width of the fracture at a depth of 3700 m (depth of the 2 °C isotherm) is 4.5 km, while at the bottom it is less than 2 km. The walls of the fracture are higher than this isotherm and keep the entire AABW flow below 2 °C in the fracture. Profiles of potential temperature, salinity, and velocity in the fracture are shown in Fig. 6.17a, b. The easterly flow occupies the bottom layer deeper than 3250 m. The maximum velocities reach 12–14 cm s^{-1} at the bottom. The AABW transport below 2 °C is 0.100 ± 0.010 Sv, while deeper than 1.9 °C the transport is 0.040 ± 0.005 Sv. The channel becomes narrower at a depth of 3750 m (1.95 °C isotherm). We do not know about any previous CTD/LADCP measurements in this fracture as in many others.

6.2.2 *Bogdanov Fracture Zone (7°10′ N)*

The Bogdanov Fracture Zone is the second fracture from the south, in which we made measurements in 2015 and repeated in 2016. The width of the fracture at a depth of 3800 m (depth of the 2 °C isotherm) is only 2 km. The walls of the fracture are higher than this isotherm. In 2016, the minimum measured potential temperature was 1.992°C over a depth of 3820 m, and the minimum salinity was 34.880 PSU (station 2577). However, the station was occupied not at the deepest place of the fracture, which is 3960 m. In 2015, the measured current with a velocity of 8 cm s^{-1} was directed to the east. The AABW transport below the 2 °C isotherm based on the measurements at one station in 2015 was 0.020 ± 0.005 Sv.

When repeating the station in 2016 (station 2596) we slightly displaced the measurements to reach the deepest place and occupied a station at a depth of 3934 m, 6 m above the bottom (see Table 6.4). The potential temperature at the bottom in 2016 was 1.971 °C, and the minimum salinity was 34.878 PSU. The profiles are shown in Fig. 6.17c. Profiles of potential temperature and salinity are also very similar here. The velocities of the bottom water flow (Fig. 6.17d) were up to 15 cm s^{-1}, which results in a transport of 0.040 ± 0.005 Sv. The easterly transport was observed below 3350 m. The maximum velocities were at 3800 m.

6.2.3 *Nameless Fracture Zone (7°28′ N)*

In 2014, one CTD/LADCP station (2558) was occupied in the Nameless Fracture Zone (7°28′ N). We made a cast in the western part of the fracture at 07°28.7′ N, 36°59.1′ W. Detailed bottom topography in this region has been poorly studied. A

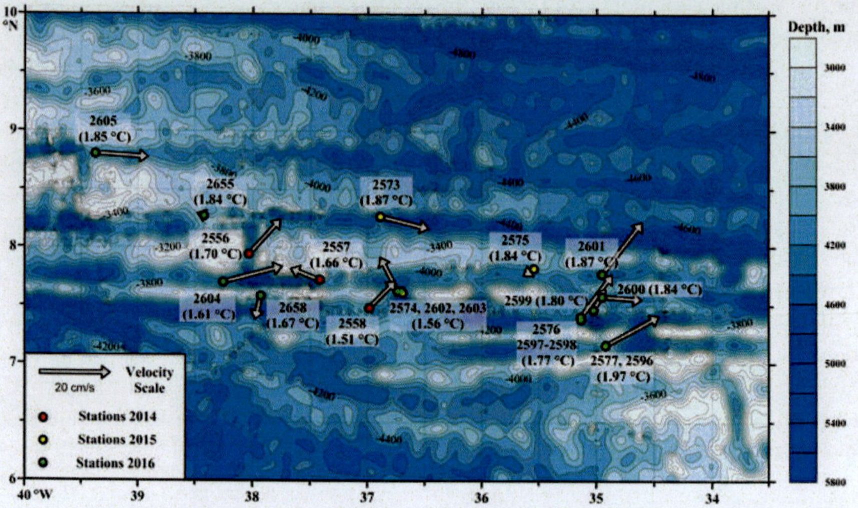

Fig. 6.18 Map of the southern fracture zones. Red, yellow, and green circles indicate our stations in 2014, 2015, and 2016, respectively. Gray arrows indicate the vectors of current velocities at the bottom. The bottom potential temperatures (°C) at each station are written in parentheses. The bottom topography is based on (Smith and Sandwell 1997). (Modified from (Morozov et al. 2018))

chart of the topography is shown below (Fig. 6.18). The topography is complicated with channels connecting the fracture zones. The fracture valley is formed by two channels divided by a high median ridge whose crest is located shallower than 3500 m. The station was occupied in the deeper northern channel, whose width at a level of the 2 °C isotherm (3700 m) is 20 km, but the fracture narrows with depth, and at a depth of 4000 m its width is only 8 km. The minimum potential temperature at the bottom (4650 m) was 1.509 °C, while the salinity was 34.838 PSU (Fig. 6.19a). The eastward velocities 3–9 cm s^{-1} were measured in the entire AABW layer to the bottom below 3750 m (Fig. 6.19b). The transport to the east was 0.170 ± 0.030 Sv.

In 2015, station 2576 was occupied in the eastern part of the Nameless Fracture Zone at 07°23.1′ N, 35°08.0′ W (Figs. 6.3, 6.18, and 6.19c, d). The eastern part of the fracture similarly to its western part consists of two channels divided by a high median ridge with the depth of this crest shallower than 3600 m. The station was occupied in the southern channel, which is approximately 9 km wide at a depth of 3800 m (the depth of 2.0 °C isotherm). The eastward velocities in the entire AABW layer were 2–3 cm s^{-1}. The minimum potential temperature at the bottom was 1.759 °C over a depth of 4529 m. The salinity at the bottom was 34.859 PSU. The transport below the 2 °C isotherm was 0.090 ± 0.010 Sv. Our echo-sounder measurements showed that the depth over the northern wall of the channel is close to 3600 m; the width of the northern channel is 6 km (at a depth of 3800 m), and the depth is approximately the same as in the southern channel.

Antarctic Bottom Water flows to the Vernadsky Fracture Zone from the south as a northward current from the Nameless Fracture Zone through a gap 11 km wide at a

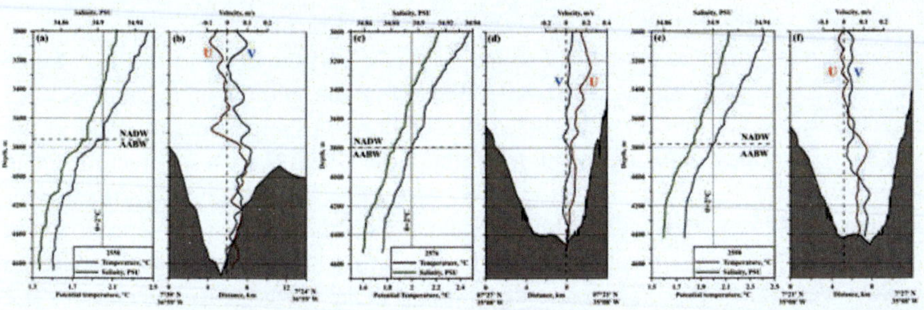

Fig. 6.19 Profiles of potential temperature, salinity, and velocity *U* (red) and *V* (blue) components in the Nameless Fracture Zone. The vertical lines at 2 °C (**a**, **c**) show the boundary of AABW with North Atlantic Deep Water (NADW): at station 2558 (**a**, **b**), the boundary of AABW is at 3750 m; at station 2576 (**c**, **d**), the boundary of AABW is at 3800 m; and at station 2598 (**e**, **f**), the boundary of AABW is at 3780 m. The vertical dashed line in the figures indicates the zero velocity and also the location of the station. The bottom is shown with gray color. (Modified from (Morozov et al. 2018))

level of 3900 m. A CTD/LADCP station (2574) was occupied in 2015 in this gap at 07°36.5′ N, 36°41.6′ W. The minimum potential temperature at the bottom was 1.572 °C, and the salinity was 34.839 PSU. Below, we shall see that this potential temperature is lower than the potential temperature at the two stations in the Vernadsky Fracture Zone east and west of the gap, but it is greater than at the station in the western part of the Nameless Fracture Zone. A decrease in the potential temperature is in some sense a signature of the AABW pathway if this decrease agrees with the measured direction of the currents. The velocities were not high and did not exceed 3 cm s^{-1}. Since the gap is quite wide, we are not sure that this station was made in the core of the flow with the maximum velocities.

In 2016, we made four CTD/LADCP casts in the eastern part of the Nameless Fracture Zone. Station 2597 was occupied at 07°23.0′ N, 35°08.0′ W. This station repeated station 2576 in 2015. The depth at the station is 4530 m and the potential temperature at the bottom was 1.767 °C (warmer by 0.08 °C than in 2015), while the salinity was the same 34.860 PSU. The velocities of the bottom current in 2016 were 10 cm s^{-1}, which is greater than in 2015 (2–3 cm s^{-1}). Another station (2598) was occupied here in 2016 (07°24.5′ N, 35°08.0′ W) close to station 2597. Hence, these two stations formed a transversal section across the fracture. The bottom potential temperature at this station was the same (1.766 °C) but the depth was 100 m shallower (4429 m); the salinity was also the same, 34.860 PSU. Profiles of potential temperature, salinity, and velocity at station 2598 are shown in Fig. 6.19e, f. The heights of the fracture walls are higher than the 2 °C potential isotherm. The maximum bottom velocities (13 cm s^{-1}) were recorded at the northern station (2598), while at the southern station (2597) the bottom velocities were lower (5.5 cm s^{-1}), which indicates that the core of the maximum velocities is displaced to the north or to the left relative to the transport direction. At both stations, the bottom velocities were directed to the northeast. The transport estimated from the data of these two stations was 0.230 ± 0.015 Sv.

In this eastern region (35° W), the Nameless Fracture Zone splits into two channels. One channel is directed to the northeast and the other to the east. Since the velocities at two stations (2597 and 2598) before the splitting of the channel were directed to the northeast, we occupied two stations in the channel directed to the northeast to study the continuation of the flow. The Nameless Fracture Zone becomes wider to the east of stations 2597 and 2598. Station 2599 was made over a depth of 4375 m. The width of the channel at a depth of 3800 m is approximately 5.5 km. The potential temperature at the bottom was 1.800 °C, and the salinity at the bottom was 34.864 PSU. The velocities directed to the northeast were 8 cm s^{-1}. The walls keep the flow below 1.9 °C in the channel, but the upper part of the flow spreads wider than the channel. Station 2600 was located downstream at 07°34.8′ N, 34°57.0′ W at a depth of 4222 m. Here, the width of the channel is approximately 9 km. It is the place, in which the Nameless Fracture Zone merges with the Vernadsky Fracture Zone. The potential temperature at the bottom was higher than upstream and was equal to 1.840 °C, and the salinity was 34.867 PSU. The eastward velocities at the bottom were 2 cm s^{-1}. The velocities at a depth of 3870 m reached 12 cm s^{-1}. The transport from the Nameless to Vernadsky fracture was estimated at 0.140 ± 0.015 Sv. We previously mentioned that the northeastern bottom current measured at stations 2597–2598 (0.230 Sv) splits in two. Part of the flow (0.090 ± 0.006 Sv) transports bottom water to the east and the other part (0.140 ± 0.009 Sv) is transported to the Vernadsky Fracture Zone.

In October 2016, station 2658 was occupied in a gap between the Vernadsky and Nameless fractures at 07°35.0′ N, 37°55.7′ W. The depth at the station was 4083 m; the minimum potential temperature was 1.672 °C; the salinity was 34.852 PSU. The velocities within 5–6 cm s^{-1} with variable directions in the bottom layer indicate that in this region there was no significant transport through this gap.

6.2.4 Vernadsky Fracture Zone (7°49′ N)

A chart of bottom topography based on a multibeam survey in the region of the Vernadsky Fracture Zone is shown in Fig. 6.20.

Several CTD-casts were made in the Vernadsky Fracture Zone. One station (2575) in 2015 was in the eastern part and another (2557) in 2014 in the western part. The stations were located at both sides of a gap in the southern wall of this fracture that allows the flow of AABW from the Nameless Fracture Zone. Station 2575 in 2015 was located at 07°49.2′ N, 35°32.5′ W. Here, the fracture consists of two quite wide zonal channels divided by a median ridge. The station was occupied in the northern channel over a depth of 4072 m; the bottom potential temperature was 1.836 °C and the salinity was 34.865 PSU. The current was negligible: the velocities directed to the southwest were 2–3 cm s^{-1}. Since there are no measurements in the southern channel it is difficult to judge about the full transport of abyssal waters through the Vernadsky Fracture Zone in this region.

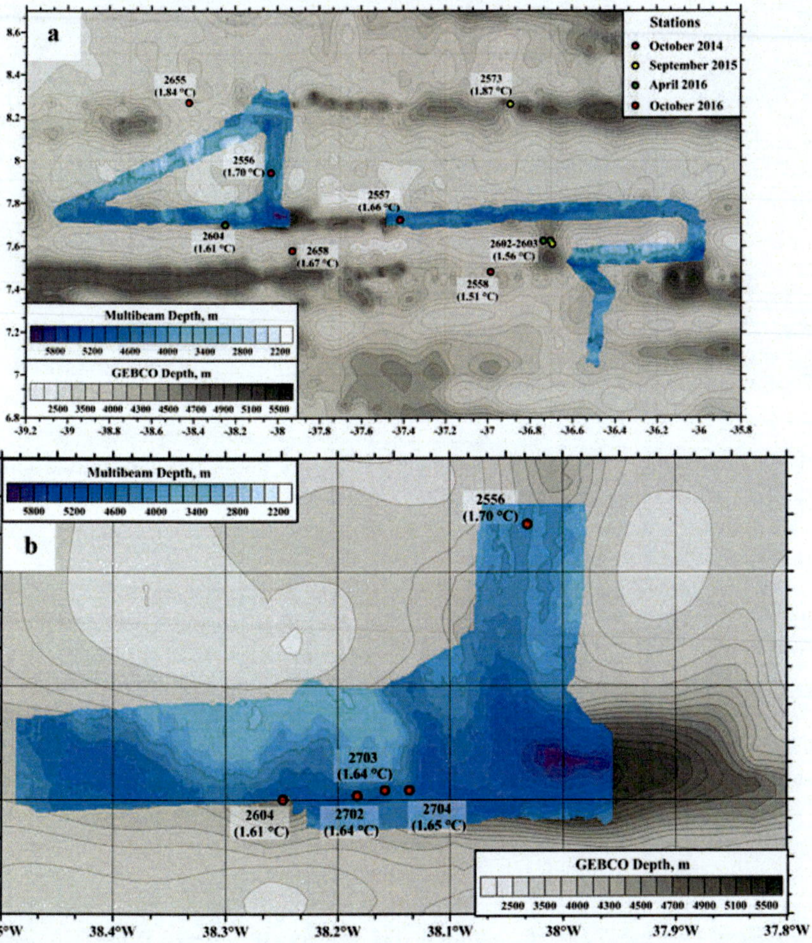

Fig. 6.20 Bottom topography in the region of the Vernadsky Fracture Zone based on a multi-beam survey from the R/V *Akademik Nikolaj Strakhov* (Courtesy of M. Kapustina, I. Dudkov, and L. Bashirova). A larger region is shown in Fig. 6.20a, and fragment of the chart on a larger scale (near the rift valley between the Vernadsky and Doldrums fracture zones) is shown in Fig. 6.20b. Stations with CTD-casts are shown

In 2014, station 2557 was occupied in the western part of the Vernadsky Fracture Zone (07°43.3′ N, 37°25.0′ W) over a depth of 4628 m. The minimum potential temperature in the western part was 1.661 °C, and the salinity was 34.854 PSU (Fig. 6.21a). The width of the channel at a level of 3700 m is 9 km. The tops of the walls are at 3000 m. The bottom current with velocities up to 9 cm s^{-1} was directed to the west (negative zonal velocity at the bottom) heading to the rift zone connecting the Vernadsky and Doldrums fracture zones (Fig. 6.21b). The AABW transport below 2 °C was estimated at 0.120 ± 0.010 Sv. The bottom water seems to flow from

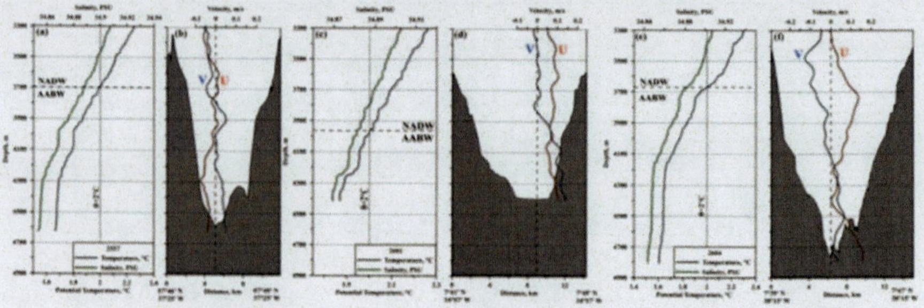

Fig. 6.21 Same as in Fig. 6.19, but for the Vernadsky Fracture Zone: station 2557 (**a**, **b**), the boundary of AABW is at 3690 m; station 2601 (**c**, **d**), the boundary of AABW is at 3960 m; station 2604 (**e**, **f**), the boundary of AABW is at 3680 m. (Modified from (Morozov et al. 2018))

the Nameless Fracture Zone through a gap to the Vernadsky Fracture Zone and then the flow splits into two. Part of the flow directs to the north through the rift zone to the Doldrums Fracture Zone and the other slowly heads to the East Atlantic along the fracture.

In 2016, several stations were occupied in the Vernadsky Fracture Zone. Station 2601 was in the eastern part of the fracture close to the East Atlantic basin (07°46.4′ N, 34°57.1′ W) at a depth of 4415 m (Figs. 6.18 and 6.21c, d). The potential temperature at the bottom was 1.870 °C, and the salinity was 34.870 PSU. The width at a depth of 3800 m (depth of the 2 °C isotherm) is approximately 9 km. The bottom velocities directed to the northeast were relatively high (17 cm s^{-1}). This station was located east of the gap, through which bottom water can flow from the Nameless Fracture Zone to the Vernadsky Fracture Zone (in the previous section, we estimated the transport at 0.140 ± 0.015 Sv). The eastward transport at station 2601 was estimated at 0.120 ± 0.012 Sv. Hence, it is likely that all the bottom water transported from the Vernadsky Fracture Zone to the east has the origin from the Nameless Fracture Zone. In addition, the previously discussed measurements at station 2575 show almost zero transport west of the inflow of AABW from the Nameless Fracture Zone.

Stations 2602 (07°37.5′ N, 36°44.2′ W, depth 4624 m, bottom potential temperature 1.565 °C, salinity 34.839 PSU) and 2603 (7°37.5′ N, 36°42.2′ W, depth 4631 m, bottom potential temperature 1.560 °C, salinity 34.839 PSU) were occupied in a wide gap (17 km wide at a depth of 3700 m) between the Nameless and Vernadsky fracture zones. In 2015, station 2574 was occupied in this gap in the deepest place (Fig. 6.18). The bottom potential temperature was 1.572 °C, and the salinity was 34.839 PSU at a depth of 4830 m. Bottom velocities at station 2602 below 4500 m were 10 cm s^{-1} directed to the north; velocities at station 2603 were close to zero. In 2015, weak currents were directed to the northeast.

Station 2604 was occupied close to the southern end of the rift valley that connects this fracture with the Doldrums Fracture Zone (07°42.0′ N, 38°15.0′ W). The depth at this station is 4810 m, the bottom potential temperature was 1.607 °C, and the salinity was 34.844 PSU. The width of the fracture at a depth of 3680 m

($\theta = 2$ °C isotherm) is approximately 13 km. Profiles of potential temperature, salinity, and velocity are shown in Fig. 6.21e, f. The eastward current was recorded below 3300 m. A maximum of the eastern velocity component (20 cm s^{-1}) was at 3760 m; a decrease in velocity was recorded between 4200–4400 m. The bottom velocities measured at this station were as high as 16 cm s^{-1}. We think that such a high-velocity jet does not occupy the entire width of the fracture but it is confined to its wall. Unfortunately, we made only one station here. A few more stations are needed in a channel 13 km wide. The vertical structure of currents with two cores corresponds to the structure at the bottom of underwater spillways (Morozov et al. 2010; Morozov et al. 2012).

6.2.5 *Doldrums Fracture Zone (8° N) and a Rift Valley South of It*

In 2014, we occupied a CTD/LADCP station (2556) at a depth of 4330 m in a rift valley connecting the Vernadsky and Doldrums fracture zones at 07°56.5′ N, 38°01.9′ W. The valley has almost a meridional orientation (Fig. 6.20b). The width of the rift valley at a level of the 3750 m isobath is 7 km. The minimum potential temperature in the valley was 1.701 °C, and the salinity was 34.859 PSU (Fig. 6.22a). The velocities of the northward current of AABW were up to 15 cm s^{-1} (Fig. 6.22b). The transport below the $\theta = 2$ °C isotherm was estimated at 0.280 ± 0.030 Sv, which indicates that the bottom water propagates to the north from the Vernadsky Fracture Zone to the Doldrums Fracture Zone. Hence, after the bottom water reaches the Doldrums Fracture Zone it should turn to the East Atlantic

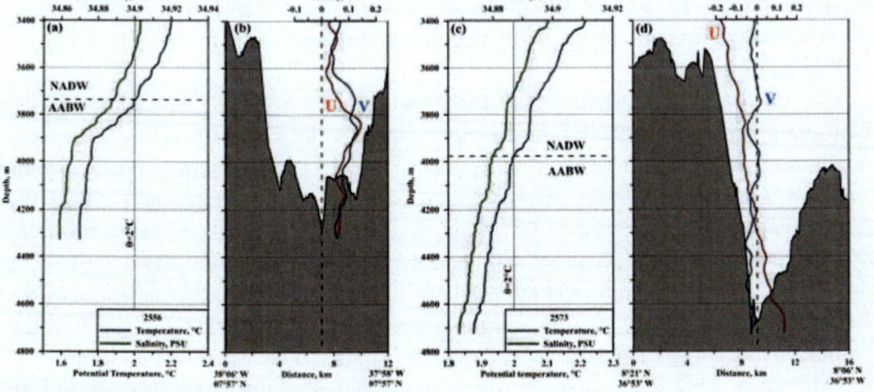

Fig. 6.22 Profiles of potential temperature, salinity, and velocity *U* (red) and *V* (blue) components in the Doldrums Fracture Zone: station 2556 (**a**, **b**), the boundary of AABW is at 3740 m; station 2573 (**c**, **d**), the boundary of AABW is at 3980 m. (Modified from (Morozov et al. 2018))

because the sills in the western part of the fracture (west of the rift zone) are shallower than in the eastern part of the Doldrums Fracture Zone.

In 2015, we occupied a station in the eastern part of the fracture at 08°15.8′ N, 36°53.4′ W (station 2573). The fracture valley at this section is very wide (20 km at a depth of 3900 m), it forms two channels divided by a median ridge. The station was located in the northern (deeper) channel. The minimum potential temperature at this station was 1.872 °C, which is higher than the potential temperature of the bottom water flowing to the fracture from the south through the rift zone. The salinity at the bottom was 34.868 PSU, which is higher than in the rift valley. Profiles of potential temperature, salinity, and velocity are shown in Fig. 6.22c, d. The eastward current occupies a layer below 4300 m, which means that the upper layer of AABW is not transported here to the east. The eastward velocities reach a maximum of 14 cm s^{-1} at 4700 m (Fig. 6.22d). The depth of the northern channel of the Doldrums Fracture Zone at the location of the station reached by the CTD probe was 4856 m, but the depth measured by the echo-sounder before the beginning of the CTD/LADCP profile was only 4700 m. Most likely, the instrument was displaced down to a greater depth by a possible flow descending over the bottom slope. This is also confirmed by the fact that the maximum velocity was found not at the bottom but at 4700 m at the level of the sill, which is usually found in the spillways in the Romanche Fracture Zone at the equator and other regions (Morozov et al. 2010; Morozov et al. 2012). These are strong and relatively narrow bottom gravity currents descending over underwater slopes. The transport of AABW below $\theta = 2$ °C isotherm was estimated at 0.045 ± 0.005 Sv. We did not occupy a station in the southern (shallower) channel. The depth there is 4320 m.

The transport of bottom water strongly decreases as the bottom water flows from the rift valley at 38° W to the eastern part of the fracture zone where we occupied station 2573 (36°53.4′ W). Potential temperature increased from 1.701 °C to 1.872 °C. The main cause of this warming is mixing with the overlying water. In addition, the bottom water can be transported to the north and south through numerous small passages in the walls of the Doldrums Fracture Zone and through the shallower southern channel, in which we did not occupy a station south of station 2573.

Our measurements clarified that the flow from the rift zone that connects the Doldrums and Vernadsky fracture zones enters the Doldrums Fracture Zone and continues to the east. In order to make clear whether the flow along the Doldrums Fracture Zone is continuous from west to east we occupied station 2655 in the fracture west of the rift zone at 08°16′ N, 38°25′ W. The depth at this station was 4038 m; the minimum potential temperature was 1.837 °C, and the salinity was 34.865 PSU. The potential temperature was higher than in the rift zone south of the fracture and the salinity was also higher. Velocity measurements showed that the velocities did not exceed 1–2 cm s^{-1}, which is close to the error of measurements; the velocities were even directed to the west. Hence, there is no throughflow in the Doldrums Fracture Zone from west to east because the sill at 38°25′ W blocks the transport. The bottom water is transported to the Doldrums Fracture Zone from the

Vernadsky Fracture Zone through the rift valley and then the current turns to the east transporting water to the Atlantic.

6.2.6 *Arkhangelsky Fracture Zone (9° N)*

In 2016, station 2605 was occupied in the Arkhangelsky Fracture Zone at a depth of 3968 m. The potential temperature at the bottom was 1.849 °C, and the salinity was 34.865 PSU; the bottom velocities directed to the east exceeded 10 cm s^{-1}. A maximum of 15 cm s^{-1} was at a depth of 3700 m. Profiles of potential temperature, salinity, and velocity are shown in Fig. 6.23a, b. The upper boundary of AABW was at 3650 m. A thermocline and a halocline (hence, also a pycnocline) are just above this boundary. The width of the channel at a depth of 3750 m (1.9 °C) is 8 km. The walls of the fracture keep the bottom water flow below 1.9 °C isotherm in the channel. The flow above 1.9 °C isotherm spreads wider. The transport below 1.9 °C isotherm was 0.088 ± 0.009 Sv. The eastward transport was found below 3580 m, which is above the upper boundary of AABW; hence, part of North Atlantic Deep Water is involved in the eastward current.

6.2.7 *Ten Degree Fracture Zone (9°57′ N)*

The Ten Degree Fracture Zone is located south of the Vema Fracture Zone. The fracture is shallow. Station 2652 was occupied here at 09°57.0′ N, 40°27.0′ W at a

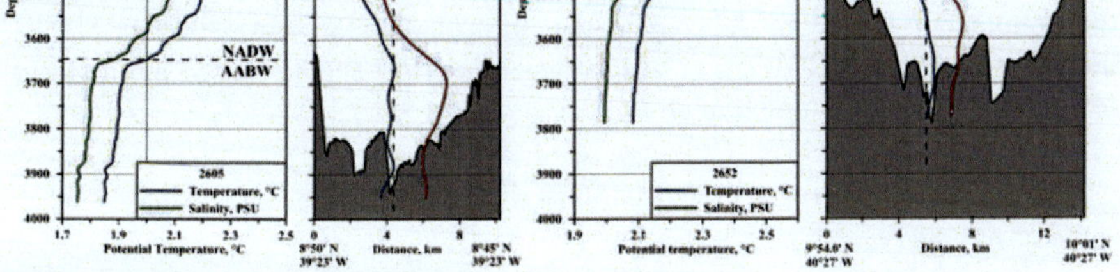

Fig. 6.23 Profiles of potential temperature, salinity, and velocity *U* (red) and *V* (blue) components in the Arkhangelsky Fracture Zone: station 2605 (**a, b**), the boundary of AABW is at 3640 m; and the Ten Degree Fracture Zone: station 2652 (**c, d**). (Modified from (Morozov et al. 2018))

depth of 3790 m. The minimum potential temperature was 2.082 °C, and the salinity was 34.888 PSU. Profiles of potential temperature, salinity, and velocity are shown in Fig. 6.23c, d. The velocities were directed to the east (11 cm s^{-1} at a depth of 3560 m). Since the potential temperature in the bottom layer is warmer than 2 °C, this fracture is not a pathway for AABW to the east, but warmer water in the lower layer 800 m thick is transported to the east with velocities within 5–10 cm s^{-1}.

6.2.8 *Rift Valley South of the Vema Fracture Zone (10°21′ N)*

One station (2651) was occupied in the deepest place (3836 m) of the rift valley south of the Vema Fracture Zone at 10°21.0′ N, 40°54.5′ W. The minimum potential temperature here was 2.014 °C and the salinity was 34.883 PSU. The properties of the seawater at the bottom indicate that AABW does not spread into the rift valley or it is strongly mixed there with the water from the overlying layer; the bottom velocities did not exceed 5 cm s^{-1}, which indicates that there is no strong transport through this rift valley, which is filled with stagnant water.

6.2.9 *Marathon Fracture Zone (12°40′ N)*

The Marathon Fracture Zone is very narrow (Figs. 6.24a, 6.25a, b). Its width at a level of 3700 m is 4.5 km. We occupied only one CTD/LADCP station there (2549) at 12°37.4′ N, 44°18.0′ W over a depth of 4230 m. The minimum recorded potential temperature was 1.920 °C and the salinity was 34.879 PSU, which is almost the limit for AABW (2 °C). The depth of the $\theta = 2$ °C isotherm was 3700 m, which is deeper than the crests of the fracture walls that are in the range 2500–3000 m. Profiles of potential temperature, salinity, and velocity components are shown in Fig. 6.25a, b. A thick layer more than 200 m thick of almost homogeneous water with a potential temperature of 2 °C exists here between 3800 and 4000 m.

A sharp vertical gradient of potential temperature and salinity exists at a depth of 4050 m where the fracture becomes very narrow (less than 1 km). It is surprising that the bottom currents were directed to the north across the narrow fracture. The northward velocity was 10 cm s^{-1}. However, this effect can be caused by the local features of the bottom topography.

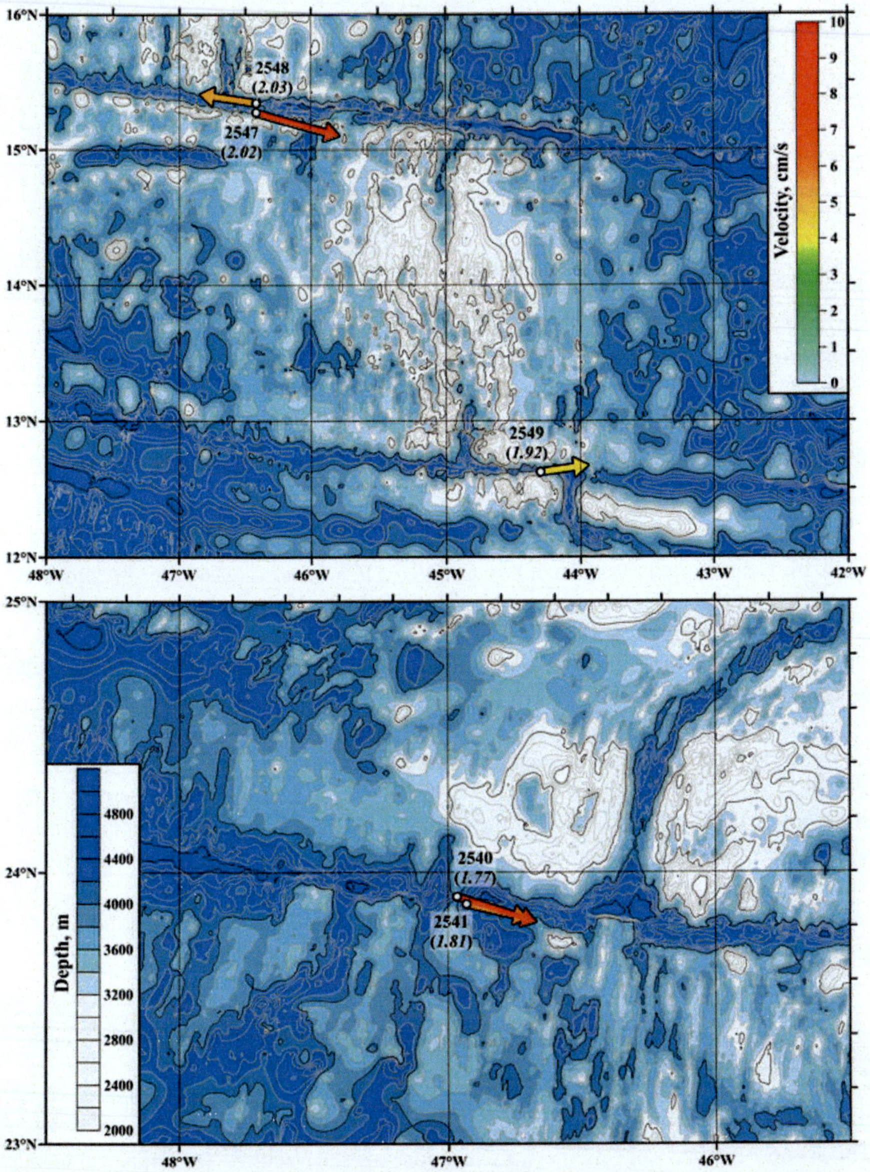

Fig. 6.24 Map of the Fifteen Twenty (Cabo Verde) and Marathon fracture zones (**a**). Map of the Kane Fracture Zone (**b**). Locations of our stations in 2014 and 2015 are shown with the circles; the arrows at the stations indicate the vectors of current velocities at the bottom. The numbers of stations according to Table 6.4 are indicated. The bottom potential temperatures at each station (°C) are written in parentheses. The bottom topography is based on (Smith and Sandwell 1997). (Modified from (Morozov et al. 2018))

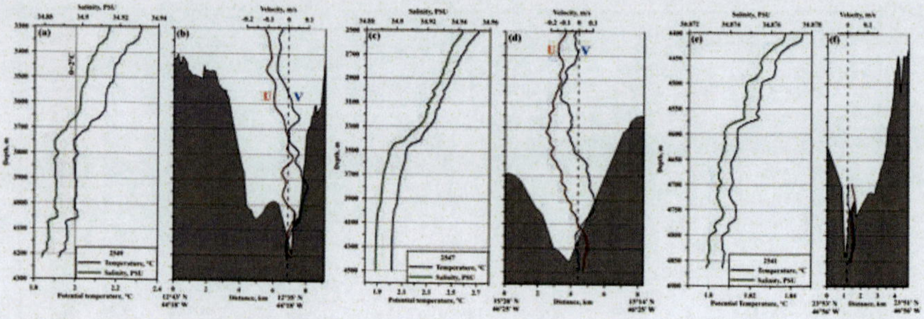

Fig. 6.25 Profiles of potential temperature, salinity, and velocity *U* (red) and *V* (blue) components in the Marathon Fracture Zone: station 2549; Fifteen Twenty (Cabo Verde) Fracture Zone: station 2547; and Kane Fracture Zone: station 2541. (Modified from (Morozov et al. 2018))

6.2.10 Fifteen Twenty Fracture Zone (Cabo Verde Fracture Zone) (15°16′ N)

In 2014, we occupied a section of two CTD/LADCP stations (2547, 2548) across the fracture near the sill located at 15°16.3′ N, 46°25.0′ W (Figs. 6.24a and 6.25c, d). The depth of the ridge crest along the walls of the fracture is 3400 m. The width of the fracture at the depth of 3400 m isobath is 16 km. A median ridge divides the fracture into two channels. The crest of the median ridge is at 3800 m. The southern channel is deeper; its depth is 4500 m compared to the northern channel with a depth of 4250 m. Profiles of potential temperature, salinity, and velocity are shown in Fig. 6.25c, d. The minimum potential temperature measured at the bottom of the southern channel was 2.018 °C, while in the northern channel the potential temperature was 2.030 °C. Bottom salinities at the two stations were 34.887. The current velocity in the southern channel below 4100 m was 7–8 cm s^{-1} to the east. Thus, we conclude that no transport of AABW occurs in this fracture because the bottom water potential temperature exceeds 2 °C.

6.2.11 Kane Fracture Zone (24° N)

The Kane Fracture Zone is the northernmost of the fractures we analyzed in our research in 2014–2016. We occupied two CTD/LADCP stations (2540, 2541) close to one of the sills of the fracture in its western part (23°54.6′ N, 46°58.0′ W) (Fig. 6.24b). The Kane Fracture Zone is more than 1000 km long and can be traced from 52° W to 40° W. In the region of our stations the depth of the sill is 4350 m and the upper line of its walls is at 3850 m. The width of the fracture at the level of 3850 m isobath exceeds 7 km. We measured the minimum potential temperature equal to 1.768 °C, and the minimum salinity was 34.870 PSU. Based on the LADCP measurements the maximum eastward current velocity of the AABW flow was

12 cm s^{-1}. The concentration of suspension that reflects ADCP beams in the bottom water in this region is very low; therefore, our velocity estimates are based only on the bottom track data. Batteries of higher voltage are needed for the LADCP here. At the time of the experiment, we had only the standard batteries. Profiles of potential temperature, salinity, and velocity components (bottom track) are shown in Fig. 6.25e, f. The velocity measurements above the 4250 m are not reliable and we cannot estimate the transport in the entire section of the channel.

6.2.12 Pathways and AABW Transport Through the Northern Part of the MAR

A general scheme of the bottom water transport through the fractures in the southern part of the ridge is shown in Fig. 6.18. At latitudes of 7–8° N, the bottom water first flows to the east through the Nameless Fracture Zone (7°28′ N); then the flow partly turns to the north along a gap at 36°42′ W and splits after passing this short wide passage. The flow turns to the west and to the east in the Vernadsky Fracture Zone (7°49′ N). The westerly flow possibly merges with the flow from the West Atlantic along the Vernadsky Fracture Zone and then turns to the north along the rift valley at 38°02′ W, and eventually flows to the East Atlantic through the Doldrums Fracture Zone (8°15′ N). One more flow can propagate from the West Atlantic along the entire Nameless Fracture Zone (7°28′ N), which splits into two parts directed to the Vernadsky Fracture Zone and to the east. The bottom water flowing along the Bogdanov Fracture Zone (7°10′ N) possibly merges with the above-mentioned flow through the rift valley that connects this fracture with the Nameless Fracture Zone (7°28′ N). The direct pathway from west to east along the entire Doldrums Fracture Zone (8°15′ N) seemingly does not exist. Thus, the AABW flows from the Doldrums (8°15′ N), Vernadsky (7°49′ N), and Nameless (7°28′ N) fracture zones most likely merge with the bottom water from the Vema (10°48′ N) and Arkhangelsky (8°48′ N) fracture zones and fill the abyssal depths of the Northeast Atlantic.

The transport through many smaller and shallower fracture zones differs from one fracture to another. There is no doubt that the coldest water flows through the Vema Fracture Zone. Among the others, excluding the Vema Fracture Zone, the Strakhov Fracture Zone is the pathway for the coldest water, because it is the southernmost fracture among the fractures with significant transport. The minimum measured potential temperature in the Strakhov Fracture Zone was 1.742 °C, while in the Vema Fracture Zone it was 1.358 °C. There are sills in the Strakhov Fracture Zone that prevent the flow of the coldest waters to the east. The minimum salinity in the Vema Fracture Zone is 34.818 PSU. Salinity gradually increases as AABW flows through the fractures. The fractures, in which we recorded significant transports, are: Strakhov, Bogdanov, Nameless (07°28′ N), Vernadsky, Doldrums, and Arkhangelsky. The transports and the minimum potential temperatures and salinities measured in the fractures are given in Table 6.7. We have to clarify that the total

Table 6.7 Transport through different fracture zones of the North Mid-Atlantic Ridge below $\theta = 2.0$ °C and the minimum potential temperature measured in these fractures

Fracture zone	Stations	Transport below the 2 °C isotherm, Sv	Minimum θ, °C and salinities, PSU at the easternmost point
Strakhov	2595	0.100 ± 0.010	1.742 °C; 34.855
Bogdanov	2596	0.040 ± 0.005	1.971 °C; 34.878
Nameless	2597–2598	0.090 ± 0.006	1.767 °C; 34.860
Vernadsky	2601	0.120 ± 0.012	1.870 °C; 34.870
Doldrums	2573	0.045 ± 0.005	1.872 °C; 34.868
Arkhangelsky	2605	0.090 ± 0.009	1.849 °C; 34.865
Sum of transports		0.485 ± 0.047	

northeastern transport from the section in the Nameless Fracture Zone (stations 2597–2598), which is 0.23 ± 0.015 Sv, splits into two parts: 0.090 ± 0.006 Sv of AABW is transported to the east and 0.140 ± 0.009 Sv goes to the Vernadsky Fracture Zone, of which 0.120 Sv is shown in Table 6.7.

The transports through the fracture zones are variable; however, their sum is not negligible and we can conclude that approximately a half of the volume of Antarctic Bottom Water is transported through many fractures in the Mid-Atlantic Ridge compared to the Vema Fracture Zone. Nevertheless, the densest and coldest water of low salinity flows through the central channel of the Vema Fracture Zone.

Individual measurements of velocities do not reveal variations of different scales. Repeated measurements with further averaging can result in the estimates of the mean flows. In our expeditions, we did our best to repeat the measurements in the Vema Fracture Zone, where the strongest transport exists. We also repeated some casts in the other fractures. In addition, we covered a large area with fractures to obtain a spatial pattern of AABW transport through the tropical part of the Mid-Atlantic Ridge. In our previous work in the Kane Gap (9° N, 19° W), we made five sections across the gap in different years and deployed a mooring with current meters (Morozov et al. 2013). A similar work was made together with our German colleagues in the Vema Channel. These results were summarized in (Morozov et al. 2010). We made the first field measurements in the channels, where no measurements were previously made to reveal the fractures most important for the bottom water exchange between the East and West Atlantic.

The main AABW transport to the northern basins of the East Atlantic occurs through the Vema Fracture Zone. Along with this transport the bottom water also flows through a number of smaller fracture zones: Bogdanov, Doldrums, Vernadsky, Strakhov, Arkhangelsky, and Nameless (7°28′ N). The estimates of the total AABW transport through this group of fractures based on the measurements in 2014–2016 are within 0.48 ± 0.05 Sv. Not all the possible pathways of AABW transport in this region have been studied by present; hence, the transport through the subequatorial fractures can increase this proportion. However, without doubts, the coldest bottom water is transported through the Vema Fracture Zone.

The bottom water transport to the East Atlantic through the fractures south of the Vema Fracture Zone occurs in a complex system of overflows from one fracture to another.

The transport of bottom waters (including AABW) to the east through the fractures north of the Vema Fracture Zone is most likely insignificant. The Kane Fracture Zone is an exception, but the potential temperature of bottom water in this fracture is close to 2 °C.

6.3 Modeling of the Flow Through the Northern MAR

We used the INMOM model (Dianskii 2013; Frey et al. 2018) for modeling circulation in the Vema Fracture Zone and selected a model domain between 10.5° and 11.25° N and between 42.7° and 39.5° W for studying the structure variability of the bottom flows along the fracture. The horizontal resolution was 0.002° (222 m) for describing the current in the narrowest sections of the channels. The bottom topography here was studied by a multi-beam echo sounder; hence, we used the bottom topography from the GMRT database (Ryan et al. 2009). The main topographic obstacle for the AABW flow is located at 41°10′–40°50′ W (Fig. 6.26a); the

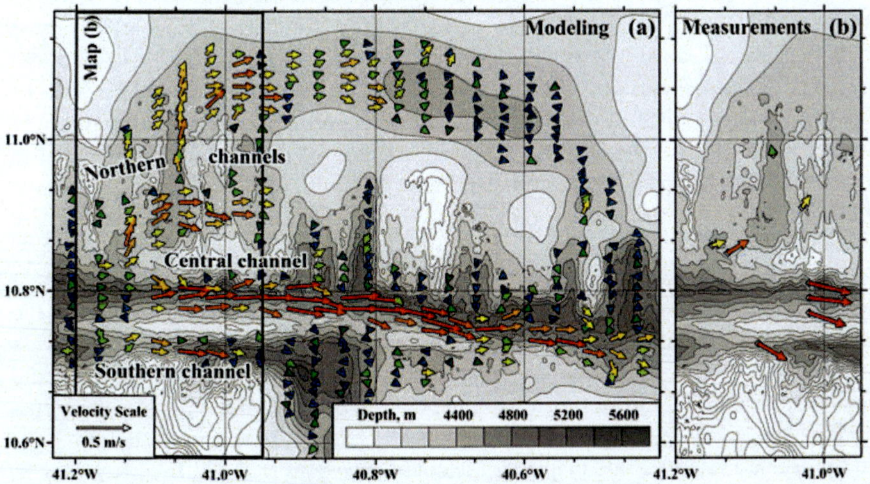

Fig. 6.26 Simulated (**a**) and measured (**b**) velocities in the bottom layer of the Vema Fracture Zone at a height of 50 m above the seafloor. The data presented here correspond to the deepest σ-level in the regions of the fracture deeper than 4250 m. The bottom topography is based on the GMRT database (Ryan et al. 2009), which assimilates multi-beam echo sounder measurements. The velocity scales are the same for the measured and simulated values. (Reproduced from (Frey et al. 2019))

Table 6.8 Maximum velocities and AABW transports in three parts of the Vema Fracture Zone

	Channels of the Vema FZ	Measurements in 2014	Measurements in 2015	Measurements in 2016	Modeling
Maximum velocity, cm/s	Central	34	46	37	38
	Southern	25	40	29	42
	Northern	21	–	21	19
AABW transport, Sv	Central	0.86	0.50	0.52	0.43
	Southern	0.26	0.30	0.14	0.16
	Northern	0.08	–	0.05	0.15
Total transport, Sv		1.20	>0.80	0.71	0.61

flow splits into three parts here. The main and coldest part of AABW propagates through the central channel. Its maximum depth at the sill is 4690 m; the width of the channel at a depth of 3850 m (the $\theta = 2$ °C isotherm) is 6.5 km. The southern channel is shallower and narrower than the central one, its depth is 4500 m and the width is 4 km. The northern channels are relatively wide and shallow; the depths of their main sills do not exceed 4350 m.

Simulated velocities and the bottom topography of the fracture are shown in Fig. 6.26a. The velocities reach 40 cm s^{-1} in the central and southern channels and 20 cm s^{-1} in the northern channels (the estimates are given in Table 6.8). Acceleration of the flow occurs in a narrow region of the central channel between 41°08′ W and 41°00′ W. The maximum velocities are gained between 41°00′ W and 40°55′ W; then the flow gradually slows down. The velocities decrease by a factor of two over the 30-mile part of the channel. High velocities in the southern channel are observed only near the sill at 41°05′ W. The channel leads to a depression with low bottom currents. A warmer part of AABW can propagate further to the east over the walls of the depression at higher levels. The currents in the northern channels are directed to the east and north depending on the local features of topography. The velocities reach 20 cm s^{-1} only in the western part of the area with relatively narrow channels; further propagation to the east occurs through the wide channel and velocities here become very low.

The transversal structure of the bottom flow was studied at the sill of the main (central) channel (Fig. 6.27). The temperature and velocities were measured over the meridional section at 41°01′ W. The simulated temperature and velocity fields were interpolated to the same section. The effect of the displacement of the potential temperature minimum is observed near the bottom of the channel. Unlike the Vema Channel located in the Southern Hemisphere, the Ekman friction in the bottom layer in the Vema Fracture Zone displaces the cold dense core to the left.

Our model also allowed us to simulate the transformation of thermohaline properties of the waters along the fracture (Fig. 6.28). White dots indicate locations of all stations with measurements up to the seafloor, the values of measured bottom potential temperature are given in Fig. 6.28. The temperature and salinity increase from west to east; there are two different mechanisms of this increase. First, the coldest and densest bottom waters cannot overflow the topographic obstacles at the

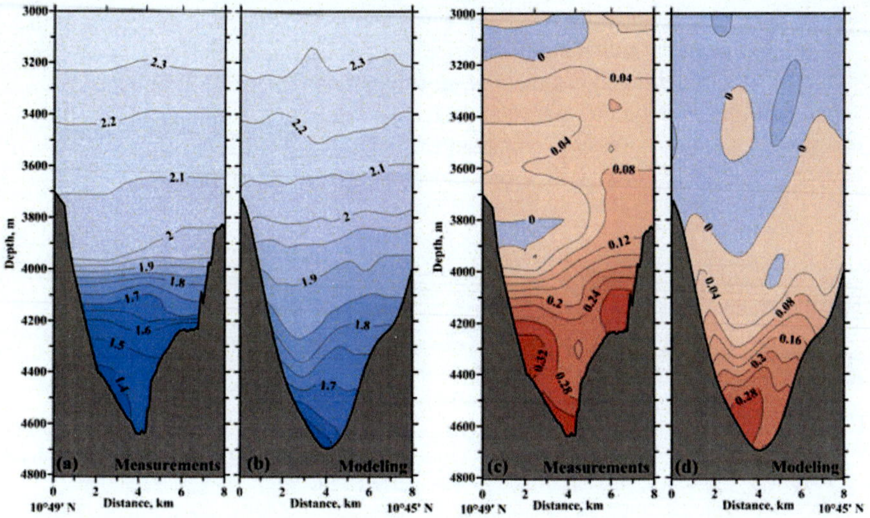

Fig. 6.27 Measured (**a**) and modeled (**b**) potential temperature, and measured (**c**) and modeled (**d**) zonal velocity at the section across the main channel of the Vema Fracture Zone at 41°01′ W. (Reproduced from (Frey et al. 2019))

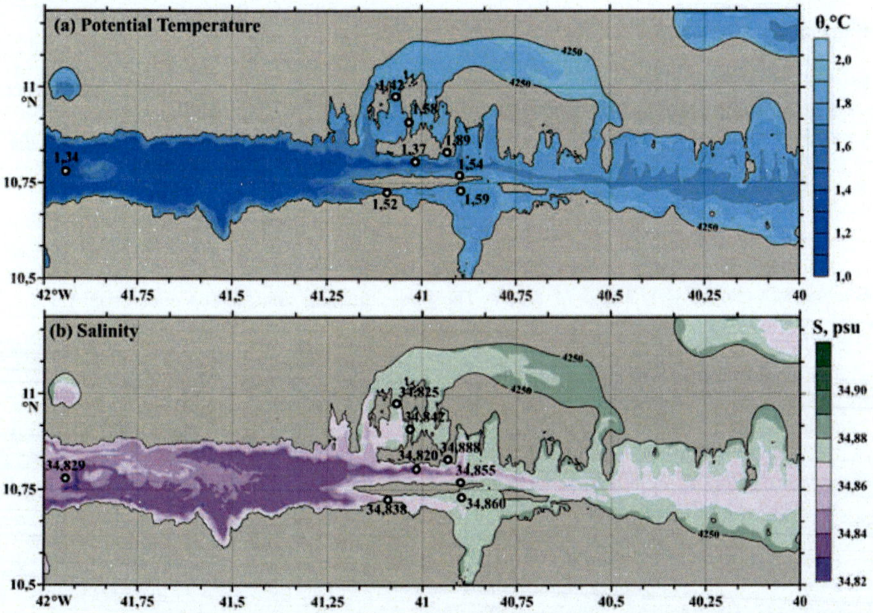

Fig. 6.28 Simulated distributions of potential temperature (**a**) and salinity (**b**) distributions in the bottom layer of the Vema Fracture Zone at the ocean depths greater than 4250 m. The measured values are indicated at the locations of stations shown with white dots. The bottom is shown with gray color. (Reproduced from (Frey et al. 2019))

sills of the channel, and only relatively warm and more saline waters spread further. Second, temperature and salinity increase due to mixing of the jet with overlying waters. Our simulation shows that the most significant increase of temperature and salinity occurs east of the main sill in the jet. Hence, the mixing of the flow is manifested in the increasing temperature and salinity. The simulated values of temperature increase were 0.4 °C and those of salinity were 0.05 PSU.

References

Arhan M, Mercier H, Bourlès B, Gouriou Y (1998) Hydrographic section across the Atlantic at 7°30′ N and 4°30′ S. Deep-Sea Res I 45:829–872

Boyer TP, Baranova OK, Coleman C, Garcia HE, Grodsky A, Locarnini RA, Mishonov AV, Paver CR, Reagan JR, Seidov D, Smolyar IV, Weathers K, Zweng MM (2018) World Ocean Database 2018. AV Mishonov, Technical Ed., NOAA Atlas NESDIS 87. https://www.ncei.noaa.gov/sites/default/files/2020-04/wod_intro_0.pdf

Broecker WS, Takahashi T, Li YH (1976) Hydrography of the Central Atlantic (I). The two-degree discontinuity. Deep-Sea Res I 23:1083–1104

Demidov AN, Dobrolyubov SA, Morozov EG, Tarakanov RY (2007) Transport of bottom waters through the Vema fracture zone in the Mid-Atlantic Ridge. Dokl Earth Sci 416:1120–1124

Dianskii NA (2013) Simulation of ocean circulation and investigation of its reaction on short-term and long-term atmospheric impacts (Fizmatlit, Moscow, 2013) [in Russian]

Eittreim SL, Biscaye PE, Jacobs SS (1983) Bottom-water observations in the Vema fracture zone. J Geophys Res Oceans 88:2609–2614

Fischer J, Rhein M, Schott F, Stramma L (1996) Deep water masses and transports in the Vema fracture zone. Deep-Sea Res I 43(7):1067–1074

Frey DI, Morozov EG, Fomin VV, Diansky NA (2018) Spatial structure of the Antarctic water flow in the Vema fracture zone of the mid-Atlantic ridge. Izvestia Atmos Oceanic Phys 54(6):621–625

Frey DI, Morozov EG, Fomin VV, Diansky NA, Tarakanov RY (2019) Regional modeling of Antarctic bottom water flows in the key passages of the Atlantic. J Geophys Res Oceans 124(11):8414. https://doi.org/10.1029/2019JC015315

Heezen BC, Gerard RD, Tharp M (1964) The Vema fracture zone in the equatorial Atlantic. J Geophys Res Oceans 69:733–739. https://doi.org/10.1029/JZ069i004p00733

Hekinian R (1982) Petrology of the ocean floor, Elsevier oceanography series, 33, Chap. 7, Oceanic Fracture Zones. Elsevier, Amsterdam, p 393

Johnson DA, McDowell SE, Sullivan LG, Biscaye PE (1976) Abyssal topography, nephelometry, currents, and benthic boundary layer structure in the Vema Channel. J Geophys Res Oceans 81:5771–5786. https://doi.org/10.1029/JC081i033p05771

Mantyla AW, Reid JL (1983) Abyssal characteristics of the World Ocean waters. Deep-Sea Res I 30(8):805–833. https://doi.org/10.1016/0198-0149(83)90002-X

McCartney MS, Bennet SL, Woodgate-Jones ME (1991) Eastward flow through the mid-Atlantic ridge at 11° N and its influence on the abyss of the eastern basin. J Phys Oceanogr 21(8):1089–1121. https://doi.org/10.1175/1520-0485(1991)021<1089:EFTTMA>2.0.CO;2

Morozov EG, Demidov AN, Tarakanov RY, Zenk W (2010) Abyssal channels in the Atlantic Ocean: water structure and flows. Springer. 266 p, Dordrecht. https://doi.org/10.1007/978-90-481-9358-5

Morozov EG, Tarakanov RY, Lyapidevskii VY, Makarenko NI (2012) Abyssal cataracts in the Romanche and chain fracture zones. Doklady Earth Science 446(6):1211–1214. https://doi.org/10.1134/S0001437013050147

Morozov EG, Tarakanov RY, van Haren H (2013) Transport of AABW through the Kane gap, tropical NE Atlantic Ocean. Ocean Sci 9:825–835. https://doi.org/10.5194/os-9-825-2013

Morozov EG, Tarakanov RY, Frey DI, Demidova TA, Makarenko NI (2018) Bottom water flows in the tropical fractures of the Northern Mid-Atlantic Ridge. J Oceanogr 74(2):147–167. https://doi.org/10.1007/s10872-017-0445-x

Müller RD, Roest WR (1992) Fracture zones in the North Atlantic from combined Geosat and Seasat data. J Geophys Res 97(B3):3337–3350

Rhein M, Stramma L, Krahmann G (1998) The spreading of Antarctic bottom water in the tropical Atlantic. Deep-Sea Res I 45:507–527

Roest WR, Collette BJ (1986) The fifteen twenty fracture zone and the north American-south American plate boundary. J Geol Soc Lond 143:833–843

Ryan WBF, Carbotte S, Coplan JO, O'Hara S, Melkonian A, Arko R, et al. (2009) Global multi-resolution topography synthesis. Geochem Geophys Geosyst 10:Q03014. https://doi.org/10.1029/2008gc002332

Schlitzer R (1987) Renewal rates of East Atlantic deep water estimated by inversion of 14C data. J Geophys Res Oceans 92:2953–2980

Smith WHF, Sandwell DT (1997) Global Sea floor topography from satellite altimetry and ship depth soundings. Science 277:1956–1962. http://topex.ucsd.edu/cgi-bin/get_data.cgi. https://doi.org/10.1126/science.277.5334.1956

Smoot NC (1989) North Atlantic fracture zone distribution and patterns shown by multibeam sonar. Geology 17:1119–1122

Stephens JC, Marshall DP (2000) Dynamical pathways of Antarctic bottom water in the Atlantic. J Phys Oceanogr 30(3):622–640

Tarakanov RY, Makarenko NI, Morozov EG (2013) Antarctic bottom water flow in the western part of the Romanche fracture zone based on the measurements in October of 2011. Oceanology 53(6):655–667. https://doi.org/10.1134/S0001437013050147

Vangriesheim A (1980) Antarctic bottom water flow through the Vema fracture zone. Oceanol Acta 3:199–207

Whitehead JA (1998) Topographic control of oceanic flows in deep passages and straits. Rev Geophys 36(3):423–440. https://doi.org/10.1029/98RG01014

WOD18 (2018) World Ocean Database 2018. https://www.nodc.noaa.gov/OC5/WOD/pr_wod.html

Wüst G (1936) Schichtung und Zirkulation des Atlantischen Ozeans, Das Bodenwasser und die Stratosphäre. In: A. Defant (Ed) Wissenschaftliche Ergebnisse, Deutsche Atlantische Expedition auf dem Forschungs - und Vermessungsschiff „Meteor" 1925–1927, 6(1), Walter de Gruyter & Co, Berlin, 411pg

Chapter 7
Eastern Basin Pathways and Further Propagation of Antarctic Bottom Water in the East Atlantic

Abstract We analyze the inflow of bottom waters to the Northeast Atlantic through the Vema, Romanche, and Chain fracture zones and their further propagation in the basin. In 1983, Mantyla and Reid wrote that bottom waters propagating through the Romanche Fracture Zone influence only the equatorial and southeastern part of the Atlantic Ocean and do not spread to the north through the Kane Gap, whereas the Vema Fracture Zone is the main pathway for bottom waters into the northeastern Atlantic. This concept was confirmed by our measurements. According to the CTD casts made in the last 20 years, waters with potential temperature less than $\theta = 1.85$ °C from the Romanche and Vema fracture zones merge in the region around the Kane Gap. Thus, the large basin of the Northeast Atlantic including the Gambia Abyssal Plain and Canary Basin is filled with bottom water that propagated through the Vema Fracture Zone. The bottom water that propagated through the Romanche Fracture Zone is localized in relatively small basins: Sierra Leone and Guinea basins with a possible insignificant outflow to the Angola Basin.

Keywords East Atlantic · Kane Gap · Angola Basin · Water transport

7.1 General Description

Antarctic Bottom Water flows to the Northeast Atlantic through the Vema, Romanche, and Chain fracture zones. Many years had passed since the *Meteor* expedition in 1927, when Wüst (1936) found the coldest waters at the bottom ($\theta = 1.74$ °C) in the Northeast Atlantic near 12° N, 34° W. He thought that cold waters spread to the East Atlantic through the Romanche Fracture Zone, as well as through an unknown passage in the Mid-Atlantic Ridge at 8–11° N. The Vema Fracture Zone remained unknown and was discovered only in 1956. During the long history of analysis of bottom water properties, all researchers agreed that these waters are of the Antarctic origin as was initially suggested by Wüst. However, there were different concepts about the inflow of these waters through the fractures in the Mid-Atlantic Ridge. Alternative opinions about possible candidates among the Vema and Romanche fracture zones were published in (Heezen et al. 1964;

E. G. Morozov et al., *Bottom Gravity Currents and Overflows in Deep Channels of the Atlantic Ocean*, https://doi.org/10.1007/978-3-030-83074-8_7

Worthington and Wright 1970a, b; Mantyla and Reid 1983; Tarakanov et al. 2017; Morozov et al. 2018). The present-day concept was for the first time suggested in (Mantyla and Reid 1983). They wrote that bottom waters propagating through the Romanche Fracture Zone influence only the equatorial and southeastern part of the Atlantic Ocean and do not spread to the north through the Kane Gap, whereas the Vema Fracture Zone is the main pathway for bottom waters into the northeastern Atlantic.

A scheme of Antarctic Bottom Water (with potential temperature $\theta < 2$ °C) spreading in the Northeast Atlantic was suggested in (McCartney et al. 1991) on the basis of a large amount of measurements before 1991. They based their analysis on the distribution of potential temperature at the bottom, which is a good indicator for Antarctic Bottom Water transport over different sections in the Eastern Basin. In particular, they revealed an eastward geostrophic transport of bottom water ranging from 1.3 to 3.0 Sv relative to the zero surfaces at potential temperatures levels of 2.0° and 2.5 °C, respectively. According to this scheme, Antarctic Bottom Water that propagated to the Cape Verde Basin in the East Atlantic through the Vema Fracture Zone splits into two branches spreading in different directions (Fig. 7.1). Their total transport significantly exceeds the transport of bottom water through the Vema Fracture Zone, which suggests the presence of a closed circulation in the basin east of the fracture zone. Here, we suggest a new scheme of the distribution of potential temperature at the bottom based on recent measurements, which appeared after 1991, with inclusion of all available historical data from the WOD18 database (Fig. 7.2) (WOD18 2018; Boyer et al. 2018).

One branch of the bottom water flow is directed from the Vema Fracture Zone to the southeast; the slopes of the Mid-Atlantic Ridge and Sierra Leone Rise are constraints for this flow from the south. Then, the flow returns to the west in the northern loop of the cyclonic circulation in the Cape Verde Basin. This flow transports not more than one half of the Antarctic Bottom Water flow that passed through the Vema Fracture Zone (McCartney et al. 1991). This branch reaches the Kane Gap near the coast of Guinea. According to the CTD casts made in the last 20 years, waters with potential temperature less than $\theta = 1.85$ °C from the Romanche and Vema fracture zones merge in the region around the Kane Gap. Waters with $\theta = 1.8$ °C are located north and south of the Kane Gap. At the same time, isothermal surfaces $\theta = 2.0$ °C and $\theta = 1.9$ °C are not separated over the Kane Gap (Fig. 7.2), which indicates possible Antarctic Bottom Water exchange through this passage. However, according to (McCartney et al. 1991) the southern part of the Cape Verde Basin (the Kane Gap region) is occupied by a cyclonic gyre; therefore, it is a cul-de-sac for the eastward flow of bottom waters from the Vema Fracture Zone. Thus, the large basin of the Northeast Atlantic including the Cape Verde and Canary basins is filled with bottom water that propagated through the Vema Fracture Zone. The bottom water that propagated through the Romanche Fracture Zone is localized in relatively small basins: Sierra Leone and Guinea basins with a possible insignificant outflow to the Angola Basin.

Such localization seems surprising because Antarctic Bottom Water transports through the Romanche and Chain fracture zones exceed 1 Sv, which is almost the

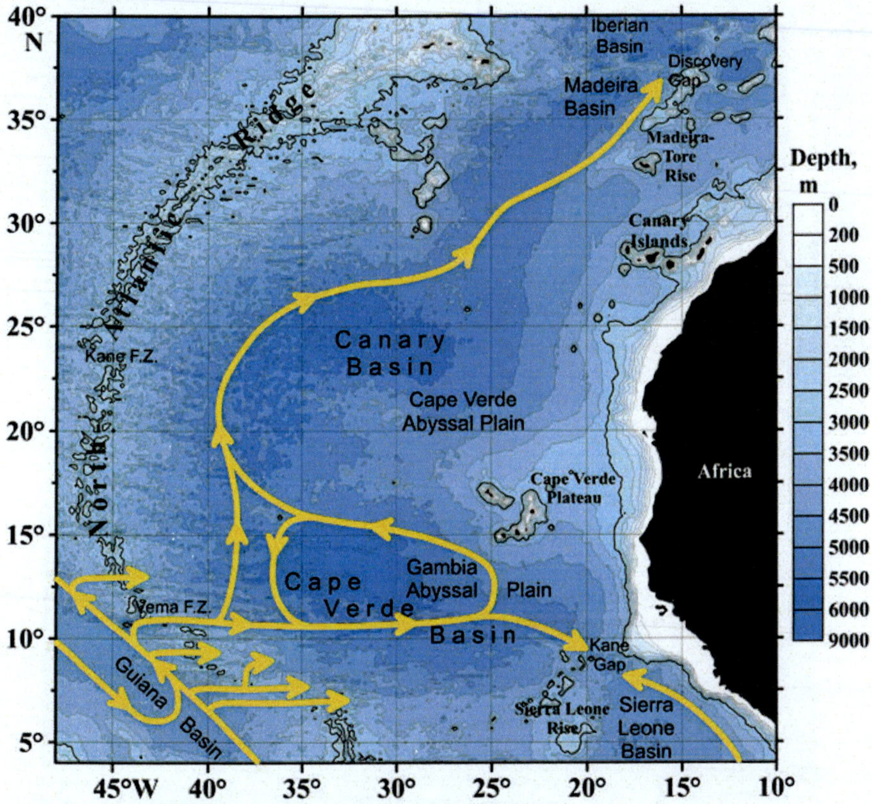

Fig. 7.1 Bottom circulation in the Northeastern Basin (Cape Verde and Canary basins) of the Atlantic Ocean. Modified after (McCartney et al. 1991; Morozov et al. 2013)

same as the water transport through the Vema Fracture Zone. We believe that it may be explained by stronger mixing in the Romanche and Chain Fracture Zones compared to the Vema Fracture Zone. This can be seen from enhanced transformation of bottom water properties in the equatorial channels. In particular, before the entrance to the Romanche and Vema fracture zones, the minimum values of potential temperature at the bottom are 0.50° and 1.33 °C, respectively (McCartney et al. 1991). The difference between the lowest values of potential temperature in these fracture zones becomes insignificant after they outflow to the East Atlantic from the fracture zones: $\theta = 1.66°$ (from Romanche) and $\theta = 1.69°$ (from Vema) (McCartney et al. 1991). In this case warming of bottom water during its propagation through the equatorial channels is not related to the geothermal heat flux (see Chap. 5, section about the Romanche Fracture Zone). Mercier and Morin (1997) showed that benthic thermocline disappears east of the Romanche and Chain Fracture zones. In the region of the Vema Fracture Zone, benthic thermocline is found only at the stations near the sills. The benthic thermocline disappears east of the fracture zone. We

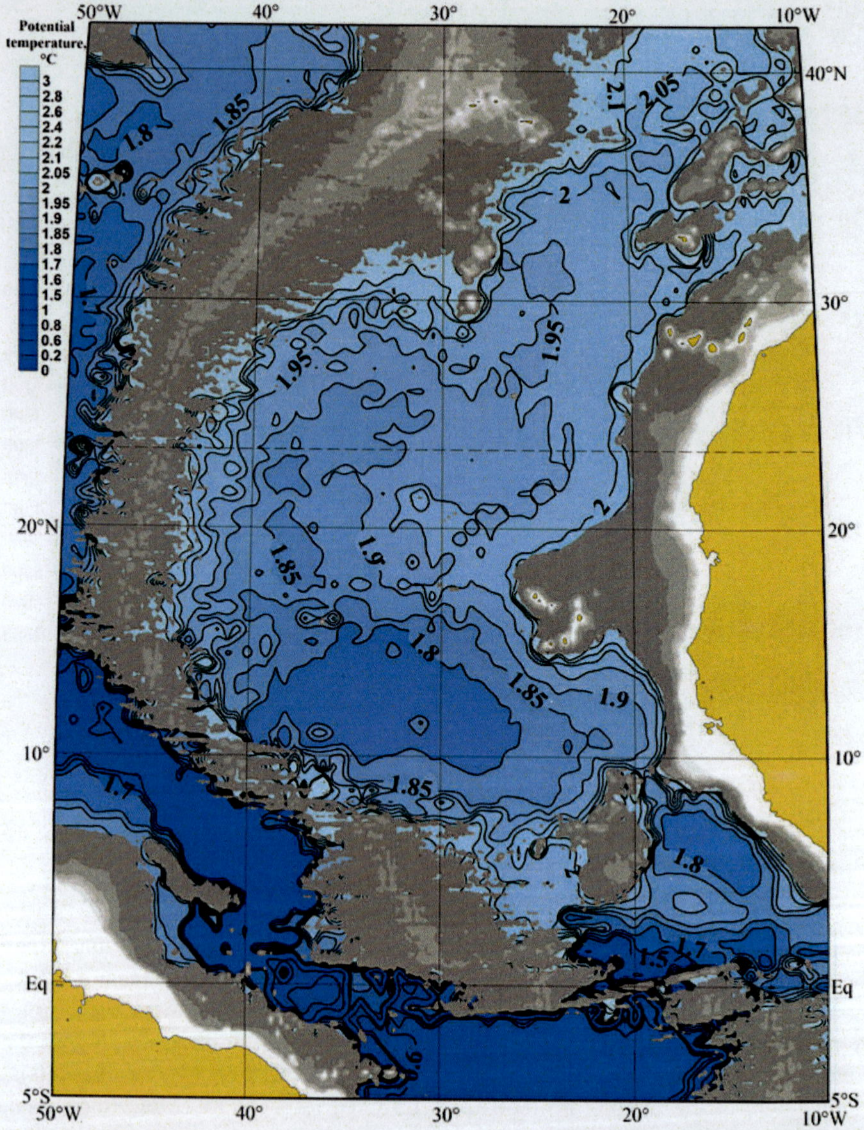

Fig. 7.2 Scheme of Antarctic waters spreading in the bottom layer of the Atlantic. Distribution of potential temperature (°C) at the bottom in the eastern part of the North Atlantic based on the WOD18 data. Only the stations deeper than 4000 m were used. Gray shade shows the location of isobaths above 4000 m

believe that this indicates ceasing of strong mixing near the bottom in the remote regions from the Mid-Atlantic Ridge.

The volume of the Antarctic Bottom Water spreading region strongly depends on the method of determining its upper boundary. This effect is strongest in the basins east of the Mid-Atlantic Ridge. If the boundary of this water propagation is determined on the basis of isopycnals $\sigma_2 = 37.11$ ($\sigma_4 = 45.92$), the spreading region of Antarctic Bottom Water in the eastern basin is limited to a small space near the Romanche and Vema fracture zones. If we take isotherm $\theta = 2$ °C as the boundary, Antarctic Bottom Water occupies the major part of the East Atlantic (Fig. 7.2) up to the Iberian Basin (van Aken 2000). Even more, if we use as a criterion $\sigma_4 = 45.87$ ($\gamma^n = 28.11$), which is equivalent to $\theta = 2$ °C in the region of the Brazil Basin, the spreading region of Antarctic Bottom Water reaches 52° N (Fig. 2.6 in Chap. 2 of this book).

According to (McCartney et al. 1991), another branch of the Bottom Water flow from the Vema Fracture Zone in the Cape Verde Basin is directed to the north and waters of the Antarctic origin fill the deepest parts of the Northeast Atlantic basins including the Canary Basin. The transport of this flow does not exceed the transport of Antarctic Bottom Water that passed through the Vema Fracture Zone. The propagation of bottom waters in the Eastern Basin occurs mainly in its western part, as shown in Fig. 7.1. According to (Saunders 1987), the core of the coldest water in the Madeira Basin, north of the Canary Basin, occurs along the foot of the Madeira Tore Rise. The presence of such a core corresponds to a deviation of the denser bottom flow to the right within the geostrophic balance. The flow reaches the ridge in the East Azores Fracture Zone at 37° N (hereinafter East Azores Ridge) dividing the Madeira and Iberian abyssal basins.

There are two deep-water passages in the East Azores Ridge at 37°20′ N, 15°40′ W and 37°15′ N, 19°25′ W providing the flow of abyssal waters into the Iberian Basin (Saunders 1987; Tarakanov et al. 2013). Our previous knowledge of the flow through these passages is based on the research described in (Saunders 1987) who named the eastern passage the 'Discovery Gap'. The passage is 150 km long. Its narrowest place is located at 37°20′ N, 15°40′ W. The width of the narrowest gap is 10 km and the depth of the sill is 4700 m. The flow was measured from six moorings and supplemented by tracking floats at a depth of 4700 m. The measured mean velocities were 5 cm s^{-1}. The flux of bottom water colder than potential temperature $\theta = 2.05$ °C was estimated at 0.2 Sv. Numerous CTD measurements around the Discovery Gap indicate that water with potential temperature below 2 °C does not propagate through this passage or the amount of water that propagates to the north is very small here.

Analysis of the historical CTD dataset and digital maps of bottom topography around the East Azores Ridge (Tarakanov et al. 2013) showed that the contribution of the second passage in the ridge, called Western Gap in this book, to the flow of abyssal waters into the Iberian Basin may be underestimated; however, it can be deeper than the eastern passage (Discovery Gap), and the colder bottom water can flow through it. Based on the results of this work, we carried out field studies in the Western Gap, the results of which will be presented in more detail in Chap. 8 of this book.

Most numerical estimates of the transport in the East Atlantic are below 1 Sv (Table 7.1) (McCartney et al. 1991; Saunders 1994; Lavin et al. 2003; Lherminier et al. 2007). Calculation of the amounts of Antarctic waters in (van Bennekom 1985)

Table 7.1 Transport in the Eastern Basin

Authors	AABW boundary	Transport, Sv	Comments
Saunders 1987	$\theta = 2.05$ °C	0.7	Geostrophic transport 36° N; reference surface $p = 3500$ db
Lavin et al. 2003	4900 m	0.4	Geostrophic transport at 24° N; variable reference level based on the a priori constraints
Álvarez et al. 2004	$\sigma_4 = 45.85$	1.6	Geostrophic transport at A25 section at 42° N with inverse correction of barotropic transport
Lherminier et al. 2007	$\sigma_4 = 45.85$	0.8	Geostrophic transport at A25 section at 42° N with inverse correction of barotropic transport + SADCP data
		1.5 ± 1.0	Geostrophic transport at OVIDE section at 42° N with inverse correction of barotropic transport + SADCP and LADCP data

revealed that bottom waters in the Northeast Atlantic contain 29% of the Antarctic waters at 50° N, 9–11% at 57° N, and only a few percent of Antarctic waters in the Norwegian Sea. Table 7.1 summarizes the results of different publications about the transport of bottom waters in the Northeast Atlantic (Canary Basin and Gambia Abyssal Plane).

We note that the northeastern part of the Atlantic Ocean is a good example that the physical processes stand aside from terminology. Although the bottom water that propagated through the Vema Fracture Zone flows further to the north, gradually transforming and losing the extreme values of its properties, many authors give other names to the bottom water in the Northeast Atlantic. Antarctic Bottom Water east of the Mid-Atlantic Ridge is called Eastern Basin Bottom Water, Lower Deep Water (van Aken 2000), or Abyssal Water of the East Basin (Stephens and Marshall 2000), which is a product of strong transformation of Antarctic Bottom Water. Arhan et al. (1998) distinguished Antarctic waters that passed the Romanche and Chain fracture zones as specific water mass and identified it as Guinea Basin Bottom Water. These waters conserve all distinguishing indicators of Antarctic Bottom Water: low salinity, low concentration of dissolved oxygen, and high concentration of nutrients (van Aken 2000). However, some authors use the term Antarctic Bottom Water for the bottom waters in the Iberian Basin (Lherminier et al. 2007).

7.2 Mixing Caused by the Barotropic Tide

Here, we make an attempt to give a physical explanation why strong mixing occurs in the equatorial channels, which results in the isolation of different basins east of the Mid-Atlantic Ridge. We believe that this fact is explained by different intensities of mixing of Antarctic waters with the overlying North Atlantic Deep Water due to internal tides generated over the slopes of the Mid-Atlantic Ridge. Another cause of different mixing in the Romanche and Vema fracture zones is the roughness of the bottom topography in these fractures. The bottom of the Vema Fracture Zone is smoother; hence, mixing is not as strong as in the Romanche Fracture Zone.

Morozov (1995) calculated the amplitudes of internal tidal waves in the World Ocean based on the integrated results of measurements and model calculations. According to these estimates, the amplitude of internal tidal waves reaches almost 50 m in the Romanche Fracture Zone region at the equator and only slightly exceeds 20 m in the Vema Fracture Zone region at 11° N.

Figure 7.3 shows a chart of amplitudes of internal tides in the Atlantic Ocean. Due to such strong difference in the amplitudes of waves in these regions, mixing of deep-water masses must differ strongly. The most intense mixing occurs over the slopes of the submarine ridge, where amplitudes of internal waves are the greatest. Internal tides are intensely generated in the regions of strong barotropic tides if the currents are normal to the ridge. Generation is intensified if the inclination of the bottom coincides with the inclinations of characteristic curves of internal tides, which depend on stratification (Morozov 2018). The conditions that favor strong generation of internal tides are much better in the region of equatorial channels than in the Vema Fracture Zone.

Velocity ellipses of the barotropic tide currents in the region of the bottom water outflow from the Romanche and Chain fracture zones were calculated based on the TPXO9 model. The ellipses in Fig. 7.4 show two remarkable properties: (i) the velocities are quite high (5 cm s^{-1}) compared to usual 1–2 cm s^{-1} in the open ocean; (ii) the ellipse is strongly elongated in the E-W direction, which is close to the direction of the Romanche and Chain fracture zones. It is essential that the ellipses were calculated from the data of surface elevations and they represent the entire tidal structure in the water column, the largest part of which is located above the deep channel. Thus, such barotropic tides can induce strong generation of internal tides.

Characteristic curves for internal tides, where amplitudes of internal waves are maximal, are inclined near the slopes of the ridges. The results of numerical calculation of the field of horizontal velocities of particle motion in internal tidal waves over the slopes of the Mid-Atlantic Ridge in the Romanche Fracture Zone region are shown in Fig. 7.5. The velocity field is presented for the phase of the barotropic tide when the velocities of the barotropic tide currents are zero. The calculation was made using the numerical model described in (Vlasenko and Hutter 2002). Strong mixing in the Romanche Fracture Zone region transforms properties of the Antarctic Bottom Water.

A similar calculation was performed for the Vema Fracture Zone region (Fig. 7.6). Velocities of internal tide are lower. Hence, internal waves provoke lesser mixing in this region.

We summarize that mixing in the equatorial region of the East Atlantic is greater than in the region of the Vema Fracture Zone. This strong mixing is caused by a strong barotropic tide in the equatorial region of the East Atlantic compared to the region of the Mid-Atlantic Ridge near the Vema Fracture Zone. Another cause of strong mixing is related to the overflows of numerous sills in the Romanche and Chain Fracture zones. The bottom in the Vema Fracture Zone is much smoother.

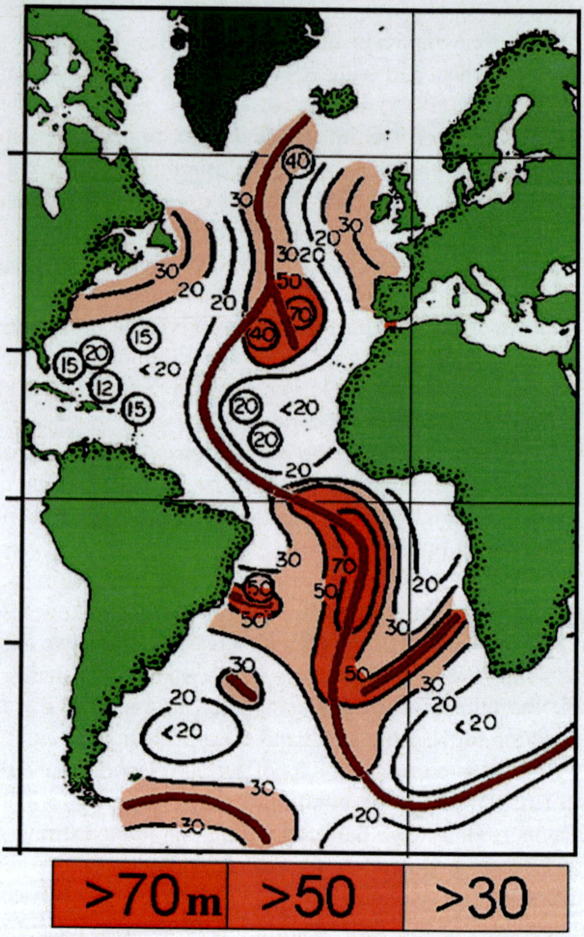

Fig. 7.3 Contour lines of amplitudes of internal tide in meters (Morozov 2018). Thick brown lines show submarine ridges. Numerals in circles denote actual measurements of amplitudes of internal tide. The internal tide amplitude exceeds 50 m in the Romanche Fracture Zone region and only slightly exceeds 20 m in the Vema Fracture Zone region

Strong mixing explains the difference in the contribution of the equatorial channels and Vema Fracture Zone to the bottom water mass composition in the Northeast Atlantic. Without strong mixing in the equatorial region, the influence of the Romanche and Vema fracture zones on the Northeast Atlantic seems to be equal:

(1) The transport of Antarctic Bottom Water through the Romanche and Chain fracture zones (both) is of the order of 1 Sv, which is almost the same as the water transport through the Vema Fracture Zone.
(2) The difference between the smallest values of potential temperature of bottom water is insignificant after the water outflows to the East Atlantic: $\theta = 1.66°$ (from the Romanche Fracture Zone) and $\theta = 1.69°$ (from the Vema Fracture Zone).
(3) The depths of the main sills in the channels do not differ greatly: 4359 m (Romanche Fracture Zone) and 4660 m (Vema Fracture Zone).

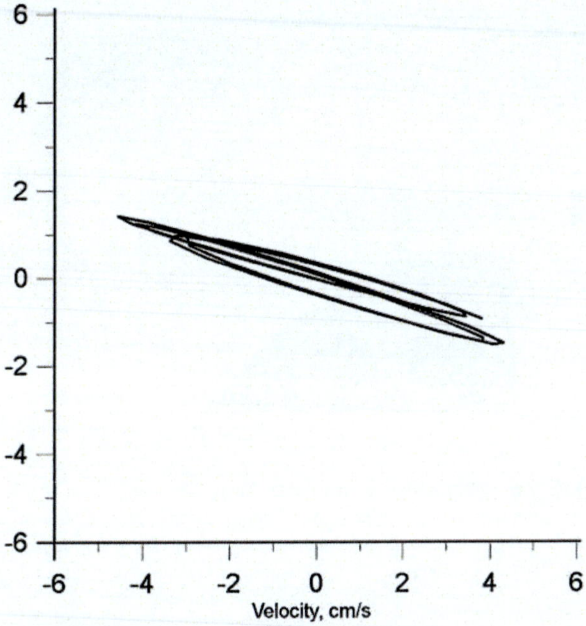

Fig. 7.4 Ellipses of M_2 barotropic tide (cm s^{-1}) in the region east of the Chain Fracture Zone based on TPXO9 model data. The ellipses are calculated for the time of measurements in the region (April, 2009)

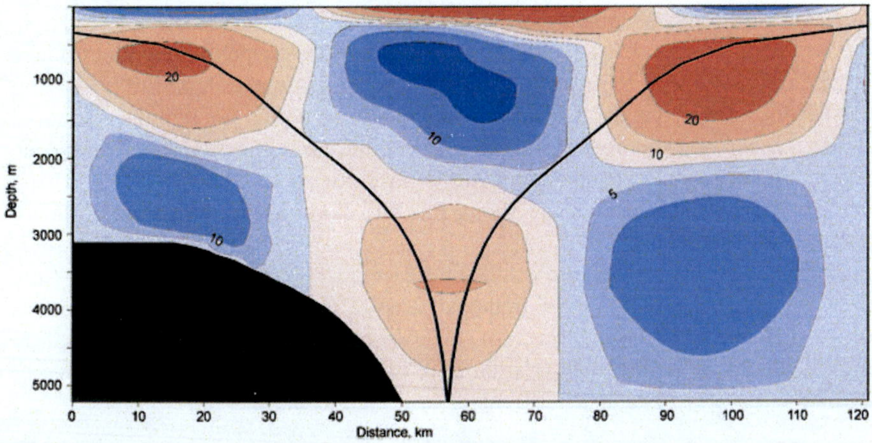

Fig. 7.5 Distribution of velocities during generation of internal tide over the slope of the Mid-Atlantic Ridge in the Romanche Fracture Zone region. Contour lines of velocities are given in cm s^{-1}. The black color denotes the ocean bottom. The thick black line shows the characteristic curve for internal tide M_2

Despite the fact that the depths of the Kane Gap allow propagation of Antarctic Bottom Water and only the $\theta = 1.8$ °C isotherm is separated over the passage, the transport of bottom water through this channel is not strong, although the depth of the sill is 4560 m.

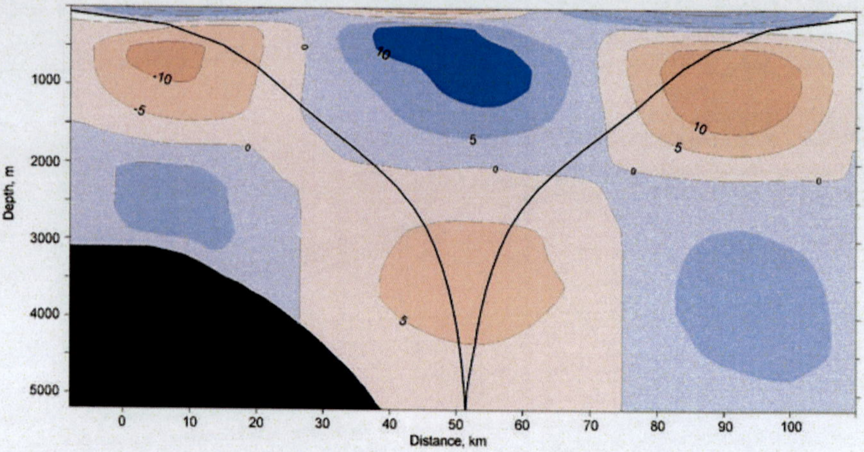

Fig. 7.6 Distribution of velocities during the generation of internal tide over the slope of the Mid-Atlantic Ridge in the Vema Fracture Zone region. Contour lines of velocities are given in cm s^{-1}. The black color denotes the ocean bottom. The thick black line shows the characteristic curve for internal tide M_2

Strong mixing with overlying waters results in strong transformation of properties of Antarctic Bottom Water after its outflow from the equatorial channels. In addition, the region of strong mixing includes also the Kane Gap, which makes this passage almost impossible for the strong northerly flow of bottom waters. Strong mixing closes this pathway for the flow of bottom water transported through the equatorial fracture zones. On the other hand, mixing conditions in the Vema Fracture Zone region are not as strong as in the equatorial East Atlantic and bottom water inflows are less transformed through the Vema Fracture Zone filling the entire Northeast Atlantic abyssal depths.

Different intensity of mixing results in different stratification at the outflow from the channels. The Brunt-Väisälä frequency east of the Romanche Fracture Zone is $N = 0.14 \cdot 10^{-3}\ s^{-1}$, while east of the Vema Fracture Zone it is equal to $N = 0.80 \cdot 10^{-3}\ s^{-1}$. A simple relation in Gargett (1984) gives the dependence of vertical diffusivity k_z on stratification:

$$k_z = \frac{10^7 m^2 s^2}{N(z)}. \tag{7.1}$$

According to this relation diffusivity in the deep layers of the Mid-Atlantic Ridge east of the Vema Fracture Zone is $k_z = 1.2 \cdot 10^{-4}\ m^2\ s^{-1}$, and east of the Romanche Fracture Zone it is 7 times greater ($k_z = 7.1 \cdot 10^{-4}\ m^2\ s^{-1}$).

7.3 Kane Gap

The Kane Gap is located between the Grimaldi Mountains, which are a part of the Sierra Leone Rise and the Guinea Plateau near the African continent (Fig. 7.7). The gap connects the Gambia Abyssal Plain and Sierra Leone Basin. The sill depth in the gap is 4566 m based on our measurements. Another gap is located east of the main passage with depths exceeding 4100 m (at longitude 18°40′ W). A chart of bottom topography based on a multibeam survey in 2016 is shown in Fig. 7.8. This chart is based on the data collected during cruise 44 of the R/V *Akademik Nikolaj Strakhov* in October 2019. A description of the bottom topography is given in (Dudkov et al. 2020).

Antarctic Bottom Water propagates to the East Atlantic Ocean through two main passages in the Mid-Atlantic Ridge: the Vema Fracture Zone (at 11° N) and Romanche and Chain fracture zones (at the equator) and then reaches the Kane Gap west of Guinea. It is seen from Fig. 7.7 that the minimum potential temperature at the bottom close to the Kane Gap ever measured by CTD casts is 1.85 °C (at depths of 4560 m that correspond to the depth of the sill). This indicates that bottom waters with temperatures lower than this value transported to the region of the Kane Gap from the two sources are separated by the sill in the gap, while warmer mixed waters

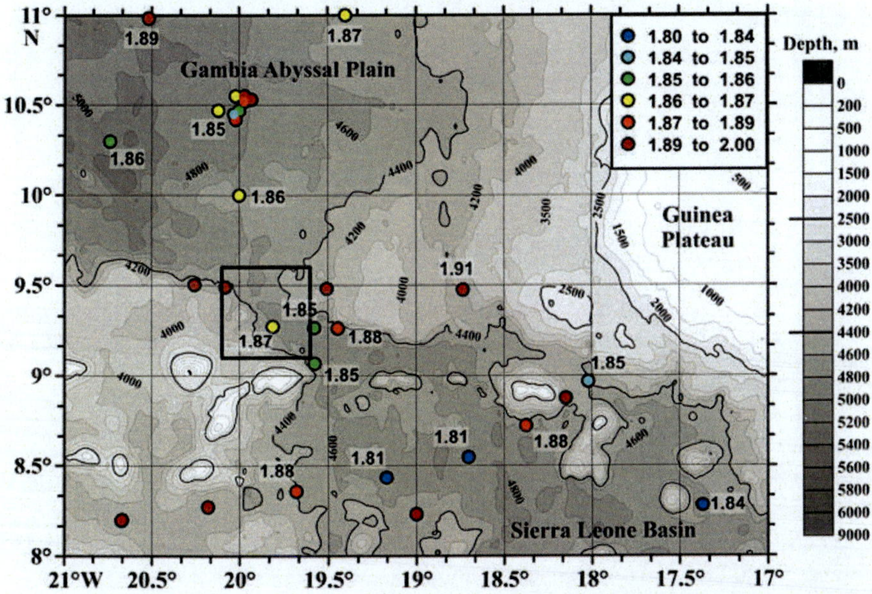

Fig. 7.7 Bathymetry of the Kane Gap region with observations of near-bottom potential temperature (°C) according to WOD18 database. Historical stations are shown with dots of different color depending on the temperature (see notations in the inset; our study region is shown with a black square). Bottom potential temperatures (°C) are indicated at the deepest stations. Reproduced from (Morozov et al. 2013)

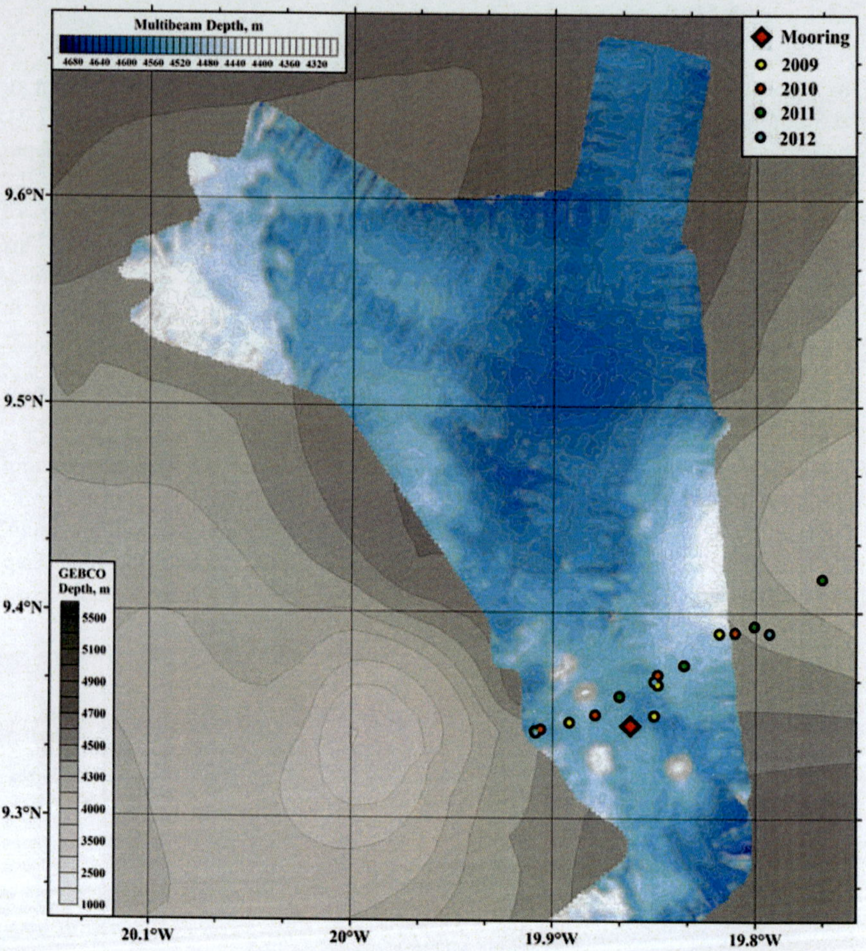

Fig. 7.8 Chart of the bottom topography in the Kane Gap based on a multibeam survey in 2019 (Dudkov et al. 2020). Locations of stations in 2009–2012 are indicated. The mooring position is shown with a rhomb

can flow over the sill. Thus, only the warmest part of Antarctic Bottom Water (2.00 °C > θ > 1.85 °C) can exchange over the sill of the Kane Gap. Historical CTD stations near the Kane Gap are shown in a map in Fig. 7.7. We note that Fig. 7.7 was constructed on the basis of water samples data and CTD-data from the WOD18 database (WOD18 2018) supplemented with the results of Hobart et al. (1975).

In the previous section, we discussed a scheme of Antarctic Bottom Water (with potential temperature θ < 2 °C) spreading in the Northeast Atlantic. According to this scheme, the Kane Gap occupies the middle position between Antarctic Bottom Water that propagated to the East Atlantic through the Vema Fracture Zone and Antarctic Bottom Water that propagated through the Romanche and Chain fracture zones. Waters with potential temperature less than θ = 1.85 °C from these two sources merge in the region around the Kane Gap. At the same time, isotherm θ = 1.90 °C is not separated over the Kane Gap (Fig. 7.2).

The echo sounder survey in October of 2012 revealed local hills at the southwestern slope of the Kane Gap. These hills are clearly seen in the map of bottom topography based on the multibeam survey (Fig. 7.8). The further analysis demonstrated significant influence of these hills on the direction and structure of the bottom currents in the entire section and at the location of the mooring. The bathymetry measured in our cruises was used to plot the bathymetry (Fig. 7.9) over all the sections occupied across the Kane Gap (Figs. 7.11, 7.12, 7.13, 7.14, and 7.15). We used the digital bottom topography of the ocean (Smith and Sandwell 1997) and recent data available in the Internet as the basis for this chart (http://topex.ucsd.edu/cgi-bin/get_data.cgi). The differences in the bottom topography in the sections are related to slightly different lines of sections in each year.

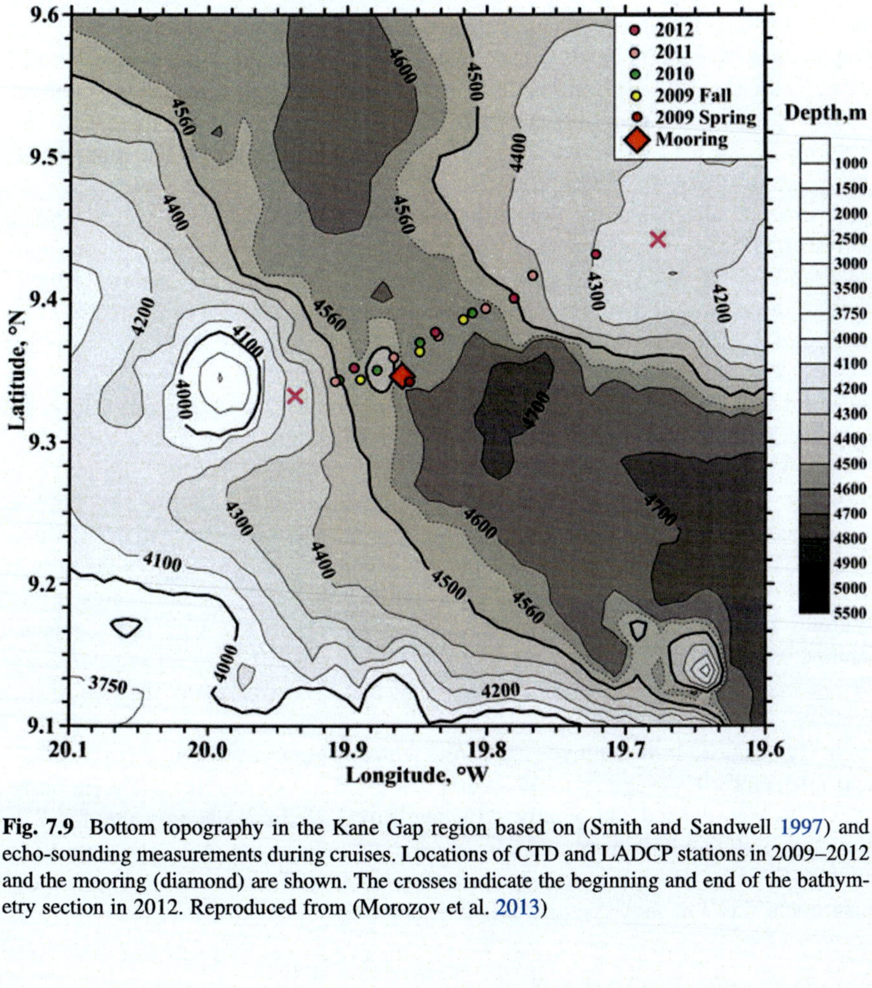

Fig. 7.9 Bottom topography in the Kane Gap region based on (Smith and Sandwell 1997) and echo-sounding measurements during cruises. Locations of CTD and LADCP stations in 2009–2012 and the mooring (diamond) are shown. The crosses indicate the beginning and end of the bathymetry section in 2012. Reproduced from (Morozov et al. 2013)

According to World Ocean Database (WOD18 2018), the first hydrographic stations in this region were occupied during the cruise of the R/V *Trident* in the Kane Gap region in 1963. They were located mostly over the adjacent seamounts. Later, bottom sediments were studied during a cruise of the R/V *Chain*. Several CTD casts were also made in the region (Hobart et al. 1975). In 1983, CTD stations were made from the R/V *Knorr*. A few hydrographic sections were made close to the gap. In 1993, WOCE hydrographic section A06 was occupied along 7°30′ N south of the gap. In 2000, a hydrographic section was made from the R/V *Akademik Ioffe* along 8° N close to the WOCE A06 section. During cruise 27 of the R/V *Akademik Ioffe* in May 2009, a CTD station was occupied at the main sill of the gap at a depth of 4552 m with the first LADCP measurements in the gap. Later, in October 2009, a section of three stations was made across the Kane Gap in the region of the sill. All available hydrographic stations in the regions of the Kane Gap and historical data from (Hobart et al. 1975), which were not included in WOD18, are shown in Fig. 7.7. The bottom temperatures at the deepest stations are also indicated.

In 2009–2012, the scientists from the Shirshov Institute of Oceanology significantly supplemented the available data set of temperature and salinity in the Kane Gap. Before, even the direction of bottom water flow through this passage was not clear. The only source of information about the currents was based on the oriented underwater photos. According to these observations, the current was directed to the northwest from the Sierra Leone Basin to the Gambia Abyssal Plain (Hobart et al. 1975).

In order to understand the flow of bottom water in the Kane Gap, we performed series of CTD casts and long-term mooring measurements at the sill of the gap (9°20′ N, 19°51′ W, approximate depth 4560 m). This program started in 2009. During the period 2009–2012, we carried out five series of CTD (SBE-19) casts combined with LADCP (RDI WHS 300 kHz) measurements (Figs. 7.8 and 7.9, Table 7.2).

We started measurements in the Kane Gap in May 2009 with only one station. In October 2009 (cruise AI-29), we continued the measurements in the Kane Gap (Fig. 7.9). A CTD-section of three stations was occupied combined with LADCP profiling. In October 2010 (cruise 32 of the R/V *Akademik Ioffe,* AI-32), another CTD-section with LADCP measurements at four stations was occupied in the Kane Gap (Fig. 7.9). In October 2011, during an expedition onboard the R/V *Akademik Sergey Vavilov* (cruise ASV-34) a CTD-section with LADCP measurements at five stations, and in October 2012 (cruise ASV-36), a CTD-section with LADCP measurements at four stations was carried out. In October of 2012, we also carried out a detailed echo sounder survey in the region of the Kane Gap along the line of the section.

In October 2010, we deployed a mooring with three current-meters (Figs. 7.8, 7.9, and 7.10) at 9°20.7′ N, 19°51.7′ W. It was recovered in October 2011. The mooring operation was carried out jointly with the Royal Netherlands Institute for Sea Research (NIOZ). The instruments measured currents at the operation frequency equal to 2 MHz. The instruments were fixed at 4352, 4477, and 4562 m. The deepest instrument was 7 m above the bottom. The sampling period was 600 s (10 min). The

Table 7.2 CTD stations at different cruises of the vessels Akademik Ioffe (AI) and Akademik Sergey Vavilov (ASV) in the Kane Gap

Date	Station, cruise	Coordinates	Lowest measurement/ ocean depth, m
01.05.2009	2084 AI-27	09°20.6′ N, 19°51.3′ W	4553/4565
30.10.2009	2159 AI-29	09°23.4′ N, 19°49.1′ W	4567/4579
30.10.2009	2160 AI-29	09°21.9′ N, 19°50.9′ W	4566/4572
31.10.2009	2161 AI-29	09°20.8′ N, 19°53.5′ W	4536/4549
20.10.2010	2423 AI-32	09°20.6′ N, 19°54.3′ W	4502/4512
20.10.2010	2424 AI-32	09°21.0′ N, 19°52.7′ W	4444/4454
21.10.2010	2425 AI-32	09°22.2′ N, 19°50.9′ W	4550/4550
21.10.2010	2426 AI-32	09°23.4′ N, 19°48.6′ W	4545/4555
18.10.2011	2461 ASV-34	09°25.0′ N, 19°42.5′ W	4387/4397
18.10.2011	2462 ASV-34	09°23.6′ N, 19°48.1′ W	4549/4559
19.10.2011	2463 ASV-34	09°22.5′ N, 19°50.1′ W	4566/4576
19.10.2011	2464 ASV-34	09°21.6′ N, 19°52.0′ W	4535/4545
19.10.2011	2465 ASV-34	09°20.5′ N, 19°54.5′ W	454,964,506
24.10.2012	2474 ASV-36	09°25.3′ N, 19°44.0′ W	4284/4292
24.10.2012	2476 ASV-36	09°23.4′ N, 19°47.6′ W	4543/4551
24.10.2012	2477 ASV-36	09°22.0′ N, 19°51.0′ W	4558/4566
24.10.2012	2478 ASV-36	09°20.5′ N, 19°54.5′ W	4498/4506

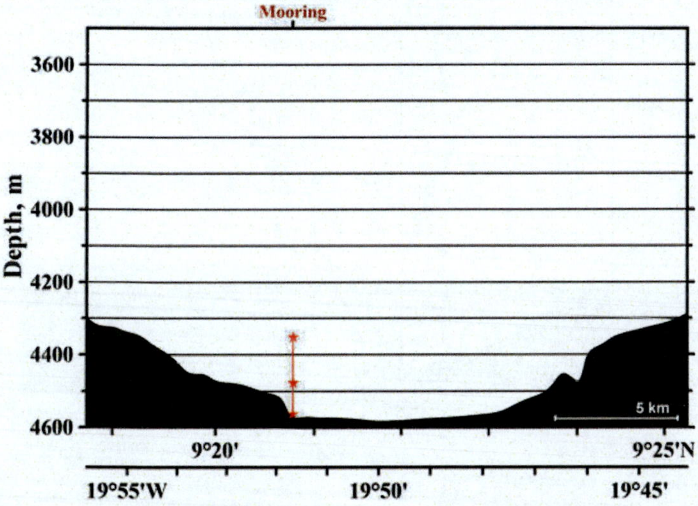

Fig. 7.10 A scheme of the mooring location over the topographic section across the Kane Gap. Crosses indicate the depths of current meters. The horizontal scale is shown in the bottom right angle. Reproduced from (Morozov et al. 2013)

sections in 2010 and 2011 were occupied immediately before and after the deployment and recovery of the mooring.

The series of five measurements in 2009–2012 allowed us to analyze the evolution of the near-bottom thermohaline structure and velocity in the Kane Gap. The measurements in May 2009 recorded potential temperature θ = 1.857 °C near the bottom in the Kane Gap, which agrees well with the previous measurements in the region. In October 2009, even lower temperatures were measured at the western slope of the gap (1.846 °C). Potential temperatures at the bottom are cooler both south and north of the Kane Gap. We suppose that an abyssal front exists near the Kane Gap, where bottom waters from two different sources encounter. The temperature and velocity profiles in May 2009 are shown in Fig. 7.11 and four temperature, salinity, and potential density σ_4 sections in 2009–2012 are shown in Figs. 7.12 and 7.13. The distributions of the velocity component in the abyssal layer normal to the sections are shown in Figs. 7.14a–d. Figure 7.15 shows θ, *S*-curves in the abyssal layer of these sections.

As seen from the distributions of potential temperature over sections and bottom topography charts, the abyssal channel of the Kane Gap oriented from southeast to northwest is a narrow passage only for the bottom waters with potential temperature

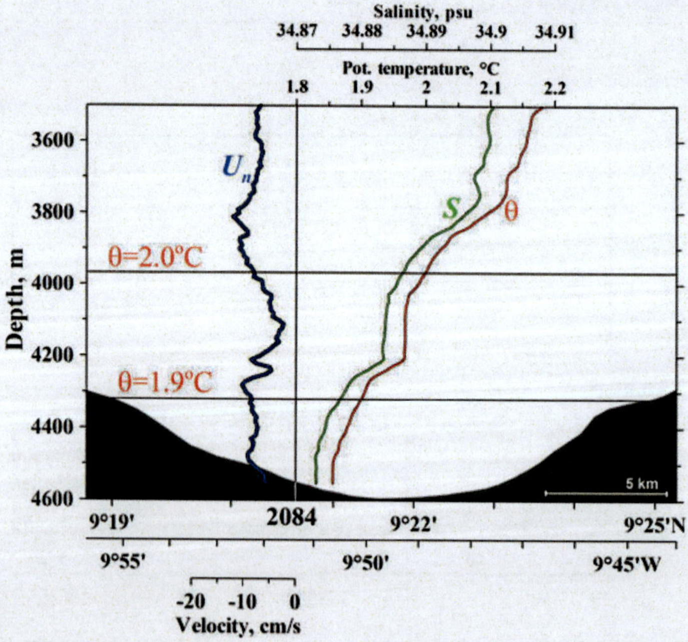

Fig. 7.11 Profiles of potential temperature (°C), Salinity, and normal-to-section velocity component (cm s^{-1}) measured in May 2009 at station AI 2084 over the sill of the Kane Gap. Positive velocities of the flow are directed to the northwest. The horizontal scale is shown in the bottom right angle. The hills in the bottom topography shown in Fig. 7.8 are located to the west of the station in 2009. Reproduced from (Morozov et al. 2013)

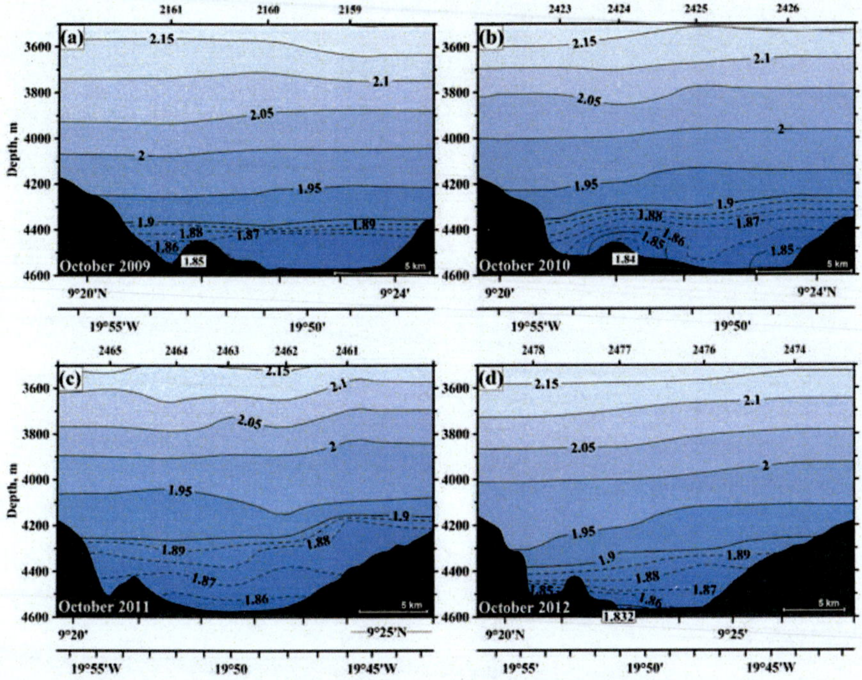

Fig. 7.12 Sections of potential temperature (°C), across the Kane Gap in October 2009 (**a**), October 2010 (**b**), October 2011 (**c**), and October 2012 (**d**). Numbers of stations are shown at the top axis and coordinates are below. The horizontal scale is shown in the bottom right angle. Reproduced from (Morozov et al. 2013)

approximately in the range 1.85° < θ < 1.90 °C. The warmer waters propagate through a significantly wider section. Hence, we cannot use the measurements in 2009–2012 to study in detail the properties and structure of water exchange (and neither calculate the transports) in the Kane Gap in the classical understanding of AABW (below the 2 °C isotherm). Therefore, in this study we shall consider only the layer of cold waters with θ < 1.9 °C.

During the first visit to the Kane Gap in May 2009, when only one CTD cast with LADCP instrument was carried out near the main sill of the Gap, we unexpectedly revealed a southeasterly abyssal current from the Gambia Abyssal Plain to the Sierra Leone Basin with a velocity of about 10 cm s^{-1}. The potential temperature at the bottom was 1.850 °C. According to (McCartney et al. 1991), bottom waters in the Kane Gap have more similar properties with the waters in the Gambia Abyssal Plain than with the waters in the Sierra Leone Basin, but the southeastern part of the Gambia Abyssal Plain is considered a cul-de-sac for further propagation of bottom waters to the south.

In October 2009, 2010, and 2012, almost unidirectional (in the general direction to the northwest) currents with velocities up to 10 cm s^{-1} in 2009, 13 cm s^{-1} in 2010, and 8 cm s^{-1} in 2012 were recorded during three of the four sections in the AABW

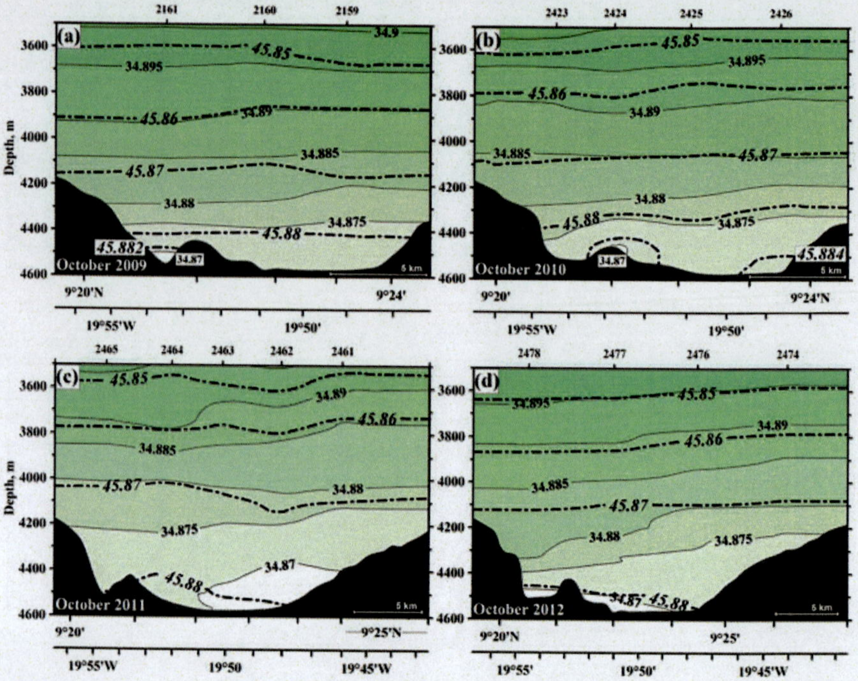

Fig. 7.13 Sections of salinity, and potential density σ_4 across the Kane Gap in October 2009 (**a**), October 2010 (**b**), October 2011 (**c**), and October 2012 (**d**). Contour lines of potential density are shown with thick dash-dotted lines. Numbers of stations are shown at the top axis and coordinates are below. The horizontal scale is shown in the bottom right angle. Reproduced from (Morozov et al. 2013)

layer (Fig. 7.14a, b, d). The cores of the current were observed in different places of the section approximately 200 m over the bottom. In October 2009, the core of the current was located close to the northeastern slope of the Kane Gap; in 2010 and 2012 it was found near the local hill at the southwestern slope. In addition, in 2009 and 2010, we found additional weaker cores of the currents at the slopes opposite to the locations of the main cores. In October 2011, we measured very weak differently oriented (southeasterly and northwesterly) currents (in the central part of the gap) whose velocities reached only 5 cm s^{-1} near the bottom (Fig. 7.14c).

It is seen from Figs. 7.12 and 7.13 that on all four sections, the densest water was displaced either to the southwestern slope of the Kane Gap or to the local hill near this slope. In October 2010, we recorded one more core of dense water near the northeastern slope (Fig. 7.13b). We also note elevation of isopycnals detected in all years of measurements over the northeastern slope of the channel for the water layer warmer than $\theta = 1.9$ °C. During the measurements in October 2009 and October 2012, we observed horizontal thermohaline inhomogeneity in the AABW layer: the colder and lower saline (along the isopycnals) water was located close to the southwestern slope of the Kane Gap (Figs. 7.12 and 7.13). The maximum

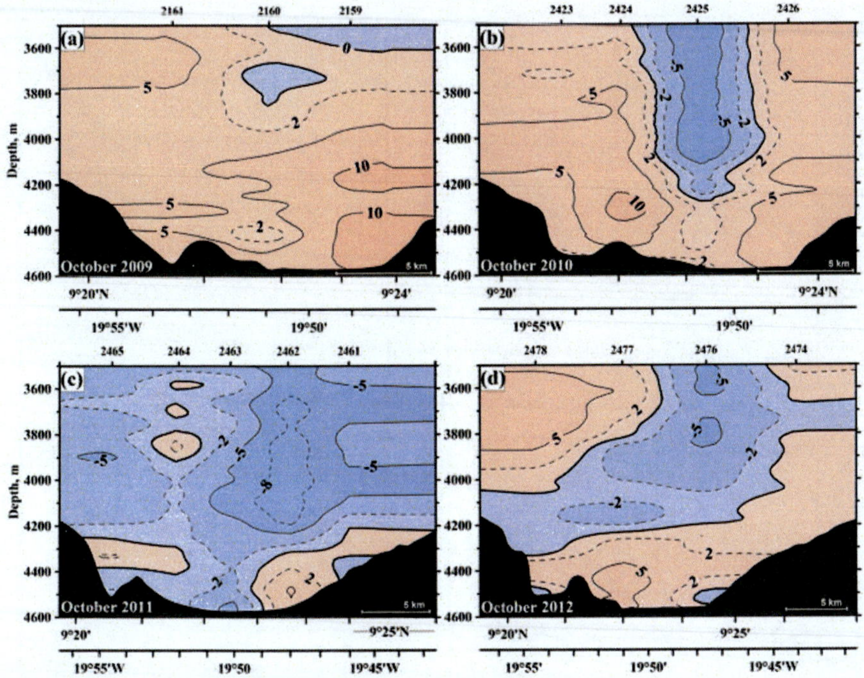

Fig. 7.14 Sections of velocity (normal to the section, positive to the northwest) (cm s^{-1}) (below) across the Kane Gap in October 2009 (**a**), October 2010 (**b**), October 2011 (**c**), and October 2012 (**d**). Numbers of stations are shown at the top axis and coordinates are below. The horizontal scale is shown in the bottom right angle. Reproduced from (Morozov et al. 2013)

differences in salinity along isopycnals reached 0.003 PSU. In October 2010 and 2011, such inhomogeneity was not found. In addition, the existence of different intrusions of more and less saline water types was recorded, for example near the bottom at station 2463 in 2011 (Fig. 7.15c). One can also see such intrusions at the stations of section in 2012 (Fig. 7.13d). In October 2009, we observed a temperature inversion, which was not density compensated (a loop in the bottom layer at station 2160 in Fig. 7.15a). We also indicate minimum potential temperatures measured during the sections in different years: 1.846 °C in 2009, 1.836 °C in 2010, 1.855 °C in 2011, and 1.832 °C in 2012.

The estimates of transports in the layer $\theta < 1.9$ °C were 0.27 Sv (to the southeast) in May 2009; 0.11 Sv, 0.19 Sv, and 0.12 Sv in October of 2009, in 2010, and in 2012, respectively. In October 2011, the transport estimates in two directions were 0.03 Sv to the northwest and 0.05 Sv to the southeast, which jointly gave an almost zero transport of 0.02 Sv to the southeast. The transport was estimated by interpolating the LADCP velocities onto a grid over the section across the channel using 10-m layers. The transport estimates through the Kane Gap even when the transport was the highest are almost two times smaller than the transport of bottom waters

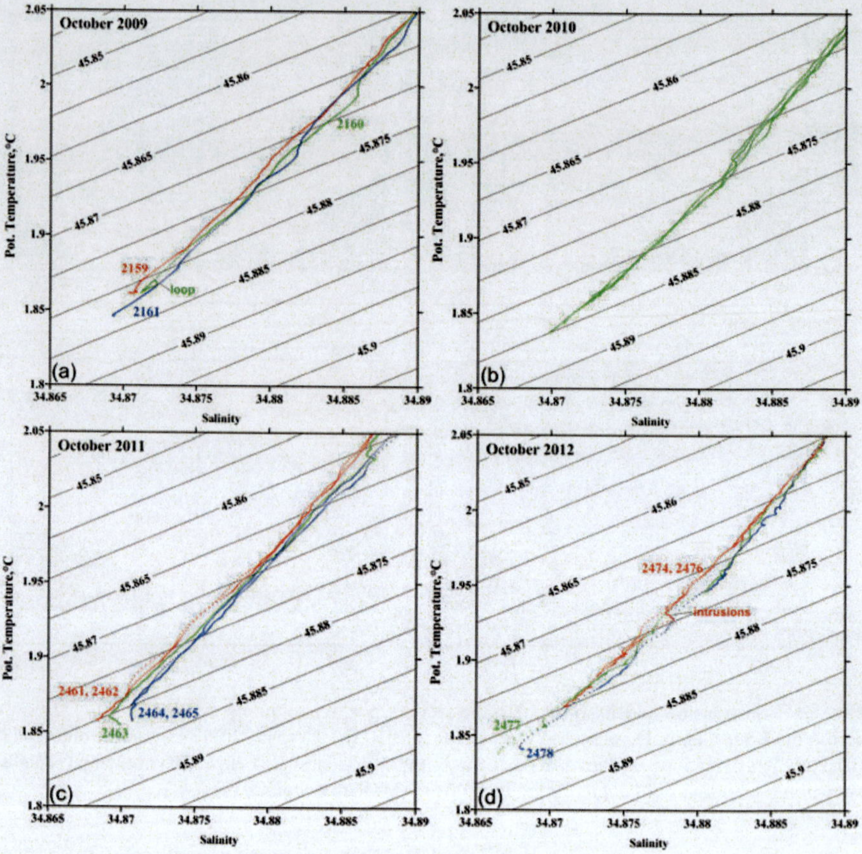

Fig. 7.15 Curves of θ, *S*-distributions of the bottom layer data from the sections across the Kane Gap in October 2009 (**a**), October 2010 (**b**), October 2011 (**c**), and October 2012 (**d**). Isopleths of potential density σ_4 are also shown. Reproduced from (Morozov et al. 2013)

through the Vema Fracture Zone, the source of bottom waters in the Northeast Atlantic Ocean.

In this research we used the data of only one mooring. We estimate the accuracy of calculating the transport based only on the current meters on this mooring and one CTD/LADCP cast before deployment of the mooring. With this in mind, we calculated the transport based on the data of sections using only one LADCP profile with extreme profiles of velocity. The estimated intervals of transports were 0.06–0.20 Sv in October 2009, 0.03–0.22 Sv in 2010, −0.15-0.08 Sv in 2011, and 0.03–0.19 Sv in 2012. Thus, the estimate of transport based on one station fluctuates approximately within ±0.1 Sv relative to the calculations based on the whole set of data along the section. The bottom topography is another factor influencing the estimate of transport. For example, the previously published transport in May 2009 (0.16 Sv) (Morozov et al. 2013) is almost 1.5 times smaller than the estimate of

0.27 Sv owing only to the new data that appeared in the digital bottom topography database in the Internet (http://topex.ucsd.edu/cgi-bin/get_data.cgi) and also owing to the measurements during our echo sounder survey in 2012.

In 2009 (may and October), we found oppositely directed flows in the Kane Gap in May and October based on the LADCP measurements. This finding gave us an idea to carry out long-term measurements of currents and temperature in the Kane Gap. We deployed a mooring for a year to investigate the time variation of the flow. The mooring with three current meters started to operate in the bottom layer in October 2010.

Figure 7.16 shows the transport rose for all three current meters. The rose (similar to the wind rose graph) was calculated in two steps. In the first step, we performed summing of all velocity measurements in each of the 5-degree sector. In the second step, the transport estimates in each of the sector were normalized by the total transport over the 360-degree circle and related to the sector angle in degrees. In our case, the transport rose almost repeats the rose of currents, i.e., the distribution of currents by directions. However, since the transport rose takes into account the intensity of the current, we present only the transport rose here.

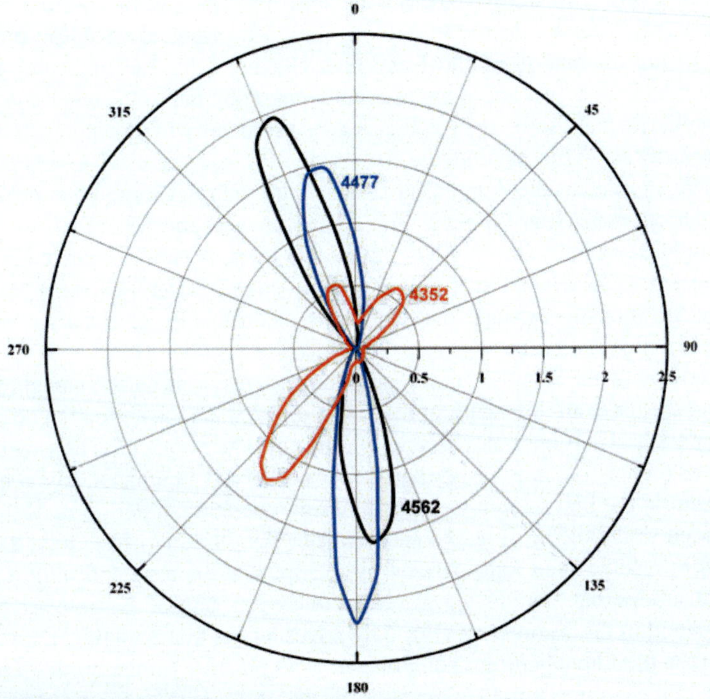

Fig. 7.16 Rose of the flow transports (percent per degree angle) over a period of the year based on the data of three current meters located at 4352, 4477, and 4562 m. Reproduced from (Morozov et al. 2013)

It is seen from Fig. 7.16 that the currents at depths of 4562 and 4477 m (at the lower and middle levels) have a bimodal structure. There are only two directions of transport. The transport at 4562 m is directed to the south with a slight deviation to the east or to the north-northwest. The transport at 4477 m is slightly different. The general alternating directions of the flow are to the south or to the north with a slight deviation to the east. The mean velocities in the direction of the maximum transport were 14 cm s^{-1} to the south and 11 cm s^{-1} to the north at the lower level; 17 cm s^{-1} and 12 cm s^{-1}, respectively at the middle level. The maximum daily average velocities were 20 cm s^{-1} to the south and 15 cm s^{-1} to the north at the lower level and 22 cm s^{-1} and 17 cm s^{-1}, respectively at the middle level. These directions of the currents do not correspond to the general northwest–southeast orientation of the Kane Gap channel. We think that it is presumably caused by the local orientation of the underwater hill slope, near which the mooring was deployed.

Four directions were distinguished at a level of 4352 m. Two of them approximately correspond to the directions of the currents at the lower levels. The mean velocities in these directions were 8 cm s^{-1} to the south and 10 cm s^{-1} to the north; the maximum daily average values were 9 cm s^{-1} to the south and 14 cm s^{-1} to the north. The two other directions are east-southeast and northwest. The main transport at this level corresponds to these pairs of directions unlike the transport at the middle and lower levels. The mean velocities in both directions were 14 cm s^{-1} to the southwest and 9 cm s^{-1} to the northeast. These directions are actually transverse relative to the orientation of the Kane Gap channel. This fact indicates that the currents at 4352 m adjust not to the channel topography but to the other constraints.

A peculiarity that can be seen in Fig. 7.16 is asymmetry of transport directions at two lower levels. This asymmetry determines the existence of a small current component normal to the Kane Gap channel. Its value average over time at the bottom current meter was 1.6 cm s^{-1} strictly to the west and 1.3 cm s^{-1} to the west with a slight deviation to the south at the middle level. A similar calculation for the upper current meter resulted in a velocity of 1.6 cm s^{-1} strictly to the west. This is just a vector sum of year-long measurements of differently oriented currents that hardly has a physical basis.

The graphs of the meridional velocity components with daily averaging of the measurements over all three current meters (Fig. 7.17) demonstrate high correlation of these curves at sub-inertial frequencies. The currents of one direction with minor exceptions (either southerly or northerly) were observed simultaneously at all three levels. During the first 6 months of the mooring operation, a period of 180 days is clearly seen in the record (Fig. 7.17). However, the duration of the record did not allow us to resolve such long periods by the spectral analysis and only a 90-day period of sub-inertial motions was revealed in the first half of the time series and a 44-day period in the second half (Fig. 7.18). During the first 3 months (November–January) the meridional current components were positive (to the north). Then, the direction of the currents turned to the opposite and from February to the beginning of April the meridional velocity components were negative. In May 2011, the regime of the currents changed to predominant oscillations with a higher (sub-inertial) frequency. We estimate the average period of these fluctuations of shorter period at

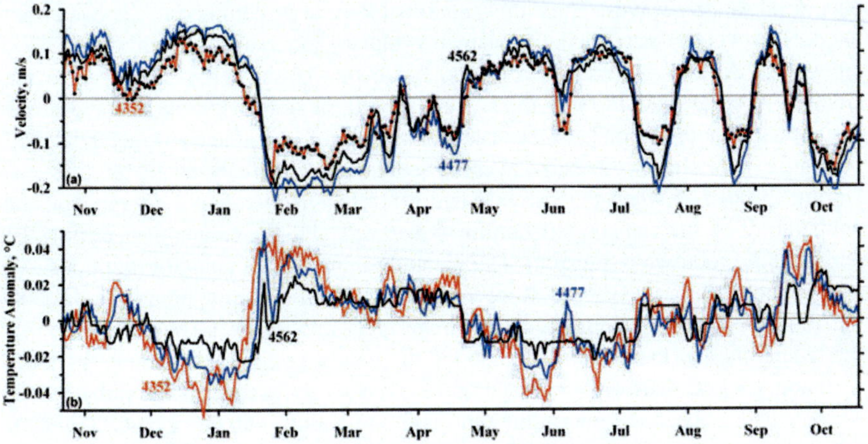

Fig. 7.17 Graphs of the current component along the main axis of the Kane Gap (north to south direction, positive to the north) (**a**) and temperature anomaly (**b**). The bottom axis shows the sequential days and months of measurements. Reproduced from (Morozov et al. 2013)

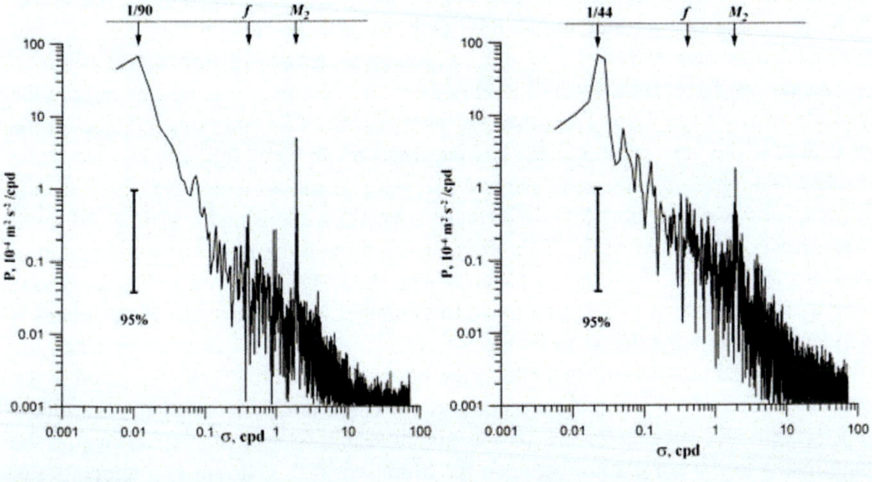

Fig. 7.18 Nearly raw spectra of kinetic energy from currents at 7 m above the bottom for the periods between November 2010 and May 2011 (left) and between May and October 2011 (right). 95% confidence intervals are shown. Reproduced from (Morozov et al. 2013)

44 days (Figs. 7.17 and 7.18). We note that such effect was not seen in the zonal velocity component, which is related to the presence of four transport directions at the upper level unlike two at the middle and lower levels (Fig. 7.16).

The seasonal periodicity of current directions in the Kane Gap can be related to the seasonal variability of transport in the Romanche Fracture Zone found in (Mercier and Speer 1998). Thierry et al. (2006) showed that the seasonal signal, which is observed in the Romanche Fracture Zone, was caused by the propagation of equatorial Rossby waves. Thus, this signal is likely to be confined to the equatorial band. However, an increase in the eastward velocity and transport could increase the amount of AABW in the Sierra Leone Basin that could partly discharge to the Gambia Abyssal Plain through the Kane Gap and the northerly flow in the passage could be a remote response to the seasonal signal in the Romanche Fracture Zone.

Figure 7.17 shows the daily average temperature anomaly at all three levels. During the period when the currents had a northward meridional component, the temperature of the transported bottom water was slightly cooler than during the period of the southward transport. The greatest temperature anomalies (both positive and negative) up to 0.05 °C were found at the middle and upper levels. The maximum temperature anomaly at the lower level was 0.025 °C. We also note that the temperature at the bottom (7 m above the bottom) varied in the range between 1.80 and 1.92 °C. The daily average temperatures vary between 1.82 and 1.88 °C. These values exceed the minimum and maximum values measured by CTD instruments. We emphasize that the minimum temperature recorded by CTD measurements was $\theta = 1.832$ °C in October 2012, (Fig. 7.12d).

The correlation of the southward (northward) transport with warming (cooling) of the bottom flow is most likely explained by the fact that bottom waters from the southern source (Romanche Fracture Zone) are cooler than those from the northwestern source (Vema Fracture Zone). This can also be seen in Fig. 7.17. Cooler bottom temperatures (relative to the temperatures at the sill) are found south of the Kane Gap rather than north of it. However, this statement can be true only in the vicinity of the Kane Gap. The other explanation of temperature dependence on the direction of the flow related to the Ekman friction can be as follows. The jet of the coldest water was displaced to the southwestern slope, which is explained by the Ekman frictional boundary layer (Northern Hemisphere) during the northward water flow (Speer and Zenk 1993). This effect is seen on three of the four temperature sections (Fig. 7.12a, b, d) and not seen in Fig. 7.13c when the velocities were low. The densest (coldest) water flowing in the channel in the Northern Hemisphere is displaced to the left relative to its motion, i.e., to the southwestern slope when the flow is directed to the northwest and to the northeast when the flow is directed to the southeast. Considering that the temperature measurements were made on the mooring, which was located closer to the southwestern slope, this effect would be manifested in the correlation between the direction of the current and temperature. However, this effect seems not likely because the temperature difference measured on the mooring during the year is approximately 0.1 °C, which is almost one order of magnitude greater than the horizontal temperature difference observed over the CTD-sections.

In order to estimate the transport based on the data of moored current meters, we plotted a section similar to the one at the station in May 2009 (Fig. 7.11). The transport calculation was made on the basis of three current meters in the layer from 4352 m to the bottom. It is seen from Figs. 7.11 and 7.12 that this depth is close to the $\theta = 1.9$ °C isotherm. The data were interpolated on a vertical grid with a step of

10 m. Then, the calculation was similar to the calculations from the LADCP data over the sections.

The maximum daily average velocities in the southern direction reached 20 cm s^{-1}, while the maximum north-northwesterly currents were as high as 15 cm s^{-1}. The daily mean transport varied in this layer between 0.36 Sv to the south and 0.33 Sv to the north. The mean transport over the period when the current was directed to the south was 0.20 Sv and 0.23 Sv when the current was directed to the north. The total transport over the entire period of observations (1 year) was 0.014 Sv to the north. This transport is determined by the previously mentioned westerly drift component of the current and seems to be quasi-constant. We emphasize that this total transport does not characterize the long-term transport through the Kane Gap (which is actually zero). In some sense, this small value is random depending on the specific point of the mooring location and line of the section across the Kane Gap.

It is interesting to compare the transports based on the LADCP data over sections in 2010 and 2011 with the transports based on the mooring data in the beginning and in the end of its operation. The transport over the period from October 21, 2010 14:00 to October 22, 2010 14:00 (beginning of the mooring operation) was 0.23 Sv to the north and from 17 October, 2011 9:30 to 18 October, 2011 9:30 (last day) was 0.16 Sv to the south. The former is close to the estimate of the transport during the section occupied in the closest time (0.19 Sv to the north in 2010); the latter is strongly different (0.02 Sv to the south in 2011). We note that the variation in the mean velocity over 1 day can reach 10 cm s^{-1}, which is especially characteristic of the period of current reconstruction. This is clearly seen in Fig. 7.17a, in which every daily record is denoted with a dot. The indications of such a situation are observed over the section in October 2011, which was occupied during the next day after recovery of the mooring. This is manifested first of all in the different directions of currents over the section.

We estimated the volume of bottom water transported through the Kane Gap during unidirectional flow and the square of the water basin territory occupied by the water transported through the channel. The mean transport during the longest period of the permanent flow to the southeast (February–April 2011) was 0.20 Sv. During this period, 1300 km^3 of AABW was transported to the south of the Kane Gap. If we assume that the mean depth south of the Kane Gap is 4650 m and the layer of AABW is located below 4000–4200 m, we estimate that the region filled with the bottom water transported from the north is limited by a distance of approximately 50 km from the sill of the Kane Gap. This estimate is valid also to the north of the Kane Gap.

We compared our direct measurements of velocity in the Kane Gap with the geostrophic estimates based on the results of other expeditions. We consider previous CTD-sections occupied close to the Kane Gap from the north and south of the passage. McCartney et al. (1991) demonstrated almost zero westerly geostrophic transport of bottom waters with $\theta < 2.0$ °C north of the Kane Gap along 29° W. In

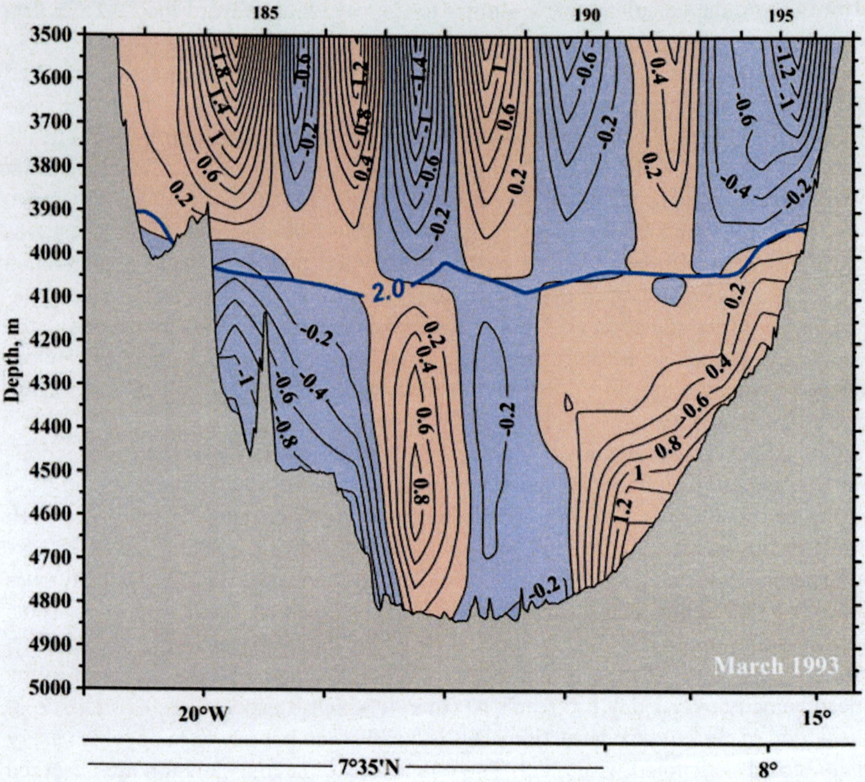

Fig. 7.19 Geostrophic velocities (cm s^{-1}) over section A6 in 1993

1993 and 2000, two hydrographic sections were occupied close to the Kane Gap. The WOCE A6 section was made by French oceanographers in 1993 and the section in 2000 was made by Russian scientists. These sections crossed the Sierra Leone Basin approximately at 8° N. Geostrophic velocities of bottom water across these sections were negligible. The sections are shown in Figs. 7.19 and 7.20.

If we take $\theta = 1.9$ °C as the upper boundary, the resulting transport would be almost zero. Calculation with a boundary at isotherm $\theta = 2.0$ °C yields a northerly transport equal to 0.06 Sv in 1993 and 0.14 Sv in 2000, which are both close to zero. We can conclude that no significant transport exists here, because the estimates are within the limits of error for geostrophic calculations.

The geostrophic section in 2000 demonstrates a northerly flow along 18° W. The salinity section shown below in Fig. 7.21 indicates that salinity of the water transported with this flow is different from bottom salinities in the other part of the section. Salinity and density contour lines are elevated above the background level (passage at 18° W) by almost 300 m, indicating the existence of a southward bottom slope flow whose core is located in this passage.

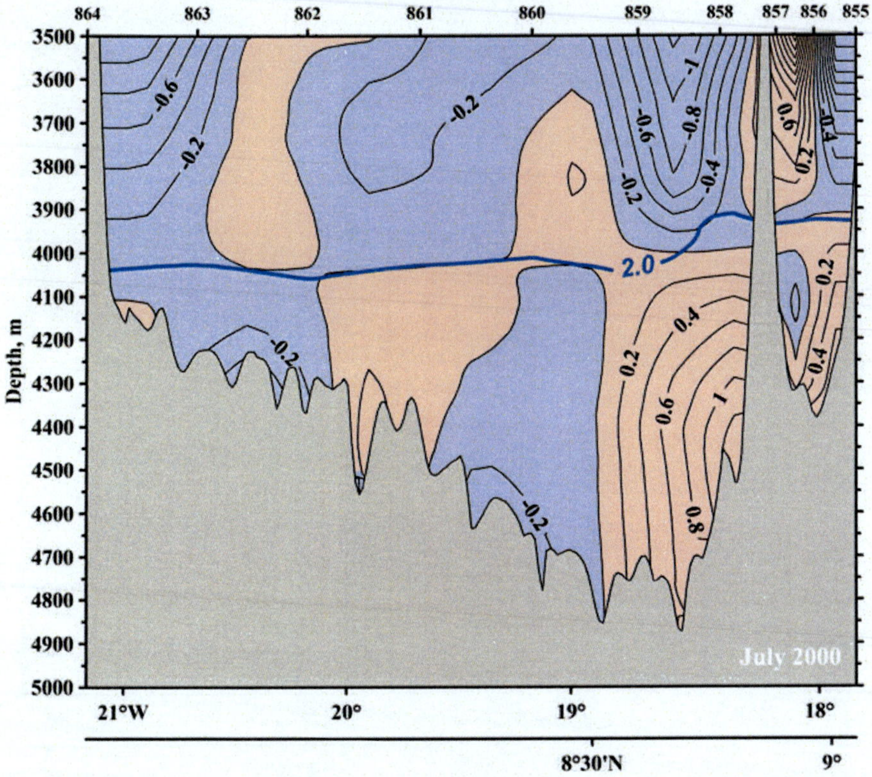

Fig. 7.20 Geostrophic velocities (cm s^{-1}) over the section in 2000

7.4 Angola Basin

The Angola Basin located in the Southeast Atlantic is a deep basin. The major part of the basin is deeper than 5000 m while the maximum depth exceeds 5700 m (Fig. 7.22). It is bounded from the west by the Mid-Atlantic Ridge. Its crest is crosscut by many narrow fractures not deeper than 3500 m (Warren and Speer 1991). From the south the Angola Basin is bounded by the Walvis Ridge. The Guinea Rise bounding the basin from the north is the deepest barrier. The depths of the sills in the passages of the Guinea Rise exceed 4000 m.

It is seen from the distribution of potential temperature near the bottom in the Atlantic Ocean (Fig. 2.3) that bottom water with potential temperature less than 2 °C penetrates through the passages in the Guinea Rise. However, ridges of the Guinea Rise almost block the water with potential temperature below 1.89 °C from propagating south into the Angola Basin. The depth of this isotherm over WOCE section

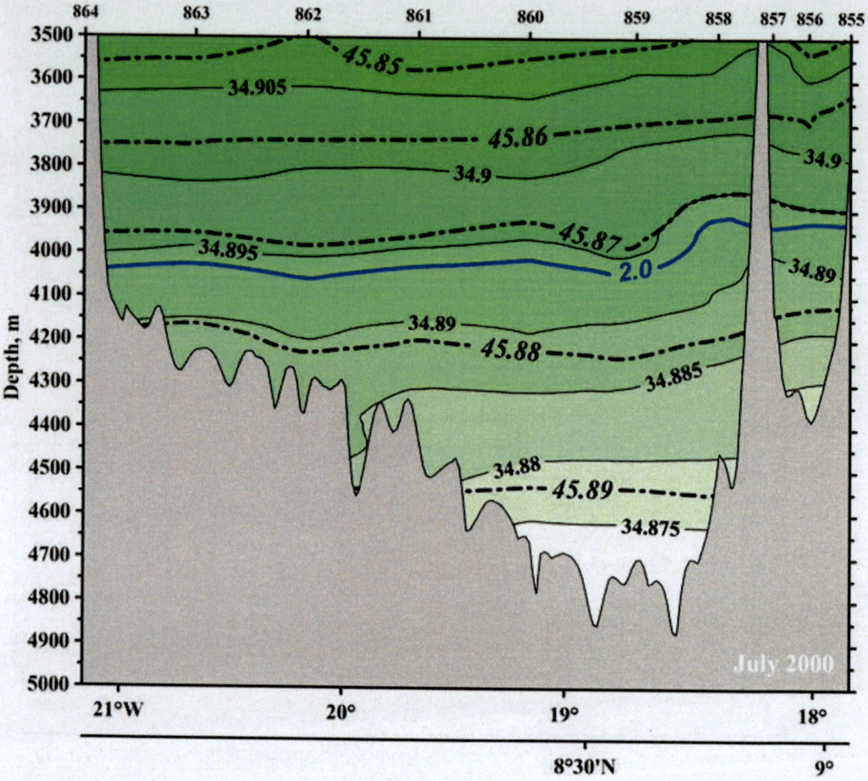

Fig. 7.21 Salinity (PSU) distribution over the section in 2000. Potential isotherm θ = 2.0 °C is shown with a blue line. Isopycnals referenced to 4000 m (σ_4 in kg/m^3) are shown

A13 (R/V *L'Atalante*, France, 1995) in the northern part of the Angola Basin suggests that the sill depth is about 4300 m (Fig. 7.23).

Connary and Ewing (1974) showed the existence of a narrow passage (5–10 miles wide) in the Walvis Ridge in the southern part of the Angola Basin near the junction of the Walvis and Mid-Atlantic ridges at 36° S, 7° W. The depth of the sill is 4120 m. This passage ('Walvis Passage' according to (Connary and Ewing 1974)) is the main pathway for Antarctic Bottom Water (in terms of Wüst (1936)) transport from the Cape Basin to the Angola Basin. Figure 7.24 shows the potential temperature distribution based on the WOCE A14 meridional CTD section approximately along 9° W (R/V *L'Atalante*, France, 1995). The coldest bottom water in the Angola Basin is found here south of 33° S. However, additional spreading of Antarctic Bottom Water to the Angola Basin through a small break in the Walvis Ridge at 31–32° S, 2° E was reported in (Shannon and Rijswijck 1969; Shannon and Chapman 1991).

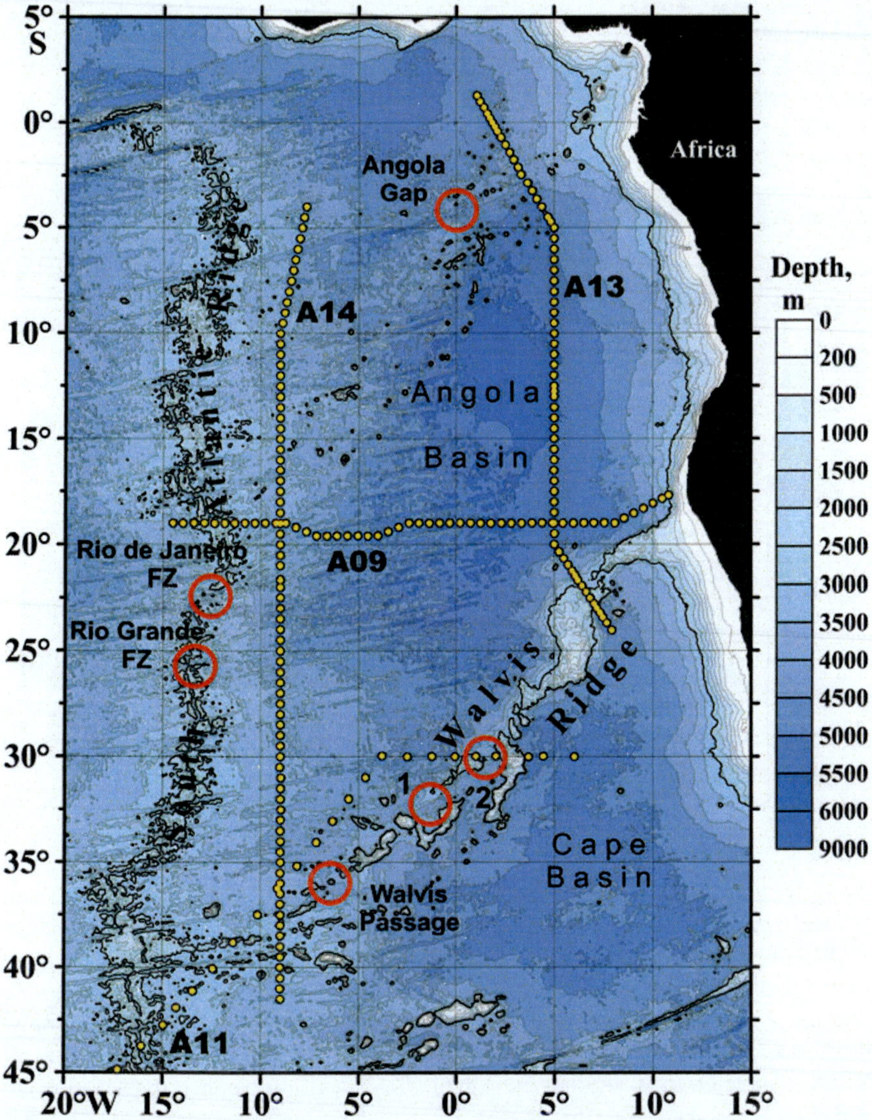

Fig. 7.22 Bathymetric chart of the Angola Basin and locations of WOCE stations. Possible passages for Antarctic Bottom Water inflow to the basin are indicated

On the basis of the chart published in (Needham et al. 1986), Warren and Speer (1991) reported the existence of a passage near 32°40′ S, 2°20′ W with a sill depth between 4000 and 4250 m, which can be the deepest corridor. However, the charts of

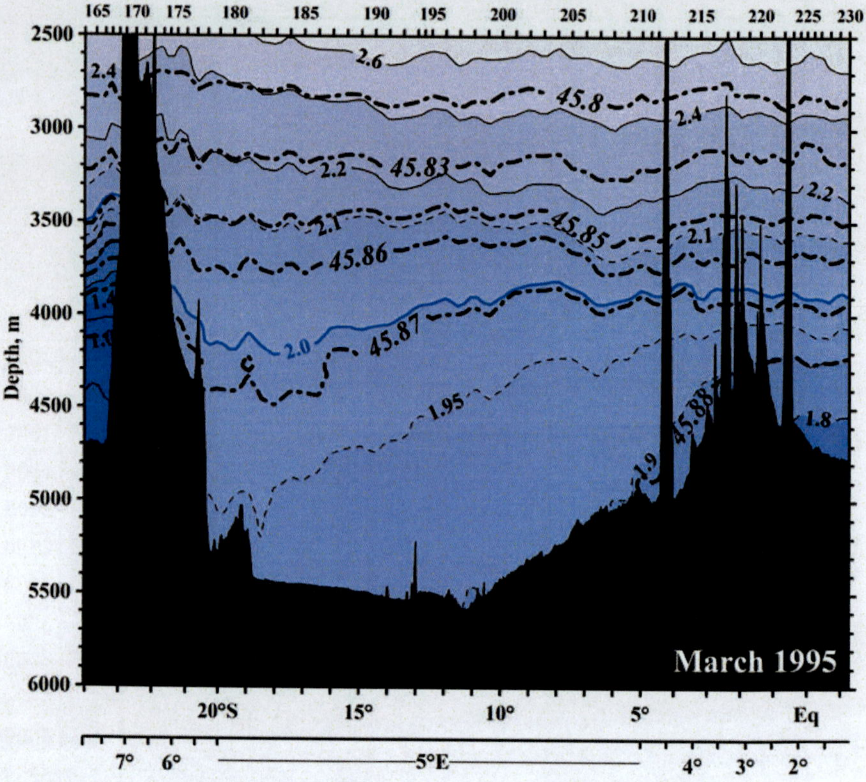

Fig. 7.23 Distribution of potential temperature (°C) along WOCE section A13

the distribution of different parameters at the bottom show that this flow does not influence strongly the water properties in the Angola Basin. This is confirmed by the WOCE A11 section (R/V *Discovery* in 1993). Potential temperature distribution shows that the inflow of cold waters occurs near 7–8° W (Fig. 7.25). Namib Col passage (22° S, 7° E) (Speer et al. 1995) is not a very deep passage that allows only the flow of North Atlantic Deep Water into or from the Angola Basin. The flow of North Atlantic Deep Water into the Angola Basin was also studied in (Arhan et al. 2003).

The depths of deep passages in the Mid-Atlantic Ridge south of the equatorial zone prevent the propagation of Antarctic waters. Analysis performed in (Mercier et al. 2000; Sandoval and Weatherly 2001) suggests that Antarctic waters propagate to the Southeast Atlantic through the Rio de Janeiro Fracture Zone (22° S) in the Mid-Atlantic Ridge. Mercier et al. (2000) analyzed the signatures of through-flows at the Rio de Janeiro Fracture Zone (22° S) and the Rio Grande Fracture Zone (26° S)

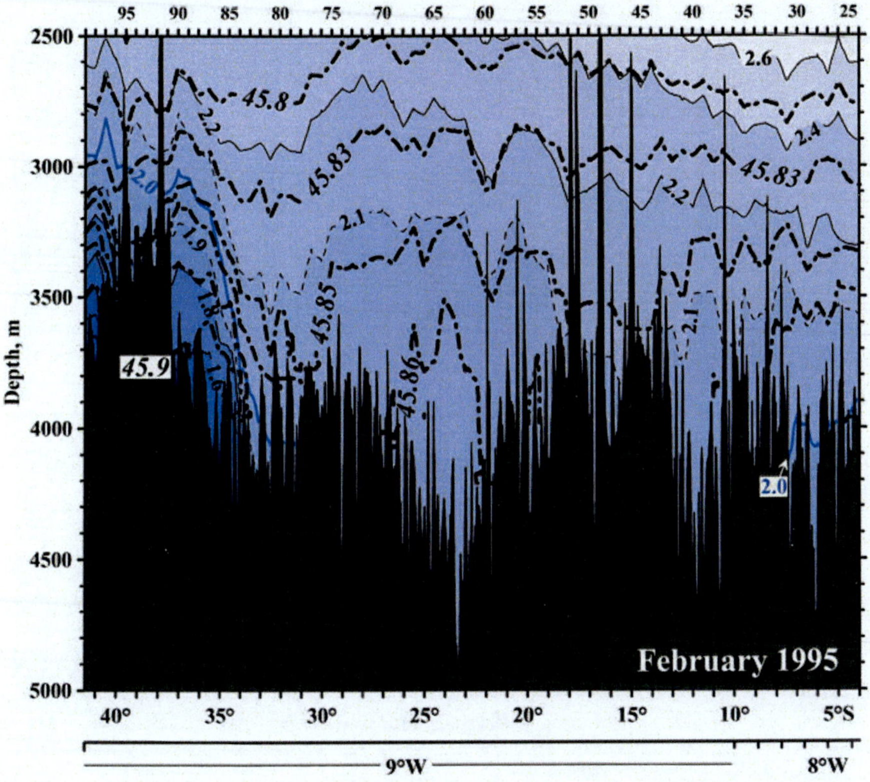

Fig. 7.24 Distribution of potential temperature (°C) along WOCE section A14

in the Mid-Atlantic Ridge in the tracer fields over a hydrographic section along 9° W. Bottom water flows are identified from anomalies of water properties in the Angola Basin. The through-flow water is supplied by a meridional band of cold and fresh water located against the western flank of the Mid-Atlantic Ridge. The authors calculated the geostrophic transport, which appeared as high as 0.5 Sv. The Tristan de Cunha Fracture Zone remains unstudied as a possible pathway for bottom waters. Figure 7.26 shows the potential temperature distribution based on the WOCE A9 CTD section across the Angola Basin occupied along 18°-19° S (R/V *Meteor*, Germany, 1991). The figure demonstrates that there is no strong inflow of bottom waters from the west through the Mid-Atlantic Ridge at 18–19° S. Only a small deep region near the bottom is occupied with cold water with potential temperature 1.95 °C.

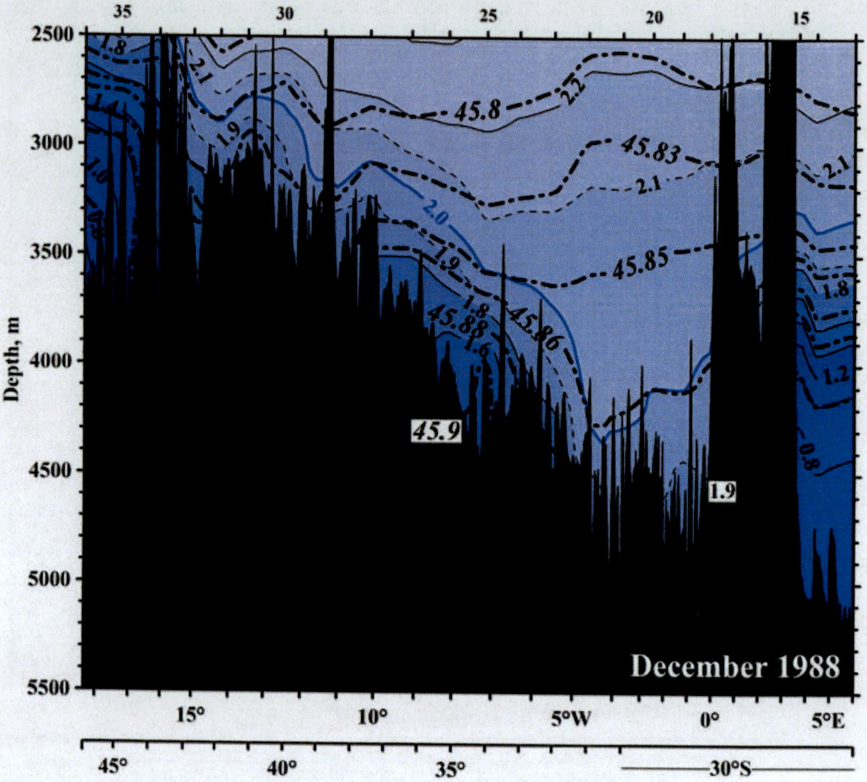

Fig. 7.25 Distribution of potential temperature (°C) along WOCE section A11

We can conclude that at depths exceeding 4000 m, the bottom water that propagates to the Angola Basin from the north across the Guinea Rise flows southward along the slopes of the Mid-Atlantic Ridge. In the southern part of the Angola Basin, a northward boundary current is supplied by the water propagating through the Walvis Passage.

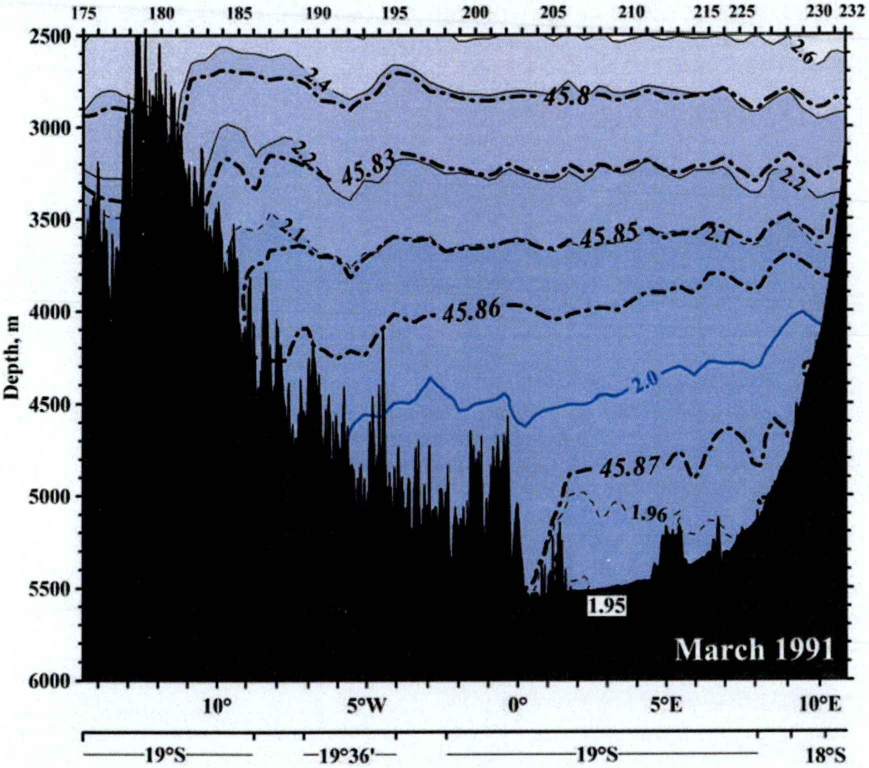

Fig. 7.26 Distribution of potential temperature (°C) along WOCE section A09

References

Álvarez M, Pérez FF, Bryden H, Ríos AF (2004) Physical and biogeochemical transports structure in the North Atlantic subpolar gyre. J Geophys Res 109:C03027. https://doi.org/10.1029/2003JC002015

Arhan M, Mercier H, Bourlès B, Gouriou Y (1998) Hydrographic section across the Atlantic at 7°30′ N and 4°30′ S. Deep-Sea Res I 45:829–872

Arhan M, Mercier H, Park Y-H (2003) On the deep water circulation of the eastern South Atlantic Ocean. Deep-Sea Res I 50:889–916

Boyer TP, Baranova OK, Coleman C, Garcia HE, Grodsky A, Locarnini RA, Mishonov AV, Paver CR, Reagan JR, Seidov D, Smolyar IV, Weathers K, Zweng MM (2018) World Ocean database 2018. AV Mishonov, technical Ed., NOAA Atlas NESDIS 87. https://www.ncei.noaa.gov/sites/default/files/2020-04/wod_intro_0.pdf

Connary SD, Ewing M (1974) Penetration of Antarctic bottom water from the Cape Basin into the Angola Basin. J Geophys Res Oceans 79:463–469

Dudkov I, Sivkov VV, Dorokhov DV, Bashirova LD (2020) Multibeam bathymetry data from the Kane Gap and south-eastern part of the Canary Basin (eastern tropical Atlantic). Data Brief 32:106055. https://doi.org/10.1016/j.dib.2020.106055

Gargett AE (1984) Vertical eddy diffusivity in the ocean interior. J Mar Res 42:359–393

Heezen BC, Gerard RD, Tharp M (1964) The Vema fracture zone in the equatorial Atlantic. J Geophys Res Oceans 69:733–739. https://doi.org/10.1029/JZ069i004p00733

Hobart MA, Bunce ET, Sclater JG (1975) Bottom water flow through the Kane Gap, Sierra Leone rise, Atlantic Ocean. J Geophys Res Oceans 80:5083–5088

Lavin AM, Bryden HL, Parrilla G (2003) Mechanisms of heat, freshwater, oxygen and nutrient transports and budgets at 24.5° N in the subtropical North Atlantic. Deep-Sea Res I 50:1099–1128

Lherminier P, Mercier H, Gourcuff C, Alvarez M, Bacon S, Kermabon C (2007) Transports across the 2002 Greenland-Portugal Ovide section and comparison with 1997. J Geophys Res Oceans 112:C07003. https://doi.org/10.1029/2006JC003716

Mantyla AW, Reid JL (1983) Abyssal characteristics of the World Ocean waters. Deep-Sea Res I 30(8):805–833. https://doi.org/10.1016/0198-0149(83)90002-X

McCartney MS, Bennet SL, Woodgate-Jones ME (1991) Eastward flow through the mid-Atlantic ridge at 11° N and its influence on the abyss of the eastern basin. J Phys Oceanogr 21(8):1089–1121. https://doi.org/10.1175/1520-0485(1991)021<1089:EFTTMA>2.0.CO;2

Mercier H, Morin P (1997) Hydrography of the Romanche and chain fracture zones. J Geophys Res Oceans 102(C5):10,373–10,389. https://doi.org/10.1029/97JC00229

Mercier H, Speer KG (1998) Transport of bottom water in the Romanche fracture zone and the chain fracture zone. J Phys Oceanogr 28(5):779–790. https://doi.org/10.1175/1520-0485(1998)028<0779:TOBWIT>2.0.CO;2

Mercier H, Weatherly G, Arhan M (2000) Bottom water throughflows at the Rio de Janeiro and Rio Grande fracture zones. Geophys Res Lett 27:1503–1506

Morozov EG (1995) Semidiurnal internal wave global field. Deep-Sea Res I 42(1):135–148

Morozov EG (2018) Oceanic internal tides. Observations, analysis and modeling. A global view. Springer, 316p

Morozov EG, Tarakanov RY, van Haren H (2013) Transport of AABW through the Kane Gap, tropical NE Atlantic Ocean. Ocean Sci 9:825–835. https://doi.org/10.5194/os-9-825-2013

Morozov EG, Tarakanov RY, Frey DI, Demidova TA, Makarenko NI (2018) Bottom water flows in the tropical fractures of the Northern Mid-Atlantic Ridge. J Oceanogr 74(2):147–167. https://doi.org/10.1007/s10872-017-0445-x

Needham HD, Carré D, Sibuet J-C (1986) Carte Bathymétrique de la Ride de Walvis: Océan Atlantique Sud. Echelle 1/4. 382, 832. IFREMER, Direction de l'Environnement et des Recherches Océaniques, Département Géosciences Marines

Sandoval FJ, Weatherly GL (2001) Evolution of the deep western boundary current of Antarctic bottom water in the Brazil Basin. J Phys Oceanogr 31(6):1440–1460. https://doi.org/10.1175/1520-0485(2001)031<1440:EOTDWB>2.0.CO;2

Saunders PM (1987) Flow through discovery gap. J Phys Oceanogr 17(5):631–643

Saunders PM (1994) The flux of overflow water through the Charlie-Gibbs Fracture Zone. J Geophys Res Oceans 99(C6):12343–12355

Shannon LV, Chapman P (1991) Evidence of Antarctic bottom water in the Angola Basin at 32°S. Deep-Sea Res I 38(10):1299–1304

Shannon LV, van Rijswijck M (1969) Physical oceanography of the Walvis Ridge region. Investigational report no 70, Division of Sea Fisheries of RSA [Available from the library of the University of Cape Town]

Smith WHF, Sandwell DT (1997) Global Sea floor topography from satellite altimetry and ship depth soundings. Science 277:1956–1962. http://topex.ucsd.edu/cgi-bin/get_data.cgi. https://doi.org/10.1126/science.277.5334.1956

Speer KG, Zenk W (1993) The flow of Antarctic bottom water into the Brazil Basin. J Phys Oceanogr 23:2667–2682

Speer KG, Siedler G, Talley L (1995) The Namib Col current. Deep-Sea Res I 42:1933–1950

Stephens JC, Marshall DP (2000) Dynamical pathways of Antarctic bottom water in the Atlantic. J Phys Oceanogr 30(3):622–640

Tarakanov RY, Morozov EG, Gritsenko AM, Demidova TA, Makarenko NI (2013) Transport of Antarctic bottom water through passages in the East Azores Ridge (37° N) in the East Atlantic. Oceanology 53(4):432–441

Tarakanov RY, Morozov EG, Makarenko NI, Frey DI, Demidova TA (2017) Measurements of bottom currents in fractures of the southern part of the mid-Atlantic Ridge on cruises 39, 40, and 41 of the R/V *Akademik Sergey Vavilov*. Oceanology 57(5):753–755

Thierry V, Mercier H, Treguier AM (2006) Seasonal fluctuations in the deep central equatorial Atlantic Ocean: a data–model comparison. Ocean Dyn 56:581–593. https://doi.org/10.1007/s10236-005-0045-y

van Aken HM (2000) The hydrography of the mid-latitude Northeast Atlantic Ocean I: the deep-water masses. Deep-Sea Res I 47:757–788

van Bennekom AJ (1985) Dissolved silica as an indicator of Antarctic bottom water penetration, and the variability in the bottom layers of the Norwegian and Iceland Basins. Rit Fiskideildar No 9, pp 101–109

Vlasenko V, Hutter K (2002) Numerical experiments on the breaking of solitary internal waves over a slope-shelf topography. J Phys Oceanogr 32:1779–1793

Warren BA, Speer KG (1991) Deep circulation in the eastern South Atlantic Ocean. Deep-Sea Res I 38(Suppl 1):281–322

WOD18 (2018) World Ocean Database 2018. https://www.nodc.noaa.gov/OC5/WOD/pr_wod.html

Worthington LV, Wright WR (1970a) North Atlantic Ocean Atlas of potential temperature, salinity, and oxygen profiles from the Erika Dan Cruise of 1962. Woods Hole Oceanographic Institution Atlas Series 2, pp 1–82

Worthington LV, Wright WR (1970b) North Atlantic Ocean Atlas of potential temperature and salinity in the deep water. Woods Hole Oceanographic Institution Atlas Series 2, 24 pp

Wüst G (1936) Schichtung und Zirkulation des Atlantischen Ozeans, Das Bodenwasser und die Stratosphäre. In: Defant A (ed) Wissenschaftliche Ergebnisse, Deutsche Atlantische Expedition auf dem Forschungs – und Vermessungsschiff "Meteor" 1925-1927, 6(1), Walter de Gruyter, Berlin, 411p

Chapter 8
Passages in the East Azores Ridge

Abstract The structure of northerly overflow of Antarctic Bottom Water (AABW) in the East Atlantic through passages in the East Azores Ridge (37° N) from the Madeira Basin to the Iberian Basin is studied on the basis of hydrographic measurements in October 2011 and in September 2013, historical data from the WOD18 database, and recent data on bottom topography. The overflow of the coldest water occurs through two passages with close depths at 16° W (Discovery Gap) and at 19°30′ W (Western Gap). It is shown that it is likely that the role of the latter passage in bottom water transport was underestimated in earlier publications because the water ($\theta > 2.01$ °C) found in the region north of the Western Gap was cooler than in the region north of the Discovery Gap ($\theta > 2.03$ °C). In 2011, we found a decrease by 0.01 °C in Antarctic Bottom Water temperature near the bottom in the Discovery Gap compared to previous measurements in 1982 (from 2.011 °C to 2.002 °C). The potential temperature at the bottom measured in 2013 in the Western Gap was 1.995 °C. The transport of bottom water through the Western Gap was not less than through the Discovery Gap.

The Discovery Gap located at 37° N, 16° W was previously considered the boundary of Antarctic Bottom Water spreading to the north in the sense that this is the water with potential temperature below $\theta = 2$ °C. In 2011, the transport of bottom water with potential temperatures below 2.05 °C was 0.24 Sv. The measurements in 2013 in the Western Gap (19° W) showed the presence of a pronounced current of bottom water to the north at speeds of up to 30 cm s^{-1}, which is much stronger than through the Discovery Gap at 16° W. Bottom potential temperatures below 2 °C were found only in a close vicinity north of the passages not exceeding 50 miles.

Keywords East Azores ridge · Discovery gap · Western gap · Terminal point for Antarctic bottom water

E. G. Morozov et al., *Bottom Gravity Currents and Overflows in Deep Channels of the Atlantic Ocean*, https://doi.org/10.1007/978-3-030-83074-8_8

8.1 General Description

The main flow of Antarctic waters in the bottom layer of the Atlantic Ocean to the north from the Weddell Sea occurs through the Scotia Sea, Argentine Basin, and Brazil Basin. As the flow reaches the equatorial zone, part of the flow crosses the Mid-Atlantic Ridge through the Romanche and Chain fracture zones, while the other part flows further to the north in the West Atlantic (Fig. 8.1). Approximately at 11° N, this flow crosses the Mid-Atlantic Ridge through the Vema Fracture Zone and other smaller fractures. This branch is considered the main source of cold bottom waters in the northern part of the East Atlantic. These waters follow to the north generally in the eastern part of the Northeast Atlantic (Saunders 1987), and reach the ridge in the East Azores Fracture Zone (hereinafter the East Azores Ridge).

The East Azores Ridge, located along 37° N, divides the Madeira Basin from the south and the Iberian Basin from the north (Fig. 8.2). The depths of both basins exceed 5300 m. The characteristic depths over the ridge crest are 3500–4500 m (Fig. 8.2). From the viewpoint of a formal definition of Antarctic Bottom Water (AABW) as water with $\theta < 2.0$ °C, the East Azores Ridge is the northern boundary of its spreading in the East Atlantic, because north of this ridge no water with such temperatures has been observed (Saunders 1987). However, many properties of

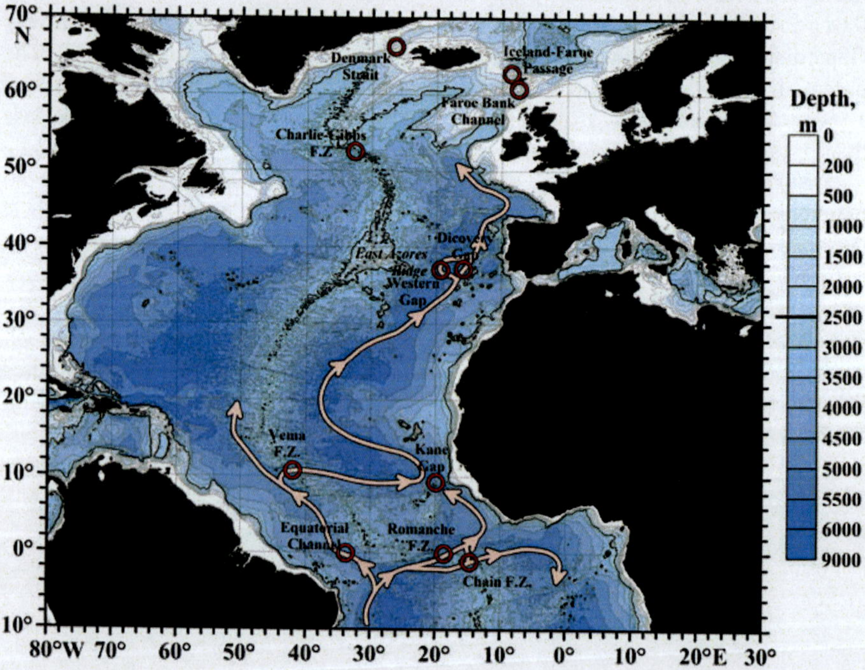

Fig. 8.1 Scheme of Antarctic Bottom Water flow in the Northeast Atlantic directed to the East Azores Ridge. The circles mark the most significant channels and passages connecting individual ocean basins. This scheme also shows the bottom flow after passing the East Azores Ridge

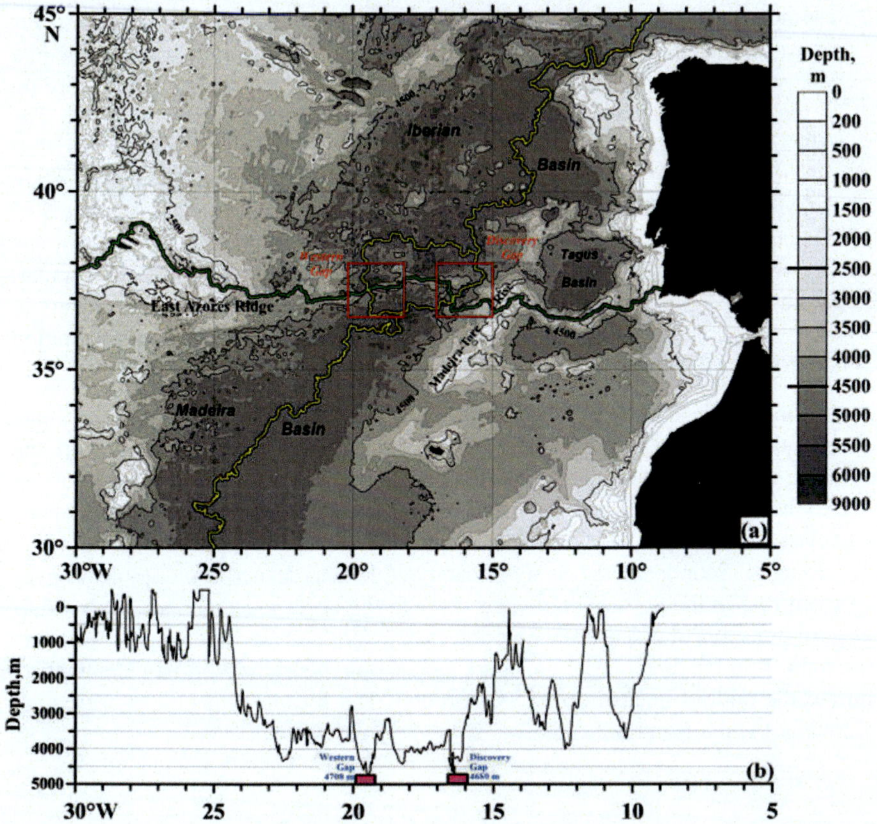

Fig. 8.2 Bottom topography in the region of the Iberian and Madeira basins based on (Smith and Sandwell 1997) digital topography (version V18.1). The yellow and green lines are the thalweg (deepest line) of the East Atlantic and the watershed of the East Azores Ridge, respectively. The red squares mark the regions of the Discovery Gap and Western Gap, shown in Figs. 8.3 and 8.5a. Depth profile (m) along the watershed line as function of longitude (b). (Reproduced from Tarakanov et al. (2013) with permission from Pleiades Publishers)

AABW, for example, high concentration of silicates are conserved even north of this ridge.

At the same time, in a number of works, e.g., (Lherminier et al. 2007) the bottom water in the northern basins of the East Atlantic (Iberian, Biscay, and West European basins) with temperatures $\theta < 2.3$ °C was interpreted as AABW. The water of this temperature range undoubtedly has an Antarctic component propagating from the south through the East Azores Ridge. The causes of this idea are the following:

1. The Mid-Atlantic Ridge, the Faroe-Shetland and Faroe-Iceland thresholds (Fig. 8.1) significantly hamper the penetration of cold water from the northern basins of the West Atlantic and from the Arctic. Cold Arctic water flowing over these thresholds (obviously lower than 2.3 °C) mixes with the surrounding

warmer waters (Tomczak and Godfrey 1994). As a result, the intensity of this flow is sufficient to maintain a cold trace of Arctic waters at the bottom only in a relatively small region of the East Atlantic adjacent to the thresholds. According to the WOD18 database, no such traces have been found south of 58° N. The water inflow with temperatures $\theta < 2.3$ °C to the northern basins of the East Atlantic from the west was found in the Charlie-Gibbs Fracture Zone (53° N) in the Mid-Atlantic Ridge. However, according to the same database, the traces of this outflow are localized in the immediate vicinity of the fracture.

2. The cold bottom water in the basins north of the East Azores Ridge, in particular, in the Iberian Basin, exhibits properties characteristic of the Antarctic waters in the Atlantic: an increased content of silicates and a reduced oxygen concentration (McCartney 1992).

According to (Saunders 1987), two passages exist in the East Azores Ridge (deeper than 4500 m) through which the coldest abyssal waters can flow from the Madeira Basin to the Iberian Basin. This is the Discovery Gap located at the eastern end of the East Azores Ridge and named by the author of the mentioned work, and a nameless western gap at 19° W (hereinafter, the Western Gap). The first of them according to (Saunders 1987) is significantly wider and deeper, and thus, determines its main role in the abyssal cold-water flow to the north together with the trend of abyssal transport of dense AABW to the eastern slope of the Madeira Basin (i.e., to the right from the direction of transport, as it should be in the Northern Hemisphere due to the bottom intensification of currents). The locations of both passages are shown in Figs. 8.1 and 8.2.

8.2 Discovery Gap

The Discovery Gap is located in the eastern part of the East Azores Ridge (approximately at 16° W); this passage is close to the Madeira–Tore Rise (Fig. 8.2). The channel is approximately 150 km long and 15–50 km wide at a depth of the 4500 m isobath (Fig. 8.3). According to the bottom topography database (Smith and Sandwell 1997) (version V18.1), available at http://topex.ucsd.edu/cgi-bin/get_data.cgi, and the multibeam survey in 2019 (Dudkov and Dorokhova 2016), three sequential sills are located in the passage (from southwest to northeast) with the depths of the sills from 4680 to 4711 m. Their location is shown in Fig. 8.3; coordinates of the sills are given in Table 8.1.

In the region of the outflow from the Discovery Gap, a decrease in depths from 4700 m to about 5000 m occurs. The slope of the gap walls exceeds 1/5 (Saunders 1987). Watershed line and thalweg (line of lowest depths in the basin) are shown in Fig. 8.2. Even more detailed map of the region is shown in Fig. 8.3c based on the multibeam survey in October 2019 during cruise 43 of the R/V *Akademik Nikolay*

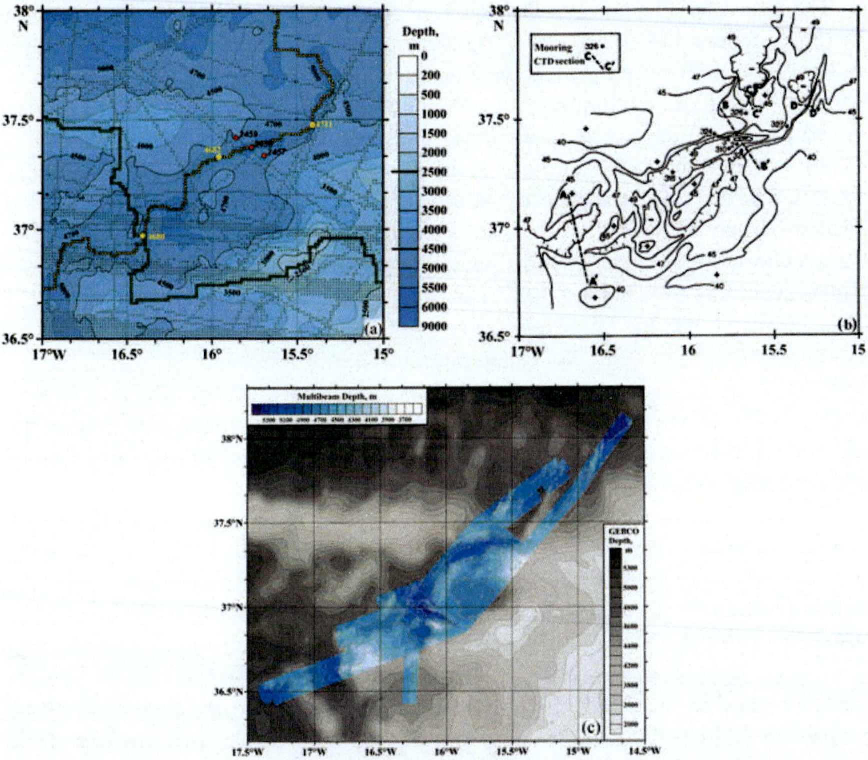

Fig. 8.3 Bottom topography around the Discovery Gap based on the digital data (Smith and Sandwell, 1997) (V.18.1); red circles with numbers near them denote the numbers of stations in 2011, the dots indicate grid nodes of the bottom topography data marked as ship echo sounder measurements. The locations of three sills in the Discovery Gap with depths are marked in yellow color; the depths in meters are indicated. The rest is as in Fig. 8.2. (Reproduced from Tarakanov et al. (2013) with permission from Pleiades Publishers (**a**). Bottom topography as shown in (Saunders 1987); the depths are in 10^2 m; station numbers are indicated: 310, 318, 321, 326, 327 (**b**); Bathymetric chart of the Discovery Gap based on the multi-beam echo-sounder survey in 2019 (Dudkov and Dorokhova 2016) superimposed on the chart based on (Smith and Sandwell 1997) (**c**))

Table 8.1 Locations and depths of the sills in the Discovery Gap

Sill	Depth, m	Coordinates
Sill at the southwestern entrance	4675	36°54′ N, 16°40′ W
Eastern sill at the exit	4780	37°30′ N, 15°24′ W
Western sill at the exit	4620	37°29′ N, 15°37′ W

Strakhov (Dudkov and Dorokhova 2016). The distance between the entrance and exit sills is about 150 km (Fig. 8.3 and Table 8.1). There is an internal pool between the two sills, in which internal circulation is possible.

The first detailed studies in the region of the Discovery Gap were carried out in 1982 (Saunders 1987). Several CTD-sections were occupied and six long-term moorings with 10 current meters were deployed during these studies. Neutral floats were released in the bottom layer at a depth of approximately 4700 m, which were traced during 10 days. The mean water transport based on the data in (Saunders 1987), was directed to the northeast along the channel. The velocities were approximately 5 cm s^{-1}. The transport of water with temperatures below 2.05 °C from the Madeira Basin to the Iberian Basin was estimated in the paper at 0.21 ± 0.04 Sv. Such a choice of the upper boundary of the cold-water flow was because the warmer waters according to (Saunders 1987) can overflow the crest of the East Azores Ridge in other places than the Discovery Gap.

In October 2011 during the expedition on the R/V *Akademik Sergey Vavilov*, a new set of investigations was performed in the Discovery Gap. The section was located in the narrowest place of the Discovery Gap close to section B-B′ west of the eastern sill to have a possibility of comparing the data with the results of previous measurements (Fig. 8.3a,b).

The CTD-casts over a section 20 km long were made using an SBE-19 plus profiler combined with lowered current profiler RDI WHS (300 kHz). Processing the velocity measurements revealed low current velocities comparable with the barotropic current velocities. A correction to eliminate the tide was performed using the TPXO9 inverse global tide model (http://www.oce.orst.edu/po/research/tide/) (Egbert and Erofeeva 2002). On October 10, 2011, the ellipse of tidal currents was elongated in the direction from northeast to southwest; the large axis of the ellipse was 3.5 cm s^{-1} and the small one was 0.8 cm s^{-1}. The distributions of the potential temperature, salinity, and velocity component normal to the section in 2011 are shown in Fig. 8.4. A survey of bottom topography was performed before the measurements in October 2011 to locate the profiling stations more carefully. The results of the multibeam survey were used to plot the section of bottom topography shown in Fig. 8.4.

The potential temperature and salinity distributions over the section in 2011 across the Discovery Gap reveal displacement of the cold fresh core to the southeastern slope of the gap similarly to the sections in 1982 (Fig. 8.4a). The vertical temperature gradient at depths of 4000–4200 m changes to the mixed layer in the deeper layers of the gap. However, note the decrease in the potential temperature compared to 1982. The minimum temperature in 1982 equal to 2.013 °C ± 0.001 °C was measured on moorings 327 (south of section D–D′ in Fig. 8.3b) and 323 (west of section B–B′ in Fig. 8.3b nearly on the line of the 2011 section) approximately 15 m above the bottom (Saunders 1987). In 2011, the minimum temperature (2.002 °C) was recorded simultaneously at two stations, 2457 and 2458 (Figs. 8.3a and 8.4a). Thus, the cooling was nearly 0.01 °C over a period of 29 years.

According to the reports of M. Kapustina and V. Krechik based on the CTD measurements in 2019, potential temperature at 37°25′ N, 15°51′ W east of this section was 2.003 °C (Kapustina and Krechik 2021). This is even cooler by 0.01 °C than in 1982. It is worth noting that in 2011, the potential temperature at the bottom at the third station 2459 (2.011 °C) was lower than the minimum temperature in

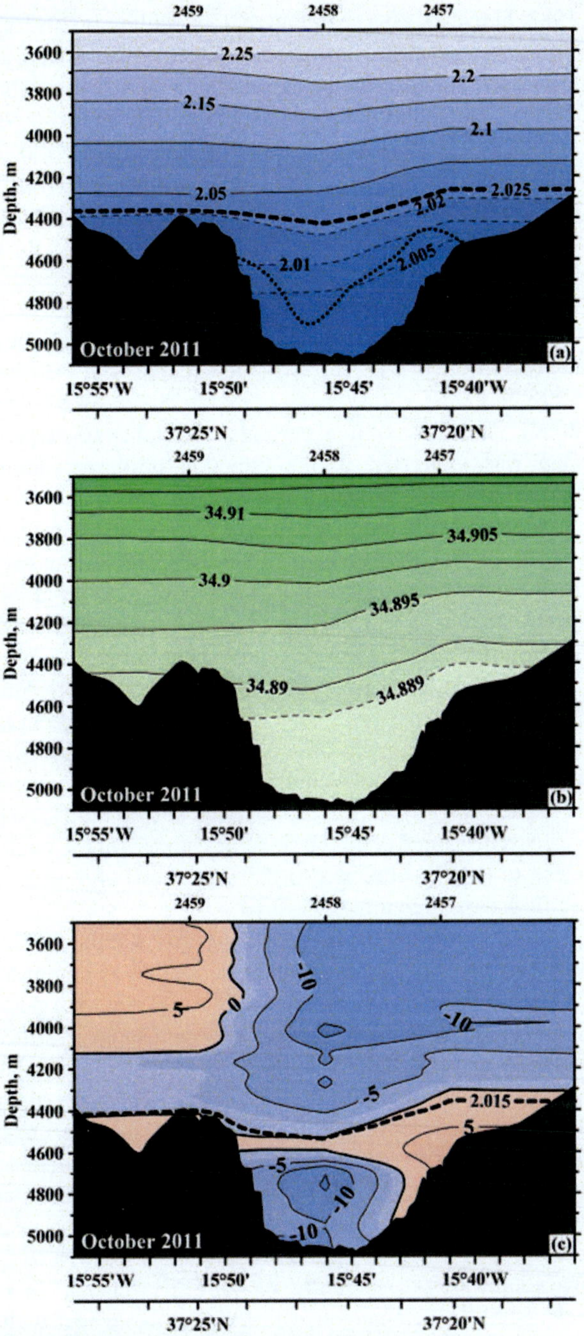

Fig. 8.4 Potential temperature (°C) (**a**), salinity (PSU) (**b**), current velocity component normal to the section (cm s^{-1}) based on the LADCP data in the Discovery Gap in 2011 (**c**). Negative current velocities (blue colors) correspond to the current directed to the southwest, positive ones (red tones) are to the northeast. The bold dashed line in Fig. (8.4a) is potential temperature isotherm 2.025 °C, the same bold dashed line in in Fig. (8.4c) is potential temperature isotherm 2.015 °C. The dots in Fig. 8.4a show the 2.025 °C isotherm based on the 1982 data. The bold dashed line in Fig. 8.4c shows potential temperature isotherm 2.015 °C. The tics and numbers above show the locations and numbers of the stations. (Reproduced from Tarakanov et al. (2013) with permission from Pleiades Publishers)

1982. However, this seemingly slight cooling in 2011 and warming in 2019 is very significant because it affects the bottom layer with a thickness of more than one kilometer. Against the background of a very high uniformity of the bottom layer in the study site, this cooling in 2011 is well illustrated by the elevation of isotherms by almost 200 m in 2011 compared to 1982 (Fig. 8.4a). Salinity is not discussed in (Saunders 1987); therefore, it is impossible to compare the data. However, we note that the minimum salinity in 2011 was 34.888 PSU at stations 2457 and 2458 and 34.890 PSU at station 2459 (Fig. 8.4b). As in 1982, the cold core is displaced to the eastern slope. The salinity distribution demonstrates a similar displacement of fresher water to the eastern slope.

The LADCP data revealed low current velocities compared with the velocities of the barotropic tide. A correction to exclude tidal velocities was performed using the TPXO9 model (http://www.oce.orst.edu/po/research/tide/) (Egbert and Erofeeva 2002). The ellipse of tidal currents on October 10, 2011 at the point of the Discovery Gap was elongated in the direction northeast-southwest. The large axis was 3.5 cm s^{-1} and the small one was 0.8 cm s^{-1}. A section of the normal velocity component after subtracting the barotropic tide is shown in Fig. 8.4c.

The current measurements over the 2011 section revealed an unexpected result. At the most remote stations of the section, 2457 and 2459, the bottom current was directed to the northeast. The velocity of the northeasterly currents at station 2457 was approximately 6 cm s^{-1} and the current at station 2459 was close to 1 cm s^{-1}, respectively (Fig. 8.4c), which agrees with measurements (Saunders, 1987). Unexpectedly strong opposing flow to the southwest exceeding 15 cm s^{-1} was recorded at the central station 2458 (Fig. 8.4c); hence, the current was directed from the Iberian to the Madeira Basin. Its core was located approximately at a depth of 4750 m, which is slightly lower than the crests of the sills in the Discovery Gap. Due to the strong currents measured at this station, the estimate of the total transport appeared negative close to 0.4 Sv (directed to the southwest) compared to a flow of 0.1 Sv in the northeastern direction.

It is noteworthy that the location of this station (37°22.4′ N, 15°46.0′ W) was less than two miles from the site of the deployment of mooring 323 (37°23.9′ N, 15°47.2′ W) in 1982 (Fig. 8.3b), which also showed a reverse flow of water, only with a much lower speed of 2 cm s^{-1} on the average (Saunders 1987). Apparently, in the basin between the sills of the passage, a local cyclonic circulation forms, the existence of which was indicated in (Saunders 1987). However, the intensity of this circulation was either underestimated due to insufficient spatial resolution of measurements in 1982, or in 2011 we recorded its local intensification in time. In addition, it is obvious that in 2011 the resolution of measurements in the narrowest place with only three stations was completely insufficient. The resulting estimate of the transport of abyssal waters 0.28 Sv (below the $\theta < 2.015$ °C isotherm) in the direction to the southwest is apparently random. The change in the upper boundary of the calculation in comparison with 1982 is associated with the previously mentioned general significant elevation of isotherms. High correlation is found between the 2.015 °C isotherm and the zero isotach in Fig. 8.4c.

8.3 Western Gap

The Western Gap in the East Azores Ridge is located at approximately 19°30′ W (Figs. 8.2 and 8.5). According to the bottom topography database (Smith and Sandwell 1997) (version V18.1), this passage actually consists of two individual depressions in the ridge (Fig. 8.5). The sill of the eastern depression is at a depth of 4708 m; its width is greater than 15 km (along the 4500 m isobath); the depth of the sill of the western depression is 4596 m and its width is 5 km.

The points around the eastern sill (4707 m) are marked in the database (Smith and Sandwell 1997) as ship echo sounder measurements (Fig. 8.5), so that its position and depth can be considered determined reliably, although no doubt, a more detailed study of the bottom topography of the area is required. Immediately on the northern side of the ridge at the eastern sill, there is a sharp deepening of depths from 4700 m to more than 5500 m. Such characteristics of the Western Gap notably differ from the modest description given in (Saunders 1987). The width of the passage is almost the same as the Discovery Gap in its narrowest place, and the depth of the sill is even

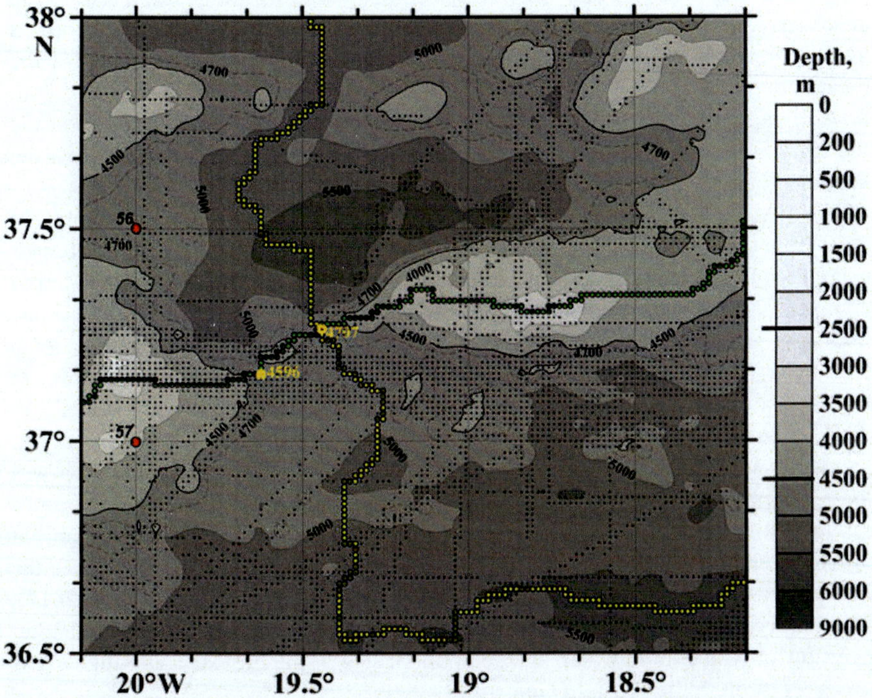

Fig. 8.5 Bottom topography in the region of the Western Gap based on (Smith and Sandwell 1997, version V18.1). The sills of the passage with depth marks are highlighted. The yellow and green lines are the thalweg (deepest line) of the East Atlantic and the watershed of the East Azores Ridge, respectively. The locations of two deepest places in the Western Gap with depths are marked in yellow color. Locations of stations 56 and 57 are shown

larger. The above description of the Western Gap indicates a possible underestimation of its role in the transport of cold abyssal waters from the Madeira Basin to the Iberian Basin reported in (Saunders 1987). According to indirect indicators, the intensity of such a flow from the Discovery Gap to the west along the southern slope of the East Azores Ridge to the Western Gap and back along the northern slope was estimated by Saunders to be half the intensity of the flow through the Discovery Gap.

To understand the relative role of the Western Gap and Discovery Gap in the transport of cold abyssal waters from the Madeira Basin to the Iberian Basin, CTD measurements to the bottom were sampled from the historical data array WOD18. The sample was limited to the ocean sector 34°30′ N– 40°30′ N, 9° W –26° W. In addition, we performed the measurements in the Western Gap in 2013. A chart of the distribution of points with measurements of the bottom potential temperature is shown in Fig. 8.6, from which it is seen that traces of the flow of water with $\theta < 2.04$ °C on the northern side of the East Azores Ridge really separate in the vicinities of both of these passages. Four points (with the designation of the observed temperature near them) in the southeastern part of the Iberian Basin (15° W) including the channel at the entrance to the Tagus Basin (Fig. 8.6a, b) are obviously associated with the Discovery Gap. Traces of the overflow through the Western Gap are observed at two points at 20° W (Fig. 8.6), the measurements have been repeated over 30 years of observations. These two points are located near the line of the greatest depth (thalweg) of the Iberian Basin, to which the coldest water is likely to flow. Hence, the point at 18° W (with temperature designation; Fig. 8.6a) in the Iberian Basin should also be attributed to the traces of the overflow through the Western Gap.

The LADCP data of velocity over the WOCE A16N section along 20° W in July 2003 from the R/V *Ronald H. Brown* allow us to judge about the intensity of cold-water flow to the north through the Western Gap. These data are available at http://currents.soest.hawaii.edu/clivar/ladcp/. On the northern side of the ridge at the station of the section close to the main sill of the Western Gap (station 56 at 37°30′ N, 20° W in Fig. 8.5 and the station with the temperature indication in Fig. 8.6), the measured current velocities in the near bottom layer reach 5 cm s^{-1} in the northern direction. Intensification of the current at the bottom was observed, i.e., an increase in velocity toward the ocean floor (Fig. 8.7). Note that the station was located in a wide overflow along the line of the thalweg at a rather large distance (55 km) from the main sill of the gap (Fig. 8.6). The velocity of the abyssal current measured under these conditions appeared approximately equal to the velocity of the northeasterly current in the narrows of the Discovery Gap (both in 1982 and 2011). This indirectly indicates that a potentially more intense than it was considered earlier overflow of cold bottom waters on the northern side of the East Azores Ridge exists precisely through the Western Gap.

It is noteworthy that over the entire period of observations, the lowest bottom temperatures were recorded precisely at the exit from the Western Gap (Fig. 8.5). This fact has two not mutually exclusive explanations:

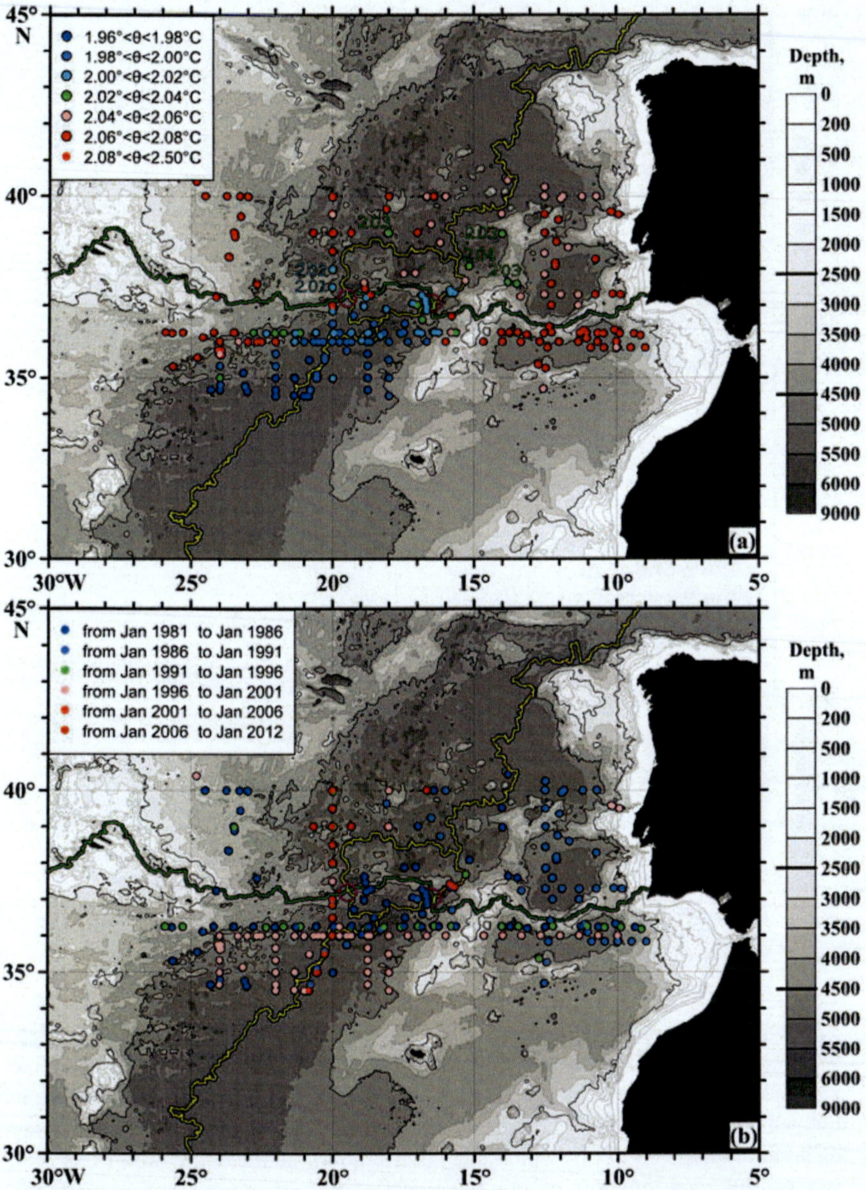

Fig. 8.6 Distribution of potential temperature at the ocean floor in the region of the East Azores Ridge based on the historical data WOD18 (**a**); Chart of the years of measurements; the purple circles indicate the Western Gap and Discovery Gap (**b**). The rest is as in Fig. 8.2. (Reproduced from Tarakanov et al. (2013) with permission from Pleiades Publishers)

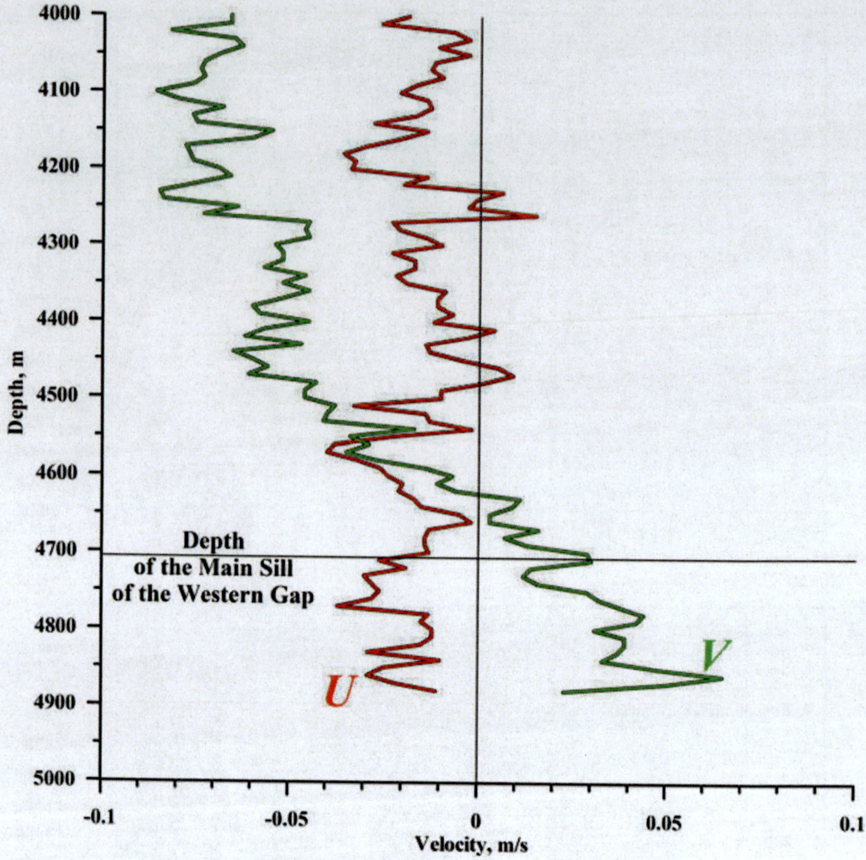

Fig. 8.7 Profiles of the zonal *U* and meridional *V* components of the current velocity in the bottom layer at station 56 (37°30′ N, 20° W) of the A16N section in 2003. The position of station 56 is shown in Fig. 8.5

1. The depth of the sill in the Western Gap is greater than in the Discovery Gap, which ensures the flow of colder water through the first one. In addition, unlike the Western Gap, the Discovery Gap is a rather long channel with strongly uneven bottom topography (in particular, there are several sills along the pathway of bottom water (Fig. 8.3), which can contribute to the intensification of abyssal mixing and, consequently, to a notable increase in the bottom temperature at the exit from the channel compared to the entrance.
2. As can be seen in Fig. 8.6a and b, extremely low temperatures (below 2.02 °C) at the outflow from the Western Gap were measured in the period after 1995, while the measurements at the exit from the Discovery Gap were carried out before 1996. Thus, lower temperatures near the Western Gap can be a manifestation of the cooling trend of bottom waters in the northern basins of the East Atlantic. In this relation, we add that in 2011 and 2019, the potential temperatures recorded

over the section across the Discovery Gap were lower by 0.01 °C than those in 1982.

Measurements of the Antarctic Bottom Water flow in the Western Gap were performed on September 12–13, 2013, during cruise 38 of the R/V *Akademik Sergey Vavilov.* A section of three stations (1501, 2502, 203) was made across the gap (Fig. 8.8) similar to the one in the Discovery Gap in 2011. The distance between the stations was about 3 km. It was assumed that the section line would be directed along the crest of the ridge. The sections of potential temperature, salinity and the component of the current velocity normal to the 2013 section are shown in Fig. 8.9. Figure 8.10 shows the velocity vectors in the bottom layer based on the measurements in 2013. In addition to CTD profiling, an echo sounding was performed in the sill area (Fig. 8.8). The maximum depth along the survey route was 4790 m, the sea depth at the central station was 4753 m. The depth profile based on the echo sounder survey is shown in Fig. 8.9. A strong cyclone was approaching while measuring, and we had to stop the casts without completing the station directly over the eastern

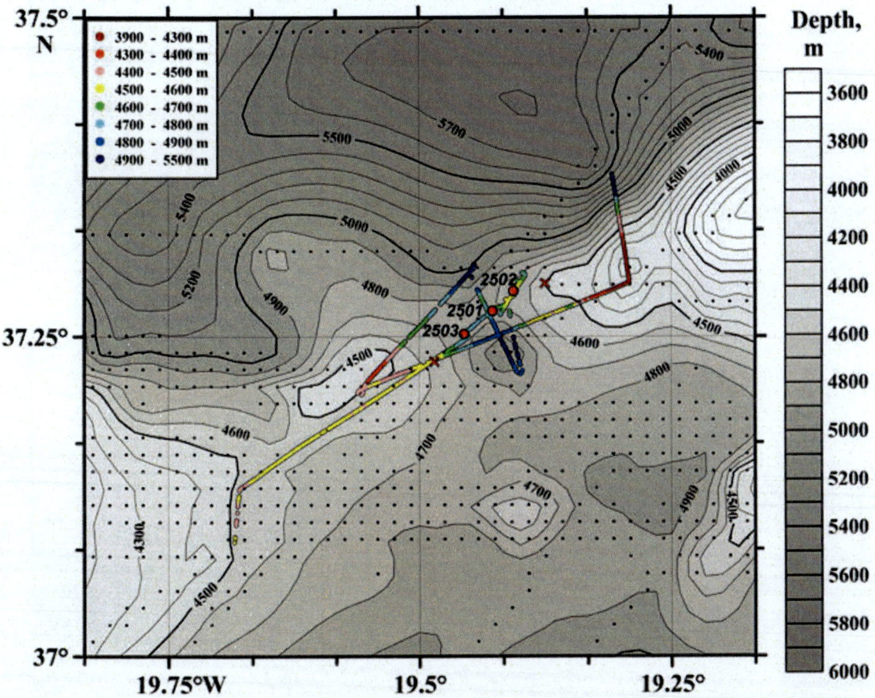

Fig. 8.8 Bottom topography in the Western Gap region based on the data of digital topography (Smith and Sandwell 1997, version V18.1), corrected by the echo sounding survey in 2013. Colors shown in the inset denote the depths. Red circles with numbers show stations and their numbers. Crosses indicate the beginning and end of the section shown in Fig. 8.9. (Reproduced from Tarakanov et al. (2013) with permission from Pleiades Publishers)

Fig. 8.9 Potential temperature (°C) (**a**), salinity (PSU) (**b**), and normal component of current velocity (cm s^{-1}) measured by LADCP (**c**) in the Western Gap in 2013. Negative current velocities correspond to the current to the south-southeast, positive are to the north-northwest. The bold dashed line in Fig. (c) shows potential temperature isotherm 2.05 °C. Tics and numbers on the top axis show position and station numbers. (Reproduced from Tarakanov et al. (2013) with permission from Pleiades Publishers)

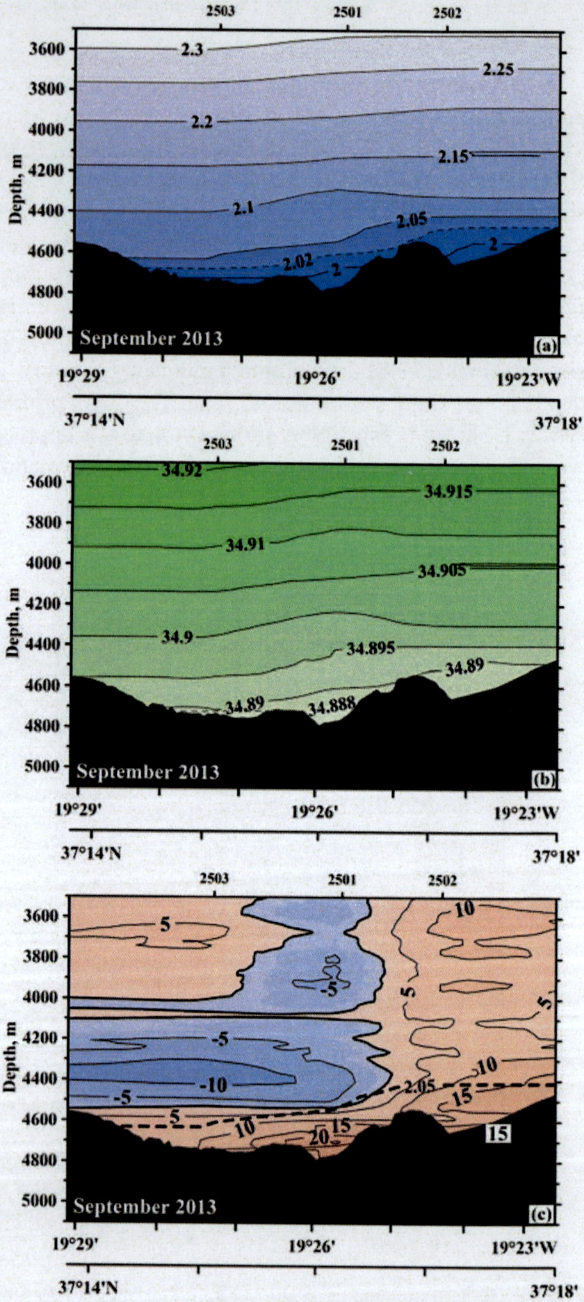

slope. After completing the three stations, we continued the echo-sounder survey in the southwestern direction escaping from the storm.

The sections of potential temperature and salinity show the displacement of cold, fresher waters in the bottom layer towards the eastern slope of the gap (Fig. 8.9a, b). However, the minimum potential temperature (1.995 °C) and salinity (34.8875 PSU) were recorded at the bottom at the deepest central station of the section. The reported temperature is lower than that measured in 2011 in the Discovery Gap (2.011 °C). The temperatures at two other stations were lower than those in the Discovery Gap: 1.997 °C at the eastern station and 2.005 °C at the western one. The measurements of current velocity show (Figs. 8.9c and 8.10), that a flow with such characteristics is directed from the Madeira Basin to the Iberian Basin. Note that the difference in the bottom potential temperatures between the Western Gap in 2013 and the Discovery Gap in 2011 was 0.007 °C. This value is very close to the value of the time trend towards cooling of bottom waters (0.009 °C, accumulated over 29 years) recorded in the Discovery Gap. To our opinion, this difference indicates that coldest water flows to the north through the Western Gap rather than through the Discovery Gap.

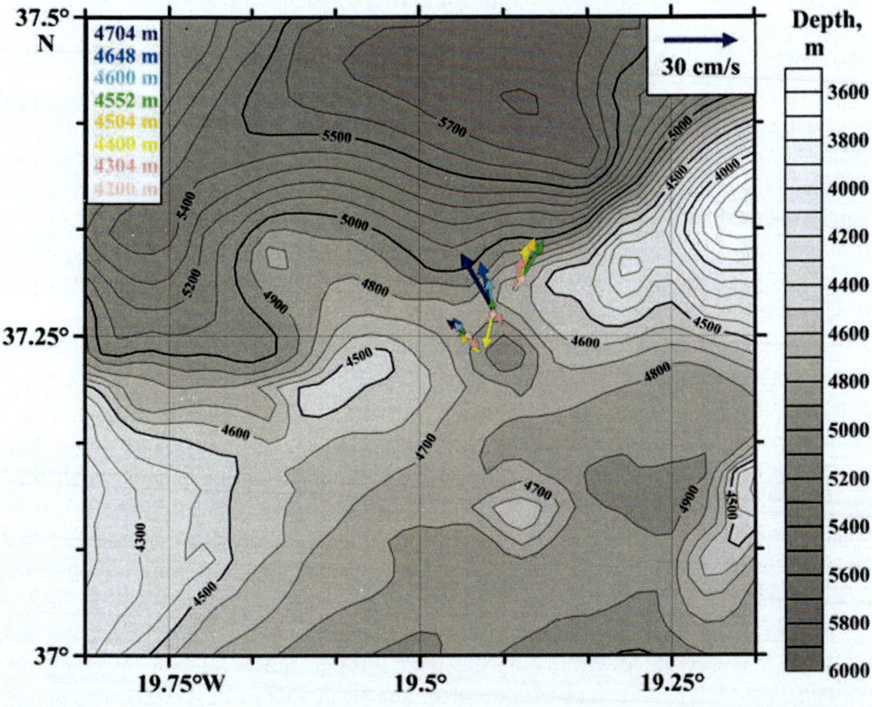

Fig. 8.10 Vectors of the current velocity in the bottom layer at different depths (see legend) in the area of the sill of the Western Gap based on the LADCP measurements in 2013. The bottom topography is as in Fig. 8.8. (Reproduced from Tarakanov et al. (2013) with permission from Pleiades Publishers)

The maximum current velocities on the Iberian Basin side of the Western Gap, based on the data in 2013 reached 27 cm s^{-1} at two stations, the central and eastern ones, approximately 40 and 80 m from the bottom, respectively (Figs. 8.9c and 8.10). At the central station, the flow into the Iberian Basin covered a layer from 4550 m to the bottom and was directed almost normal to the section. At the eastern station, the flow covered a much larger part of the bottom layer (Fig. 8.9c) and accelerated significantly to the bottom, starting approximately from 4300 m. The latter is clearly seen in the chart of velocity vectors at different depths (Fig. 8.10). The second core with velocities up to 21 cm s^{-1} was located directly at the bottom at this station (Fig. 8.9c). The bottom flow at the eastern station was directed almost parallel to the main section line (Fig. 8.10), which indicates that the station point was shifted slightly to the north of the ridge crest. At the western station, the current velocities in the direction to the Iberian Basin were low: up to 12 cm s^{-1} at 20 m above the bottom (Fig. 8.9c, 8.10). At the central and western stations, the core of the return flow with velocities up to 15 cm s^{-1} was located approximately at a depth of 4400 m (Figs. 8.9c and 8.10).

The just described two-core structure of the current velocity over the section across the Western Gap in 2013 is most likely associated with the difference in the balance of forces supporting these currents. The upper core, apparently, is of quasi-geostrophic nature and is a continuation of the deep boundary current propagating northward along the foot of the Madeira-Tore Rise and then westward along the southern slope of the East Azores Ridge. The lower core is obviously a hydraulically controlled flow of water accumulating in the abyssal of the Madeira Basin, and is maintained by the so-called semi-geostrophic balance, when the balance of forces across the channel is geostrophic and non-geostrophic in the longitudinal direction. It is the non-geostrophicity of the lower current core in the Western Gap that manifests itself in the formation of a cold tongue at the bottom of the Iberian Basin north of the Western Gap (Fig. 8.6a).

The transport of the abyssal flow through the Western Gap from the Madeira Basin to the Iberian Basin below the 2.05 °C potential isotherm based on the 2013 data is estimated at 0.22 Sv. This value is even slightly higher than that in the Discovery Gap in 1982–1983 (Saunders, 1987). However, it should be noted that due to the presence of a cooling trend in time in the East Azores Ridge, which was previously discussed, a negative adjustment of the transport through the Western Gap is required for a correct comparison of the transports through the passages. To estimate the transport through the Western Gap, we take the limitation for isotherms 2.015–2.025 °C, which is the closest in the Discovery Gap in depth in 2011 to the 2.05 °C isotherm in 1982, then the transport is estimated at 0.12–0.15 Sv. At the same time, it is quite possible that the flow through the Western Gap could be underestimated due to the lack of stations on the eastern slope of the gap. The depth of the 2.05 °C isotherm in the Discovery Gap in 2011 is at least 150 m higher compared to the Western Gap (Figs. 8.4a and 8.9a), allowing the presence of a velocity core in the latter even higher up the slope than this was revealed by measurements in 2013.

8.4 Modeling

The deepest passages in the East Azores Ridge at 37°20′ N are the Discovery Gap in the eastern part of the ridge at 15°40′ W (Saunders 1987) and the Western Gap at 19°30′ (Tarakanov et al. 2013). The narrowest part in the passage Discovery Gap is located at 37°20′ N and 15°40′ W. The main sill of the Western Gap is located at approximately 37°17′ N, 19°25′ W (Fig. 8.11a). Our measurements in 2013 at three stations across the channel revealed a bottom northward current with velocities up to 25 cm s^{-1}; the maximum velocities were found near the bottom.

We used the INMOM model (Dianskii 2013) for numerical simulation of the currents in the passages. The model was adjusted to the domain from 34.5° to 40.5° N and from 13° to 23° W; the horizontal resolution was 0.01°. The domain completely covers both the Discovery Gap and the Western Gap. The simulated velocities in the bottom layer are shown in Fig. 8.11b. Overall, the results of modeling show a complicated pattern of bottom water propagation in these regions. Velocities in the deepest part of the Discovery Gap are low and do not exceed 5 cm s^{-1}. These modeling results are generally consistent with the existing measurements. According to the model, the main current here is observed along the northern wall of the Discovery Gap; the velocities here are up to 12 cm s^{-1}. There

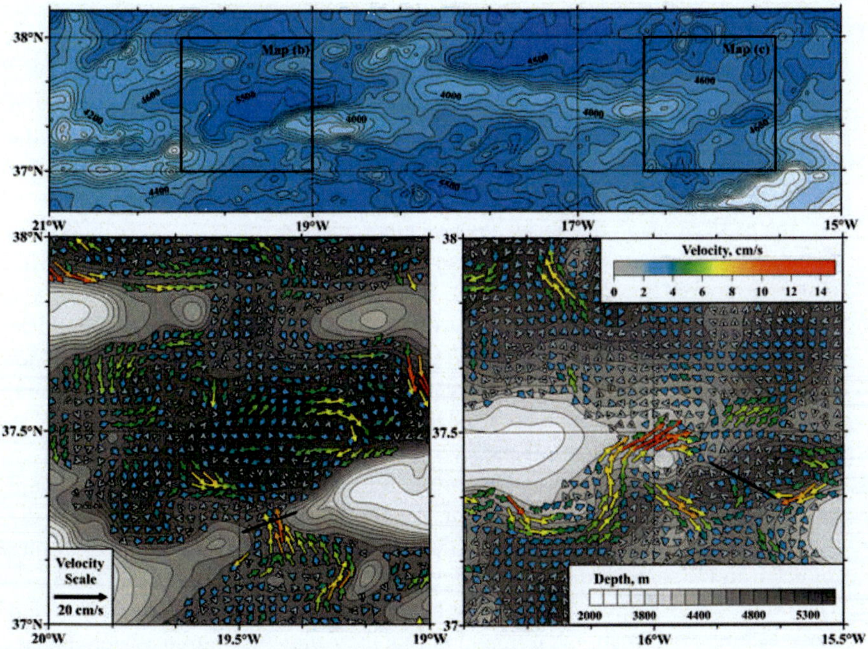

Fig. 8.11 Topography of the East Azores Ridge (top panel, (**a**), modeled velocities in the bottom layer in the area of the Western Gap (left bottom panel, (**b**) and Discovery Gap (right bottom panel, (**c**). Black solid lines in (**b**) and (**c**) show our LADCP sections in 2013 and 2011, respectively

Table 8.2 Parameters of the flows through passages of the East Azores Ridge

	Passages of the East Azores Ridge	Measurements	Modeling
Maximum velocity, cm/s	Discovery Gap	5.0	12
	Western Gap	27.0	11
AABW transport, Sv	Discovery Gap	0.28	
	Western Gap	0.15	
Total transport, Sv		0.43	

are other slow currents in the gap along local features of the bottom topography (Fig 8.11c). Parameters of the flows in both passages through the East Azores Ridge are summarized in Table 8.2.

The simulated velocities at the main sill of the Western Gap do not exceed 11 cm s^{-1}, which is notably lower than the measured values. We suppose that the current here is not steady; unfortunately, there are no long-term measurements of the bottom currents at the sill. Simulated velocity fields do not show strong currents to the north of the main sill (Fig. 8.11c). We calculated the transports through the East Azores Ridge for waters with potential temperature less than 2.3 °C. These estimates as well as the maximum velocities at different sections are presented in Table 8.2.

As a result of investigations in the passages in the East Azores Ridge we conclude:

1. A decrease in the potential temperature of the coldest water in the Discovery Gap compared to 1982 was found amounting to about 0.01 °C from 2.011 °C in 1982 to 2.002 °C in 2011. Analysis of the historical data shows that this decrease is most likely a consequence of the trend towards cooling of abyssal waters in the northern basins of the East Atlantic, but can also be related to the fluctuations of the flow.
2. An intensification of the local abyssal circulation in the Discovery Gap was found, which manifested itself in an increase in bottom velocities in 2011 compared to 1982.
3. It was found that the water that flows from the Madeira Basin to the Iberian Basin through the Western Gap in the East Azores Range, located at 19°30′ W is colder than that flowing through the Discovery Gap. The minimum measured potential temperature in the Western Gap in 2013 was 1.995 °C compared to 2.002 °C in the Discovery Gap in 2011.
4. In the Western Gap, high velocities of the bottom current up to 27 cm s^{-1} were revealed. According to the 2013 data, the water transport was 0.22 Sv for the water colder than 2.05 °C, which is comparable with the estimate of the water transport in the same temperature range through the Discovery Gap in 1982.
5. The properties of bottom water allow us to confirm that the Discovery Gap and the Western Gap remain the terminal points for Antarctic Bottom Water spreading in the sense that this is the water with potential temperature cooler than 2 °C since the transport of bottom water cooler than 2 °C to the north is negligibly small.

References

Boyer TP, Baranova OK, Coleman C, Garcia HE, Grodsky A, Locarnini RA, Mishonov AV, Paver CR, Reagan JR, Seidov D, Smolyar IV, Weathers K, Zweng MM (2018) World Ocean Database 2018. AV Mishonov, Technical Ed., NOAA Atlas NESDIS 87. https://www.ncei.noaa.gov/sites/default/files/2020-04/wod_intro_0.pdf

Dianskii NA (2013) Simulation of ocean circulation and investigation of its reaction on short-term and long-term atmospheric impacts (Fizmatlit, Moscow, 2013) [in Russian]

Dudkov I, Dorokhova E (2016) Multibeam bathymetry data of discovery gap in the eastern North Atlantic. Data Brief 31:105679. https://doi.org/10.1016/j.dib.2020.105679

Egbert GD, Erofeeva SY (2002) Efficient inverse modeling of barotropic ocean tides. J Atmos Ocean Tech 19:183–204. https://doi.org/10.1175/1520-0426(2002)019<0183:EIMOBO>2.0.CO;2

Kapustina MV, Krechik VA (2021) Propagation of Antarctic bottom water in the discovery gap (Northeast Atlantic) based on the field data in 2019. Oceanology 61(5)

Lherminier P, Mercier H, Gourcuff C, Alvarez M, Bacon S, Kermabon C (2007) Transports across the 2002 Greenland-Portugal Ovide section and comparison with 1997. J Geophys Res Oceans 112:C07003. https://doi.org/10.1029/2006JC003716

McCartney MS (1992) Recirculating components to the deep boundary current of the northern North Atlantic. Prog Oceanogr 29:283–382

Saunders PM (1987) Flow through discovery gap. J Phys Oceanogr 17(5):631–643

Smith WHF, Sandwell DT (1997) Global Sea floor topography from satellite altimetry and ship depth soundings. Science 277:1956–1962. http://topex.ucsd.edu/cgi-bin/get_data.cgi; https://doi.org/10.1126/science.277.5334.1956

Tarakanov RY, Morozov EG, Gritsenko AM, Demidova TA, Makarenko NI (2013) Transport of Antarctic bottom water through passages in the East Azores ridge (37° N) in the East Atlantic. Oceanology 53(4):432–441

Tomczak M, Godfrey SR (1994) Regional oceanography: an introduction. Pergamon Press, p 422

WOD18 (2018) World Ocean Database 2018. https://www.nodc.noaa.gov/OC5/WOD/pr_wod.html

Chapter 9
Flows Through the Northern Channels in the North Atlantic

Abstract We study the flows in the Charlie Gibbs Fracture Zone and Bight Fracture Zone in the Mid-Atlantic Ridge, the overflows in the Strait of Gibraltar, in the Strait of Denmark, Faroe Bank Channel, and in the Bear Island Trough. The flows are investigated on the basis of CTD/LADCP casts and moorings and review of previous publications. The Charlie Gibbs Fracture Zone is the main pathway for deep water exchange across the Reykjanes Ridge, while the Bight, Maxwell, and Faraday fractures play secondary roles. We also investigate other northern regions. The cold dense water that is formed in the Greenland Sea overflows the ridge between Greenland and Iceland. It gives origin to the lower limb of the Atlantic Meridional Overturning Circulation. The transport of Arctic waters through the Denmark Strait ranges between 2.0 and 3.5 Sv. No strong seasonal or interannual variability of water transport exists in this region, but large fluctuations were found on short time scales of a few days. The overflow over the Denmark Strait occurs in the form of large and cold lenses. These lenses are called boluses. This form of overflows of cold lenses water alternates with usual overflow currents. We analyzed CTD and LADCP measurements in the western Barents Sea in July–August 2017. We studied the water structure over a meridional section between Norway and Svalbard. The velocities of the bottom flow were measured on a mooring deployed in the deepest part of the Bear Island Trough. Velocity measurements were compared with the tidal velocities from satellite altimetry data. Despite strong barotropic tides in the region, the mean bottom flow was permanently directed from the Barents to the Norwegian Sea, which corresponds to the measured thermohaline structure of the deep waters.

Keywords Deep channels in the Northern Hemisphere · Charlie Gibbs Fracture Zone · Bear Island Trough · Strait of Gibraltar · Danmark Strait · Faroe Shetland Channel

---

With a contribution by Walter Zenk, GEOMAR Helmholtz Centre for Ocean Research Kiel, Germany

E. G. Morozov et al., *Bottom Gravity Currents and Overflows in Deep Channels of the Atlantic Ocean*, https://doi.org/10.1007/978-3-030-83074-8_9

9.1 Charlie Gibbs Fracture Zone

In the winter period, cold dense water is formed in the Norwegian and Greenland seas due to severe cooling and intense heat release to the atmosphere. The water mass from the Greenland Sea overflows the shallow threshold between Greenland and Iceland and flows into the Irminger Basin. It gives origin to the Denmark Strait Overflow Water (DSOW). The water mass from the Norwegian Sea overflows the threshold between Iceland, Faeroe, and Shetland Islands and flows into the Iceland Basin. The water mass, which is formed during the overflow over the latter thresholds, is called Iceland Scotland Overflow Water (ISOW). In the Iceland Basin, this water is mixed with the significantly warmer and more saline waters of the Northeast Atlantic, which are subject to the influence of Mediterranean Water. As a result, Iceland Scotland Overflow Water appears more saline and warmer than the deep waters propagating into the Atlantic through the Denmark Strait between Greenland and Iceland (DSOW). Iceland Scotland Overflow Water flows first to the west with the Deep Northern Boundary Current (McCartney 1992) and then to the south along the eastern slope of the Reykjanes Ridge. A circulation scheme of different layers of the North Atlantic is shown in Fig. 1.5.

Large portions of Iceland Scotland Overflow Water penetrate to the West Atlantic at 53° N through the deep Charlie Gibbs Fracture Zone and then flow predominantly to the north along the western slope of the Reykjanes Ridge. Following the 2500-m isobath, Iceland Scotland Overflow Water reaches the northern boundary of the Labrador Basin, where it becomes part of the Deep Western Boundary Current (DWBC). Iceland Scotland Overflow Water in the North Atlantic is traced by a local salinity and silicate maximum between the cores of the Deep Western Boundary Current and Labrador Water. In the literature, Iceland Scotland Overflow Water is sometimes called Iceland Shetland Water (in the eastern North Atlantic) or Charlie Gibbs Water (in the western basin). This issue was discussed in (Saunders 1994; Dickson et al. 1980). McCartney (1992) assumed that the flow through the Charlie Gibbs Fracture Zone should contain a component of ancient deep water from the eastern basin, which is of the southern origin. The through-flow should consist of 40% of this water and 60% of overflow water of the northern origin.

The Charlie Gibbs Fracture Zone is located between 52–53° N and 30–35° W in the Mid-Atlantic Ridge. A bathymetric chart of this region plotted from the data of multibeam survey during cruise S55 of the R/V *Akademik Nikolay Strakhov* is shown in Fig. 9.1. Details of the bottom topography are described in (Skolotnev et al. 2021). The fracture zone comprises two channels separated by a median ridge. The fracture zone is more than 400 km long in the west-east direction. Two transversal sills are located in each of the channels at depths of 3600–3700 m. The sill is located at 35° W in the northern channel and at 30° W in the southern channel.

In 2002, during cruise 48 of the R/V *Akademik Mstislav Keldysh,* a CTD section was occupied along 52–53° N, which crossed the Charlie Gibbs Fracture Zone. Water structure and transport were studied at 15 stations located in this region. A scheme of the CTD stations occupied in 2002 in the cruise is shown in Fig. 9.2.

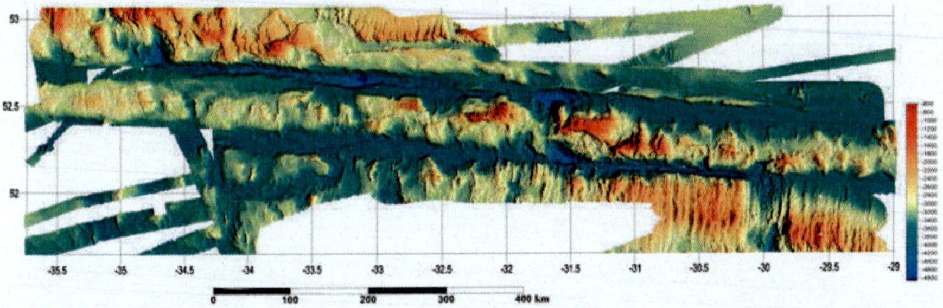

Fig. 9.1 Chart of the Charlie Gibbs Fracture Zone based on the data of multibeam survey data during cruise S55 of the R/V *Akademik Nikolay Strakhov* collected by Skolotnev et al. (2021). (Courtesy of S.G. Skolotnev)

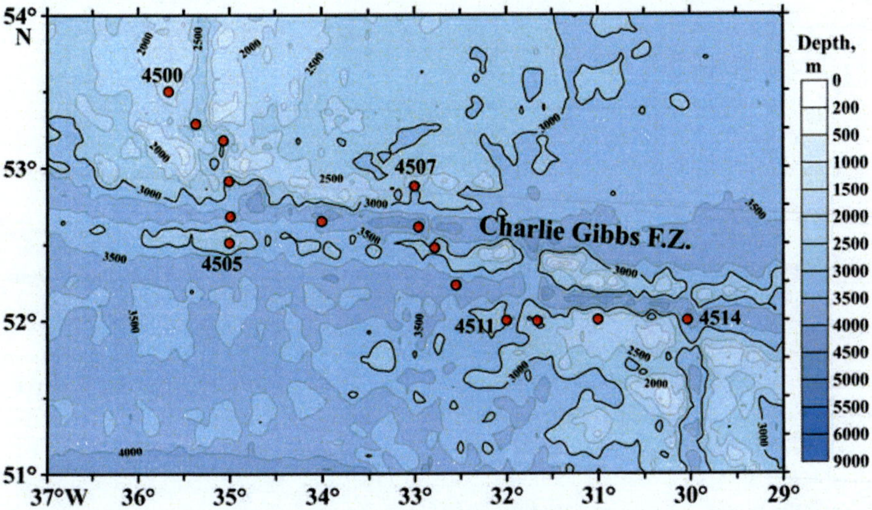

Fig. 9.2 Scheme of CTD-stations occupied in 2002 near the Charlie Gibbs Fracture Zone (4500–4511)

The line of stations was positioned to cross the channels in the fracture zone twice. The first section was located along the rift valley and across the northern channel approximately along 35° W (stations 4500–4505); the second section crossed the northern and southern channels along 33–32° W (stations 4507–4511).

The salinity section approximately along 34° W is shown in Fig. 9.3. Investigations in 2001 during cruise 9 of the R/V *Akademik Ioffe* (Dobrolyubov and Falina 2002) revealed that the main salinity maximum (> 34.97 PSU) in the Iceland Scotland Overflow Water core is not located above the deepest channels (western section at 53°10′ N). Figure 9.4 shows θ,*S*-curves for the three deepest stations in the Charlie Gibbs Fracture Zone (4504, 4508, and 4510) and curves for the stations located west of the fracture zone (with the typical Labrador Sea Water and deep

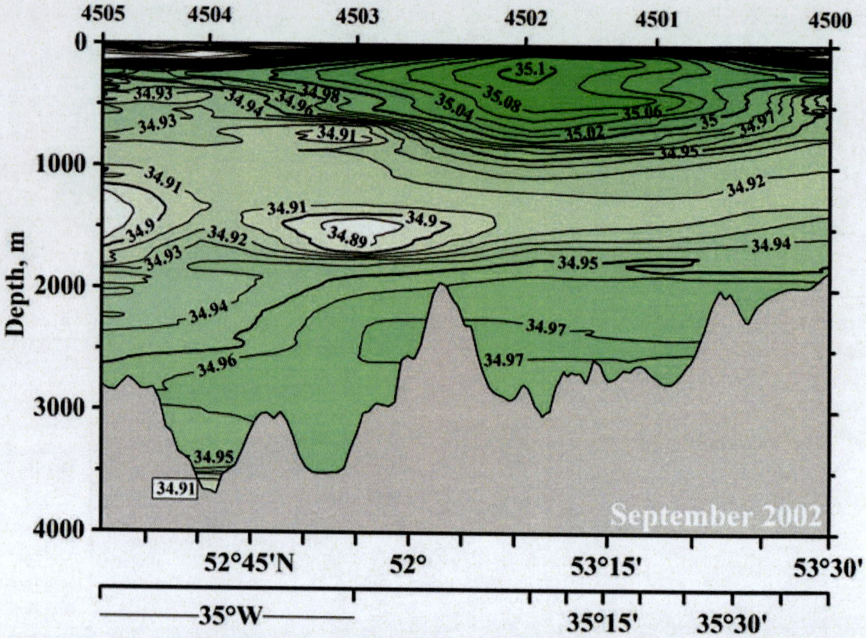

Fig. 9.3 Salinity section (PSU) across the Charlie Gibbs Fracture Zone (western section approximately along 35–34° W). Triangles on the top axis indicate stations. Reproduced from Dobrolyubov et al. (2003b) with permission from Pleiades Publishers

layers of Denmark Strait Overflow Water). East of the fracture zone, one can see typical curves of Iceland Scotland Overflow Water. The shape of θ,S-curves indicates that the water flowing west through the Charlie Gibbs Fracture Zone is a product of mixing between Iceland Scotland Overflow Water and Denmark Strait Overflow Water. Thus, recirculation and intense mixing of deep waters of different origin are observed in this region.

Figure 9.5 shows a scheme of the geostrophic water transport through the northern and southern channels in the Charlie Gibbs Fracture Zone. Geostrophic calculations showed that the western water transport below isopycnal $\sigma_0 = 27.82$ (approximately between 1800 m) and bottom is 1.2 Sv over the section along 35° W above the rift valley. The westerly flow over the same section in the northern channel is 1.2 Sv. On the section along 33° W, the westerly flow transport along the northern channel is 2.5 Sv, whereas the easterly flow transport along the southern channel (52°50′ N) is 3.2 Sv.

Let us compare the estimates of water transport in the Charlie Gibbs Fracture Zone in 2002 with the instrumental measurements made in the previous years. Dickson et al. (1980) analyzed the results of instrumental measurements in the southern channel carried out from October 1977 to July 1978 and found a westerly transport along the northern slope of the southern channel and easterly transport along the southern slope. Both easterly and westerly mean transports were

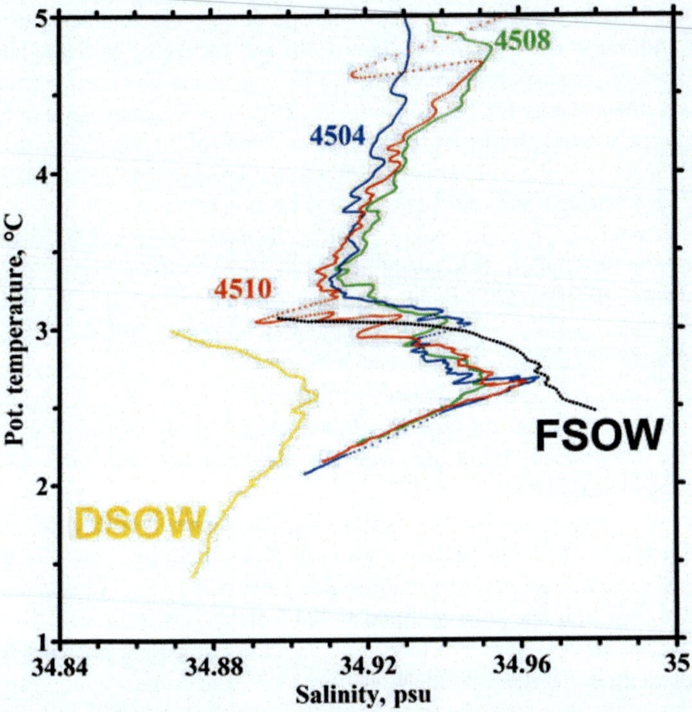

Fig. 9.4 Curves of θ,*S*-distributions for the Charlie Gibbs Fracture Zone (stations 4504, 4508, 4510) and curves for stations with a typical Iceland Scotland Overflow Water and Denmark Strait Overflow Water. Reproduced from Dobrolyubov et al. (2003b) with permission from Pleiades Publishers

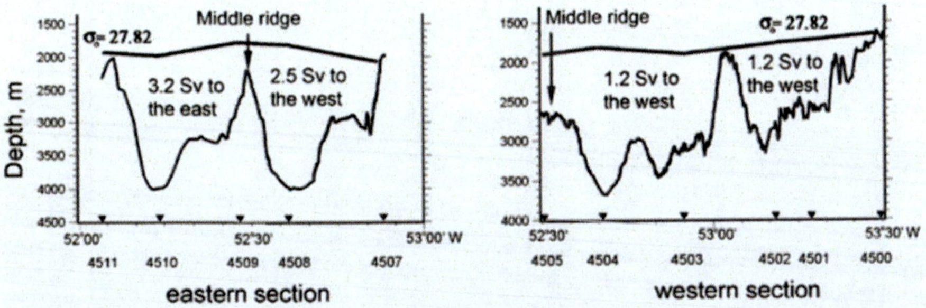

Fig. 9.5 Water transport through the Charlie Gibbs Fracture Zone. Lines in the upper parts of the graph denote the upper boundary of Iceland Scotland Overflow Water based on the position of potential density $\sigma_0 = 27.82$. Eastern section (left panel), western section (right panel). Stations are indicated by black triangles on the bottom axis. Reproduced from Dobrolyubov et al. (2003b) with permission from Pleiades Publishers

approximately 0.24 Sv. Saunders (1994) carried out the measurements using CTD profiling and moorings deployed in June 1988 and recovered in September 1989. Only the westerly transport estimated at 2.4 Sv was found in the northern channel. The measurements carried out during cruise 9 of the R/V *Akademik Ioffe* in 2001 also found only the westerly transport in the northern channel. In this cruise, the number of stations located in the southern channel was not enough for the calculations.

Thus, the westerly water transport was always found. The easterly transport in the southern channel was revealed only in 1978 and 2002. These are the years of low North Atlantic Oscillation (NAO) and increased Arctic water transport through the Strait of Denmark, Faeroe-Shetland, and Faeroe-Iceland thresholds. In 1988–1989, the measurements were carried out during the high North Atlantic Oscillation index, decreased water transport from the Arctic Ocean (Denmark Strait Overflow Water and Iceland Scotland Overflow Water), and intense Labrador Sea Water formation. In 2002, the Iceland Scotland Overflow Water branch at 60° N was also weakened west of the Reykjanes Ridge and was not observed near the Greenland slope (Dobrolyubov et al. 2003a).

The flow through the Charlie Gibbs Fracture Zone in the depth range of 1500–1600 m was also studied using RAFOS floats 463, 469, and 474 that were deployed at a depth of 1530–1560 m in August 1998. All of them passed the ridge in the western direction and then returned to the Iceland Basin through the northern channel of the Charlie Gibbs Fracture Zone with the Labrador Sea Water flow. The floats popped up in February 2000, November 1999, and August 2000, respectively. Trajectories of the floats are shown in Fig. 9.6.

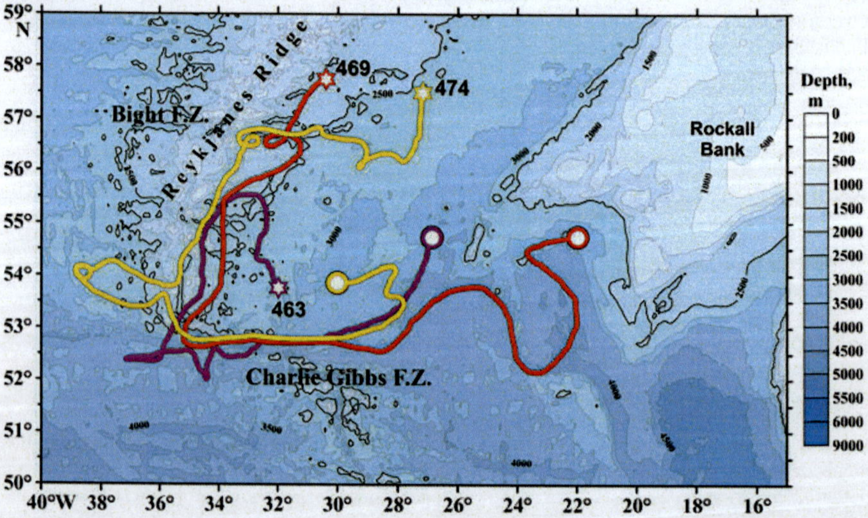

Fig. 9.6 Bathymetry of the Iceland Basin with the Mid-Atlantic Ridge on the western side and the Rockall Bank and Porcupine Bank on the eastern side. Trajectories of three RAFOS floats (IfM tracking numbers 463, 469, and 474) are shown. The launching position and last fix before surfacing are marked with stars and circles, respectively

Other fractures in the Reykjanes Ridge also provide water transport. The Bight Fracture Zone is likely the second most important channel after the Charlie Gibbs Fracture Zone. The Charlie Gibbs Fracture Zone undoubtedly plays the leading role for exchange of overflow waters between the eastern and western rims of the Mid-Atlantic Ridge. Nevertheless, Lagrangian observations by miscellaneous floats have demonstrated the importance of several other (secondary) channels. This category includes the Faraday, Maxwell, and Bight fracture zones, as was shown in (Bower et al. 2002; Lankhorst and Zenk 2006).

The drifts of RAFOS floats through the Bight Fracture Zone are shown in Fig. 9.7. We compiled an ensemble of three eddy resolving RAFOS trajectories (with IFM tracking numbers 411, 476, 478) that all crossed the Mid-Atlantic Ridge at ~57° N through the Bight Fracture Zone. The instruments were launched in the northern part of the Iceland Basin (star symbols). They were ballasted for a nominal pressure of 1500 dbar, corresponding to a depth of ~1500 m. Return positions are shown as circles. The eastern slope of the Mid-Atlantic Ridge is marked by the high-

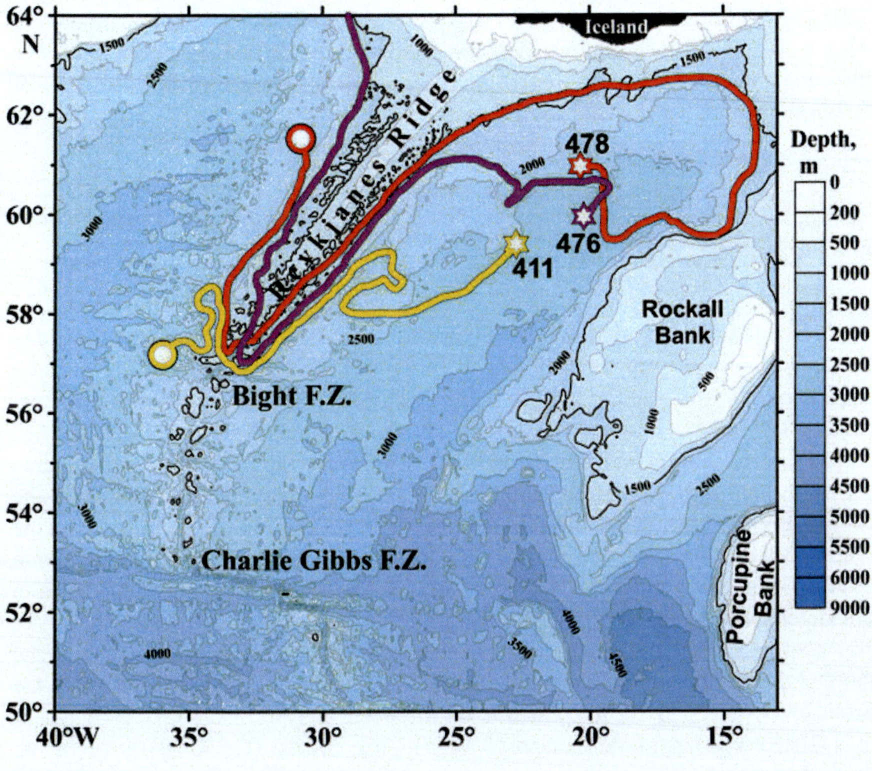

Fig. 9.7 Bathymetry of the Iceland Basin with the Mid-Atlantic Ridge on the western side and the Rockall and Porcupine banks on the eastern side. Trajectories of three RAFOS floats are shown. The launching position and last fix before surfacing are marked with stars and circles, respectively. Contour lines of 500, 1000, 1500, 2000, 2500, and 3000 m are shown

speed path of Iceland Scotland Overflow Water (Lankhorst and Zenk 2006). On the western side, all floats reduce their drift speed on a northeasterly course. The image emphasizes the importance of this additional pathway through the ridge. In the older literature, this pathway was assumed as impermeable for Iceland Scotland Overflow Water north of the Charlie Gibbs Fracture Zone at ~52° N.

Statistics of floats crossing the Reykjanes Ridge in both directions is given in (Bower et al. 2002) and depth of the ridge crest as function of latitude. Floats transported by the North Atlantic Current at the upper level crossed the ridge eastbound preferentially over the Charlie Gibbs Fracture Zone (31 of the total 61), the Faraday Fracture Zone (11/61), and, to a lesser extent, the Maxwell Fracture Zone (4/61) (float depths 200–800 m). The distribution of float deployment latitudes is significantly different from the distribution of crossing latitudes. Floats crossed the Reykjanes Ridge from east to west north of 53° N mainly through the Bight Fracture Zone and two unnamed gaps near 53.5° N and 55.0° N. North of 53° N, the gaps are not aligned east–west and the distribution of float crossing latitudes is therefore somewhat wider around these gaps.

Figure 9.8 displays the drift of an Argo float over almost 3 years from August 2002 to May 2005 at a density level $\sigma_\theta \sim 27.8$ corresponding to a nominal depth of 1500 m. The instrument pathway is strongly guided by the bottom topography. The instrument was launched in the Iceland Basin. Then, it crossed the Mid-Atlantic Ridge at a latitude of the Bight Fracture Zone (57° N) and reached the Irminger Sea, where it headed toward the Denmark Strait.

9.2 Overflow in the Denmark Strait

The cold dense water that is formed in the Greenland Sea overflows the ridge between Greenland and Iceland. It gives origin to the lower limb of the Atlantic Meridional Overturning Circulation as the flow of Lower North Atlantic Deep Water in the Deep Western Boundary Current. The dense waters from the Greenland Sea form Denmark Strait Overflow Water (DSOW) flow to the Denmark Strait as three jets: the shelfbreak East Greenland Current over the shelfbreak of Greenland, the separated East Greenland Current in the eastern part of the Blosseville Basin, and the North Icelandic Jet over the northern shelfbreak of Iceland (Fig. 9.9) (Mauritzen 1996). The sources of these waters are water masses with different characteristics in the Greenland Sea (Köhl 2010; Mastropole et al. 2017).

According to the model estimates of the transport of bottom dense water to the North Atlantic (the bottom water layer with potential density $\sigma_0 > 27.80$) through the Strait of Denmark in 1958–2006 was on the average 1.86 ± 0.96 Sv (Filyushkin et al. 2013). The maximum transport was as high as 3.7 Sv. (Haine 2010; Kohl et al. 2007; Macrander et al. 2007).

Investigations with moorings in this region showed that there is no strong seasonal or interannual variability (Jochumsen et al. 2012), but large fluctuations were found on short time scales of a few days. Cooper (1955) was the first to reveal

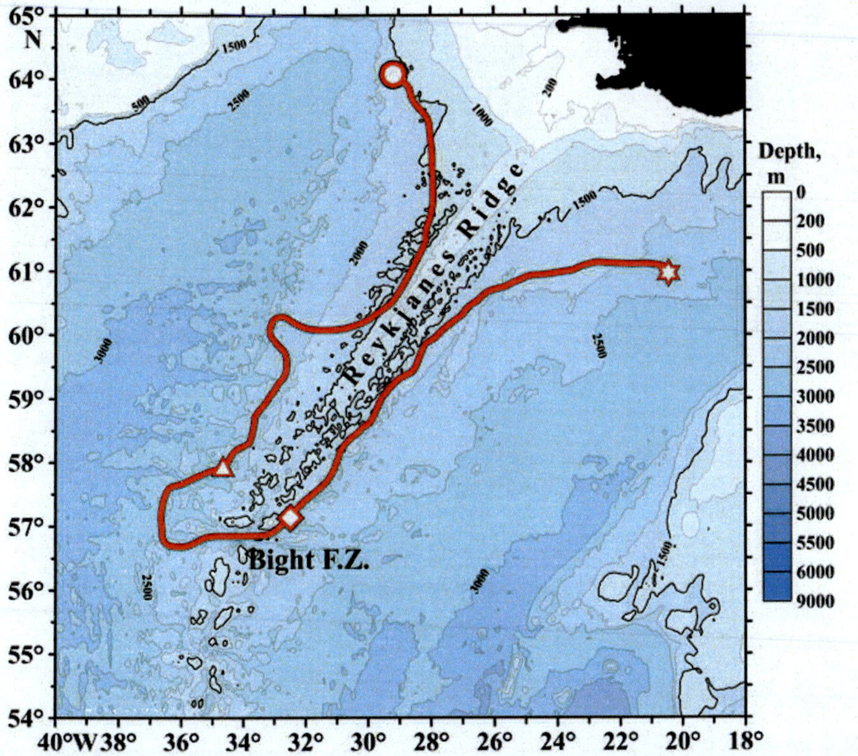

Fig. 9.8 Bathymetry of the Iceland Basin. Trajectory of an Argo float (No. 6900186) crossing the Reykjanes Ridge through the Bight Fracture Zone (57° N, 34°W) is shown. The launching position is marked with an asterisk. The last fix is marked with an open circle. The white triangle and black circle denote positions of selected profiles. Reproduced from (Lankhorst and Zenk 2006)

large cold lenses overflowing the ridge of the Denmark Strait on the basis of hydrographic sections; these lenses were called boluses. Later, Worthington (1969) confirmed the existence of boluses as overflows of cold water at high velocity alternating with warmer slow currents. High-frequency (with a period of several days) variability of the bottom flow in the Denmark Strait may be a response to the similar variability of meteorological impact as shown in model calculations in (Haine et al. 2009).

It was reported in (von Appen et al. 2017) on the basis of measurements on the mooring station and CTD-casts across the strait that there are two alternating dominant structures in the form of a bolus and pulse. The authors present the bolus as a large lens of water with weak stratification and cyclonic rotation. When a bolus passes through the strait, the interface depth with the overlying water layer displaces up in the vertical direction by about 60 m, and the bottom temperature drops by 0.48 °C. During the pulse phase, the rotation changes to anticyclonic, and the interface deepens by 90 m. The overflow speed increases, but there is no notable

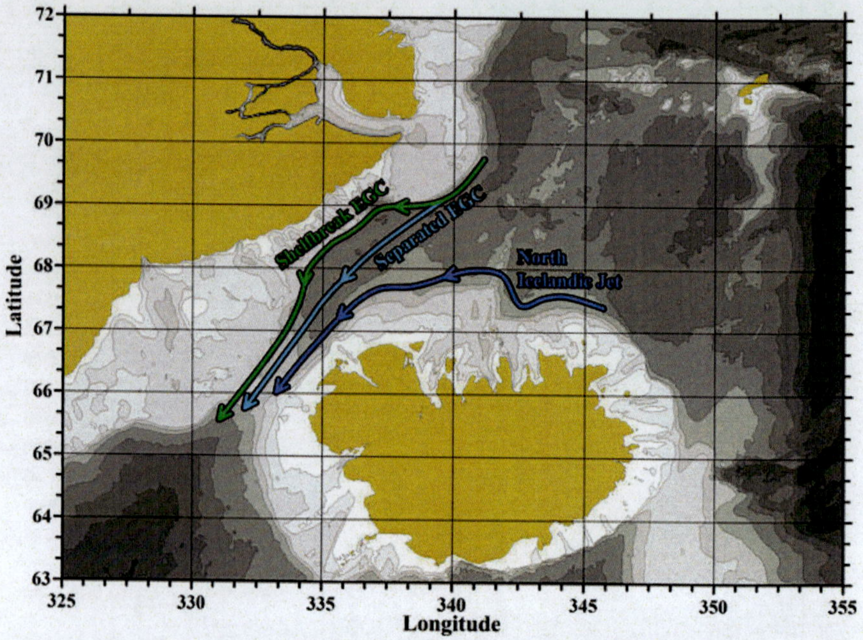

Fig. 9.9 Currents in the Strait of Denmark

change in the bottom temperature. On average, boluses (and pulses) appear with an interval of 3–6 days without seasonal variability.

According to (Mastropole et al. 2017), boluses transport the coldest, saltiest, and densest portion of DSOW. The water in boluses is on the average 0.049 ± 0.003 kg/m^3 denser than the overflow water without boluses. The average cross-section of boluses is approximately 4 km^2. Boluses contain mostly Arctic-Origin Water with smaller percentage of other water. The following percentage of different waters is reported in (Mastropole et al. 2017): $80 \pm 9\%$ of Arctic-Origin Water; $14 \pm 9\%$ of Atlantic-Origin Water; $4 \pm 3\%$ of Polar Surface Water; and $2 \pm 2\%$ of Irminger Water. The percentage in the non-bolus DSOW is: $63 \pm 9\%$ of Arctic-Origin Water; of $23 \pm 9\%$ Atlantic-Origin Water; $8 \pm 3\%$ of Polar Surface Water; and of $6 \pm 1\%$ Irminger Water.

Since the variability of water transport on a small time scale is very high in this region, some authors attribute the appearance of boluses to the influence of nonlinear internal waves of large amplitude incident onto the underwater slope. This idea is supported by some authors on the basis of observations, laboratory and numerical modeling (Bourgault et al. 2008; Moore et al. 2016; Venayagamoorthy and Fringer 2007). In this case, the appearance of boluses should correspond to the periods of internal waves or intervals of amplification of their amplitudes.

The variations in the overflow water are significant for the large-scale circulation. The existing variations in the water properties are related both to the time variability and spatial variability. These variations have two possible causes related to the changes in the properties of the Nordic water mass sources and to the changes in the composition of the overflowing water. The changes in wind and temperature conditions in the source regions cause changes in convection; hence, DSOW is in its turn is subjected to the changes in its composition, thus influencing the properties of North Atlantic Deep Water.

The typical profiles of temperature in the narrows of the Denmark Strait almost on the sill are shown in Fig. 9.10.

Since the flow of Denmark Strait Overflow Water consists of the water from at least two sources (Mauritzen 1996; Köhl 2010), we made an attempt to compare the temperature relations in the two source regions and in the strait where we found strongly modified and mixed water masses. We used the data of CTD casts from the World Ocean Database 2018 (WOD18 2018). The source of warmer water is north of Iceland where the North Icelandic Jet is formed over the shelfbreak approximately at 650 m depth. We used a CTD cast in this region at station at 67.79° N, 19.69° on April 7, 2004 at a depth of 797 m. The source of colder water is in the Greenland Sea, east of the Greenland shelf. A typical CTD-cast from this region was at station at 68.78° N, 22.38° on August 30, 1973 at a depth of 1509 m.

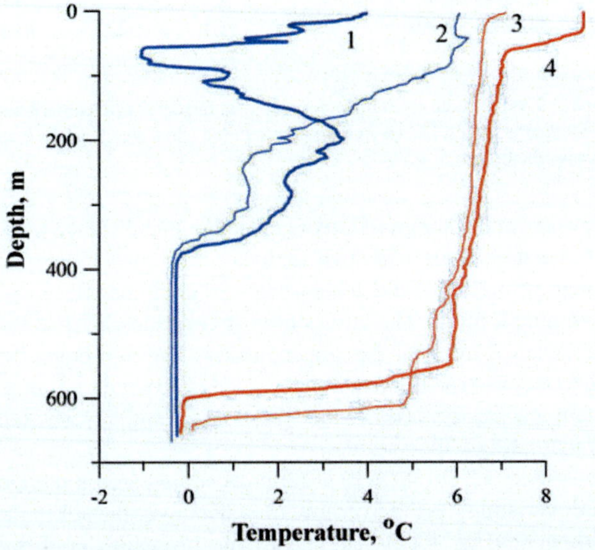

Fig. 9.10 Vertical distribution of potential temperature of different waters in the Strait of Denmark. Curves 1, 2 correspond to the bolus overflow; curves 3, 4 correspond to the usual flow of water: station from the cruise of R/V *Akademik Mstislav Keldysh* 5894 at 66.076° N, 27.079° E on July 29, 2018 at a depth of 658 m (curve 1); station at 66.08° N, 27.05° on May 18, 2003 at a depth of 657 m (curve 2, typical for bolus); station at 66.08° N, 27.05° E on May 17, 2002 at a depth of 646 m (curve 3, typical for non-bolus overflow), station from the cruise of R/V *Akademik Mstislav Keldysh* 5837 at 66.076° N, 27.080° E on July 25, 2018 at a depth of 654 m (curve 4, typical for non-bolus overflow). Reproduced from Morozov et al. (2020) with permission from Pleiades Publishers

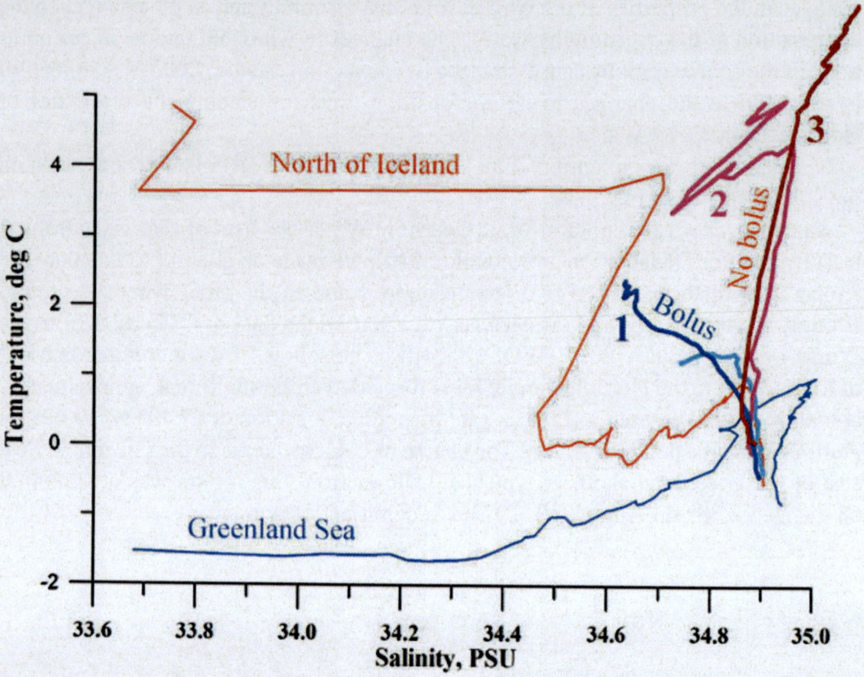

Fig. 9.11 Curves of θ,*S*-distributions of different waters in the Strait of Denmark and to the north of it in the regions of water sources. Curve *1* corresponds to the usual flow; curves *2* and *3* correspond to the distribution in the bolus

The θ,*S*-curves over the slopes of Greenland and Iceland shown in Fig. 9.11 differ strongly. As the waters from these two regions arrive to the Denmark Strait their properties change strongly, but the general features could be traced even on the θ,*S*-curves in the Denmark Strait. The curves almost coincide closer to the bottom, but diverge as the distance from the bottom increases. The non-bolus "warm" curves corresponding to the water with greater percentage of Icelandic water tend to greater salinities as temperature increases, while the bolus "cold" curves tend to smaller salinities with increasing temperature.

Two hydrographic (CTD/LADCP) sections occupied across the Denmark Strait reveal very high variations in the overflow of waters over the sill in the presence of bolus and without it (Fig. 9.12). These variable properties confirm the existing knowledge of the water masses and currents over the sill (Mastropole et al. 2017). The warm and salty water (Irminger Water) of the subtropical origin occupies the Iceland shelf. Denmark Strait Overflow Water is located farther to the west of the section. A sharp temperature and salinity front separates these waters. The section without bolus is very similar to the one shown in (von Appen et al. 2014). In this work, the authors also distinguish a separate jet (spill jet) at a distance of 28 km from the Greenland shelf break in the density range 27.6–27.8. Figure 9.12 shows the

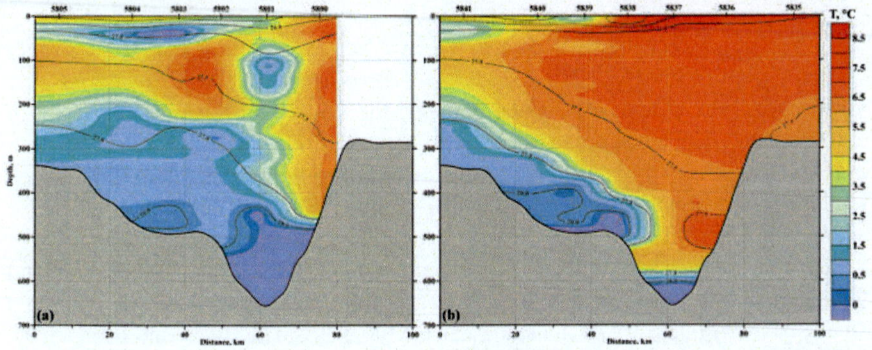

Fig. 9.12 Two vertical sections of potential temperature showing the (**a**) presence and (**b**) absence of a bolus of overflow water from cruise of R/V *Akademik Mstislav Keldysh*: section occupied on July 23, 2018 (**a**); and (**b**) section occupied on July 25, 2018 (**b**). The isopycnals of potential density (boundary of the bolus: 28.0 kg/m^3 and boundary of the bottom flow: 27.8 kg/m^3) are denoted by solid lines. Reproduced from Morozov et al. (2020) with permission from Pleiades Publishers

main major properties of the temperature distribution revealed from the CTD section across the strait in the bolus regime and without it.

It has been reported in many publications that short-period variability exists in the overflow transport and velocity (Spall and Price 1998; Girton and Sanford 2003). Variations in the overflow are manifested in the alternating appearance of boluses of colder water. Since then, boluses have been identified in various observations and numerical models (Girton and Sanford 2003; Käse et al. 2003; Koszalka et al. 2013). The spectra of velocity fluctuations reveal a prominent peak at a period of 4–5 days. This peak is much stronger than the tidal M_2 peak (Morozov et al. 2019).

Internal seiches can be a possible cause of five-day variations in the bottom flow. The period of seiches is determined by the Merian formula (Merian 1828):

$$T_{seich} = {}^{2L}\!/_{n\sqrt{gH_{mean}}}, \tag{9.1}$$

here, T is the period of seiches, L is characteristic size of the basin, H_{mean} is the mean depth, g is acceleration due to gravity, n is the mode number. The seiches are natural oscillations of the basin. The lowest mode ($n = 1$) is the fundamental mode. The period of surface seiches in the basin north of the Denmark Strait is a few hours. Internal seiches, which are the seiches of the fluctuating bottom layer, are important for regulating the bottom flow. Characteristic variation in the thickness of the bottom flow with bolus and without bolus is as large as 150–300 m, which is possibly related to internal seiches. We consider that the boluses are formed in the basin north of the strait at a distance of 150–200 km from the sill. The characteristic transversal size of the basin at a depth of the 200 m isobath is 250 km. To analyze internal seiches we have to substitute the reduced gravity instead of gravity

$$g' = g\frac{\Delta\rho}{\rho},$$

where $\Delta\rho$ is the density difference between the upper and lower layers equal to $2{\cdot}10^{-3}$ g/cm^3. The mean depth of the basin between 200 m isobaths across the strait is 60 m. Then the period of seiches of the lower layer is estimated as 5.3 days, which is close to the periodicity of the overflow regimes. Since all the initial values for estimating this period are approximate, the result seems reasonable.

We consider vertical profiles of temperature at several stations on the Greenland Continental slope north of the sill in the strait in September in different years in a small square from the WOD18 database: from 68°08′ N to 68°14′ N and from 25°05′ N to 25°15′ W. The graphs of vertical profiles are shown in Fig. 9.13.

A similar law of temperature decrease is seen in the lower part of each curve; however, the curves are vertically shifted relative to the others by 100–300 m, which is caused by seiches of the lower layer. Mixing and vertical homogeneity in the bolus is caused by cascading of cold water on the Greenland shelf down the slope. This descent of cold waters cools water in the bolus and makes it homogeneous. There is no similar process on the Iceland shelf where shelf water is warmer.

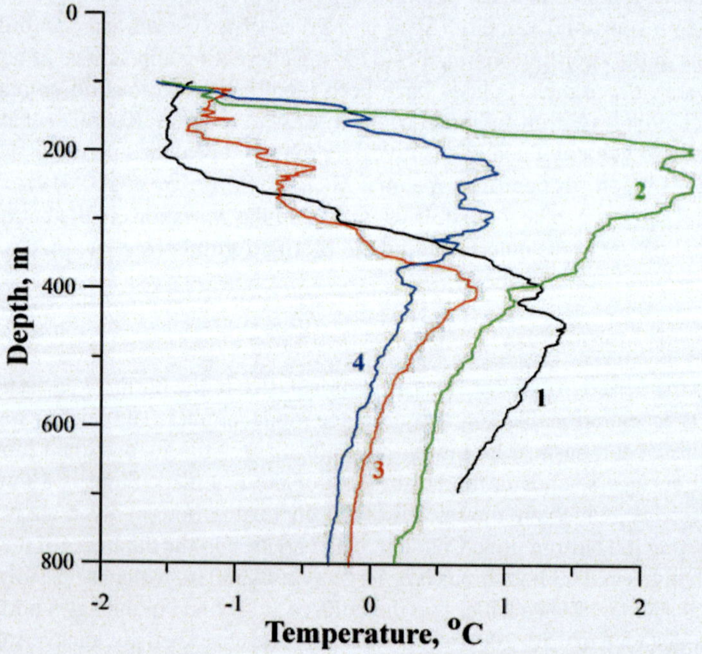

Fig. 9.13 Vertical profiles of temperature at stations in September in different years north of the sill: station *1* on 3.09.1997 (black line), station *2* on 26.09.2007 (green), station *3* on 7.09.1998 (red), and station *4* on 11.09.1995 (blue). Reproduced from Morozov et al. (2020) with permission from Pleiades Publishers

Another mechanism that leads to the formation of bolus as a mixed volume of water is the following. It is shown by numerous measurements that the southern velocity component increases during the bolus type of overflow. Usually it exceeds 0.7 m s^{-1}. High vertical velocities appear when the current overflows the sill of the Denmark Strait. A scheme in Fig. 2.2 shows the mechanism of mixing near the sill. It was reported in the publications (Thurnherr and Speer 2003; Thurnherr et al. 2005) that hydraulically controlled flow over the sills is responsible for bottom mixing above the slope. A local overturning circulation cell is formed over the slope with increase vertical velocities that mixes the bottom flow (Thurnherr and Speer 2003; Thurnherr et al. 2020). The flow regime changes as the velocity increases and the Froude number exceeds unity. The critical regime appears when the Froude number is $Fr > 1$. When the current overflows the sill of the Denmark Strait the Froude number is calculated as

$$Fr = v/\sqrt{g'H},$$

where v is the meridional component of velocity, $\sqrt{g'H}$ is the velocity of gravity waves in the shallow flow; $H = 100$ m is the thickness of the flow; $g' = g\Delta\rho/\rho$ is reduced acceleration due to gravity; $\rho = 1$ g/cm^3 is the in situ density; $g = 9.8$ m s^{-1}; $\Delta\rho = 0.0005$ g/cm^3 in the overflowing current. The critical regime of overflowing appears when the bottom current exceeds $v = 0.7$ m s^{-1}:

$$Fr = \frac{0.7}{\sqrt{\frac{9.8\times100\times0.0005}{1}}} \approx 1.0.$$

We made CTD/LADCP sections across the bolus from south to north and from west to east. We crossed the bolus from south to north before arriving to the narrows of the strait. This allowed us to study the spatial structure of the bolus.

These data allowed us to construct a map of the location of bottom isopycnals that outline the bolus. A map of the location of the bolus in the strait on July 21–23, 2018 and the grid of stations covering the bolus are shown below in Fig. 9.14. The temperature section along the bolus from south to north is shown in Fig. 9.15.

Bottom water in the presence of a bolus flows from the strait and rolls down the slope under the influence of gravity. In this case, the flow core is displaced to the Greenland slope under the influence of the Coriolis force (Girton and Sanford 2003). The lowest part of the bottom water up to 50 m thick flows forward down the slope with a high velocity and penetrates under a layer of less dense water. This forcing elongates the bolus in the southwestern direction, which is seen on the map in Fig. 9.14. This thin layer of the bolus front is followed by the entire body of the bolus with a height exceeding 150 m.

The bolus we crossed with stations is characterized by increased velocities and mixed bottom layer up to 300 m thick. The velocity section is shown in Fig. 9.16. The maximum velocities exceeding 50 cm s^{-1} were recorded at the leading edge of the bolus. The velocity section along the bolus shows that there is a front of high

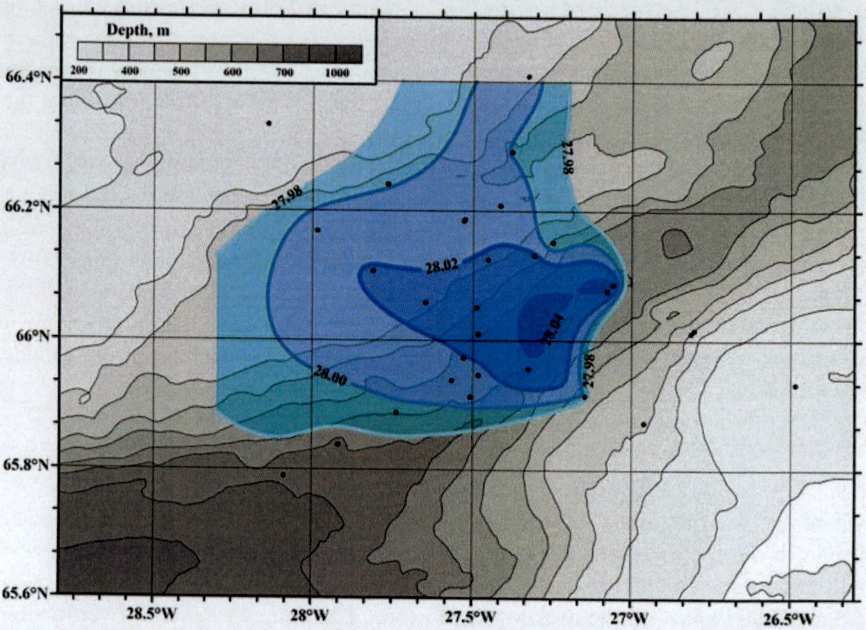

Fig. 9.14 Map of the location of the bolus in the Strait of Denmark (July 21–23, 2018). The map shows the points of the stations and the lines of four isopycnal contours of the bolus: 1.02798, 1.02800 (bold), 1.02802, 1.02804 kg/m^3. Reproduced from Morozov et al. (2020) with permission from Pleiades Publishers

speed located in its leading part. The meridional velocity section across the Strait of Denmark in the presence of a bolus is shown in Fig. 9.17. The velocities in the bolus core exceed 60 cm s^{-1}, which is not observed in the absence of a bolus when the measured speeds do not reach 40 cm s^{-1}.

The data of measurements on two sections along and across the bolus made it possible to construct a velocity map in the bolus. It turned out that in its middle part there is a minimum of speed, while higher speeds are noted in the frontal part and in the narrowness of the strait through which the bolus passed (Fig. 9.18).

The leading part of the bolus escapes from its center because the inclination of the slope increases to the south. The velocities are also high in the rear part of the bolus because it passes through the narrows of the strait. The velocities decrease after passing the narrows and then again increase in the spillway down the slope.

We performed a numerical simulation of the bottom flow in the Strait of Denmark using the modern three-dimensional INMOM ocean dynamics model developed at the Institute of Numerical Mathematics, Russian Academy of Sciences (Dianskii 2013). The model is based on the complete system of primitive equations of hydrothermodynamics in spherical coordinates in the approximations of hydrostatics and Boussinesq (Dianskii 2013). The INMOM model is well tested in the underwater channels of the Atlantic Ocean (Frey et al. 2019; Morozov et al. 2019).

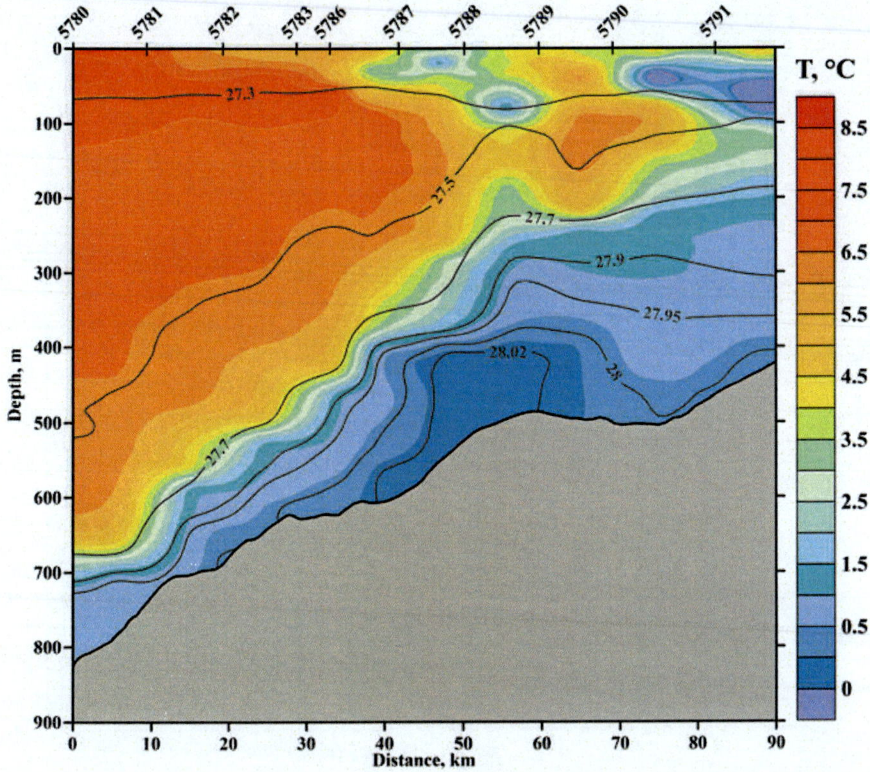

Fig. 9.15 Temperature section along the bolus from south to north. Stations are shown at the top axis. Reproduced from Morozov et al. (2020) with permission from Pleiades Publishers

We selected for modeling a rectangular region in the strait between 65 and 67° N, and between 26 and 29° W. The seabed topography data with an initial resolution of 30′ was taken from database (Smith and Sandwell 1997). The temperature and salinity fields from the monthly average climatic data of the WOA2013 (WOA2013 2013) atlas were set as initial conditions (Antonov et al. 2010); the initial velocity field was assumed zero.

The fields of horizontal velocities, potential temperature and salinity at the sigma levels, sea level deviation from the unperturbed surface are the forecasting parameters of the model. The horizontal resolution is 0.02° by latitude and longitude. The vertical sigma levels are set unevenly; the best resolution is specified near the bottom. The total number of sigma levels is 33, with a depth at 4650 m, the deepest 10 levels are set at the bottom from 4600 to 4150 m with a step of 50 m. The total duration of the calculations was 30 days.

The density distribution in the bolus was set based on the data of our CTD-casts. The velocities north and south of the transverse ridge exceed 100 cm s^{-1}. A chart of simulated velocities is shown in Fig. 9.19. It was shown above based on the

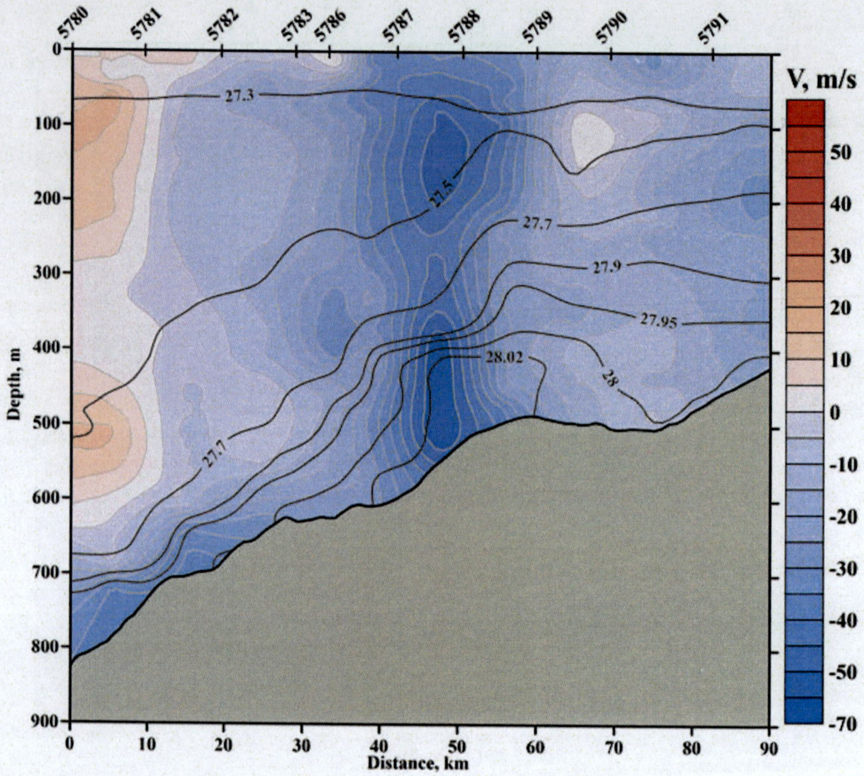

Fig. 9.16 Section of meridional velocity along the bolus. Blue colors are related to the south direction of velocities. Stations are shown at the top axis. Reproduced from Morozov et al. (2020) with permission from Pleiades Publishers

measurements that the core of the flow with maximum velocities is displaced closer to the Greenland slope. It was shown above that, according to the CTD/LADCP casts a minimum of velocities is located in the center of the bolus. A similar minimum of velocity in the center of the bolus is also noted in the numerical simulation of the flow in the strait. A similar deceleration of the flow is observed with a small rotation of the flow according to the curvature of the topography in (Käse and Oschilies 2000). The maximum velocities were found at the leading front of the bolus, which causes elongation of its form in the direction of propagation to the southwest.

Thus, we studied the spatial structure of the overflow of bottom water in the Denmark Strait in the presence of a bolus. Intermittent overflow over the transversal ridge from two sources in the Arctic located north of the strait is observed in the strait. A unique natural hydrodynamic phenomenon is noted in the overflow. A large volume of cold strongly mixed water flows in the form of a lens, which in this case is identified by the term bolus, periodically flows through the strait (with a period of about 5 days). It is shown that the transverse size of the bolus bounded by the 1.028 kg/m^3 isopycnal has a scale of about 70 km, while its thickness reaches 300 m.

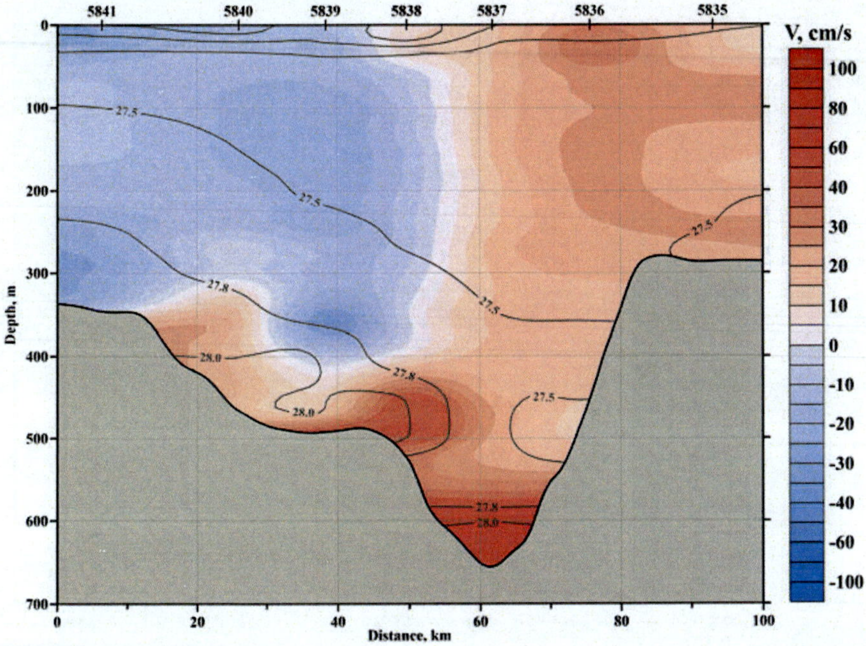

Fig. 9.17 Velocity section across the strait based on the LADCP profiling. The solid lines in the section show the density contour lines. Stations are shown at the top axis. Reproduced from Morozov et al. (2020) with permission from Pleiades Publishers

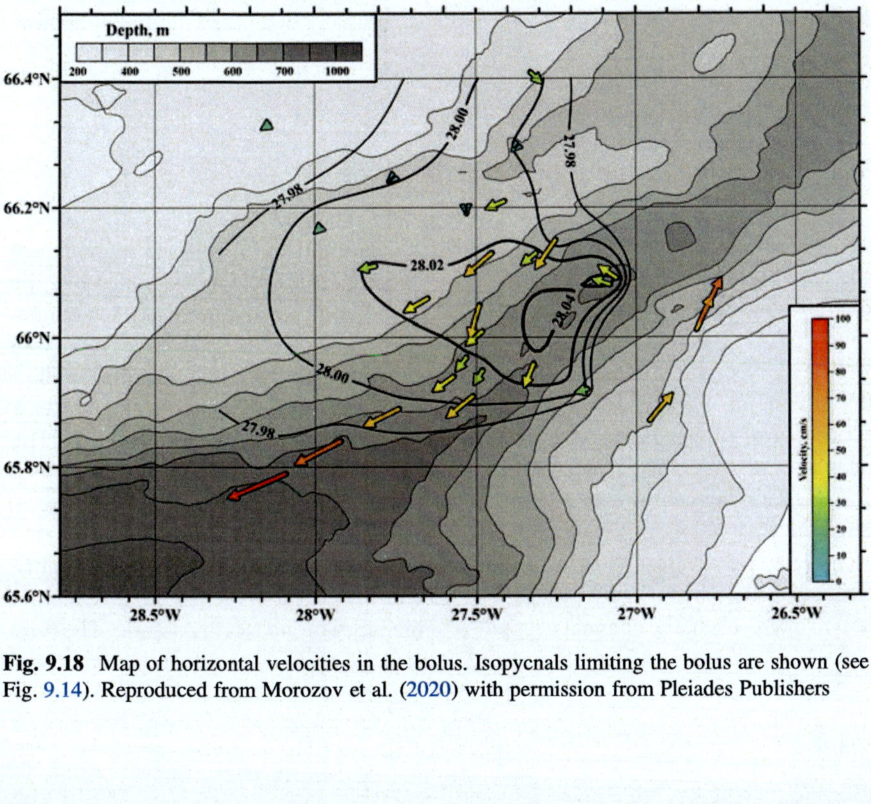

Fig. 9.18 Map of horizontal velocities in the bolus. Isopycnals limiting the bolus are shown (see Fig. 9.14). Reproduced from Morozov et al. (2020) with permission from Pleiades Publishers

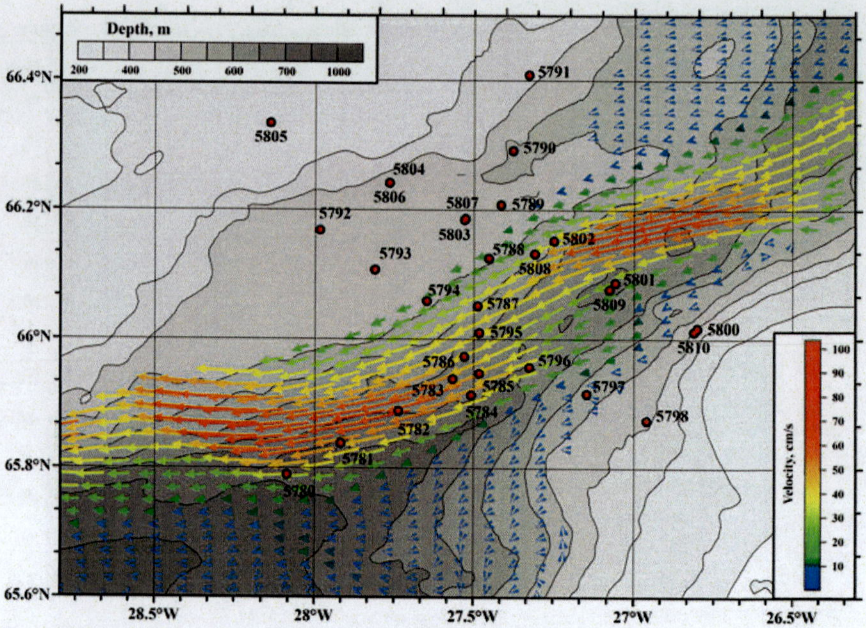

Fig. 9.19 Map of simulated horizontal velocities in the bolus. Locations of stations and their numbers are showns. Reproduced from Morozov et al. (2020) with permission from Pleiades Publishers

This stratified form of overflow with a bolus alternates with the "normal" overflow over the submarine ridge.

9.3 Overflow in the Faroe-Shetland Channel

The Faroe-Shetland region has been well studied and the results of research are presented in many publications. In this section, we review the measurements of the bottom flow from the Arctic to the Atlantic through the Faroe Bank Channel (Fig. 9.20).

The depths over the Iceland Faroe Ridge do not exceed 500 m, while the depth of the sill in the Faroe Bank Channel is approximately 850 m. This channel is a continuation of the Faroe-Shetland Channel, whose depths exceed 1000 m. The main flow of Arctic waters occurs through these channels located between the Faroes and Shetland Islands. A map of the bottom topography (Fig. 9.20) shows the Faroe–Shetland Channel narrowing into the Faroe Bank Channel. The bottom flow from the Norwegian Sea to the southwest follows the shallowing channel towards the southwest; then the channel makes a turn to the northwest and reaches a sill, which is 850 m deep. Then, the channel deepens into the basin of the North Atlantic. The flow

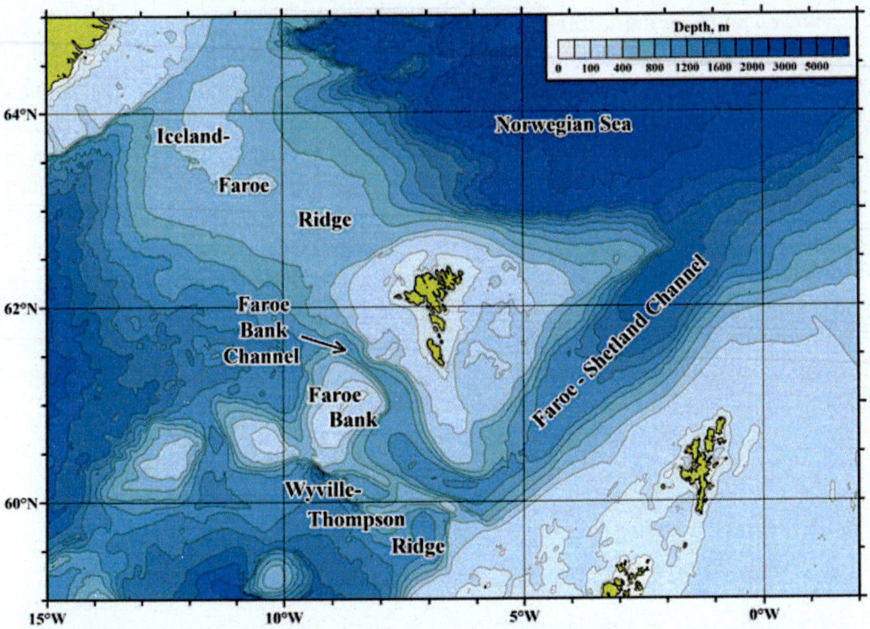

Fig. 9.20 Bathymetry of the region and the main passages for the bottom flows from the Arctic region

of cold water is cascading down the slope into the North Atlantic. This flow occurs under the influence of pressure gradient, bottom friction, and the Earth's rotation. The overflow descends the Iceland-Faroe slope and mixes with overlaying water (Seim and Fer 2011; Saunders 2001; Mauritzen et al. 2005; Fer et al. 2010). Our analysis is based on several review publications.

Hansen and Østerhus (2000) write that a bottom flow water colder than 3 °C and salinity below 35.0 PSU occupies a layer below 500 m to the bottom. The overflow water is much colder and less saline than the upper Atlantic water.

The water in the Faroe-Shetland Channel originates from the waters north of the channel: Modified East Icelandic Water (MEIW), whose temperature and salinity values range within 1.0°-3 °C and 34.70–34.90 PSU; Norwegian Sea Arctic Intermediate Water (NSAIW) with temperature and salinity variations from −0.5 to +0.5 °C and 34.87–34.90 PSU; and Norwegian Sea Deep Water (NSDW) below the −0.5 °C potential isotherm whose salinity is approximately 34.91 PSU. The properties of the deepest water in the Faroe–Shetland Channel are slightly different from its origin in the Norwegian Sea. The properties of the two initial intermediate water masses MEIW and NSAIW are less pronounced in the channel, which is possibly caused by mixing. Generally speaking, only the coldest and low saline waters of the Norwegian Sea basin reach the sill of the Faroe Bank Channel.

Hansen and Østerhus (2007) analyzed the ADCP data from moorings in the Faroe Bank Channel over a period of 1995–2005. The mean overflow transport based on the velocity field was estimated at 2.1 ± 0.2 Sv with seasonal and inter-annual variations of approximately 10%, but without any trend over the whole period. The bottom current transport with potential density exceeding 27.8 kg m^{-3} was 1.9 ± 0.3 Sv with average properties: potential temperature $\theta = 0.25$ °C, salinity S = 34.93 PSU.

These estimates exceed the previous values reported in (van Aken 1988) who estimated the overflow at 1.1 Sv. However, (Saunders 1990) gives an estimate of cold water (<3 °C) overflow at 1.9 Sv. The dense water transport over the Wyville-Thomson Ridge is approximately 0.3 Sv (Saunders 1990). The depths of the ridge are approximately 600 m. Velocities of the bottom current over the ridge sometimes exceed 100 cm s^{-1}. Approximately the same overflow volume was estimated in (Sherwin et al. 2008). They found that a significant part of the overflow is channelled through a newly discovered canyon that leads southward down the southern flank of the Faroe Bank into the Ellett gully. The overflow waters are sometimes identified as Faroe Shetland Channel Bottom Water (FSCBW) with a nominal temperature of 0 °C (Sherwin et al. 2008).

The isotherms in the channel are inclined so that they are deeper over the southwestern wall of the channel on Faroe Bank and shallower over the Faroe Plateau on the northeastern side. The difference in the deepening of the 0–7 °C potential isotherms is approximately 100 m. The channel width at the depth of these isotherms is approximately 16 km. The 3 °C isotherm divides the two water masses in the channel: the overflow and upper flow.

A sharp pycnocline is 50–100 m thick. It is located at depths of 500–600 m between the 27.55 and the 27.95 kg m^{-3} isopycnals. Borenäs et al. (2001) found pitching of isotherms at the southwestern wall of the Faroe Bank Channel, which indicates close locations of isotherms 1–7 °C at the wall of the channel compared to their wider position at the opposite wall.

The core of bottom current of overflow waters is approximately 200–300 m thick with a sharp gradient in the depth interval 200–300 m above the bottom. The interface is approximately at 250 m above the bottom. The bottom layer approximately 70 m thick is well mixed. The bottom layer is separated from the upper layer by an interface 120–180 m thick. The typical speed of the overflow current is 50–100 cm s^{-1}, the maximum currents reach 135 cm s^{-1} when the current accelerates down the slope of the Faroe Bank Channel to the ocean (Fer et al. 2010). The mean direction of the flow along the Faroe Bank Channel at the sill is to the northwest (304°).

McKenna et al. (2016) analyzed propagation of Arctic waters to the North Atlantic and indicate that only Norwegian Sea Arctic Intermediate Water (NSAIW) and Norwegian Sea Deep Water (NSDW) propagate through the conduit of the Faroe Bank Channel to participate in the Atlantic thermohaline circulation. Mixing processes in the Faroe-Shetland Channel result in an exchange of properties between the inflow and outflow in the channel. The flow of colder Norwegian Sea Deep Water occurs closer to the Faroe Plateau, thus causing the previously mentioned inclination of isotherms.

9.4 Overflow in the Gibraltar Strait

It is well known that a two-layer system of opposite flows exists in the Strait of Gibraltar. These flows are caused by strong evaporation in the Mediterranean Sea and low freshwater discharge from the rivers. Thus, a strong difference in water density is formed between the Atlantic Ocean and the Mediterranean Sea, which is mostly caused by salinity difference. Due to strong evaporation, the Mediterranean Sea level gradually decreases. A strong surface current of relatively fresher water from the Atlantic compensates intense evaporation and keeps the sea level balance. Salinity and density of surface water increase due to evaporation; thus, it descends to the deeper layers. Hence, water densities at equal depths in the sea and in the ocean become different and a deep-water current of more saline Mediterranean Water flows into the ocean. A topography chart of the strait is shown in Fig. 9.21.

Strong semidiurnal tidal waves propagate in the Strait. The tidal velocities are of the same order of magnitude as the mean flow at the surface and sometimes exceed the mean current. Thus, an unsteady flow is formed. The tide generates an internal wave when the tidal currents flow over uneven topography. The strongest generation of internal tide occurs over the slopes of the Camarinal Sill.

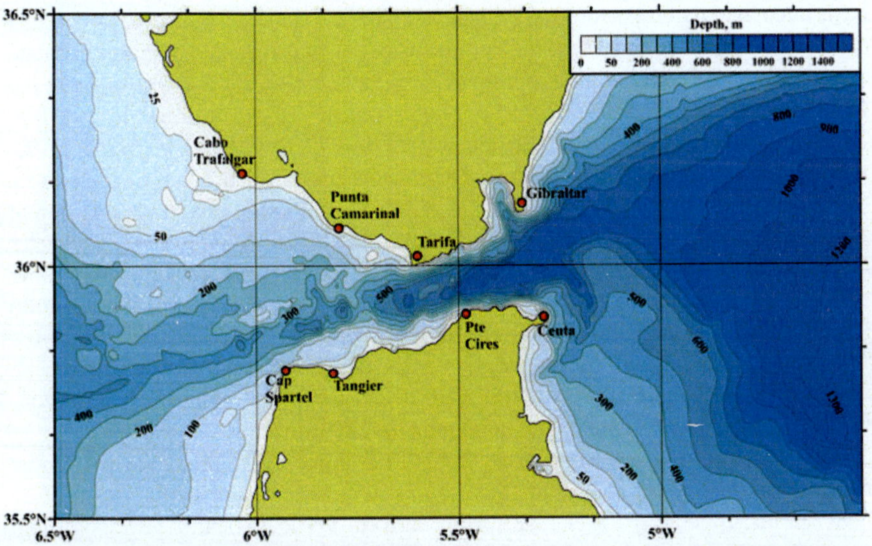

Fig. 9.21 Topography of the Gibraltar Strait

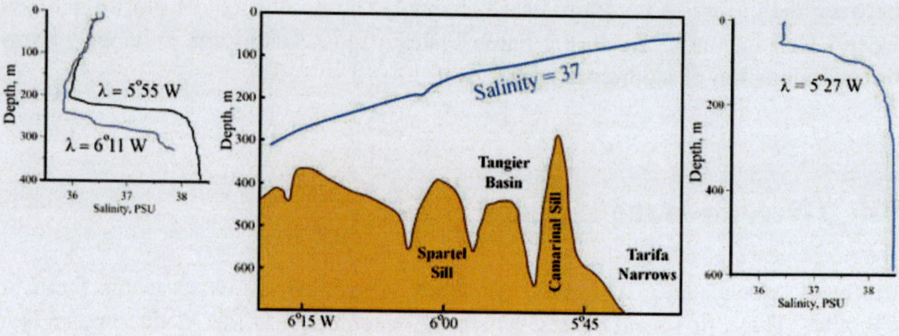

Fig. 9.22 Bottom profile along the Strait of Gibraltar (middle). Isohaline of salinity 37 PSU is shown schematically. Profiles of salinity west (left) and east (right) of the Camarinal Sill in the strait

Generally, the velocity of inflowing Atlantic waters is approximately two times smaller than the surface current outflowing from the Mediterranean Sea. The upper part of the strait is wider than the deeper part of the strait. However, the velocities strongly depend on the location of measurements because the interface between the currents in the eastern part of the strait is located at 50 m depth and deepens in the western direction to 200–250 m (Fig. 9.22). The interface between Atlantic and Mediterranean waters is clearly seen from the location of the halocline. The halocline does not generally coincide with the thermocline, because a strong seasonal thermocline exists under the surface Atlantic water during almost the whole year. As a result, there are two pycnoclines, one for the seasonal thermocline and another for the halocline.

The flow regime and high amplitude internal waves in the Strait of Gibraltar have been intensively studied for many centuries. The estimates of transport by Bryden et al. (1994) based on the direct measurements of currents revealed that the exchange across Gibraltar Strait is due in nearly equal parts to the mean currents and to tidal fluctuations. They estimate that the Mediterranean Water outflow is equal to −0.68 Sv, while the flow of Atlantic waters is 0.72 Sv. Velocity profile of the zonal velocity component based on ADCP data from an instrument at the bottom (274 m) at 35°55′ N, 5°45′ W is shown in Fig. 9.23. Velocities with a vertical step of 10 m were averaged over a period from October 20, 1994 to April 3, 1995. The water flowing into the Mediterranean Sea compensates for the outflow and evaporation, which is approximately 52 cm of the sea level per year. In a later publication in 2000 (Tsimplis and Bryden 2000) the authors estimate based on ADCP measurements that the transports are 0.78 Sv for the Atlantic inflow and 0.67 Sv for the Mediterranean outflow. High velocities of the surface flow into the Mediterranean Sea are confirmed by radar measurements in (Lorente et al. 2019).

In the publication by Bray et al. (1995) the authors extended the traditional two-layer model of exchange to three layers assuming that there is an interface between the upper and lower layers with uniform vertical shear. They found that approximately half of the inflowing and outflowing water transports occurs in the interface layer. The vertical exchange between layers in the interface is about half of

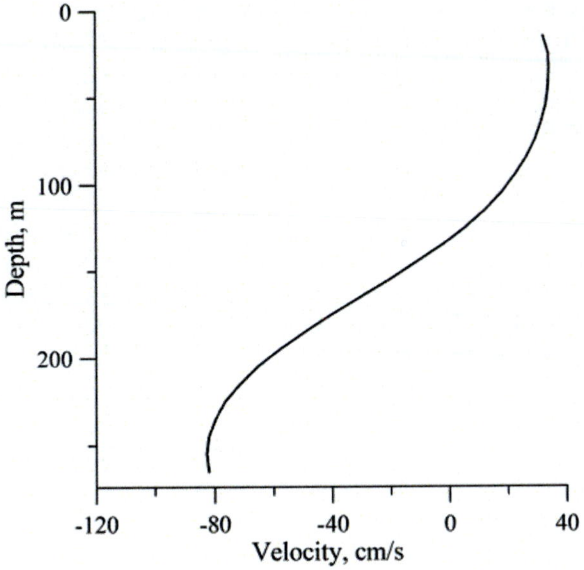

Fig. 9.23 Vertical profile of the zonal component of velocity over the Camarinal Sill in the Strait of Gibraltar

the horizontal exchange. Internal waves of high amplitude contribute to the vertical exchange in the strait.

Horizontal transport within the interface layer strongly changes along the strait; thus, substantial vertical exchange exists between the interface and the upper and lower layers. On the eastern end of the Strait, roughly half of the transport into the Mediterranean is found in the interface layer, while on the west end an equivalent outflowing transport occurs in the interface layer. In the middle of the strait low transport within the interface layer is observed near the Camarinal Sill. The interface is deeper, thicker, fresher, and cooler on the west end of the strait than in the eastern part in the region of the Tarifa Narrows. Here, a minimum in thickness and a maximum in salinity gradient are found. La Violette and Arnone (1988) found that the high amplitude internal tides and turbulence resulting from a hydraulic jump over the Camarinal Sill might be the major cause of mixing between the upper and lower waters in the strait.

The flows in the strait were studied by Armi and Farmer (1988). The analysis reported in their publication is related to the internal hydraulics of the strait and to the existing of hydraulic control regions, in which strong influence on the exchange between the upper and lower layers occurs. They revealed periodical changes in the flow regime from subcritical to supercritical depending on the tidal phase. The overflow process is hydraulically controlled over Camarinal and Spartel sills. Two basin reservoirs: the Tangiers Basin and the basin east of the Tarifa Narrows accumulate the transported water. If the speed of the current exceeds the speed of the gravity waves, a "critical flow" appears and the waves cannot propagate upstream. In this case, the motion is cut-off from the influence of the downstream areas, because the waves cannot propagate upstream. This regime of the flow is hydraulically controlled.

A very interesting property of the flow regime is observed in the strait (Armi and Farmer 1988). The interface rapidly deepens in-phase with the tide west of the Camarinal Sill. It is followed by sudden eastward acceleration of the upper layer flow and formation of an internal bore and a velocity front. It is reported in Lacombe and Richez (1982) that the interface deepens by more than 200 m in 1 h; at the same time, the surface currents increase by 1 m s^{-1}. This phenomenon is known as the tidal bore, usually observed when a tidal wave propagates upstream in a river. Internal tides generated over the slopes of the Camarinal Sill are extremely strong. The vertical excursions of water particles exceed 150–200 m (Bockel 1962; Boyce 1975; Morozov et al. 2002).

Sailing directions in the Strait of Gibraltar indicate that strong currents and tidal streams may be experienced in the strait. The maximum velocities of the surface flow, which may be expected are about 100 cm s^{-1} in the western direction caused by tides and up to 350 cm s^{-1} in the eastern direction when the mean current and tidal current coincide. In the middle of the strait near Tarifa, the strong eastward stream commences at the time of high water in Tarifa. The direction of the flow changes to the opposite 6 h later.

After passing the Strait of Gibraltar and partial mixing with the Atlantic oceanic waters at the interface layer, Mediterranean Water descends to the level of the corresponding density in the Atlantic Ocean at 1000–1200 m. The downslope descent of saline waters from the Strait of Gibraltar is sometimes called a giant underwater cataract (Whitehead 1989). These waters spread in the entire eastern Atlantic Ocean reaching 50° N and 20° S. Initially, lenses of high salinity waters are formed. These lenses are circular with a thickness of about 400 m and a diameter of about 50 km with clockwise rotation. After the discovery of these lenses in 1981 they were called Meddies (Mediterranean Eddies). The temperature in the Meddies is by 4 °C greater than in the surrounding waters and the salinity is by 0.9 PSU higher. These lenses exist for several years gradually mixing with the surrounding waters. Finally, they decay and form a layer of Mediterranean Water.

9.5 Gravity Current in the Bear Island Trough

The Norwegian Current transports relatively warm, saline waters into the Arctic Ocean. This current has offshore and coastal (North Cape Current) branches (Bjerke and Torsethaugen 1989). The former branch warms the west coast of Svalbard, while the latter transports warm water to the Barents Sea.

The interaction between cold and low-saline Arctic waters and relatively warmer and more saline Atlantic waters is a key feature of the entire North Atlantic region. Strong interaction between these waters occurs in the Barents Sea. Together with the Fram Strait, it contributes to the water mass exchange between the Arctic and Atlantic oceans (Giraudeau et al. 2016). The shelf slopes define the western and

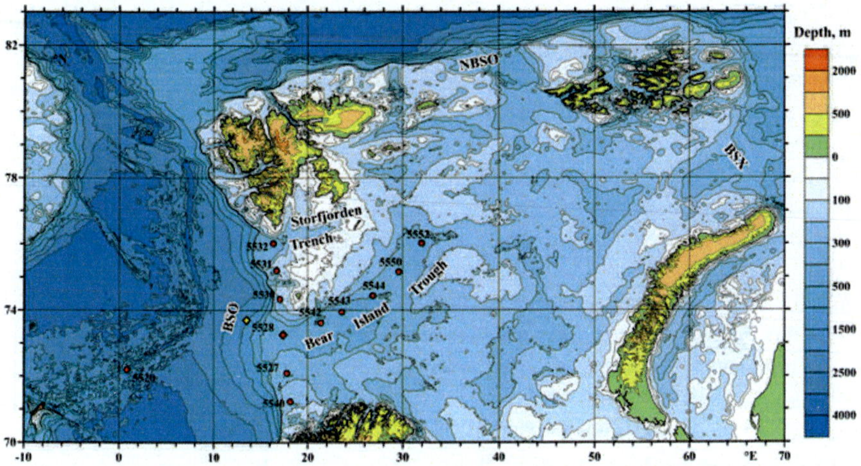

Fig. 9.24 Locations of the measurements in the western Barents Sea in July–August 2017. Red circles show our CTD stations; red diamond is the site of mooring deployment in the Bear Island Trough; yellow diamond is the mooring location in 1995. Bathymetry is based on GEBCO 2014 data. Abbreviations: Northern Barents Sea Opening (NBSO); Barents Sea Opening (BSO); Barents Sea Exit (BSX)

northern margins of the Barents Sea (Fig. 9.24). The western boundary is called the Barents Sea Opening (BSO) (Smedsrud et al. 2013); the northern boundary is the Northern Barents Sea Opening (NBSO) (Lind and Ingvaldsen 2012). The eastern strait between Franz Josef Land and Novaya Zemlya is called the Barents Sea Exit (BSX) (Gammelsrød et al. 2009). The outflow of cold deep waters from the Barents Sea occurs through all three boundaries (Årthun et al. 2011). The first oceanographic measurements in this region were performed in the beginning of the twentieth century (Nansen 1906).

The deepest pathway for the water exchange between the Norwegian and Barents seas is the Bear Island Trough. It is located in the southwestern part of the Barents Sea; the characteristic depths of the trough are up to 500 m; the background depths are 100–200 m shallower. The length of the trough exceeds 500 km and the width is approximately 100 km. The bottom current in the Bear Island Trough is directed to the Norwegian Sea (Lukashin and Shcherbinin 2007).

We analyzed CTD and ADCP measurements in the western Barents Sea carried out onboard the Russian R/V *Akademik Mstislav Keldysh* (cruise 68) in July–August 2017. We studied the water structure over a meridional section between Norway and Svalbard. The velocities of the bottom flow were measured on a mooring deployed in the deepest part of the trough. We compared direct velocity measurements with the tidal velocities calculated from satellite altimetry data. Despite strong barotropic tides in the region, the mean bottom flow was permanently directed from the Barents to the Norwegian Sea, which corresponds to the measured thermohaline structure of the deep waters.

The goal of this section is to analyze the hydrographic conditions in the region of the Bear Island Trough in July–August 2017. We analyzed our direct velocity measurements in the deepest part of the trough and thermohaline structure of the bottom waters in the western Barents Sea.

On July 23–August 3, 2017, we occupied two CTD-sections across and along the Bear Island Trough in the western Barents Sea. Each section included six stations. We deployed a mooring with Acoustic Current Doppler Profiler (ADCP) close to the bottom at the point of intersection (station 5528) between these two sections in the deepest part of the Bear Island Trough. The bottom topography along the sections was measured by the Kongsberg EA600 12 kHz single beam echo sounder. The locations of CTD stations and moorings are shown in Fig. 9.24.

Temperature and salinity profiles at station 5528 are shown in Fig. 9.27. Salinity decreases strongly (from 35.11 to 35.03 PSU) from 200 m to the bottom. The bottom layer about 30 m thick between depths of 450 and 480 m is well mixed. The bottom potential temperature at the westernmost station in the trough was 3.52 °C, and salinity was 35.04 PSU. The ADCP velocity measurements were carried out near the upper boundary of this layer. The θ,*S*-curves for four stations along the Bear Island Trough are shown in Fig. 9.28.

Observed potential temperature and salinity over both sections are shown in Figs. 9.25 and 9.27. The surface temperature was approximately 7–8 °C along the entire quasi-meridional section from Norway to 76° N near Svalbard slowly decreasing to the north. Unlike the surface temperatures, the bottom temperatures decreased strongly to the north forming a strong density gradient that drives the geostrophic currents. Despite the fact that the bottom water temperature in the deepest part of the Bear Island Trough was quite low (3.52 °C), it was not the coldest water over this section. The Storfjorden Trench located immediately south of Svalbard to the north

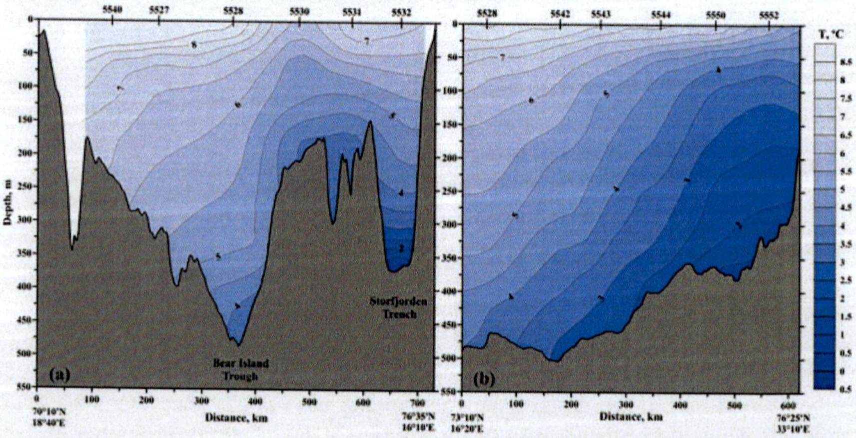

Fig. 9.25 Potential temperature sections across (**a**) and along (**b**) the Bear Island Trough measured on July 23 – August 3, 2017. The mooring was deployed at station 5528 in the deepest part of the quasi-meridional section. The vertical scale and temperature levels are the same in both figures. The grey color shows the ocean bottom (Frey et al. 2017)

of Bear Island is 150 m shallower but the bottom temperature there was much lower (1.05 °C), which is approximately 2.5 °C lower than in the Bear Island Trough. The second section along the trough (Fig. 9.25b) shows significant increase in the bottom temperature in the southwestern direction along this trough. The minimum temperature here was 1.53 °C at a depth of 400 m in the northeastern part; the temperature near the exit to the Bear Island Trough (station 5528) was 3.52 °C at a depth of 500 m. The temperature increase can be caused by the influence of warm Atlantic waters in the western part of the Barents Sea and by mixing of cold bottom water with warmer waters above the bottom layer during its southwestward propagation.

The salinity section along the trough reveals a tongue of low saline waters descending down the slope from northeast to southwest (Fig. 9.26). Salinity increases in the course of the propagation of this water downslope the Bear Island Trough. The θ,*S*-curves at stations along the trough, show that the coldest water was found at the bottom of the shallowest station in the northeastern part of the trough. As the water descends down the trough it becomes warmer and the cold "tails" on the θ,*S*-curves are sequentially truncated (stations 5552, 5544, 5528) (Fig. 9.28). The curves differ only above the density level 27.85, which corresponds to 200 m. The θ, *S*-curve for station 5520 in the western part of the Norwegian Sea is different from the diagrams in the Bear Island Trough, which indicates that the water structure in the ocean and in the trough are different. The bottom water in the Bear Island Trough is formed predominantly by descending of cold low saline water from the slopes that becomes warmer and more saline as it flows. The bottom water in the region west of the trough is generally of the oceanic origin. The difference between them drives the geostrophic current to the north.

Mooring observations of the velocity time series were carried out at 73°15.6′ N, 17°24.6′ E in 25 m above the seafloor using downward looking Teledyne RD

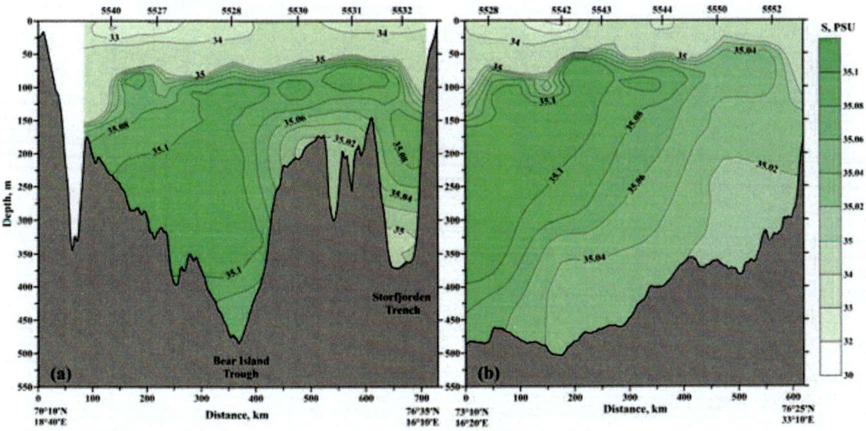

Fig. 9.26 Salinity sections across (**a**) and along (**b**) the Bear Island Trough measured on July 23 – August 3, 2017. The vertical scale and salinity levels are the same in both figures. The grey color shows the ocean bottom (Frey et al. 2017)

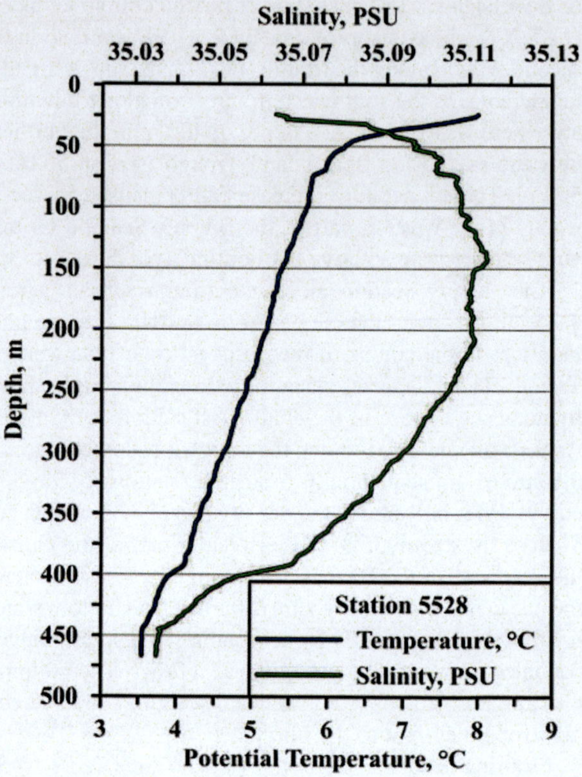

Fig. 9.27 Vertical profiles of potential temperature and salinity at the location of mooring in the Bear Island Trough (station 5528) (Frey et al. 2017)

Instruments Doppler Volume Sampler (DVS) 2400 kHz ADCP. It was deployed for 4 days (July 24–28, 2017) with a 10 min sampling interval. The range of measurements was 1 m from the ADCP with a blank distance of 30 cm.

The velocity measurements reveal a westward bottom flow in the Bear Island Trough (Fig. 9.29). Strong tides in this region are responsible for the variations in the bottom currents. We calculated tidal velocities by the TPXO9 model using satellite altimetry data (Egbert and Erofeeva 2002). A comparison of measured and calculated tidal velocities shows that the observed change of the bottom flow direction is caused by the semidiurnal tides. Nevertheless, the mean current is directed to the northwest. The four-day mean westward component was 4.1 cm s^{-1}; the northward component was 6.6 cm s^{-1}.

Thus, the thermohaline structure and velocity data in the Bear Island Trough showed that the bottom water spreads along the trough from the Barents to the Norwegian Sea. The four-day mean velocity at the exit from the trough was 7.8 cm s^{-1} directed to the northwest. This flow of warm Norwegian Atlantic Current waters transports heat to the northwestern and western parts of the Svalbard Archipelago, which does not freeze in winter. The bottom temperature in the Bear Island Trough was 3.52 °C, while it was lower (1.05 °C) in the Storfjorden Trench, which is

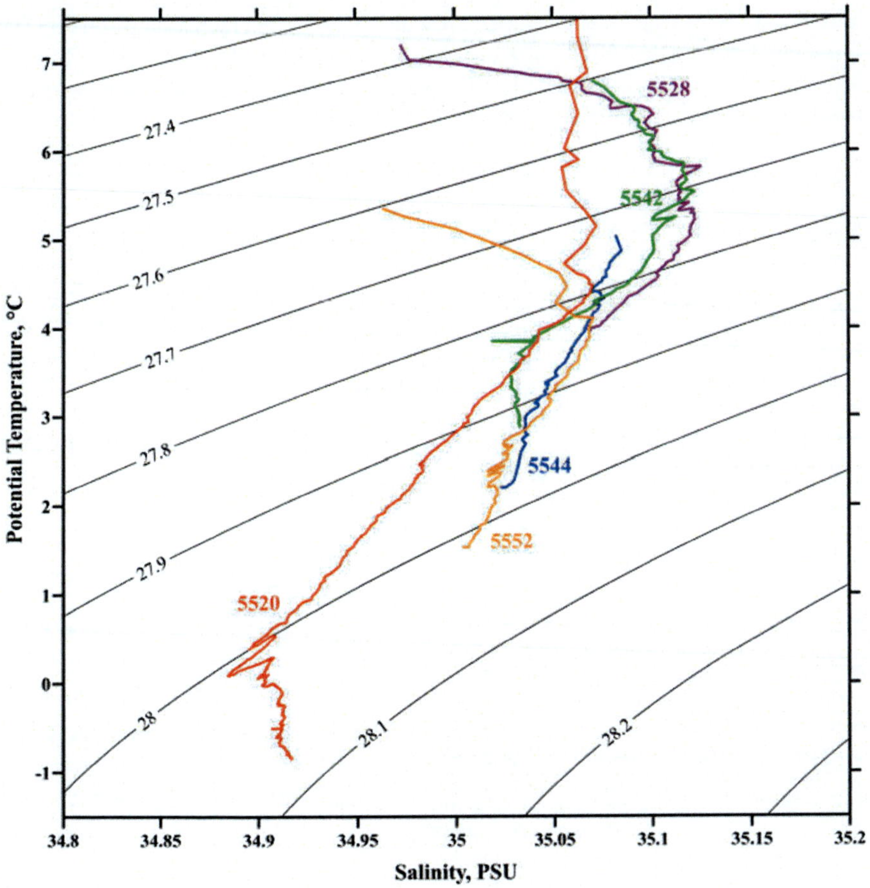

Fig. 9.28 Curves of θ,*S*-distributions for four stations along the trough. Stations 5528, 5544, and 5552 are in the Bear Island Trough, station 5516 is southwest of the trough near Jan Mayen Island (Frey et al. 2017)

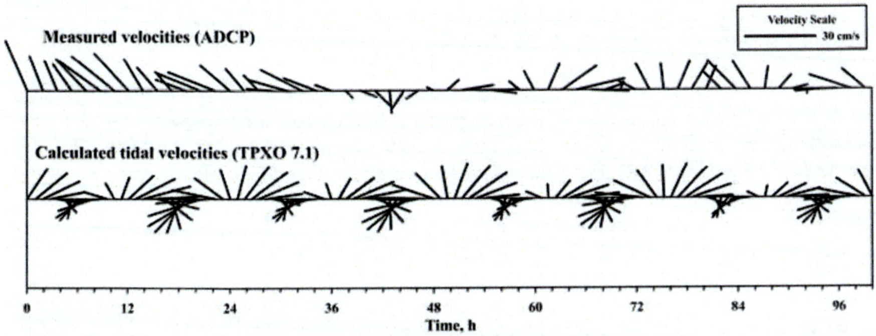

Fig. 9.29 Velocity vectors based on four-day mooring measurements (station 5528) of the bottom flow in the Bear Island Trough on July 24–28, 2017 at 73°15.6′ N, 17°24.6′ E (top); and calculated tidal velocities for the same period (bottom) (Frey et al. 2017)

150 m shallower. The bottom temperatures along the trough increased from 1.53 °C in its eastern part to 3.52 °C in the western boundary of the Barents Sea. Strong semidiurnal tides in this region influence bottom currents significantly, but the mean flow is directed from the Barents Sea to the Norwegian Sea.

References

Antonov JI, Seidov D, Boyer TP, et al. (2010) World Ocean Atlas 2009, vol. 2: Salinity. (Ed. Levitus S). U.S. Government Print Office, Washington, DC, 184 p

Armi L, Farmer DM (1988) The flow of Mediterranean water through the Strait of Gibraltar; Farmer DM, Armi L (1988) The flow of Atlantic water through the Strait of Gibraltar. Prog Oceanogr 21:1–105

Årthun M, Ingvaldsen RB, Smedsrud LH, Schrum C (2011) Dense water formation and circulation in the Barents Sea. Deep-Sea Res I 58:801–817

Bjerke PL, Torsethaugen K (1989) Environmental conditions on the Norwegian Continental Shelf, Barents Sea. Report no STF60 A89052. Norwegian Hydrotechnical Laboratory, SINTEF, Trondheim

Bockel M (1962) Travaux Ooceanographiques de l' "Origny" a Gibraltar. Cahiers Oceanogr 14:325–329

Borenäs KM, Lake IL, Lundberg PA (2001) On the intermediate water masses of the Faroe–Bank Channel overflow. J Phys Oceanogr 31:1904–1914

Bourgault D, Kelley DE, Galbraith PS (2008) Turbulence and boluses on an internal beach. J Mar Res 66(5):563–588

Bower AS, le Cann B, Rossby T, Zenk W, Gould J, Speer K, Richardson PL, Prater MD, Zhang H-M (2002) Directly measured mid-depth circulation in the northeastern North Atlantic Ocean. Nature 419:603–607

Boyce FM (1975) Internal waves in the Strait of Gibraltar. Deep Sea Res I 22:597–610

Bray N, Ochoa J, Kinder TH (1995) The role of the interface in exchange through the Strait of Gibraltar. J Geophys Res 100:10755–10776

Bryden HL, Candela J, Kinder TH (1994) Exchange through the Strait of Gibraltar. Prog Oceanogr 33:201–248

Cooper LHN (1955) Deep water movements in the North Atlantic as a link between climatic changes around Iceland and biological productivity of the English Channel and Celtic Sea. J Mar Res 14:347–362

Dianskii NA (2013) Simulation of ocean circulation and investigation of its reaction on short-term and long-term atmospheric impacts (Fizmatlit, Moscow, 2013) (in Russian)

Dickson RR, Gurbutt PA, Medler KJ (1980) Long-term water movements in the southern trough of the Charlie-Gibbs Fracture Zone. J Mar Res 18(3):571–583

Dobrolyubov SA, Falina AS (2002) Interannual variability of the intermediate North Atlantic water circulation from a multiparameter analysis of the water masses. Oceanology 42:619–627

Dobrolyubov SA, Lappo SS, Morozov EG, Pisarev SV, Sokov AV (2003a) Variability of water masses in the North Atlantic based on hydrographic sections along 60° N. Dokl Earth Sci 390(2):566–570

Dobrolyubov SA, Lappo SS, Morozov EG, Sokov AV (2003b) Water transport through the Charlie–Gibbs Fracture Zone. Dokl Earth Sci 391(6):903–905

Egbert GD, Erofeeva SY (2002) Efficient inverse modeling of barotropic ocean tides. J Atmos Ocean Technol 19:183–204. https://doi.org/10.1175/1520-0426(2002)019<0183:EIMOBO>2.0.CO;2

Fer I, Voet G, Seim KS, Rudels B, Latarius K (2010) Intense mixing of the Faroe Bank Channel overflow. Geophys Res Lett 37:L02604. https://doi.org/10.1029/2009GL041924

Filyushkin BN, Moshonkin SN, Myslenkov SA, Zalesnyi VB (2013) Simulation of the interannual and seasonal variability of the overflow transport through the Denmark Strait. Oceanology 53(6):643–654

Frey DI, Novigatsky AN, Kravchishina MD, Morozov EG (2017) Water structure and currents in the Bear Island Trough in July–August. Russ J Earth Sci 17:ES3003. https://doi.org/10.2205/2017ES000602

Frey DI, Morozov EG, Fomin VV, Diansky NA, Tarakanov RY (2019) Regional modeling of Antarctic bottom water flows in the key passages of the Atlantic. J Geophys Res Oceans 124:8414–8428. https://doi.org/10.1029/2019JC015315

Gammelsrød T, Leikvin Ø, Lien V, Budgell WP, Loeng H, Maslowski W (2009) Mass and heat transports in the NE Barents Sea: observations and models. J Mar Syst 75(1):56–69

Giraudeau J, Hulot V, Hanquiez V, Devaux L, Howa H, Garlan T (2016) A survey of the summer coccolithophore community in the western Barents Sea. J Mar Syst 158:93–105

Girton J, Sanford T (2003) Descent and modification of the overflow plume in the Denmark Strait. J Phys Oceanogr 33(7):1351–1364

Haine TWN (2010) High-frequency fluctuations in Denmark Strait transport. Geophys Res Lett 37: L14601. https://doi.org/10.1029/2010GL043272

Haine TWN, Zhang S, Moore GWK, Renfrew IA (2009) On the impact of high-resolution, high-frequency meteorological forcing on Denmark Strait ocean circulation. Q J R Meteorol Soc 135:2067–2085. https://doi.org/10.1002/qj.505

Hansen B, Østerhus S (2000) North Atlantic–Nordic Seas exchanges. Prog Oceanogr 45:109–208

Hansen B, Østerhus S (2007) Faroe Bank Channel overflow 1995–2005. Prog Oceanogr 75:817–856

Jochumsen K, Quadfasel D, Valdimarsson H, Jónsson S (2012) Variability of the Denmark Strait overflow: moored time series from 1996–2011. J Geophys Res Oceans 117:C12003. https://doi.org/10.1029/2012JC008244

Käse RH, Oschlies A (2000) Flow through Denmark Strait. J Geophys Res Oceans 105(C12): 28527–28546. https://doi.org/10.1029/2000JC900111

Käse R, Girton JB, Sanford TB (2003) Structure and variability of the Denmark Strait overflow: model and observations. J Geophys Res Oceans 108(C6):3181. https://doi.org/10.1029/2002JC001548

Kinder TH, Bryden HL (1990) The aspiration of deep waters through Straits. The physical oceanography of Sea Straits. Pratt LJ (Ed). Kluwer Academic Publishing, Dordrecht, 295–319

Köhl A (2010) Variable source regions of Denmark Strait and Faroe Bank Channel overflow waters. Tellus 62A:551–568. https://doi.org/10.1111/j.1600-0870.2010.00454.x

Köhl A, Käse RH, Stammer D (2007) Causes of changes in the Denmark Strait overflow. J Phys Oceanogr 37(6):1678–1696

Koszalka IM, Haine T, Magaldi M (2013) Fates and travel times of Denmark Strait overflow water in the Irminger Basin. J Phys Oceanogr 43(12):2611–2628. https://doi.org/10.1175/jpo-d-13-023.1

La Violette PE, Arnone RA (1988) A tide-generated internal waveform in the western approaches to the strait of Gibraltar. J Geophys Res Oceans 93:15,653–15,667

Lacombe H, Richez C (1982) The regime of the Strait of Gibraltar. In: Nihoul JCJ (ed) Hydrodynamics of semi-enclosed seas. Elsivier, Amsterdam, pp 13–74

Lankhorst M, Zenk W (2006) Lagrangian observations of the mid-depth and deep velocity fields of the northeastern Atlantic Ocean. J Phys Oceanogr 36:43–63

Lind S, Ingvaldsen RB (2012) Variability and impacts of Atlantic water entering the Barents Sea from the north. Deep-Sea Res I 62:70–88. https://doi.org/10.1016/j.dsr.2011.12.007

Lorente P, Piedracoba S, Sotillo MG, Álvarez-Fanjul E (2019) Long-term monitoring of the Atlantic yet through the strait of Gibraltar with HF radar observations. J Mar Sci Eng 7(1): 3–19. https://doi.org/10.3390/jmse7010003

Lukashin VN, Shcherbinin AD (2007) Nepheloid layer and horizontal flux of the sedimentary material in the Norwegian Sea. Oceanology 47(6):894–908

Macrander A, Käse RH, Send U, Valdimarsson H, Jónsson S (2007) Spatial and temporal structure of the Denmark Strait overflow revealed by acoustic observations. Ocean Dyn 57:75–89. https://doi.org/10.1007/s10236-007-0101-x

Mastropole D, Pickart RS, Valdimarsson H, Våge K, Jochumsen K, Girton J (2017) On the hydrography of Denmark Strait. J Geophys Res Oceans 122:306–321. https://doi.org/10.1002/2016JC012007

Mauritzen C (1996) Production of dense overflow waters feeding the North Atlantic across the Greenland-Scotland ridge. Pt 1: evidence for a revised circulation scheme. Deep-Sea Res I 43(6):769–806. https://doi.org/10.1016/0967-0637(96)00037-4

Mauritzen C, Price J, Sanford T, Torres D (2005) Circulation and mixing in the Faroese channels. Deep-Sea Res I 52:883–913

McCartney MS (1992) Recirculating components to the deep boundary current of the northern North Atlantic. Prog Oceanogr 29:283–382

McKenna C, Berx B, Austin WEN (2016) The decomposition of the Faroe-Shetland Channel water masses using parametric optimum multi-parameter analysis. Deep-Sea Res I 107:9–21

Merian JR (1828) Ueber die Bewegung tropfbarer Fl¨ussigkeiten in Gef˙Assen [on the motion of drippable liquids in containers] (thesis). Schweighauser, Basel (in German)

Moore CD, Koseff JR, Hult EL (2016) Characteristics of bolus formation and propagation from breaking internal waves on shelf slopes. J Fluid Mech 791:260–283. https://doi.org/10.1017/jfm.2016.58

Morozov EG, Trulsen K, Velarde MG, Vlasenko VI (2002) Internal tides in the Strait of Gibraltar. J Phys Oceanogr 32:3193–3206

Morozov EG, Frey DI, Gladyshev SV, Klyuvitkin AA, Novigatsky AN (2019) Internal tides in the Denmark Strait. Izvestia Atmos Oceanic Phys 55(3):295–302

Morozov EG, Frey DI, Gladyshev SV, Gladyshev VS (2020) Hydrodynamics of the bottom-water flow from the Arctic to the Atlantic through the strait of Denmark, Izvestiya. Atmos Oceanic Phys 56(5):479–487

Nansen F (1906) Northernwaters: captain Roald Amundsen's oceanographic observations in the Arctic seas in 1901. With a discussion of the origin of the Bottom-Waters of the Northern Seas. No. 3 (in Commission by Jacob Dybwad)

Saunders PM (1990) Cold outflow from the Faroe Bank Channel. J Phys Oceanogr 20:28–43

Saunders PM (1994) The flux of overflow water through the Charlie-Gibbs Fracture Zone. J Geophys Res Oceans 99(C6):12343–12355

Saunders PM (2001) The dense northern overflows. In: Siedler G, Church J, Gould J (eds) Ocean circulation and climate. Academic, San Diego, pp 401–417

Seim KS, Fer I (2011) Mixing in the stratified interface of the Faroe Bank Channel overflow: the role of transverse circulation and internal waves. J Geophys Res Oceans 116:C07022. https://doi.org/10.1029/2010JC006805

Sherwin TJ, Griffiths CR, Inall ME, Turrell WR (2008) Quantifying the overflow across the Wyville Thomson Ridge into the Rockall Trough. Deep-Sea Res I 55(4):396–404

Skolotnev SG, Sanfilippo A, Peyve AA, Nestola Y et al (2021) Seafloor spreading and tectonics at the Charlie Gibbs transform system (52-53°N, Mid-Atlantic Ridge): preliminary results from R/V A. N. Strakhov expedition S50. Ofioliti 46(1):83–101. https://doi.org/10.4454/ofioliti.v46i1.539

Smedsrud LH, Esau I, Ingvaldsen R, Eldevik T, Haugan PM, Li C, Lien VS, Olsen A, Omar AM, Otteraa OH (2013) The role of the Barents Sea in the Arctic climate system. Rev Geophys 51(3):415–449

Smith WHF, Sandwell DT (1997) Global Sea floor topography from satellite altimetry and ship depth soundings. Science 277:1956–1962. http://topex.ucsd.edu/cgi-bin/get_data.cgi. https://doi.org/10.1126/science.277.5334.1956

Spall M, Price J (1998) Mesoscale variability in Denmark Strait: the PV outflow hypothesis. J Phys Oceanogr 28(8):1598–1623

Weddell Gyre that leads to trapping of the coldest water south of the South Scotia Rise and periodical release of this trapping. This process is responsible for temperature fluctuations of Weddell Sea Deep Water volumes in the Argentine Basin, which later flow through the Vema Channel and others.

9. We studied variability of the flow along the Vema Channel. The temperature in the jet increases due to mixing with the overlying North Atlantic Deep Water while the flow propagates from south to north in the Vema Channel.
10. A cold jet of Weddell Sea Deep Water with a potential temperature of -0.1 °C was observed above the western slope of the Vema Channel at a depth of 4200 m. This flow was observed at the topography break and can be characterized as "coastal" intensification. The coldest type of AABW ($\theta < 0.0$ °C) flows to the north in the Vema Channel as two jets one of which is located in the deep channel bed and the other approximately 300 m above over the western slope of the channel. We traced the latter jet over a distance exceeding 250 km. This flow eventually fills the deep basins of the Northeast Atlantic. The deep channel bed of the Vema Channel does not end in the region 26°40′ S, 34°00′ W as a widening to the north in the Brazil Basin but continues to the north as flows in smaller abyssal channels. The flow along the deep bed of the Vema Channel gradually disappears approximately at 25°30′ S while flowing along these small channels.
11. The Antarctic Bottom Water ($\theta < 2$ °C) flow through the Romanche and Chain fracture zones is estimated at 0.5 Sv in each channel. The mean velocities are 10–20 cm s^{-1}. The bottom water propagating through the Romanche and Chain fracture zones spreads only to the southeastern and equatorial parts of the Atlantic.
12. We found a deep spillway (acceleration of the abyssal flow when it passes the narrows of the deep channel and descends the slope after overflowing the sills) in the western part of the Romanche Fracture Zone (22°27′–22°32′W). This is the main inflow of Antarctic Bottom Water into the Romanche Fracture Zone. The spillway consists of two flows and exists in the layer $\theta < 1.20$ °C (deeper than 4150 m). The vertical velocities in the spillway can reach 15 cm s^{-1}.
13. Spillways were found in the eastern parts of the Chain and Romanche fracture zones. Bottom water descends from a sill approximately from 4000 m down to 4500 m. The flow splits into a dense water downstream along the bottom and a mixed layer flow continuation at a depth of 4000 m. This phenomenon is similar to laboratory experiments.
14. The Antarctic Bottom Water ($\theta < 2$ °C) flow to the East Atlantic through the Vema Fracture Zone (11° N) is estimated at 1.0 Sv (ranging from 0.7 to 1.2 Sv in different years). The greatest velocity reaches 45 cm s^{-1}. The Vema Fracture Zone is the main pathway for Antarctic Bottom Water to the Northeast Atlantic. A spillway with the descent of water by 170 m was found in the southern channel of the Vema Fracture Zone.
15. The main AABW transport to the northern basins of the East Atlantic occurs through the Vema Fracture Zone. Along with this transport, the bottom water also flows through a number of smaller fracture zones: Bogdanov, Doldrums,

Vernadsky, Strakhov, Arkhangelsky, and Nameless (7°28′ N). The estimates of the total AABW transport through this group of fractures based on the measurements in 2014–2016 were equal to 0.48 ± 0.05 Sv. However, without doubts, the coldest bottom water is transported through the Vema Fracture Zone.

16. The bottom water transport to the East Atlantic through the fractures south of the Vema Fracture Zone occurs in a complex system of overflows from one fracture to another.
17. The transport of bottom waters (including AABW) to the east through the fractures north of the Vema Fracture Zone is most likely insignificant. The Kane Fracture Zone is an exception, but the potential temperature of bottom water in this fracture is close to 2 °C.
18. The further propagation of Antarctic Bottom Water to the north in the East Atlantic is almost limited by the Kane Gap. We found an alternating regime of flow, which changes direction several times during a year. The seasonal signal seems to dominate. The maximum daily average values of southerly velocities reach 0.20 m s^{-1}. The velocity and transport at the bottom are aligned along the slope of a local hill near the southwestern side of the Kane Gap. The distribution of velocity directions at the upper boundary of AABW is wider. The transport of AABW ($\theta < 1.9$ °C) based on the mooring and LADCP data varies approximately within ±0.35 Sv in the northern and southern directions. The annual mean AABW transport through the Kane Gap is almost zero.
19. The Discovery Gap is the terminal point of Antarctic Bottom Water Spreading. The Discovery Gap located at 37° N, 16° W is known as the boundary of Antarctic Bottom Water spreading in the sense that this is the water with potential temperature cooler than $\theta = 2$ °C. A decrease in the potential temperature of the coldest water in the Discovery Gap compared with 1982 was found. We found a northerly flow of Antarctic Bottom Water whose temperature is only slightly warmer than $\theta = 2$ °C. The transport of bottom water with a temperature cooler than $\theta = 2.05$ °C is 0.24 Sv. The measurements in 2013 in the Western Gap (37° N, 19° W) showed the presence of a pronounced current of bottom water to the north at speeds of up to 30 cm s^{-1}, which is much stronger than through the Discovery Gap at 16° W. Bottom temperatures below $\theta = 2$ °C are found only in a close vicinity not exceeding 50 miles north of the passage.
20. Data analysis of the flows across the Mid-Atlantic Ridge in the North Atlantic demonstrates that the Charlie Gibbs Fracture Zone and also the Bight Fracture Zone are important channels for bottom water flow.
21. The cold dense water that is formed in the Greenland Sea overflows the ridge between Greenland and Iceland. The transport of Arctic waters through the Denmark Strait ranges between 2.0–3.5 Sv. No strong seasonal or interannual variability of water transport exists in this region, but large fluctuations were found on short time scales of a few days. The overflow over the Denmark Strait occurs as a spillway in the form of large cold lenses. These lenses are called boluses. This form of overflows of cold lenses water alternates with usual slower currents. The period of variability is approximately 3–6 days without seasonal